# REACTOR DOSIMETRY

# PROCEEDINGS OF THE

# Seventh ASTM-EURATOM Symposium on Reactor Dosimetry

## STRASBOURG, FRANCE
## 27-31 AUGUST 1990

edited by


**G. TSOTRIDIS**
*Commission of the European Communities,*
*Joint Research Centre, Institute for Advanced Materials, Petten, The Netherlands*

**R. DIERCKX**
*Commission of the European Communities,*
*Joint Research Centre, Institute for Safety Technology, Ispra, Italy*

**P. D'HONDT**
*SCK/CEN, Mol, Belgium*



SPRINGER-SCIENCE+BUSINESS MEDIA, B.V.

**Library of Congress Cataloging-in-Publication Data**

ASTM-Euratom Symposium on Reactor Dosimetry (7th : 1990 : Strasbourg, France)
Proceedings of the Seventh ASTM-Euratom Symposium on Reactor Dosimetry, Strasbourg, France, 27-31 August 1990 / edited by G. Tsotridis, R. Dierckx, P. d'Hondt.
p. cm.
English and French.
Includes index.
ISBN 978-0-7923-1792-0
1. Nuclear reactors--Materials--Effect of radiation on--Congresses. 2. Radiation dosimetry--Congresses. I. Tsotridis, G. II. Dierckx, R. III. Hondt, P. d'. IV. Title. V. Title: Proceedings of the 7th ASTM-Euratom Symposium on Reactor Dosimetry, Strasbourg, France, 27-31 August 1990.
TK9185.A1A88 1990
621.48'33--dc20 92-15198
ISBN 978-0-7923-1792-0 ISBN 978-94-011-2781-3 (eBook)
DOI 10.1007/978-94-011-2781-3

Publication arrangements by
Commission of the European Communities
Directorate-General Telecommunications, Information Industries and Innovation,
Dissemination of Scientific and Technical Knowledge Unit, Luxembourg

EUR 14356 EN

Originally published by Kluwer Academic Publishers in 1992

# PREFACE

## THE EURATOM WORKING GROUP ON REACTOR DOSIMETRY AND THE ASTM-EURATOM SYMPOSIA

The Euratom Working Group on Reactor Dosimetry (EWGRD) started around 1960 with members having been nominated by the governments, from each European laboratory working in reactor physics and technology. The goal was to exchange directly experience and know-how in reactor dosimetry and related programmes.

A need for normalisation was felt in order to guarantee that: the same nuclear data is used; measurements in different laboratories give the same results (need for intercalibration experiments and standards); results are expressed such that a comparison with results from other laboratories is possible.
In sub-groups, specific arguments were discussed resulting in final recommendations. These final recommendations were then discussed in a plenary meeting and accepted as a recommendation for European usage. Several of these recommendations were published, e.g. radiation damage dosimetry guidebooks, and a nuclear data guidebook. Also a programme, executed by the BCMN GEEL, for the production and selling of Reference Materials for Neutron Dosimetry is sponsored by the EWGRD. Workshops in the field of radiation damage and on the pressure vessel steels programme in Europe were successfully organised.

The group works in close contact with ASTM (American Society for Testing of Materials). Altogether seven symposia were jointly organized, and held, alternatively in Europe and USA. The next symposium, the eighth, will be organized by ASTM in 1993 in the USA.

The first ASTM-Euratom symposium, which took place in 1975 at the Joint Research Centre, Petten, the Netherlands, generally defined the status of neutron dosimetry and damage analysis programs and identified the needs of the emerging nuclear industry. The second symposium, hosted in 1977 by the Electric Power Research Institute, Palo Alto, California, emphasized data and techniques used to characterize neutron and gamma ray environments , including the use of well-characterized benchmark studies. The third meeting, which took place in 1979 at the Joint Research Centre, Ispra, Italy, focussed on the interactions between materials experts and dosimetry metrologists. The thrust of the next two symposia, which were held in 1982 at the National Bureau of Standards, Washington, D.C., and in 1984 at the GKSS Research Centre, Geesthacht, Federal Republic of Germany, was on radiation metrology techniques, data bases, damage correlations analysis, and standardization.

However, a growing number of the papers and discussions were concerned with reactor pressure vessel surveillance techniques and correlations of the resulting data, thus reflecting a progression from test reactor research to applications in a maturing power reactor industry. The sixth symposium, held in 1987 at Jackson Hole USA, was designed to provide and update the complete field of reactor physics and dosimetry. This present and seventh symposium was held in Strasbourg, France. As in the sixth symposium benchmarking to standardize physics dosimetry and materials was a major topic. There was much interest in light water reactor pressure vessel surveillance work and in Plant life extension work. Several papers reported on the results of multi-national joint research projects designed to improve the accuracy of pressure vessel dosimetry and materials property trend curves by performing interrelated experiments in a number of well-characterized neutron fields.

**R. Dierckx**
Commission of the European Communities
Joint Research Centre
Institute for Safety Technology
Ispra, Italy

**G. Tsotridis**
Commission of the European Communities
Joint Research Centre
Institute for Advanced Materials
Petten, The Netherlands

## ORGANIZATION OF THE
## 7th ASTM-EURATOM SYMPOSIUM ON REACTOR DOSIMETRY

### SPONSORED BY:

Commission of the European Communities
Institute for Advanced Materials

and the

ASTM Committee E-10 on Nuclear Technology and Applications

### Co-SPONSORS:

U.S. National Institute of Standards and Technology

U.S. Nuclear Regulatory Commission

U.S. Department of Energy

U.S. Electric Power Research Institute

American Society of Mechanical Engineers

American Nuclear Society

### IN COOPERATION WITH:

International Atomic Energy Agency

Council of Europe (European Networks of Scientific Cooperation)

Centre de Recherches Nucleaires de Strasbourg


### ACKNOWLEDGEMENTS

This symposium was part of the efforts and support of many individuals and organizations from all parts of the world, and it would be difficult to name everyone who contributed to the Symposium's success.

Special recognition should be given to Mrs. M. Cobut and Mrs. Debeauvais, Centre de Recherche Nucleaire - CRN, Strasbourg, for their support in organizing the Symposium.

## CHAIRMEN OF THE 7th ASTM-EURATOM SYMPOSIUM COMMITTEE

| | |
|---|---|
| **EURATOM Co-Chairman:** | **R. Dierckx**<br>Commission of the European Communities<br>Joint Research Centre, ISPRA<br>Italy |
| **ASTM Co-Chairman:** | **W.N. McElroy**<br>Consultants and Technical Services<br>Richland, WA |

## EURATOM PROGRAMME COMMITTEE

| | |
|---|---|
| **Chairman:** | **R. Dierckx**<br>JRC, Ispra, Italy |
| **Scientific Secretary:** | **G. Tsotridis**<br>JRC, Petten, The Netherlands |
| **Members:** | **A. Alberman**<br>CEN Saclay, France |
| | **W.G. Alberts**<br>PTB, Braunschweig, Federal Republic of Germany |
| | **P. D'hondt**<br>CEN/SCK Laboratories, Belgium |
| | **A.J. Fudge**<br>AERE, Harwell, UK |
| | **F. Hegedüs**<br>PSI, Würenlingen, Switzerland |
| | **M. Petilli**<br>ENEA, Casaccia, Italy |
| | **H.J. Nolthenius**<br>ECN, Petten, The Netherlands |

## ASTM SYMPOSIUM COMMITTEE

| | |
|---|---|
| Executive Chairman: | **W.N. McElroy**<br>Consultants and Technical Services<br>Richland, Washington |
| General Chairman: | **H. Farrar IV**<br>Rockwell, Los Angeles, California |
| Vice-Chairman: | **F.H. Ruddy**<br>Westinghouse, Pittsburgh, Pennsylvania |
| Program Chairman: | **E.P. Lippincott**<br>Westinghouse, Pittsburgh, Pennsylvania |
| Program Vice-Chairman: | **J.G. Williams**<br>University of Illinois, Urbana, Illinois |
| Scientific Secretary: | **D.W. Vehar**<br>Sandia, Alburquerque, New Mexico |
| Members: | **S.L. Anderson**<br>Westinghouse, Pittsburgh, Pennsylvania<br>**L.R. Greenwood**<br>Battelle Northwest Lab., Richland, Washington<br>**F.B.K. Kam**<br>Oak Ridge National Laboratory, Oak Ridge, Tennessee<br>**G.P. Lamaze**<br>NIST, Gaithersburg, Maryland<br>**A.L. Lowe**<br>Babcock & Wilcox, Lynchburg, Virginia<br>**G.C. Martin**<br>General Electric, Pleasanton, California<br>**E.D. McGarry**<br>NIST, Gaithersburg, Maryland<br>**M. Nakazawa**<br>University of Tokyo, Tokyo, Japan<br>**J.W. Rogers**<br>EG & G, Idaho Falls, Idaho<br>**W.L. Server**<br>Tenera (EPRI), Berkeley, California<br>**F.W. Stallmann**<br>Martin Marietta Engineering Systems, Oak Ridge, Tennessee<br>**A. Taboada**<br>NRC, Washington, DC<br>**J.J. Wagschal**<br>Hebrew University, Jerusalem, Israel |

## LIST OF CONTENTS

## COMPUTER CODES AND METHODS

## NUCLEAR DATA



## FUSION

## RADIATION FIELDS CHARACTERISTICS

## GENERAL INTEREST

## SUMMARIES OF WORKSHOP MEETINGS

# REACTOR LIFE ASSESSMENT

# SURVEILLANCE FOR LIFE ATTAINMENT
# AND LICENSE RENEWAL

S. T. Byrne
ABB Combustion Engineering
Nuclear Power
Windsor, Connecticut 06095

W. L. Server
Tenera Engineering Services
Berkeley, CA 94704

## ABSTRACT

The reactor vessel surveillance program was originally designed as a means of monitoring neutron irradiation damage to verify initial design estimates. However, current needs have changed the traditional perspective towards the surveillance program. These needs come about because of two major issues facing nuclear reactor vessel owners; they are life attainment and license renewal. Some of the early reactor vessels are subject to relatively high embrittlement rates such that attainment of the original design life of the vessel requires various actions to manage the embrittlement rate. Conversely irradiation embrittlement in numerous other reactor vessels is relatively benign such that it can be demonstrated that the power plant operating license can be extended while maintaining adequate margins of safety for the vessel.

Decisions involving both life attainment and license renewal place new demands on the surveillance program. In some cases, the new demands can be met by minor program adjustments; in other cases, major adjustments are needed to obtain the requisite data. In either case, the situations and potential solutions are many and varied, such that there is a definite need for a set of guidelines on how a reactor vessel owner can adapt an existing surveillance program to meet current needs. This paper discusses various approaches for managing irradiation embrittlement in support of life attainment and license renewal decisions. Strategies such as integrated programs, replacement surveillance capsules, and supplemental dosimetry measurements are addressed.

## INTRODUCTION

The reactor vessel surveillance program was established originally as a means of monitoring neutron irradiation damage to verify initial design estimates. However, present needs have changed this traditional perspective toward the surveillance program. These needs have come about because of two major issues facing nuclear reactor vessel owners; they are life attainment and license renewal. Both issues require confirmation that neutron embrittlement will not exceed limits related to the vessel design safety margin. The most viable source of confirmatory data is the reactor vessel surveillance program. However, the surveillance program was not designed for this purpose, and the explicit data required may not be immediately available.

Through modifications to an existing surveillance program, it may be possible to obtain information currently required. The following sections describe embrittlement management needs and discuss several approaches for managing irradiation embrittlement via modifications to the surveillance program for obtaining data important to life attainment and license renewal.

## BACKGROUND

Neutron irradiation embrittlement is addressed in the design for the reactor vessel beltline (that region which surrounds the active height of the reactor core). Most reactor vessels have a surveillance program designed based on ASTM E185 (1) which is intended to monitor the degree of embrittlement for the beltline materials. The program includes base metal and weld region materials which were selected to represent the vessel beltline. Selection criteria reflected the then current understanding of irradiation embrittlement trends. Surveillance capsules installed within the vessel contain mechanical test specimens to measure changes in strength and toughness and monitors to measure peak temperatures and neutron fluence rate during the service life of the vessel. The surveillance capsules are removed periodically during the service life in order to evaluate the test specimens and monitors. The results are then compared to original design predictions to verify their accuracy.

Over the past thirty years, the ability to predict the degree of irradiation embrittlement has advanced significantly as more and more data became available from surveillance and experimental programs. (2-7) The residual elements copper and phosphorus were identified (2,3) as important variables affecting radiation sensitivity of reactor vessel base and weld materials. It was subsequently determined (4,5) that nickel content was also a significant variable, especially for reactor vessel weldments. These observations eventually led to the development of irradiation damage correlations (6,7) based on copper and nickel content and neutron fluence; separate correlations (6,7) based on copper and nickel content and neutron fluence; separate correlations were developed for plates and welds. Those correlations formed the basis for Revision 2 to USNRC Regulatory Guide 1.99. (8)

Reactor vessel surveillance program data are presently evaluated using the Regulatory Guide. It provides a set of criteria for judging the credibility of the data and a prescription for applying credible data sets in vessel integrity analyses. The prescriptive approach enables the user to adjust the radiation damage prediction based on the post-irradiation measurements and to reduce the amount of margin which is added to the mean predicted transition temperature shift. Therefore, availability of credible surveillance data can be very beneficial in terms of reducing the predicted shift for the most limiting reactor vessel beltline material.

Direct realization of these benefits, however, requires that the limiting material is one of the materials originally selected for the surveillance program. In many programs, this is not the case. Surveillance materials selected two or three decades ago can be different from what is presently considered to be the limiting vessel material. Lacking the limiting material in the surveillance program will, therefore, limit the benefits available from having the program unless an alternative approach can be employed.

## EMBRITTLEMENT MANAGEMENT NEEDS

The reactor vessel surveillance program is one resource which can be used to help manage embrittlement in support of vessel life attainment and license renewal. Four general embrittlement management needs which can be supported using the surveillance program are identified below:

1. Reactor Vessel Operating Limits - Heat-up and cool-down of the reactor vessel is controlled within a range of pressures and temperatures to prevent brittle fracture of the vessel. Limits on pressure and temperature are adjusted based on predictions of the irradiation induced shift in the ductile-to-brittle transition temperature of the limiting vessel material. Availability of surveillance data for the limiting material increases confidence in the predicted shift and, under certain conditions, permits the use of less restrictive operating limits. This approach provides greater operating flexibility in the near term and a longer period of time in which the vessel can be operated.

2. Vessel Integrity - Pressure or temperature transients (e.g., pressurized thermal shock) which could lead to brittle failure must be considered in the vessel design.

Furthermore, a transient could occur during operation, a flaw could be discovered during an in-service inspection, or a part of the vessel could have been fabricated with a low upper-shelf toughness material. Analysis of vessel integrity in such situations relies in part on predictions of irradiation embrittlement. Surveillance data for the vessel beltline materials can be used to demonstrate the conservatism of the embrittlement predictions and thereby support operation of the vessel up to its intended service life or beyond (e.g., for license renewal).

3. Vessel Annealing - In cases where irradiation embrittlement is predicted to limit the service life of the vessel, in-situ vessel annealing may provide a means of restoring the toughness of the beltline materials such that the vessel can continue to operate safely. ASTM Standard E509 (9) presents guidelines for demonstrating the effectiveness of an in-situ anneal, including "development of a surveillance program to monitor the effects of subsequent irradiation on the annealed vessel beltline materials."

4. License Renewal Monitoring - One license renewal consideration is provision for surveillance monitoring beyond the original service life of the reactor vessel. Explicit requirements have not been established, but ASTM Standard E185 (1) guidelines are assumed to apply. Information on the neutron fluence and irradiation embrittlement of the vessel beltline materials should be obtained for the license renewal period using the existing or a modified surveillance program. In some cases, the program will have matured such that none of the original surveillance capsules will be left in the vessel or the accumulated neutron fluence of the remaining capsules will be too great to be meaningful during the license renewal period. In other cases, a sufficient number of additional capsules will be available and can be used for monitoring during the renewal period.

## APPROACHES

Various approaches are available for using the surveillance program in support of life attainment and license renewal such as the three discussed below. The strategy adopted will, of course, have to be tailored to the plant-specific needs identified above and to the limitations of the existing program. Limitations include those resulting from surveillance capsule design, availability of the limiting beltline material, irradiation environment factors, and time constraints with respect to meeting specific needs.

1. Management of Existing Programs - The original surveillance program provides several potential resources

- o number of remaining capsules
- o surveillance materials and their relation to the limiting beltline materials
- o correlation monitor material
- o capsule lead factor

The number of remaining capsules determines in part the capability for monitoring over the license renewal period or for vessel annealing. The encapsulated surveillance materials provide the means for determining the irradiation sensitivity of the vessel beltline, depending on how closely the surveillance materials match the limiting vessel beltline materials. Availability of a correlation monitor material in one or more capsules provides the means for satisfying one of the criteria in Regulatory Guide 1.99, Revision 2 (8) to establish credibility of the surveillance data. The capsule lead factor (the ratio of neutron fluence rate at the capsule to the rate at the vessel inside surface) influences the capsule withdrawal schedule; a lead factor of five (5) with an eight year capsule exposure will yield a neutron fluence equivalent to forty years at the vessel inside surface.

In cases where the limiting material or the correlation monitor material are included in only a fraction of the capsules, the capsules designated for withdrawal and evaluation can be changed to provide data on those specific materials on a more timely basis. This approach would produce the data required to establish credibility, as defined by Regulatory Guide 1.99, (8) and potentially provide a benefit in reducing shift predictions for use in setting vessel operating limits or in performing vessel integrity analyses. In cases where the controlling material is not included in any of the capsules, the surveillance program can still fulfill the long term monitoring function for the license renewal period and, potentially, provide a near term benefit by reducing uncertainty in radiation damage predictions. In either case, capsule lead factor must be considered such that available capsules are used efficiently; for example, with high lead factor capsules it may be beneficial to move them to low lead factor positions or to remove and store those capsules, and then reinsert them at a later date to provide monitoring in support of license renewal or vessel annealing.

Available resources should also be managed in terms of types and quantities of test specimens. For capsules which contain fracture toughness (e.g., compact tension) specimens, the test method can be selected based on plant specific needs, such as upper shelf toughness, crack initiation toughness, or crack arrest toughness. For capsules with a limited number of Charpy impact specimens, the broken specimen halves can be reconstituted into new specimens and tested to acquire additional data for that material and irradiation exposure.

2. Integrated Surveillance Program - The integrated program concept was first used in a situation where plant specific capsules had to be removed; the capsules were installed in another, or host reactor to provide surveillance monitoring (10). This concept would also be advantageous if extended to the case where surveillance data from a host reactor represents the limiting vessel material in another reactor which has its own surveillance capsules but not its limiting material. This situation occurs frequently with vessels fabricated at the same facility; typically several vessels were fabricated using the same heats of submerged arc weld wire and the surveillance weld for each vessel was selected from one of those weld wire heats. If the host reactor and the reactor of interest are of the same design, the host reactor's surveillance data are readily transferable to the other reactor vessel. Conversely, if the two vessel designs are different, additional analysis is needed to demonstrate that the differences in irradiation environment are inconsequential before the host reactor data can be applied to the reactor of interest.

   If an integrated surveillance program can be established, surveillance data from the host reactor can be used in conjunction with data from the plant specific surveillance capsules to yield credible data on the limiting beltline material. Once credibility has been established, the data can be used to improve margin for vessel operating limits and vessel integrity analysis and to provide support for vessel license renewal. Data from an integrated surveillance program would also provide support to decisions concerning vessel annealing and, potentially, to conducting the post-anneal surveillance program. (Note: Acceptance criteria for use of data from an integrated surveillance program are not addressed explicitly in Regulatory Guide 1.99 (8).

3. Replacement Surveillance Capsule Program - It is not always possible to obtain critical information from an existing program or through an integrated program; however, in situations where remote installation of a new capsule in the vessel can be performed, it may be possible to obtain that critical information via a replacement capsule program. This approach is useful for obtaining data in support of vessel integrity analyses, increased margin for vessel operating limits, license renewal, and vessel annealing where the replacement capsules could contain the following:

   - specimens representing the limiting beltline material
   - supplemental dosimetry
   - supplemental fracture toughness specimens

   Potential sources of the limiting material or representative material include surveillance material archive blocks, reconstituted Charpy specimens, vessel archive material, and NDE calibration blocks. Supplemental dosimetry can include the same flux monitors used in the existing program or new state-of-the-art monitors, and they can be installed in vacant capsule holders inside the vessel and in the cavity outside the reactor vessel. Fracture toughness specimens machined from the limiting or representative material can be irradiated in replacement capsules and used to obtain desired toughness properties. Specimen size, however, will be limited by the surveillance capsule dimensions.

   The replacement capsule approach is advantageous in that the data will be obtained from specimens irradiated in the reactor vessel of interest, so the data can be applied directly using established methods. Major considerations influencing the effectiveness of replacement capsules include:
   - availability of vacant capsule holder
   - capability for remote installation
   - availability of representative material
   - time constraints for obtaining the required data versus plant outage schedule and capsule lead factor.

## SUMMARY

The surveillance program was originally intended to provide a means for monitoring irradiation response of the reactor vessel. Furthermore, technology advancements over the intervening decades and changes in regulatory requirements have limited the effectiveness of many surveillance programs in meeting present day needs specific to vessel life attainment and license renewal.

Those needs related to embrittlement management include data required for:

- o increased vessel operating flexibility
- o vessel integrity analysis
- o vessel annealing
- o license renewal period surveillance monitoring.

In order to meet these demands, three approaches have been established for tailoring an existing surveillance program to provide relevant data on a timely basis:

1. Management of Existing Programs - Utilize existing surveillance program resources, and modify the capsule withdrawal schedule, the capsules designated for withdrawal, or the test methods to meet plant specific needs.

2. Integrated Surveillance Program - Utilize surveillance data from a host reactor in conjunction with the existing plant specific surveillance capsule data to establish irradiation sensitivity of the limiting vessel beltline material.

3. Replacement Surveillance Capsule Program - Install new capsules containing the limiting vessel material, supplemental dosimetry, or supplemental fracture toughness specimens to obtain data required to meet plant specific needs.

Each approach has advantages and limitations which must be considered for a given application. One or more approaches can be useful to assist in the management of irradiation embrittlement.

## REFERENCES


1. American Society of Testing and Materials, "Standard Practice for Conducting Surveillance Tests for Light-Water Cooled Nuclear Power Reactor Vessels," ASTM E185 (Current edition approved July 1982; Originally published in 1961).

2. U. Potapovs and J. R. Hawthorne, "The Effect of Residual Elements on 550°F Irradiation Response of Selected Pressure Vessel Steels and Weldments," Nuclear Applications, Vol. 6, No. 1, 1969, pp. 27-46.

3. J. R. Hawthorne, J. J. Koziol, and R. C. Groeschel, "Evaluation of Commercial Production A533-B Plates and Weld Deposits Tailored for Improved Radiation Embrittlement Resistance," Properties of Reactor STructural Alloys After Neutron or Particle Irradiation, ASTM STP570, American Society for Testing and Materials, 1975, pp. 83-103.

4. E. C. Biemiller and S. T. Byrne, "Evaluation of the Effect of Chemical Composition on the Irradiation Sensitivity of Reactor Vessel Weld Metal," Irradiation Effects on the Microstructure and Properties of Metals, ASTM STP611, American Society for Testing and Materials, 1976, pp. 418-433.

5. J. D. Varsik and S. T. Byrne, "An Empirical Evaluation of the Irradiation Sensitivity of Reactor Pressure Vessel Materials," Effects of Radiation on Structural Materials, ASTM STP683, American Society for Testing and Materials, 1979, pp. 252-266.

6. G. R. Odette, et.al., "Physically Based Regression Correlations of Embrittlement Data from Reactor Pressure Vessel Surveillance Programs," Electric Power Research Institute, NP-3319, January 1984.

7. G. L. Guthrie, "Charpy Trend Curves Based on 177 PWR Data Points," in "LWR Pressure Vessel Surveillance Dosimetry Improvement Program," NUREG/CR-3391, Vol. 2, prepared by Hanford Engineering Development Laboratory, HEDL-TME 83-22, April 1984.

8. "Radiation Embrittlement of Reactor Vessel Materials," U.S. Nuclear Regulatory Commission, Regulatory Guide 1.99, Revision 2, May 1988.

9. American Society for Testing and Materials, "Standard Guide for In-Service Annealing of Light-Water Cooled Nuclear Reactor Vessels," ASTM E509-86, May 30, 1986.

10. A. L. Lowe, et.al., "Integrated Reactor Vessel Material Surveillance Program," Babcock & Wilcox, BAW-1543A, Rev. 2, May 1985.

# MAGNOX REACTOR PRESSURE VESSEL DOSIMETRY - A NEW ASSESSMENT

P.J.H. Heffer, L.T. Jones, D.A. Thornton
A.F. Avery*, N.R. Smith* and A.K. Ziver*
Nuclear Electric, plc,
Berkeley Nuclear Laboratories,
Berkeley, Gloucestershire, GL13 9PB UK

## ABSTRACT

This paper describes a major study undertaken to produce new dpa dose assessments for all Magnox RPVs operated by Nuclear Electric plc. The work combines complex calculations based on the Monte Carlo methodology with extensive plant measurements. Wide variations in neutron flux and spectrum are found at the vessel surfaces which contrast with those in similar PWR studies. The results form part of a full review of the structural integrity of Magnox RPVs undertaken by Nuclear Electric plc to ensure their continued safe operation.

## INTRODUCTION

Nuclear Electric plc operates ten Magnox reactors with pressure circuits fabricated entirely from steel. Magnox Reactor Pressure Vessels (RPVs) are large spheres, typically 20m in diameter and fabricated from carbon manganese steel plate ranging in thickness from 7.5 to 11.4cm. Approximately half of the plate welds are of the automatic submerged-arc type, with high copper content. This material exhibits the largest properties change with irradiation and reactor safety cases concentrate upon it.

The original steels surveillance scheme instigated for Magnox plant

*AEA, Winfrith Technology Centre, Dorchester, Dorset, DT2 8DH, UK

## Side-Core Calculations

Neutrons reaching the RPV at the side of the core do so by bulk penetration of the side reflector, which comprises a solid graphite structure supported by a steel restraint. A typical structure is shown in Figure 2. This is readily amenable to modelling in a single step calculation based on a 45° symmetric wedge of the core. The geometries and sources in the inner parts of the active core are fully smeared, but the fuel channels at the core periphery are modelled in detail.

The reflector is represented explicitly and the core restraint is modelled as a series of steel tubes and girders. Minor approximations are made, such as the smearing of concentric tubes to one cylinder, but none has a significant effect on the bulk neutron transport.

The reactor vessel is modelled explicitly, as are the main vessel attachments, such as gas seals and baffle plates. Outside the vessel, a metre thickness of the primary bio-shield is represented to ensure correct boundary conditions to the neutron transport calculation.

Dose rates are calculated on the inner and outer surfaces of the RPV and at intermediate thicknesses through the steel. The vessel is divided meridionally into zones of typically 5° spacing, but close to important engineering features or in regions of rapidly varying flux the zones are reduced to 1° or less. In cores with irregular outer boundaries to fuelled regions, azimuthal sub-divisions are also imposed at 15° intervals.

## Above- and Sub-Core Calculations

Axial neutron leakage from Magnox cores is characterised by neutron streaming behaviour. At the top of the core, the fuel channels are entirely open, giving free line of sight from the fuel to the vessel. Above the fuel is a reflector region varying in thickness from 45cm to 100cm and a steel structure, known as a charge pan, rests on top of the reflector. The charge pans comprise either one or two horizontal steel plates and their effective thickness varies from 1.5cm to 14cm. A typical top-core geometry is shown in Figure 3. Fuel is loaded into and removed from the core via steel tubes which penetrate from outside the RPV through a series of massive nozzles. Each tube services a number of fuel channels, and there are equivalent tubes for each control rod. The general arrangement can be seen in Figure 1.

Because streaming is a significant feature of neutron leakage from the core, it is not acceptable to approximate the upper core

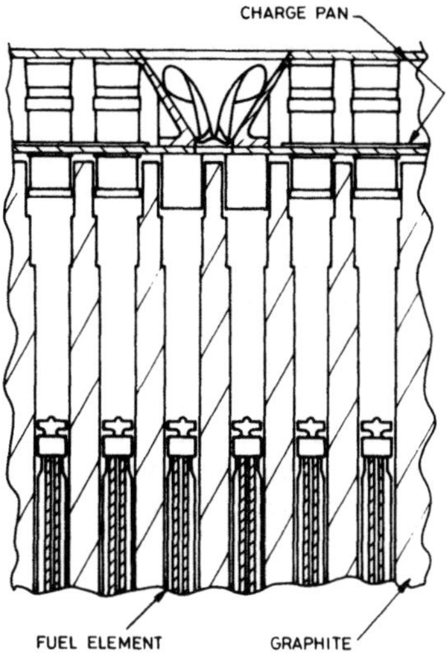


FIG.3. TOP CORE GEOMETRY.

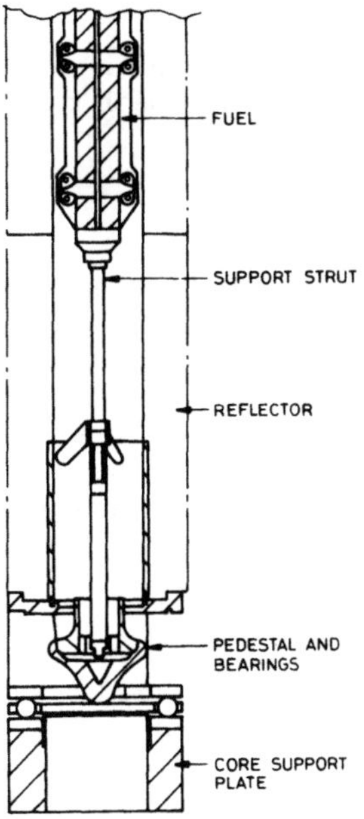


FIG.4. BOTTOM CORE GEOMETRY.

geometry to any significant extent. However, the large number of fuel channels, up to 4,500 in this type of Magnox design, makes explicit modelling of each channel intractable. To overcome this, the calculation is split into two stages. In the first, a detailed model of the top of the active core, reflector and charge pan are prepared, but limited to the smallest symmetrical zone of the core, up to 36 fuelled channels. This is used to calculate the energy and angular dependent neutron leakage from the surface of the reactor. This spatially invariant leakage form is folded with the core radial power profile to generate a surface source for the second stage of the calculation, which represents explicitly all the above-core steelwork and vessel penetrations. At this stage, the core is represented approximately as a series of smeared zones which are adequate to model back-scattered neutrons.

The below-core region is treated in an analogous way. In this case the core geometries are more complex; the fuel rests on steel supports which limit neutron streaming and the core itself is supported on massive steel plates up to 15cm thick mounted on a grid of beams. A typical bottom-core geometry is shown in Figure 4. The grid is fixed to extensions of the main RPV supports which rise from the concrete foundations. Interspersed in these supports is the complex arrangement of steel plates of the debris fairing. This is illustrated in Figure 1.

RPV doses beneath the core have been calculated using the same two-stage approach described above.

## RESULTS

RPV dose estimates are based on a combination of measured and calculated results. Measured fluxes are not sufficiently comprehensive to allow direct dose assessment over the complete RPV and spectrum determinations are relatively sparse. Measurements also require translation from the experimental location (often ex-vessel) to the RPV inner wall. Therefore, calculations are relied upon to interpret direct measurements, to supply data away from measurement locations and to supply spectrum data at all locations.

The quality of the calculations can be assessed by comparison with direct measurements where these are available. Results to date, concentrated on one plant, give a high degree of confidence in the theoretical routes. At the side of the core, the absolute nickel activation rate as measured by an ex-vessel foil is predicted to within 8% of measurement. Above the core, nickel foil measurements inside a surveillance tube at the RPV inner surface are predicted to within 5%. Beneath the core, the ex-vessel measurements are predicted less well; discrepancies up to 28% are found. However, measurements inside the RPV from the debris plate samples are predicted to within 10%; these are much more directly related to the RPV inner wall flux.

The quality of neutron spectrum predictions can be judged from the analysis of the MFA packs, which has been carried out by AEA Technology at Harwell. The results are assessed by comparing the initial predicted ratio of dpa rate to fission equivalent flux to that generated from the MFA analysis and spectrum adjustment. To date, only one pack shows a ratio difference in excess of 10%, and in this case there was considerable damage to the pack foils arising from the high exposure temperatures. The absolute values of these ratios, as measured by packs in extreme locations, have varied by nearly an order of magnitude.

The accuracy of these assessments is position dependent, a function of the degree of experimental support in a given region, the modelling difficulties and the rate of change of flux in the locality. At the best locations, uncertainties of order 13% (one standard deviation) are estimated, rising to 30% in regions of rapid local flux perturbations. Extensive sensitivity calculations have been carried out to identify the dominant sources of modelling uncertainty and to help quantify their magnitude.

## DISCUSSION

The results of the assessments confirm the complex spatial variations of neutron flux and spectrum anticipated from the design of these plants. Macroscopic systematic changes in flux are found following broadly the variations of power distribution within the reactor core, but these are subject to strong local perturbations. In the vicinity of major engineering features, such as the main core supports, fluxes have been found to change by up to a factor four over 25cm of vessel height. There are equally marked changes in neutron spectrum, which varies much more at different locations of the RPV wall than in PWR plant.

The importance of the work can be seen from the comparison of the two RPV dose assessments presented in Figure 5. Using fission equivalent flux as the parameter, the maximum RPV doses are found at the vessel equator, entirely in keeping with dose maxima at the beltline of PWRs. Doses above and below the core are relatively low. Using a dpa assessment, the dose maximum shifts to the sub-core regions and the above-core region increases in relative

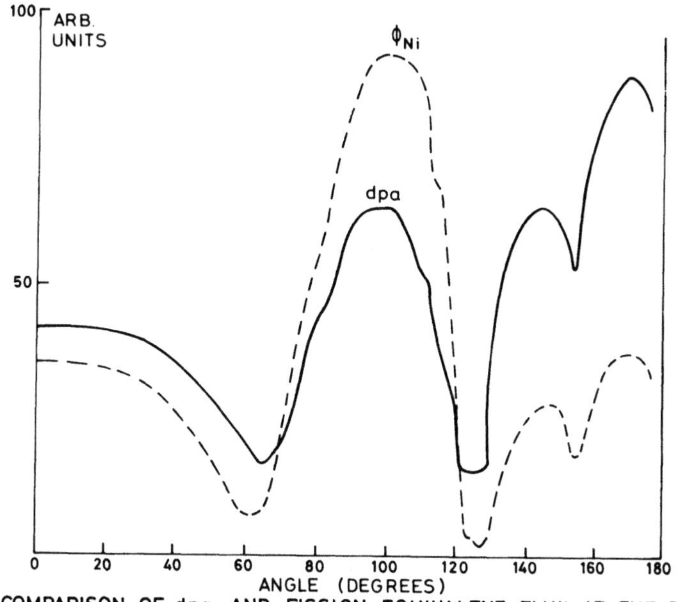


FIG.5. COMPARISON OF dpa AND FISSION EQUIVALENT FLUX AT THE PRESSURE VESSEL INNER SURFACE.

importance. This general pattern of change occurs for all Magnox plant, though the detail varies considerably for different designs.

There are, of course, parallel changes in the dosimetry of the Magnox materials database. These have changed markedly the form of the mechanical properties trend curves and have improved the interpretation of the experimental data (3). All this work is being used by Nuclear Electric plc in a full review of the structural integrity of Magnox RPVs to ensure their continued safe operation.

## CONCLUSIONS

(1) A new assessment of the dosimetry of all Magnox reactor steel RPVs operated by Nuclear Electric plc is nearing completion, based on dpa and combining experimental and theoretical data.

(2) Comparison of theoretical results with measurement confirm that dose predictions can be made to high accuracy, with total uncertainties ranging between 13-30%.

(3) The assessments show that wide spatial variations in neutron flux and spectrum occur at the surface of Magnox RPVs. These are associated both with changes in the core power density and with the localised effects of major engineering features.

(4) RPV dose distributions based on dpa are significantly different from those based on fission equivalent flux. These changes are much more marked than those experienced in high flux regions of PWR vessels.

(5) The assessments are being used by Nuclear Electric plc in a major programme reviewing the structural integrity of their Magnox RPVs to ensure their continued safe operation.

## ACKNOWLEDGEMENT

This paper is published by permission of Nuclear Electric plc and AEA Technology.

## REFERENCES

(1) Miller P C, ANSWERS Users' Guide (MCBEND), 1989.
(2) ASTM, Annual Book of Standards, PCN:01-120228-35, 1988.
(3) Heffer P J H, Buswell J T, Jones, R B, and Bolton C J, ibid.
(4) Miller, P.C., ANSWERS Users' Guide (SNAP), 1988.
(5) Miller, P.C., ANSWERS Users' Guide (FENDER), 1988.

# A COMPARISON OF FRACTURE TOUGHNESS DATA ON A PRESSURE VESSEL WITH THE ASME $K_{IR}$ CURVE

Michael P. Manahan, Sr.

Nuclear Engineering, Penn State University

University Park, PA 16802, USA

## ABSTRACT

Calculation of the Pressure-Temperature (P-T) operating curves for a nuclear power plant requires an assessment of the neutron embrittlement effects on the static and dynamic fracture resistance of the pressure vessel. The current practice used by the nuclear industry is to shift the ASME reference stress intensity factor ($K_{IR}$) curve by an amount equal to the Charpy 41-Joule transition temperature shift. While it can be argued on a qualitative basis that some correlation between notched impact data and dynamic, plane-strain-fracture toughness data is reasonable and quantitative, demonstration is lacking. Further, recently measured static, plane-strain-fracture toughness measurements made using an ASTM A212B steel indicate that, for this case, the ASME $K_{IC}$ and $K_{IR}$ curves, when shifted by the Charpy 41-Joule transition temperature shift, are non-conservative by about 44°C.

In response to this concern and to the need within the nuclear industry's plant life extension (PLEX) program for in-service materials data, several new test techniques have been developed. These test methods include miniature Charpy and miniature fracture toughness testing. These new test methods are described and benchmark data presented. In addition, a procedure for measuring the $K_{IR}$ curve after neutron irradiation is described the method. Further, the new specimen test techniques described below require a method which results in a much smaller HAZ. Therefore, research at Penn State is currently underway which focuses on the laser welding approach. This method is capable of producing welds with very low porosity and with HAZ regions of less than 0.2 mm. The results of this work will be published elsewhere in the near future.

The measured fracture toughness data is shown in figure 1. It was not possible to satisfy the ligament requirements at all test temperatures. However, the thickness requirement was satisfied in all cases. It was not possible to fit the fracture toughness data versus temperature because of the sparsity of data. However, a comparison with the trend curves from refs. [18] and [19] is provided in Fig. 1. It is important to note that shifting the $K_{Ic}$ and $K_{IR}$ curves by the Charpy $\Delta T_{41}$ is nonconservative by at least 44°C for this particular steel. Even if the margin term in ref. [20] is added to the adjusted $RT_{NDT}(ART_{NDT})$, the $K_{Ic}$ and $K_{IR}$ curves would be nonconservative by at least 22°C.

References [21, 22] compare the Charpy shift and $K_{Ic}/K_{ID}$ shift for an A508 forging, A533B plate, an A302B plate, and several welds. The average underprediction of the base metal $K_{Ic}$ shift using Charpy data is 22°C. Approximately 75% of the base metal data indicated that the Charpy shift underpredicted the $K_{Ic}$ shift, and individual data points varied by as much as 75°C. Although the data base is limited, dynamic toughness is less scattered and has smaller shifts. The dynamic toughness shifts are reasonably well represented by the RG1.99(2) $\Delta RT_{NDT}$ (plus margin).

While several researchers have successfully correlated Charpy and plane strain fracture toughness for specific heats and classes of steel, a general correlation which is applicable to all ferritic steels has not been developed. As discussed in [4], several authors have presented comprehensive discussions on the differences between Charpy and $K_{IC}$ testing in terms of stress state, notch acuity and other critical variables [23, 24, 25]. A comparison between notched bar data for several specimen sizes is not valid because the fracture mode is different at various Charpy energy indices (ex. at 41J). As discussed in [4], in order to relate notched bar data from specimens of various sizes using an energy based parameter, it was necessary to remove the crack formation energy from the analysis and consider only the energy associated with stable crack growth and subsequent cleavage. Based on this observation, one would not expect the shift measured using total absorbed energy in a notched bar to accurately represent the shift in the measured static and dynamic fracture toughness. There are, of course, cases where the standard Charpy specimen does represent the $K_{IR}$ curve shift, but such agreement would be fortuitous in general.

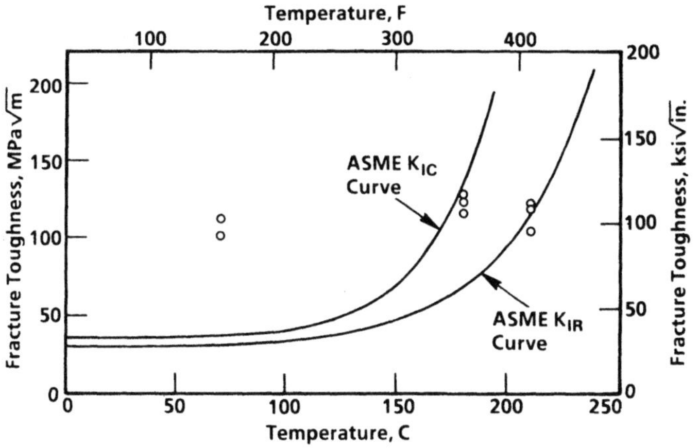


Figure 1. Comparison of $K_i$ data with the ASME $K_{Ic}$ and $K_{IR}$ curves.

## 3.0 MINIATURIZED FRACTURE TESTING

An in-depth discussion of the development of the Miniaturized Notch Test[(1)] (MNT) is provided in the literature [4, 10, 11]. The goal of this work was to develop an energy based parameter which can

---

(1) The processes described are explained in part in U.S. patent no. 4,567,774 dated February 4, 1986. A patent application on further improvements of the methodology is pending.

be used to relate data from specimens with non-standard dimensions to data obtained using conventional, dynamic Charpy specimens. Early studies lead to the conclusion that total absorbed energy, whether normalized or not, is not a generally applicable MNT parameter. While it has been established that shear fracture appearance can serve as a useful parameter for any specimen size, accurate measurement is tedious. In addition, in the nuclear industry, current regulations require the conventional Charpy 41-J index transition temperature. Therefore, the % post-maximum load energy was developed to serve as an energy based parameter for miniature Charpy testing. The test has been shown to be capable of yielding data of the same accuracy as that obtained with conventional Charpy testing.

A new approach(1) to plane-strain fracture toughness testing has also been developed. The method enables the testing of specimens which are substantially thinner than currently believed necessary by the technical community to obtain valid $K_{Ic}$ data. The method developed is a direct, experimental approach and does not involve the use of analytical "correction factors." The basic idea is to modify the stress field in the vicinity of the crack tip to produce plane-strain conditions. There are a variety of experimental procedures that can be used to modify the stress field. Details concerning the modified stress-field fractue toughness theory validation is reported in [12], and the experimental demonstration is given in reference [1].

The miniature specimen data are compared to data obtained from thermal shock of a 15.24-cm pressure vessel in Figure 2. The fracture modes are identical and the miniature $K_{Ic}$ data lie within the experimental scatter of the thick vessel data. Future applications include weld reconstitution of broken Charpy specimens to yield a miniaturized fracture toughness specimen.

---

(1) The processes described are explained in part in U.S. patent no. 4,567,774 dated February 4, 1986. A patent application on further improvements of the methodology is pending.

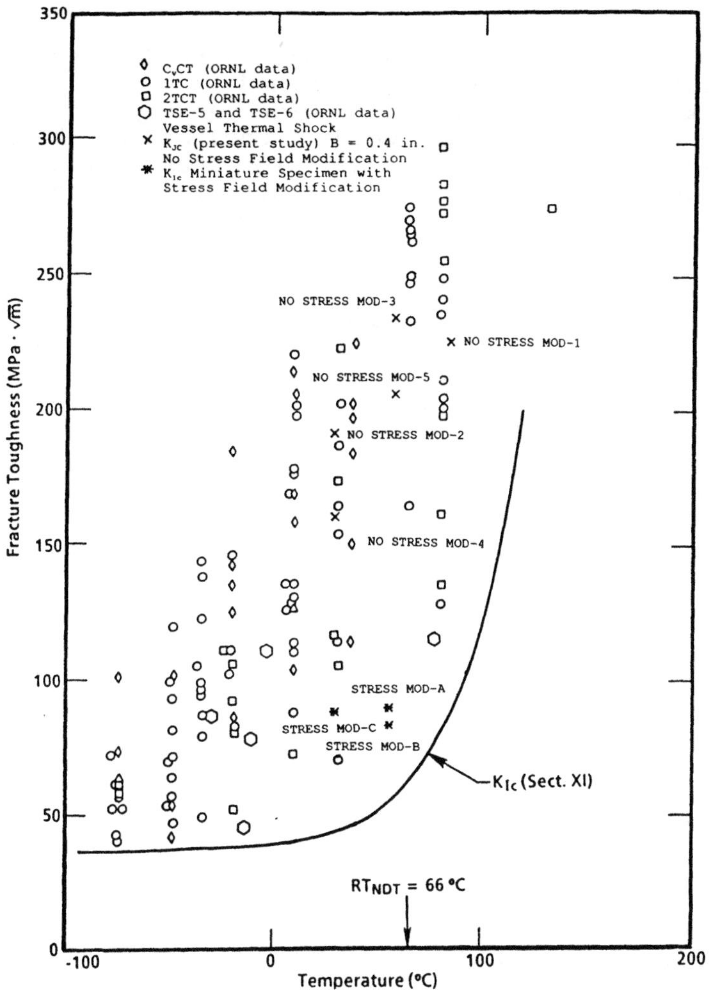


Figure 2. Modified Stress Field Data Compared with ORNL Thick-Section Test Results for TSE-5 and TSE-6 Material

## 5.0 SUMMARY

In order to ensure safe operation to end of-license and during the PLEX period, utilities and regulators desire plant-specific fracture toughness and Charpy data. MST is a cost-effective means by which to provide these data. The proof-of-principle experiments conducted over the past several years demonstrate the efficacy of the technical approach of stress-field modification. What is needed now is a comprehensive research effort aimed at benchmarking this approach on a wide variety of pressure vessel plates and welds. Once benchmarked, these data can serve as the basis for change of the applicable ASTM standards. It is anticipated that this effort will set the stage for regulatory acceptance and future use by the nuclear utilities.

## 6.0 REFERENCES


1. M.P. Manahan, "Plane-Strain Fracture Toughness Determination Using Stress Field Modified Miniature Specimens", submitted to the International Journal of Fracture Mechanics (August 1990).

2. M.P. Manahan, A.S. Argon, and O.K. Harling, "The Development of a Miniaturized Disk Bend Test for the Determination of Postirradiation Mechanical Properties", Journal of Nuclear Materials 103 and 104 (August 1981) 1545-1550.

3. M.P. Manahan, "Mechanical Behavior of Magnetite from the Oconee-2 and TMI-1 Steam Generators Using Miniaturized Specimen Technology", Journal of Materials Science accepted for publication March 1989, in press.

4. M.P. Manahan and C. Charles, "A Generalized Methodology for Obtaining Quantitative Charpy Data from Test Specimens of Nonstandard Dimensions", Journal of Nuclear Technology, Vol. 90, pp 245-259, May 1990.

5. B. Majumdar, C.E. Jaske and M.P. Manahan, "Determining Creep-Crack-Growth Behavior Using Miniature Specimens", International Journal of Fracture Mechanics accepted for publication January 1989, in press.

6. M.P. Manahan, R. Kohli, J. Santucci, and P. Sipush, "A Phenomenological Investigation of In-Reactor Cracking of Type 304 Stainless Steel Control Rod Cladding", Invited Paper Journal of Nuclear Engineering and Design, 113 (1989) 297-321.

7. M.P. Manahan, et al., "Fracture Toughness of Irradiated ASTM A212B Pressure Vessel steel" Final Report, Battelle Memorial Institute, (September 1988).

8. M.P. Manahan, "Surveillance Capsules A' and C' for Nine Mile Point Unit 1", Final Report from Battelle Memorial Institute to Niagara Mohawk Power Corporation (September 1987).

9. M.P. Manahan, "A New Postirradiation Mechanical Behavior Test--The Miniaturized Disk Bend Test", Invited Paper at the 1982 American Nuclear Society Winter Meeting in Washington, DC Journal of Nuclear Technology 63 (November 1983) 295-315.

10. M.P. Manahan, "Determination of Fracture Behavior of Ferritic Steel Using Miniature Specimens", Journal of Nuclear Materials, 166 (1989) 321-330.

11. M.P. Manahan and C. Charles, "Determination of Charpy Transition Temperature of Ferritic Steels Using Miniaturized Specimens", Journal of Materials Science, accepted for publication March 1989, in press.

12. M.P. Manahan and M. Nakagaki, "Modified Stress-Field Fracture-Toughness Theory Validation", submitted to International Journal of Fracture Mechanics (August 1990).

13. M.P. Manahan, A.R. Rosenfield, and J.A. Vanecho, "Final Report on Miniaturized Fracture Mechanics Feasibility Study to Corporate Technical Development", Battelle Columbus Laboratories (1985).

14. M.P. Manahan, "Oyster Creek Re-Insertion Capsule Design and Fabrication", Final Report to General Public Utilities (1990).

15. ASTM Standard E813, Test Method for $J_{IC}$, A Measure of Fracture Toughness, ASTM Philadelphia 03.01.

16. ASTM Standard E1253, "Standard Guide for the Reconstitution of Irradiated Charpy Specimens" (1987).

17. Perrin, et al., "Reconstituted Charpy Impact Specimen", EPRI NP2759 (1982).

18. WRC Bulletin 175, PVRC Recommendations on Toughness Requirements for Ferritic Materials, August 1972.

19. ASEM Section XI, Division 1, Figure A-4200-1.

20. Regulatory Guide 1.99, Revision 2, "Radiation Embrittlement of Reactor Vessel Materials" (1988).

21. A.L. Hiser, "Comparison of Irradiation-Induced Transition Temperature Increases ($\Delta T$) from Notch Ductility and Fracture Toughness Test", Presentation to ASTM Committee E10.02.10, January, 1987.

22. W.L. Server, "Fracture Toughness Shifts and Bounds Due to Radiation Embrittlement", Presentation to ASTM Committee E10.02.10, January, 1987.

23. D. Broek, "Elementary Engineering Fracture Mechanics" 4th rev. ed. Dordrecht: Martinus Nijhoff 184 (1986) 116.

24. R.W. Hertzberg, Deformation and Fracture Mechanics of Engineering Materials, John Wiley & Sons, New York (1983).

25. S.T. Rolfe and J.M. Barsom, Fracture and Fatigue Control in Structures, Prentice-Hall, Englewood Cliffs, New Jersey (1977).

26. R.D. Cheverton, et al., "Pressure Vessel Fracture Studies Pertaining to the PWR Thermal-Shock Issue: Experiments TSE-5, TSE-5A, and TSE-6", ORNL-6163, Oak Ridge National Laboratory (1985).

# ETUDE DE L'INFLUENCE DU SPECTRE NEUTRONIQUE SUR LA FRAGILISATION DES ACIERS DE CUVE DE REACTEURS A EAU PRESSURISEE

A. Alberman*, D. Beretz**, L. Bourdet*, F. Hedin***, C. Rieg***, P. Soulat*

* Commissariat à l'Energie Atomique, CEN/SACLAY
91191 Gif sur Yvette Cedex

** Commissariat à l'Energie Atomique, CEN/GRENOBLE,
B.P. 85 X, 38041 Grenoble Cedex

*** Electricité de France, SEPTEN
12-14, avenue Dutrievoz, 69628 Villeurbanne Cedex

## ABSTRACT

Pressure vessel neutron embrittlement is a major concern with regard to PWR lifetime. In order to achieve a better assessment of the vessel end-of-life, a collaborative CEA-EDF research program has been established. The goal is a better monitoring of vessel embrittlement from surveillance capsule data. This program should establish the best correlation between damage to vessel steels and neutron fluence and spectrum. A specific irradiation location has been designed in the reflector of SILOE research reactor (Grenoble), in such a manner that the neutron spectrum, comparable to a surveillance capsule spectrum, is very different from a "standard" irradiation location in OSIRIS research reactor (Saclay). A dpa/$\phi$ (E> 1 MeV) ratio differing by a factor about 2 is foreseen in order to establish the influence of the neutron spectrum on steel embrittlement. Main design options for this program (irradiation devices, neutron spectra, dosimetry) are presented.

RESUME

La fragilisation de la cuve sous irradiation constitue un élément fondamental de la durée de vie d'un réacteur électronucléaire à eau pressurisée. Afin de mieux évaluer cette durée, un programme de recherches a été défini entre CEA et EDF visant à améliorer l'exploitation du programme de surveillance de la cuve. Il s'agit d'obtenir la meilleure corrélation expérimentale entre dommage d'irradiation sur l'acier de cuve et fluence d'irradiation. Pour cela, un emplacement spécial d'irradiation a été étudié dans le réflecteur du réacteur de recherche SILOE (Grenoble), de sorte à obtenir un spectre très différent de celui d'un emplacement "standard" OSIRIS (Saclay), tout en restant représentatif de celui d'une capsule de surveillance. On vise un rapport dpa/$\phi$ (E>1 MeV) sensiblement double d'un spectre à l'autre, afin de discriminer l'influence du spectre neutronique sur la fragilisation de l'acier. La communication présente les principaux éléments de dimensionnement retenus pour ce programme : dispositifs d'irradiation, spectres neutroniques, dosimétrie.

## 1 - INTRODUCTION

Dans le cadre des réflexions actuelles de l'Electricité de France sur la durée de vie des réacteurs à eau pressurisée (REP), l'évaluation des marges de sûreté quant à la rupture fragile de la cuve est de première importance. En effet, contrairement aux autres équipements lourds (générateur de vapeur ...), il n'est guère envisageable de changer une cuve lorsque sa résistance à la rupture fragile aura été jugée trop faible.

De même que pour les différents réacteurs électronucléaires à eau pressurisée, dans le monde, le parc français fait l'objet d'un programme de surveillance (1). L'état de la cuve est apprécié à l'aide des résultats des essais mécaniques sur éprouvettes d'irradiation. Le facteur d'accélération éprouvettes/cuve (ou facteur de flux) est déterminé par calcul. Faute de corrélation suffisamment bien établie, ce facteur n'intègre pas pour l'instant la variation du spectre neutronique entre les positions considérées.

Pour préciser ce point, un programme expérimental, appelé ESTEREL, a été engagé par le CEA et EDF afin de rechercher le meilleur indicateur du dommage d'irradiation des capsules d'irradiation et de la cuve sachant que celui qui est aujourd'hui utilisé est le flux de neutrons d'énergie supérieure à 1 MeV.

## 2. PRESENTATION DU PROGRAMME ESTEREL

Ce programme consiste à interpréter, de manière explicite et qualifiée, l'irradiation d'un même matériau de cuve REP dans des environnements neutroniques très différents.

Pour cela, on envisage deux irradiations d'éprouvettes dans deux spectres de réacteur de recherche à même fluence de dommages, telle que le nombre de déplacements par atome (dpa) par exemple, pour des niveaux instantanés (dpa/s) assez voisins. Cette fluence de dommages doit correspondre à des fluences de neutrons (E > 1 MeV) très différentes dans les deux spectres :

1) Emplacement standard A d'irradiation d'aciers de cuve en réflecteur de pile-piscine (OSIRIS),

2) Emplacement à spectre "perturbé" B.

## 2.1. Spécifications neutroniques visées de l'emplacement spécial B

Il s'agit de déterminer les caractéristiques principales d'un emplacement d'irradiation spécial B en réflecteur de la pile-piscine SILOE, choisie pour cette expérience, positionné derrière un écran neutronique (de façon sensiblement analogue à une capsule de surveillance en géométrie REP du palier 900 MWe), pour obtenir un spectre neutronique très dégradé vers les énergies de neutrons intermédiaires. Cet emplacement spécial B doit répondre à trois exigences :

- niveau de flux rapide (E > 1 MeV) tel que :

  $(\phi > 1\ \text{MeV})\ A \sim 2\ (\phi > 1\ \text{MeV})\ B$

  pour une même fluence de dommage exprimée en dpa,

- niveau de flux suffisant : 2,5 à $4.10^{12}$ $n.cm^{-2}.s^{-1}$ (E > 1 MeV) en B, pour garder une durée d'irradiation raisonnable.

- même vitesse d'irradiation (en dpa/s) visée pour s'affranchir d'un éventuel effet de niveau de flux. A notre connaissance, cette dernière condition n'a jamais pu être réalisée dans les programmes d'étude réalisés jusqu'à présent.

## 2.2. Spécifications du dispositif d'irradiation

Dans le cadre de ce programme, un nouveau dispositif d'irradiation - régulé électriquement - et de grande capacité (~ 80 à 100 éprouvettes Kcv) a été développé. Il a été étudié de sorte à pouvoir fonctionner indifféremment dans les deux environnements nucléaires : A et B.

La température requise est de 288±5°C. Ceci implique de limiter l'échauffement gamma sur les éprouvettes.

## 2.3. Spécifications des éprouvettes d'irradiation

Afin de mieux cerner l'influence du spectre neutronique, il a été décidé d'irradier deux matériaux de cuve différents dans la même nuance. Ils sont réalisés à l'aide de joints soudés et sont ajustés sur des teneurs en éléments fragilisants (Cu ; P ; Ni) distinctes.

Pour améliorer la statistique de mesures, on a choisi d'irradier 30 éprouvettes Charpy-V et 10 éprouvettes de traction par matériau. Bien entendu, le chargement est identique pour les deux irradiations.

## 2.4. Dosimétrie et calculs de référence

Afin de bien caractériser les deux emplacements, la dosimétrie comporte des :

| | |
|---|---|
| - détecteurs de dommages | : GAMIN, Tungstène, Saphirs, |
| - dosimètres d'activation | : cuivre, nickel, indium, cobalt, |
| - dosimètres fissiles | : neptunium 237, uranium 238. |

Les schémas de calcul retenus pour l'exploitation de la dosimétrie dans l'emplacement spécial sont dérivés de ceux effectués pour la dosimétrie de dommages DOMPAC (2). On effectuera notamment un calcul de référence Monte Carlo (TRIPOLI 2).

# 3. OPTIMISATION DU PROGRAMME

L'influence du spectre neutronique sur la fragilisation des aciers de cuve implique de pouvoir apprécier une différence de décalage de température de transition fragile ductile (ΔTT) entre les deux emplacements d'irradiation.

Cette différence (éventuelle) doit être mesurée d'autant plus précisément que les spectres ont des caractéristiques voisines.

C'est pour cette raison que l'on recherche :

1) des spectres suffisamment différents dans la gamme des énergies supérieures à 0.1 MeV, seuil inférieur en pratique des modèles de fluences de dommages couramment proposés pour les aciers,

2) un décalage suffisant de ΔTT permettant de limiter l'incertitude relative sur la mesure de la température de transition. Ceci conduisant, pour un acier ou une soudure représentatifs du parc électronucléaire français, à envisager une fluence d'irradiation élevée.

### 3.1. Critère de discrimination du spectre neutronique

Une fluence d'irradiation s'analyse comme un paramètre intégral de corrélation permettant de qualifier l'irradiation. Le programme ESTEREL consiste à tenter de préciser la fluence d'irradiation la mieux adaptée :

$\phi > 1$ MeV *(ou $\phi$ 1)*, $\phi > 0.1$ MeV *(ou $\phi$ 0.1)* ou dpa.

Aussi, on analysera les deux spectres visés (standard A et spécial B) en terme d'indices de spectre, c'est-à-dire quotient de paramètres intégraux tels que :

$\frac{\phi\ 1}{\phi\ 0.1}$, $\frac{\text{dpa}}{\phi\ 1}$, etc

On choisit à priori l'indice $\frac{\text{dpa}}{\phi\ 1}$ comme critère de discrimination.

L'écart "spectral" entre les deux emplacements pourra être qualifié par le rapport de dommages "R" :

$$R = \frac{(\text{dpa} / \phi\ 1)\ B}{(\text{dpa} / \phi\ 1) A}$$

Ce rapport étant, par définition, le rapport de dpa (emplacement spécial/ emplacement standard) obtenu lors d'une irradiation à même fluence rapide ($\phi$ 1) ou, à contrario, le rapport des fluences rapides (emplacement standard/emplacement spécial) à même dpa.

Un rapport R le plus différent de 1 est bien entendu l'objectif recherché.

### 3.2. Critères métallurgiques

Pour la faisabilité du programme ESTEREL, il est proposé de choisir une valeur de R que l'on peut tenter de préciser en fonction de la précision des mesures de $\Delta$TT. On admet généralement une loi de variation $\Delta$TT fonction de la fluence rapide $\phi$ 1 du type :

$$\Delta TT = C(\phi\ 1)^N \qquad \text{avec } 0.3 < N < 0.5$$

Si le dpa est le meilleur paramètre d'irradiation, les valeurs de $\Delta$TT seront très voisines dans les deux emplacements A et B. Par contre, si c'est la fluence rapide ($\phi$ 1), la mesure de $\Delta$TT en B sera inférieure à celle obtenue en A.

Pour la faisabilité de ce programme, il faut donc s'assurer d'avoir une valeur de R suffisante en regard de l'incertitude de mesure.

Si $\delta$ est l'incertitude de mesure (à un niveau de confiance à préciser) de $\Delta$TT, on pourra discriminer le meilleur paramètre d'irradiation selon le schéma de la figure 1.

L'incertitude sur ΔTT est sensiblement constante en valeur absolue. Il faut donc déterminer un décalage minimum pour diminuer l'incertitude relative. On admettra que ce décalage minimum sera obtenu dès que les incertitudes associées aux mesures métallurgiques en A et B ne se recouvrent pas :

$$\Delta TT_A - \delta = \Delta TT_B + \delta$$

On obtient le critère de discrimination du décalage minimum en fonction de l'incertitude de mesure et de R :

$$\Delta TT_A = \frac{2\delta}{1 - \frac{1}{R^N}}$$

La figure 2 montre les abaques correspondant au cas N = 0.4, valeur moyenne généralement constatée dans l'exploitation de programmes d'irradiation.

Les matériaux choisis dans le programme ESTEREL sont deux joints soudés représentatifs des joints soudés des viroles de coeur des cuves 900 MWe d'EDF :

| | P | Cu | Ni |
|---|---|---|---|
| M1 : joint soudé à bas résiduels | 0.010 % | 0.04 % | 0.70 % |
| M2 : joint soudé, plus sensible à l'irradiation | 0.019 % | 0.10 % | 0.70 % |

On se fixe une fluence rapide de $\phi\ 1 = 6$ à $7\ 10^{19}$ $n.cm^{-2}$ pour obtenir un ΔTT élevé, de l'ordre de ~ 45°C (M1) et ~ 85°C (M2).

## 4. CARACTERISTIQUES NEUTRONIQUES

On peut augmenter le rapport R dans un emplacement d'irradiation spécial en favorisant un décalage de spectre des neutrons de fission vers les énergies intermédiaires (≅ 0.1 MeV) par le biais notamment de la diffusion inélastique sur les atomes de fer. D'où l'idée d'implanter un large bloc d'acier ferritique en périphérie du réacteur SILOE contenant en son sein le dispositif d'irradiation (figure 3). Les calculs neutroniques préliminaires ont montré qu'une distance d'environ 20 cm entre le combustible et l'axe du dispositif est adéquate. Le bloc d'acier est protégé de la corrosion par un dépôt de nickel, l'avant est muni de canaux pour assurer le refroidissement.

Le tableau I résume les caractéristiques neutroniques attendues des deux emplacements :

| | dpa/s | $\phi$ 1 ($n.cm^{-2}.s^{-1}$) | dpa/$\phi$ 1 (barn/neutron) | $\frac{\phi\ 0.1}{\phi\ 1}$ |
|---|---|---|---|---|
| Spécial B | $6.10^{-9}$ | $3.10^{12}$ | 2200-2400 | 5 |
| Standard A | $6.10^{-9}$ | $5.10^{12}$ | 1300-1400 | 2 |
| Rapport | 1 | 1,7 | 1,7 | 2,5 |

Tableau I

CARACTERISTIQUES NEUTRONIQUES DES EMPLACEMENTS D'IRRADIATION DU PROGRAMME ESTEREL

## 5. CONCLUSION

Le programme ESTEREL visant à rechercher le meilleur indicateur de dommage d'irradiation des éprouvettes de surveillance de la cuve a fait l'objet d'une étude d'optimisation dont les principaux résultats sont résumés dans ce papier. Il met en oeuvre un nouveau dispositif d'irradiation de grande capacité, dans les réacteurs OSIRIS et SILOE. Ce dispositif est en cours de qualification.

Un emplacement spécial à spectre perturbé a été conçu dans SILOE de sorte à obtenir environ 1,7 fois plus de dpa qu'à OSIRIS, pour une même fluence neutronique (E>1 MeV). Un important programme de dosimétrie permettra de qualifier cet emplacement spécial.

Le début des irradiations est prévu en 1991 ; les analyses métallurgiques devraient être réalisées courant 1992.

## REFERENCES


1. J.C. Nimal, L. Bourdet, R. Lloret, A. Bévilacqua, F. Hédin, C. Brillaud, Détermination des Caractéristiques Neutroniques du Programme de Surveillance des REP de 900 MW en France, ce symposium

2. A. Alberman, O. Hoclet et Coll., Expérience de dosimétrie DOMPAC, Rapport CEA R 5217 (1983)

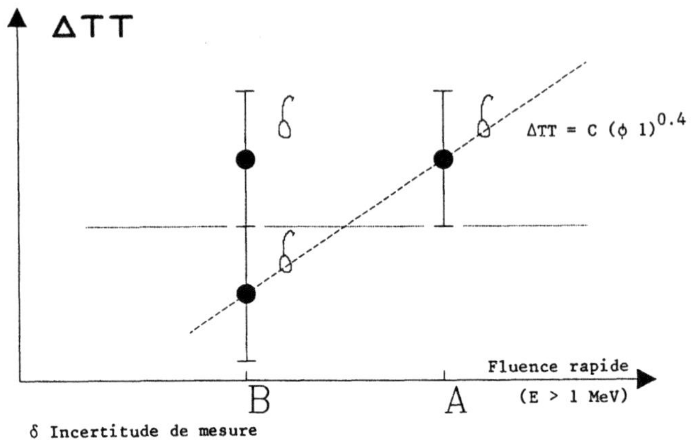


FIG. 1 - Décalage de température de transition fragile/ductile prévu dans les deux emplacements d'irradiation A et B pour une même fluence de dommages (p. ex. en dpa)

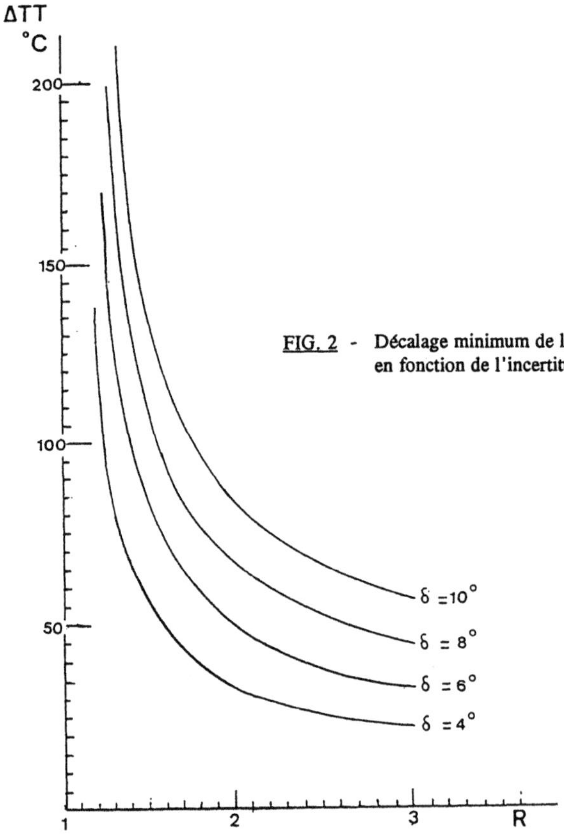


FIG. 2 - Décalage minimum de la température de transition fragile-ductile en fonction de l'incertitude et de "R"

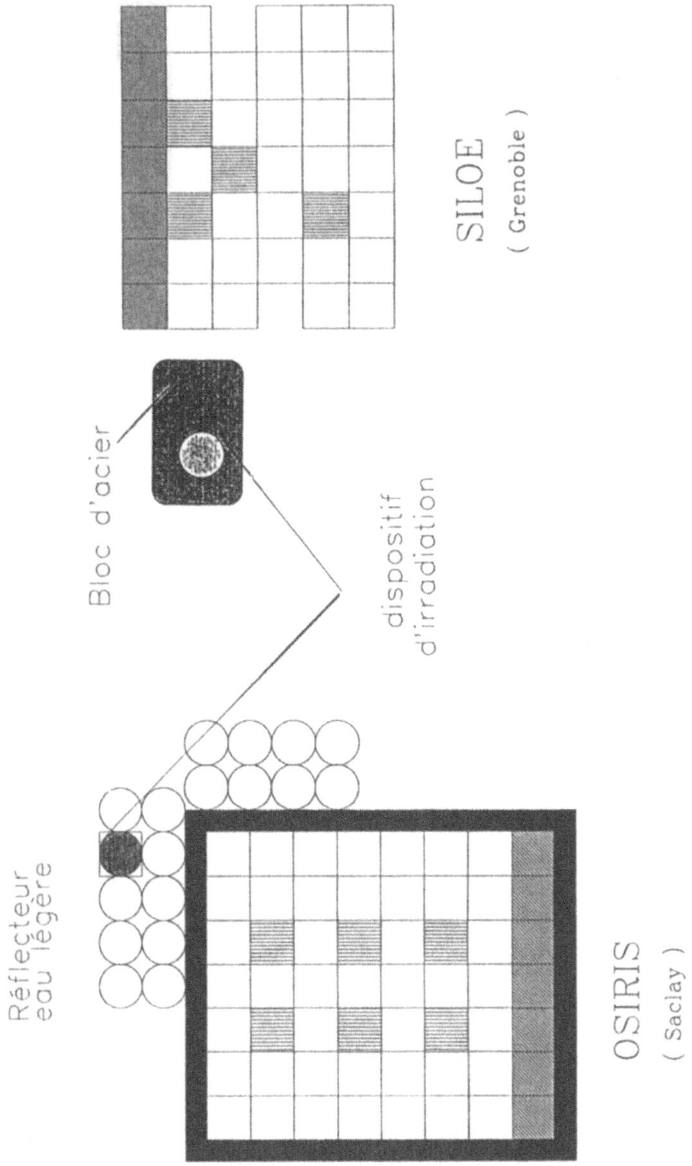


**<u>FIG. 3</u>** - **Emplacements d'irradiation du programe ESTEREL**

# POWER REACTOR EMBRITTLEMENT DATA BASE

F. B. K. Kam, F. W. Stallmann, and J. A. Wang

Oak Ridge National Laboratory
Oak Ridge, Tennessee 37831 USA

## ABSTRACT

Regulatory and research evaluations of embrittlement prediction models and of pressure vessel integrity can be greatly expedited by the use of a well-designed, computerized data base. The Power Reactor Embrittlement Data Base (PR_EDB) is such a comprehensive collection of data for U.S. commercial nuclear reactors(1). The Nuclear Regulatory Commission (NRC) has provided financial support, and the Electric Power Research Institute (EPRI) has provided technical assistance in the quality assurance (QA) of the data to establish an industry-wide data base that will be maintained and updated on a long-term basis. Successful applications of the data base to several of NRC's evaluations have received favorable responses and support for its continuation. The future direction of the data base has been designed to include the test reactor and other types of data of interest to the regulators and the researchers.

---

## BACKGROUND

The PR_EDB is a collection of computerized data files that contain information about the irradiated and unirradiated properties of steels used in U.S. power reactor pressure vessels. These data were accumulated directly from surveillance reports and other published reports provided, for the most part, by the NRC, EPRI and the Materials Property Council.

This data base is funded by the NRC and has been designed and generated for research evaluations of embrittlement prediction models and pressure vessel integrity issues. Based on the favorable comments from data evaluations completed, NRC has indicated continued future support for the data base. The elapsed time to complete an evaluation and the QA of the data were critical factors in their decision.

The EPRI Reactor Vessel Embrittlement Management Program has adopted PR_EDB as the basis for an industry-wide data base. Under EPRI subcontracts, the reactor vendors are providing assistance in the QA of this data base and will provide additional data and documentation. The QA checks by the vendors are in progress; the anticipated date for completion of this phase is September 30, 1990.

## ARCHITECTURE, DATA COLLECTION, AND SOFTWARE

In keeping with current data base technology, the information in the PR_EDB is contained in a collection of data files. Only similar or logically related data are stored in the same file, and the information from several single files can be combined by means of "key identifiers" that code in a unique manner the reactor (PLANT_ID) and surveillance capsule (CAPSULE), the material (HEAT_ID) and specimen orientation (SPEC_ORI), and the reference to the document (REF_ID). For instance, chemistry information is contained in the file CHEM_PR, neutron fluences and related information are in REAC_PR, and the Charpy transition temperature shifts and the upper-shelf energies are in SHFT_PR. Tables relating transition temperature shifts with fluences and chemistry can be constructed by combining these three files using the key identifiers PLANT_ID and CAPSULE for the irradiation environments and HEAT_ID for the materials. Each record of every file is also connected to its source document via the REF_ID so that every data item can be traced unambiguously to its source. Because the structure and content of each file is independent of the others, new data types can be added easily without disturbing the existing files. The same features, which are characteristic of relational data bases, make it relatively simple to update, maintain, and perform quality control on the data files.

All data are collected first into "raw data files," and all information, including numbers, is entered as characters to preserve as much as feasible the details of the original documentation of the experimental results.

NRC and EPRI groups have decided temporarily against having evaluated files until reliable and consistent data sets can be confirmed. As a result, in the raw data files, multiple determinations of the same quantity are given in different records, each with its proper references. All different determinations are kept in PR-EDB, and it is up to the user as to which determination to use for a particular application or whether averages should be obtained from several of them. Hence, to analyze the data, the user must make some decisions reflecting on the credibility of the data in the raw data files and prepare a processed data file for the analysis.

The current compilation contains data from 92 reactors and consists of 79 different weld materials (161 data points) and 110 different base materials (381 data points). The different types of data that are implemented or planned for this data base are shown in Fig. 1.

The PR_EDB can be used with any personal computer that runs under a disk operating system (DOS) above Version 2.0. A fixed disk with at least 5 Mbytes of free space and 512 kbytes of random access memory (RAM) is required for efficient use of the data and software. The data files are given in dBASE format, which allows the use of dBASE III Plus, and higher versions as well, as any compatible software (e.g., CLIPPER) to inspect, manipulate, and process the data. However, many routine tasks can be greatly expedited and simplified with the use of special-purpose software. For this reason, a menu-driven software program, EDB-Utilities, has been written to perform many processing steps, such as the retrieval and selection of data, data-manipulation calculations, saving data in files, displaying data to the screen or printer, and the fitting and plotting of Charpy impact data. The computer programs for all the data evaluations performed for the NRC and other researchers are maintained in a file as a guide to the future direction of the software development.

## APPLICATIONS

A comparison of the PR_EDB with the Regulatory Guide 1.99 (Rev. 2) data base showed a larger spread for the base material, but the scatter appears to be random (Fig. 2). The residual parameter shown on the figures is defined to be the observed value of transition temperature shift ($\Delta RT_{NDT}$) minus the value of transition temperature shift calculated by the Regulatory Guide 1.99 (Rev. 2). For the weld materials, the scatter of the new data is of the same magnitude as that for the Regulatory Guide (Fig. 3). A re-evaluation of the two standard deviation values for the base and weld materials in PR_EDB did not show a need to assign separate safety margins for the two materials (Figs. 4 and 5).

An analysis was made to see if there was any bias in the Charpy specimen orientation (TL vs. LT) to the transition temperature shift and to the decrease in upper-shelf energy curves. No bias was observed for the Charpy specimen orientation in the transition temperature shift curve (Fig. 6). However a bias can be clearly seen in the TL-orientation as compared with the LT-orientation (Fig. 7). The data points contained only pairs with the same fluence in the same capsule.

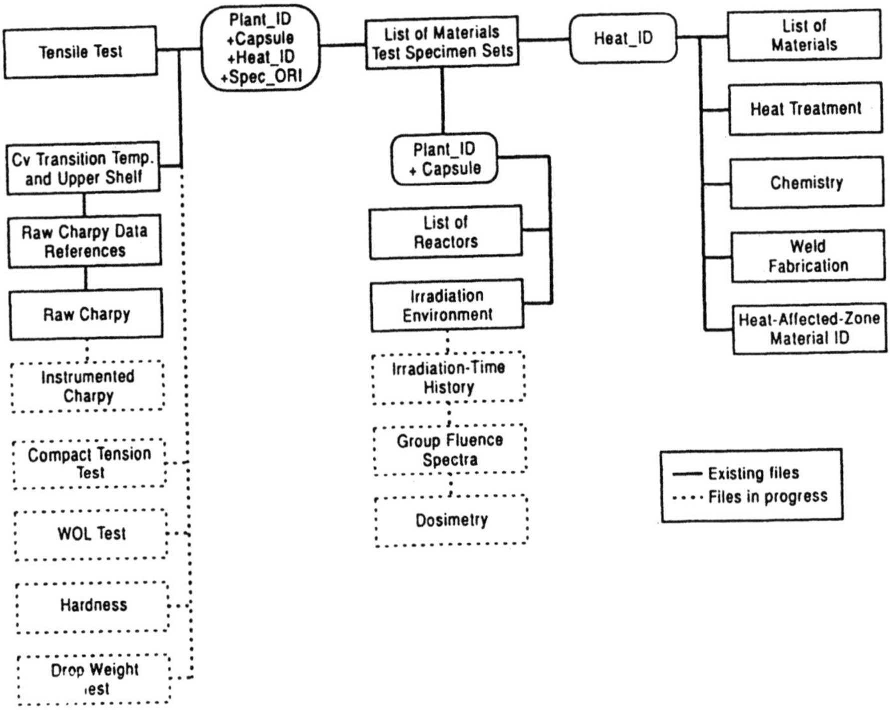


Fig. 1. Architect of the Power Reactor Embrittlement Data Base (PR_EDB).

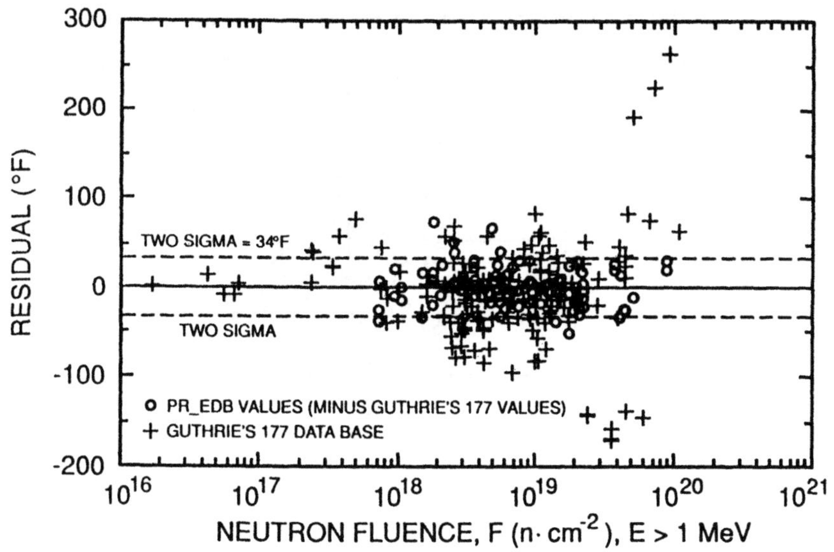


Fig. 2. Comparison of the data for base materials in PR_EDB with Regulatory Guide 1.99 (Rev. 2).

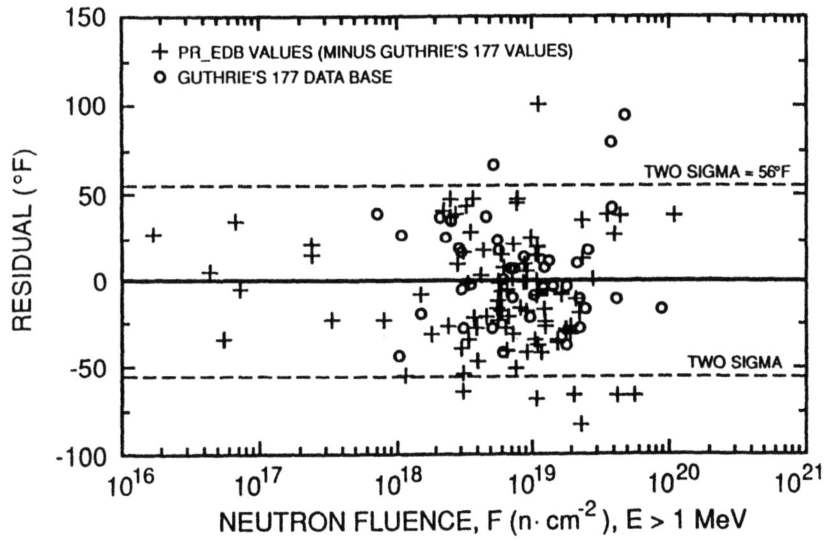


Fig. 3. Comparison of the data for weld materials in PR_EDB with Regulatory Guide 1.99 (Rev. 2).

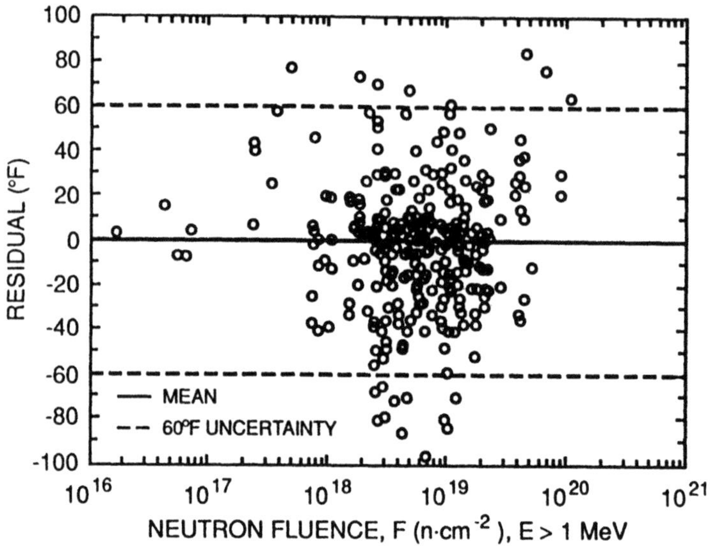


Fig. 4. Comparison of the data for base materials in PR_EDB with Regulatory Guide 1.99 (Rev. 2).

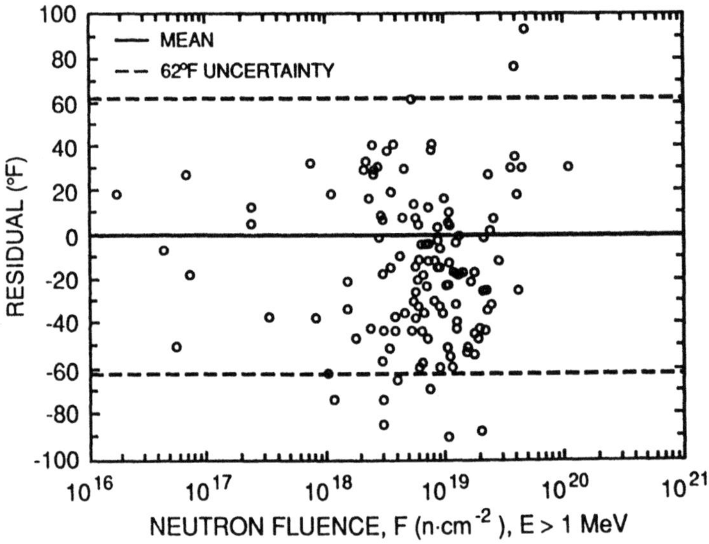


Fig. 5. Comparison of the data for weld materials in PR_EDB with Regulatory Guide 1.99 (Rev. 2).

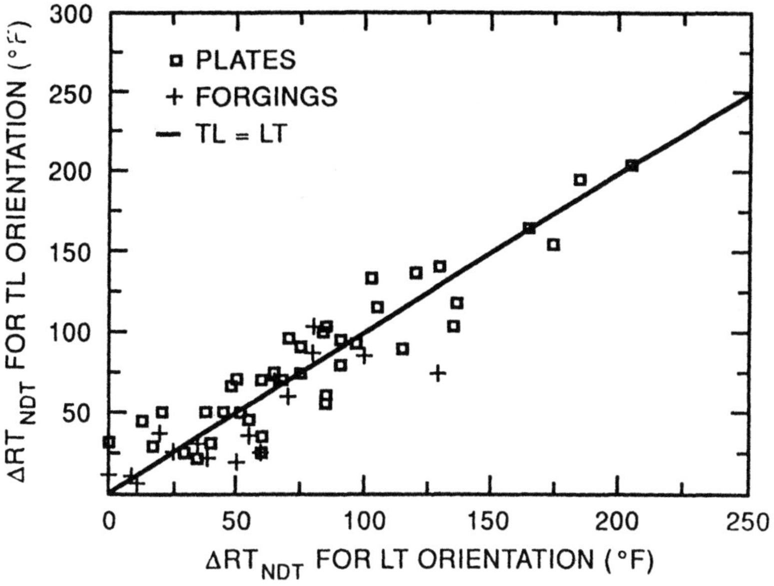


Fig. 6. Comparison of transition temperature shifts for TL and LT orientations.

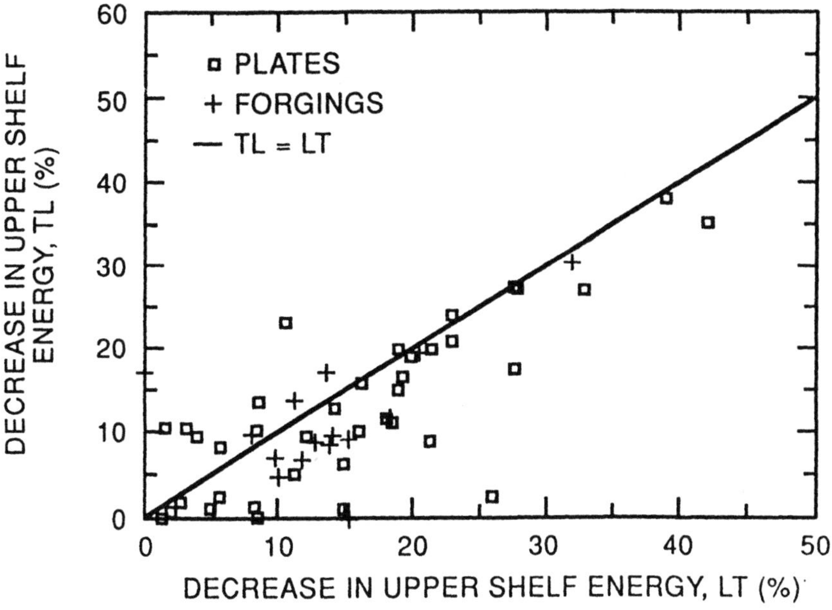


Fig. 7. Decrease in upper-shelf energy for TL and orientations (%).

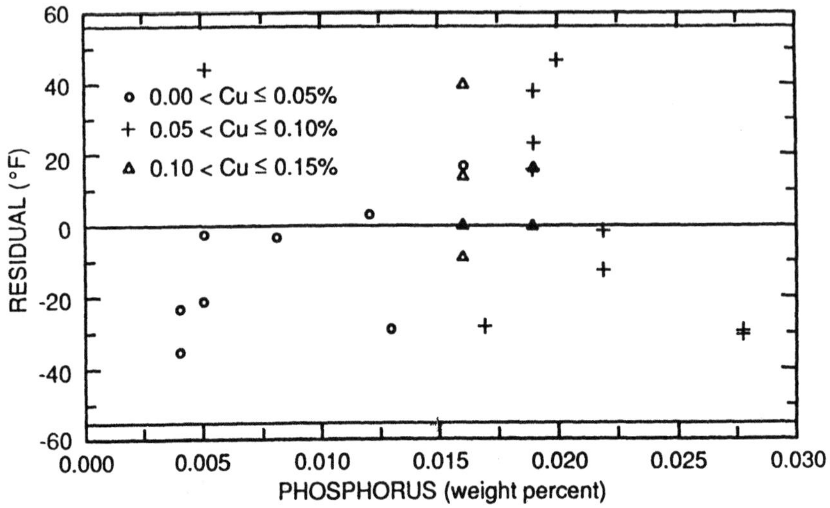


Fig. 8. Residual as a function of phosphorus fow low-copper welds.

Table 1

## COMPARISON OF THE UPPER SHELF ENERGIES FOR THE TL- AND LT- ORIENTED CHARPY V-NOTCH SPECIMENS AND THE EFFECT OF SULFUR -- A302B

| UNIRRADIATED SPECIMENS | | IRRADIATED SPECIMENS | | SULFUR |
|---|---|---|---|---|
| TL VALUES (ft-lbf) | 65% OF LT VALUES (ft-lbf) | TL VALUES (ft-lbf) | 65% OF LT VALUES (ft-lbf) | (%) |
| 67 | 68.3 | 68 | 62.4 | 0.024 |
| 75 | 67.6 | | | 0.026/0.020* |
| 84 | 71.5 | | | 0.020/0.018* |
| 86 | 62.4 | | | 0.025/0.016* |
| 98 | 85.2 | 88 | 78.0 | 0.017 |
| 102 | 101.0 | 84 | 72.8 | 0.019 |
| 105 | 107.3 | 68 | 61.8 | 0.019 |
| 108 | 91.0 | | | 0.015 |
| 108 | 91.7 | 106 | 89.1 | 0.015 |
| 108 | 91.7 | 107 | 89.7 | 0.015 |
| 112 | 118.0 | | | 0.016 |
| 115 | 76.5 | | | 0.017 |

*VALUES FROM TWO DIFFERENT LABORATORIES (LUKENS/KAWIN)

A discussion of some work by J. R. Hawthorne showed a significant phosphorus effect in low-copper material (about 0.01% copper), but not for high-copper content (0.30% copper). The safety issue for regulatory was: "Does Regulatory Guide 1.99 (Rev. 2) underpredict the Charpy shift for low-copper material when the phosphorus is high?" A preliminary study indicated that some of the surveillance weld data did show a phosphorus effect (Fig. 8). Unfortunately, the large scatter of the residuals plus the fact that a plot of the residuals versus copper showed a similar bias prevented any definitive conclusion.

Lastly, a data evaluation was made to check if test results from longitudinally oriented (LT) specimens reduced to 65% of their value provided conservative estimates of the expected values from transversely oriented (TL) specimens. Also, the effect of sulfur was investigated. Generally, the results in Table 1 indicated that conservative estimates were obtained, and the higher sulfur content showed lower upper shelf energies.

These four data evaluations were just a few of the applications that were performed for the NRC. Completion of the quality assurance task for the PR_EDB will enhance its use by industry and researchers.

## FUTURE DIRECTION

Data collection and software development will be a continuing process. A move is under way to establish a program in cooperation with the Nuclear Management and Resource Council (NUMARC) and NRC for long-term maintenance of PR_EDB to ensure that the data base does not become outdated as have previous data bases.

A vast amount of test reactor data has been collected and will be integrated with the PR_EDB to form a larger Embrittlement Data Base (EDB). This data base will contain many more types of data and will be used to explore and test a variety of models for irradiation damage mechanisms and predictions. PR_EDB, however, will be maintained as a separate data base as requested by EPRI who will provide the quality assurance for added power reactor data.

Data collection anticipated in the coming years will include:

1. data from instrumented Charpy impact tests, including digitized graphics representations of these tests;
2. test results from CT and WOL surveillance specimens;
3. calculated group fluences and fluence rates;
4. dosimetry data;

5. power-time history;
6. pressure-temperature data; and
7. beltline materials and their reported properties for all U.S. power reactors.

Simultaneous analytical studies will be made to identify weaknesses and inconsistencies in the structure and content of the data base.

## DISTRIBUTION

The distribution of PR_EDB will be made to all users by the Oak Ridge National Laboratory with the approval of the NRC.

## ACKNOWLEDGEMENTS

Research sponsored by the Office of Nuclear Regulatory Research, U.S. Nuclear Regulatory Commission, under Interagency Agreement DOE 0551-0551-A1 with the U.S. Department of Energy under contract DE-AC05-84OR21400 with Martin Marietta Energy Systems, Inc.

## REFERENCES

1. F. W. Stallmann et al., "PR_EDB, Version 1, Power Reactor Embrittlement Data Base: Program Description," USNRC Report NUREG/CR-4816, June 1990.

# LIGHT WATER REACTOR SURVEILLANCE

# AN ANALYSIS OF THE H.B.ROBINSON UNIT 2 PWR

# USING THE MONTE-CARLO CODE MCBEND

S W Power

Radiation Physics, Shielding and Criticality Department,

AEA Technology Winfrith, Dorchester, Dorset, DT2 8DH

**ABSTRACT**

H.B Robinson Unit 2 is a three-loop Westinghouse PWR. To assess a low-leakage core configuration, measurements were made in the downcomer annulus and the cavity, and these have now been compared with calculations.

Neutron calculations were performed using the Monte-Carlo code McBEND. As the reactor has eight-fold symmetry, it was modelled, in three dimensions, by a 45$^{o}$ sector and reflective boundaries. The shielding was modelled exactly, but the core composition smeared. Response rates were calculated for ten detectors, for a position oppposite the flats of the core at mid-height, and the calculation was biased to produce good statistics there.

Calculated reaction rates were compared with measurements (after correction for various factors). A slight underprediction (within the estimated uncertainty) was found for most reactions. The $^{63}$Cu(n,$\alpha$) rate was overpredicted (within the estimated uncertainty). The $^{237}$Np(n,fx) detector reaction rate was underpredicted by 35%, but this result seems anomalous.

## INTRODUCTION

The H.B Robinson Unit 2 (HBR-2) station is a three-loop 665 MW(e) Westinghouse PWR owned by the Carolina Power & Light Company

At the start of fuel cycle 9 it was loaded with previously burned fuel in the locations on the flats of the core. In order t measure the effect of this low-leakage core arrangement, special dosimetry was introduced into the surveillance position in the downcomer annulus, and into the reactor cavity.

The extensive range of measurements carried out provided an opportunity to evaluate the Monte-Carlo code McBEND [1] for radial shield calculations on PWRs.

## THE MCBEND MODEL

### Geometry and Materials

The reactor core and shield was modelled in three dimensions using combinatorial geometry. The reactor has eight-fold symmetry so was modelled by a 45° slice, with reflective boundaries at its sides, as shown in figure 1. The full height of the reactor was modelled, although approximations were made at regions away from the core mid-height. The shield, baffles, and pressure vessel wer modelled explicitly, using exact material compositions. The core was modelled as a single region consisting of fuel ($UO_2$), fuel pin cladding (Zircaloy), fuel assembly supports and sheathing (Inconel and SST304), and boronated water smeared together. The nuclear data used was the 8220 group DICE dataset, derived from the UKNDL.

### Source

The calculation was performed for a single time interval around half way through the cycle, with a burn-up of 5,445 MWD/Te. This should provide a good estimate of the measured activities [3] The source intensity [4] was specified as two components:

(a) Relative powers for the 4116 pins in the 45° sector, calculated using PDQ7, were re-normalized to assembly averaged powers measured at 1782.9MW(t) and processed using INCORE. The pi powers were converted to source densities using 8.167E16 neutrons $s^{-1}$ $MW^{-1}$. To reduce the amount of data required, groups of 3 by 3 pins were combined into larger sources.

(b) The axial profile was assumed to be the same for all pins, an specified in 19 vertical intervals over the height of the core.

The source spectrum used was the Watt-Cranberg expression for $^{235}U$ fission, as this is built into McBEND, enabling a relatively simple source definition. Sampling of the source was weighted automatically, using the importance map.

### Scoring

Reaction rates were scored at the $0^o$ dosimetry position at the core mid-height in the reactor cavity as shown in figure 1. The scoring region was 82cm high, and covered the full width of the cavity. There was less than 10% variation in flux across this region. Reaction rates were also scored at other positions, including the surveillance capsule in the downcomer annulus. The detector responses scored were : $^{46}Ti(n,p)$, $^{54}Fe(n,p)$, $^{58}Fe(n,\gamma)$, $^{58}Ni(n,p)$, $^{59}Co(n,\gamma)$, $^{63}Cu(n,\alpha)$, $^{235}U(n,fx)$, $^{238}U(n,fx)$, $^{237}Np(n,fx)$, and $^{45}Sc(n,\gamma)$. The cross-sections used were from the International Reactor Dosimetry File 1985 [5].

### Acceleration of the Monte Carlo Calculations.

The finite difference diffusion code SNAPSH [6] was used to solve the adjoint equations in order to derive spatial and energy dependent importance functions for use in accelerating the McBEND calculation. Two SNAPSH calculations were performed: the first provided the X, Y, and energy components of the importance function; the second provided the axial component only.

The geometry was generated automatically from the McBEND input, and exactly the same material definitions were used. The adjoint source was located at the $0^o$ dosimetry position in the cavity to produce an importance map optimized for that scoring region. An adjoint source spectrum with equal flux in each energy group was used to give equal importance at all energies. The importance function was in 10 energy groups with 17 regions in the X direction, 31 in the Y direction, and 14 in the axial direction.

## MAIN MONTE CARLO CALCULATION

The code was run on 4 Transputers, and 8 million particles were started. Good statistics were only obtained for the $0^o$ dosimetry position, because the adjoint source for the importance map calculation was in this region. The results were adjusted from 1782.9MW(t) to the design full power of 2300MW(t).

## CORRECTIONS TO THE MAIN CALCULATION

The effects of of factors not included in the main calculation were identified in separate studies, and are summarised in Table 2.

These sensitivity calculations were carried out using a small source consisting only of the one and a half assemblies nearest the $0^{o}$ dosimetry position (the reflection plane simulates of another one and a half). Calculations showed that about 65% of the contribution to the responses comes from these assemblies.

**Plutonium Fission**

The main case used was the Watt-Cranberg representation of a $^{235}U$ fission spectrum incorporated in McBEND. However in H.B.Robinson the high burn-up assemblies contain substantial proportions of plutonium. The different number of neutrons emitted per fission by plutonium had already been taken into account in the derivation of the neutron source strengths.

The nuclide inventory code FISPIN [9] was used to calculate fission fractions at 22.2GWd/Te (the mean burnup over the whole core), and 27.35GWd/Te (the mean burnup in the 3 assemblies nearest the $0^{o}$ dosimetry position). As the 3 assemblies contribute 65% of the detector activations the fission rates were combined assuming 65% 27.35GWd/Te and 35% 22.2GWd/Te. The resulting fractions were: $^{235}U$: 40.2±1.1%, $^{238}U$: 8.6±0.1%, $^{239}Pu$:42.5±0.6%, $^{241}Pu$: 8.6±0.4%. As $^{238}U$ and $^{241}Pu$ make only a small contribution to the total flux they were included as $^{239}Pu$ (which has a similar fission spectrum). The source was therefore taken to be 40.2±1.1% $^{235}U$ and 59.8±1.1% $^{239}Pu$.

A case was run using a $^{239}Pu$ source spectrum, entered in 33 energy groups. This was compared with a case using a $^{235}U$ source spectrum also entered in group form, and showed that a pure plutonium source would produce responses 1.06 to 1.29 times that from a pure uranium source. These results were combined with the fission fractions to produce the corrections in Table 2.

**Burnup**

As the burnup increases during the cycle, the proportion of Plutonium will increase, and so the spectrum will change. For the assemblies nearest the $0^{o}$ dosimetry position the Pu content will increase by about 10%, which would have an effect of only 1-2% on the reaction rates. As the calculation was performed for a point near the middle of the cycle, the effect of burnup will be less than 1%, and has been ignored.

**Variation in Source Distribution with Time**

During the cycle the source distribution changes, more power being produced by the peripheral assemblies towards the cycle end. The measured results have been corrected, to the cycle-averaged power distribution. A factor of 0.95 should be applied to the calculated results to correct them to that distribution from the power distribution at the 5,445 MWD/Te of the calculation.

**Core Asymmetry**

The calculation was for one eighth of the core only, and could not take account of asymmetry in the source distribution. The $0^{o}$ dosimetry position is on the reflection plane in the McBEND model, so half the flux reaching it will be due to the eighth that was not modelled. By considering the three assemblies nearest the measurement position, a correction of 0.989 was derived.

## UNCERTAINTIES IN THE MAIN CALCULATION

The statistical uncertainties on the calculated reaction rates were better than ±5% (one standard deviation) for all reactions and close to ±1% for most. There are additional considerations which increase the overall uncertainty of the calculation, they are summarised in Table 2.

**Cross-Sections**

There are uncertainties in the materials cross-sections, and in the reaction cross-sections of the dosimeters. Of particular importance is the inelastic cross-section of iron. The effect of cross-section uncertainties on threshold reaction rates in a PWR cavity is considered in [8]; it gives the following standard deviations: $^{103}Rh(n,n')$: 9.5%, $^{115}In(n,n')$: 18.3%, $^{32}S(n,p)$: 27.5%; this provides a good estimate of the range of uncertainties expected for the threshold detectors used in H.B.Robinson. The uncertainty value for $^{32}S(n,p)$ is appropriate to the $^{46}Ti(n,p)$, $^{54}Fe(n,p)$, $^{58}Ni(n,p)$, $^{63}Cu(n,\alpha)$, and $^{238}U(n,fx)$ reactions; the value for $^{103}Rh(n,n')$ is appropriate to the other reactions.

**Dimensions and Materials**

The calculation was performed using design dimensions and materials. [2] examines the effect of varying the dimensions within the given manufacturing tolerances, and concludes that the maximum effect on the neutron fluence at the surveillance position is ±2.5%. It also examines the plant-specific atomic number densities for the steels in the reactor, and concludes that using these would have no significant effect on the neutron fluencies. [7] considers the effect of uncertainties in the coolant density and ascribes a standard deviation of approximately ±6% for flux predictions in the cavity.

**Core Smearing**

The core components (fuel, clad, supports, and water) were smeared in the model as including all fuel pins explicitly would be impractical. This may result in some inaccuracies in its effective attenuation of neutrons. As most of the flux received at the $0^{o}$ dosimetry position is from those assemblies on the edge of the core, where there has been only a short penetration through the core material, any error in the attenuation in the core is assumed to have negligible effect.

### Core Source Data

The factor used to convert pin powers to source rate densities was taken from [4]. The FISPIN calculations confirmed that factor, but suggest that an uncertainty (1$\sigma$) of ±1.3% should be applied to it.

In [2] the uncertainties on the measured power distributions from INCORE are considered: the assembly averaged powers are ascribed a standard deviation of ±2.3% and the axial distributions a standard deviation of ±1.0%. The accuracy of the PDQ7 calculated pin powers is estimated as ±1.5%.

## COMPARISON OF CALCULATION AND MEASUREMENT

The calculated reaction rates, corrected by the factors considered in above, are compared with the measurements in table 1. It can be seen that McBEND underpredicts for all reactions except $^{63}Cu(n,\alpha)$ (average C/M ratio is 0.94), but is within one standard deviation of unity for all reactions except $^{237}Np(n,fx)$.

The $^{63}Cu(n,a)$ detector, with a very small cross-section below 5.5MeV, has the highest energy response of all those included in the calculation, and its reaction rate was overpredicted by a factor of 1.22, although this is still within one standard deviation of unity. The accuracy of the correction factor for this reaction is poor as few particles of high energy that were tracked, so the anomalous result may be simply a product of the statistical uncertainty. The $^{237}Np(n,fx)$ detector reaction rate was underpredicted by a factor of 0.65, with the main contribution to the reaction rate coming from neutrons with energies between about 0.5 and 1.0 MeV This seems anomalous as reactions with both higher and lower energy responses are predicted much more accurately.

The largest source of uncertainty on the calculated response rates is due to uncertainties in the cross-section, which are estimated [8] as ±27.5% (1$\sigma$) for high threshold reactions, and 9.5% for lower energy reactions. This may be pessimistic as examination of the results of this calculation show that the standard deviations due to cross-sections may be reduced to ±15% and ±0%, for high and low energy reactions respectively, and C/M ratios for all reactions except $^{237}Np(n,fx)$ are still unity to within one standard deviation.

| Reaction | MCBEND Calculation | | Measurement | | $\frac{\text{Calculation}}{\text{Measurement}}$ |
|---|---|---|---|---|---|
| | (dps/atom) | % $\sigma$ | (dps/atom) | % $\sigma$ | $1\sigma$ |
| $^{46}Ti(n,p)$ | 5.76E-18 | 30.2 | 6.21E-18 | 10.0 | 0.93 ±0.32 |
| $^{54}Fe(n,p)$ | 3.31E-17 | 29.1 | 3.59E-17 | 5.0 | 0.92 ±0.30 |
| $^{58}Fe(n,\gamma)$ | 2.63E-15 | 12.4 | 2.68E-15 | 8.0 | 0.98 ±0.15 |
| $^{58}Ni(n,p)$ | 5.13E-17 | 28.9 | 5.44E-17 | 10.0 | 0.94 ±0.31 |
| $^{59}Co(n,\gamma)$ | 1.18E-13 | 12.9 | 1.20E-13 | 5.0 | 0.99 ±0.14 |
| $^{63}Cu(n,\alpha)$ | 4.81E-19 | 31.8 | 3.95E-19 | 5.0 | 1.22 ±0.32 |
| $^{235}U(n,fx)$ | 1.07E-12 | 12.4 | 1.10E-12 | 5.0 | 0.98 ±0.13 |
| $^{238}U(n,fx)$ | 2.38E-16 | 28.6 | 2.90E-16 | 5.0 | 0.82 ±0.29 |
| $^{237}Np(n,fx)$ | 4.59E-15 | 12.0 | 7.06E-15 | 5.0 | 0.65 ±0.13 |
| $^{45}Sc(n,\gamma)$ | 5.01E-14 | 12.5 | 5.41E-14 | 10.0 | 0.93 ±0.16 |

**Table 1: Comparison of Calculated and Measured Responses**

## REFERENCES


1. McBEND User Guide to Version 6, ANSWERS, Winfrith Technology Centre, 1990

2. Anderson S. L. Summary of H.B.Robinson PTS Analysis, RSAC-CPL-170, Westinghouse Nuclear Technology Division, 1983.

3. Lippincott E.P. et al, Evaluation of Surveillance Capsule and Reactor Cavity Dosimetry from H.B.Robinson Unit 2 Cycle 9, WCAP-11104, Carolina Power & Light Company, NUREG/CR-4576, U.S. Nuclear Regulation Commission, 1987.

4. Maerker R.E., Oak Ridge National Laboratory, Private communication, 1987.

5. Cullen D.E. & McLaughlin P.K., The International Reactor Dosimetry File (IRDF-85). IAEA-NDS-41, International Atomic Energy Agency, 1985.

6 SNAPSH User Guide to Version 2, ANSWERS, Winfrith Technology Centre, 1990

7. Maerker R.E., Application of the LEPRICON Methodology to LWR Pressure Vessel Surveillance Dosimetry. NRC FIN No. B0415. Oak Ridge National Laboratory, 1986.

8. McCracken A.K., Winfrith Technology Centre, Private Communication, 1985.

9. Burstall R.F., FISPIN - A Computer Code for Nuclide Inventory Calculations ND-R-328(R), Risley Technology Centre, 1979.

| Reason for correction or uncertainty | Correction factor | Uncertainty (1σ) |
|---|---|---|
| **Corrections** | | |
| Plutonium Fission | | |
| for $^{46}$Ti(n,p) | 1.11 | ±0.10 |
| for $^{54}$Fe(n,p) | 1.07 | ±0.06 |
| for $^{58}$Fe(n,γ) | 1.03 | ±0.03 |
| for $^{58}$Ni(n,p) | 1.07 | ±0.05 |
| for $^{59}$Co(n,γ) | 1.04 | ±0.05 |
| for $^{63}$Cu(n,α) | 1.18 | ±0.14 |
| for $^{235}$U(n,fx) | 1.04 | ±0.03 |
| for $^{238}$U(n,fx) | 1.06 | ±0.03 |
| for $^{237}$Np(n,fx) | 1.05 | ±0.01 |
| for $^{45}$Sc(n,γ) | 1.03 | ±0.03 |
| Source distribution | 0.95 | |
| Core Asymmetry | 0.99 | |
| **Uncertainties** | | |
| Cross-sections | | |
| for threshold reactions | | ±27.5% |
| for others | | ±9.5% |
| Dimensions | | ±2.5% |
| Coolant density | | ±6.0% |
| Source Intensity | | ±1.3% |
| Core Source data | | ±2.9% |
| **Total** | | |
| for $^{46}$Ti(n,p) | 1.04 | ±0.31 |
| for $^{54}$Fe(n,p) | 1.00 | ±0.29 |
| for $^{58}$Fe(n,γ) | 0.97 | ±0.12 |
| for $^{58}$Ni(n,p) | 1.00 | ±0.29 |
| for $^{59}$Co(n,γ) | 0.98 | ±0.12 |
| for $^{63}$Cu(n,α) | 1.10 | ±0.34 |
| for $^{235}$U(n,fx) | 0.97 | ±0.12 |
| for $^{238}$U(n,fx) | 1.00 | ±0.29 |
| for $^{237}$Np(n,fx) | 0.99 | ±0.12 |
| for $^{45}$Sc(n,γ) | 0.97 | ±0.12 |

**Table 2: Correction Factors and Uncertainties**

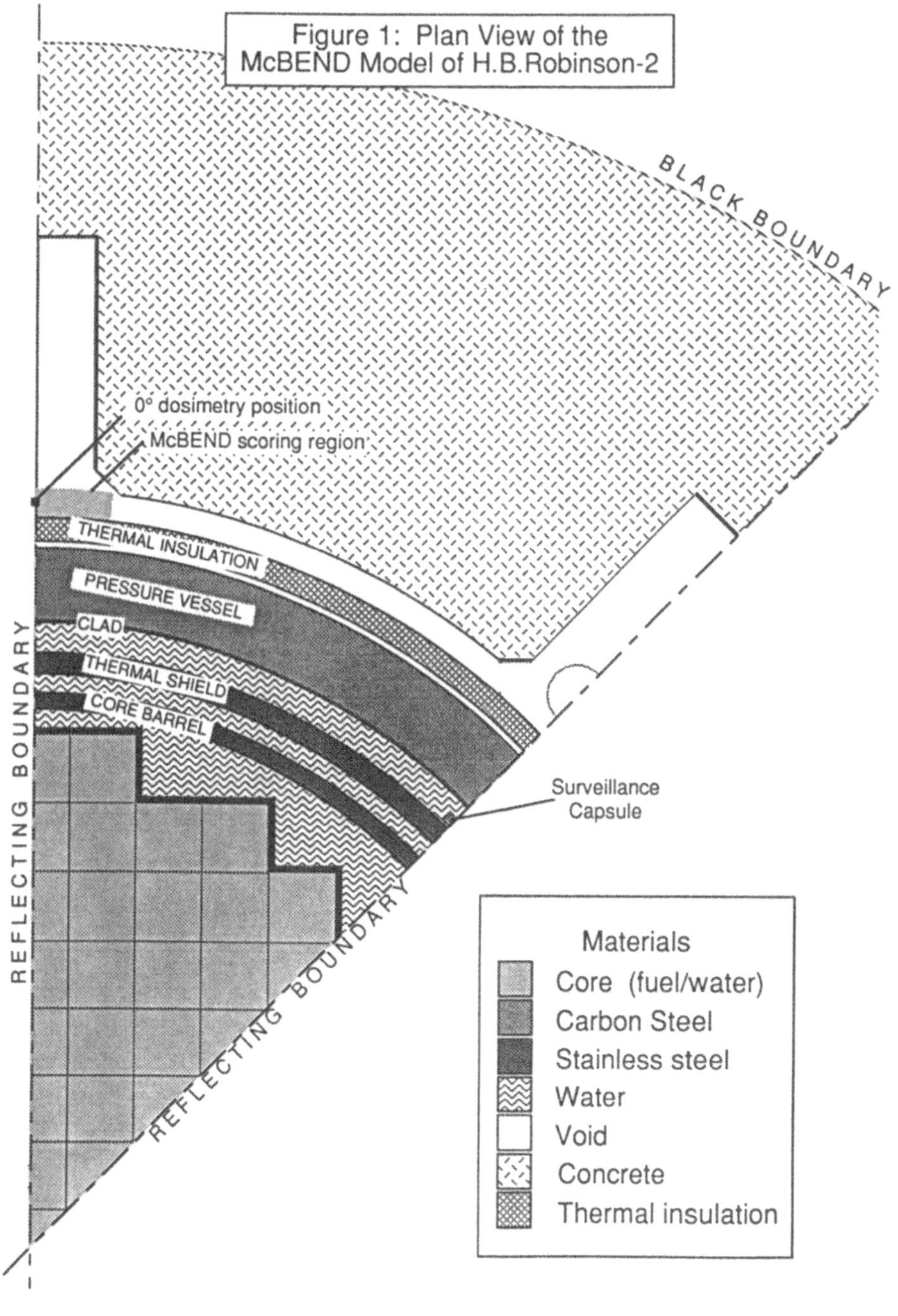


Figure 1: Plan View of the McBEND Model of H.B.Robinson-2

# ANALYSIS OF PRESSURE VESSEL CAVITY AND SURVEILLANCE CAPSULE DOSIMETRY FROM A TWO LOOP PWR

**I.Remec and M.Najžer**

**"J.Stefan" Institute, Ljubljana**

**Slovenia, Yugoslavia**

**Abstract**

At Krško Nuclear Power Plant cavity dosimetry experiment was performed. Measured activities from the cavity and simultaneously withdrawn capsule were used to adjust calculated neutron spectrum and determine neutron exposure parameters. Obtained good consistency of results confirmed the validity of cavity dosimetry. For several cycles, fuel loading schemes designed to reduce the maximum pressure vessel flux, were used. Analysis, indicating that approximately 25% reduction was obtained, is presented.

# Introduction

This paper describes the dosimetry and neutron transport calculation part of the continuation and improvement of the Pressure Vessel Surveillance Program at the Krško Nuclear Power Plant (NPP). A companion paper [1] discusses the results of surveillance capsule testing.

Krško NPP is a 632 MWe Westinghouse pressurized water reactor, which started operation in 1981. In the analysis of the first surveillance capsule, it was found that pronounced peak exists in the fast flux distribution at the pressure vessel (PV) inner wall. This peak gave rise to considerable projected end-of-life fluence of app. $6\,10^{19}\ n\,cm^{-2}$. Additional considerable increase above the given value was expected in the following cycles due to higher enrichment of reload fuel batches which when peripheral fuel loading scheme is used results in increased PV fast flux. For this reason the possibility to reduce PV maximum exposure was studied [2]. It was found that if for example once or twice burned fuel elements are placed on core flat edge (see Fig. 1) instead of fresh ones, app. 30% reduction of maximum PV flux can be obtained. Loading schemes meeting this requirement represent subclass of low leakage loading patterns. They were used at Krško NPP for the $5^{th}$ and subsequent fuel cycles.

To verify the effect of loading pattern modification cavity dosimetry was installed at the beginning of the $7^{th}$ cycle. It was withdrawn during the refueling one year later in the fall of 1989, simultaneously with the second surveillance capsule. In the following analysis of the fuel pattern modification effect and dosimetry from both locations is presented.

# Calculations

Two dimensional DOT-4.2 [3] and one dimensional ANISN [4] discrete ordinates transport codes were used for neutron flux calculations. Cross sections were taken from the SAILOR library [5]. 47 energy groups (with 19 groups above 1MeV ), $P_3$ expansion of scattering cross sections and $S_8$ angular quadrature was used throughout. Part of the reactor horizontal cross section shown in Fig. 1 was modelled in $R - \theta$ geometry using DOT 4 variable mesh with 84 to 97 radial and 60 azimuthal intervals. Axial model for R-Z calculations covered the height from app. 1m below to app. 2m above the core with 67 intervals, while 86 intervals in radial mesh described the geometry at $0^o$ azimuth. The same radial mesh was used for 1 dimensional radial calculations.

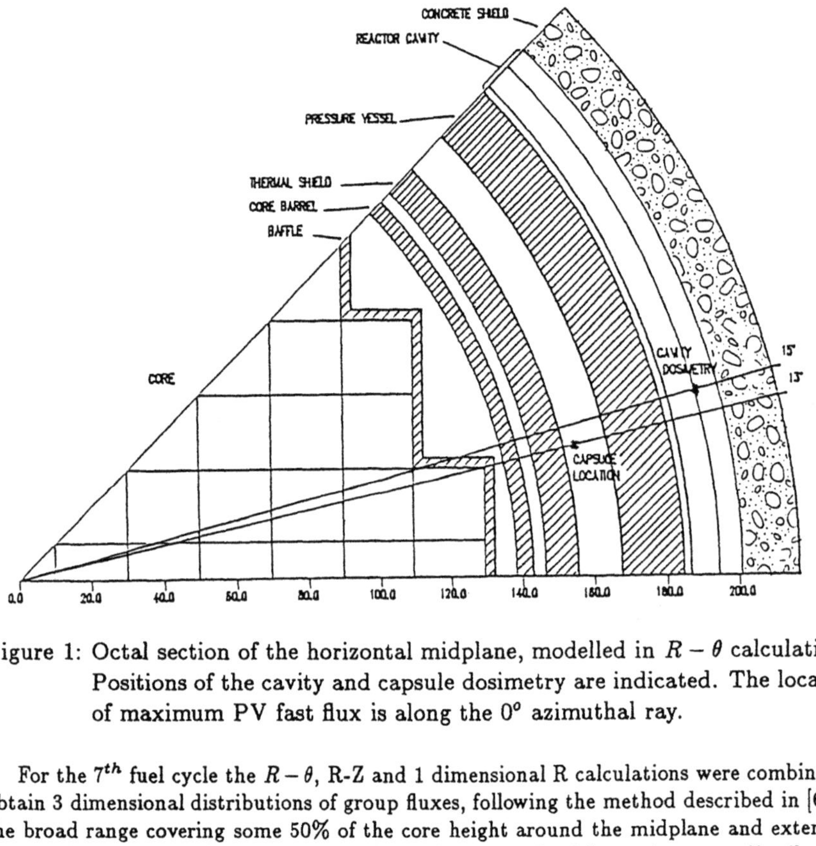


Figure 1: Octal section of the horizontal midplane, modelled in $R - \theta$ calculations. Positions of the cavity and capsule dosimetry are indicated. The location of maximum PV fast flux is along the $0^o$ azimuthal ray.

For the $7^{th}$ fuel cycle the $R-\theta$, R-Z and 1 dimensional R calculations were combined to obtain 3 dimensional distributions of group fluxes, following the method described in [6]. In the broad range covering some 50% of the core height around the midplane and extending radially beyond the PV outer wall the obtained shapes of axial reaction rates distributions were found to coincide closely with axial distribution of the power in the core (as used in R-Z calculation). Since surveillance capsule and maximum PV flux lay within this axial limits axial variations for the other cycles were approximated with the axial power distribution in the core. Measured axial distributions, obtained from the plant, were used. For the $R - \theta$ calculations the neutron source was prepared from the two dimensional power distributions obtained from core design diffusion calculations.

To take into account time variations of neutron flux for each of the 7 fuel cycles 2 direct transport calculations were done: one for the beginning and one for the end of cycle. Typically, the beginning to end of cycle differences for the threshold reactions, $\phi > 1MeV$ and displacement per atom rate (dpa/s) were below 10%. Linear variation of reaction rates with burnup was then assumed during the cycle. To account for quick power changes, it was assumed that flux is proportional to the core power and power variations obtained from the plant daily operation history were superimposed on the above described gradual changes due to fuel depletion.

## Surveillance capsule and cavity dosimetry

The surveillance capsule removed at the end of cycle 7 was standard Westinghouse surveillance capsule, which was attached at the thermal shield at the azimuth of $13^o$ (see Fig. 1).

Cavity dosimetry experiment consisted of three sets of radiometric monitors, mounted on the stainless steel wire. Each set included Fe,Ni,Cu,Ti,bare Co, Cd shielded Co and $^{238}U$ or $^{232}Th$ fission monitor. Monitors were placed in small cylindrical holders fixed successively one to another. Medium set was attached to the wire at the height of the core midplane with top and bottom sets 1m above and 1m below, respectively. The experiment was lowered into the cavity from the bottom of the reactor pit and fixed at the top. After the irradiation the activities were measured with high resolution Ge-Li detector/computer system.

## Results and discussion

Comparison of calculated cycle averaged maximum PV fast fluxes is given in Table 1. It can be seen that modified loading schemes resulted in app. 25% reduction in maximum PV flux, from the average value of $3.6\ 10^{10}\ n\ cm^{-2}s^{-1}$ in the first 4 cycles to the average of $2.7\,10^{10}\ ncm^{-2}s^{-1}$ in the last 3 cycles. Figure 2 shows that greatest reduction was obtained at the peak position as it was required. Location of the maximum remained unchanged, at $0^o$ azimuth and app. 40 cm below the core midplane. In the last row of Table 1 the ratios of cycle averaged core midplane power densities of the outermost three fuel elements on the core flat edge are given. Obviously these ratios, which can be easily calculated from data in standard core design report, give good estimates of cycle to cycle variations of maximum PV flux.

Calculated and measured activities at the end of irradiation are given in Table 2. In general the agreement is good, thus confirming the validity of calculations and measurements. For the downcomer location the agreement is to within 10%. Calculated values with the exception of Np, are somewhat lower than measured. In the cavity calculated activities exceed the measured by app. 16% on the average. Calculated to measured ratios at the off midplane locations do not differ significantly from values at the midplane. This indicates that in the axial region covered by the measurements, neutron streaming in the cavity is not important and the symmetric $S_8$ quadrature, used in R-Z calculation, is adequate. However all measurements were still done relatively close to the core midplane where flatness of power axial distribution probably greatly reduces possible streaming effects.

Table 1: Comparison of Maximum $\phi > 1\,MeV$ in the Pressure Vessel, at the 1/4 Thickness Position. Cycle Averaged Values are Shown.

| | Cy.1 | Cy.2 | Cy.3 | Cy.4 | Cy.5 | Cy.6 | Cy.7 |
|---|---|---|---|---|---|---|---|
| $\phi > 1\,MeV^{a}$ | 3.54 | 3.67 | 3.65 | 3.43 | 2.68 | 2.77 | 2.71 |
| $\phi\,/\,\phi_{cy1}{}^{b}$ | 1.00 | 1.04 | 1.03 | 0.97 | 0.76 | 0.78 | 0.77 |
| $P/P_{cy1}{}^{c}$ | 1.00 | 1.03 | 1.03 | 0.97 | 0.74 | 0.75 | 0.73 |

[a] Fast flux in $10^{10} n\,cm^{-2}\,s^{-1}$.
[b] Fast flux in the cycle divided by cycle 1 fast flux.
[c] Cycle averaged core midplane power density for the outermost three fuel elements on the core flat adge at $\theta = 0^{o}$, divided by the corresponding value for the cycle 1.

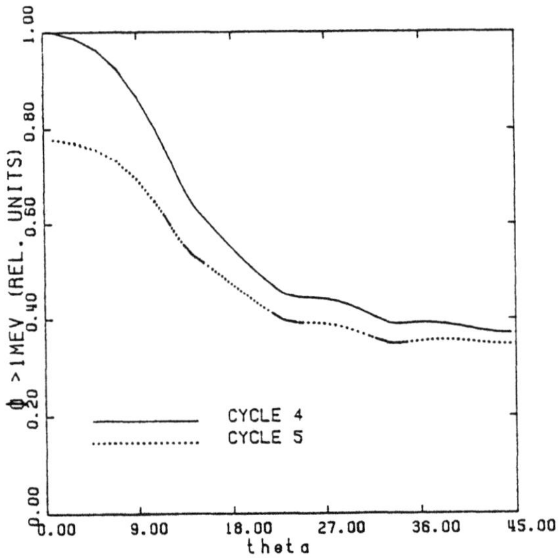


Figure 2: Azimuthal distribution of $\phi > 1MeV$ for the cycle 4 (out-in-in) and cycle 5 (modified fuel loading pattern).

Table 2: Comparison of calculated and measured end-of-irradiation activities.

| Reaction | H[b] | Activity[a] Calculated dis $s^{-1}$ $atom^{-1}$ | Measured | C/M |
|---|---|---|---|---|
| Cavity location | | | | |
| $^{54}Fe(n,p)\ ^{54}Mn$ | 106 | $4.69\ 10^{-17} \pm 20$[c] | $4.32\ 10^{-17} \pm 4$ | $1.09 \pm 0.22$[d] |
| $^{232}Th(n,f)\ ^{137}Cs$ | 100 | $1.48\ 10^{-19} \pm 22$ | $1.16\ 10^{-19} \pm 23$ | $1.27 \pm 0.40$ |
| $^{58}Ni(n,p)\ ^{58}Co$ | 94 | $1.47\ 10^{-16} \pm 21$ | $1.24\ 10^{-16} \pm 5$ | $1.19 \pm 0.26$ |
| $^{46}Ti(n,p)\ ^{46}Sc$ | 88 | $1.59\ 10^{-17} \pm 23$ | $1.38\ 10^{-17} \pm 5$ | $1.16 \pm 0.27$ |
| $^{63}Cu(n,\alpha)\ ^{60}Co$ | 82 | $1.32\ 10^{-19} \pm 21$ | $1.21\ 10^{-19} \pm 5$ | $1.09 \pm 0.24$ |
| $^{54}Fe(n,p)\ ^{54}Mn$ | 6 | $5.33\ 10^{-17} \pm 20$ | $4.57\ 10^{-17} \pm 4$ | $1.16 \pm 0.24$ |
| $^{238}U(n,f)\ ^{137}Cs$ | 0 | $7.79\ 10^{-19} \pm 21$ | $6.52\ 10^{-19} \pm 18$ | $1.19 \pm 0.33$ |
| $^{58}Ni(n,p)\ ^{58}Co$ | -6 | $1.61\ 10^{-16} \pm 21$ | $1.33\ 10^{-16} \pm 5$ | $1.21 \pm 0.26$ |
| $^{46}Ti(n,p)\ ^{46}Sc$ | -12 | $1.72\ 10^{-17} \pm 23$ | $1.54\ 10^{-17} \pm 5$ | $1.11 \pm 0.26$ |
| $^{63}Cu(n,\alpha)\ ^{60}Co$ | -18 | $1.41\ 10^{-19} \pm 21$ | $1.30\ 10^{-19} \pm 5$ | $1.08 \pm 0.23$ |
| $^{63}Cu(n,\alpha)\ ^{60}Co$ | -82 | $1.39\ 10^{-19} \pm 21$ | $1.26\ 10^{-19} \pm 5$ | $1.10 \pm 0.24$ |
| $^{46}Ti(n,p)\ ^{46}Sc$ | -88 | $1.69\ 10^{-17} \pm 23$ | $1.56\ 10^{-17} \pm 5$ | $1.08 \pm 0.25$ |
| $^{58}Ni(n,p)\ ^{58}Co$ | -94 | $1.56\ 10^{-16} \pm 21$ | $1.34\ 10^{-16} \pm 5$ | $1.16 \pm 0.25$ |
| $^{232}Th(n,f)\ ^{137}Cs$ | -100 | $1.59\ 10^{-19} \pm 22$ | $1.21\ 10^{-19} \pm 22$ | $1.31 \pm 0.41$ |
| $^{54}Fe(n,p)\ ^{54}Mn$ | -106 | $5.09\ 10^{-17} \pm 20$ | $4.37\ 10^{-17} \pm 5$ | $1.16 \pm 0.24$ |
| Downcomer location | | | | |
| $^{63}Cu(n,\alpha)\ ^{60}Co$ | | $3.80\ 10^{-17} \pm 21$ | $4.15\ 10^{-17} \pm 7$ | $0.92 \pm 0.20$ |
| $^{54}Fe(n,p)\ ^{54}Mn$ | | $6.48\ 10^{-15} \pm 20$ | $6.57\ 10^{-15} \pm 4$ | $0.99 \pm 0.20$ |
| $^{58}Ni(n,p)\ ^{58}Co$ | | $1.02\ 10^{-14} \pm 21$ | $1.05\ 10^{-14} \pm 7$ | $0.97 \pm 0.21$ |
| $^{237}Np(n,f)\ ^{137}Cs$ | | $1.41\ 10^{-15} \pm 24$ | $1.32\ 10^{-15} \pm 8$ | $1.07 \pm 0.27$ |
| $^{59}Co(n,\gamma)\ ^{60}Co$ | | $3.77\ 10^{-12} \pm 40$ | $3.79\ 10^{-12} \pm 9$ | $1.00 \pm 0.41$ |
| $^{59}Co(n,\gamma)\ ^{60}Co(Cd)$ | | $2.05\ 10^{-12} \pm 40$ | $2.15\ 10^{-12} \pm 11$ | $0.95 \pm 0.39$ |

[a] In disintegrations per second per target atom.
[b] Distance above core midplane in cm.
[c] Read as activity 4.69 $10^{-17}$ with standard deviation of 20 %.
[d] Read as calculated to measured activity ratio 1.09 with standard deviation of 0.22.

Uncertainties given in Table 2 for the calculated activities take into account cross sections and group fluxes uncertainties. Besides fission yield errors were added for fission detectors, and 5% error was joined to account for uncertainties in reaction-rate variations with time. For the Th foils the C/M ratio is considerably above the average. This probably indicates that the photofission correction of 20% applied to the measured activities, was too high. Since gamma field wasn't calculated, this correction was estimated from previous calculations and other data [7]. However, 20% uncertainty was conservatively attributed to the correction, resulting in high Th activity uncertainty. Similarly, photofission and correction for $^{235}U$ content contributed to high uncertainty quoted for $^{238}U$ foil. Other contributions to the measured activity errors taken into account are: $\gamma$-ray counter efficiency calibration uncertainty, random counting error, $\gamma$-ray attenuation correction error (for the cavity Ni and Cu dosimeters, and capsule Np dosimeter) and coincidence correction error ( for detectors where $^{60}Co$ activity is measured).

# Adjustment results

For the unfolding calculations LSL-M2 [8] least-squares logarithmic neutron spectra adjustment computer code was used. Activation cross sections and covariances were taken from [9].

Measured activities were converted to reaction rates at the end of irradiation (end of cycle 7) with the use of calculated conversion coefficients. Also, capsule activities, irradiated at different positions, were extrapolated to the center of the capsule using calculated gradients. Correlation matrix of so obtained "measured" reaction rates is shown in Table 3, together with their standard deviations.

Since complete uncertainty analysis was not available, the group fluxes autocorrelation matrix ( correlation matrix of group fluxes at the same location) was generated following the suggestions in ASTM E944-83 [10] and [11]. Number of groups was reduced to 33 and autocorrelation of the form:

$$C_{ij} = e^{-|E_i - E_j|/31.5\,MeV}$$

where $E_i$ and $E_j$ denote mean energy of the group i and j, was taken. Standard deviation of 20% and 40% was attributed to the group fluxes with energy above 0.1 MeV and below 0.1 MeV, respectively. For the correlation of group fluxes at any two different locations, for example between surveillance position and PV quarter thickness, autocorrelations given above, multiplied by cross-correlation factor of 0.8 were assumed. Series of calculations, performed for different autocorrelation matrices generated or taken from the literature [12], [13], showed, that adjusted exposure parameters are quite insensitive to the detailed matrix shape, and indicated that the above assumption is probably satisfactory. However, adjustments at the locations without dosimetry strongly depend on the cross-correlation factor. Therefore more realistic estimates are desired to improve the reliability and accuracy of adjustment results.

Table 3: Correlation Coefficients and Standard Deviations of Measured Reaction Rates.[a]

| | σ % | $^{238}U$ | $^{58}Ni$ | $^{54}Fe$ | $^{46}Ti$ | $^{63}Cu$ | $^{237}Np$ | $^{58}Ni$ | $^{54}Fe$ | $^{59}Co$ | $^{59}Co^b$ | $^{63}Cu$ |
|---|---|---|---|---|---|---|---|---|---|---|---|---|
| Cavity | | | | | | | | | | | | |
| $^{238}U$ | 19 | 100 | | | | | | | | | | |
| $^{58}Ni$ | 7 | 23 | 100 | | | | | | | | | |
| $^{54}Fe$ | 6 | 24 | 84 | 100 | | | | | | | | |
| $^{46}Ti$ | 7 | 21 | 70 | 80 | 100 | | | | | | | |
| $^{63}Cu$ | 7 | 22 | 60 | 67 | 57 | 100 | | | | | | |
| *Downcomer* | | | | | | | | | | | | |
| $^{237}Np$ | 11 | 16 | 35 | 37 | 31 | 33 | 100 | | | | | |
| $^{58}Ni$ | 9 | 16 | 57 | 60 | 49 | 42 | 28 | 100 | | | | |
| $^{54}Fe$ | 6 | 21 | 74 | 86 | 70 | 57 | 36 | 67 | 100 | | | |
| $^{59}Co$ | 11 | 12 | 34 | 38 | 32 | 56 | 21 | 32 | 43 | 100 | | |
| $^{59}Co^b$ | 12 | 11 | 30 | 33 | 28 | 49 | 19 | 28 | 38 | 35 | 100 | |
| $^{63}Cu$ | 9 | 15 | 42 | 47 | 40 | 69 | 27 | 39 | 53 | 49 | 43 | 100 |

[a]Matrix is symmetric,the lower half is given. Coefficients are multiplied by 100.
[b]Co is cadmium shielded.

Table 4: Flux and dpa Rate Adjustments and Standard Deviations at Downcomer, Cavity and 1/4 Pressure Vessel Thickness as Functions of Measured Reaction Rates used in Unfolding.

| Location | | Before Adjustment | Downcomer Dosimetry | Cavity Dosimetry | Downcomer + Cavity |
|---|---|---|---|---|---|
| downcomer | $\phi > 1MeV$ | $0 \pm 20$[a] | $1 \pm 6$ | $-11 \pm 13$ | $1 \pm 6$ |
| | $\phi > 0.1MeV$ | $0 \pm 20$ | $1 \pm 5$ | $-11 \pm 14$ | $1 \pm 5$ |
| | $dpa/s$ | $0 \pm 20$ | $1 \pm 5$ | $-11 \pm 14$ | $1 \pm 5$ |
| cavity | $\phi > 1MeV$ | $0 \pm 20$ | $1 \pm 13$ | $-14 \pm 8$ | $-14 \pm 7$ |
| | $\phi > 0.1MeV$ | $0 \pm 20$ | $1 \pm 13$ | $-14 \pm 10$ | $-14 \pm 7$ |
| | $dpa/s$ | $0 \pm 20$ | $1 \pm 13$ | $-14 \pm 9$ | $-14 \pm 7$ |
| 1/4 Pressure Vessel | $\phi > 1MeV$ | $0 \pm 20$ | $1 \pm 13$ | $-11 \pm 13$ | $-6 \pm 12$ |
| | $\phi > 0.1MeV$ | $0 \pm 20$ | $1 \pm 13$ | $-11 \pm 14$ | $-6 \pm 12$ |
| | $dpa/s$ | $0 \pm 20$ | $1 \pm 13$ | $-11 \pm 14$ | $-6 \pm 12$ |

[a]Read as adjustment of 0% with standard deviation of 20%.

# SPECIAL MEASUREMENTS IN PWR : SAINT LAURENT B1 POWER PLANT

R. LLORET*, A. BEVILACQUA*, F. HEDIN**, C. RIEG**

* CEA-DRN-DRE-SRS-SAPR - Centre d'Etude Nucléaires
85 X - 38041 GRENOBLE Cédex - FRANCE

** Electricité de France-SEPTEN-Division Réacteur
12,14 avenue Dutrievoz - 69628 VILLEURBANNE FRANCE

ABSTRACT

The present trend in the operation of PWR 900 MW plants is characterized in particular by refuelling with mixed fuel elements. Special studies are being carried out in France to ascertain the consequences this entails on the irradiation of the vessel and surveillance capsules. More generally, to reduce uncertainties in routine surveillance dosimetry, other investigations have been performed, especially studies of the disturbances on dosimeter reaction rates, on fluxes and damage, caused within the volume of the surveillance capsules by certain internal structures.

To validate these studies, a specially instrumented capsule will equip the Saint Laurent B1 plant unit. This plant unit was chosen as typical of a MOX fuel at equilibrium. The equipment irradiated over the 1991 cycle, should furnish details of local flux disturbances in the capsule due to neighbouring metal structures, determine maximum operating temperatures, and confirm the neutron spectrum calculated in the case of mixed fuel. It will also provide data for the study of correlations between the dosimetry in the surveillance capsule and in the cavity, along the generatrices for three characteristic azimuths where the neutron parameters have been calculated.

The experimental programme is described. The results will be available in 1992.

## DEFINITION OF THE SCOPE OF INVESTIGATION AND AIMS

The present trend in the operation of PWR 900 MW power plants is characterized in particular by refuelling with mixed fuel including MOX fuel elements.

The routine surveillance dosimetry programme for the PWR reactors is based on the exclusive use of flux integrators (activation and fissile dosimeters) that obviously cannot serve to measure evolutions of the neutron spectrum during capsule irradiation. Consequently a special measurement was planned in the St Laurent B1 plant unit (1) during the 4th operating cycle with mixed refuelling (1990-1991)*. The plant unit was selected as being typical of a MOX refuelling that had reached a state of equilibrium and also because of the availability of a U capsule (4 years irradiation in an "all-uranium" core) with its measurement results that could be compared with those of the special measurement.

Given this main objective linked with MOX refuelling we took advantage of this situation to set a number of other aims, namely :

- an investigation of the consequences of MOX refuelling on irradiation of the reactor vessel and the surveillance capsules;
- the measurement of disturbances with respect to dosimeter reaction rates, fluxes and damage caused in the volume of the capsules by certain internal structures (at present 3-D computations are the only means available to determine these effects), which can be included in the general objective of reducing uncertainties in the routine surveillance dosimetry;
- the search for correlations between dosimetry in the internal capsule and the cavity while using a second measured point in order to adjust the radial distributions of the calculated fluxes;
- the measurement of axial and azimuthal flux distributions in the cavity by means of measurement systems placed appropriately in 3 characteristic positions;
- the determination of the irradiation temperature of metallurgical specimens with improved accuracy;
- an attempt to reduce overall uncertainties.

If the experimental programme runs according to plan, results should be available by mid 1992.

## EXPERIMENTAL RIGS - LAY-OUT IN THE REACTOR

For the whole of the special measurements, it is planned to use 2 types of instrumentation that will be installed in the St Laurent B1 reactor throughout the cycle (1990-1991).

---

*The irradiation was initially planned to take place during the 3rd MOX cycle (1989-1990) but had to be postponed for one year due to problems of operability between cycles.

## Internal instrumentation

This comprises a standard surveillance capsule with a filling specially designed to satisfy the set objectives. It will be installed in the location of the U capsule ($\theta$ = 20°, vacant since the programmed widhdrawal of this capsule) during dismantling of the internal eqquipments on stand at the end of 1990. There are 23 measurement units distributed over the whole of the capsule from axial level PM* - 81.6 to level PM* + 50.7 cm.

*PM = Core Mid- Plane

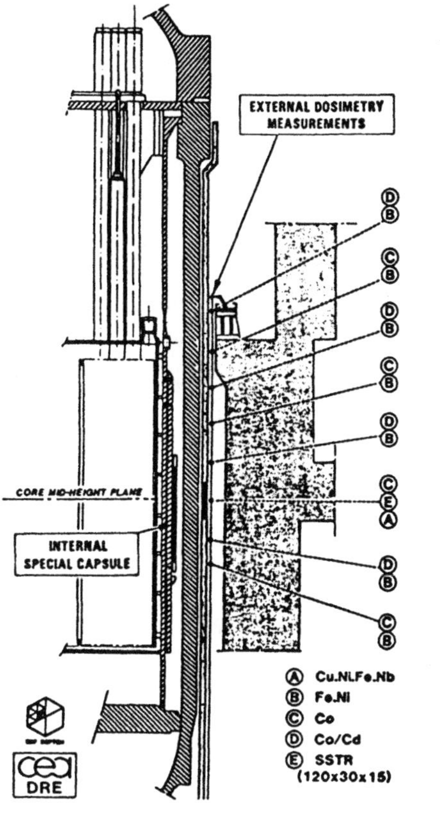


Fig.1 : Special dosimetry in a 900 MW PWR.
Axial cross-section. Instrumentation Lay-out

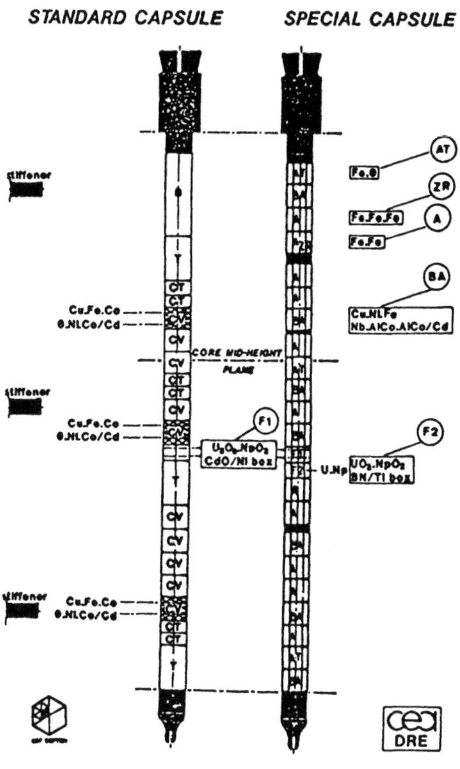


Fig.2 : Filling of the capsules

## External Instrumentation

- Lay-out in the reactor

It comprises 3 identical measurement systems with azimuthal positions (Fig.3) selected in order to obtain :

. a measurement of minimum flux at $\underline{\theta = 50°}$
. a measurement of maximum flux at $\underline{\theta = 180°}$
. an external measurement point at $\underline{\theta = 70°}$ (special internal capsule equivalent azimuthal position).

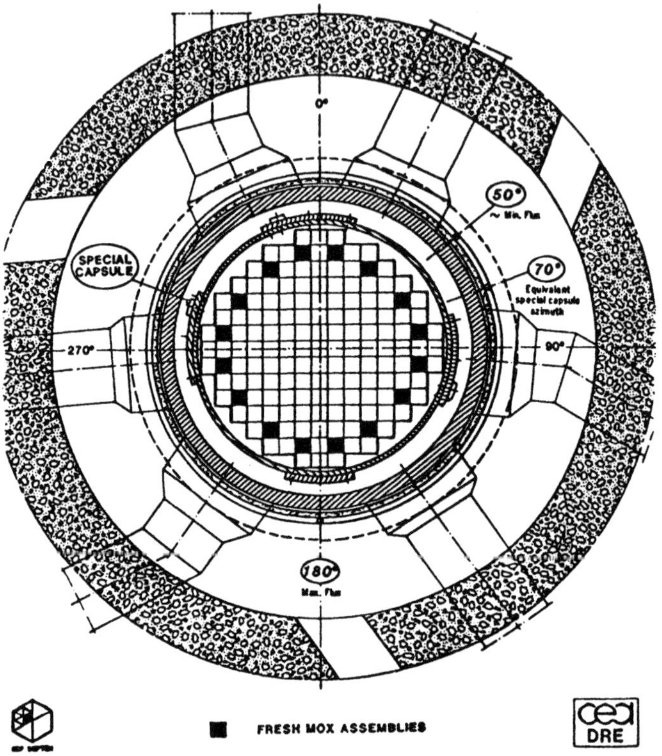


**Fig.3 : Special dosimetry in a 900 MW PWR.**
**Azimuthal lay-out of the external instrumentation**

Each of the 3 measurement systems has 8 measurement points (Fig. 1 regularly distributed over a height of 333.5 cm between axial levels PM*-85 cm and PM*+250 cm. Consequently measurements in the cavity cove an area situated between the axial levels of the weld of C2-C1 shell-plates and 80 cm above the top of the core.

- <u>Description of the experimental rigs</u>

The rigs comprise 3 main components :

· <u>Measurement rack</u>
The dosimeter-holder boxes are fixed onto 2 parallel cables 100 mm apart, held taut (50 daN tension) along the generatrix of the reactor vessel at about 2 cm from the jacketing. The dosimeter-holders are fitted with a polypropylene damper to eliminate any noise due to vibrations of the rack during reactor operation;

. upper support (fixed on the vessel support ring)
. lower support (fixed on the footbridge of the vessel hole)

## DESCRIPTION OF DOSIMETER FILLING

### Special internal capsule

The filling selected is described below with the details of the objectives explaining this choice.

The drawing shows a comparison of the filling of a standard capsule against that of a special capsule (Fig. 2).

. There are 23 measurement points over the whole height, obtained by means of 23 standard 900 MW PWR dosimeter units having 2 to 6 housings for dosimeters depending on the type of unit. As in a standard capsule, these units are located along the axis, the rest of the volume being filled with dummy CHARPY test-specimens and steel shims of the same grades as those in the routine surveillance programme.

- Dosimeters

Each unit is equipped depending on its type (Fig. 2) with dosimeters as described in tables 2 and 3 page 6.

- Temperature monitors

Table 1 describes the temperature indicators selected.

| Temperature monitors | Melting Point °C |
|---|---|
| Pb-Ag 1,7 - Sb 6 | 263 |
| Pb-Ag 2 - Sb 4 | 280 |
| Pb-Ag 2 - Sb 3 | 288 |
| Pb Ag 2 - Sb 2 | 293 |
| Pb-Ag 2,5 | 304 |

Table 1

- Special Objectives

Measurement of the effect of MOX refuelling on the neutron spectra in the capsule : obtained by comparing the 3 BA and F1 units with the corresponding units (in the same axial position) of the U capsule (withdrawn in August 1986).

Measurement of the variations of neutron spectrum depending on the different positions of the BA units in relation to the stiffeners of the core barrel (adjustment of calculated spectrum).
Comparison of F1 fissile materials used in the standard programme and the F2 fissile materials certified by EURATOM.

| Dosimeter | Reaction | Dimensions (mm) dia. | Dimensions (mm) length | Remarks : |
|---|---|---|---|---|
| Iron | $Fe^{54}$ (n,p) | 1 | 10 | Dosimeters supplied by CERCA (Romans) equipment used in surveillance capsules PWR 900 standardized |
| Nickel | $Ni^{58}$ (n,p) | 1.3 | | |
| Copper | $Cu^{63}$ (n,α) | 0.81 | 10 | |
| Aluminium-Cobalt 0.474 % Co (bare) | $Co^{59}$ (n,γ) | 1 | 10 | |
| Aluminium-Cobalt 0.474 % Co (Cd cover) | " | 1 | 10 | |
| Cadmium cladding | | 3.2 | 13 | |
| Niobium (quartz clad) | $Nb^{93}$ (n,n') | 2.1 | 10 | Dosimeter supplied by (CERCA (Romans) Originating from CEA/IRDI/DERPE/DPER /SIDIN |

Table 2 : Activation Dosimeters

| Dosimeter | Reaction | Material-form | Mass mg | Origin Remarks | Encapsulage |
|---|---|---|---|---|---|
| $Np\ O_2$ | $^{237}Np$(n,F) | Mixture MgO + $Np\ O_2$ | ≈ 20 | Origin CEA Supplied CERCA | Double(*) Nickel casing (CERCA) |
| $U_3O_8$ | $^{238}U$(n,F) | $U_3O_8$ powder | ≈ 20 | - Supplied CERCA | |
| $Np\ O_2$ | $^{237}Np$(n,F) | Spheres Ø 0.8 mm | ≈ 20 | Supplied CEA IRDI/DERPE/DPER /SIDIN | Single(*) Titane casing (CERCA) |
| $UO_2$ | $^{238}U$(n,F) | Spheres Ø 0.5 mm ou Ø 1 mm | ≈ 20 | Origin CBNM | |

(*) High purity specified.

Table 3 : Fissile Dosimeters

Monitoring over the whole capsule of the fast flux axial distribution by means of the Fe dosimeter (42 points over 1324 mm).

Measurement by Fe dosimeters of the azimuthal variation (3 Fe over 3 cm) and the radial attenuation of the fast flux (2 Fe over 1.5 cm) in two typical positions in relation to the stiffeners.

Measurement of temperatures in 3 characteristic positions : low, middle, high.

Cavity external measurement system

The 3 identical measurement systems comprise 1 or 2 dosimeter-holder boxes (Fig.4) in each of the 8 measurement positions (Fig. 1).

Fig. 4 : Dosimeter-holder box

The filling of the dosimeters in each box is listed in table 4 page 8. (dimensions of the box (cm) : 10 x 10 x 2)

- Special objectives

Envelope measurement of the neutron field in the cavity of a 900 PWR (thermal, epithermal and fast fluxés).

Validation and adjustment of the calculated spectrum by the measurement (activation dosimeters, SSTR) in the core mid-plane in 3 characteristic azimuthal positions.

Comparison of internal capsule and external measurements for the same azimuthal position (equivalent to capsule U), in the case of cavity measurements in addition to the internal measurements.

Table 4

| Dosimeter Reaction | Dimensions<br>Ø<br>mm | <br>Thickness<br>mm | <br>Masse<br>mg<br>(approx) | Typ of dosimeter-holder box |
|---|---|---|---|---|
| Cu 63 (n,α)<br>Ni 58 (n,p)<br>Fe 54 (n,p)<br>Nb 93 (n,n') | 10<br>10<br>10<br>10<br>and 4 | 1<br>1/10<br>0.5<br><br>20E-3 | 700<br>70<br>310<br>2.8<br>6.5 | A |
| Co 59 (n,γ)<br>(AlCo bare) | 3 | 1/10 | 6.5 | C |
| Fe 54 (n,p)<br>Ni 58 (n,p) | 10<br>10 | 0.5<br>1/10 | 310<br>70 | B |
| Co 59 (n,γ)<br>(AlCo Cd cover | 3 | 1/10 | 6.5 | D |
| U 235 (n,F) bare<br>Pu239 (n,F) " | Thin deposit of fissile material $10^{-9}$ to $10^{-12}$ g/cm² depending on fluence to be measured | | | SSTR<br>Sealed boxes<br>Origin Westinghouse<br>Energy System Inc.<br>Pittsburg USA) |
| U 238 (n,F) Cd cover<br>Np237 (n,F) Cd "<br>U 235 (n,F) Cd "<br>Pu239 (n,F) Cd " | Support and mica (traces)<br><br>Cd Ø 17 mm | | | |

CALCULATIONS OF NEUTRON SPECTRA

The calculations will be performed using the TRIPOLI Code (2) that solves the transport equation by the polykinetic Monte-Carlo method in a 3-D geometry. For the special internal capsule, the neutron spectrum and its variance-covariance matrix will be provided at the core mid-plane level with a view to using the STAY'SL adjustment procedures (3), the data being completed by the axial distributions over the full height, the fast neutron flux and the cross-sections of the dosimeters employed. For the dosimetry in the cavity, the same calculations will be carried out in the core mid-plane for the 3 azimuthal positions chosen.

General comment on the installation of the externam measurement systems :

To fix the upper and lower supports will entail working in an amber classified zone. This implies a strict organisation of operations and a carreful selection of access times.

## REFERENCES


1. St Laurent Bl - Projet d'équipement de la dosimétrie spéciale.
   1. Mesures internes : capsule de surveillance.
   2. Mesures en cavité : paroi externe de la cüve.
   A. BEVILACQUA - IRDI/DERPE/DPER/SIAAR/SIDIN 228-97.

2. Programme de Monte-Carlo polycinétique TRIPOLI-2. Tomes I à III. Tiré à part.
   J.C. NIMAL et Al.

3. F.G. PEREY - Least Squares Dosimetry Unfolding.
   The programme STAY'SL - ORNL/TM-6062 - ENDF-254.

# EFFECT OF AXIAL FLUX DENSITY VARIATIONS ON THE DETERMINATION OF NEUTRON FLUENCES FOR LWR-PV DOSIMETRY

H.-C. Mehner

Zentralinstitut für Kernforschung Rossendorf

Germany

## ABSTRACT

Long term irradiations for pressure vessel dosimetry usually involve reactor operation at various neutron intensity levels. The calculation of neutron fluences is commonly based on the assumption that the neutron density is directly proportional to the thermal reactor power, where by the total irradiation time is be broken, into periods of uniform power levels. However, this assumption is not in general justified for applications in LWR-PV environment (surveillance probes, ex-vessel dosimetry).

This study includes the effect of flux density variations caused by control rod movements and fuel burn-up in the analysis of dosimetric experiments. It is shown that these variations involve corrections of the measured reaction rates up to 25 % for nuclides with short half-lives ($^{59}Fe$, $^{58}Co$, $^{46}Sc$). The corrections for nuclides with longer half-lives such as $^{54}Fe$, $^{60}Co$, $^{93m}$ do not exceed 3 %.

## INTRODUCTION

To assess the structural integrity of reactor pressure vessel (RPV), surveillance probes made of pressure vessel material are irradiated, together with neutron fluence monitors, in capsules located behind the thermal shield. This in-vessel dosimetry has been sometimes supplemented by ex-vessel dosimetry in order to test transport calculations, to follow core changes at a wide variety of axial and azimuthal locations and to provide a measure of redundancy in the event of loss of some in-vessel dosimeters. Furthermore, in some cases probes have been scraped from the inner surface of pressure vessels to measure directly the neutron exposures on the pressure vessel inner vall.

All these experimental work requires the knowledge of neutron exposures at the locations of interest. Most of these experiments involve long-lived activation detectors such as Fe, Ni, Ti, Cu, Nb, and Co. The first step in the determination of neutron fluence is the calculation of the reaction rate. It is usual to assume, in general, that the neutron flux density is directly proportional to the total reactor power. However, this assumption is not a priori justified for applications in LWR-PV-dosimetry. Therefore, this study includes the effect of axial flux density variations caused by control rod movements and fuel burn-up in the evaluation of activation detectors.

## FORMALISM

Long irradiation in power reactors involve operations at various reactor power levels as well as shut down periods. In addition, changes in isotopic content of fuel by burn-up and control rod movements lead to non-uniform neutron flux density at detector location during the irradiation. Therefore, the total irradiation period is divided into continuous series of periods during each of which the flux level is constant.

Taking into account multiple irradiations at different neutron flux levels, the reaction rate R can be written as follows:

$$R = A/[N_0\ (1 - e^{-\lambda t_i})\ e^{-\lambda t_d} \cdot f] \qquad (1)$$

where:

$A$ - activity of the detector
$N_0$ - number of parent nuclei in the activation detector
$\lambda$ - decay constant
$t_i$ - irradiation time, $t_i = \sum_{k=1}^{n} \Delta t_k$
$t_d$ - decay time (time elapsed between the end of the irradiation and the start of the activity counting)

with f as a "correction factor for reactor history":

$$f = \frac{T_{eff} \cdot \sum_{j=1}^{n} f_j\ (1 - e^{-\lambda \Delta t_j})\ e^{-\lambda (t_i - t_j)}}{(1 - e^{-\lambda t_i}) \sum_{j=1}^{n} f_j\ \Delta t_j} \qquad (2)$$

where:

$T_{eff}$ - irradiation time in effective full-power days
$f_j$ - function proportionally to the neutron flux density in the j-th irradiation period
$\Delta t_j$ - duration of the j-th irradiation period
$t_j$ - time to the end of irradiation period j, $t_j = \sum_{k=1}^{n} \Delta t_k$

if it is assumed that the neutron spectrum is constant at all periods.

The correction factor for reactor history is so defined that f=1 is for one irradiation period at uniform full power level.

The quantities $f_j$ are differently considered within the standard rule or quidelines of various countries. In ASTM E 261 /1/, it is assumed that the quantities $f_j$ are directly proportional to reactor power, while Ref. /2/ involves deviations at longer irradiations, but the quantities $f_j$ are also considered to be proportional to reactor power at one operation cycle.

The importance of local power variations due to control rod configuration was taken into account by G. Alloggio et al. /3/. The effect of local power history was investigated by comparing calculations with measurements carried out at the cavity core midplane level of a BWR.

## FLUX DENSITY VARIATIONS DESCRIPTION

Fig. 1 shows a section of a WWER-440 power reactor. Measurements of neutron fluence can be performed at survaillance locations (between barrel and pressure vessel), behind the pressure vessel as well as from samples of the inner surface or plate of RPV. To analyze the effect of flux density variations on the correction factor of reactor his-

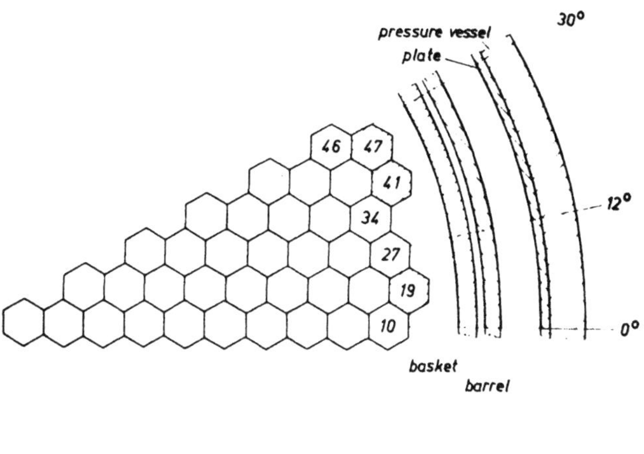


Fig. 1

Lay-out of WWER-440. Numbers indicate places for fuel assemblies or steel dummies.

tory, both fission source distributions from core border places and flux density calculations at the pressure vessel outer wall (RPV/OW) have been used. Especially, the study is focused on the question, if it is sufficient to use the source distribution, only, or if calculations at RPV must be taken.

The fission source distribution was determined on the basis of 3-dimensional power density and burn-up distributions computed with the help of the code PYTHIA /4/ for time periods of 20 effective days. The calculation of flux density at RPV/OW was carried out with the 3 dimensional Monte-Carlo multigroup transport code SMO /5/ by using the fission source distribution. Three different loading configurations have been taken into consideration:

case 1: normal loaded core with fresh fuel assemblies at the border fuel assembly places no. 19, 41, 47

case 2: low leakage core, where highly burned fuel assemblies are inserted in the locations no. 19, 41, 47

case 3: steel dummy assemblies at places no 19, 41, 47.

For the purpose of investigating the effect of flux density variations on the correction factor for reactor history, three representative azimuthal positions have been considered: calculations with source distributions at the locations no. 10, 27 and 47 (for dummies 46) and the corresponding angles at 0°, 12° and 30° for the flux density at RPV/OW; in each case 10 axial positions were considered.

As an example, Fig. 2 shows the normalized axial source distribution at the core border in comparison with the normalized

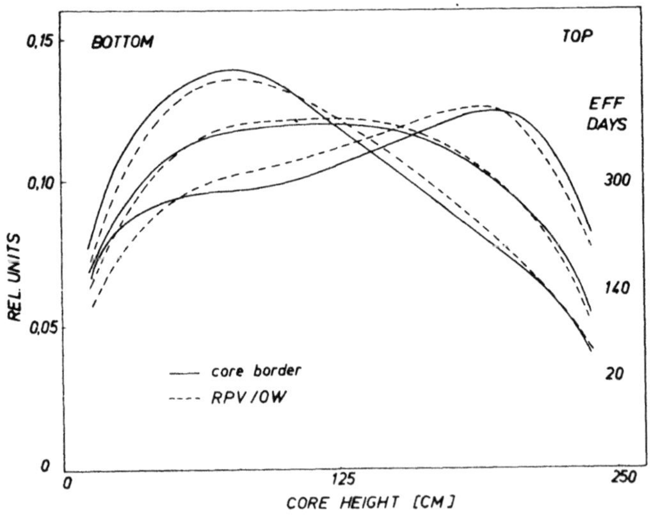


Fig. 2

Comparison of normalized source distribution from the core border with normalized flux density (E > 1 MeV) at RPV/OW for a low leakage core (case 2) flux density (E > 1 MeV) at RPV/OW for 3 different states of a low leakage core (case 2) at the beginning, in the middle and at the end of a cycle. The curves represent average values over the 3 azimuthal locations. The averaging is justified because the differences between various azimuthal locations are smaller than 3 % both at the core border and at RPV/OW. Apparently, the axial source distribution is not determined by the burn-up of the single assemblies but by the macro flux of the whole core. Furthermore, the curves at the core border are widely consistent with the curves at the RPV/OW for all three states. It is seen that the maximum of the distribution is shifted from the lower to the upper part of the core during one cycle. Similar strong variations can also be observed at other core configurations. These time-depending axial variations cause the effects described in the following.

## RESULTS

On the basis of the above mentioned source distribution or flux density at RPV/OW, correction factors for reactor history have been calculated by using eq. (2) for the three cases and for one cycle. The cycle was divided into periods of 40 effective days. The results are plotted in Fig. 3-5. Only axial variations have been taken into account, while periods of reduced power or zero-power have been excluded. Thus, the correction factor for reactor history f only involves flux density variations during reactor operation. Values of f were computed for the often used long-lived nuclides $^{59}$Fe,

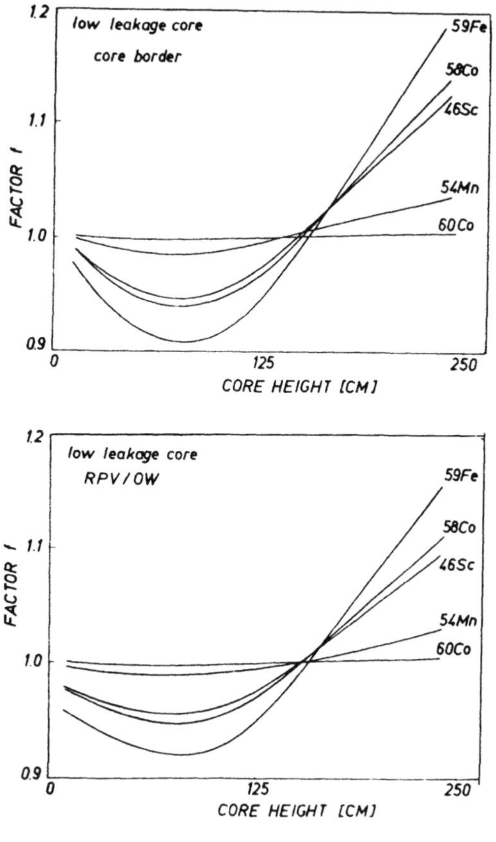


Fig. 3

The correction factor for reactor history f for a low leakage core (case 2) at core border and RPV/OW

$^{58}$Co, $^{46}$Sc, $^{54}$Mn, $^{60}$Co, and $^{93m}$Nb with half-lives between 44 days and 16 years. The differences between various azimuthal locations are smaller than 2 %. Therefore, in Fig. 3-5 the average values over the 3 azimuthal locations are plotted, exclusively.

Fig. 3 shows the comparison between core border and RPV/OW for a low leakage core (case 2). Significant differences can only be observed in axial border regions, but they do not exceed 5 % in any case. It is worth commenting here that fission source distributions from the core border produce values of f which are in very good agreement with those directly calculated by the flux density distribution at the pressure vessel outer wall. This fact cannot be overemphasized. The calculated curves of f in Fig. 3-5 differ from each other due to different time-depending axial flux density variations at the three investigated cases. The deviations from f=1 associated with the short-lived nuclides $^{59}$Fe, $^{58}$Co, and $^{46}$Sc are large and reach up to 25 %. The deviations do not exceed 3 % for

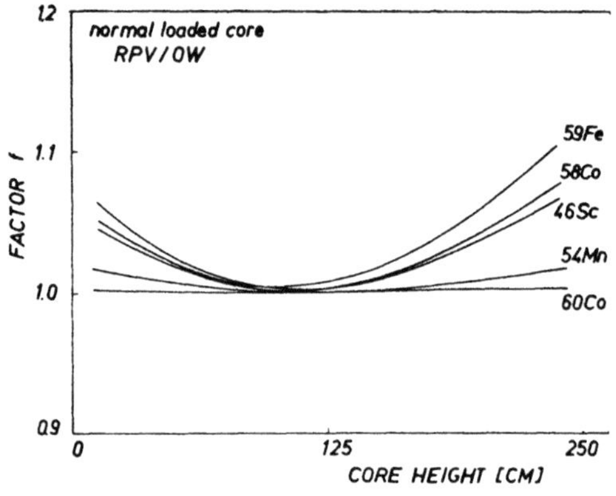


Fig. 4

The correction factor for reactor history f for a normal loaded core (case 1)

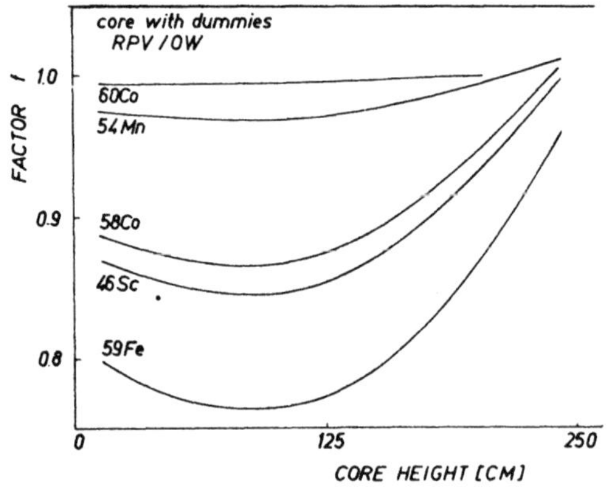


Fig. 5

The correction factor for reactor history f for a core with dummy assemblies (case 3)

$^{54}Mn$, while the effect is negligible for the longer-lived products $^{60}Co$ and $^{93m}Nb$.

In the case of a normal loaded core the deviations from f=1 are small in the middle core region for all nuclides. On the other hand, great deviations occur at cores with dummy assemblies as well as low leakage cores. WWER-440 reactors are provided with a welding joint in a distance of about 30 cm from the lower core boundary. Experiments at this location have to take into account the axial flux density variations, in any case.

## CONCLUSIONS

The commonly used methods of determining neutron fluences in power reactor surveillance dosimetry are examined and found to be unsatisfactory in regard to corrections of reactor history. In general, it is not sufficient to take into account only power reactor history and to neglect axial flux density variations due to fuel burn-up and control rod movements.

The effect of these axial flux variations strongly depends on the core configuration and the reactor operation regime. For short-lived nuclides, corrections can be necessary up to 25 % while the effect can be neglected for longer-lived products such as $^{60}Co$ or $^{93m}Nb$. In any case, the effect of axial flux density variations should be estimated to avoid systematic errors during evaluation of reactor dosimetry measurements.

For the core configurations studied, only small azimuthal and radial dependence have been found. Therefore, it is sufficient to study the axial variations at a single azimuthal angle. Fission source distributions from the core border region can be used, if no axial calculation from the experimental locations (surveillance places, pressure vessel inner or outer wall) are available. To characterize the time-depending axial flux density variations monthly steps are sufficient. Due to small influence of axial flux density variations, the reaction $^{93}Nb(n,n')^{93m}Nb$ is especially favourable to reactor dosimetry measurements.

## ACKNOWLEDGEMENTS

The author wishes to express his appreciations to H.-U. Barz, Rossendorf, G. Suschowk, K.F. Pötter, Rheinsberg, for performing the computations of the neutron source distribution as well as the fluxes at pressure vessel.

## REFERENCES


/1/ ASTM E 261, "Standard Method for Determining Neutron Flux, Fluence, and Spectra by Radioactivation Techniques", ASTM Annual Book of Standards, Vol. 12.02.83

/2/ DIN 25456, Teil 1, "Neutronenfluenzmessung", Mai 1983

/3/ G. Alloggio et al., "Power Reactor Neutron Dosimetry: Experimental Estimate of the Influences Due to Local Changes in Core Configuration and Gamma-Induced Reactions", Reactor Dosimetry: Methods, Applications, and Standardization, ASTM STP 1001, Philadelphia 1989, pp. 281-292

/4/ S. Thomas, Kernenergie 29 (1986) p. 413

/5/ H.-U. Barz, K.-F. Pötter, G. Suschowk, Kernenergie 31 (1988) p. 133

# ACCURACY DETERMINATION IN NEUTRON FLUX DENSITY MONITORING

B.Ošmera, V.Štěpánek

Nuclear Research Institute

250 68 Řež, Czechoslovakia

## ABSTRACT

The neutron flux density and fluence monitoring method in irradiation experiments is based on the knowledge of the neutron spectrum in the irradiation facility and on the cross section of the employed detector. The accuracy of monitoring is predominantly determined by the accuracy of these quantities and by the experimental technique. Supposing that complete covariance matrices of the neutron spectrum and of the neutron cross sections are available, the variances of the monitored quantities have been derived in terms of variances of the entering variables. The variance of the sum in an arbitrary group has been derived too and the size of the covariance of two group sums has been found. Both quantities, the neutron flux densities and the cross sections, are treated as random variables, possibly correlated. The method is distribution-free and it differs from attempts to introduce computational algorithms with incorporated apriori information.

## ACCURACY OF THE FLUX DENSITY

The requirements upon monitoring a netron field emerge from irradiation experiments in research reactors and from irradiation damage of construction materials due to irradiation in nuclear power reactors. This monitoring is carried out usually in a steady state operation of a nuclear reactor. The multigroup approximation is characteristic for the majority of practical computations.

In a multigroup approach, the flux density $\phi$ integrated over all neutron energies appears as [1]

$$\phi = R_M / \bar{\sigma}_M \tag{1}$$

where $R_M$ is the reaction rate of the monitor in a particular site of the reactor and $\bar{\sigma}_M$ is the cross section of the monitor.

It can be seen that the accuracy and the correctness of mnitoring the neutron flux density are given by accuracy of the cross section $\bar{\sigma}_M$ and by unbiasedness of the reaction rate $R_M$ where $R_M$ is a measured quantity and $\bar{\sigma}_M$ is obtained by computation from a multigroup approximation of the neutron flux density and of the cross section according to

$$\bar{\sigma}_M = \frac{\sum_{j=1}^{N} \sigma_M^j \varphi^j}{\sum_{j=1}^{N} \varphi^j} \tag{2}$$

where $\varphi^j$ is the neutron group flux density and N is the number of the groups.

The variance of the monitored quantity $\phi$ in (1) is obtained in such a way that $\phi$ is expanded into a two-dimensional Taylor series and the variance of this approximation is calculated. Under assumption of stochastic independence of $R_M$ and $\bar{\sigma}_M$ we obtain

$$V[\phi] = \frac{1}{\bar{\sigma}_M^2} V[R_M] + \frac{R_M^2}{\bar{\sigma}_M^4} V[\bar{\sigma}_M] \tag{3}$$

where V stands for the operator of the variance.

(Remark: the expected values that were to appear in (3) are replaced by measured or computed values of $R_M$ or $\overline{\sigma}_M$.)

Thus we have obtained a dispersion measure of the monitored quantity in terms of variances of the entering variables. In the same time we replace the variance of $R_M$ by the empirical (sample) variance of this measured random variable whereas we are able to compute the variance of $\overline{\sigma}_M$. Regarding that $\overline{\sigma}_M$ is a weighted scalar product of quantities $\sigma_M^j$ and $\varphi^j$ (with a weight $(\sum_j \varphi^j)^{-1}$) whose covariance matrices are known and where $(\sigma_M^j, \varphi^j)$ is a couple of independent components, it can be proved that

$$V[\overline{\sigma}_M] = \frac{1}{\left(\sum_{j=1}^{N} \varphi^j\right)^2} \left\{ \sum_j V[\sigma_M^j \varphi^j] + \sum_{i \neq j} \sum_j cov(\sigma_M^i \varphi^i, \sigma_M^j \varphi^j) \right\} \quad (4)$$

The formulae for the variance and the covariance of the products $\sigma_M^j \varphi^j$ evaluated by the means of parameters of $\sigma_M^j$ and $\varphi^j$ are shown in Appendix 1.

Whereas the measurement uncertainty of the cross section is by rule shown in recommended data collections [2] and is frequently accompanied by a complete covariance matrix, the neutron spectrum, which is most relevant for reliable monitoring, is obtained either from computations or from other measurements. If the proper problem of determining a suitable spectrum of the weight function is successfully solved [4] there remains a need of unifying the group formats of both fundamental quantities $\varphi$ and $\sigma$ including their covariance matrices (if given) to the purpose of evaluating the accuracies of the monitoring.

## ACCURACY WITHIN GROUPS

In multigroup approximation the number N of random variables is reduced to a number M (M < N) in such a way that the groups involve $n_k$ components (k=1,2,...,M), in which $\sum_{k=1}^{M} n_k = N$. In the k-th group we obtain the variance of the sum $S_k = \sum_{j=1}^{n_k} \sigma_D^j \varphi^j$ in the form

$$V[S_k] = \sum_{j=1}^{n_k} V[\sigma_D^j \varphi^j] + \sum_{\substack{i=1 \\ i \neq j}}^{n_k} \sum_{j=1}^{n_l} cov(\sigma_D^i \varphi^i, \sigma_D^j \varphi^j) \quad (5)$$

and, moreover, we get for the covariance of two groups

$$cov(S_k, S_\ell) = \sum_{i=1}^{m_k} \sum_{j=1}^{m_\ell} cov(\sigma_D^i \varphi^i, \sigma_D^j \varphi^j) \tag{6}$$

(the results are derived in Appendix 2). In these formulae the superscripts apply to quantities belonging to the particular groups in concern.

## APPENDIX 1

$\bar{\sigma}_M$ is a weighted scalar product of $\sigma_M^j$ and $\varphi^j$

$$\bar{\sigma}_M = w \sum_{j=1}^{N} \sigma_M^j \varphi^j \quad , \quad w = \left( \sum_{j=1}^{N} \varphi^j \right)^{-1}$$

The components in pairs ( $\sigma_M^j, \varphi^j$) are stochastically independent, the covariance matrices of $\sigma_M^j$ and of $\varphi^j$ are known. Thus according to the theorem for the variance of a sum of random variables [3]

$$V[\bar{\sigma}_M] = w^2 \{ \sum_j V[\sigma_M^j \varphi^j] + \sum_{i \neq j} \sum_j cov(\sigma_M^i \varphi^i, \sigma_M^j \varphi^j)$$

However,

$$V[\sigma_M^j \varphi^j] = E[(\sigma_M^j \varphi^j)^2] - E^2[\sigma_M^j \varphi^j]$$

where E is the operator of the expectation. Hence, introducing the independence within the couples

$$V[\sigma_M^j \varphi^j] = E[(\sigma_M^j)^2] \cdot E[(\varphi^j)^2] - E^2[\sigma_M^j] \cdot E^2[\varphi^j]$$

and evaluating the second general moments in terms of variances of the single components

$$V[\sigma_M^j \varphi^j] = V[\sigma_M^j] \cdot V[\varphi^j] + V[\sigma_M^j] \cdot E^2[\varphi^j] + V[\varphi^j] \cdot E^2[\sigma_M^j]$$

Similarly we obtain for the covariances

$$cov(\sigma_M^i \varphi^i, \sigma_M^j \varphi^j) = cov(\sigma_M^i, \sigma_M^j)\, cov(\varphi^i, \varphi^j) +$$

$$+ cov(\sigma_M^i, \sigma_M^j) \cdot E[\varphi^i] \cdot E[\varphi^j] + cov(\varphi^i, \varphi^j)\, E[\sigma_M^i$$

Thus the variance of $\bar{\sigma}_M$ is expressed in terms of the parameters of entering quantities $\sigma_M^j$ and $\varphi^j$.

## APPENDIX 2

Since the sum $S_k$ in the k-th group has been introduced as

$$S_k = \sum_{j=1}^{n_k} \sigma_D^j \varphi^j$$

we obtain its variance from Eq.(4) omitting the weight and replacing N and $\sigma_M^j$ with $n_k$ and $\sigma_D^j$ respectively. The changes result in (5).

Applying the definition of covariance to a covariance of two sums $S_k, S_l$ we derive Eq.(6) in a similar manner as above.

## REFERENCES


1. B.Ošmera, "The Monitoring of Irradiation Experiments" in Neutron Spectrometry and Reactor Dosimetry, ČSKAE Praha, 1985 (in Czech).

2. D.E.Cullen, N.P.Kocherov, and P.M.McLaughlin, "Comparison of Experimental and Calculated Spectrum-Averaged Neutron Cross Sections for the IAEA International Reactor Dosimetres File", Nucl.Sci.Eng. 83, (1983).

3. N.L.Johnson, F.C.Leone, Statistics and Experimental Design, Wiley, New York, 1964.

4. B.Ošmera, M.Holman, "Surveillance Neutron Dosimetry and Cavity Neutron Flux Monitoring at Czechoslovak VVER-440 Power Reactors", this symposium.

# PRESSURE VESSEL NEUTRON DOSIMETRY AT ARKANSAS NUCLEAR ONE

## PAST, PRESENT, AND FUTURE


| S.C. Apple, L.M. Hodgson, R.R. Culp | C.O. Cogburn | J.N. Miller |
|---|---|---|
| Arkansas Tech University | University of Arkansas | Entergy Inc. |
| Russellville, AR 72801 USA | Fayetteville, AR 72701 | Russellville, AR 72801 |



## ABSTRACT

Pressure vessel cavity neuton dosimetry experiments have been performed at the Arkansas Power & Light Co.'s ANO facility since 1977. The experiments have included several conventional foil sets, as well as several sets of commercially available niobium foils. The foil set irradiations represent several fuel cycles, operational conditions, and different axial and azimuthal positions for each reactor. EPRI Report NP-5733, PRESSURE VESSEL NEUTRON DOSIMETRY at THREE PWR's, summarizes the work through April, 1988.

Current foil sets are being analyzed to determine reaction rates, spectral indices, fluence, and displacements per atom in iron (dpa). Experiments II-10, II-11 and II-12 were irradiated utilizing a low-leakage fuel management configuration. The analysis of these experiments indicates a reduction in the neutron flux. Also, niobium foils have been irradiated concurrently with the conventional foil sets in the experiments since 1984. Development of the Nb monitor is important since its neutron cross-section is similar to the damage cross-section of iron. Also, niobium monitors may be used for irradiation periods ranging from half of a fuel cycle to several fuel cycles because of its long half-life and activation sensitivity.

Future experiments will use both conventional foil sets and Nb foils. The large data base developed over the last 12 years will continue to be expanded and is expected to provide valuable information for plant life extention and future reactor designs.


---

## IRRADIATION EXPERIMENTS

Experiments at ANO have included foil sets placed in both units, ANO-I (B&W) and ANO-II (C.E.). The experiments in Unit-I are currently placed in a 1" I.D. ex-core instrumentation calibration source tube at 155°, as indicated in Figure 1. Experiments in Unit-II are placed in the uncompensated ion-chamber (U.I.C.) well #10, as indicated in Figure 2. The foil sets contain aluminum-covered Ag, Ni, and Ti foils, cadmium-covered Co, Cu, Fe, Nb, and Sc foils, and cd-covered $^{237}Np$, $^{235}U$, and $^{238}U$ capsules. Table 1 contains neutron monitor information. The monitors are suspended at the horizontal midplane of each reactor by a stainless steel beaded chain which can be used to provide axial flux information. The beads and steel pins connecting them contain cobalt, chromium, iron, and nickel in sufficient quantities to allow the counting of $^{58}Co$, $^{60}Co$, $^{51}Cr$, $^{59}Fe$, and $^{54}Mn$.

## CALIBRATION

The foil sets were counted on two gamma spectrometry systems. Each detection system consists of a high resolution germanium detector and associated electronics interfaced to a multi-channel buffer/mca. The systems are calibrated for both energy and efficiency using certified standard radiation sources traceable to the National Institute of Standards and Technology (NIST). The system calibration conforms to established criteria in ASTM Standard E-181.

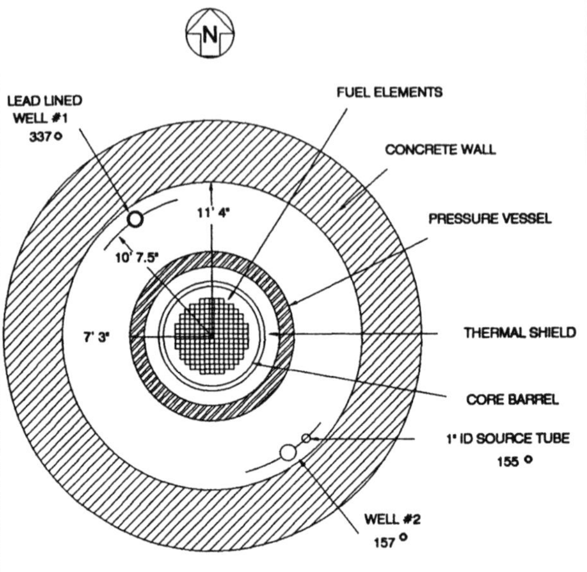


**FIGURE 1. UNIT-I Location diagram for radiometric monitors. Currently in 1" I.D. source tube.**

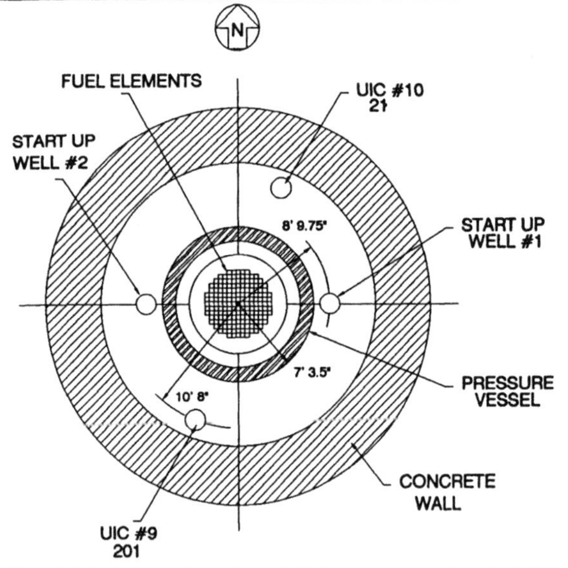


**FIGURE 2. UNIT-II Location diagram for radiometric monitors. Currently in U.I.C. #10.**

## ANALYSIS

Neutron cross-sections and d.p.a. cross-sections for Unit-I were obtained from transport calculations performed in the unshielded detector well at 157°, using core composition values corresponding to the middle of fuel cycle 4. The cross-sections are listed in Table 2.

Neutron cross-sections and d.p.a. cross-sections for Unit - II were obtained from transport calculations performed in both the 21° and 201° azimuths (which were similar), using core composition values corresponding to fuel cycle 2. The cross-sections are listed in Table 3.

Saturated reaction rates for the isotopes used were calculated using

$$RR_{sat}\,(\text{dps/atom}) = A' \cdot \frac{e^{\lambda T}}{N} \cdot k_{sat}$$

where

$$A' = \frac{\dot{c}}{\varepsilon \cdot P_{\gamma}}$$

and

$$N = \frac{w \cdot N_A \cdot P \cdot A}{M}$$

$\lambda$ is the decay constant for the isotope in days $^{-1}$
A is the isotopic abundance ( atom fraction )
T is the time from EOI to mid-count in days
$\dot{c}$ is the count rate in counts per second
$\varepsilon$ is the detector efficiency
$P_{\gamma}$ is the $\gamma$ probability ( gammas per disintegration )
w is the dosimeter weight (g)
$N_A$ is Avogadro's number ( 6.0225 E+23 atoms/g−atom )
M is the atomic weight
P is the dosimeter purity ( weight fraction ).

The saturation factor, $k_{sat}$, was calculated over the irradiation period by summing the irradiation components during which the reactor power level could be considered constant, using the equation

$$k_{sat} = \left\{ \sum_{i=1}^{I} [P_i\,(1 - e^{-\lambda t_i}) \cdot e^{-\lambda T_i}] \right\}^{-1}$$

where

$P_i$ is the fraction of full power during the ith interval
$t_i$ is the duration of the ith interval
$T_i$ is the time from the end of ith interval to EOI
$\lambda$ is the decay constant for the isotope in days $^{-1}$.

| Element | I.D. | Description | Weight (g) | Purity | Vendor |
|---|---|---|---|---|---|
| Cobalt | 'A' | 1/2" foil | 0.0442 | 0.54% | Reactor Exp |
| Copper | 'J' | 1/2" foi | l0.2944 | 99.995% | Reactor Exp. |
| Iron | 'AD' | 1/2" foil | 0.1161 | 99.984% | Reactor Exp. |
| Niobium | 'C' | 1/2' foil | 0.1440 | 167 ppm Ta | Reactor Exp. |
| Nickel | 'H' | 1/2" foil | 0.2841 | 99.981% | Reactor Exp. |
| Scandium | 'A' | 1/4" foil | 0.0104 | 99.83% | Reactor Exp. |
| Silver | '1' | 1/2" foil | 0.04184 | 4.65% | ORNL |
| Titanium | '2' | 1/4" foil | 0.0361 | 99.876% | Reactor Exp. |
| Neptunium | '23' | Capsule | 0.009176 | 100% | ORNL |
| Uranium-235 | 6 | Capsule | 0.1105 | 99.89% | ORNL |
| Uranium-238 | '19' | Capsule | 0.092647 | 100% | ORNL |

**Table 1. Neutron monitor information.**

(Units: cm$^2$)

| Reaction | E > 1.0 MeV | E > 0.1MeV | Total |
|---|---|---|---|
| 63Cu(n,$\alpha$) | 9.980E-28 | 1.086E-28 | 4.362E-29 |
| 54Fe(n,p) | 6.424E-26 | 6.989E-27 | 2.807E-27 |
| 58Ni(n,p) | 8.914E-26 | 9.698E-27 | 3.896E-27 |
| 46Ti(n,p) | 1.314E-26 | 1.430E-27 | 5.742E-28 |
| 237Np(n,f) | 4.404E-24 | 4.791E-25 | 1.925E-25 |
| 235U(n,f) | | | 7.108E-24 |
| 238U(n,f) | 3.391E-25 | 3.690E-26 | 1.482E-26 |

D.P.A. Cross-sections for E> 1 MeV (barns/neutron)

| Position | D.P.A. Cross section |
|---|---|
| Inside PV | 1577 |
| T/4 of PV | 1750 |
| 3T/4 of PV | 2821 |
| Cavity | 3436 |

**Table 2. Unit I reaction cross-sections.**

(Units: cm$^2$)

| Reaction | E > 1.0 MeV | E > 0.1MeV | Total |
|---|---|---|---|
| 63Cu(n,$\alpha$) | 7.421E-28 | 8.548E-29 | 3.222E-29 |
| 54Fe(n,p) | 6.134E-26 | 7.065E-27 | 2.663E-27 |
| 58Ni(n,p) | 8.668E-26 | 9.983E-27 | 3.764E-27 |
| 46Ti(n,p) | 1.117E-26 | 1.287E-27 | 4.852E-28 |
| 237Np(n,f) | 4.373E-24 | 5.037E-25 | 1.899E-25 |
| 235U(n,f) | | | 7.219E-24 |
| 238U(n,f) | 3.472E-25 | 3.999E-26 | 1.508E-26 |

D.P.A. Cross-sections for E> 1 MeV (barns/neutron)

| Position | D.P.A. Cross section |
|---|---|
| Inside PV | 1564 |
| T/4 of PV | 1776 |
| 3T/4 of PV | 2720 |
| Cavity | 3314 |

**Table 3. Unit II reaction cross-sections.**

Reaction rates were compared to the $^{58}Ni(n,p)$ reaction rate for ANO-I experiments 13 through 15 and ANO-II experiments 9 through 12. The spectral index values, relative to ANO I-5 and ANO II-2 indices, respectively, show small variations over the experiments. This supports the assumption that the neutron energy spectrum in the pressure vessel cavity remains fairly constant. The results of the spectral indices for Unit-I and Unit-II are shown graphically in Figures 3 and 4, respectively. The results are consistent with calculations performed in previous experiments, as summarized in EPRI Report NP-5733.

The average cavity fluence for each reactor was determined using the reaction rates and cross sections for the seven reactions listed in Table 2 and 3 , and the effective full power days in the equation

$$F_{ave} = \underset{i=1}{\overset{7}{AVG.}} \left[ \frac{RR_{sat_i} \cdot EFPD \cdot 86400}{CS_i} \right]$$

where

$RR_{sat}$ corresponds to the saturated reaction rate
EFPD is the effective full power days
$CS_i$ reaction cross sections from Table 2 or 3.

Results of fluence calculations for Experiments I-11 through I-15 and II-8 through II-12 are included in Table 4 and the *total cavity fluence* values vs EFPD are shown graphically in Figure 5. Note the change in slope due to the new low-leakage fuel management core configuration placed into operation at the beginning of fuel cycle 6, experiment II-10.

The displaced atoms in iron (d.p.a.) were calculated using the average fluence, the displaced cross-sections (Table 2 or 3), and the geometry factors in Table 5 by

$$DPA = \frac{F_{ave} \cdot CS_{dpa} \cdot GF}{1 \times 10^{-24}}$$

where

$CS_{dpa}$ dpa cross sections from Table 2 or 3
$GF_{is}$ from Table 5.

The results of the d.p.a. calculations are included in Table 6. The values for the 1/4 thickness position of the reactor vessel, for each reactor, are shown graphically in Figure 6. Again, for Unit-II, note the change in slope at the beginning of fuel cycle 6, Experiment II-10.

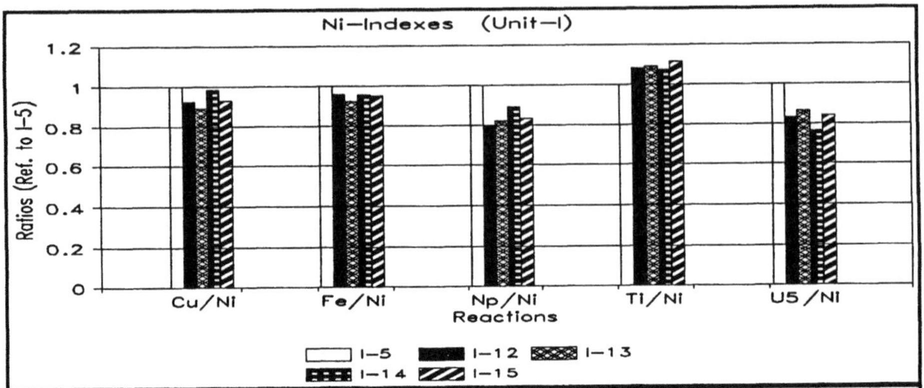


**Figure 3. Unit-I Ni-indices.**

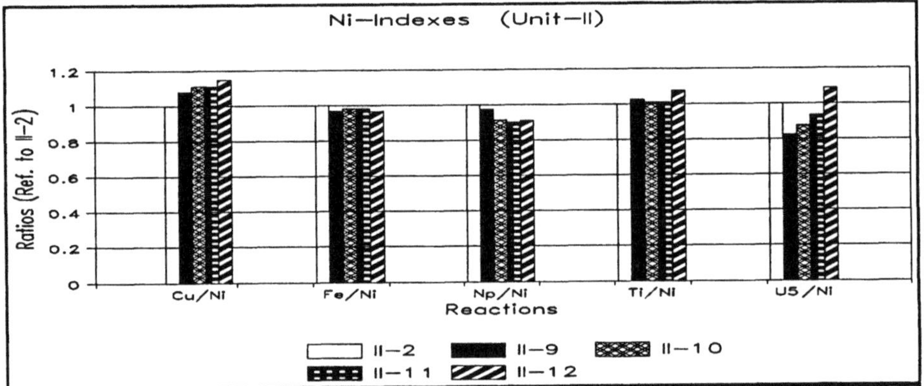


**Figure 4.Unit-II Ni-indices.**

| Experiment | E > 1.0 MeV | E > 0.1 MeV | Total |
|---|---|---|---|
| (Unit -1) | | | |
| I-11 | 3.52E+15 | 3.23E+16 | 8.02E+16 |
| I-12 | 2.58E+15 | 2.36E+16 | 5.83E+16 |
| I-13 | 7.90E+15 | 7.25E+16 | 1.79E+17 |
| I-14 | 3.25E+15 | 2.98E+16 | 7.20E+16 |
| I-15 | 3.89E+15 | 3.58E+16 | 8.80E+16 |
| ((Unit-2) | | | |
| II-8 | 3.60E+16 | 3.13E+17 | 8.07E+17 |
| II-9 | 3.14E+16 | 2.73E+17 | 6.93E+17 |
| II-10 | 3.70E+16 | 3.21E+17 | 8.21E+17 |
| II-11 | 2.21E+16 | 1.92E+17 | 4.85E+17 |
| II-12 | 8.55E+15 | 7.42E+16 | 1.94E+17 |

**Table 4. Cavity fluence at horizontal mid-plane.**

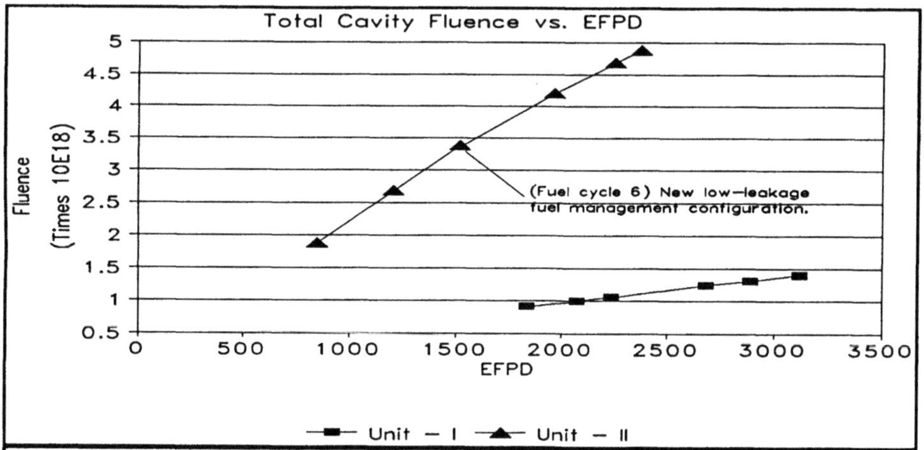


**Figure 5. Total Cavity Fluence vs EFPD.**

| Location | E > 1.0 MeV | E > 0.1 MeV | Total |
|---|---|---|---|
| (Unit I) | | | |
| Inside PV Cavity | 33.6 | 7.27 | 7.89 |
| T/4 PV Cavity | 20.3 | 7.33 | 4.67 |
| 3T/4 PV Cavity | 4.16 | 3.23 | 2.14 |
| (Unit II) | | | |
| Inside PV Cavity | 33.1 | 8.58 | 7.83 |
| T/4 PV Cavity | 18.3 | 7.43 | 4.40 |
| 3T/4 PV Cavity | 4.06 | 3.35 | 2.00 |

**Table 5. Geometry factors.**

| Experiment | Inner | T/4 | 3T/4 | Cavity |
|---|---|---|---|---|
| (Unit I) | | | | |
| I-13 | 0.4186 | 0.2806 | 0.0927 | 0.0271 |
| I-14 | 0.1722 | 0.1155 | 0.0381 | 0.0112 |
| I-15 | 0.2061 | 0.1382 | 0.0457 | 0.0134 |
| TOTAL* | 3.3893 | 2.2801 | 0.7459 | 0.2114 |
| (Unit II) | | | | |
| II-9 | 1.627 | 1.021 | 0.347 | 0.104 |
| II-10 | 1.916 | 1.203 | 0.409 | 0.126 |
| II-11 | 1.124 | 0.706 | 0.240 | 0.072 |
| II-12 | 0.442 | 0.277 | 0.094 | 0.028 |
| TOTAL* | 11.226 | 7.054 | 2.394 | 0.718 |

*** Includes previously reported data.**

**Table 6. Displaced atoms per 1000 atoms.**

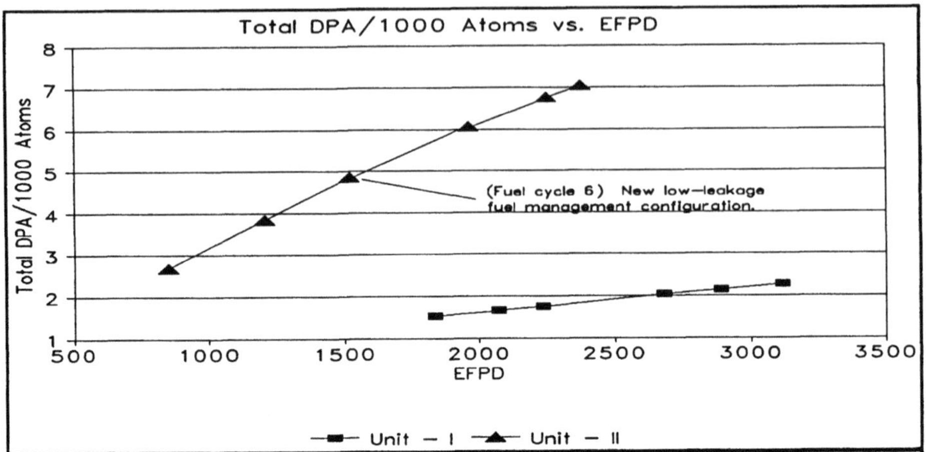


**Figure 6. DPA vs EFPD.**

## CONCLUSIONS

The pressure vessel cavity monitoring program at Arkansas Nuclear One has provided, through measurement of the neutron fields, additional confirmation that the neutron energy spectrum in the cavity remains fairly constant from one fuel cycle to the next. The agreement of the Ni-Indices is generally $\pm$ 20%. The measurements have also provided information as to the reduction in flux due to the new low-leakage fuel management configuration (Unit-II).

The results furthur indicate that cavity dosimetry is a viable means of supplementing in-vessel dosimetry, due to the ease of placement and retrieval, varied geometrical configurations, and the lower cost. Continuation of the ex-vessel dosimetry program is planned for ANO, with an increased effort to the development of new dosimeters.

Current work on the development of Nb as a neutron monitor is focusing on analysis techniques, problems associated with dissolution and deposition, and comparison of fluence calculations with conventional foil data (See L.M. Hodgson et.al., "Niobium as an Ex-Vessel Neutron Dosimeter for PWR's", 7th ASTM-EURATOM Symposium on Reactor Dosimetry).

Future work includes ex-vessel monitoring using conventional foil sets and Nb foils. Also, core composition information from later fuel cycles will be analyzed to provide new cross-section information. The large data base developed over the last 12 years will continue to be expanded and is expected to provide valuable information for plant life extension and future reactor designs.

## ACKNOWLEDGEMENTS

The authors would like to acknowledge Entergy Inc. for their assistance and support.

# ASTM STANDARDS IN SUPPORT OF CODES AND REGULATIONS ASSOCIATED WITH LWR POWER PLANT LICENSING, OPERATION AND SURVEILLANCE

**W. N. McElroy and P. D. Hedgecock**

Chairman
ASTM Subcommittee E10.05

Consultants & Technology Services
Richland, WA, USA

Chairman
ASTM Subcommittee E10.02

Nutech Engineers, Inc.
San Jose, CA, USA

**E. P. Lippincott and A. L. Lowe, Jr**

ASTM Subcommittee E10.05

Westinghouse Electric Corp.
Pittsburgh, PA, USA

ASTM Subcommittee E10.02

B&W Nuclear Service Co.
Lynchburg, VA, USA

## ABSTRACT

In this review paper , consideration is given to United States "Regulatory Instruments" (e.g., Codes, Regulations, and Standards) that are available for use in support of Light Water Reactor (LWR) power plant licensing, operation, surveillance and life attainment. The "Regulatory Instruments" and a number of "Aging Issues" believed to be most important and applicable to the management of radiation and age-related degradation of the pressure boundary aspects of current nuclear power plant components are identified and briefly discussed. The need for national and international coordination of "Standards Development Organization's (SDO)" codes and standards development work is also addressed; particularly, in regards to the current activities of the ASTM E10.02 and E10.05 Subcommittees and Task Groups E10.05.04, E10.05.11, E10.02.11, and E10.02.12 associated with 1) ASTM E706 Standard on "Master Matrix for LWR Pressure Vessel Surveillance Standards, E706(0)" and 2) Nuclear Plant Life Extension (NUPLEX) issues and the "ASTM Standards--NUPLEX Matrix" of approximately 450 of the 8500 existing ASTM standards that have some relevance to the issue of plant life extension (PLEX).

## INTRODUCTION

As stated by Frederick Seitz (1),

"America's electricity demands are soaring, up 5% a year -- twice what anyone forecast even 5 years ago. Rolling blackouts were frequent at my Florida home last winter. Brownouts were anticipated in parts of the nation this summer. In fact, our demand for electricity is climbing so fast that over the next decade U.S. generating capacity must increase by a third. Fossil fuels supply nearly 3/4 of this energy. But the smoke-belching stacks of coal-, gas-, and oil-fired plants are also responsible for about 1/2 of our air pollution. That, we used to think, is a small price to pay for progress. But there is an alternative, one that produces no smoke and can actually create more fuel than it consumes. In many regions it is even cheaper than coal-fired electricity: nuclear power. Already nuclear power is the second largest source of our electricity, and a new family of "fail-safe" nuclear reactors -- some now under construction in Japan -- may one day make nuclear power even cheaper and more plentiful. But before these can be built in the United States, major changes must be made in the way nuclear plants are financed, licensed and operated. For that to happen, all of us need to understand the truth about issues long clouded by misinformation."

In this review paper, consideration will be given to the U.S. "Regulatory Instruments" (e.g., Codes, Regulations, and Standards) that support the licensing, operation, surveillance, and life attainment of Light Water Reactor (LWR) nuclear power plants.

The reader is referred to Reference (2) for an overview of the basic principles, techniques, and methodology to be used in the design, implementation, analysis and interpretation of neutron exposure and materials property change data used to evaluate the present and future condition of the pressure vessel (PV) and its support structures. The "Methodology" is summarized in the ANNEX to ASTM Committee E10 Standard E853 on "Analysis and Interpretation of Light-Water Reactor Surveillance Results". ASTM Standard E853 is the umbrella standard for the entire set of 20 ASTM E706 Standards on "Master Matrix for LWR Pressure Vessel Surveillance Standards, E706(0)". Copies of the ASTM E706, E853 and other ASTM Committee A-1, E10, E24, and E28 standards are provided in Appendix D of Ref. (2).

Background information is provided in Ref. (2) on United States "Regulatory Instruments" (3) with extensive referencing to the literature. This information is associated with understanding and managing "radiation and age related degradation" of a nuclear power plant's reactor pressure vessel (RPV) and its support structures. Included are discussions of 1) the assessment, monitoring, and evaluation of the impact of steel property changes on plant safety and operation, 2) a review of the design and implementation of in- and ex-vessel surveillance programs and 3) plant life extension (PLEX) issues; much of this information has been extracted from

References 3 through 23.

A comprehensive "Reference Listing of Supporting Technical Documentation" is also provided in Ref. (2). This documentation (over 200 references) is grouped under the following categories:

- o <u>Licensing, Regulation, Codes and Standards</u>
- o <u>Nuclear Plant Design, Life Attainment, Aging, and License Renewal</u>
- o <u>Surveillance Program Studies</u>
- o <u>Plant Specific Surveillance Program Studies</u>
- o <u>Research Program Studies</u>
- o <u>Calculational Benchmark Program Studies</u>
- o <u>Experimental Benchmark Program Studies</u>
- o <u>Measurement Methods, Reference Data, Sensor Materials, and Calibration Field Studies</u>

## CODES, REGULATIONS, AND STANDARDS

Next to be considered is a review of the "Regulatory Instruments" that currently exist (or are under revision and development) for regulatory bodies and utilities to use to assure nuclear plant safety and the continued operability of reactor pressure vessels and their support structures. That is, what regulatory instrument guidance is available, under revision, and under development for use by regulatory agencies, utilities, research organizations, and others to accurately predict, monitor, and manage nuclear power plant materials radiation and age related degradation.

Over the lifetime of a reactor plant, irradiation of key structural components produces radiation induced changes to material properties. Assessment of these changes is required to ensure continued safe operation of the plant. For example, material property degradation may impose more restrictive pressure/temperature (P-T) limits for normal heatup and cooldown of the reactor coolant system or for emergency response guidelines. Of even more concern in many cases is the response of the reactor vessel to emergency conditions such as pressurized thermal shock (PTS).

A guide to these and other regulatory issues associated with understanding and managing aging for pressurized water reactor (PWR) and boiling water reactor (BWR) pressure vessels (PV) is provided in Werry's "Guide to Regulatory Instruments for LWR Pressure Vessels: Aging and License Renewal Considerations" (3).

Wechsler's "Featured Overview" article on "Radiation Embrittlement in Pressure-Vessel Steels of Nuclear Power Plants" reviews the development of technology, codes, regulations, and standards related to the radiation embrittlement of LWR nuclear power plants' PV steels

(5). Wechsler states:

> "The growing knowledge base on radiation's influence on matter is enabling researchers to better predict radiation embrittlement over the lifetime of a reactor. However, while much has been learned about the nature of radiation embrittlement, vexing questions remain that have a direct bearing on regulatory policy and, indeed, on the viability of older nuclear power plants."

Katz has summarized the activities of a number of code writing bodies [Standards Development Organizations (SDOs) such as: American Society of Mechanical Engineers (ASME), ASME Board of Nuclear Codes and Standards (BNCS), American Society for Testing and Materials (ASTM), and Institute of Electrical and Electronics Engineers (IEEE)] that have formed groups to address the issue of codes and standards for mechanical and electrical components pertaining to PLEX (6). His review of existing nuclear codes and standards indicates many shortcomings in addressing nuclear plant life extension PLEX. The Katz (6), Hedgecock and Perrin (4), McElroy, Lippincott, Lowe, and Hedgecock (2), and Wechsler (5) review documents are recommended reading for information on current SDO initiatives to support PLEX.

## Regulatory Instruments and Aging Issues

As just discussed, a guide to understanding and managing aging for PWR and BWR pressure vessels is provided by Werry (3). United States "Regulatory Instruments" believed to be most important and applicable to the understanding and management of radiation and age related degradation of the reactor pressure vessel of nuclear power plants are (3):

1. The Standard Review Plan (SRP)
2. The Code of Federal Regulation (CFR)
3. The NRC Regulatory Guides (RG)
4. The ASME Boiler and Pressure Vessel Code (BPVC) [Sections (SC) III and XI]
5. The Generic Safety Issues (GSI)
6. The Technical Specifications (TS)

and supporting the above are:

7. The American Society for Testing and Materials (ASTM) Standards

[It is noted that Instrument 7 was not included in the Ref. (3) listing.]

For other plant components and materials discussed in Refs. (2 and 4), a more diverse instrument list could include, for example, American Nuclear Society (ANS), IEEE, American Concrete Institute (ACI), American National Standards Institute (ANSI) and National Fire

Code (NFC) Codes and Standards.

In considering codes and standards development needs for the RPV and its support structures, each instrument needs to be reviewed by comparing the content of an instrument to each of, say, "Six Aging Issues":

1) Examination, Inspection, and Surveillance.
2) Irradiation Embrittlement, Fracture Toughness.
3) Fatigue.
4) Low-flux Long-term Irradiation of Vessel and External Supports.
5) Crack Initiation, Stud Failure and Flaw Propagation.
6) Corrosion.

These specific aging issues were derived from the general guidance found in Section 2.1 of NUREG-1144, Rev. 1 (19). In this regard, Werry states:

"Typical degradation causes are neutron embrittlement, fatigue, erosion, corrosion, oxidation, thermal embrittlement, and chemical reactions. Among the other factors of aging are service wear, testing and improper installation, application, or maintenance."

In relation to the above aging issues, most of the instruments were developed for plant design and construction, for plant pre-operational stages and for safety analysis reports; i.e., the instruments may discuss all of the pertinent issues, but they were not, necessarily, written to solve aging problems or manage the aging issues.

Werry's (3) descriptions for instruments 1) through 6) and McElroy's description for instrument 7) follow:

SRPs -- are published by the NRC and provide guidance for preliminary Nuclear Power Plant (NPP) Safety Analysis Reviews (SARs) and Final Safety Analysis Review (FSAR) for plant operations.

CFR -- These regulations codify general and permanent rules published in the U.S. Federal Register by the Executive departments and agencies of the U.S. Federal Government. For the NRC and this Guide (3), the significant CFR is Title 10 Part 50, Domestic Licensing of Production and Utilization Facilities which includes the General Design Criteria for NPPs.

RGs -- These guides are published and made available to the public and the NPP industry to describe methods acceptable to the NRC to implement specific parts of the Commission's regulations and to delineate techniques used by NRC staff to evaluate specific problems, postulated accidents, and provide general guidance to applicants.

ASME-BPVC-SCIII & SCXI -- The ASME Codes establish rules of safety governing design, fabrication and inspection during the construction of boilers and pressure vessels; i.e., Section III for protection against nonductile failure of NPP components and Section XI for inservice inspection of NPP components. The ASME Code Editions, Addena and Code Cases used by the owners of NPPs are subject to the acceptance of regulatory and enforcement authorities; i.e., the NRC.

GSI -- These are published by the NRC to identify safety issues generic to NPPs. Solutions to GSIs usually result in NRC Generic Letters or NRC NUREG reports.

TS -- Technical Specifications are mandatory for NPP operational specifications and provide instructions for limiting conditions and surveillance requirements for the plant. Each plant's specifications are adaptations of Standard Technical Specifications and are mandated in the CFRs.

ASTM -- By reference, these standards support the above regulatory instruments. More specifically, ASTM Standard E185 on "Practice for Conducting Surveillance Tests for Light-Water Cooled Nuclear Power Reactor Vessels, E706(IF)" is identified as a part of Appendix H to 10 CFR 50. As previously stated, the ASTM Standard E853 on "Practice for Analysis and Interpretation of Light-Water Reactor Surveillance Results, E706(IA)" is the umbrella standard for the entire set of 20 ASTM E706 Master Matrix standards. From a reactor safety and surveillance program perspective, ASTM Standards E853, E185, E482, E560 and the other 16 supporting standards provide combined physics-dosimetry-metallurgy guidelines for NPP surveillance programs. These include recommended procedures and methods for 1) calculational and experimental approaches for materials and environmental condition assessment and 2) verification of the correctness of the conservatism of predictions of fluence limits for RPVs and their support structures.

A review of these instruments leads to the conclusion that further revisions are required to include more specific regulatory guidance for utility proof of performance for the in-situ radiation and age related degradation for materials and components to attain maximum life. Consequently, one might anticipate a need for improved codes and standards in specific areas where there is a lack of regulatory guidance associated with environmental and material condition assessment.

## ASTM & ASTM COMMITTEE E10 STANDARDS DEVELOPMENT ACTIVITIES

Information on the organization and scope of the activities of ASTM and Committee E10, as well as listings of the names of the subcommittees, task groups, and standards (now under study, revision, or development) are given in Appendix C of Ref. (2).

ASTM-SDO activities have demonstrated that new technical issues are constantly addressed related to the revision and applicability of existing ASTM standards for the condition assessment of the environment, materials and system components. For Committee E10, there is a continuing need to provide direction and coordination among SDOs (nationally and internationally) for specific codes and standards activities that are uniquely applicable to the listed aging issues as they are related to nuclear plant life attainment and license renewal!

As an example, there is a continuing need for national and international coordination of the activities of the ASTM E10.02 and E10.05 Subcommittees. The many "Task Groups" associated with these two subcommittees have the direct responsibility for the standards work associated with the required "5 year" updating and revision of each of the ASTM E706 "Master Matrix" set of 20 LWR pressure vessel and support structure surveillance standards (20).

For instance, an important aspect of the current work associated with the three ASTM E10 Plant Life Extension (PLEX) related Task Groups [E10.02.11 (on Metallurgy), E10.05.11 (on Physics-Dosimetry) and E10.02.12 (on Surveillance for Life Attainment and License Renewal)] was the completion by Hedgecock of a NUPLEX Matrix of approximately 450 ASTM Standards believed to be of relevance to PLEX. This matrix is presented and discussed in Refs. (2 and 4).

With reference to Steve Byrne's new ASTM E10.02.12 Task Group (TG), it is stated:

> "Its purpose and potential scope is described in the Proposed New Standard on RPV Surveillance for Life Attainment and License Renewal. The end product of this TG could be an appendix to ASTM E185, a unique standard, or some combination thereof. The primary impetus is to provide timely guidance to the industry for addressing life attainment and license renewal goals beyond that provided in E185."

Task Group E10.05.11 has been responsible for the development and revision of the ASTM Standard E1035 on "Determining Radiation Exposures for Nuclear Reactor Vessel Support Structures, E706(IG)". This standard provides guidance for the assessment of the radiation environment for support structures (20-23).

The development of ASTM Committee E10 Standards and the Supporting Documentation has and continues to be a very large international undertaking. An on-going and periodic updated "workshop" review of the material contained in Ref. (2) should help to improve the overall coordination of the current ASTM E706 Master Matrix, ASTM Standards -- NUPLEX MATRIX, and other related standards development work. This certainly is one of the more important goals of the "One-Day ASTM Standards Technology Training Course Workshops" (2) as well as interesting new people to join in this most important and rewarding standards work. That is, "ASTM International Standards" work which is directed towards helping to improve the safety, reliability, and operation of existing and future LWR nuclear power plants!

## REFERENCES


1. F. Seitz, "Must We Have Nuclear Power", Reader's Digest, p. 113, August 1990.

2. W. N. McElroy, E. P. Lippincott, A. L. Lowe, Jr., and P. D. Hedgecock, Condition Assessment and Surveillance of Nuclear Reactor Pressure Vessel Steels, CASNRPVS/90-1, ASTM Standards Technology Training Course Workbook, American Society for Testing and Materials, 1916 Race St., Philadelphia, PA 19103-1187, USA, May 1990, First Edition.

3. E. V. Werry, Guide to Regulatory Instruments for LWR Reactor Pressure Vessels: Aging and License Renewal Considerations, PNL-6848, Battelle Pacific Northwest Laboratories, March 1989.

4. P. D. Hedgecock and J. S. Perrin, "The Role of ASTM Standards in Nuclear Power Plant Life Extension", Proc. of the June 1990 ASME Pressure Vessel and Piping Conference, Nashville, TN, June 18-22, 1990.

5. M. S. Wechsler, "Radiation Embrittlement in the Pressure-Vessel of Nuclear Power Plants", Journal of Metals, p. 7, December 1989.

6. L. R. Katz, "The Codes and Standards Aspects of Nuclear Plant Life Extension", Proc. of Nuclear Power Plant Life Extension, Topical Meeting, July 31-August 3, 1988, Snowbird, Utah, 1988

7. W. N. McElroy et al., "Surveillance Dosimetry of Operating Power Plants", Proc. of the 4th ASTM-EURATOM Symposium on Reactor Dosimetry, Gaithersburg, MD, NUREG/CP-0029, Nuclear Regulatory Commission, Vol. 1, p. 3, July 1982.

8. F. B. K. Kam et al., Pressure Vessel Fluence Analysis and Neutron Dosimetry, NUREG/CR-5049, ORNL/TM-10651, Oak Ridge National Laboratory, December 1987.

9. W. N. McElroy and L. S. Kellogg, "Dosimetry Adjusted Reactor Physics Parameters for Pressure Vessel Neutron Exposure Assessment", Transactions of ANS, 57, p. 223, November 1988.

10. A. Taboada, P. N. Randall, and C. Z. Serpan, Jr., " Status of Regulatory Issues and Research in Light Water Reactor Surveillance Dosimetry in the U. S.", Proc. of the 5th ASTM-EURATOM Symposium on Reactor Dosimetry, Geesthacht, Federal Republic of Germany, September 24-28, 1984, EUR 9869, Vol. 1, p. 185, Commission of the European Communities, 1985.

11. C. Z. Serpan, Jr., "Impact of USA-Euratom Cooperative Dosimetry Research on NRC Regulation of Light Water Reactors", Reactor Dosimetry: Methods, Applications, and Standardization, 6th ASTM-EURATOM Symposium on Reactor Dosimetry, Jackson Hole, WY, ASTM STP 1001, p. 7, May 1989.

12. L. Lois, "Reactor Dosimetry and Nuclear Reactor Regulation", Reactor Dosimetry: Methods, Applications, and Standardization, 6th ASTM-EURATOM Symposium on Reactor Dosimetry, Jackson Hole, WY, ASTM STP 1001, p. 12, May 1989.

13. R. Gold and W. N. McElroy, "The LWR-PV-SDIP: Past Accomplishments, Recent Developments, and Future Directions", Reactor Dosimetry: Methods, Applications, and Standardization, 6th ASTM-EURATOM Symposium on Reactor Dosimetry, Jackson Hole, WY, ASTM STP 1001, p. 44, May 1989.

14. A. Taboada, P. N. Randall, and C. Z. Serpan, Jr., "Overview of U.S. Research and Regulatory

Activities on Neutron Radiation Embrittlement of Pressure Vessel Steel", Radiation Embrittlement and Surveillance of Nuclear Reactor Pressure Vessel Steels: An International Review, Third Volume, ASTM STP 1011, p. 27, February, 1989.

15. L. E. Steele, Ed., Radiation Embrittlement and Surveillance of Nuclear Reactor Pressure Vessel Steels: An International Review, Third Volume, ASTM STP 1011, February 1989.

16. F. A. Garner, C. H. Henager, Jr., and N. Igata, Eds., Proc. of the 13th International Symposium on Effects of Radiation on Materials, ASTM STP 956, Part II, Seattle, WA, Dec. 1987.

17. H. Farrar IV and E. P. Lippincott, Eds., Proc. of the 6th ASTM-EURATOM Symposium on Reactor Dosimetry on Reactor Dosimetry: Methods, Applications, and Standardization, ASTM STP 1001, Jackson Hole, WY, May 1989.

18. N. H. Packan, R. E. Stoller and A. S. Kumar, Eds., Proc. of the 14th International Symposium on Effects of Radiation on Materials, ASTM STP 1046, Vol. II, Andover, MA, May 1990.

19. Nuclear Plant Aging Research (NPAR) Program Plan", NUREG-1144, Rev. 1, U. S. Nuclear Regulatory Commission, September 1987.

20. W. N. McElroy, Selected ASTM E10.05/E10.02 Subcommittees and Task Group Reports, CTS-ASTM Letter Report to Members of Subcommittee E10.05 on Nuclear Radiation Metrology, Consultants and Technology Services, Richland, WA, January 5, 1990.

21. W. C. Hopkins and W. L. Grove, "A Study of the Embrittlement of Reactor Vessel Steel Supports", Proc. of the 5th ASTM-EURATOM Symposium on Reactor Dosimetry, Geesthacht, Federal Republic of Germany, September 24-28, 1984, EUR 9869, Vol. 1, p. 621, Commission of the European Communities, 1985.

22. W. C. Hopkins, "Reactor Pressure Vessel Supports for PWRs and BWRs", in N. Shah and P. E. MacDonald, Eds., Residual Life Assessment of Major Light Water Reactor Components - Overview, NUREG/CR-4731, Vol. 1, Nuclear Regulatory Commission, June 1987.

23. R. E. Johnson, " Comments on ASTM E1035 Standard Practice for Determining Radiation Exposures for Nuclear Reactor Vessel Support Structures", Attachment to ASTM April (1990) Ballot Card, No. 4-90, May 1990; Johnson is a member of ASTM Committee E24.

# Contribution of the VENUS-Engineering Mock-up Experiment to the LWR-PV Surveillance

P. D'hondt, A. Fabry, G. Minsart, H. Aït Abderrahim, L. Leenders, R. Menil, S. De Leeuw, G. De Leeuw-Gierts
SCK/CEN, Mol, Belgium

F.B.K. Kam, R.E. Maerker
ORNL, Oak Ridge, TN, USA

E.D. McGarry
NIST, Gaithersburg, MD, USA

W.N. McElroy
CTS, Richland, WA, USA

ABSTRACT: This paper highlights selected accomplishments within the "Belgian PWR-Pressure Vessel Surveillance Programme" sponsored by the Belgian Utilities and its contribution to the "LWR-PV Surveillance Dosimetry Improvement Programme" supported by the U.S. Nuclear Regulatory Commission.

Major conclusions drawn from the VENUS PWR-Engineering Mock-up experiments are presented. Observed trends of theory- versus-experiment comparisons are summarized and compared to observations in PCA and in a Belgian Power Plant.

Further work planned to improve the theory-versus-experiment comparisons is outlined.

KEY WORDS: pressure vessel surveillance, light water reactor, neutron fluence measurement and calculation, benchmark field

The safety evaluation of a reactor pressure vessel requires the consideration of the vessel embrittlement resulting from the irradiation by fast neutrons produced in the reactor and propagating through the vessel. The determination of the related mechanical property changes over the whole operation period relies upon a surveillance programme and accurate calculation of the neutron damage exposure at any location in the vessel wall is needed to make proper use of the surveillance dosimetry measurements.

The VENUS PWR Engineering Mock-up experiment is part of the "Belgian PWR-Pressure Vessel Surveillance Programme" sponsored by the Belgian Utilities. Through a cooperation agreement, it contributes also to the "LWR-PV Surveillance Dosimetry Improvement Programme" supported by the U.S. Nuclear Regulatory Commission. Its objective is to validate the neutron transport calculation methods applied in the surveillance programmes; this objective is mostly addressed by systematic intercomparisons between calculations and measurements performed in research reactors (PCA and PSF at Oak Ridge Nat.Lab., VENUS at the SCK/CEN Mol) and in real power plants where a sufficient amount of measurement results is available.

The PCA (and the associated PSF) experiment was focused on the "radial" fast neutron and gamma propagation in a one-dimensional (slab) model corrected for transverse and vertical leakage, or in a X-Y model corrected for vertical leakage.

The VENUS PWR Engineering Mock-up is an additional step towards a more precise modelling of actual PWR ex-core configurations (see fig. 1) :

- core made of fuel pins having the same diameter and similar U-235 enrichment, with the same pitch as the 17x17 fuel assemblies of the modern PWR's,
- staircase shaped core boundary, with a steel baffle of the same thickness as in actual PWR's,
- barrel and thermal shield simulator (PAD) at distances from the core representative of the actual ex-core geometries.

The VENUS mock-up allows to study both radial and azimuthal neutron and photon distributions in an environment representing a generic 3-loop power plant.

The VENUS programme was established in three parts, all referring to the same modelisation of a PWR :

VENUS-1 : the core was made of a regular pattern of fresh $UO_2$ fuel pins; due to fuel availability, the central zone was made of 3.3 W/o U-235 enriched Zr4-clad oxide fuel whereas the outer zone was made of 4.0 W/o U-235 enriched rods with SS-304 cladding.

VENUS-2 : is the same as VENUS-1 but the 8 outer rows of $UO_2$ fuel pins in the branches of the cross have been replaced by MOX (2.0 W/o U-235 and 2.7 W/o Pu) fuel pins having the same diameter and clad. This modification was a simulation of a "low-leakage" loading defined in order to investigate how far major conclusions from the clean $UO_2$ core configuration (VENUS-1) could be affected by significant amounts of Pu-239 in the outer part of the core loading.

VENUS-3 : the core shape and dimensions, the baffle and other ex-core components were kept unchanged, but the core loading has been modified in order to simulate the PLSA (Partial Length Shield Assembly) concept, aiming at a strong decrease of the neutron fluence in the circumferential welds of a vessel. The PLSA zones in VENUS-3 encompass five fuel rows in two opposite branches of the cross-shaped core. To build these zones, the 4.0 W/o SS-clad fuel has been put in the central part of the loading and the 3.3 W/o Zr4-clad pins, which are dismountable, were located in the branches of the core; the pins of the two PLSA zones have been dismounted and the UO2 pellets have been replaced by stainless steel rods over half the core height.

In the VENUS configurations, the investigated parameters were the pin-to-pin power distribution, the propagation of the neutrons outside the core, the damage exposure and the gamma heating in the steel structures. Most important in a surveillance programme is the fast neutron fluence accumulated in the vessel at the maximum of the azimuthal distribution; the experimental effort was thus concentrated on this aspect of the computer code validation.

The experimental techniques used for determination of the core power distribution, the absolute core power, the axial buckling and fast neutron propagation were very similar for all of the three benchmark experiments. All equivalent fission flux measurements have been benchmark-field referenced to the MOL cavity uranium-235 fission spectrum standard neutron field and are traceable to a flux transfer from the NIST californium-252 standard neutron source. These techniques and the comparisons between measurements and results computed at MOL, have been reported in previous papers 1, 2, 3, 4, 5] for the VENUS-1 and -2 campaigns. For VENUS-3, the core description and absolute power measurements were reported by SCK/CEN [6,7], and the theoretical analysis has been performed by ORNL [8,9].

In parallel, the calculational procedure adopted for the analysis of ex-core neutron propagation and gamma field in PCA/PSF and VENUS, has been applied to power reactors (BR3 and a Belgian power plant). An important difference between the analysis of benchmark experiments and the study of power reactors is that the pin-to-pin power distribution is not measured in actual PWR's and this distribution is varying during core cycles; one has thus to rely on computations confirmed by sparse measurements in thimbles for defining the detailed fast neutron source distribution across the core, and particularly in the outer assemblies of the loading. This remains a sensitive point of the study, since the assemblies of the outer row of a power reactor contribute for more than 80 % to the fluence at the vessel (maximum azimuthal). It should be noted also that only few ex-core measurements are performed in power reactors : in accelerated locations such as surveillance capsules and sometimes beyond the vessel, in the cavity between the vessel thermal insulation and the concrete shield; this makes more difficult to cross-check calculations and measurements and to normalize them to the core power.

Several major conclusions can be drawn from the VENUS-1 and -2 benchmark exercices, from the previous PCA experiment and from the PWR surveillance data.

- the procedure adopted by SCK/CEN for computing pin-to-pin power distribution across the core (transport code DOT 3.5 with either the SCK/CEN 40-group cross section library MOL-BR2-40GR or the 27-group US set derived from the DLC43/CSRL 218-group reference cross section library) yields an excellent agreement with the experimental results : for VENUS-1 ($UO_2$ core) generally better than 0.6 %, with local deviations not exceeding - 3% close to absorber pins (pyrex) and + 5 % along the outer baffle; for VENUS-2 (with a MOX outer zone), the power in the central zone is slightly underestimated (- 2 to - 3 % on the average) and the fission density in the MOX zone is overestimated (+ 4 to + 5 %) with the same, local, larger deviations as in VENUS-1.
- the calculational scheme, based on the transport code DOT 3.5 and the VITAMIN-C 171-group cross section set (derived from ENDF/B-IV) and elaborated by the SCK/CEN for the analysis of ex-core fast neutron propagation, produces relative azimuthal distributions of threshold detector responses within the experimental uncertainty margins.
- an underestimation of these responses is observed in the radial direction : the C/E values, while azimuthally uniform, decrease when the distance from the core boundary increases, and the corresponding slope is higher for the highest thresholds (see table 1). This trend is similar both for VENUS-1 and VENUS-2; it had already been observed in the PCA results (see fig. 2).
- one could expect that the above-mentioned trend can be extrapolated for larger distances from the core; however, the analysis of measurements performed in a power reactor does not show this effect for any detector when comparing C/E results for surveillance capsules and for ex-vessel locations (see Table 2). The presence of the concrete shield just beyond the ex-vessel measurement points could have some influence (insufficient modelling of the concrete in the calculations ?).
- the procedure for analysing the gamma heating and gamma doses outside the core appears to be very satisfactory (VENUS-1 campaign).

Tentative improvements are already envisaged :

- the radial slope observed for the C/E values could be partially removed when using updated inelastic scattering cross sections for iron, as evaluated by C.Y.FU [10]. On the basis of a comparative study performed by M.L. WILLIAMS (LSU, Oak Ridge) [11], the corresponding improvement has been estimated and is indicated in table 1 (values between brackets).

- the high energy group cross sections of oxygen and hydrogen, and especially the anisotropy of the scattering, should be scrutinized. Legendre expansion up to P3 could be not sufficient to represent the marked "forward peaking" in the scattering of fast neutrons by water. If a P5 set has to be produced, it should be qualified by specific benchmarks.
- improved data for the anisotropy of scattering in water could be also beneficial when computing the back-scattering of concrete, thus leading to a more consistent pattern of C/E for ex-vessel locations. Here also, benchmark studies of a concrete environment should be useful.

It is hoped that, after such additional steps for refining the analysis are taken, the theoretical predictions, which presently already agree with the experimental results within 15 % for most sensors, will show deviations of less than 10 %; most importantly the present inconsistent pattern should be removed.

REFERENCES


1. A. FABRY et al., "VENUS PWR Engineering Mock-Up : Core Qualification, Neutron and Gamma Field Characterization," Proceedings of the 5th ASTM-EURATOM Symposium on Reactor Dosimetry, Geesthacht (F.R.G.), September 24-28, 1984.

2. A. FABRY et al., "LWR-PV-SDIP : VENUS PWR Core Source and Azimuthal Lead Factor Experiments and Calculational Tests," BLG 586/NUREG CR-3323 report, to be published.

3. A. FABRY et al., "Improvement of LWR Pressure Vessel Steel Embrittlement Surveillance : 1984-1986, Progress report on Belgian Activities in cooperation with USNRC and other R&D Programs", Proceedings of the 6th ASTM-EURATOM symposium on Reactor Dosimetry, Jackson Hole, STP 1001, p. 17.

4. G. MINSART et al., "Activities of the SCK/CEN in the frame of LWR pressure vessel dosimetry", Proceedings of the 7th International Conference on Radiation Shielding, Bournemouth (U.K.) 12-16 September 1988.

5. H. AÏT ABDERRAHIM, "Implementation and Experimental Validation of a Computational Model for Neutron Dosimetry and Gamma Heating Analysis in PWR Reactors," PhD Thesis (in French) (University of ORSAY, France), Ed. by ANRT-Grenoble 90/PA11/2061, March 1990.

6. L. LEENDERS, "LWR-PVS-Benchmark Experiment VENUS-3 : Description and Qualification," SCK/CEN Report nr FCP/VEN/01, September 1988.

7. L. VERBRUGGEN, "LWR-PVS-Benchmark Experiment VENUS-3 : Absolute Core Power," SCK/CEN Report nr FCP/VEN/03, December 1989.

8. R.E. MAERKER, "Analysis of the VENUS-3 Experiments," NUREG/CR-5338 ORNL/TM-11106, August 1989.

9. R.E. MAERKER et al., "Analysis of the VENUS-3 Experiments", This conference (paper 120, poster session I).

10. C.Y. FU and D.M. HETRICK, "Update of ENDF/BV Mod-3 Iron : Neutron producing reaction cross-sections and angle correlations," ORNL/TM-9964, July 1986.

11. M.L. WILLIAMS and F.B.K. KAM, private communication.

TABLE 1 : Azimuthal Averaged C/E-values observed in VENUS-2

| Detector | Threshold (MeV) | Outer Baffle | Barrel | Water Gap II | Neutron Pad |
|---|---|---|---|---|---|
| Np-237(n,f) | 0.60 | 1.04 (1.05) | 0.96 (0.98) | 0.88 (0.89) | 0.85 (0.89) |
| Rh-103(n,n') | 0.80 | 1.00 (1.01) | 0.93 (0.95) | 0.90 (0.92) | 0.83 (0.87) |
| In-115(n,n') | 1.20 | 1.01 (1.02) | 0.93 (0.96) | 0.91 (0.94) | 0.81 (0.86) |
| Ni-58(n,p) | 2.80 | 1.01 (1.03) | 0.94 (1.02) | - - | 0.86 (0.90) |
| Al-27(n,α) | 7.20 | 0.96 (0.99) | 0.86 (0.91) | 0.88 (0.94) | 0.70 (0.78) |

TABLE 2 : Range of C/E values obtained for threshold detectors

| Dosimeters (threshold) | Surveillance Capsules | Ex-vessel locations (PWR) | |
|---|---|---|---|
| | | 39 & 321° | 170 & 350° |
| $^{237}$Np(0.6 MeV) | 0.91 ± 0.05 | 0.86 ± 0.10 | 0.71 ± 0.08 |
| $^{238}$U (1.5 MeV) | 0.70 ± 0.06 | 0.71 ± 0.04 | 0.64 ± 0.08 |
| $^{58}$Ni (2.8 MeV) | 0.70 ± 0.05 | 0.80 ± 0.05 | 0.72 ± 0.10 |
| $^{54}$Fe (2.9 MeV) | 0.74 ± 0.03 | | |
| $^{63}$Cu (6.6 MeV) | 0.88 ± 0.07 | 0.93 ± 0.07 | 0.89 ± 0.10 |

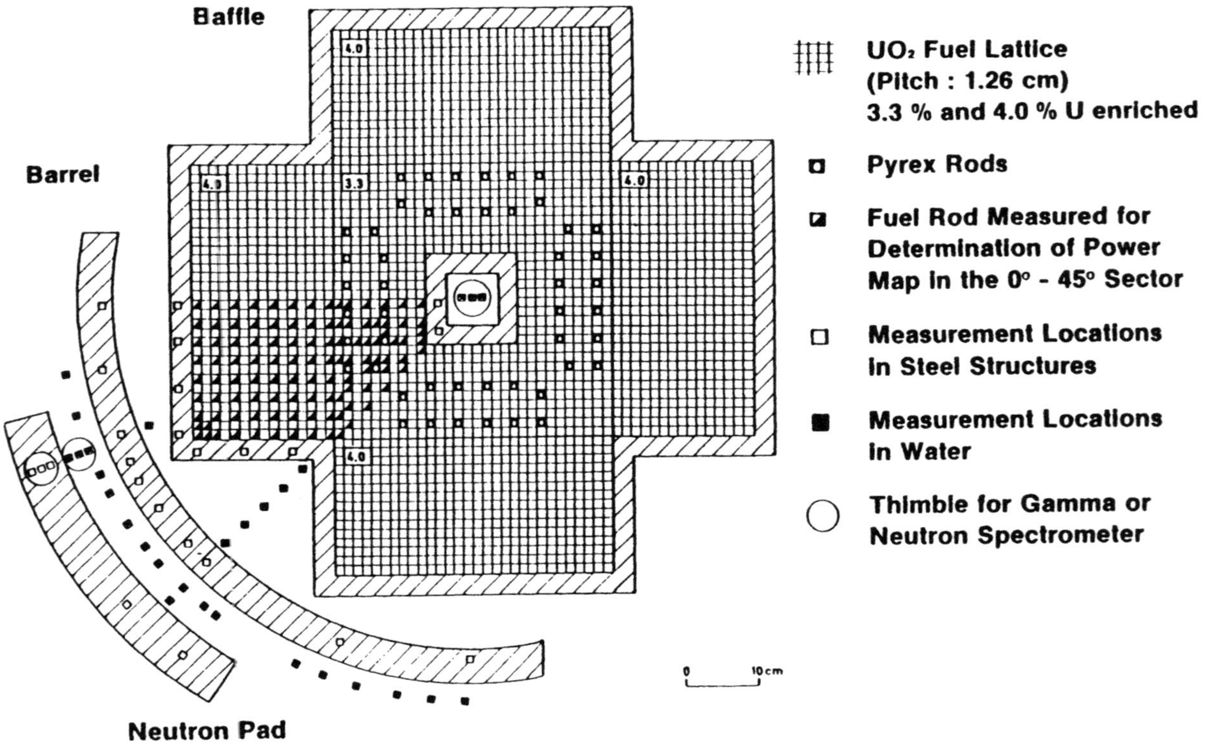


Fig. 1. GEOMETRY OF THE VENUS CORE CONFIGURATION FOR THE REACTOR PRESSURE VESSEL PROGRAM

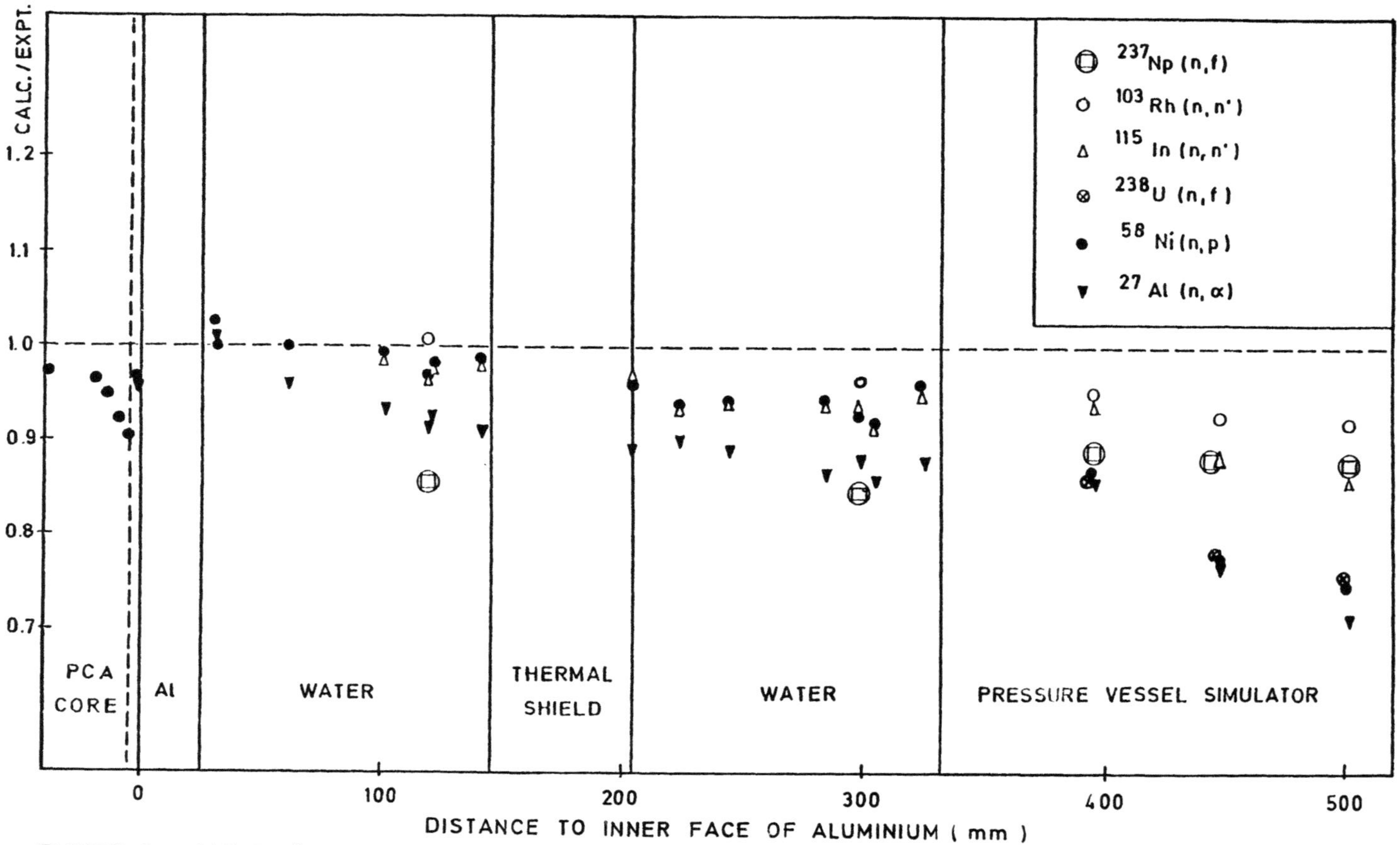


FIGURE 2. C/E Ratios for Absolute Equivalent Fission Fluxes per Unit Core Source Strength in PCA 12/13 LWR-PV Benchmark.

# BENCHMARK FIELD STUDY OF DEEP NEUTRON PENETRATION

Raymond Gold, James H. Roberts and
Christopher C. Preston
Metrology Control Corporation
Richland, Washington, USA 99352
and
James F. Morgan and Kenneth Sale
Lawrence Livermore National Laboratory
Livermore, California, USA 94550

## ABSTRACT

A unique benchmark neutron field has been established at the Lawrence Livermore National Laboratory (LLNL) to study deep penetration neutron transport. At LLNL, a tandem accelerator is used to generate a monoenergetic neutron source that permits investigation of deep neutron penetration under conditions that are virtually ideal to model, namely the transport of monoenergetic neutrons through a single material in a simple geometry. General features of the Lawrence Tandem (LATAN) benchmark field are described with emphasis on neutron source characteristics and room return background. The single material chosen for the first benchmark, LATAN - 1, is a steel representative of Light Water Reactor (LWR) Pressure Vessels (PV).

## 1.0 Introduction

In recognition of significant safety and economic issues associated with radiation-induced embrittlement of reactor pressure vessels, the U.S. Nuclear Regulatory Commission (NRC) sponsored the light water reactor pressure vessel surveillance dosimetry improvement program (LWR-PV-SDIP) over the last decade or more to improve, maintain, and standardize neutron dosimetry, damage correlation, and the associated reactor analysis procedures used to predict the integrated effect of neutron exposure to LWR-PV. Owing to the complexity of this embrittlement phenomenon, experimental and calculational strategies formulated in the LWR-PV-SDIP were subsequently implemented in a multi-laboratory international program.

Four principal laboratories collaborated in the conduct of this program: Centre D'Etude De L'Energie Nucleaire /Studiecentrum Voor Kernenergie (CEN/SCK), Mol, Belgium; the Hanford Engineering Development Laboratory (HEDL), Richland, Washington; the National Bureau of Standards (NBS), Gaithersburg, Maryland; and Oak Ridge National Laboratory (ORNL), Oak Ridge, Tennessee. Subsequently, collaboration with additional European laboratories was established: Rolls-Royce & Associates Ltd.

(RRA), Derby, U.K.; the United Kingdom Atomic Energy Estabishment Winfrith (UKAEEW), Winfrith, U.K.; and the Kernforschungsanlage, Julich, GmbH, (KFA), Julich, Federal Republic of Germany. Through the complementary capabilities of personnel at these seven laboratories, it has been possible to carry out a productive and highly cost-effective program. The major benefit of LWR-PV-SDIP has been and continues to be a significant improvement in the accuracy of the assessment of the current metallurgical condition and the remaining safe operating lifetime of LWR-PV.[1] These efforts have been extensively reported in the ASTM-EURATOM series of symposia on Reactor Dosimetry and a review of the status and accomplishments of the LWR-PV-SDIP was presented at the very last symposia in Jackson Hole.[2]

Benchmark fields have served as important elements in implementing experimental and calculational strategies adopted in the LWR-PV-SDIP. In the broadest sense, a benchmark neutron field represents a practical test bed for the testing, validation, and/or confrontation of different methods, be they experimental or theoretical. Progress by trial and error is, perhaps, the essence of benchmark field work. One essential characteristic of a benchmark field is that all environmental variables are well controlled and accurately defined. Benchmark fields can be used primarily for calculational purposes. In many instances benchmark fields are used for both experimental and calculational purposes simultaneously.

Independent contributions from US and EURATOM laboratories participating in LWR-PV-SDIP benchmark field studies have confirmed the existence of a serious and persistent discrepancy between neutron field calculations and experiment. The calculated-to-experimental (C/E) ratio is considerably less than unity in LWR-PV environments. Moreover this discrepancy increases, i.e. C/E decreases farther from unity, both with increasing neutron energy and with increasing depth of penetration into the PV.

This discrepancy is the most significant issue in current LWR-PVS neutron dosimetry efforts. Since the PV is non-replaceable, radiation induced embrittlement of the PV controls the effective lifetime of a nuclear power plant. Because of the critical nature of the PV, successful resolution of this discrepancy would have a major impact on current plant life extension (PLEX) efforts for LWR.

It must be emphasized that the same general discrepancy has arisen in another international neutron dosimetry program conducted for completely different reasons. Within nearly the same time frame that LWR-PV-SDIP was conducted, a joint US-Japan program was carried out to re-evaluate radiation exposures of atom bomb survivors. A final report describing the results and conclusions of this program, which has been designated Dosimetry System 1986 (DS86), has been issued.[3] In general, studies of Hiroshima and Nagasaki atom bomb survivors have provided valuable data for quantifying the effects of radiation on man. These data have been used by the United Nations Scientific Committee on the Effects of Atomic Radiation (UNSCEAR) for comprehensive assessments of the effects and risks of ionizing radiation. Conclusions of UNSCEAR reports[4,5] often form the basis that governments use for policy decisions on radiation protection standards.

An essential part of this joint US-Japan reassessment program for evaluating radiation doses at Hiroshima and Nagasaki has been the replication of Little-Boy in the form of a critical reactor assembly. Replication of Little-Boy, which was the atom bomb exploded at Hiroshima, was faithfully accomplished at Los Alamos National Laboratory (LANL). Available specifications together with spare or excess parts were used to set up and operate the Little-Boy replica as a low power critical reactor assembly at LANL. Replication of Little-Boy included the thick steel casing of the bomb. The thickness of this casing varied with polar angle, with a minimum thickness at the waist of the assembly (90 ), which is comparable to the nominal 8 inch thickness of a LWR-PV. The steel casing thickness increases with decreasing polar angle to a maximum thickness near the top of the assembly (0 ) that is considerably larger than the nominal LWR-PV thickness. Consequently, the Little-Boy replica shares one significant feature with all LWR-PV-SDIP benchmarks, namely deep penetration neutron transport in steel.

In view of the sparsity of neutron dosimetry data that actually exists from the Hiroshima and Nagasaki sites[6], calculations have been relied upon to generate A-bomb radiation exposures. Hence, this Little-Boy replica constitutes a benchmark radiation field for the testing and intercomparisons of experimental and calculational radiation metrology methods. Rigorous testing of these calculational methods using the Little-Boy benchmark was a crucial aspect of this program. Extensive experimental and calculational efforts have been expended to characterize the radiation field emitted by the Little-Boy replica.[7-9] Both the neutron and gamma-ray components of this mixed radiation field have been addressed. These Little-Boy replica characterization efforts are reviewed in Chapter 2 of the final report on DS86[10], where it is concluded that the C/E ratio is acceptable at the waist of the critical assembly (90 ), but C/E decreases and can be as low as 0.7 at the top of the critical assembly (0 ).

This C/E trend is evident in nuclear research emulsion (NRE) neutron spectrometry carried out at the Little-Boy replica.[11,12] However, the more detailed absolute power normalization experiments of the Little-Boy replica conducted with solid state track recorders(SSTR)[13] yield a C/E behavior for these NRE data that is qualitatively identical to LWR-PV-SDIP benchmark results. More specifically, one finds that C/E decreases with increasing penetration from approximately 0.7 at 90 to 0.5 at 0 . Moreover, these NRE data also confirm the trend for C/E to decrease with increasing neutron energy.

## 2.0 Speculative Ad-Hoc Solutions

The persistent nature of this discrepancy in different LWR-PV-SDIP benchmark fields, such as the PCA, VENUS and NESDIP, has fueled the speculation that deficiencies exist in the neutron cross section of iron. In fact, two different groups have introduced different adjustments of the inelastic cross section of iron to obtain improved C/E ratios.[14-17] However, these adjusted iron cross sections do not provide consistent improvements, but rather results that vary from one LWR-PV environment to another.[18]

Concern about the uniqueness of such cross section adjustments has already been raised.[2] Indeed, arbitrary changes in the iron inelastic cross section can reduce the magnitude of this discrepancy for certain cases, but there is no definitive experimental basis for making these adjustments. Since a given cross section is essentially a continuous function of neutron energy, it is an element of function space. Consequently, there exist more than a denumerable infinity of adjustments in the iron cross section that can be introduced to improve the C/E ratio attained in a particular LWR-PV environment. It is therefore not surprising to subsequently learn that this (non-unique) adjustment does not consistently provide improved C/E ratios in other environments.

Perhaps an even simpler ad-hoc solution can be obtained by adjusting the high energy portion of the fission neutron spectrum. Indeed, deep penetration transport in reactor environments places considerable credence on the high energy portion of the fission spectrum. Moreover for deep penetration neutron transport, emphasis is not placed on a fixed energy region of the fission neutron spectrum, but rather shifts upward in energy as the depth of penetration increases. Deficiencies in this high energy region, in particular use of a source term that provides too few high energy neutrons, could qualitatively explain both the energy and penetration dependent trends of the discrepancy obtained in the LWR-PV-SDIP and the Little-Boy benchmarks.

Examination of the experimental data defining the higher energy region of the fission neutron spectrum, especially above 8 MeV, reveals very poor statistical precision. Moreover, the latest comparison between theory and experiment[19] suggests significant uncertainties in this high energy region of the fission neutron spectrum. However, since the fission neutron spectrum is also an element of function space, arbitrary adjustments of the high energy region of this spectrum will yield solutions that possess the same non-uniqueness obtained from cross section adjustments.

## 3.0 Deep Penetration Neutron Benchmark Field

Benchmark neutron fields established for both LWR-PV-SDIP and Little-Boy have addressed specific real world issues. In so doing, these benchmarks have had to simulate the actual transport of fission spectrum neutrons through composite materials in complex geometries. While these elements were necessary to the basic purpose of these benchmarks, they present serious difficulties for the resolution of the observed discrepancy.

Indeed, it is difficult if not impossible, to resolve this significant discrepancy in neutron dosimetry let alone attain more stringent accuracy goals required to implement PLEX or establish personnel radiation protection standards with a "business as usual" approach. Use of the same experimental and theoretical methods, whether in power reactor PV environments or LWR-PV benchmark field tests, will in all likelihood repeat past history rather than provide new insights. Results from the series of dedicated LWR-PV benchmarks at VENUS and NESDIP verify this conclusion. What is needed is a new and independent benchmark field that provides simpler and more ideal conditions for the comparison of theory and experiment, so that a higher plane of understanding of deep penetration neutron transport can be achieved.

To satisfy this need for a higher plane of understanding, a unique benchmark field has been set up at the Lawrence Livermore National Laboratory (LLNL). At LLNL, a tandem accelerator facility is used to study neutron transport from a monoenergetic source through a single material in a simple geometry. Some of the more significant advantages of the Lawrence Tandem (LATAN) benchmark field are:

1. Simplified Geometry
   (a) Single Material
   (b) Point Source
   (c) Cylindrical Symmetry

2. Monoenergetic Neutron Source
   (a) Energy of the monoenergetic source neutrons can be varied
   (b) Independence of the fission neutron spectrum

The single material chosen for the first benchmark, LATAN-1, is a steel representative of LWR-PV. Because it has been demonstrated that active neutron detectors, such as fission chambers and proportional counters, produce non-negligible perturbations, measurements in LATAN-1 will initially emphasize passive dosimeters. Passive neutron dosimeters of high sensitivity and absolute accuracy will be used, such as SSTR and NRE. In contrast with conducting a sequence or series of irradiations, LATAN-1 irradiations have used a three dimensional array of these passive dosimeters embedded at different mean-free-path locations from the point source. This approach reduces the need for normalization between irradiations with a concomitant reduction of systematic uncertainty.

## 4.0 <u>The LATAN Benchmark Neutron Field</u>

Monoenergetic neutrons for the irradiations are generated by nuclear reactions in a gas target system. The deuteron beam from the LLNL Tandem is incident on a newly designed gas cell located at the end of the 0 beam line. By filling the cell with various gases, such as deuterium or tritium, monoenergetic neutrons are produced with energies from 2 to 20 MeV.[20] For a typical gas fill of 5 atm of deuterium, a neutron beam of $5x10^7$ neutrons/sec, strongly peaked around zero degrees, is produced. For LATAN-1, these neutrons are then incident on the center of a cylindrically symmetric assembly of the representative steel with appropriately placed neutron dosimeters, such as SSTR, located throughout the assembly. The absolute magnitude of the neutron flux is measured by the associated particle technique. A solid state detector is used to count the number of charged particles produced in the gas cell and this result is used to determine the absolute number of emitted neutrons.

One of the most important limitations of this kind of benchmark neutron field can be the room return background, i.e. the neutrons which scatter in the material surrounding the experimental facility and then find their way back into the assembly. The experimental hall at the LLNL Tandem was designed to reduce this problem. The walls are thin and most of the neutrons are allowed to pass out of the experimental area and to interact at a large distance in an earthen berm, where their effect is reduced by $1/R^2$. However, the floor of the experimental hall is heavy concrete, and the hall contains heavy experimental equipment such as switching and quadrupoles magnets, which can act as effective neutron scattering sources. Consequently, a combination of experimental and computational methods are used to understand and quantify the magnitude of this room return effect.

Experimentally the magnitude of the room return can be studied by using a pulsed deuteron beam and active neutron detectors. A short pulse ( 1ns) of deuterons is incident on the gas cell and active neutron detectors located within the assembly neutron transport calculations, which mock-up most of the significant neutron scattering sources. The agreement between measured and calculated values of the room return increases our confidence in the values assigned to this background effect.

The LATAN benchmark facility provides ideal conditions to study deep neutron penetration. In particular, LATAN-1 is a new and completely different benchmark field that can be used to investigate the current systematic discrepancies that plague calculations in LWR-PV environments. The unique characteristics of LATAN-1 will, hopefully, affect progress with these current difficulties.

## References


1. R. Gold and W. N. McElroy, "Radiation-Induced Embrittlement in Light Water Reactor Pressure Vessels," Nucl. Eng. and Design 104, 155-174 (1987).

2. R. Gold and W. N. McElroy, "The light water pressure vessel surveillance dosimetry improvement program (LWR-PV-SDIP): Past accomplishments, recent developments and future directions," in: Proc. of the Sixth ASTM-EURATOM Symposium on Reactor Dosimetry, June 1-5, 1987, Jackson Hole, WY, STP 1001, 44-61 (1989).

3. Radiation Effects Research Foundation. Dosimetry system 1986, U.S.-Japan joint reassessment of atomic bomb radiation dosimetry in Hiroshima and Nagasaki. Hiroshima, Japan: Radiation Effects Research Foundation; 1987.

4. United Nations Scientific Committee on the Effects of Atomic Radiation. Genetic and somatic effects of ionizing radiation. 1986 report to the General Assembly, with annexes. New York: United Nations; 1986.

5. United Nations Scientific Committee on the Effects of Atomic Radiation. Sources and effects of ionizing radiation. 1988 report to the General Assembly, with annexes. New York: United Nations; 1988.

6. W. E. Loewe, E. Mendelsohn, T. Hamada, T. Maruyama, S. Okajima, J. V. III Pace, M. Sakanoue, S. Kondo, T. Hashizume, J. Marcum, and W. A. Woolson, "Measurements of Neutron Fluences", Chapter 5 in US-Japan Joint Reassessment of Atomic Bomb Radiation Dosimetry in Hiroshima and Nagasaki, Final Report, Vol 1, Radiation Effects Research Foundation, 5-2 Hijiyama Park, Minami-Ku, Hiroshima, 732, Japan (1987).

7. "Little-Boy Replica, A Seminar to Discuss Construction, Operation, Measurements and Calculation", Los Alamos National Laboratory, Los Alamos, NM, September 1983.

8. "Measurements on a True Replica of Little-Boy", Special Session 8, Computer Applications in Health Physics, Proceedings of the 17th Midyear Topical Symposium of the Health Physics Society, Pasco, WA, February 1984.

9. "Replication of Little-Boy", Special Sessions I and II, 29th Annual Meeting of the Health Physics Society, New Orleans, LA, June 1984.

10. P. P. Whalen, "Calculation and Verification of Source Terms", Chapter 2 in US-Japan Joint Reassessment of Atomic Bomb Radiation Dosimetry in Hiroshima and Nagasaki, Final Report, Vol 1, Radiation Effects Research Foundation, 5-2 Hijiyama Park, Minami-ku, Hiroshima, 732, Japan (1987).

11. R. Gold, J. H. Roberts, and C. C. Preston, "Nuclear Research Emulsion Neutron Spectrometry at the Little-Boy Replica", HEDL-7559, October 1985.

12. J. H. Roberts, R. Gold, and C. C. Preston, "Measurements of the Absolute Neutron Fluence Spectrum Emitted at 0° and 90° from the Little-Boy Replica", 13th International Conference on Solid State Nuclear Track Detectors, Rome, September 23-27, 1985, Nucl. Track and Radiation Measurements 12, 553 (1986).

13. R. Gold, J. H. Roberts, and C. C. Preston, "Little-Boy Replica Power Normalization," 34th Annual Meeting of the Health Physics Society, Albuquerque, 25-29 June 1989.

14. J. Butler, M. D. Carter, I. J. Curl, M. R. March, A. K. McCracken, M. F. Murphy, and A. Packwood, "The PCA Replica Experiment: Part I, Winfrith Measurements and Calibrations, NUREG/CR-3324, AEEW-R 1736, Part I, NRC, Washington, D.C., Jan. 1984.

15. M. D. Carter, et al., "The ASPIS Iron Benchmark Experiment: Results and Calculational Model," in Proceedings, OECD Specialists' Meeting on Nuclear Data and Benchmarks for Shielding, Paris, 1980.

16. R. E. Maerker, "LEPRICON Analysis of Pressure Vessel Surveillance Dosimetry Inserted into H. B. Robinson-2 During Cycle 9," NUREG/CR-4439, ORNL/TM-1032, 1986.

17. R. E. Maerker, "Application of LEPRICON Methodology to LWR Pressure Vessel Surveillance Dosimetry," Sixth ASTM-EURATOM Symposium on Reactor Dosimetry, June 1-5, 1987, Jackson Hole, WY, STP 1001, 405-414 (1989).

18. S.P. Grant and A. Packwood, "Workshop on the NESDIP Benchmark", in: Proc. of the Sixth ASTM-EURATOM Symposium on Reactor Dosimetry, June 1-5, 1987, Jackson Hole, WY, STP

1001, 775-777 (1989).

19. D. Madland, and J. Rayford Nix, "New Calculations of Prompt Fission Neutron Spectra and Average Prompt Neutron Multiplicities", LANL Report LA-UR-81-2968.

20. J. E. Brolly and J. L. Fowler, "Monoenergetic Neutron Sources: Reactions with Light Nuclei" in Fast Neutron Physics Part 1. J. B. Marion & J. L. Fowler ed. Interscience Publishers Inc., London, 1960.

# ANALYSIS OF PRESSURE VESSEL CAVITY AND SURVEILLANCE CAPSULE DOSIMETRY FROM A TWO LOOP PWR

I.Remec and M.Najžer

"J.Stefan" Institute, Ljubljana

Slovenia, Yugoslavia

Abstract

At Krško Nuclear Power Plant cavity dosimetry experiment was performed. Measured activities from the cavity and simultaneously withdrawn capsule were used to adjust calculated neutron spectrum and determine neutron exposure parameters. Obtained good consistency of results confirmed the validity of cavity dosimetry. For several cycles, fuel loading schemes designed to reduce the maximum pressure vessel flux, were used. Analysis, indicating that approximately 25% reduction was obtained, is presented.

# Introduction

This paper describes the dosimetry and neutron transport calculation part of the continuation and improvement of the Pressure Vessel Surveillance Program at the Krško Nuclear Power Plant (NPP). A companion paper [1] discusses the results of surveillance capsule testing.

Krško NPP is a 632 MWe Westinghouse pressurized water reactor, which started operation in 1981. In the analysis of the first surveillance capsule, it was found that pronounced peak exists in the fast flux distribution at the pressure vessel (PV) inner wall. This peak gave rise to considerable projected end-of-life fluence of app. $6\,10^{19}\ ncm^{-2}$. Additional considerable increase above the given value was expected in the following cycles due to higher enrichment of reload fuel batches which when peripheral fuel loading scheme is used results in increased PV fast flux. For this reason the possibility to reduce PV maximum exposure was studied [2]. It was found that if for example once or twice burned fuel elements are placed on core flat edge (see Fig. 1) instead of fresh ones, app. 30% reduction of maximum PV flux can be obtained. Loading schemes meeting this requirement represent subclass of low leakage loading patterns. They were used at Krško NPP for the $5^{th}$ and subsequent fuel cycles.

To verify the effect of loading pattern modification cavity dosimetry was installed at the beginning of the $7^{th}$ cycle. It was withdrawn during the refueling one year later in the fall of 1989, simultaneously with the second surveillance capsule. In the following analysis of the fuel pattern modification effect and dosimetry from both locations is presented.

## Calculations

Two dimensional DOT-4.2 [3] and one dimensional ANISN [4] discrete ordinates transport codes were used for neutron flux calculations. Cross sections were taken from the SAILOR library [5]. 47 energy groups (with 19 groups above 1MeV ), $P_3$ expansion of scattering cross sections and $S_8$ angular quadrature was used throughout. Part of the reactor horizontal cross section shown in Fig. 1 was modelled in $R-\theta$ geometry using DOT 4 variable mesh with 84 to 97 radial and 60 azimuthal intervals. Axial model for R-Z calculations covered the height from app. 1m below to app. 2m above the core with 67 intervals, while 86 intervals in radial mesh described the geometry at $0^o$ azimuth. The same radial mesh was used for 1 dimensional radial calculations.

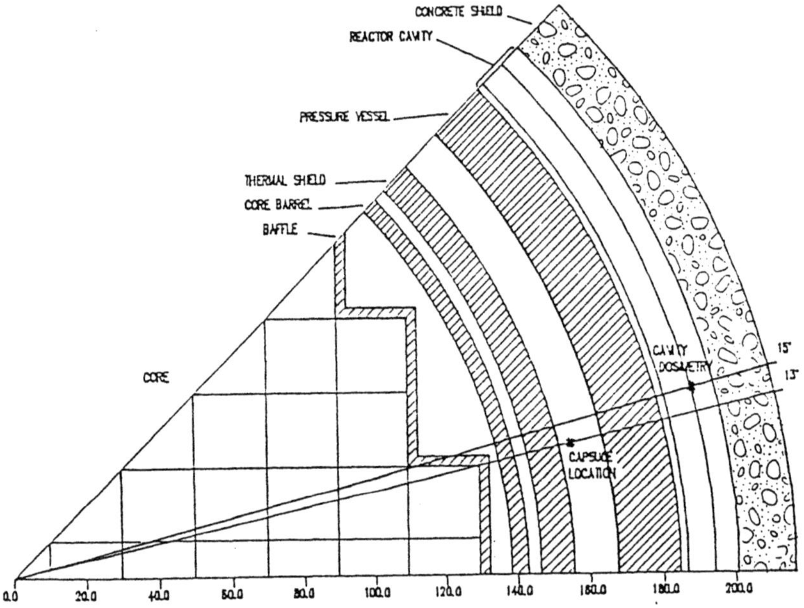


Figure 1: Octal section of the horizontal midplane, modelled in $R-\theta$ calculations. Positions of the cavity and capsule dosimetry are indicated. The location of maximum PV fast flux is along the $0^o$ azimuthal ray.

For the $7^{th}$ fuel cycle the $R-\theta$, R-Z and 1 dimensional R calculations were combined to obtain 3 dimensional distributions of group fluxes, following the method described in [6]. In the broad range covering some 50% of the core height around the midplane and extending radially beyond the PV outer wall the obtained shapes of axial reaction rates distributions were found to coincide closely with axial distribution of the power in the core (as used in R-Z calculation). Since surveillance capsule and maximum PV flux lay within this axial limits axial variations for the other cycles were approximated with the axial power distribution in the core. Measured axial distributions, obtained from the plant, were used. For the $R-\theta$ calculations the neutron source was prepared from the two dimensional power distributions obtained from core design diffusion calculations.

To take into account time variations of neutron flux for each of the 7 fuel cycles 2 direct transport calculations were done: one for the beginning and one for the end of cycle. Typically, the beginning to end of cycle differences for the threshold reactions, $\phi > 1MeV$ and displacement per atom rate (dpa/s) were below 10%. Linear variation of reaction rates with burnup was then assumed during the cycle. To account for quick power changes, it was assumed that flux is proportional to the core power and power variations obtained from the plant daily operation history were superimposed on the above described gradual changes due to fuel depletion.

# Surveillance capsule and cavity dosimetry

The surveillance capsule removed at the end of cycle 7 was standard Westinghouse surveillance capsule, which was attached at the thermal shield at the azimuth of 13° (see Fig. 1).

Cavity dosimetry experiment consisted of three sets of radiometric monitors, mounted on the stainless steel wire. Each set included Fe,Ni,Cu,Ti,bare Co, Cd shielded Co and $^{238}U$ or $^{232}Th$ fission monitor. Monitors were placed in small cylindrical holders fixed successively one to another. Medium set was attached to the wire at the height of the core midplane with top and bottom sets 1m above and 1m below, respectively. The experiment was lowered into the cavity from the bottom of the reactor pit and fixed at the top. After the irradiation the activities were measured with high resolution Ge-Li detector/computer system.

# Results and discussion

Comparison of calculated cycle averaged maximum PV fast fluxes is given in Table 1. It can be seen that modified loading schemes resulted in app. 25% reduction in maximum PV flux, from the average value of 3.6 $10^{10}$ $n\,cm^{-2}s^{-1}$ in the first 4 cycles to the average of 2.7 $10^{10}$ $n\,cm^{-2}s^{-1}$ in the last 3 cycles. Figure 2 shows that greatest reduction was obtained at the peak position as it was required. Location of the maximum remained unchanged, at 0° azimuth and app. 40 cm below the core midplane. In the last row of Table 1 the ratios of cycle averaged core midplane power densities of the outermost three fuel elements on the core flat edge are given. Obviously these ratios, which can be easily calculated from data in standard core design report, give good estimates of cycle to cycle variations of maximum PV flux.

Calculated and measured activities at the end of irradiation are given in Table 2. In general the agreement is good, thus confirming the validity of calculations and measurements. For the downcomer location the agreement is to within 10%. Calculated values with the exception of Np, are somewhat lower than measured. In the cavity calculated activities exceed the measured by app. 16% on the average. Calculated to measured ratios at the off midplane locations do not differ significantly from values at the midplane. This indicates that in the axial region covered by the measurements, neutron streaming in the cavity is not important and the symmetric $S_8$ quadrature, used in R-Z calculation, is adequate. However all measurements were still done relatively close to the core midplane where flatness of power axial distribution probably greatly reduces possible streaming effects.

Table 1: Comparison of Maximum $\phi > 1\,MeV$ in the Pressure Vessel, at the 1/4 Thickness Position. Cycle Averaged Values are Shown.

| | Cy.1 | Cy.2 | Cy.3 | Cy.4 | Cy.5 | Cy.6 | Cy.7 |
|---|---|---|---|---|---|---|---|
| $\phi > 1\,MeV^a$ | 3.54 | 3.67 | 3.65 | 3.43 | 2.68 | 2.77 | 2.71 |
| $\phi\,/\,\phi_{cy1}{}^b$ | 1.00 | 1.04 | 1.03 | 0.97 | 0.76 | 0.78 | 0.77 |
| $P/P_{cy1}{}^c$ | 1.00 | 1.03 | 1.03 | 0.97 | 0.74 | 0.75 | 0.73 |

[a] Fast flux in $10^{10} n\ cm^{-2}\ s^{-1}$.
[b] Fast flux in the cycle divided by cycle 1 fast flux.
[c] Cycle averaged core midplane power density for the outermost three fuel elements on the core flat adge at $\theta = 0°$, divided by the corresponding value for the cycle 1.

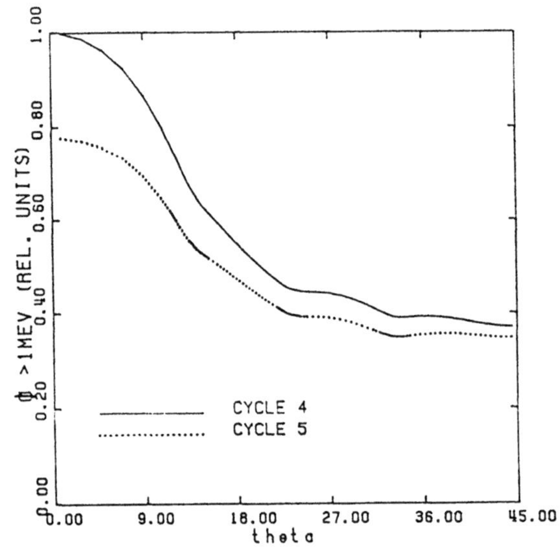


Figure 2: Azimuthal distribution of $\phi > 1MeV$ for the cycle 4 (out-in-in) and cycle 5 (modified fuel loading pattern).

Table 2: Comparison of calculated and measured end-of-irradiation activities.

| Reaction | H[b] | Activity [a] Calculated (dis $s^{-1}$ $atom^{-1}$) | Activity [a] Measured (dis $s^{-1}$ $atom^{-1}$) | C/M |
|---|---|---|---|---|
| Cavity location | | | | |
| $^{54}Fe(n,p)\,^{54}Mn$ | 106 | $4.69\ 10^{-17} \pm 20$[c] | $4.32\ 10^{-17} \pm 4$ | $1.09 \pm 0.22$[d] |
| $^{232}Th(n,f)\,^{137}Cs$ | 100 | $1.48\ 10^{-19} \pm 22$ | $1.16\ 10^{-19} \pm 23$ | $1.27 \pm 0.40$ |
| $^{58}Ni(n,p)\,^{58}Co$ | 94 | $1.47\ 10^{-16} \pm 21$ | $1.24\ 10^{-16} \pm 5$ | $1.19 \pm 0.26$ |
| $^{46}Ti(n,p)\,^{46}Sc$ | 88 | $1.59\ 10^{-17} \pm 23$ | $1.38\ 10^{-17} \pm 5$ | $1.16 \pm 0.27$ |
| $^{63}Cu(n,\alpha)\,^{60}Co$ | 82 | $1.32\ 10^{-19} \pm 21$ | $1.21\ 10^{-19} \pm 5$ | $1.09 \pm 0.24$ |
| $^{54}Fe(n,p)\,^{54}Mn$ | 6 | $5.33\ 10^{-17} \pm 20$ | $4.57\ 10^{-17} \pm 4$ | $1.16 \pm 0.24$ |
| $^{238}U(n,f)\,^{137}Cs$ | 0 | $7.79\ 10^{-19} \pm 21$ | $6.52\ 10^{-19} \pm 18$ | $1.19 \pm 0.33$ |
| $^{58}Ni(n,p)\,^{58}Co$ | -6 | $1.61\ 10^{-16} \pm 21$ | $1.33\ 10^{-16} \pm 5$ | $1.21 \pm 0.26$ |
| $^{46}Ti(n,p)\,^{46}Sc$ | -12 | $1.72\ 10^{-17} \pm 23$ | $1.54\ 10^{-17} \pm 5$ | $1.11 \pm 0.26$ |
| $^{63}Cu(n,\alpha)\,^{60}Co$ | -18 | $1.41\ 10^{-19} \pm 21$ | $1.30\ 10^{-19} \pm 5$ | $1.08 \pm 0.23$ |
| $^{63}Cu(n,\alpha)\,^{60}Co$ | -82 | $1.39\ 10^{-19} \pm 21$ | $1.26\ 10^{-19} \pm 5$ | $1.10 \pm 0.24$ |
| $^{46}Ti(n,p)\,^{46}Sc$ | -88 | $1.69\ 10^{-17} \pm 23$ | $1.56\ 10^{-17} \pm 5$ | $1.08 \pm 0.25$ |
| $^{58}Ni(n,p)\,^{58}Co$ | -94 | $1.56\ 10^{-16} \pm 21$ | $1.34\ 10^{-16} \pm 5$ | $1.16 \pm 0.25$ |
| $^{232}Th(n,f)\,^{137}Cs$ | -100 | $1.59\ 10^{-19} \pm 22$ | $1.21\ 10^{-19} \pm 22$ | $1.31 \pm 0.41$ |
| $^{54}Fe(n,p)\,^{54}Mn$ | -106 | $5.09\ 10^{-17} \pm 20$ | $4.37\ 10^{-17} \pm 5$ | $1.16 \pm 0.24$ |
| Downcomer location | | | | |
| $^{63}Cu(n,\alpha)\,^{60}Co$ | | $3.80\ 10^{-17} \pm 21$ | $4.15\ 10^{-17} \pm 7$ | $0.92 \pm 0.20$ |
| $^{54}Fe(n,p)\,^{54}Mn$ | | $6.48\ 10^{-15} \pm 20$ | $6.57\ 10^{-15} \pm 4$ | $0.99 \pm 0.20$ |
| $^{58}Ni(n,p)\,^{58}Co$ | | $1.02\ 10^{-14} \pm 21$ | $1.05\ 10^{-14} \pm 7$ | $0.97 \pm 0.21$ |
| $^{237}Np(n,f)\,^{137}Cs$ | | $1.41\ 10^{-15} \pm 24$ | $1.32\ 10^{-15} \pm 8$ | $1.07 \pm 0.27$ |
| $^{59}Co(n,\gamma)\,^{60}Co$ | | $3.77\ 10^{-12} \pm 40$ | $3.79\ 10^{-12} \pm 9$ | $1.00 \pm 0.41$ |
| $^{59}Co(n,\gamma)\,^{60}Co(Cd)$ | | $2.05\ 10^{-12} \pm 40$ | $2.15\ 10^{-12} \pm 11$ | $0.95 \pm 0.39$ |

[a] In disintegrations per second per target atom.
[b] Distance above core midplane in cm.
[c] Read as activity 4.69 $10^{-17}$ with standard deviation of 20 %.
[d] Read as calculated to measured activity ratio 1.09 with standard deviation of 0.22.

Uncertainties given in Table 2 for the calculated activities take into account cross sections and group fluxes uncertainties. Besides fission yield errors were added for fission detectors, and 5% error was joined to account for uncertainties in reaction-rate variations with time. For the Th foils the C/M ratio is considerably above the average. This probably indicates that the photofission correction of 20% applied to the measured activities, was too high. Since gamma field wasn't calculated, this correction was estimated from previous calculations and other data [7]. However, 20% uncertainty was conservatively attributed to the correction, resulting in high Th activity uncertainty. Similarly, photofission and correction for $^{235}U$ content contributed to high uncertainty quoted for $^{238}U$ foil. Other contributions to the measured activity errors taken into account are: $\gamma$-ray counter efficiency calibration uncertainty, random counting error, $\gamma$-ray attenuation correction error (for the cavity Ni and Cu dosimeters, and capsule Np dosimeter) and coincidence correction error ( for detectors where $^{60}Co$ activity is measured).

# Adjustment results

For the unfolding calculations LSL-M2 [8] least-squares logarithmic neutron spectra adjustment computer code was used. Activation cross sections and covariances were taken from [9].

Measured activities were converted to reaction rates at the end of irradiation (end of cycle 7) with the use of calculated conversion coefficients. Also, capsule activities, irradiated at different positions, were extrapolated to the center of the capsule using calculated gradients. Correlation matrix of so obtained "measured" reaction rates is shown in Table 3, together with their standard deviations.

Since complete uncertainty analysis was not available, the group fluxes autocorrelation matrix ( correlation matrix of group fluxes at the same location) was generated following the suggestions in ASTM E944-83 [10] and [11]. Number of groups was reduced to 33 and autocorrelation of the form:

$$C_{ij} = e^{-|E_i - E_j|/31.5\,MeV}$$

where $E_i$ and $E_j$ denote mean energy of the group i and j, was taken. Standard deviation of 20% and 40% was attributed to the group fluxes with energy above 0.1 MeV and below 0.1 MeV, respectively. For the correlation of group fluxes at any two different locations, for example between surveillance position and PV quarter thickness, autocorrelations given above, multiplied by cross-correlation factor of 0.8 were assumed. Series of calculations, performed for different autocorrelation matrices generated or taken from the literature [12], [13], showed, that adjusted exposure parameters are quite insensitive to the detailed matrix shape, and indicated that the above assumption is probably satisfactory. However, adjustments at the locations without dosimetry strongly depend on the cross-correlation factor. Therefore more realistic estimates are desired to improve the reliability and accuracy of adjustment results.

Table 3: Correlation Coefficients and Standard Deviations of Measured Reaction Rates.[a]

| | σ % | $^{238}U$ | $^{58}Ni$ | $^{54}Fe$ | $^{46}Ti$ | $^{63}Cu$ | $^{237}Np$ | $^{58}Ni$ | $^{54}Fe$ | $^{59}Co$ | $^{59}Co^b$ | $^{63}Cu$ |
|---|---|---|---|---|---|---|---|---|---|---|---|---|
| Cavity | | | | | | | | | | | | |
| $^{238}U$ | 19 | 100 | | | | | | | | | | |
| $^{58}Ni$ | 7 | 23 | 100 | | | | | | | | | |
| $^{54}Fe$ | 6 | 24 | 84 | 100 | | | | | | | | |
| $^{46}Ti$ | 7 | 21 | 70 | 80 | 100 | | | | | | | |
| $^{63}Cu$ | 7 | 22 | 60 | 67 | 57 | 100 | | | | | | |
| *Downcomer* | | | | | | | | | | | | |
| $^{237}Np$ | 11 | 16 | 35 | 37 | 31 | 33 | 100 | | | | | |
| $^{58}Ni$ | 9 | 16 | 57 | 60 | 49 | 42 | 28 | 100 | | | | |
| $^{54}Fe$ | 6 | 21 | 74 | 86 | 70 | 57 | 36 | 67 | 100 | | | |
| $^{59}Co$ | 11 | 12 | 34 | 38 | 32 | 56 | 21 | 32 | 43 | 100 | | |
| $^{59}Co^b$ | 12 | 11 | 30 | 33 | 28 | 49 | 19 | 28 | 38 | 35 | 100 | |
| $^{63}Cu$ | 9 | 15 | 42 | 47 | 40 | 69 | 27 | 39 | 53 | 49 | 43 | 100 |

[a] Matrix is symmetric, the lower half is given. Coefficients are multiplied by 100.
[b] Co is cadmium shielded.

Table 4: Flux and dpa Rate Adjustments and Standard Deviations at Downcomer, Cavity and 1/4 Pressure Vessel Thickness as Functions of Measured Reaction Rates used in Unfolding.

| Location | | Before Adjustment | Downcomer Dosimetry | Cavity Dosimetry | Downcomer + Cavity |
|---|---|---|---|---|---|
| downcomer | $\phi > 1MeV$ | $0 \pm 20$[a] | $1 \pm 6$ | $-11 \pm 13$ | $1 \pm 6$ |
| | $\phi > 0.1MeV$ | $0 \pm 20$ | $1 \pm 5$ | $-11 \pm 14$ | $1 \pm 5$ |
| | $dpa/s$ | $0 \pm 20$ | $1 \pm 5$ | $-11 \pm 14$ | $1 \pm 5$ |
| cavity | $\phi > 1MeV$ | $0 \pm 20$ | $1 \pm 13$ | $-14 \pm 8$ | $-14 \pm 7$ |
| | $\phi > 0.1MeV$ | $0 \pm 20$ | $1 \pm 13$ | $-14 \pm 10$ | $-14 \pm 7$ |
| | $dpa/s$ | $0 \pm 20$ | $1 \pm 13$ | $-14 \pm 9$ | $-14 \pm 7$ |
| 1/4 Pressure Vessel | $\phi > 1MeV$ | $0 \pm 20$ | $1 \pm 13$ | $-11 \pm 13$ | $-6 \pm 12$ |
| | $\phi > 0.1MeV$ | $0 \pm 20$ | $1 \pm 13$ | $-11 \pm 14$ | $-6 \pm 12$ |
| | $dpa/s$ | $0 \pm 20$ | $1 \pm 13$ | $-11 \pm 14$ | $-6 \pm 12$ |

[a] Read as adjustment of 0% with standard deviation of 20%.

In the Table 4 adjusted values of $\phi > 1MeV$ , $\phi > 0.1MeV$ and dpa/s, obtained from different adjustment runs, based on capsule dosimetry only, cavity dosimetry only and both used simultaneously, are compared. Adjustment based on capsule dosimetry ( column 4 in Table 4) gives marginal increase in exposure parameters, while efficiently reduces uncertainties to some 65% of the initial value in the cavity and PV and even to 25% to 30% at the capsule (dosimetry) location. Cavity dosimetry resulted in 11% to 14% decrease in values with standard deviation reduced to 40% to 50% in the cavity (dosimetry) location and 65% to 70% of initial uncertainty at other locations. Last column of Table 4 gives adjustments based on simultaneous use of cavity and capsule dosimetry. It shows that the adjustment at the position of dosimetry (either cavity or capsule) is almost unaffected by addition of dosimetry at the remote location, while adjustments in the PV are close to the average of the previous two runs. Some additional decrease in standard deviations is observed also.

Exposure parameters, obtained from the adjustment runs based on capsule only and cavity dosimetry only, are in good agreement. In most cases the differences remain within the one standard deviation error band. This leads to the conclusion that cavity dosimetry provides equally useful means for calculated flux verification and adjustment as capsule dosimetry.

## Conclusion

Modified loading schemes, with burned fuel elements located on the core flat edges resulted in approx. 25% reduction of maximum pressure vessel fast flux. This value is probably typical for small two-loop PWRs (121 fuel elements) of similar design, while higher reductions have been reported for three and four loop reactors.

Results for exposure parameters obtained from the surveillance capsule dosimetry and simultaneously withdrawn cavity dosimetry indicate good consistency, therefore confirming the validity and adequacy of cavity dosimetry used.

## Acknowledgement

Thanks are due to Messrs. Jenčič and Moškon for measurements of neutron monitor activities.

This paper is based in part on the work supported by the U.S.-Yugoslav Joint Fund for Scientific and Technological cooperation, in cooperation with the U.S. department of Energy, under Grant No. JF772.

## References

[1] M. Najžer,B. Glumac,A. Loose,I. Remec,"Analysis of the Second Pressure Vessel Surveillance Capsule from the Krško Nuclear Power Plant", this conference

[2] I. Remec, I. Kodeli, M. Ravnik, S. Slavič,M. Kromar, "Influence of the Fuel Loading Schemes on the Neutron Flux in the Pressure Vessel for the Two Loop PWR", Proc. ENC-4: Fourth International ENS/ANS Conference, Geneva,Switzerland, 1986

[3] W.A. Rhoades, " DOT IV, Version 4.2: Two-Dimensional Discrete Ordinates Radiation Transport Code System",RSIC Code Package CCC-320, ORNL, Oak Ridge, Tenn., 1979.

[4] W.W. Engle,Jr.," ANISN: A One-Dimensional Discrete Ordinates Transport Code with Anisotropic Scattering", K-1639,March 1967

[5] G.L. Simmons and R. Roussin, "SAILOR: Coupled,Self-Shielded, 47-Neutron, 20-Gamma Ray, P3, Cross Section Library for Light Water Reactors", RSIC-DLC-76, 3-8-1983, ORNL,RSIC

[6] R.E. Maerker, B.L. Broadhead and M.L. Williams,"Recent Progress and Developments in LWR-PV Calculational Methodology", Proc. Fifth ASTM-Euratom Symposium on Reactor Dosimetry, Sept. 1984,Vol. 2, p.639-648

[7] G. C. Martin,Jr. and C. O. Cogburn, "Special Considerations for LWR Neutron Dosimetry Experiments", Proc. Fifth ASTM-Euratom Symposium on Reactor Dosimetry, Sept. 1984, Vol. 1, p.399-406

[8] PSR-233 MICRO/LSL-M2, Least-Squares Logarithmic Adjustment of Neutron Spectra, RSIC Peripheral Shielding Routine Collection

[9] D.E. Cullen, P.K. McLaughlin, "The International Reactor Dosimetry File (IRDF-85), IAEA-NDS-41, Rev.1, 1986

[10] ASTM E 944-83, Standard Practice for Application of Neutron Spectrum Adjustment Methods in Reactor Surveillance

[11] F.W. Stallmann, Theory and Practice of General Adjustmen and Model Fitting Procedures, NUREG/CR-2222, 1981

[12] R.E. Maerker, B.L. Broadhead, B.A. Worley, M.L. Williams and J.J. Wagschal, " Application of the LEPRICON Unfolding Procedure to the Arkansas Nuclear One-Unit 1 Reactor", Nuc. Sci. Eng., 93, 137(1986)

[13] R.E. Maerker, "LEPRICON Analysis of Pressure Vessel Surveillance Dosimetry Inserted into H.B. Robinson-2 During Cycle 9 ", Nuc. Sci. Eng., 96, 263(1987)

# SURVEILLANCE NEUTRON DOSIMETRY AND CAVITY NEUTRON FLUX MONITORING AT CZECHOSLOVAK VVER-440 POWER REACTORS

B. Ošmera, M. Holman

Nuclear Research Institute, Řež
Škoda Concern, Plzeň, Czechoslovakia

ABSTRACT

The neutron flux density estimation at the surveillance chain and at the vicinity of the reactor pressure vessel is based on the results of the mock-up studies in the LR-0 experimental reactor. The neutron spectra were measured within the energy range 10 keV - 10 MeV at the surveillance chain location, the pressure vessel inner and outer surfaces in the central plane of the VVER-440 mock-up. To transfer the mock-up results to the power reactor we use the measured neutron spectra and space-energy or energy and space indices which are determined as the relation of integral neutron flux densities at the points $R_k$ and $R_l$ and denoted as

$$K\ (E_i,\ E_j,\ R_k/R_l) = \emptyset\ (E > E_i,\ R_k)\ /\ \emptyset\ (E > E_j,\ R_l).$$

## INTRODUCTION

Since the radiation embrittlement of the pressure vessel (PV) governs the PV service life-time a great attention has been paid to the investigation of the neutron field parameters in the vicinity of the PV. These parameters cannot be always calculated with sufficient accuracy and therefore it is necessary to carry out accurate measurements in benchmark fields and low power mock-up of the reactor.

In order to reach demanded quality in the space-energy neutron flux density distribution measurements, the corresponding neutron spectrometers and methods were developed and improved /2, 3/. Further progress has been reached since the Working Group on Reactor Dosimetry (WGRD) at the Czechoslovak Atomic Energy Commision has been established in 1980. Some published results associated with the WGRD activity could be found /1, 2, 3, 4/.

Since 1983 the main effort has been concentrated on the

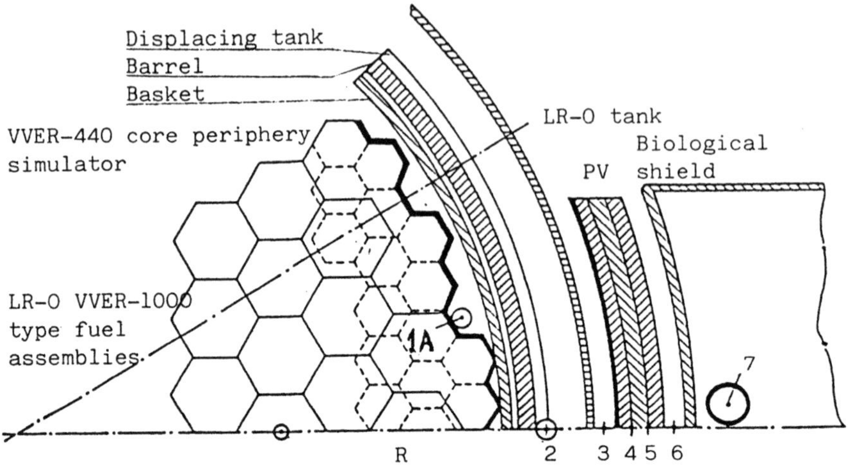


FIG. 1 - Schematic of VVER-440 mock-up.
(R2 = 1,625 m, R3 = 1,771 m, R6 = 1,92m)

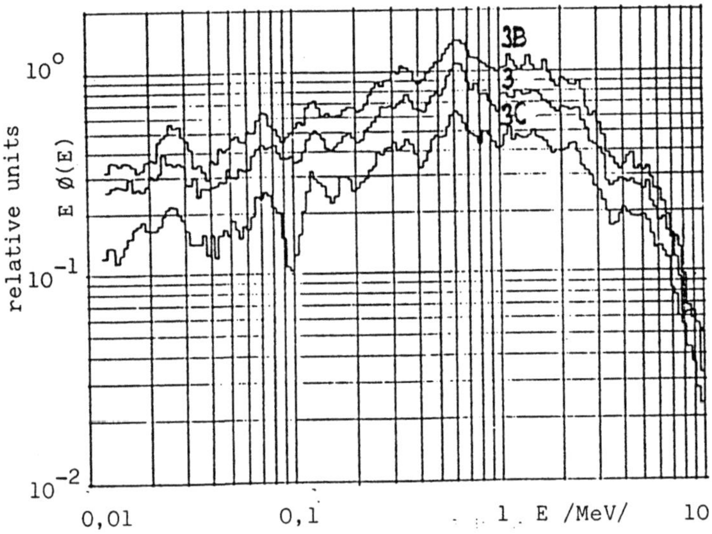


FIG. 4 - Neutron spectra measured in point 3 for 10 cm (3B), 12 cm (3) and 15,6 (3C) thickness of water layer between barrel and PV inner surface.

LR-0 reactor assembly. Suitable geometrical condition of the LR-0 experimental reactor facilitate construction of the VVER reactor type fullscale physical model in radial direction from the core to the biological shielding of the reactor.

It makes possible to obtain direct experimental estimation of the neutron spectra in the critical location of the power reactor (surveillance chain, inner and outer PV surface, atc.) and some space-energy indices for neutron exposure monitoring.

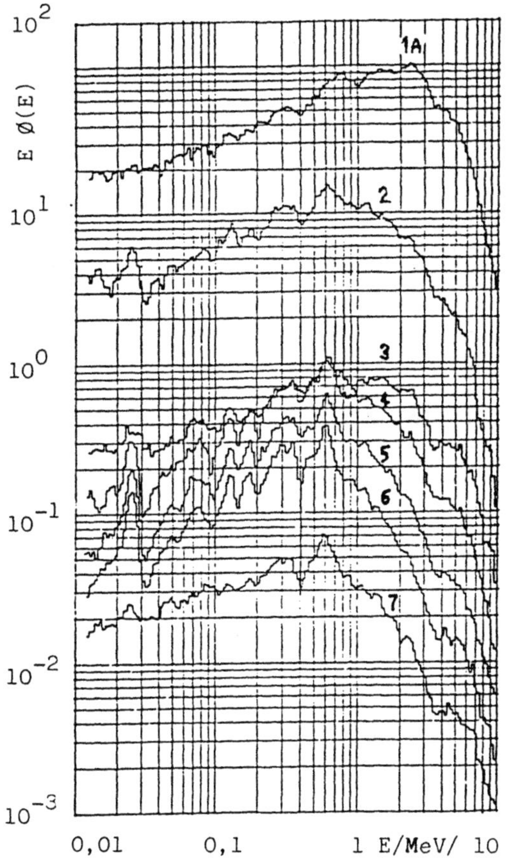


FIG. 2 - Measured neutron spectra normalized to Ø (E> 1 MeV) in point 3.

## THE VVER-440 MOCK-UP RESULTS AND PV NEUTRON EXPOSURE MONITORING

The VVER-440 symmetry sector ($60^{\circ}$) from the core periphery area to the inner region of the biological shield was modelled on the LR-0 experimental reactor (Fig. 1). The fuel composition in the model core was chosen in such a way that the radial - azimuthal power distribution within its periphery area was close as much as possible to the equilibrium power distribution in the core peripheral area of the commercial VVER-440. To simulate the water density reduction in operated VVER-440 reactor the steel displacing tank was placed in the space between the barrel an the LR-0 tank. The PV simulator consisted of three identical layers which could be successivaly shifted in radial direction in order to form an additional air gap layer for the spectrometric measurement over the PV thickness (measuring points 4,5). Further details concerning the mock-up design, experimental procedures, calculations and results could be found in /5, 6/.

The neutron spectra measured in the central plane of the mock--up core at the points numbered in the Fig. 1 are presented in Fig. 2. The neutron spectra measured in point 3 with different water layers between barrel and PV inner surface are in Fig. 4 (12 cm water layer thickness corresponds to a water density of 770 kg/m$^3$). The measurement of differential energy distribution of the fast neutron flux density were carried out within the energy range of 10 keV - 10 MeV by means of proton recoil detectors (a scintillation detector with a 10 x 10 mm stilbene crystal, two 40 mm diameter spherical proportional hydrogen filled counters, gas pressure 981 kPa and 392,4 kPa) /6/.

The measured neutron spectra with statistical errors are presented in Fig. 3 (uncertainty in the energy scale is not included).

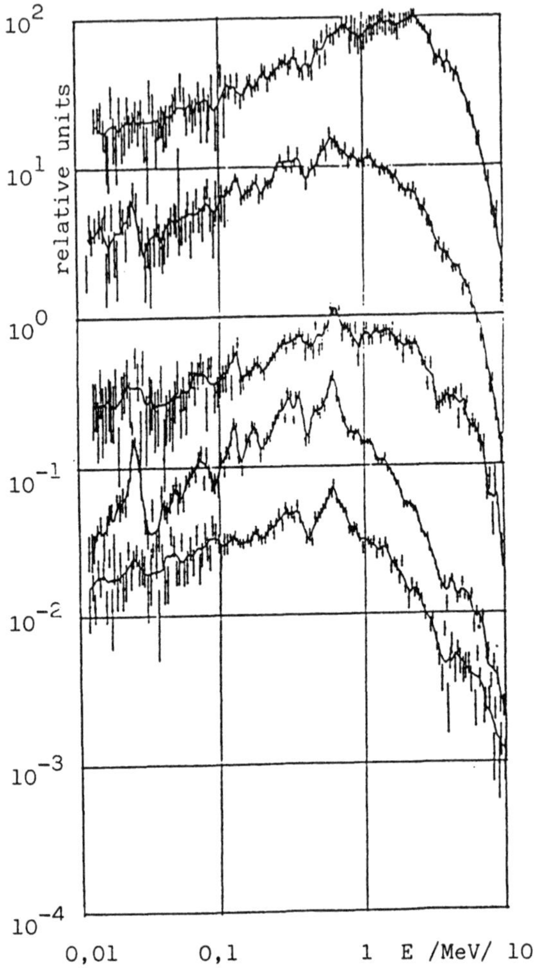


FIG. 3 - Measured neutron spectra with statistical errors. Solid curves represent smoothed spectra.

For measurement at point 2 the spectrometer was set in the displacing tank to remove the spectrum perturbation /6/. This spectrum represents the spectrum at the surveillance chain position. The measuring points 3 and 6 correspond to PV inner and outer surfaces, 1/4 PV thickness spectrum is close to that one measured at the point 4. In Table 1 some spectral indices are summarized. The integral cross sections

TABLE 1 - Relative integral energy distribution of neutron flux density in the central plane

| Point | Ø (E > 1,0 MeV) | Ø (E > 0,5 MeV) | Ø (E > 0,1 MeV) |
|---|---|---|---|
| surveillance chain | 1 | 1,8 | 3,033 |
| PV inner surface | 1 | 1,641 | 2,487 |
| 1/4 of PV thickness | 1 | 1,910 | 3,279 |
| PV outer surface | 1 | 2,593 | 5,361 |

Integral cross sections (DOSCROSS 81) $\sigma$ (E > 1,0 MeV) /m$^2$/

| Reaction | Surveill. chain | PV inner surface | 1/4 of PV | PV outer surface |
|---|---|---|---|---|
| $^{54}$Fe(n,p) | 7,491-30 | 1,035-29 | 7,709-30 | 5,863-30 |
| $^{63}$Cu(n,$\alpha$) | 5,532-32 | 1,237-31 | 8,011-32 | 6,668-32 |
| $^{93}$Nb(n,n') | 2,238-29 | 2,253-29 | 2,255-29 | 2,359-29 |

90 % response interval /MeV

| Reaction | Surveill. chain | PV inner surface | 1/4 of PV | PV outer surface |
|---|---|---|---|---|
| $^{54}$Fe(n,p) | 2,2-7,7 | 2,3-8,7 | 2,2-8,7 | 2,1-8,7 |
| $^{63}$Cu(n,$\alpha$) | 4,9-11,0 | 5,2-11,7 | 5,1-11,5 | 5,2-11,6 |
| $^{93}$Nb(n,n') | 0,63-5,2 | 0,62-6,2 | 0,575-5,5 | 0,475-5,1 |

TABLE 2 - Space - energy indices (maximum flux in the central plane)

K ($E_i$, $E_i$, surveillance position / $R_1$)

| Point $R_1$ | Energy $E_i$ /MeV/ 0,1 | 0,5 | 1 | 3 | DPA /s$^{-1}$/ E > 0,1 MeV |
|---|---|---|---|---|---|
| PV inner surface | 11,74 | 10,9 | 9,63 | 7,01 | 10,04 |
| 1/4 of PV thickness | 15,04 | 15,29 | 16,22 | 16,04 | 14,71 |

/4/ Reactor Dosimetry. Proceedings of the Czechoslovak Seminar, October 20 - 22, 1987, W .G.R.D. Prague, 1988 (in Czech., ÚJI 156 16 Praha 5, Zbraslav).

/5/ OŠMERA, B., et al.: Mock-up Experiments for the Determination of the Radiation Loading of the VVER Reactor Type Pressure Vessel. Report ÚJV 7797, Řež, 1986 (in Russian, restricted distribution).

/6/ OŠMERA, B. et al.: Experimental Validation of Neutron Calculations for VVER Pressure Vessel Radiation Load Evalution. Reactor Dosimetry: Methods, Applications and standardization. ASTM STP 1001, 1989, p. 130.

/7/ OŠMERA, B., VYCHYTIL, F.: Measurement of Neutron Flux Density at PV Outer Surface of VVER-440, ÚJV - 7582 (internal report, in Russian), 1985.

/8/ VIKHROV, V. I. et al.: Investigation of Neutron Field Parameters at PV of VVER-440 Reactors (in Russian), to be published in Voprosy atomnoi nauki i tekhnika yadernych reaktorov, 1990.

/9/ POŠTA, S.: Actual Status of Nb Monitor Evaluation, interlaboratory note, 1989 (in Czech.).

# REACTOR VESSEL FLUENCE MONITORING AND REDUCTION

E.P. Lippincott and S.L. Anderson

Westinghouse Electric Corporation

Pittsburgh, PA 15230

## Abstract

Reactor vessel fluence monitoring has been carried out in a number of reactor power plants in order to provide for increased vessel life. A data base of results is being used to provide increased reliability and to indicate experimental uncertainties. Applications of vessel monitoring to validate fluence reduction measures have been made with good results. A number of techniques have been proposed for reducing the rate of vessel neutron exposure that are being considered for future implementation.

## Introduction

In many nuclear power plants, the reactor vessel can be the limiting component with respect to plant life. The vessel is of concern because fast neutron leakage from the reactor core causes embrittlement of the carbon steel, thus allowing for possible fracture under postulated conditions of transient thermal stress. Therefore, accurate definition and minimization of fast neutron fluence to the vessel directly relate to plant lifetime.

Since the neutron fluence in the vessel cannot be measured directly, it must be inferred from measurements near the vessel combined with calculated neutron transport. The most accurate fluence determination must make use of plant specific measurements combined with transport methodology that has been benchmarked both in similar power reactors and in relevant benchmark fields such as PCA, NESDIP, and VENUS.[1] Improvement in accuracy of the vessel fluence assessment can result in increased vessel life due to the reduction in uncertainty in the point in time at which a screening criterion is passed. However, the amount of life extension available through improved vessel exposure definition is limited and much more substantial gains can be made through exposure reduction. In particular, measures can be taken to minimize the fast neutron fluence at the most critical vessel locations such as welds. Examples of such measures are discussed below after the discussion of fluence determination methods.

Surveillance Capsule Measurements

Since the inside of the vessel receives the highest exposure, measurements of neutron fluence on the inside of the vessel are most important. In most cases, measurements inside the vessel are limited to surveillance capsules which have the disadvantage of only being available at infrequent intervals and they can only be located at a few points which may not be near critical vessel locations. Thus calculations which take into account both the time and spatial variations of flux are necessary to project the fluence to the vessel.

In recent years, the accuracy of surveillance capsule dosimetry measurements has improved. Since data from larger numbers of similar plants have become available, it has been possible to establish data bases of results that can be used to identify errors in data and data uncertainties. Table 1 presents a comparison of capsule data from 8 different plants with similar geometry normalized to the Fe54 np reaction. The conclusion reached from this set is that expected relative dosimetry results should fall within an error band of 4-8%. This is in sharp contrast to some of the results observed in Simon's analysis of earlier capsule results.[2]

Reactor Cavity Dosimetry

Because of the limitations on amount of data available from capsule dosimetry, a number of plants have instituted programs to measure the neutron fluence in the cavity region outside the vessel. These measurements can be carried out at multiple locations and for each fuel cycle, but suffer from the disadvantage of being further from the inside of the vessel. A description of typical cavity dosimetry programs was presented at the last symposium.[3]

Results have demonstrated that a combination of the surveillance capsule and cavity measurements together with detailed plant-specific neutron transport calculations is the most accurate approach currently available for vessel fluence monitoring. This conclusion is supported by comparisons of measurements with calculations in benchmarks such as PCA, NESDIP, and VENUS, and in power reactors. These have produced a substantial body of data useful for validation of vessel monitoring techniques, definition of fluence uncertainty, and indications of areas of needed improvement (e.g. handling of the iron inelastic cross section).

Since the cavity dosimetry results described in reference 3 were obtained, a much larger body of experience has been gained with cavity measurements. In particular, SSTR dosimetry application to cavity measurements has progressed to where high quality results have been demonstrated in a large number of cases [4] and the cavity SSTR fission deposits have been accurately benchmarked both in standard benchmark fields [5] and in reactor cavity exposures by comparison with radiometric fission monitors. The latter comparisons indicate that the SSTR dosimetry in cavity applications produces results comparable in accuracy to those in Table 1 for surveillance capsules.

The use of experimental results from cavity dosimetry is dependent on the accuracy of the transport calculation to extrapolate from the cavity to the inside of the vessel. This makes the cavity results very dependent on the relevant cross sections and computer modeling. The transport methods currently in use by the authors typically give results that show very little difference in calculation to experiment bias between surveillance capsule and cavity measurements in cases where both measurements are available. The cross sections that give these results are the Sailor 47 group set which is based on ENDF/B-IV. Recently ORNL has been calculating some of the same cases using ENDF/B-VI cross section sets and has obtained some improvement in prediction of higher energy neutron measurements.[6] However, the ORNL calculations still cannot get much improvement in the energy range around 1 MeV. Thus the improvement in prediction of some reaction rates leads to a worse prediction of the neutron spectrum in the cavity. It is concluded therefore that further work is necessary before the updated cross section evaluations can be routinely applied in the same fashion as the older benchmarked sets.

Recently, cavity measurements have been utilized to establish the efficacy of fluence reduction measures. A variety of schemes have been studied for the lowering of vessel exposure, including the reduction of neutron leakage from the core and use of special shielding. Some of these designs introduce asymmetries in fluence patterns that cannot be calculated with the same degree of confidence as more simple configurations, thus enhancing the importance of the measured profiles. It is, therefore, expected that vessel fluence monitoring measurements will continue to play an important role in achieving maximum power plant life.

## Vessel Fluence Reduction Measures

The exposure of the reactor vessel to fast neutrons can be reduced using two basic methods: reduce the amount of fast neutrons leaking out of the core region or increase the shielding between the core and the vessel. The first of these involves changing the distribution of power generation (fissions) in the outside fuel assemblies. Although in some cases this may involve the reduction of fluence at only a specific point on the vessel azimuth, in most cases the general reduction of vessel fluence is also desirable. Several techniques have been applied to accomplish this reduction:

TABLE 1

RELATIVE RADIOMETRIC RESULTS FROM 58 DEGREE CAPSULES

REACTION RATE (rps/nucleus)

| Capsule | Cu63 na | Fe54 np | Ni58 np | U238 nf* | Np237 nf* | Co59 ng | Co59 ng* |
|---|---|---|---|---|---|---|---|
| A | 5.71E-17 | 5.81E-15 | 8.03E-15 | 3.48E-14 | 2.76E-13 | 5.20E-12 | 2.73E-12 |
| B | 6.15E-17 | 5.81E-15 | 7.17E-15 | 3.63E-14 | 3.27E-13 | 4.77E-12 | 2.25E-12 |
| C | 5.41E-17 | 5.81E-15 | 7.72E-15 | 3.26E-14 | 3.04E-13 | 5.19E-12 | 2.61E-12 |
| D | 6.04E-17 | 5.81E-15 | 7.37E-15 | 3.33E-14 | 3.11E-13 | 5.46E-12 | 2.68E-12 |
| E | 5.47E-17 | 5.81E-15 | 7.22E-15 | 3.07E-14 | 2.76E-13 | 4.80E-12 | 2.47E-12 |
| F | 6.07E-17 | 5.81E-15 | 7.67E-15 | 3.58E-14 | 3.15E-13 | 5.59E-12 | 2.78E-12 |
| G | 5.79E-17 | 5.81E-15 | 8.03E-15 | 3.09E-14 | 3.20E-13 | 5.61E-12 | 2.95E-12 |
| H | 5.97E-17 | 5.81E-15 | 7.45E-15 | 3.58E-14 | 3.05E-13 | 5.51E-12 | 2.84E-12 |
| AVERAGE | 5.83E-17 | 5.81E-15 | 7.58E-15 | 3.38E-14 | 3.04E-13 | 5.27E-12 | 2.66E-12 |
| 1 SIGMA | 4.80% | | 4.41% | 6.57% | 6.29% | 6.42% | 8.24% |

Reaction rates have been normalized to Fe(n,p) reaction rate of 5.81E-15 reactions per second.

* Indicates that sensor was cadmium covered.

1. Placement of burned fuel on the periphery of the core

2. Insertion of poison rods in peripheral assemblies

3. Replacement of enriched fuel pins at the outer edge with natural or depleted uranium pins

4. Replacement of fuel pins at the outer edge with stainless steel pins

Each of these methods either reduces the fissions occurring in the fuel itself or removes fuel from the outside edge. The first method, sometimes referred to as a low leakage core design, is easy to implement in most plants and can result in fuel cost savings. When larger flux reductions are desired, a combination of the low leakage core with one of the other methods can be used. Vessel exposure rate reductions as large as a factor of 2 can be obtained with reasonable designs.

Fluence reduction using these types of fuel management techniques are limited by limits on peaking factors and the fuel amount needed to maintain the rated power. In addition, the reconstitution of burned assemblies to replace fuel at the outer edge can be a problem to complete during a refueling outage. The alternative of specially designed assemblies is also inconvenient due to increased costs and the restriction on loading pattern flexibility.

The insertion of poison rods allows for reduction in fission rate in only part of an assembly when protection of a circumferential weld is the major concern. Use of part length poison centered on the weld axial location provides the minimum impact on fuel cycle cost while achieving almost the maximum reduction at the weld. Such part-length techniques have also been applied using natural uranium or stainless steel in specially designed fuel pins.

The other technique of vessel fluence reduction involves increasing the shielding between the neutron source in the fuel and the vessel. Shielding is presently provided in reactors by the reactor internal structures (baffle plates, core barrel, and thermal shield or neutron pads in PWRs) made of stainless steel and by the water between these. Both the water and stainless steel serve a shielding function. The hydrogen in the water elastically scatters the neutrons to lower energies. However, the hydrogen cross section falls rapidly with increasing neutron energy. Thus the shielding effectiveness of hydrogen is limited above 1 MeV but it is particularly effective below this energy. The iron and other stainless steel components do not greatly reduce the neutron energy by elastic scattering but they do cause reflection of the neutrons thereby increasing the path length to reach the vessel. Of more significance is the inelastic scattering which is effective above 1 MeV, and plays the major role in reducing higher neutron energies down to the 1 MeV region.

The relative amount of water and stainless steel in current plants is such that the vessel shielding can be improved by insertion of additional steel. Since the high energy neutrons are transported from core to vessel, the most effective place to put additional steel for shielding is close to the core, so that the neutrons in the 1-10 MeV range are scattered down by the steel and then the water can scatter the neutrons in the 0.1 to 2 MeV range. Placing steel next to the core also has the benefit of increasing the reflection of neutrons back to the core resulting in an increase in fuel efficiency. Implementation of such reactor internals modifications should be considered for existing plants with vessel embrittlement problems and also for the next generation of plants currently being designed.

Another alternative for vessel shielding involves use of other materials that can provide increased shielding per unit thickness. Examples might be tungsten which has a relatively high cross section at higher energies, or materials which combine the properties of steel and hydrogen such as titanium hydride. Although such materials have not yet been tried, the combination of shielding using these materials with the other techniques discussed above are calculated to give vessel fluence reductions more than a factor of 5 in some cases.

## Conclusions

The use of surveillance capsule and ex-vessel dosimetry for monitoring neutron exposure to the reactor vessel has been established as an accurate and reliable technique. As a result, plants with vessel embrittlement problems can use dosimetry to minimize uncertainty in vessel condition that arises from the exposure term. Measures have also been taken by a number of plants to reduce the neutron exposure rate at the vessel by introducing changes to the reactor core. Additional flux reduction can be obtained by increasing the shielding between the core and vessel and such measures are currently being considered by several plants.

## References


1. Lippincott, E.P. and McElroy, W.N., "Power Reactor Benchmark Studies", Reactor Dosimetry: Methods, Applications, and Standardization, ASTM STP 1001, 1989, pp 308-313.

2. Simons, R.L., in "LWR Power Reactor Surveillance Physics-Dosimetry Data Base Compendium", W.N. McElroy,ed., NUREG/CR-3319, 1985.

3. Lippincott, E.P., Anderson, S.L., and Fero, A.H., "Application of Ex-Vessel Neutron Dosimetry for Determination of Vessel Fluence", Reactor Dosimetry: Methods, Applications, and Standardization, ASTM STP 1001, 1989, pp 147-154.

4. Ruddy, F.H., and Seidel, J.G., "Plant Life Extension Related Applications of Solid State Track Recorder Neutron Dosimetry", this conference.

5. Ruddy, F.H., and McGarry, E.D., "Benchmark Referencing of Solid State Track Recorder Neutron Dosimeters in Standard Neutron Fields", this conference.

6. Williams, M.L., Childs, R.L., and Asgari, M., "Analysis of H.B. Robinson PWR Vessel Fluence for Cycle 10 Utilizing Partial Length Shield Assemblies", NUREG/CR-5530, 1990.

# DETERMINATION DES CARACTERISTIQUES NEUTRONIQUES DU PROGRAMME DE SURVEILLANCE DES TRANCHES FRANCAISES - REP 900 MWe

J.C. NIMAL* - L. BOURDET* - T. VERGNAUD*

R. LLORET** - A. BEVILACQUA**

F. HEDIN*** - J.C. LEFEBVRE***

* CEA/DMT/SERMA - CEN/SACLAY -
91191 GIF SUR YVETTE CEDEX - FRANCE

** CEA/DRE - CEN/GRENOBLE -
B.P.85 X
38041 GRENOBLE CEDEX - FRANCE

*** EDF/SEPTEN - VILLEURBANNE -
12-14, AVENUE DUTRIEVOZ
69628 VILLEURBANNE CEDEX - FRANCE

RESUME

Le parc électronucléaire français comporte une trentaine de centrales de 900 MW standardisées dont la puissance nominale est 2775 MW thermiques. Ces centrales sont équipées de capsules de surveillance destinées à anticiper l'endommagement de la cuve, ces capsules contiennent également des dosimètres. Lors du retrait de la capsule, on extrait les dosimètres dont on mesure l'activité. Des calculs de spectres au niveau des capsules sont effectués à l'aide du programme de Monte Carlo TRIPOLI. Le code ACTIGE calcule les activités à saturation à partir des sections efficaces calculées pour chaque dosimètre, du diagramme de fonctionnement et des activités mesurées. Une analyse d'incertitudes fournit les matrices de variance covariance sur les spectres qui, combinées aux activités, permettent au code STAY'SL d'ajuster la meilleure forme de spectre avec laquelle les dosimètres seront interprétés. Le flux moyen pondéré mesuré est comparé à celui calculé par TRIPOLI. Le facteur d'anticipation, quotient du flux sur les éprouvettes au flux au point le plus chaud de la cuve, est déduit du calcul TRIPOLI. Ce papier passe en revue les différentes phases du calcul et de l'interprétation expérimentale et donne les résultats les plus significatifs.

## DESCRIPTION DU DISPOSITIF DE SURVEILLANCE

Un support (figures 1 et 2) de capsules est fixé par soudage sur la fac externe de l'écran thermique. Ces capsules contiennent des éprouvettes d métal de base et de métal affecté thermiquement par les soudures. Outre ce éprouvettes, elles contiennent les dosimètres intégrateurs suivants $Cu^{63}(n,\alpha)Co^{60}$, $Ni^{58}(n,p)Co^{58}$, $Fe^{54}(n,p)Mn^{54}$, $U^{238}(n,f)$ et $Np^{237}(n,f)$. L dispositif contenant les éprouvettes a une hauteur d'environ un mètre. De raidisseurs horizontaux (figure 3) relient le cloisonnement à l'enveloppe d coeur. Ils sont pris en compte car ils induisent des perturbations de spectre et d'intensité de flux.

## LES CALCULS DE SPECTRES PAR TRIPOLI ET LES CALCULS DE FACTEURS D'ANTICIPATION

Le code TRIPOLI (1) effectue les calculs de spectre sur les éprouvette et les dosimètres et permet de déterminer le facteur d'anticipation du flu ($E > 1$ MeV). Ce facteur assure le passage du flux de l'emplacement d référence de la capsule au flux au point le plus chaud de la cuve.

Ce code TRIPOLI résoud l'équation du transport par une méthode d Monte Carlo polycinétique. Le choix de la méthode de Monte Carlo permet

- une représentation à trois dimensions de la géométrie. Cette dernièr est traitée exactement y compris la description des raidisseurs, de éprouvettes et porte capsules (voir figures 1, 2, 3). Seul le coeur a été traité e homogène, hypothèse licite vu que nous nous intéressons aux neutron rapides.

- une représentation fine des sections efficaces à 315 groupes, ave distinction des collisions sur chaque élément et, au sein d'un élémen distinction des diverses interactions (élastiques, inélastiques). Les lo d'anisotropie sont décrites par points et les renvois énergétiques de faço exacte.

- une représentation fine à trois dimensions des sources de neutrons.

En ce qui concerne les sources, deux types d'assemblages ont ét considérés : les assemblages périphériques ou de deuxième rangée, pou lesquels les sources sont décrites crayon par crayon, les assemblages interne ayant une distribution uniforme de sources.

REP 900 MWe - SERIE "CPY"
GEOMETRIE DES CALCULS TRIPOLI

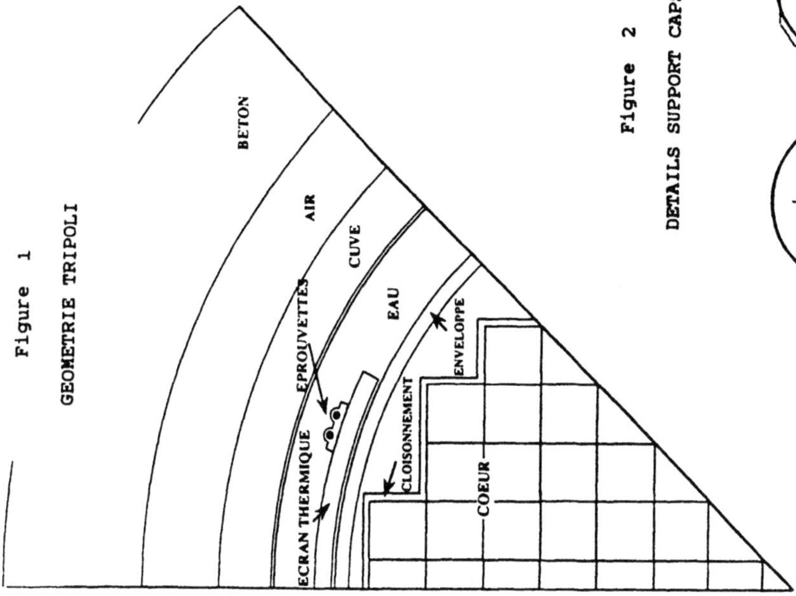


Figure 1
GEOMETRIE TRIPOLI

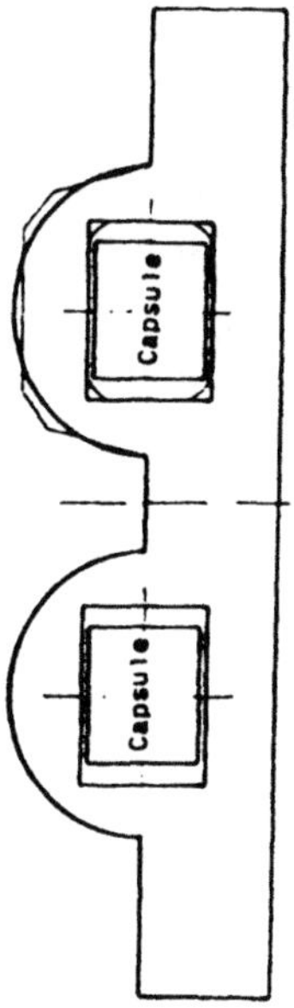


Figure 2
DETAILS SUPPORT CAPSULES

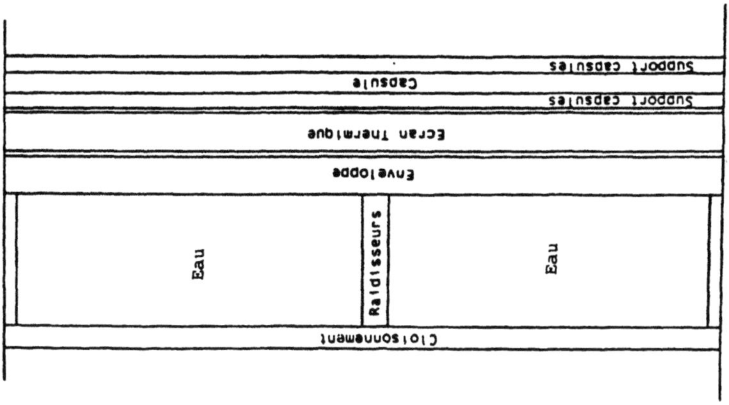


Figure 3
DETAILS DES RAIDISSEURS

Les sources de neutrons sont émises avec les spectres de $U^{235}$, $U^{23}$ $Pu^{239}$, $Pu^{241}$, chaque spectre correspondant à une distribution spatia proportionnelle à la densité de fission par isotope fissile. Cette dernière e déduite d'un calcul de cellule par APOLLO (2) en fonction du taux c combustion local moyen. La carte de puissance crayon par crayon est fourn par EDF/SEPTEN (3). La distribution axiale de puissance est déduite c calcul de coeur assemblage par assemblage. Les calculs sont normés à puissance nominale du réacteur en prenant en compte l'énergie dégagée p la capture des neutrons qui ne participent pas à la fission.

Les résultats essentiels par capsule sont :

- les spectres neutroniques
- les flux ($E > 1$ MeV soit $\phi_{1C}$, $E > 0.1$ MeV), les dpa
- les taux de réaction à saturation, soit $\tau_{KC}$, sur le dosimètre K.

On déduit de ces quantités les sections efficaces $\sigma_{1K}$ moyennes sur spectre au-dessus de 1 MeV :

$$\sigma_{1K} = \frac{\tau_{KC}}{\phi_{1C}}$$

Ces dernières quantités ne sont sensibles qu'à la forme du spectre et n à sa valeur absolue.

## L'ANALYSE D'INCERTITUDES

Les incertitudes prises en compte sont les suivantes :

1. incertitudes d'origine stochastique (Monte Carlo)
2. incertitude sur la puissance nominale du réacteur
3. incertitude sur les taux de fission par isotope fissile, l'énergie promp et le nombre de neutrons dégagés par la fission
4. incertitude sur la dimension des internes et de la température de l'ea
5. les incertitudes dues aux sections efficaces des matériaux sont en cou d'évaluation (4) de même que les incertitudes dues à la carte radiale puissance.

Nous obtenons ainsi des matrices de variance covariance pour spectre et des incertitudes sur le facteur d'anticipation du flux.

Les principaux résultats sont regroupés dans le tableau ci-desso incluant une incertitude due aux sections efficaces, estimée par ailleurs.

| | $1\sigma$ |
|---|---|
| Facteur d'anticipation du flux (2 calculs TRIPOLI indépendants) | 4.7 % |
| Flux $E > 1$ MeV dans les éprouvettes | 7.9 à 13.2 % |

## METHODOLOGIE DE LA DOSIMETRIE

### Métrologies d'activités

Les radionucléides $^{60}Co$, $^{58}Co$, $^{54}Mn$, $^{137}Cs$ sont mesurés en spectrométrie quantitative gamma au moyen d'une diode germanium hyper-pur étalonnée périodiquement sur des étalons en provenance du Laboratoire de Mesure des Rayonnements Ionisants (LMRI/SACLAY). Les corrections de linéarité, géométrie, atténuation dans les dosimètres et leurs boîtiers, pic somme, etc... sont appliquées d'une façon standardisée. Les incertitudes qu'on attribue aux activités spécifiques sont données dans le tableau 1. Elles sont très limitées et n'amènent aucune remarque particulière sauf pour les dosimètres fissiles : là, la présence de $^{60}Co$ en quantité non contrôlée en provenance des boîtiers en nickel ou en acier, exige de procéder à des corrections atteignant 25 % de l'aire du pic du $^{137}Cs$ du dosimètre $^{238}U$. En effet, l'énergie de ce pic coïncide avec celle de simple échappement de la raie 1173 keV du Cobalt. Cette correction réduit la précision à 7 % environ (1 $\sigma$).

### Méthodologie générale

Elle est donnée sur la planche 1, qui appelle quelques remarques :

- ACTIGE est un code qui permet de calibrer pour chaque dosimètre le diagramme de fonctionnement de la tranche en unité de flux. Il utilise pour ce faire les sections efficaces issues de TRIPOLI et calcule avec précision l'évolution de chaque dosimètre.

- La perturbation neutronique affectant la répartition axiale des taux de réaction et des flux, provenant de la présence d'épais renforts horizontaux en acier, est bien visible (fig. 4). Elle peut atteindre 12 %. Calculée par TRIPOLI et lissée, cette perturbation, une fois explicitée, réduit la dispersion des dosimètres d'une même capsule et permet de distribuer la fluence aux éprouvettes avec exactitude. On particularise alors dans la capsule un "point de référence", où la perturbation est nulle.

- Dans la série des réacteurs (dite "CP0") qui a précédé la série standardisée, il avait été nécessaire de tenir compte d'un déficit de puissance des éléments périphériques pour le premier cycle, le combustible alors chargé n'ayant pas atteint son équilibre. Cette correction -importante, jusqu'à 18 %- n'a pas lieu d'être dans la série standardisée.

- L'évaluation des incertitudes (variances-covariances) des activités, spectre calculé et sections efficaces dépendant de l'énergie, permet d'appliquer en routine la procédure d'ajustement STAY'SL (Réf.5). Elle fournit un spectre ajusté dont seule la forme est conservée, comme fournissant le meilleur compromis entre les informations neutroniques contenues dans le calcul TRIPOLI et celles contenues dans les taux de réaction. On redépouille alors les dosimètres sur la forme ajustée, et on fait la moyenne des flux obtenus, pondérée par les incertitudes des sections efficaces moyennes mesurées en benchmark (Réf.6 et tableau 1).

| INCERTITUDES EVALUEES EN % (1 ) | | | |
|---|---|---|---|
| | sur activités mesurées | sur sections eff mesurées en benchmark | totales sur le flux > 1 MeV mesuré |
| Np | 4.3 | 3.8 | 5.8 |
| U | 6.8 | 3.3 | 7.6 |
| Fe | 2.8 | 6.1 | 6.7 |
| Ni | 2.7 | 5.0 | 5.7 |
| Cu | 2.7 | 11.2 | 11.5 |

TABLEAU 1

La procédure STAY'SL réduit un peu les écarts à la moyenne pondéré dont la tendance générale est la suivante (indice de spectres moyens s 33 capsules) :

U/Np non significativement différent de 1,00
Ni/Fe ≈ 0.96 ± 0.03
Cu/Fe ≈ 1.13 ± 0.02

le fer fournissant la "meilleure" valeur, c'est-à-dire pratiquement la vale pondérée.

## RESULTATS

Les fluences neutroniques mesurées au moyen de la méthodologie de planche 1 sont attribuées aux éprouvettes. Par contre, les flux mesurés point de référence sont ramenés de la puissance moyenne réelle l'irradiation à la puissance nominale du réacteur, pour pouvoir êt comparés aux calculs, et celà en groupant en séries homogènes les capsul irradiées pendant la même durée et sous le même azimut. Par exemp (planche 2) :

- série S du CP0 (Fessenheim-Bugey) irradiées 1,5 an environ en déb de vie, azimut 15° ;
- série U du CPY, (900 MW standard), 4 ans environ, azimut 20° ;
- série V du CPY, 7 ans environ, azimut 20°, etc....

Ainsi regroupés, les résultats révèlent des effets de sé impressionnants : sur les 18 premières capsules de type U, l'écart type sur flux mesuré par le dosimètre Fer n'est que de 1,2 % (Fig. 5) ; sur 11 premières capsules de type V (non présentées ici), l'écart type des flux N est de 1,7 % ! Ces faits ne peuvent être attribués au hasard, ils so hautement significatifs d'une très grande fidélité dans la construction, fonctionnement des tranches, les mesures de puissances, de flux et dosimétrie. Cet écart type caractérisant la dispersion des mesures ne trad que la reproductibilité des processus dans les échantillons considérés, il donc en général inférieur à l'incertitude évaluée pour une mesure (donn dans le tableau 1 et illustrée par la figure 5), qui est la somme quadratique l'incertitude sur l'activité mesurée et de celle sur les sections efficaces.

FIGURE 4

# NEUTRONIC AXIAL PERTURBATIONS CAUSED BY STIFFENERS

They are calculated along the surveillance capsule, by means of the TRIPOLI-2 code (3-Dimensions), in term of fluxes and reaction rates used for dosimetry.

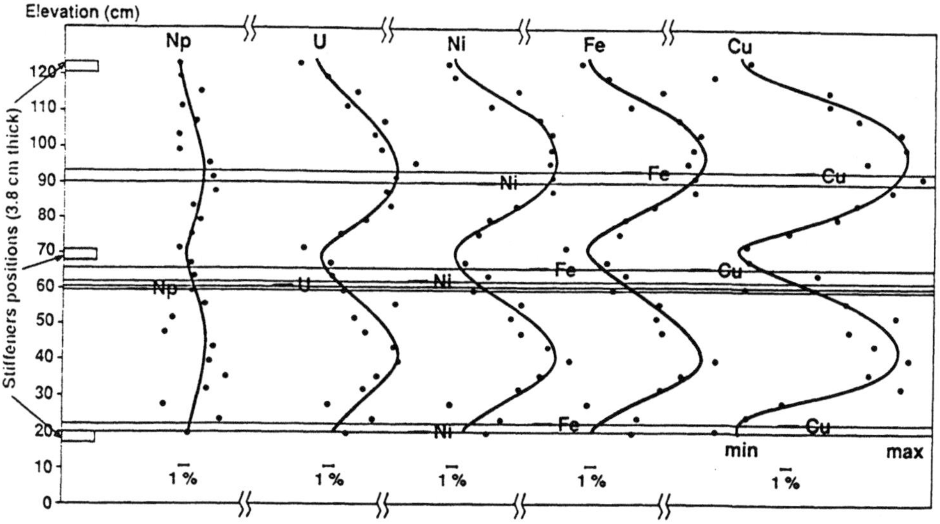


Axial distributions (smoothed) of the computed reaction rates at the 20° azimuth, and axial positions of the dosimeters (Standard 900 Mw units).

FIGURE 5

Rapport MESURE / CALCUL des $\Phi 1$ ( E > 1.0 Mev ) mesurés par les différents dosimètres,et leur moyenne pondérée

Echantillon de 18 capsules U (4 ans,20°)

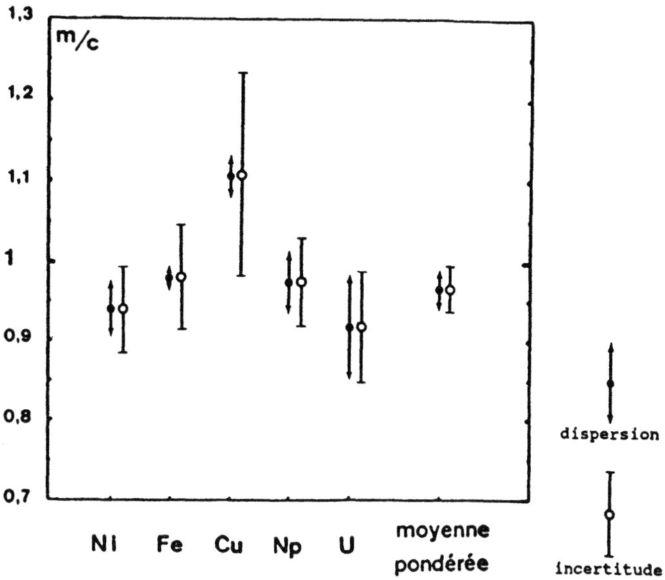

PLANCHE 1

**MÉTHODOLOGIE de la DOSIMÉTRIE du CPY**

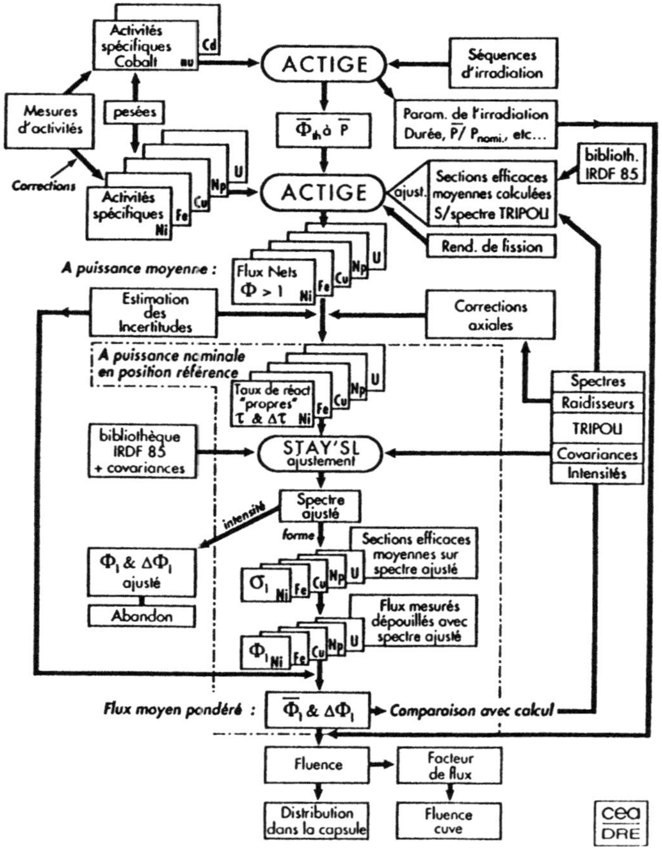


PLANCHE 2

**HISTOGRAMME DES MESURES DE SURVEILLANCE**

( Avril 1990 - 52 capsules )

**Comparaison mesure/calcul en terme de flux > 1 Mev**

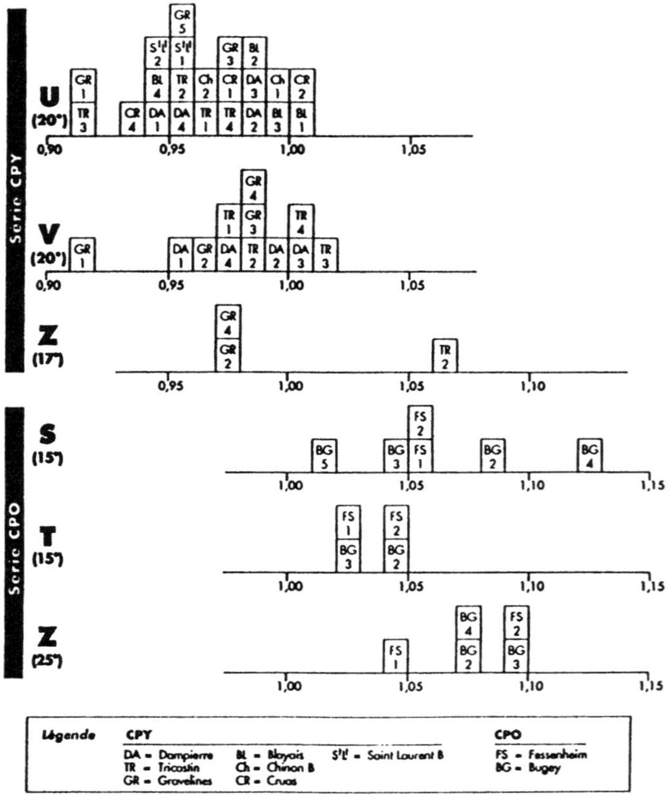

## DISCUSSION

Au cours de ce processus, on a délibérément écarté l'intensité du flux ajusté par STAY'SL. En effet elle constitue une sorte de moyenne entre calcul et mesure, son incertitude résultante ne parait pas réaliste. Nous retenons par contre la forme du spectre ajusté, mais il nous manque néanmoins le moyen d'exprimer son incertitude. Pour cette raison, les incertitudes que nous annonçons sur les moyennes pondérées nous semblent optimistes. Néanmoins l'écart calcul-mesure se situe en général à l'intérieur des incertitudes évaluées.

## REFERENCES


1. Programme de Monte Carlo polycinétique TRIPOLI-2
   Tomes I à III
   J.C. NIMAL et al.

2. APOLLO - rapport SERMA/S/1193
   A. HOFFMANN et al.

3. J.C. LEFEBVRE (EDF/SEPTEN).
   Communication privée.

4. Y. KODELI (CEA/SERMA/LEPP/SACLAY)
   Communication privée

5 F.G. PEREY - Least Squares Dosimetry Unfolding : the program STAY'SL - ORNL/TM.6062 - ENDF 254.

6 J.H. BAARD, W.L. ZIJP, H.J. NOLTHENIUS
  Nuclear Data Guide for Reactor Neutron Metrology
  EUR 12354 EN and also KLUWER Academic publishers - Dordrecht - The Netherlands

# CALCULATION AND MEASUREMENT OF NEUTRON FLUX DISTRIBUTION AT THE WWER-440 PRESSURE VESSEL

H.-C. Mehner, H.-U. Barz, B. Böhmer, U. Hagemann, I. Stephan

Zentralinstitut für Kernforschung Rossendorf

Germany

## ABSTRACT

Dosimetry experiments and neutron flux calculations have been performed for the pressure vessel of the WWER-440 reactor. The calculations were done with the 3-D Monte Carlo code TRAMO using Green's functions weighted by the concrete source distribution.

The experimental study has included analysis of several sets of activation detectors (Ti, Fe, Ni, Co, Cu, Ag, and Nb), irradiated at five azimuthal and ten axial locations at the reactor pressure vessel outer wall. The evaluation of the measured reaction rates was carried out with the adjustment code COSA based on the generalized least square method.

Comparisons between calculated and measured neutron exposure in the flux maximum indicate agreement on an absolute scale within 5 % to 15 % for different core loading schemes.

## INTRODUCTION

Neutron irradiation embrittlement of pressure vessel steels is one of the key points for the assessment of the structural integrity of reactor pressure vessels. The determination of the neutron exposure can be carried out with neutron transport calculation methods as well as by measurements with activation detectors. Calculations provide results with complete information referring to the spatial and energetic distribution of the neutron field in the environment of the pressure vessel, but the accuracy of the absolute values is limited. On the other hand, experiments promise a low degree of uncertainty, but only a limited number of energy integrals at some selected positions is accessible to measurements. So, it can be concluded that the combination of transport-theoretical calculations with dosimetry measurements is necessary to satisfy the requirements for neutron exposure determination.

Therefore, calculations with a 3-D Monte Carlo code as well as dosimetry experiments have been performed to study the neutron exposure at WWER-440 vessels. The investigations have been made for the Greifswald power reactors with a normal loaded core and with low leakage cores, where some high damage contributing fuel assemblies are replaced by burnt fuel assemblies or stainless steel dummies, respectively.

## CALCULATION METHOD

The calculation of the 3-D neutron field distribution at the pressure vessel is splitted into two steps:

- Calculation of the fission source distribution by standard burn-up calculations using diffusion approximation.
- Calculation of the neutron field at the pressure vessel by using the distribution of fission neutrons q(r). For this calculation transport methods are needed. For several reasons (e.g. description of geometrical details) the Monte Carlo method has been used.

Because of the great number of possible source distributions for all the reactor histories, partial solutions $G^i$ have been calculated for a certain grid of the core. So, the neutron transport problem has been solved for a unity source of fission neutrons in each single space element i of the core. Fixed detailed distributions over its cross-section are used for the outer assemblies of the core, otherwise equal distribution is assumed.

Presumed (as can be shown) that the shielding properties of fuel assemblies in the relevant energy range are approximately independent on the source distribution (that means in our case on the burn-up) we can represent the total fluence F depending on time for each neutron energy group j by summation over all p partial solutions to

$$F(r,j,t) = \sum_{i=0}^{p} \int_{0}^{t} q(t')\, G^{i}(r,j)\, dt'$$

($q^i$ - total number of fission neutrons in space element i).

The code TRAMO /1,2/ is a nonanalogous multigroup transport code for a given source distribution, which is able to take into account all relevant geometrical details of the reactor WWER-440. With a special given weight distribution depending on space and group oblidged for each particle, which itself can be calculated with the Monte Carlo code, one can obtain results with small variances also for great optical distances between the source and detector region. The method to calculate single partial solutions on the basis of Monte Carlo method besides has the advantage that the statistical error of the composed solution will be very small in general also if this error is considerable for the single solution. Further the accuracy is better than pure group calculation, because the elastical slowing down is calculated exactly using the scattering law taking into 20 angle group for anisotropic scattering in the center of mass system /3/.

The calculations are performed for a 30 degree symmetry sector of the core. Fig. 1 gives a scheme of this used symmetry sector. As grids we used 10 layers of 25 cm for each fuel assembly. Only grids in the 3 outer rows of assemblies are needed, because the effect of the more inner distributed assemblies can be neglected (contribution less than 1 %). The used detailed distribution for the outer assemblies (127 fuel rods) were taken for a fixed mean burn-up. As results mean values of fluxes are given for 10 groups of 26 ABBN-groups, 10 angle intervals and 10 layers in the height

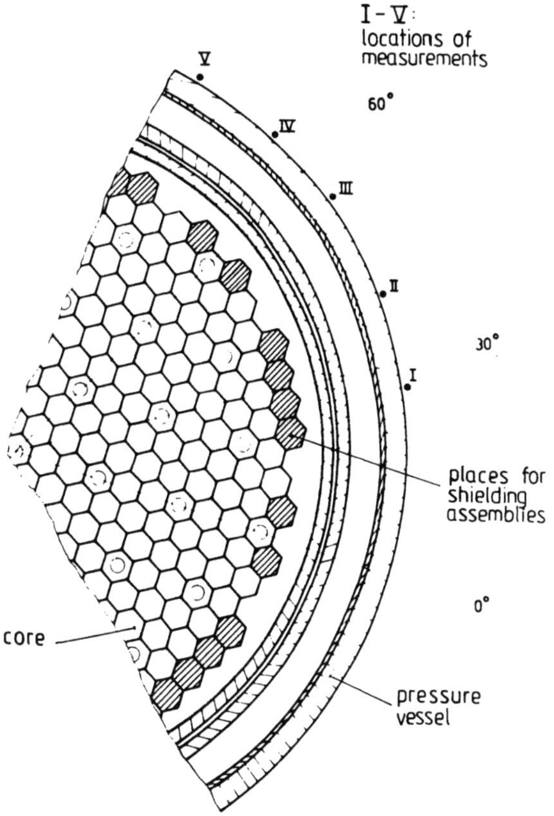


Fig. 1

Top view for the WWER-440 symmetry sector with marked core places for shielding assemblies for the inner and outer wall of the pressure vessel.

In order to obtain the influence of shielding dummy assemblies substituted instead of fuel elements and special marked in Fig. 1 a new set of function $G^i$ had to be calculated /4/, because by this dummy assemblies the partial solutions are of course essentially influenced. For the calculation of the total fluence time steps of approximately 40 days are used. Within the time intervals a linear slope is assumed for the sources in each space interval.

## MEASUREMENT AND EVALUATION

The experiments have been performed at the pressure vessel outer wall of the units 1, 3, and 4 of the Greifswald NPS using nearly the same experimental set-up. In Ref. /5/ an experiment is described in detail at unit 3 with a normal loaded core. The activation detectors Ti, Fe, Ni, Co, Cu, Ag, and Nb were installed in 5 azimuthal garlands (Fig. 1). Every garland contained 10 axial arranged sets of monitors covered by quartz glass ampoules ($6 \cdot 20$ mm). The mounting and dismounting of the garlands was performed during the reactor refueling period. Therefore, the detectors have been irradiated for a full fuel cycle.

The Nb activation detectors developed at the Zentralinstitut für Festkörperphysik und Werkstofforschung Dresden, Germany, were circular, metallic foils with thicknesses around 15 mg/cm$^2$. All Nb detectors showed a remarkably brittleness after irradiation, but no pulverizing of the foils has been observed if they were capsuled under Ar-atmosphere. The Ta-impurity was $<$ 1 ppm so that no special preparations were necessary. The self-absorption of the $K_\alpha$ and $K_\beta$ X-radiation was calculated to 0.86 and 0.89, respectively.

To generate the covariance matrix of the uncertainties the errors of the uncertainty components have been estimated. The applied procedures are specified in Ref. /5/. Correlations have mainly to be taken into account due to the efficiency calibration of the Ge(Li)-detector as well as for the correction of the power history of the reactor /6/.

A database management system REACDOS has been created to collect all primary data received from reactor dosimetry measurements of WWER /7/. Beside the recording of primary data both reaction rate and its uncertainty can be computed by special data processing routines. The system REACDOS has been implemented on a PC/AT by using the database language dBASE III.

The evaluation of the measured reaction rates has been performed with the adjustment code COSA1 and its newest version COSA2 /8/. The adjustment procedure (Fig. 2) adds the information from

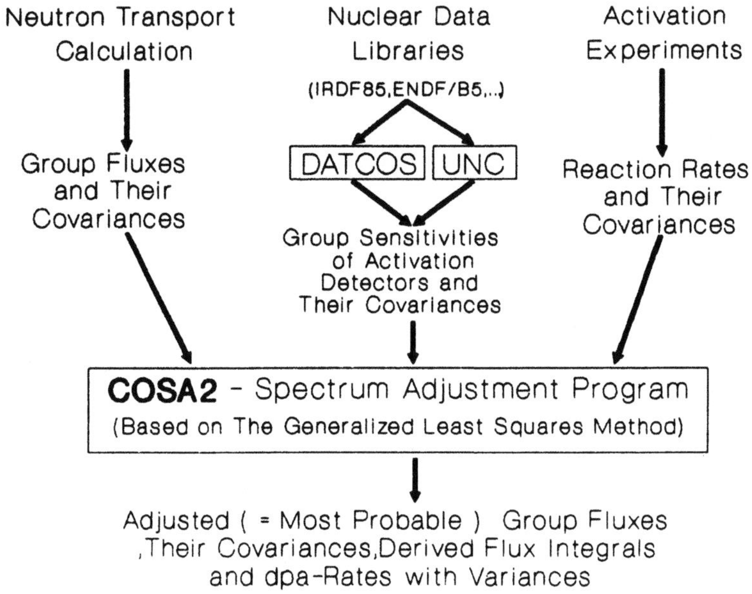


Fig. 2

Scheme of the spectrum adjustment procedure

neutron transport calculation and from activation experiments to most probable group fluxes (spectra) and derived integral quantities. (In the following these results are called experimental results.)

The input spectrum was composed from Monte Carlo calculations in the fast energy region and from removal diffusion calculation in the thermal and epithermal part. The same average unnormalized spectrum was used as approximation for all detector positions in a given reactor.

The adjustment code and most of the used cross section data and data processing codes have been tested in the REAL-84/88 exercises of the IAEA /9,10/. Additionally, the influence of different approximations and uncertainties during the adjustment procedure on the final results has been investigated /11/. From this investigation, the REAL-84/88 exercises and the experience of the reported evaluations we conclude, that the quality of available detector cross section data needs improvement. No covariance data are available for important reactions, while their quality seems to be very low for some other reactions. The situation with spectrum covariance calculation is also not satisfactory for us, because we have no access to cross section covariances of reactor materials, up to now. Therefore, we have used artificial model functions or spectrum covariances for a similar reactor available through REAL-84/88. Consequently, the error values of the final results had to be increased relatively to the least square calculation.

## RESULTS

Three different core configurations have been taken into consideration:

- a normal loaded core with fresh fuel assemblies at some outer positions
- a low leakage core, where highly burnt fuel assemblies are inserted in some outer locations
- a core configuration with steel dummies at outer places.

In Fig. 3 a comparison between calculation and measurement is plotted for the core midplane. The neutron flux density >0.5 MeV is shown, because this value is the basis for evaluating the shift of the nil-ductilty temperature in the Soviet guidelines for WWER-reactors.

The comparison shows the good agreement between calculated and experimental values in all three investigated core configurations. The average differences are within the given experimetal error, which has been estimated to about 15 - 20 %. Going into detail one can say that the average difference (experimental value - calculated

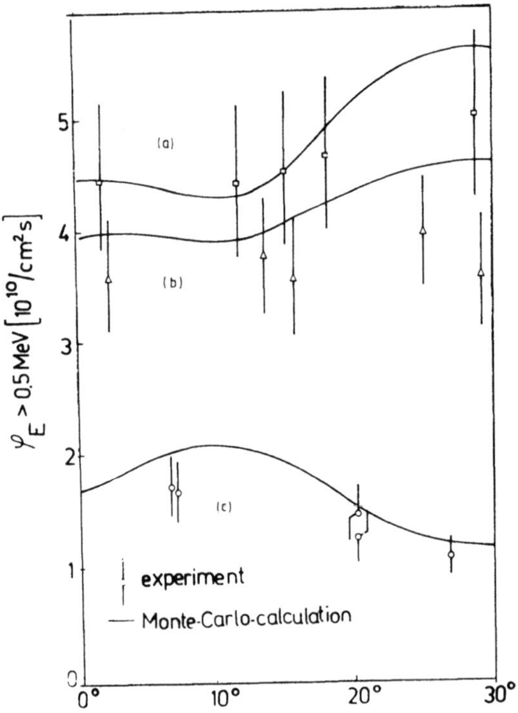


Fig. 3

Comparison of calculated with experimental neutron flux density > 0.5 MEV

(a) normal loaded core
(b) low leakage core
(c) core with dummy assemblies

value) of the neutron flux density > 0.5 MeV is about - 5 % for a normal loaded core, - 11 % for a low leakage core, and - 15 % for a core configuration with steel dummies. In any case the calculations are conservative.

Larger differences are obtained if one compares value outside the central axial region. Possible sources of these deviations are uncertainties in the arrangement of the activation detectors and also inaccuracies of the neutron source distribution nearly the axial reflectors depending on the burn-up calculations.

It can be concluded from Fig. 3 that a low leakage core reduces the neutron exposure by about 20 %, while a core with dummies leads to a decrease on about the factor 4.5 in the earlier azimuthal maximum. The reduction is only of the factor 2.1 in the new maximum (core with dummies). Both possibilities have been found to be useful methods to reduce the neutron exposure at reactor pressure vessel.

## CONCLUSIONS

For calculating the neutron fluences near the pressure vessel of WWER type reactors the described method is favourable applicable to compute reliable values in the full reactor operation time with respect of the real core loading schemes. The ex-vessel dosimetry work demonstrates possibilities and limits of experimental methods based on activation detectors and its evaluation by adjustment codes.

The comparison between adjusted experimental neutron exposures and calculations in the core midplane shows that the present computations overestimate the flux density at about 5 - 15 % depending on the core configuration. In core regions near the axial reflector the differences are greater. The results of the work are a good starting-point for estimating the EOL-fluence of all investigated reactor pressure vessels.

## REFERENCES


/1/ H.-U. Barz, Kernenergie 30 (1987) p. 233

/2/ H.-U. Barz, "TRAMO - a Flexible Multigroup Neutron Transport Code on the Basis of the Monte Carlo Method for Flux Calculations", Report ZfK-705, Zentralinstitut für Kernforschung Rossendorf, 1990

/3/ B. Böhmer, "MCANG - Eine Bibliothek mit Gruppendaten kummulativer Winkelverteilungen für Monte-Carlo-Rechnungen", Report ZfK-531, Zentralinstitut für Kernforschung Rossendorf, 1984

/4/ H.-U. Barz, K.-F. Pötter, G. Suschowk, Kernenergie 31 (1988) p. 133

/5/ H.-C. Mehner, B. Böhmer, U. Hagemann, I. Stephan, "Neutronendosimetrie an der WWER-440-Druckbehälteraußenwand", Report ZfK-623, Zentralinstitut für Kernforschung Rossendorf, 1987

/6/ H.-C. Mehner, "Effect of flux density variations on the determination of neutron fluences for LWR-PV dosimetry", paper presented to this Symposium

/7/ H.-C. Mehner, "REACDOS - a database management system for reactor pressure vessel neutron dosimetry", Report ZfK-693, Zentralinstitut für Kernforschung Rossendorf, 1990

/8/ B. Böhmer, "COSA2 - Ein Spektrumsjustierungsprogramm zur Auswertung von Aktivierungsmessungen auf der Basis der verallgemeinerten Methode der kleinsten Quadrate", Report ZfK-733, 1991

/9/ W.L. Zijp, E.M. Zsolnay, H.J. Nothenius, E.J. Szondi, "Final Report of the Real-84 Excercise", ECN-212, Petten, 1988

/10/ H.J. Nolthenius, "Results of the REAL-88 Exercise", ECN-89-095, Petten, 1989

/11/ B. Böhmer, "Accuracy of the adjustment of neutron spectra on the basis of activation measurements" (in Russian), Proceedings of the 18th Symposium to Problems of Physics of WWER, Prachatice, Czechoslovakia, September 1989, vol. 1, p. 498-515

# AN IMPROVED METHOD FOR MONITORING NEUTRON DOSE ON PWR VESSEL STEEL BY FLUX SPECTRUM MEASUREMENT WITH A FEW NONFISSIONABLE FOILS

Li Zhaohuan Wang Yongqing Li Ain

China Institute of Atomic Energy

P.O. Box 275(54), Beijing, Zip Code 102413
People's Republic of China

An improved method for monitoring fast neutron dose on reactor pressure vessel of Qin Shan PWR by the measurement of neutron spectrum with aluminium alloy foils, including Ni-Al, Ti-Al, Fe-Al, Cu-Al and Co-Al, has been developed in CIAE. The content of objective element in each alloy and interference activation due to impurities are determined. A simulation experiment has been conducted on Swimming Pool Reactor (SPR). A code, called NEUSPEC, based on SAND-II and improved by a large cycle iteration method which gives a reasonable neutron spectrum distribution, especially the distribution in energy region about from 0.5 to 2.5 Mev, in which the reaction rate information could hardly be obtained by these foils, has been developed. The error of fast neutron fluence in simulation experiment is <3.5%. Considering the nonuniformity of alloy foils and the uncertainty in census of total operation time the estimated error of fast neutron dose for Qin Shan PWR will increase to 8～10%.

## Introduction

Along with increasing of irradiated fast neutron dose above a determined limit on the reactor pressure vessel (RPV) of the nuclear plant the nil ductility transition temperature (NDTT) gradually raises. During almost whole life time of plant operation, the RPV commonly requires monitoring on irradiated fast neutron dose. This study on fast neutron dose monitor technique is arranged according to the requirement of designers of Qin Shan PWR.

For the multiple foils activation method for monitor of fast neutron dose on the vessel the elements used as detector foils are commonly chosen from fissionable and nonfissionable material. The threshold energy Ec of fissionable detector foils including U-238, Th-232 and Np-237 (about 0.5 Mev) is much lower than the Ec of available nonfissionable detector foils (about 2.5 Mev in our case). Due to the limitation of our practical conditions, such as unavailability of pure fissionable material, unavailability of sample process laboratory and etc, only nonfissionable elements can be selected. Also each of detector foils must be made in the form of aluminium alloy, not in the form of pure metal foils. The selected detectors are Ni-Al, Ti-Al, Fe-Al, Cu-Al and Co-Al alloy foils.

The damage of vessel steel by irradiated fast neutrons depends on the incident fluence of fast neutron flux spectrum. The incident neutrons with higher energy will cause more damage than the neutrons with lower energy. Consequently, it is in the course of nature to measure a neutron spectrum distribution for the monitor of the RPV. According to the RPV monitor schedule of Qin Shan PWR all well sealed monitor jars in each of which a set of detector foils is located are inserted into adequate positions in the vessel before the reactor start-up, then during each of chosen core refueling time one or two of monitor jars are removed out of reactor for test. The irradiation time for individual jar (also for individual set of foils) is different. It can vary from a period of refueling cycle to the longest, about 20 years. The half-life time of selected neutron detector foils is short, related to the irradiated time. Thus, almost all of foil activations will saturate, even for foils in such jars that are planned to irradiate for one or several refueling cycles. Consequently, the method of RPV monitor by the measurement of neutron spectrum distribution is more adequate for Qin Shan PWR, because it permits a rather long term irradiation and gives energy structure of incident neutrons.

The energy of incident neutron that affects the damage of vessel steel must be higher than a determined value, which is 1 Mev commonly used in China. The lowest threshold energy among selected nonfissionable detector foils is about 2.5 Mev, much higher than 1 Mev. The fraction of fast neutron spectrum incident upon the RPV, as theoretically estimated, is dominant in energy range from 1 to 2.5 Mev. So it is important on this study to obtain a more reasonable neutron flux spectrum distribution especially in this energy region. The informations which could be provided by such a few selected detector foils are limited. An improved method, called large cycle iteration, which could provide a reasonable neutron flux spectrum distribution with a few informations, must be developed.

The following section gives the method to determine the objective element content and interference of impurities in aluminium alloy foils. The next section gives description of the simulation experiment, then the description of the improved method and of the Code NEUSPEC. The last section gives results in the simulation experiment.

## Quantitative Analysis and Activation Determination of Alloy Foils[1]

Two tasks have to be solved in neutron spectrum measurement with alloy foils by neutron activation method. One is to determine the content of activation element (called objective element) in each kind of foils. The other is to evaluate the possible interference activation due to impurities existing in the alloy (called interference element).

The required content of objective element in each kind of detector foils is estimated according to the irradiation schedule and controlled in metallurgical process. After fabrication the content of objective element in alloy foils is determined by method of comparative activation analysis, irradiated in the well-established isotropic thermal neutron flux field[2] and coupled with pure metal foils. The results are shown in table 1.

Table 1 Contents of Objective and Main Interference Element in Alloy

| No | Alloy | Nominal Content wt% * | Objective Element Content wt% | Objective Element Error % | Interference Element Ele-ment | Activation Content ppm | Activation Error % | Spect.** Content ppm |
|---|---|---|---|---|---|---|---|---|
| 1 | Fe-Al | 1.10 | 1.048 | 0.78 | Mn | 7.42 | 2.0 | ⩽ 5 |
| | Fe-Al | 2.99 | 3.084 | 1.01 | Mn | | | < 5 |
| 2 | Ti-Al | 1.06 | 1.107 | 1.71 | Sc | <0.02 | | < 2.5 |
| | Ti-Al | 3.10 | 3.432 | 1.61 | Sc | <0.001 | | << 2.5 |
| 3 | Cu-Al | 6.54 | 6.927 | 1.90 | Co | 3.76 | 4.0 | << 6 |
| 4 | Ni-Al | 1.16 | 1.089 | 1.34 | | | | |
| | Ni-Al | 3.12 | 3.246 | 1.25 | | | | |
| | Ni-Al | 9.98 | 10.91 | 0.72 | | | | |
| 5 | Co-Al | 0.15 | 0.1289 | 1.70 | | | | |
| | Co-Al | 0.49 | 0.5304 | 1.77 | | | | |
| | Co-Al | 1.91 | 2.230 | 1.67 | | | | |

* The nominal content is provided by the producer.
** Results of spectroscopic analysis are provided by Prof Gu Guoying and Shi Meijue. They noted that the error may be rather large.

The interference elements in each kind of foils are found out by the gamma-ray spectrometry analysis with foils irradiated in the core of SPR at full power for 24 hours and in the well-established neutron field. The main interference activation is shown in the table 2. The content of main interference element contained in each foil is, independently, determined by neutron activation analysis and mass spectroscopic analysis.

The discrimination and determination method of objective activation in the existence of interference impurities is dependent on behavior of interference. The contribution of interference reaction $^{59}Co(n,\gamma)^{60}Co$ in Cu-Al alloy is, directly, determined by the comparison measurement irradiated both in the well-established thermal neutron field and in the reactor core with pure copper foils. Interference reactions, $^{55}Mn(n,2n)^{54}Mn$ and $^{45}Sc(n,\gamma)^{46}Sc$ in Fe-Al and Ti-Al alloy, respectively, depend on neutron flux spectrum. They are deducted by multiple iteration approach in the code NEUSPEC. Formulas for these corrections are shown in Ref. 1 (Eqs. 1~5).

Table 2 The Interference Element and Reaction in Foils

| No | Alloy | Objective Reaction | Main I.E. | Interference Reaction |
|---|---|---|---|---|
| 1 | Cu-Al | $^{63}Cu(n,\alpha)^{60}Co$ | Co | $^{59}Co(n,\gamma)^{60}Co$ |
| 2 | Fe-Al | $^{54}Fe(n,p)\ ^{54}Mn$ | Mn | $^{55}Mn(n,2n)^{54}Mn$ |
| 3 | Ti-Al | $^{46}Ti(n,p)\ ^{46}Sc$ | Sc | $^{45}Sc(n,\gamma)^{46}Sc$ |

## Large Cycle Iteration Method and Code NEUSPEC

The logarithm-exponent iteration method, known as SAND-II method, is well applicable, from our experiences[3][4], for the derivation of multiple-foil flux spectra in the thermal reactor, in which the neutron energy covers a wide range from $1.0\times10^{-10}$ to 20 Mev. In that cases a great number of detector foils are used. The individual foil is respondent to a specific neutron energy range, and the whole detector foils are respondent to the almost whole spectrum of neutron energy. So it may be called multiple information method. In the current case a few detector foils could be selected under the considerations for a rather long term irradiation, and a rather long time waiting between the removal from reactor and the reaction rate measurement, and for, as possible, low burn-up of objective element and its daughter isotope. The selected detector foils will provide a few informations, especially in energy region from 1 to 2.5 Mev as mentioned above. A large iteration method, also called a few informations method, must be developed for derivation of neutron flux spectrum.The idea of this method is based on the SAND-II iteration. A next large cycle iteration begins from a new modified initial spectrum obtained by spectrum deviation in previous SAND-II iteration including Monte Carlo sampling, until each side of deviations of neutron flux spectrum distribution approaches to within a set of convergent conditions. The validity and ability of large cycle iteration method were tested before the actual calculation by using the reaction rates measured in F-5 and D-11 (see the following section) and by canceling one-by-one the reaction rate of foil in each test.

The Code NEUSPEC is based on above idea for derivation of neutron flux spectrum distribution in the simulation experiment for the study on the RPV monitor of Qin Shan PWR. The Code is compiled for general use. Many switches are set for the choice of its functions, including the function of large cycle iteration, the deduction of interference activation by reaction rate of $^{55}Mn(n,2n)^{54}Mn$ and $^{45}Sc(n,\gamma)^{46}Sc$, respectively, as mentioned in the previous section. The energy group structure of neutron activation cross sections is almost the same as in Ref.5 (only extended to 640 groups). The principle for selection and evaluation of neutron reaction group cross sections is followed from Ref.6, that is "the use of consistent set of evaluated energy-dependent cross-section values". Besides, in coordinate with 640 groups neutron activation cross sections a 427 group constants of scattering and total cross section are made for foil self-shielding correction calculation. This set of neutron group cross sections is well applicable in the derivation of measured neutron spectrum in thermal reactor[3] [4]. The code gives the neutron flux spectrum distribution, two side spectrum deviations, the spectrum variance, the thermal neuron flux, the parameters of spectrum, the fast neutron fluence and the fast neutron dose. The fast neutron fluence is an integral of spectrum above certain neutron energy (and the time is unit). The fast neutron dose is an integral of spectrum above certain neutron energy weighted by neutron energy (also in unit plant operation time).

## Simulation Experiment and Results

A simulation experiment for monitoring neutron dose for RPV of Qin Shan PWR has been done on Swimming Pool Reactor in CIAE.The irradiation time of detector foils in this experiment is chosen in order that the incident fast neutron fluence is approximate to or a little more than that in first two removal monitor jars in Qin Shan PWR. The bare and Cd-covered Co-Al foils are also located as in Qin Shan irradiation condition. The experiment is in B-3 position of SPR which is on the boundary between the reactor core and reflector. The neutron flux spectrum distribution derived by the NEUSPEC is shown in Fig.1 in solid line. In this figure it can be seen that the spectrum distribution in energy region from about 0.5 to 3.0 Mev is reasonable. In order to comparison in Fig. 1 are also introduced multiple-foil-derived neutron flux spectra,which were measured in position F-5 (in the center zone of core) and D-11 (outside of beryllium reflector, in graphite reflector, about 14 cm from core boundary) in SPR, respectively (Ref. 4 ).

Spectrum parameters,fast neutron fluence and fast neutron dose are summarized in table 3. In this experiment the error of fast neutron fluence and fast neutron dose are $< 3.5\%$. Considering the nonuniformity of objective element in each foil alloy and the uncertainty in census of total operation time the estimated error of fast neutron dose in actual monitoring in Qin Shan PWR will be larger than that in simulation experiment and will increase to $8\sim10\%$.

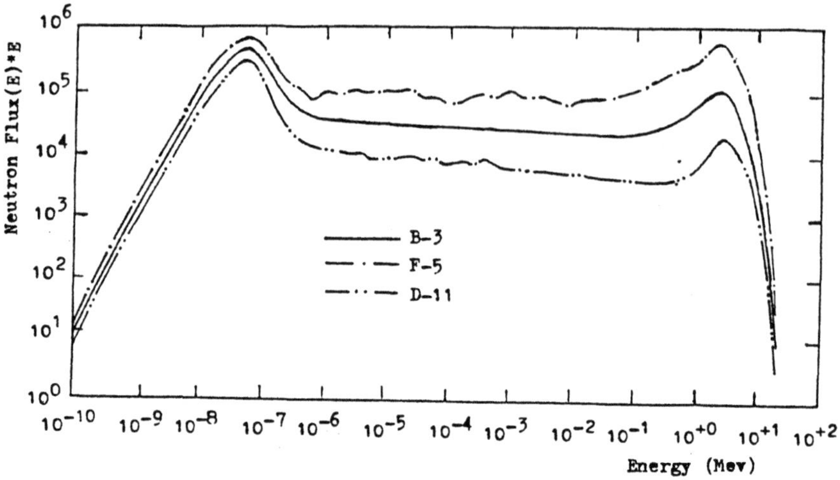


Fig. 1 The Neutron Flux Spectrum Distributions

Table 3 A Summary of Results

| No | Parameter | Unit | Value | Error * % |
|---|---|---|---|---|
| 1 | Thermal Neutron Flux | n/cm² -sec | $2.48\times10^{12}$ | 3.0 |
| 2 | Neutron Temperature | ev | 0.0268 | |
| 3 | Intermediate Flux/Thermal Flux | | 0.0273 | |
| 4 | Fermi Spectrum Deviation | | 0.0364 | |
| 5 | Fast Neutron Fluence (1 sec) | n/cm² | | |
| | $E \geqslant 0.1$ Mev | | $5.40\times10^{11}$ | 3.3 |
| | $E \geqslant 0.5$ Mev | | $4.60\times10^{11}$ | 3.3 |
| | $E \geqslant 0.8$ Mev | | $4.22\times10^{11}$ | 3.4 |
| | $E \geqslant 1.0$ Mev | | $3.98\times10^{11}$ | 3.4 |
| 6 | Fast Neutron Dose (1 sec) | n-Mev/cm² | | |
| | $E \geqslant 0.1$ Mev | | $1.27\times10^{12}$ | 3.3 |
| | $E \geqslant 0.5$ Mev | | $1.25\times10^{12}$ | 3.3 |
| | $E \geqslant 0.8$ Mev | | $1.22\times10^{12}$ | 3.4 |
| | $E \geqslant 1.0$ Mev | | $1.20\times10^{12}$ | 3.4 |

* The error including the errors in the measurement of absolute reaction rate and the uncertainty of group neutron activation cross sections is calculated by NEUSPEC.

The technique developed here is applicable to monitoring the fast neuron dose on the RPV of PWR or BWR. Its advantages are simple, low cost and without any possible contaminations.

References

1. Li Zhaohuan, Wang Yongqing, Li Ai, Chu Shaochu and Zhang Jian, Quantitative Analysis and Activity Division of Alloy for Neutron Monitoring in PWR, Chinese Journal of Nuclear Science and Engineering, vol. 9, No. 3, pp 285-288, (1989)
2. Li Zhaohuan, Zhao Yongfu and Wang Yongjing, A low Cost Isotropical Thermal Neutron Flux Field, C. J. Nucl. Sci. & Eng. vol. 7 No.3-4, pp 222-227, (1987)
3. Li Zhaohuan, Gao Jijin and Wang Yongjing, Intermediate and Fast Neutron Spectrum Measurement in Swimming Pool Reactor, C. J. Nucl. Sci. & Eng. vol. 2, No. 1, pp 22-31, (1982)
4. Li Zhauhuan and Wang Yongjing, Neutron Spectrum Measurement in Thermal Reactor by Multiple Foils Method, C. J. Nucl. Sci. & Eng. vol. 4, No. 1, pp 24-34, (1984)
5. R. L. Simons and W. N.McElroy, Evaluated Reference Cross Section Library, BNWL-1312, (1970)
6. W. N. McElroy and L. S. Kellogg, Fuels and Materials Fast-Reactor Dosimetry Data Development and Testing, Nuclear Technology, vol.25 No. 2, pp 180-223, (1975)

# B&W REACTOR DOSIMETRY PROGRAMS

L. Petrusha and J.F. Walters

B&W Nuclear Service Company

Lynchburg, Virginia, U.S.A.

ABSTRACT

Since the extensive redefinition of the Babcock & Wilcox (B&W) Reactor Vessel Surveillance Program in the late 1970s, B&W's plan for monitoring reactor vessel fluence has been developed in three phases. The first phase, the Integrated Reactor Vessel Surveillance Program (1978-present), was designed to irradiate material specimens from all B&W-designed 177 fuel assembly reactors in two host reactors and establish the relationship between neutron fluence and reactor vessel material degradation. Second (1982), a neutron flux perturbation experiment was performed to benchmark the perturbation effect due to the capsule and current fluence calculational methodology. Finally, the Cavity Dosimetry Program (1984-present) was initiated to demonstrate the effectiveness of cavity dosimetry and develop a benchmarked vessel fluence analysis methodology for future plant-specific application. This paper presents highlights of the individual B&W Reactor Dosimetry Programs and details of the Cavity Dosimetry Benchmark Experiment presently being conducted.

## INTRODUCTION

B&W initiated the Integrated Reactor Vessel Surveillance Program (IRVSP) in response to failure of the surveillance capsule holder tubes attached to the reactor vessel internals of operating plants in 1976. At that time, the capsule holders were removed, redesigned, and installed in two host reactors that had not yet begun operation. Plant-specific surveillance capsules were installed in these holders adjacent to the thermal shield in the downcomer region. This program, designed to irradiate material specimens from all the B&W-designed 177 fuel assembly (FA) reactors, was the first phase in B&W's program for monitoring reactor vessel fluence. Thus, the overall goal of the IRVSP was to characterize the effects of irradiation on reactor vessel beltline welds and materials.

The large diameter of the capsules led to the next stage of the program - the Flux Perturbation Experiment. This experiment established the perturbation effects due to the capsule itself and benchmarked the presently used fluence calculational procedures. The final phase, now being pursued, is the Cavity Dosimetry Program. The goals of this program are to establish the viability of using cavity dosimetry for future monitoring of vessel fluence and develop a benchmarked methodology for predicting this fluence.

## INTEGRATED REACTOR VESSEL SURVEILLANCE PROGRAM (1)

The IRVSP incorporates six surveillance capsules from each of the participating reactor vessels. These plant-specific capsules contain samples of weld metal, plate or forging material, and heat-affected zone material from the vessel beltline. In addition to the plant-specific capsules, six large research capsules were also installed in the two host reactors primarily to obtain fracture toughness data. These capsules were designed to study the effects of irradiation on a number of weld metals, not contained in the original capsules, which are now known to be highly sensitive to irradiation damage because of their chemistry and low initial Charpy upper shelf energies. Dosimeters were distributed throughout both types of capsules to indirectly obtain measurements of the neutron fluence levels at various specimen stack locations.

To date, 21 capsules from the IRVSP have been removed and tested. The neutron fluence has been determined for each of the tested capsules and their associated reactor vessels by normalizing neutron transport calculations to capsule dosimeter activity measurements. The average measured and calculated results from the last seven tested capsules have agreed to within ±6%. Six capsules from the original IRVSP, currently being irradiated, will be tested in the future. Eight additional capsules were fabricated in 1989 and inserted in the host reactors in 1990 to obtain irradiated weld metal data to support reactor vessel life extension.

## B&W FLUX PERTURBATION EXPERIMENT (2)

The second phase of the B&W dosimetry program was the Flux Perturbation Experiment. This experiment determined the neutron flux perturbation effect due to the relatively large capsule itself and the capsule holder tube, established a benchmark for the presently used surveillance capsule fluence calculational procedures, and allowed for a comparison of B&W dosimeters with those from other laboratories. The Flux Perturbation Experiment simulated the thermal shield, downcomer region, pressure vessel, and cavity region of a 177 fuel assembly reactor by an arrangement of steel slabs and a void box in the Oak Ridge National Laboratory Reactor Poolside Facility. Two simulated surveillance capsules, containing extensive dosimetry provided by B&W and Hanford Engineering Development Laboratory (HEDL), were then irradiated in the downcomer region. Capsule A simulated the standard 177 fuel assembly plant surveillance capsule and holder tube, while capsule B simulated the larger research capsules. Following irradiation, the dosimeters were sent to B&W and HEDL for analysis.

Flux distributions were calculated using the two-dimensional DOT4.3 neutron transport code and standard capsule fluence analysis methods, modified to reflect the geometry of the experiment. The calculated results showed that the presence of the surveillance capsule and holder tube did cause significant perturbations: Capsule A caused the flux greater than 1.0 Mev to peak by approximately 47% at the center of the capsule, while Capsule B caused a 43% increase. The capsules also created a shadowing effect on the reactor vessel. The shadow of Capsule A on the reactor vessel reduced the fast flux by approximately 5%; Capsule B reduced the fast flux by about 10%.

Calculated and measured activities were compared for over 400 dosimeters and/or fission products. B&W calculated and measured data for the group of dosimeters from the central section of the capsules, where the flux gradient was smallest, agreed within $\pm$8% for both capsules A and B. Measured activities for the B&W fission dosimeters averaged approximately 11% higher than the HEDL fission dosimeters. B&W iron (Fe) and nickel (Ni) results were about the same as HEDL in Capsule A, but 12% higher than HEDL in Capsule B. Thus, the differences in calculated and measured values indicated that the calculations adequately accounted for the perturbation introduced by the capsule and provided a valuable benchmark for the neutron transport calculational method used by B&W. The Flux Perturbation Experiment also provided valuable benchmark experience and insight for future use in cavity flux calculations.

## B&W OWNERS GROUP CAVITY DOSIMETRY PROGRAM (3)

### Program Objectives

At the present time, B&W Nuclear Service Company is focusing its efforts on developing and benchmarking a methodology by which cavity dosimetry would be used to monitor reactor vessel fluence.

By the mid-1990s, the B&W Owners Group (B&WOG) Integrated Reactor Vessel Surveillance Program will be nearing completion, with all reactor vessel surveillance capsules removed from their host reactors. At that time, it is believed that all the empirical data needed to establish the relationship between neutron fluence and material degradation for B&W-designed 177 fuel assembly plants will have been collected and reactor vessel fluence will then continue to be monitored using cavity dosimetry. Therefore, in response to Nuclear Regulatory Commission regulations in 10CFR50 Appendix H, which require reactor vessel fluence surveillance throughout the lifetime of pressurized water reactors, the B&WOG Cavity Dosimetry Program was started in 1984.

The Cavity Dosimetry Program was developed with a number of objectives. As stated earlier, the primary goal of the program is to establish a methodology to determine the vessel fluence based on dosimetry measurements in the cavity region and neutron transport calculations, allowing for long-term vessel fluence surveillance. This will be accomplished through the abundance of dosimetry data obtained from the Cavity Dosimetry Benchmark Experiment. Two irradiated standard surveillance capsules will supply data near the inside surface of the reactor vessel, while numerous dosimetry distributed throughout the cavity region will provide the dosimetry measurements near the outside surface of the vessel for the same irradiation period. Thus, using both sets of data, B&W plans to benchmark an analytically determined fluence distribution through the vessel and produce a benchmarked calculational procedure for future plant-specific cavity and vessel fluence predictions.

Another objective of the program is to establish the merits of using new types of dosimetry to monitor vessel fluence - specifically, solid state track recorders (SSTRs), helium accumulation fluence monitors (HAFMs), paired uranium detectors (PUDs), and niobium foil. Lithium fluoride (LiF) detector chips have also been incorporated into the experiment in an attempt to measure the gamma flux at various locations throughout the cavity region, to allow for benchmarking the effect of gamma-induced photofission events in the fission-type dosimeters.

Of particular interest for future cavity application is the verification of SSTR response. These detectors appear to have significant advantages over the presently used radiometric monitors (RMs). Unlike RMs, SSTRs do not decay and also leave a permanent record of their irradiation history; thus, prompt counting is not required. They also do not become significantly activated when irradiated, making shipping and handling easier and less expensive. By comparing SSTR response to RMs at the same locations, the Cavity Dosimetry Program plans to verify the accuracy of SSTRs for use in future cavity measurements.

The program also calls for statistical evaluation of the wealth of data resulting from the Cavity Benchmark Experiment to quantify the uncertainties associated with the calculated fluence values and the dosimetry measurements. Independent verifications of dosimetry measurements will be made by various participating laboratories in a "round robin" effort for comparison of like dosimetry to further reduce overall uncertainties in the data measurements. Having these numerous dosimetry measurements at various axial and azimuthal locations will also provide for mapping axial and azimuthal flux distributions in the reactor cavity, assessing the flux perturbation effects caused by large structural material in the region, and evaluating the effects of axial neutron streaming in the cavity.

Finally, the Cavity Benchmark Experiment is to be B&W's contribution to the Light Water Reactor Pressure Vessel Surveillance Dosimetry Improvement Program (LWR-PV-SDIP). The LWR-PV-SDIP has shown concern regarding the accuracy of the calculated neutron flux at deep penetrations (>4 in.) in the steel reactor vessel, with a wall thickness of 8.5 in.. This problem is presently not a concern in the standard reactor vessel surveillance capsule analyses, which focus mainly on the inside surface of the vessel. However, this issue will become important when vessel fluence predictions are made using cavity dosimetry measurements. Therefore, the Cavity Dosimetry Program will address the LWR-PV-SDIP concerns using in-vessel and ex-vessel data for deep scattering comparison.

## Status of the Cavity Dosimetry Program (4)

The Cavity Dosimetry Program objectives will be accomplished using the results of the Cavity Dosimetry Benchmark Experiment now in progress. The tasks of conceptual design (1984), cavity dosimetry hardware development for the experiment (1985-1987), purchase of dosimetry (1986-1988), installation (1988), and removal (1990) have all been successfully completed. The program is now proceeding with dosimetry testing, methodology development, and the cavity fluence analysis.

The Cavity Dosimetry Benchmark Experiment, also known as the In-Out Experiment, was a full-scale test conducted in the Davis-Besse Unit 1 B&W-designed 177 fuel assembly reactor, using both in-vessel and ex-vessel dosimetry. The dosimetry consisted of 633 RMs (286 activation foils or wires, 47 fission foils, and 300 flux mapping chain segments), 35 sets of SSTRs, 22 ultra-high purity niobium dosimeters, 40 HAFMs (30 Beryllium and 10 Lithium), 3 PUDs, and 9 LiF detector chips. The LiF chip gamma fluence detectors were specially developed by the National Institute of Standards & Technology (NIST) for this specific application to provide accurate results at the high exposure levels expected in the experiment. The six contributors of the above dosimetry to the program were - B&W Owners Group; HEDL; Center for the Study of Nuclear Energy, Mol, Belgium (CEN/SCK); NIST; Rockwell International; and the University of Arkansas. Table 1 specifies the types of dosimeters in the experiment and their locations.

The in-vessel dosimetry consisted of two standard unirradiated surveillance capsules installed in a surveillance capsule holder tube at the peak flux location. These capsules contained six standard B&W RM dosimeter sets covering incident neutron threshold energies from 0.5 ev to 2.5 Mev.

TABLE 1
*Cavity Dosimetry*

| Holder | Location[(a)] | Type of Dosimetry[(b)] |
|---|---|---|
| A | 11.5° Seal Plate Elevation | RM, HAFM, LiF |
| B, C | 11.5° Nozzle Elevation | RM, SSTR, HAFM, LiF, Niobium, Chain Samples |
| D, E | 11.5° Upper Active Fuel Elevation | RM, SSTR, HAFM, LiF, Niobium, Chain Samples |
| F, G | 11.5° Core Midplane | RM, SSTR, HAFM, LiF, PUD, Niobium |
| H | 42.5° Core Midplane | RM, HAFM, LiF, Niobium, Chain Samples |
| J, K | 11° Core Midplane | RM, HAFM, Niobium, Chain Samples |
| L, M | 6° Core Midplane | RM, SSTR, Chain Samples |
| N | 42.5° Core Midplane | RM, SSTR |
| P, Q | 26.5° Core Midplane | RM, HAFM, LiF, Niobium |
| R | 11.5° Seal Plate Elevation | RM, SSTR, LiF |
| S | 11.5° Core Midplane Source Tube A | RM, SSTR, Niobium |
| T | 11.5° Core Midplane Source Tube B | RM, SSTR, HAFM, LiF, Niobium |
| U | 11.5° Source Tube "Connector" | RM, SSTR |

NOTES: (a) Holder locations are shown in Figure 1
(b) RM — Radiometric Monitor
SSTR — Solid State Track Recorder
HAFM — Helium Accumulation Fluence Monitor
LiF — Lithium Fluoride Detector
PUD — Paired Uranium Detector

The cavity dosimetry consisted of sixteen specially fabricated aluminum dosimetry holders, each containing five sets of dosimeters. Cable assemblies containing these holders were then designed in a manner that allowed for accurately known measurements of the dosimeter locations, maintaining the dosimetry in a known direction either facing towards or away from the core, and easy installation and removal. Five cable assemblies containing dosimeter holders at various axial positions were installed in the cavity at specific azimuthal positions. The azimuthal locations were chosen to avoid possible areas of large flux gradients, which might be difficult to predict analytically. Figure 1 shows the general arrangement of the cavity dosimetry holders. The assemblies at 6°, 11°, and 11.5° were located in the region of maximum flux, while the holder at 42.5° was in the minimum flux region.

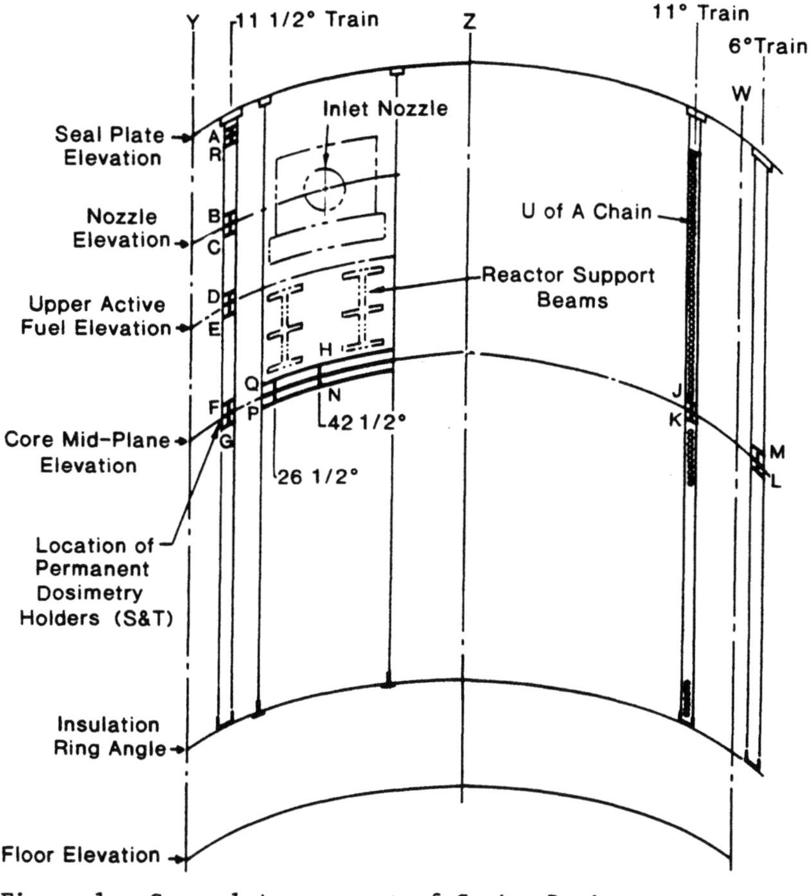


Figure 1: *General Arrangement of Cavity Dosimetry Benchmark Experiment*

Four 50-ft.-long beaded stainless steel chains were also hung in the cavity region to help achieve accurate axial flux profiles at the azimuthal positions of interest. The chain assemblies hung independently beneath Nuclear Instrumentation boxes in four of the open source check tube penetrations, one in each quadrant of the cavity. The chains were anchored with a heavy weight at the containment floor to limit lateral movement during plant operation. An additional 35-ft.-long University of Arkansas stainless steel chain was suspended from the 11° train.

All 80 sets of dosimetry, stainless steel chains, and surveillance capsules were installed for one cycle of operation in the Davis-Besse nuclear plant and removed at the completion of Cycle 6 in the spring of 1990. The dosimetry has been shipped from the site to B&W's research laboratory and is presently being analyzed. B&W is also currently organizing a "round-robin" effort for independent laboratory analysis of some of the dosimetry. The five laboratories planned to be included in this effort are B&W, CEN/SCK, NIST, University of Arkansas, and University of Virginia. The results of the independent analyses will then be compiled and statistically evaluated by B&W to quantify the uncertainties in cavity measurements that are attributable to variations in the equipment and analysis techniques used at different laboratories. The data will be refined and discrepancies will be resolved to reach a "final" set of dosimeter activity measurements.

## Analysis of the Cavity Benchmark Experiment Results

As stated earlier, the Cavity Dosimetry Benchmark Experiment was designed to allow for accurate comparisons to be made between calculated and measured flux values inside and outside of the reactor vessel at the same azimuthal position. The benchmarking of calculated results to actual dosimetry measurements will be accomplished using an iterative analysis, from which a final analytical procedure for the prediction of vessel fluence from cavity dosimetry measurements will be developed. The procedure will be based on a semi-empirical process, in which the vessel fluence will be correlated indirectly to activity measurements of the irradiated dosimeters in the reactor vessel downcomer and cavity regions.

Two-dimensional DOT4.3 R-$\theta$ and R-Z models have been developed to include extensive modelling of the in-vessel capsule assembly, the cavity dosimetry assemblies, and perturbing cavity structures. The DOT analysis will calculate the space- and energy-dependent neutron flux throughout the models and activities at the specific dosimeter locations. These DOT-calculated activities will then be compared with the final laboratory measured dosimeter activities resulting from the round-robin effort. Corrections to the calculations will be made to account for the presence of U-235 impurities in the U-238 dosimeters, photofission events in the fission-type dosimeters, and non-saturation of the specific dosimeter activities. Through an iteration process aimed at minimizing the differences, other correction factors will also be developed to account for the effects of axial neutron streaming in the cavity, the perturbing effects of large structural material in the cavity, the effect on fluence measurements of plutonium fissions in the peripheral high burnup fuel assemblies of low leakage cycles, and perhaps some factors unknown to date. Once the measured and calculated dosimeter activities are in good agreement, the results will be considered acceptable and a normalization factor will be calculated. Thus, in this manner, the final form of the semi-empirical method will be defined to calculate reactor vessel fluence by normalizing analytical calculations to measured cavity dosimetry data.

## CONCLUSIONS

From past experience and knowledge gained through the Integrated Reactor Vessel Surveillance Program and the Flux Perturbation Experiment, B&W is now successfully moving forward with its plans to monitor vessel fluence in the future with cavity dosimetry. The results from the Cavity Dosimetry Experiment are presently being analyzed and a methodology is being developed to predict vessel fluence from cavity dosimetry measurements. Note that, even though some modifications will have to be made to account for varying cavity configurations, the methodology resulting from he Cavity Benchmark Experiment will be applicable to all future B&W-177 fuel assembly reactor vessel fluence analyses.

## REFERENCES


1. S. Fyfitch, L.B. Gross, A.L. Lowe, Jr., Master Integrated Reactor Vessel Surveillance Program, Babcock & Wilcox Report No. BAW-1543, Rev.3 (September 1989).

2. N.L. Snidow, B&WOG Flux Perturbation Experiment at ORNL, Measured and Calculated Dosimeter Results, Babcock & Wilcox Report No. BAW-1886, (September 1985).

3. S.Q. King, The B&W Owners Group Cavity Dosimetry Program, Babcock & Wilcox Report No. BAW-1875-A (July 1986).

4. G.S. Carter, B&WOG Cavity Dosimetry Benchmark Program Installation Report, Babcock & Wilcox Report No. BAW-2092 (September 1989).

# DETERMINATION OF THE DOSES ABSORBED BY THE CHOOZ A PWR VESSEL.

# SUMMARY AND CONCLUSIONS OF INVESTIGATIONS.

R. Lloret*, A. Bévilacqua*, J.J. Bonnin*
J.C. Nimal**, L. Bourdet**
F. Hédin***, J.C. Guilleret****

* CEA/DRN/DRE/SRS/SAPR - Centre d'études nucléaires, 85 X, 38041 Grenoble Cedex.
** CEA/DRN/DMT/SERMA/LEPP - Centre d'études nucléaires de Saclay, 91191 Gif-sur-Yvette Cedex.
*** Electricité de France/SEPTEN/Division Réacteurs 12,14 Av[e] Dutrievoz, 69628 Villeurbanne.
**** Centrale Nucléaire des Ardennes. BP 160. 08600 Givet.

## ABSTRACT

The "CHOOZ A" reactor (300 MWe commissioned for industrial operation in 1967) is the oldest PWR nuclear power plant in service in France today. Due to problems encountered with the internal equipment, it was necessary to set up a replacement surveillance programme and this has proved to be more difficult to interpret. A vast programme of investigations was undertaken to complete the knowledge required for operating permission (1). This work, described in the proceedings of a former Symposium (2) will not be recalled here, but the original characteristics of this reactor along with the progress achieved to meet this requirement will be reported.

Firstly, the computation of neutron fluxes was completely revised using the code TRIPOLI 2. This code uses the Monte-Carlo method to solve the problem of the transport of polykinetic neutrons in any 3-dimensional geometry. In this case it furnished a precise 3-dimensional description for the calculation of specimens located beneath the core, and a precise 2-dimensional description for the vessel from behind the baffle onto the cavity.

In addition, by measuring the $^{54}$Mn reaction rate on both the specimens and on the vessel, it was possible to obtain an experimental flux lead factor. This is an original approach to solve the problem of the unusual position of the surveillance specimens. Mini-chips were removed from the vessel lining by milling and analysed to determine the reaction rates per unit mass with minimum uncertainty. Samples from the specimens were analysed in the same way.

Computations and measurements are in excellent agreement : to within 5 % on the specimens and to within 10 % on the vessel.

## INTRODUCTION

The CHOOZ A reactor (300 MWe, commissioned in 1967) is the oldest French nuclear PWR power plant in service today. The programmed shutdown of this plant unit was extended for an additional 9 months (1987-88) in order to substantiate the data records on the reactor vessel operating reliability. In fact, a complete re-evaluation of safety margins with respect to sudden failure of the vessel had been decided following a non-conformity in the results of the material embrittlement surveillance programme on the vessel, although after analysis this non-conformity proved to be only apparent. This entailed important investigations, over and above the work already carried out in the past, to determine the effect of irradiation on the embrittlement of the reactor vessel (fluence and $RT_{NDT}$) and to check by non-destructive testing that there was no defect of significant size present.

Among these investigations, an important part concerning the determination of the neutron doses absorbed by specimens and the vessel by means of the TRIPOLI code, as well as by dosimetry on samples, will be discussed in this communication. Beforehand, the problems encountered in 1986 will be briefly recalled as an introduction to the various measures decided and undertaken - non-destructive testing and reconstituted test specimens.

## NEW FACTORS AND ACTION DECIDED

After further dosimetry on test specimens and in the cavity (2) and the re-evaluation of fluences, the safety margin was evaluated in 1983 for a $RT_{NDT}$ at the end of service life not exceeding 95°C in association with otner measures such as heating of the safety injection. A new factor was introduced in 1986 with the results of measurements on the capsules that evidenced a shift in the transition temperature greater than the maximum anticipated. An assessment of all the data available pointed out the need to verify the factors governing operating reliability and to set up additional or redundant means (calculations, measurements) with a view to reducing uncertainty on such factors. Consequently :

- the $RT_{NDT}$ value forecast for the vessel was re-evaluated by re-examining the combination embrittlement ($RT_{NDT}$)/fluence of the test specimens in the surveillance programme, and vessel fluence;
- the validity of the size of defects taken into account in the fracture mechanics studies was checked by means of non-destructive testing adapted to the detection and dimensioning of possible defects located in the inner wall of the vessel. This testing, carried out over a large zone of shell-ring and welds, did not bring to light any defect over 3 mm in size, thereby conferring a significant implicit safety margin on the fracture mechanics results.

Re-evaluation of embrittlement due to irradiation of the reactor vessel consisted in :

- first, completing the embrittlement data available, in particular for the non-monitored shell-ring since it was likely to be more susceptible to embrittlement than the monitored shell-ring due to a higher phosphorus content of the order of a few thousandths. To do this, impact test specimens were reconstituted (Battelle Laboratories) to evaluate the shift in the transition temperature for the corresponding impact strength.

- secondly, in determining fluence on the test specimens and on the vessel and the uncertainty associated with those values. This reevaluation of fluences was rendered necessary because of the characteristics of the computational tools used to determine them at that time. It was suspected that the calculated fast neutron fluence had been notably underestimated and that the neutron spectrum used for the analysis of the dosimeters in the surveillance programme was not sufficiently precise.

So, it was decided :

- to determine the neutron spectrum by calculations based on the Monte-Carlo method, in 3-D geometry, for the test specimens located under the core, and to use this spectrum for the analysis of the routine and special dosimetry data; (Fig.1)
- to take activity measurements on samplings from both the stainless steel lining of the inner surface of the vessel and the test specimens in the surveillance programme, in order to compare them with those predicted by the calculation. However, a direct gamma activity measurement ($^{54}$Mn) on the inner surface of the vessel, that had been envisaged and studied, was not retained.

The results of this consistent plan of action were to govern the future operating reliability of the CHOOZ A reactor vessel.

## TRIPOLI CALCULATIONS

Calculations of neutron spectra, saturation reaction rates and flux on the vessel and test specimens were performed using the TRIPOLI program (3), that solves the problem of neutron transport in 3-D geometry by a Monte Carlo method. The cross sections are processed with 315 groups between 19 MeV and thermal energy and a lot of detail concerning in particular the anisotropy of the neutrons re-emitted by reaction on the material. The sources are deduced from fine calculations of power distribution taking into account fissions on $^{235}$U, $^{239}$Pu, $^{238}$U, $^{241}$Pu. Spatial distribution of fissions between these different fissile isotopes is calculated by the APOLLO (4) collision probability code. Reaction rates use the IRDF85 sections modified by the following $R_k$ ratios, resulting from a trend analysis in the fission spectra (5) :

| Dosimeter | Nb | Ni | Fe | Cu | Np | U |
|---|---|---|---|---|---|---|
| $R_k$ | 0.87 | 1.0493 | 1.00 | 0.9709 | 0.9747 | 0.995 |

Two computations were performed by TRIPOLI :

- one in the direction of the lower part in order to deduce the reaction rates and flux of neutrons of energy greater than 1 MeV on the test specimens suspended from the core supporting plate;
- the other radially in order to provide these same quantities on

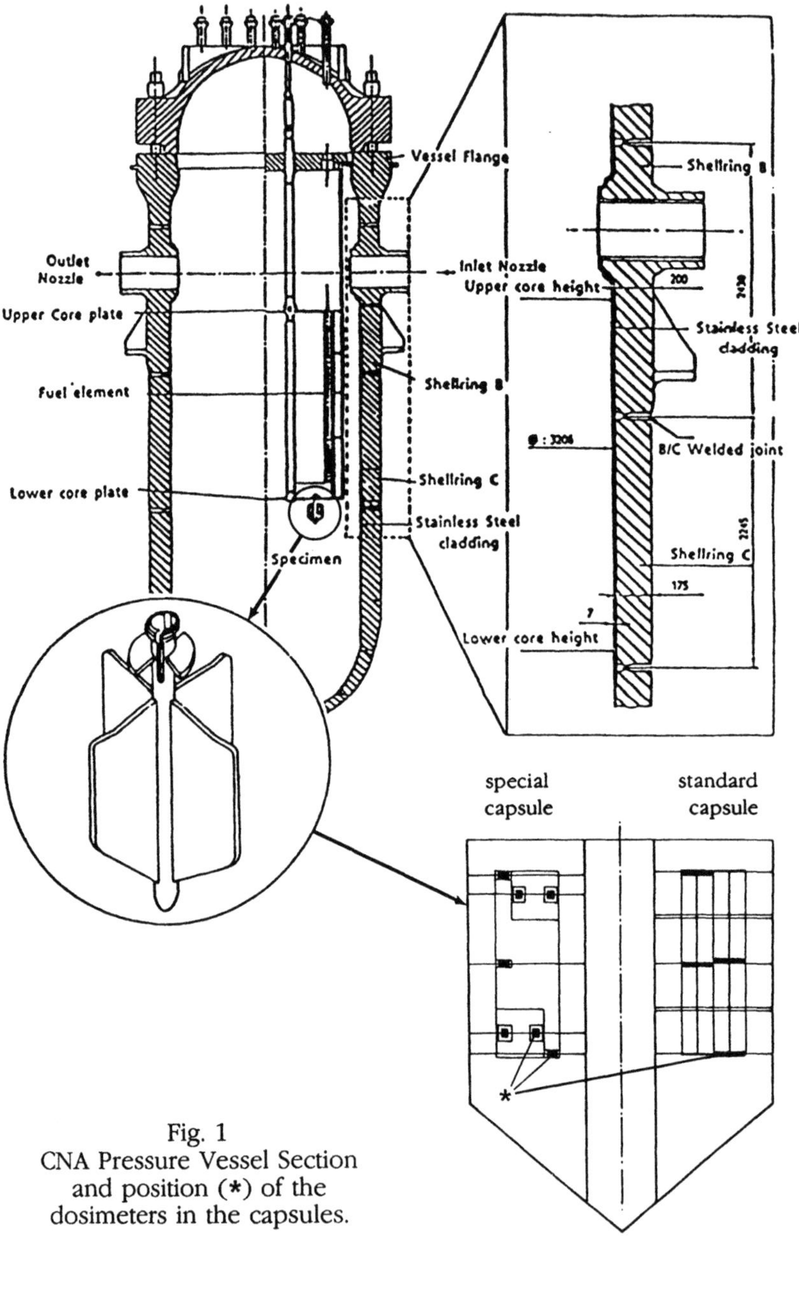


Fig. 1
CNA Pressure Vessel Section
and position (*) of the
dosimeters in the capsules.

the inner surface of the vessel (samplings) and in the cavity.

In the first calculation an infinite (periodic) network was considered for the four overhanging assemblies, with the central position coinciding with the axis of the capsule. The sources are described radially, rod after rod, with an axial distribution furnished by Belgonucleaire, the core control authority. Figure 2 is an exploded view giving some idea of the fineness of the description in the lower part. The estimated uncertainty in these calculations (sources, reactor power output and statistics) is less than 4.5 % (1 standard deviation). This calculation was used to re-interpret the measurements of the special capsule (2). A comparison with experimental results is shown on figure 4, before an adjustment by STAY'SL further improved the overall consistency.

In a second calculation, an eighth of the core was considered, symmetrical on the boundary vertical planes. The eighth of core chosen corresponds to the sector giving maximum power over the first 15 cycles. The calculation is therefore an envelope from the point of view of fluence on the vessel. The sources are described rod after rod for the peripheral fuel elements and per sub-assembly for the other fuel elements.

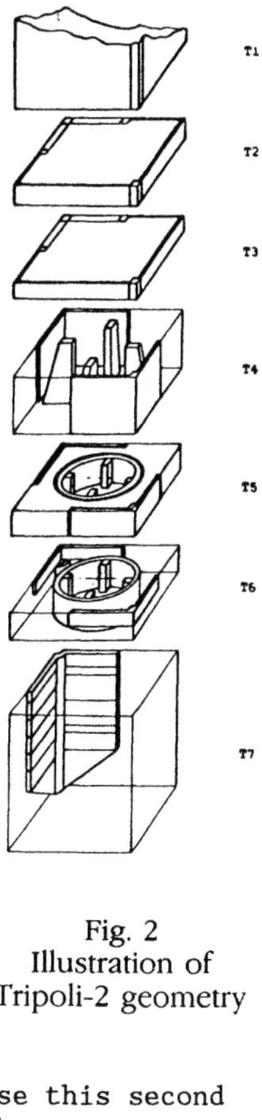

Fig. 2
Illustration of Tripoli-2 geometry

## DOSIMETRY ON SAMPLINGS

### Method

In the method of dosimetry on samplings, that uses samples of the vessel lining and from the test specimens as dosimeters, the determination of fast neutron fluence can be based on the reactions $^{54}Fe(n,p)^{54}Mn$, $^{58}Ni(n,p)^{58}Co$ and $^{93}Nb(n,n')^{93m}Nb$. Three conditions must be respected : it must be possible to measure an activity; to be able to assign this activity to a number of nuclei with reasonable certainty; to have available a "good" mean cross section.

These conditions are satisfied only by the reaction $^{54}Fe(n,p)$. The proportion of Nb, for instance, is too small (less than 30 ppm). For more guarantee, the composition of the samples was determined by two separate methods : chemical separation on ion-exchange resin, followed by weighing of the iron after separation and activation analysis. It was feared that a prohibitive activity from $^{60}Co$ might compromise this second method through a masking effect, but this was not the case.

### Sampling locations and precautions taken

On the surface of the vessel lining, the location corresponds to a maximum of flux azimuthal distribution, and one in the mid-plane (PM) and another at the weld of shell plates B and C (B/C).

The other sampling location is in the vicinity of the notches of the test specimens, on both the upper and the lower lays (Fig. 1). So it was hoped the measurement would be, if not a direct measurement of the "flux factor", at least a direct measurement of a reaction rate factor.

The tool used is a tungsten carbide milling cutter that complied with the following precautions : a) remove as little as possible of the oxide deposit that may not have the same activation history as the metal of the lining : to this end, prior brushing down is carried out and the milling pass is relatively deep, at least 0.1 mm; b) produce chips that can be handled; c) avoid generating a site susceptible to crack initiation.

Shaping and nature of the samples

The samples intended for neutron analysis are compacted in the form of pellets 4 mm in diameter and approximately 1.7 mm thick. (Fig.3). After tests, cold compacting was preferred to melting. Typical composition is : Cr : 20; Ni : 10; Co : 0.05 %. They are designated by PM and B/C.

The test specimen samplings, irradiated for 17 years, are either base metal from shell ring C or from the heat affected zone. designated by "EH" and "EB". Typical composition is : Mn : 1.7; Ni : 0.65; Cr : 0.15; Co :0,02%.

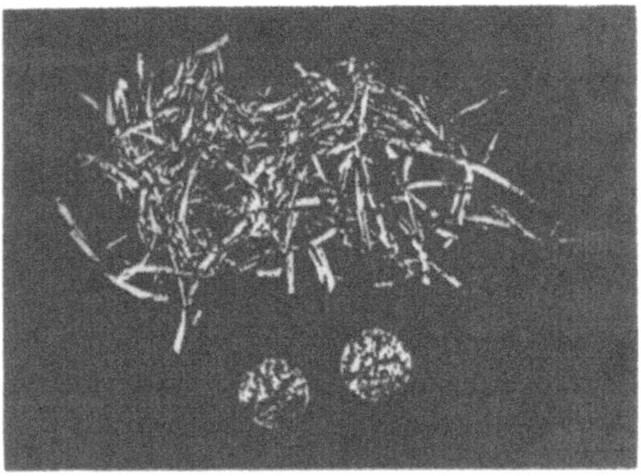

Fig. 3
Appearence of the chips before/after compacting

Chemical separation method

Chemical separation is obtained on DOWEX ion exchange resins, 1x10, particle size distribution 65-90 μ, in a hydrofluoric and hydrochloric medium, in order to separate 3 groups of elements (6).
1) Pure irons. After separation the iron oxide $Fe_2O_3$ is weighed
2) Group of Cr, Mn and Ni, including $^{51}Cr$ and $^{54}Mn$
3) Group of cobalts, 58 and 60.

In this way the number of iron atoms and the corresponding activity in $^{54}Mn$ is obtained in an eluate free from any activity in $^{60}Co$. Activity measurements before and after separation check separation yields.

Nb was also eluted, about 380 mg of lining metal were processed in this way to obtain some 70 Bq of $^{93m}Nb$ but they were not usable.

Method by nuclear weighing

The samples are reactivated with weighed control samples of pure iron and nickel. The activities acquired during re-irradiation (obtained by subtraction) correspond to mass ratios. Irradiations were performed under a boron nitride neutron filter in a continually rotating leak-tight container, in the MELUSINE research reactor, at a fluence of $3.10^{18}$ $n.cm^{-2}$ (E > 1 MeV). All activity measurements are by quantitative gamma spectrometry. The 1 % Mn present originally has not effect.

Analysis results

Iron and nickel controls were determined. For iron, the declared uncertainty is approximately 0.7 % (3 standard deviations). However the ratios :

$$\left(\frac{\text{chemical sep.}}{\text{activation}}\right)_{\text{specimen}} = 1.02 \text{ and } \left(\frac{\text{chemical sep.}}{\text{activation}}\right)_{\text{vessel}} = 0.99$$

show deviations higher than the incertainties declared. Since some imperfections were observed in the recovery of Mn groups in the chemical method, the contents measured by the activation method were chosen.

SUMMARY OF RESULTS

Compared to the values recorded in 1987, the results permit an increase in safety margins : fluence on the vessel has been reevaluated by + 16 %. This is compensated by a higher increase in the fluence on the test specimens (22 to 26 %). In addition, transposing specimen results to the vessel is facilitated by the great similarity of the calculated spectra : between the mid plane of the test specimens and the vessel, the following ratios are observed : 1.85 for Ø > 1 MeV, 1.88 for Ø > 0.1 MeV, and 1.85 for the dpa. Calculations and measurements are very close, as shown in the following table for the flux > 1 MeV, expressed in $10^9$ $n.cm^{-2}.s^{-1}$.

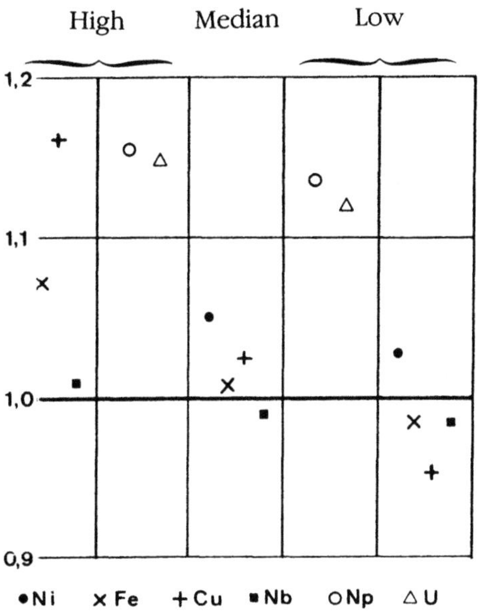


Fig. 4
Special capsule
Flux > 1 MeV : meas/calc

| Location | TRIPOLI Calculations | Special Capsule | Sampling | C/E |
|---|---|---|---|---|
| Vessel lining | 8.81 | - | 8.03 | 1.10 |
| Upper specimen | 24.7 | 26.1 | 26.0 | 0.95 |
| Lower specimen | 11.7 | 11.7 | 11.5 | 1.02 |

| Location | TRIPOLI Calculation | Special capsule | Sampling | C/E |
|---|---|---|---|---|
| Lead Factor specimens/inner wall | H : 2.77 | - | 3.17 | 0.8[illegible] |
| | L : 1.30 | - | 1.40 | 0.9[illegible] |
| Cavity : $^{54}Fe$ reaction rate | 3.14E-16 | 2.74E-16 | | 1.1[illegible] |

DISCUSSION AND CONCLUSIONS

In order to evaluate the safety margins of the Ardennes Nuclear Power Plant, important measures were undertaken in several areas. One of these key areas was neutron dosimetry. A new examination of neutron data confirmed that the values adopted earlier were underestimated, but to a greater degree for the test specimens than for the reactor vessel, and entailed an increase in the flux factor (specimens/vessel) of about 10 %. The vessel spectra and specimen spectra defined precisely by means of the TRIPOLI code, were found to be very similar which meant that the results on the test specimens could be transposed to the reactor vessel without correction. The excellent consistency of results obtained by computation, by the routine and special dosimetry, as well as by the samplings, and their relevant uncertainties, imparted high credibility to these values.

Moreover, it was found that shell-ring B, suspected of degrading more quickly, became less embrittled than shell-ring C that is closely monitored by the surveillance programme. Consequently, the value of $RT_{NDT}$ at the end of service life is now 77°C instead of the previous 95°C. In addition, the excellent results of the non-destructive testing on the inner wall of the reactor vessel and improved precision of the fracture mechanics calculations have validated safety margins authorizing operation of the power plant untill the end of its normal service life.

These investigations authorized re-start up of the plant unit in 1988. The nature of these investigations and the results obtained represent an important contribution to the overall validation of the neutron data used in running the surveillance programme. In this respect, they will be of general benefit for all the PWR vessels in France.

REFERENCES

(1) F.Hédin, S.Baugé, J.C. Guilleret, B. Barthelet - P.T.S. Evaluation in France : the case of CHOOZ A Reactor. International Nuclear Power Plant Aging Symposium. Bethesda, Md. August 1988.

(2) A.Bévilacqua, M.Campani, C.Dupont, R.Lloret, J.C.Nimal, M.Poitou, R.Riehl. Amélioration de la surveillance de la cuve d'un réacteur à eau pressurisée de la SENA. REACTOR DOSIMETRY, Fifth ASTM-Euratom Symposium, Geesthacht, 1984, D.Reidel Pub.Co.

(3) J.C.Nimal et al. Programme de Monte-Carlo polycinétique TRIPOLI 2.

(4) A.Hoffman et al. APOLLO. Rapport SERMA/S/1193.

(5) J.H.Baard, W.L. Zijp, H.J.Nolthenius. Nuclear Data Guide for Reactor Neutron Dosimetry. EUR 12354 EN. Kluwer Academic Pub. Dordrecht. NL.

(6) G.Auboin, J.Laverlochère. Séparations par résines échangeuses d'ions appliquées à l'analyse par radioactivation. R/CEA n° 2359.

(7) F.G.Perey. Least Squares Dosimetry Unfolding : the Program STAY'SL. ORNL/TM/6062 - ENDF 254.

# Comparison of Fast Fluences Determined via Fe-, Nb-, and $ThO_2$-Detectors with Theoretical Values for Several Irradiation Positions in BWR and PWR Plants

E. Polke

SIEMENS AG UB KWU

Hammerbacherstraße 12+14
D-8520 Erlangen

**Abstract**

In the pressurized- and boiling-water-reactors built by SIEMENS AG, KWU Group, the fluences E>1 MeV were measured at different positions inside and outside the reactor pessure vessel (RPV) using Nb-, Fe-, and $ThO_2$-detectors. The comparison of the fluences obtained from the Fe-, and Nb-detectors, when the latter is used with a fit to the cross sections published by Gayther, led to a good agreement with the theoretical values.

**Introduction**

SIEMENS AG, KWU Group, builds pressurized-water-reactors and boiling-water-reactors of different core configurations, thermal powers, and RPV sizes .

The RPV steel embrittlement of all Siemens nuclear power plants is monitored by several surveillance programs. For fluence determination, detectors are installed inside and outside the RPV. The reactions which were used for fluence determination above E>1 MeV are Fe54(n,p)Mn54, Nb93(n,n')Nb93m, and the fission reaction Th232(n,f)Cs137.

In the past there were sometimes discrepancies between the fluences determined via Fe-detectors and those calculated via the Nb93m activities. There is agreement on the activation cross sections of the reaction Fe45(n,p)Mn54 but still debate on which of the published cross section sets for the reaction

Nb93(n,n')Nb93m is correct. The $ThO_2$-detector can only be used at irradiation positions with low gamma flux density because Th has a noticeable photo-fission cross section.

We compared theoretical fluences calculated with the transport code DOT 4.2 /1/ with fluence values determined via Fe- and Nb-detectors using different sets of cross sections for the reaction Nb93(n,n')Nb93m.

**The Activation Reactions Fe45(n,p)Mn54 and Nb93(n,n')Nb93m**

For determination of the fluence in the energy range E>1 MeV Nb- and Fe-detectors are used in general. The shape of the cross section distribution for both activation reactions and the half life of the activated nuclides are different.

As illustrated in Fig. 1 the Fe-detector is qualified for measuring neutron fluence in the energy range E>1 MeV. These cross sections were taken from the International Reactor Dosimetry File IRDF 85 /2/.

The shape of the cross section distribution for Nb93(n,n')Nb93m, as published in 1971 by Hegedüs /3/ and based on measurements, differs markedly from that of Strohmaier et al. /4/ (IRDF 85), as based on calculations (see Fig. 1).

FIG.1-Cross section distribution for Fe54(n,p)Mn54 and Nb93(n,n')Nb93m

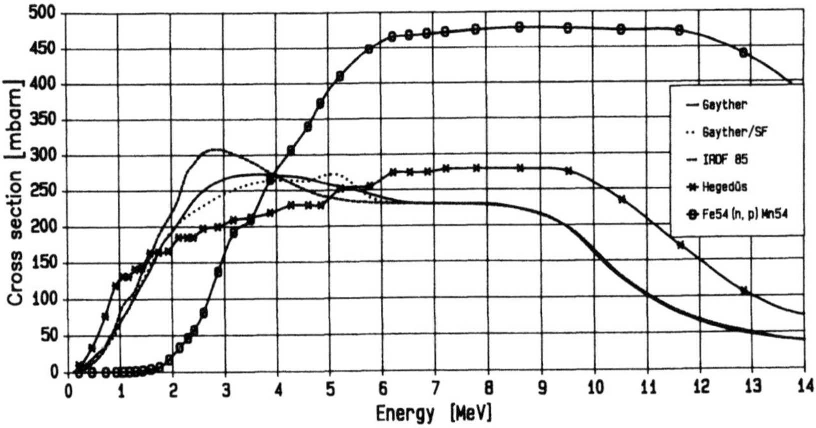


The solid line in Fig. 1 (Gayther) is a free-hand fit made by Gayther et al /5/ to their experimental data for the energy range 0.7 to 6 MeV. For the energy range 0.1 to 0.7 MeV optical-statistical model calculations by Smith /6/ were used and for energies above

6 MeV the shape is the same as that from the data of Strohmaier et al.

The dotted line in Fig.1 is a best estimate fit to the data of Gayther et al made by SIEMENS KWU for the energy range 1 to 6 MeV (Gayther/SF). For other energies the data of Strohmaier were used.

These four different cross sections distributions for Nb93(n,n')Nb93m were used to calculate fluences as based on Nb-detector activities which were then compared to theoretical fluences and fluences obtained from Fe-detectors. These comparisons were done for different types of nuclear reactors and for different irradiation positions.

**Fluence detectors in BWRs of different size inside the RPV.**

As shown in Tables 1-3 there was always good agreement between theoretical fluences and those calculated via Fe-detectors in the energy range E>1 MeV in boiling-water-reactors of different sizes analysed. The agreement with theoretical fluences for Nb-detectors depends on the cross section distribution used.

| Det. Mat. | Cross Section Source | Φ Calc. / Φ Meas. |
|---|---|---|
| Fe | IRDF 85 | 1.02 |
| Nb | IRDF 85 | 1.20 |
| Nb | Hegedüs | 1.29 |
| Nb | Gayther/SF | 1.06 |
| Nb | Gayther | 1.13 |

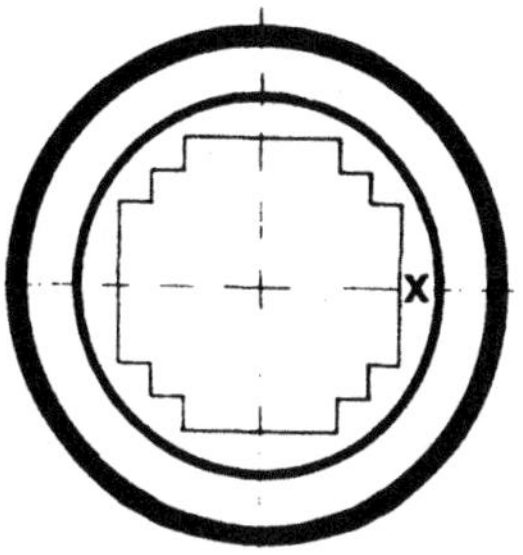


**Tab. 1-BWR 16 MWe RPV-surveillance-program Comparison of fluences with E>1 MeV calculated with transport code (Φ Calc.) for the detector position (X) and fluences determined via fluence detectors (Φ Meas.)**

Best agreement was obtained using Gayther/SF cross sections. Differences between theoretical and experimental fluences as a dependance on the size of the reactor were seen using Nb93 cross sections published by Hegedüs, the agreement being better for reactors with large core configurations than that with smaller ones (see Tables 1-3). The neutron spectrum at the irradiation position is much harder in reactors with large water layers. Altogether similar experimental fluences were obtained with all Nb93 cross sections.

| Det. Mat. | Cross Section Source | Φ Calc. / Φ Meas. |
|---|---|---|
| Fe | IRDF 85 | 0.98 |
| Nb | IRDF 85 | 1.17 |
| Nb | Hegedüs | 1.12 |
| Nb | Gayther/SF | 1.07 |
| Nb | Gayther | 1.11 |

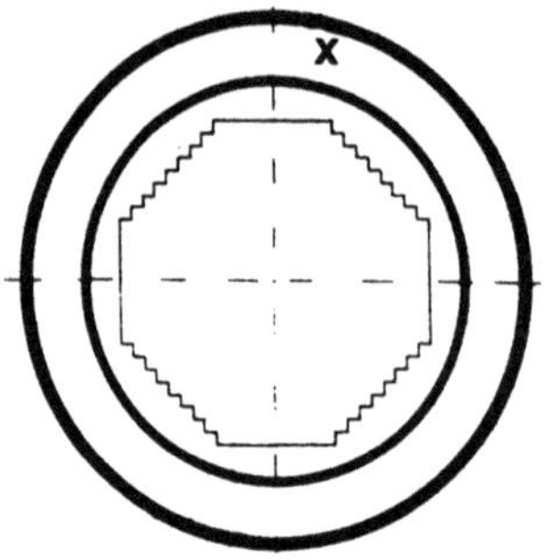


**Tab. 2-BWR 800 MWe RPV-surveillance-program Comparison of fluences with E>1 MeV calculated with transport code (Φ Calc.) for the detector position (X) and fluences determined via fluence detectors (Φ Meas.)**

In addition fluences determined from Cs137 activities are listed in tab.3. These fluences are much larger than the theoretical ones.

| Det. Mat. | Cross Section Source | Φ Calc. / Φ Meas. |
|---|---|---|
| Fe | IRDF 85 | 1.05 |
| Nb | IRDF 85 | 1.20 |
| Nb | Hegedüs | 1.15 |
| Nb | Gayther/SF | 1.09 |
| Nb | Gayther | 1.14 |
| ThO2 | IRDF 85 | 0.78 |

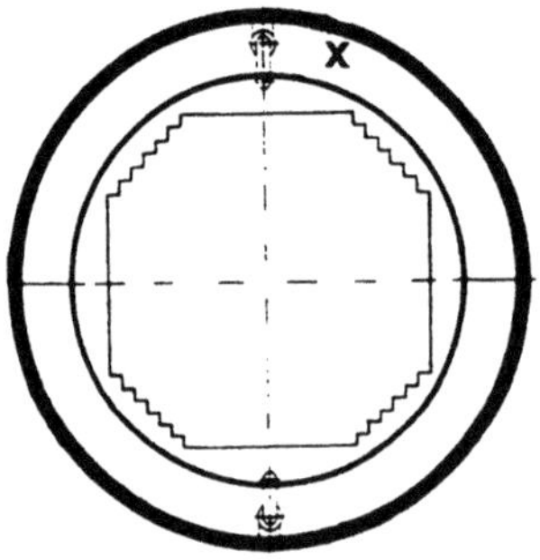


**Tab. 3-BWR 900 MWe RPV-surveillance-program Comparison of fluences with E>1 MeV calculated with transport code (Φ Calc.) for the detector position (X) and fluences determined via fluence detectors (Φ Meas.)**

**Fluence detectors in BWRs of different size outside the RPV.**

Comparison of fluences monitored by Fe- and Nb-detectors positioned outside the RPV were calculated for two boiling-water-reactors of different size. The results illustrated in Tables 4+5 indicate similar good agreement between theoretically and experimentally determined fluences as seen with fluence detectors inside the RPV (compare Tables 1-3). The exception to this are the experimentally obtained fluences using Nb cross sections published by Hegedüs The neutron spectrum is very soft outside the RPV. Again, there was best agreement between theoretically and experimentally obtained fluences for the Nb-detectors using the cross sections of Gayther/SF.

| Det. Mat. | Cross Section Source | Φ Calc. / Φ Meas. |
|---|---|---|
| Fe | IRDF 85 | 1.04 |
| Nb | IRDF 85 | 1.08 |
| Nb | Hegedüs | 1.50 |
| Nb | Gayther/SF | 0.98 |
| Nb | Gayther | 1.14 |

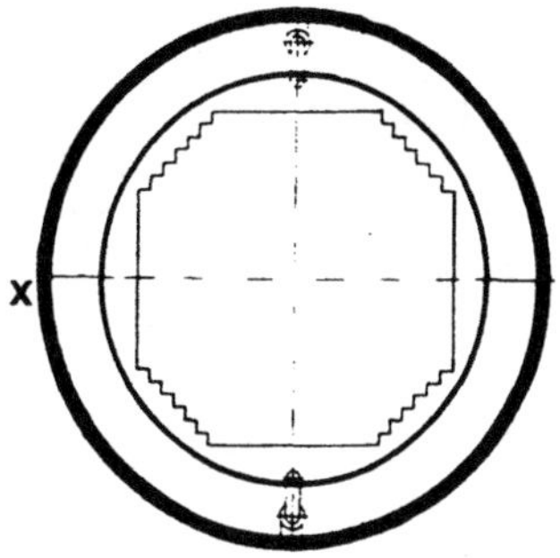


**Tab. 4-BWR 900 MWe RPV-surveillance-program Comparison of fluences with E>1 MeV calculated with transport code (Φ Calc.) for the detector position (X) and fluences determined via fluence detectors (Φ Meas.)**

| Det. Mat. | Cross Section Source | Φ Calc. / Φ Meas. |
|---|---|---|
| Fe | IRDF 85 | 1.13 |
| Nb | IRDF 85 | 1.12 |
| Nb | Hegedüs | 1.53 |
| Nb | Gayther/SF | 1.01 |
| Nb | Gayther | 1.16 |

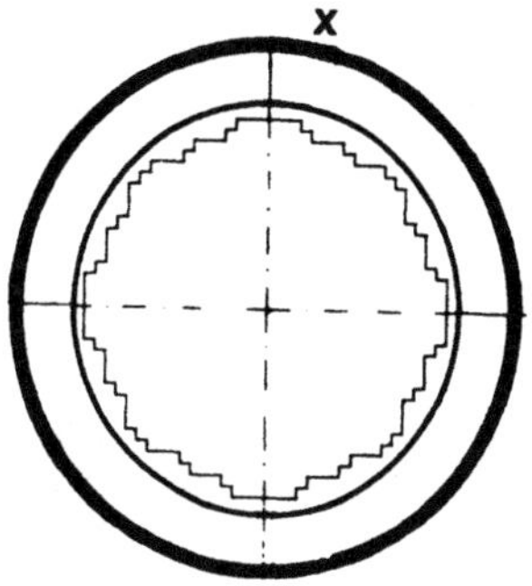


**Tab. 5-BWR 1300 MWe RPV-surveillance-program Comparison of fluences with E>1 MeV calculated with transport code (Φ Calc.) for the detector position (X) and fluences determined via fluence detectors (Φ Meas.)**

**Fluence detectors in PWRs inside the RPV.**

Many fluence detectors inside the RPV of various pressurized-water-reactors built by SIEMENS/KWU were irradiated for different time periods and subsequently analysed. For the comparison of theoretically and experimentally obtained fluences, detectors from two different pressurized-water-reactors were chosen. The results are shown in tables 6+7. Also in these reactors there is usually good agreement between fluences with E > 1 MeV obtained from Fe-detectors and theoretical fluences with E > 1 MeV. Similar good agreement with theoretical fluences was obtained for Nb-detectors, again, in general, best using Gayther/SF Nb-cross sections.

| Det. Mat. | Cross Section Source | Φ Calc. / Φ Meas. |
|---|---|---|
| Fe | IRDF 85 | 1.08 |
| Nb | IRDF 85 | 1.17 |
| Nb | Hegedüs | 1.24 |
| Nb | Gayther/SF | 1.05 |
| Nb | Gayther | 1.11 |

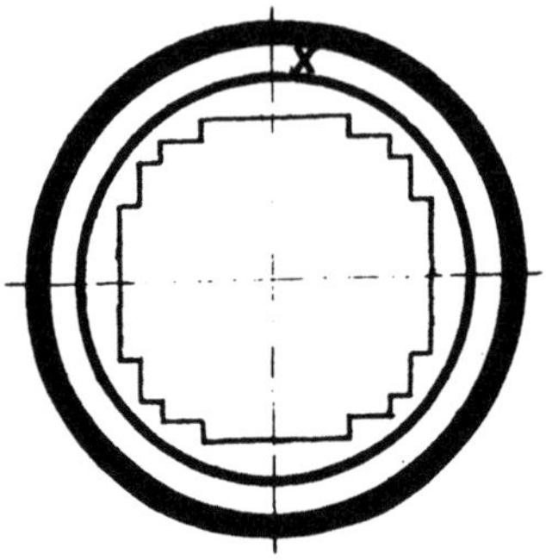


**Tab. 6-PWR 1300 MWe RPV-surveillance-program Comparison of fluences with E>1 MeV calculated with transport code (Φ Calc.) for the detector position (X) and fluences determined via fluence detectors (Φ Meas.)**

In addition fluences determined via $ThO_2$-detectors are listed in Tab.7. These fluences are in good agreement with the theoretical ones .

| Det. Mat. | Cross Section Source | Φ Calc. / Φ Meas. |
|---|---|---|
| Fe | IRDF 85 | 1.10 |
| Nb | IRDF 85 | 1.17 |
| Nb | Hegedüs | 1.23 |
| Nb | Gayther/SF | 1.03 |
| Nb | Gayther | 1.09 |
| ThO2 | IRDF 85 | 0.97 |

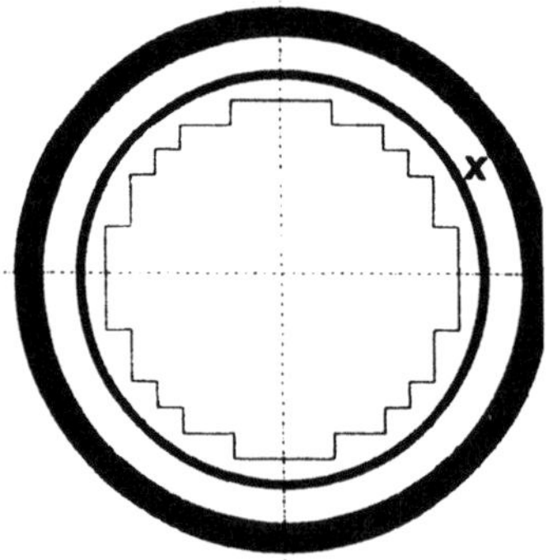


**Tab. 7-PWR 850 MWe RPV-surveillance-program Comparison of fluences with E>1 MeV calculated with transport code (Φ Calc.) for the detector position (X) and fluences determined via fluence detectors (Φ Meas.)**

**Fluence detectors in PWR outside the RPV.**

Also in pressurized-water-reactors built by SIEMENS/KWU fluence detectors at the outside of the

RPV were irradiated. This fluence comparison analysis is shown in Table 8. The agreement of the experimentally obtained fluences with the theoretical data is similar to those obtained from detectors positioned outside boiling-water-reactor vessels. Both detector types show experimental fluences in good fit with theoretical fluences, but here again the Hegedüs cross sections lead to fluences that are far below theoretical values.

| Det. Mat. | Cross Section Source | $\Phi$ Calc. / $\Phi$ Meas. |
|---|---|---|
| Fe | IRDF 85 | 0.97 |
| Nb | IRDF 85 | 1.05 |
| Nb | Hegedüs | 1.67 |
| Nb | Gayther/SF | 0.95 |
| Nb | Gayther | 1.09 |

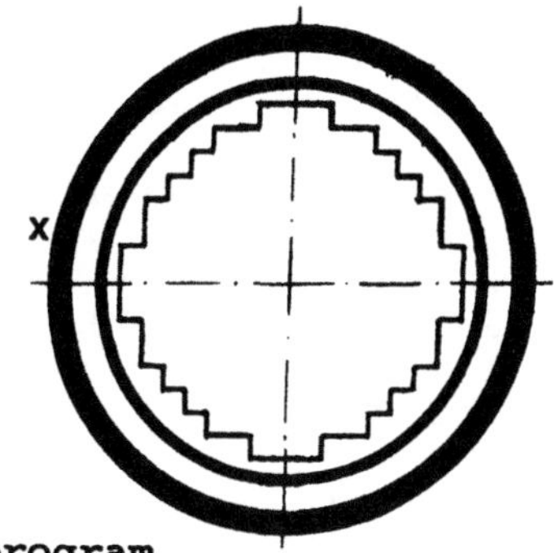


**Tab. 8-PWR 670 MWe RPV-surveillance-program Comparison of fluences with E>1 MeV calculated with transport code ($\Phi$ Calc.) for the detector position (X) and fluences determined via fluence detectors ($\Phi$ Meas.)**

## Discussion

The comparison calculations given here indicate that Fe-detector activities of fluences with E > 1 MeV give rise to fluences that are in good agreement with those obtained via neutron transport programs. Also the data obtained via Nb-detectors are excellent using the cross sections of Gayther et al.

The best agreement with theoretical fluences for Nb-detectors was obtained with Gayther/SF cross sections.

A possible reason for the good agreement of Nb-fluences (based on Hegedüs) with theoretical fluences when Nb-detectors are positioned inside the RPV and the discrepancies using his cross sections with Nb-detectors positioned outside the RPV is that the spectral shape of Hegedüs cross sections does not reflect the real curve. As the neutron spectrum can vary rapidly with irradiation position, a correct estimate can be obtained simply by chance. This happens when the wrong cross section curve is compensated within a single spectrum.

Nb detectors have two advantages on comparison with Fe-detectors.

1. Whereas the effective cross section of the reaction Fe54(n,p) Mn54 is very sensitive to changes in neutron spectrum, those changes are less relevant for the effective cross section of the Nb93(n,n)Nb83m reaction. Therefore the Nb-detectors are superior when the neutron spectrum at the irradiation position is not exactly known.

2. The activation product Mn54 has a relative short half life of 312.5 days in comparison with Nb93m with its half life of 16.1 years. At irradiation periods over a period of many years the neutron fluence can be better estimated using Nb-detectors.

In contrast to Fe- and Nb-detectors the $ThO_2$-detectors yield fluences with E>1 MeV that are in general greater than theoretical values. The differences between $ThO_2$- and theoretical fluences are greater in BWRs than in PWRs. This is referred to as the photofission-effect, because, at the irradiation position, the gamma to neutron ratio in BWRs is greater than in PWRs.

In conclusion, iron and niobium, when the latter is used with the Gayther/SF cross section data, appear to be equally valid the determination of neutron fluences. They compliment each other due to different spectral behaviour and different half lifes of their activation products. The $ThO_2$-detector is qualified for fluence determination only at irradiation positions with a small gamma field.

**References**


/1/ Rhoades, Simpson, Childs, Engle
DOT-IV Radiation Transport Code System
RSIC COMPUTER CODE COLLECTION, Package CCC-320

/2/ Cullen, Kocherov, McLaughlin
The International Reactor Dosimetry File
IAEA-NSA-41/R

/3/ Hegedüs
Détecteur de fluence de neutrons rapides basé sur reaction Nb93(n,n')Nb93m EIR-Bericht Nr. 195, 71

/4/ Strohmaier, Tagesen, Vonach
Physics Data 13-2, (1980) 62

/5/ Gayther, Murphy, Randle, Taylor, Uttley
The Nb93(n,n')Nb93m cross section
Harwell Laboratory, AERE R 12612

/6/ Smith, Smith, Howerton
ANL/NDM-88 (1985)

# COMPUTER CODES AND METHODS

# EXPERIENCES FROM AN INTERNATIONAL UNFOLDING INTERCOMPARISON WITH BONNER SPHERES

*A.V. Alevra, M. Matzke and B.R.L. Siebert*

Physikalisch-Technische Bundesanstalt

W - 3300 Braunschweig, Fed. Rep. Germany

**ABSTRACT**

EURADOS Working Group IV organized a numerical benchmark problem to intercompare unfolding codes for Bonner spheres which are often used in practical neutron spectrometry. Eight laboratories using ten different codes took part in this exercise. A summary of the results and an analysis are given here.

## 1. INTRODUCTION

A Bonner sphere (BS) spectrometer (1) consists of a set of polyethylene spheres of various diameters which contain in their centers a small detector sensitive to thermal neutrons. It is possible to extract spectrometric information from BS measurements, as the fluence response of the BS changes with the spheres' diameter. BS spectrometers have the advantage of an almost isotropic response over a very large energy range from thermal to tens of MeV, but they have a poor energy resolution which restricts the information to a few measured data from which the spectral information is derived through a few-channel unfolding.

In view of the importance of BS's in practical neutron spectrometry, and with the aim of optimizing and harmonizing their application within the European Community, EURADOS Working Group IV*) have designed a numerical benchmark problem for unfolding codes. Eight laboratories took part in this exercise and the results have been published in detail (2). This paper summarizes the results obtained and provides an analysis.

*)EURADOS (European Dosimetry Group) is a Committee founded by the EEC

## 2. THE EURADOS IV BENCHMARK PROBLEM

### 2.1 Formalism, Unfolding Codes and Participants

In a neutron field of fluence **$\Phi$** a set of $n_D$ BS's with a response matrix $\underline{\mathbf{R}}$ (**$\Phi$**, $\underline{\mathbf{R}}$ described in $n_E$ energy bins) produce the readings (3)

$$\mathbf{M} = \underline{\mathbf{R}}\ \mathbf{\Phi} \tag{2.1}$$

The fluence **$\Phi$** can be obtained as a least-squares solution to Eq. (2.1) by minimizing $X^2$:

$$X^2 = (\ \mathbf{M} - \underline{\mathbf{R}}\ \mathbf{\Phi}\ )^T\ \underline{\mathbf{C}}_M^{-1}\ (\ \mathbf{M} - \underline{\mathbf{R}}\ \mathbf{\Phi}\ ) \tag{2.2}$$

where $\underline{\mathbf{C}}_M$ is the covariance matrix of the readings **M** and $^T$ indicates transposition. Upon introducing the structure matrix $\underline{\mathbf{S}}$:

$$\underline{\mathbf{S}} = \underline{\mathbf{R}}^T\ \underline{\mathbf{C}}_M^{-1}\ \underline{\mathbf{R}} \tag{2.3}$$

one obtains the solution:

$$\mathbf{\Phi} = \underline{\mathbf{S}}^{-1}\ \underline{\mathbf{R}}^T\ \underline{\mathbf{C}}_M^{-1}\ \mathbf{M} \tag{2.4}$$

This paper deals with the ***few-channel*** unfolding problem where $n_D < n_E$. In this case $\underline{\mathbf{S}}$ is singular (4) and there is no unique solution. In order to overcome this difficulty, almost all the codes used in the unfolding intercomparison (2) have different ways of introducing additional ***a priori*** information (API). Another possibility would be to expand **$\Phi$** in a series of $n_G$ guess functions and to obtain a unique solution by requiring $n_G \leq n_D$. Table 1 lists the participants and the codes used. Details and references of the codes are given in (2).

Table 1 The participants in the few-channel unfolding intercomparison

| Scientist(s) | Institution | City | Country | Code |
|---|---|---|---|---|
| Aroua-Grecescu | IRA | Lausanne | Switzerland | SAND 5 |
| Buxerolle | DPT/SIDR/GDN CEA-CEN | Cadarache | France | MORELPA |
| Matzke | PTB | Braunschweig | Germany | SAND II$_M$<br>STAY'SL$_M$<br>MIEKE |
| Mourgues | DPT/SIDR/GDN CEA-CEN | Cadarache | France | SPEC2001 |
| Perks | Harwell Laboratory | Harwell | Great Britain | LOUHI82 |
| Schraube | GSF | Neuherberg | Germany | SAND II |
| Thomas | NPL | Teddington | Great Britain | STAY'SL |
| Zaborowski | CEA-CEN | Cadarache | France | SOHO |

## 2.2 Data Given and Quantities Requested

Equation (2.1) was used to simulate the readings of BS. The readings $M_d$, the response functions $R_d(E)$ and a fluence-to-dose-equivalent conversion function, h(E), as given by ICRP 21 (5) were given to the participants.

The energy region from 7.94 meV to 19.95 MeV was divided into 47 logarithmic equidistant bins, 5 per decade. The spectral neutron fluences used for the simulation were composed using five spectra: thermal, $1/E^{\alpha}$, $^{252}Cf$ fission (bare and moderated (5)), and a special monoenergetic contribution, $\Phi_E(E) = \delta(E-E_i)$, where i specifies a selected energy bin. The readings of an eight and four (underscored) BS spectrometer were simulated using the bare detector and the <u>2"</u>, 3", <u>5"</u>, <u>8"</u>, 10", <u>12"</u> and 18" spheres. In the latter, only simple combinations of $1/E^{\alpha}$ and $^{252}Cf$ spectra were considered, whereas all five spectra were combined in the case of the eight BS's. The participants were given the energy bin number, i, if a monoenergetic component was added.

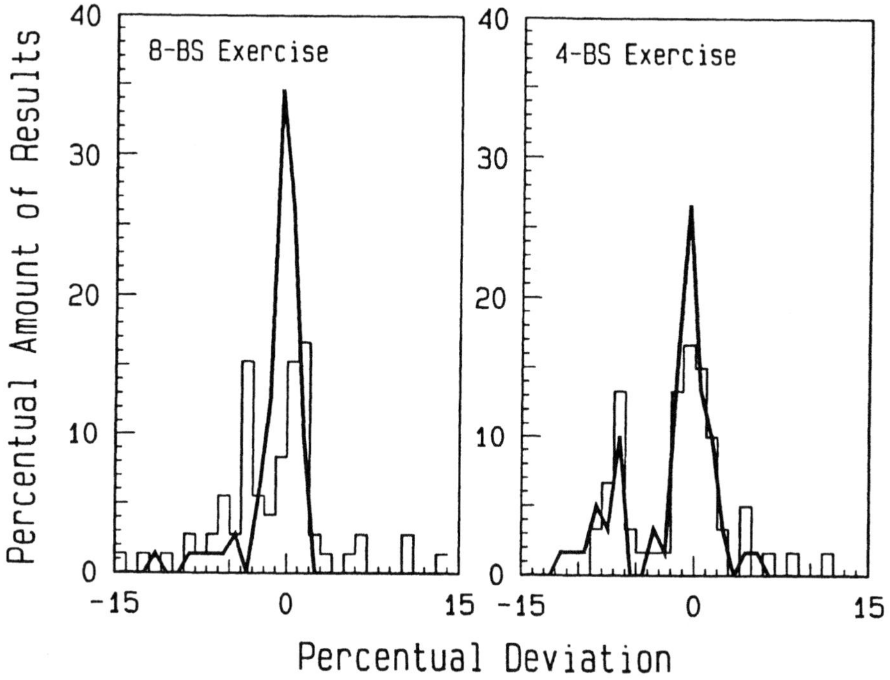


Fig. 1 Deviations of the integral results given by the participants from the original values in the 8- and 4-BS exercises. The thick, full lines indicate the results for the fluence, the histograms show the results for the dose equivalent.

Each participant received data resulting from six spectra *measured* with the 8-BS set and another six spectra *measured* with the 4-BS set, and was asked to unfold and specify the neutron fluence, $\Phi_j$, in the 47 energy groups given. This group fluence can be used to compute the integral fluence and the dose equivalent, $H = \mathbf{h}^T \mathbf{\Phi}$, where the $n_E$ components of the vector **h** represent the bin-averaged fluence-to-dose-equivalent conversion function. An uncertainty analysis was deliberately not requested, as most of the existing codes cannot provide it.

## 3. SUMMARY OF RESULTS AND ANALYSIS

The integral results obtained by the participants (71 and 60 results for the 8- and 4-BS exercises, respectively) are summarized in Fig 1, in terms of percentual deviation from the original values. Most results for the fluence (thick lines) are concentrated in a band of ± 2 % around the true values. A wider spread of the results for dose equivalent (histograms) was expected because of the strong energy dependence of the conversion function h(E), but most of the results lie in a band of ± 6 %. The secondary peak appearing in the 4-BS exercise is due to a spectrum with a somewhat unrealistic $1/E^\alpha$ contribution at neutron energies exceeding the fission spectral distribution which was not *guessed* by the participants.

Fig 2 shows spectral results for a quite simple, smooth spectrum used in the 8-BS exercise. Three participants using variants of the

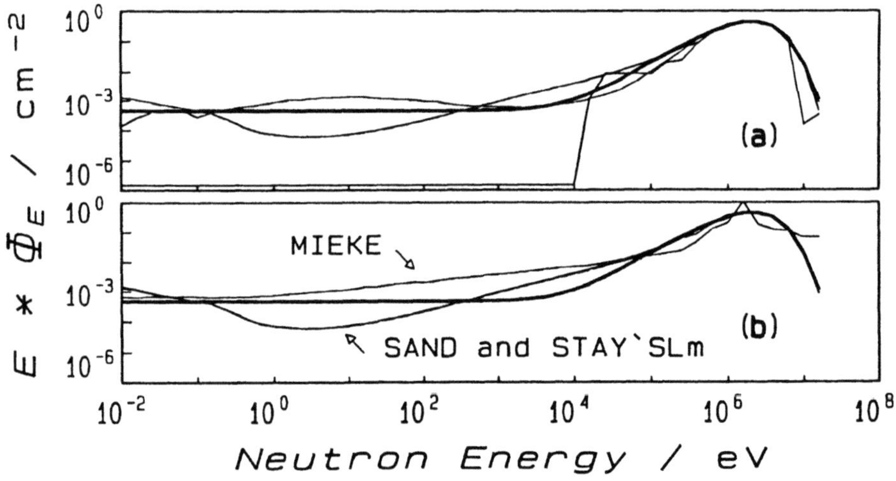


Fig. 2 One original spectrum normalized to unity integral fluence (thick line) and the corresponding solution spectra (thin lines) given by: (a) = three participants using the SAND code; (b) = one participant using three different codes.

SAND code (6) (Fig 2a) obtained different results, which demonstrates the influence of judgement in supplying the necessary API. The importance of the API is also evidenced by Fig 2b which displays results for the same spectrum obtained by one participant using three codes. The results for SAND and STAY'SLm (7) are indistinguishable in this plot, as with the latter code the result of SAND and a diagonal covariance matrix with relative standard deviations of 100 % had been used as API. MIEKE (4) did not use any API and produced a different result.

Fig 3 shows results obtained by one participant with STAY'SL for three of the spectra used in the 8-BS exercise. The general shape of the spectra has been quite well calculated. Sharp structures, however, are washed out or even ignored (Fig 3c). The monoenergetic component is well matched in Fig 3a but only seen as a broad peak in Fig 3b. This is partially explained by the energy dependence of the response functions. The response for the 18" sphere peaks at about 10 MeV (peak in Fig 3a), whereas all response functions are smooth at 25 keV (peak in Fig 3b). Another explanation is that with the STAY'SLm code, the monoenergetic component was guessed to amount to 20 %, which comes

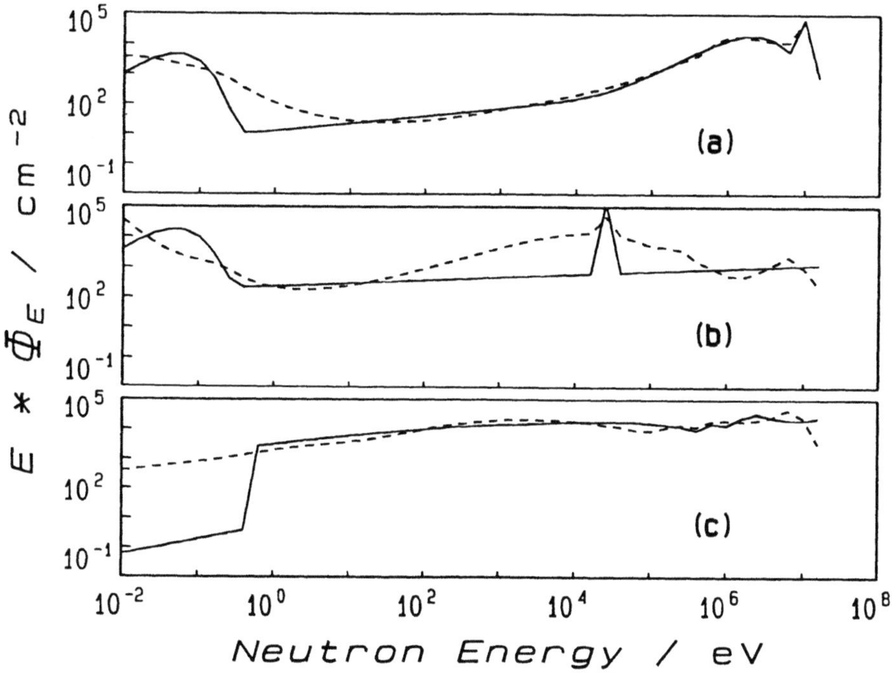


Fig. 3 Three of the solution spectra (dashed lines) obtained by one participant using STAY'SL in the 8-BS exercise compared with the corresponding original spectra (full lines)

close to its actual strength in Fig 3a, i.e. 36 %, but misses the 60 % used in the spectrum shown in Fig 3b.

Such results can be better understood and analysed by discussing the rank, r, of the structure matrix, $\underline{\mathbf{S}}$ (see Equation 2.4). The rank of $\underline{\mathbf{S}}$ is equal to its number of non-vanishing eigenvalues. It is intuitively clear that, without using API, no more than r ***pieces of information*** may be extracted from a set of measurements. The rank of $\underline{\mathbf{S}}$ is $n_D$ if the response functions are linearly independent. In the case of the BS the response functions overlap considerably, although seen strictly mathematically, they remain linearly independent. The question is whether we can still obtain $r = n_D$ good ***pieces of information***. If not, then the information supplied by some spheres could be redundant.

To illustrate this, instead of Bonner spheres, let us consider sets of four or eight ideal detectors whose response functions are assumed to be Gaussian. Their maxima are assumed to be log-equidistant covering the entire energy region considered in the intercomparison and their full-widths-at-half-maximum are chosen to be an integer number of bins, $n_W$. For a set of eight such ideal detectors with $n_W = 8$, the response functions overlap only slightly and we obtain for $\underline{\mathbf{S}}$ eight non-vanishing eigenvalues. With increasing $n_W$, the eight eigenvalues do not vanish, but some of them become very low in magnitude. The magnitudes of the eigenvalues of $\underline{\mathbf{S}}$ for various sets of ideal detectors with different energy resolutions (indicated by the values of $n_W$) are shown in Fig 4 together with the eigenvalues of $\underline{\mathbf{S}}$ obtained for a real set of Bonner

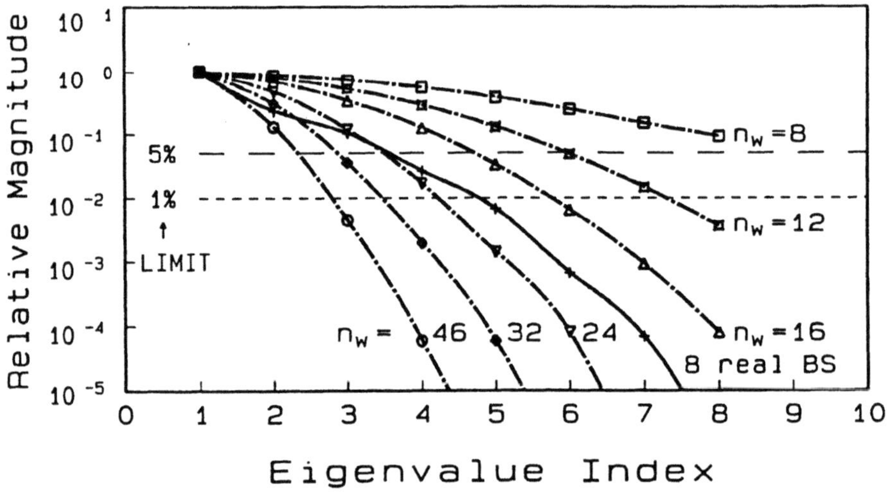


Fig. 4 Relative magnitudes of the eigenvalues of the structure matrix, $\underline{\mathbf{S}}$, for a real set of Bonner spheres, and for sets of eight *ideal* detectors (see text) with different energy resolutions indicated by $n_W$.

spheres. For each detector set the eigenvalues were normalized to the largest value in the set. If we consider a limit of 1% of the largest eigenvalue obtained, then we see that for $n_w$ = 24 for instance, four of the eigenvalues fall below this limit. It is obvious that in this case, the spectrometric information supplied by the set of eight detectors is redundant and it is reasonable to consider that only four of them are effective. Putting the limit at 5% we find that only three detectors can contribute more than 5% to the spectrometric information obtained. The number of eigenvalues whose magnitudes exceed the established limit (8) is denoted here as $r_{eff}$, meaning ***effective rank***.

These results suggest that the performances of a real BS spectrometer whose response functions are fairly broad in energy, are not as good as might be expected from the number of spheres it contains. Indeed, for the eight and four (not shown in Fig 4) sphere sets of BS used in the intercomparison, with a limit of 1% we obtain effective ranks of $\underline{S}$ of 4 and 3, respectively.

In practice the actual shape of the spectra plays a prominent role, as can be seen when comparing Figs 3a and 3b. A detailed study on the optimal selection of spheres is given in (9), where a consistent algorithm is given which allows the above mentioned criteria to be quantified.

## 4. CONCLUSIONS AND ACKNOWLEDGMENTS

No attempt was made to distinguish between ***good*** and ***bad*** unfolding codes. The most important thing was to have as many participants as possible in the exercise who are willing to learn something from it. Publication of the details of the results in (2) was intended to help all those interested in improving their ***unfolding skill***.

All the unfolding codes used in the intercomparison performed quite well in obtaining the integral quantities total fluence and dose equivalent. The spectral results are satisfactory for simple spectra, but show some larger deviations for complicated spectra. The influence of the API appeared to be important, showing that experience is necessary in using the unfolding codes.

An analysis in terms of the effective rank, $r_{eff}$, of the structure matrix provides some physical insight into the matter. The more the response functions overlap in energy, the smaller the $r_{eff}$ found.

To summarize, the good results are explained mainly by the fact that quite realistic spectra, i.e. those occurring in radiation protection practice, were used in the benchmark problem and the experience of the participants therefore resulted in good API guesses. Consequently, a study of the spectral neutron fluence in radiation protection environments in order to improve the available API is necessary.

These results are encouraging and suggest that a second stage of the intercomparison which includes a variance analysis and quantifies the influence of the *a priori information* used could be organized.

The authors wish to thank EURADOS and the participants in the intercomparison for their support and active contributions. This work was partially supported (A.V. Alevra) by the European Community under Contract No. BI6-A-012-D.

## REFERENCES

1. R.L. Bramblett, R.I. Ewing, and T.W. Bonner, "A New Type of Neutron Spectrometer", Nucl. Instr. and Meth., **9** (1960) pp 1-12

2. A.V. Alevra, B.R.L. Siebert, A. Aroua, M. Buxerolle, M. Grecescu, M. Matzke, M. Mourgues, C.A. Perks, H. Schraube, D.J. Thomas, and H.L. Zaborowski, Unfolding Bonner-Sphere Data: A European Intercomparison of Computer Codes, Laboratory Report PTB-7.22-90-1, Braunschweig (1990).

3. A.V. Alevra and B.R.L. Siebert, Influence of Neutron Spectra and Fluence Response Data on the Determination of Dose Equivalent with Bonner Spheres, PTB Report ND-28, Braunschweig (1986).

4. K. Weise and M. Matzke, "A Priori Distributions from the Principle of Maximum Entropy for Monte Carlo Unfolding of Particle Energy Spectra", Nucl. Instr. and Meth., **A280** (1989) pp 103-112

5. International Commission on Radiological Protection, Data for Protection against Ionizing Radiation from External Sources - Supplement to ICRP Publication 15, ICRP Publication **21**, Pergamon Press (1973)

6. W.N. McElroy, S. Berg, T. Crockett, and R.G. Hawkins, SAND-II, a Computer-Automated Iterative Method for Neutron Flux Spectra Determination by Foil Activation, Report AFWL-TR-67-41, U.S. Air Force Weapons Laboratory (1967).

7. F.G. Perey, Least-Squares Dosimetry Unfolding; the Program STAY'SL, Report ORNL/TM-6062, Oak Ridge National Laboratory, USA (1977).

8. E.J. Szondi, The Covariance Matrix of Neutron Spectra, Nucl. Reactor of the Techn. University of Budapest, Report BME-TR-RES-10/86 (1986)

9. K. Weise, "Optimization in Neutron Spectrometry and Dosimetry with Bonner Spheres Using a General Measure of Quality of Experiments", To be published.

# Bayesian Unfolding of Pulse-Height Spectra

Milos TICHY

Institute of Radiation Dosimetry,

Prague, Czechoslovakia

## Abstract

Utilization of a method based on the Bayes' theorem to many-channels unfolding problem is presented. The described and tested program DIFBAS, which is used for processing of spectra from NE-213 scintillator, can be also employed for any type of detector of neutrons or gamma rays whose response functions are known. It utilizes covariances of both pulse height and a priori neutron/gamma spectra to create the a posteriori covariance matrix. Tests with a moderated fusion spectrum and Gaussian correlations covering influence of input parameters and numerical stability are described. Results are compared with those gained from the FORIST code. The necessity of use of correlations for calculation of uncertainties of integral quantities is also noted.

## 1. Introduction

The unfolding of a neutron spectrum from a pulse height (p.h.) spectrum (from an organic scintillator detector) cannot be successfully performed by a usual derivative method for spectra with substantial part of neutrons above 6 MeV due to influence of products of reactions on carbon. For this case a matrix method, which is the topic of this article, should be used. In this method the unfolding means to solve the following matrix equation:

$$\mathbf{Y} = \mathbf{A}\,\mathbf{X} \qquad (1)$$

The vector **Y** containing N elements approximates the p.h. spectrum and the required neutron spectrum is approximated by a vector **X** (M elements). **A** is the response matrix (N*M type).

The size of dimensions M and N is crucial for the solution of eq.(1). N is generally greater then M but even in the case N=M the task can be hardly solved by a simple inversion of **A** because of the proximity of its columns to linear dependence. The solution of eq.(1) is thus based on minimization of the following quadratic form:

$$EX(\mathbf{X})=(\mathbf{Y}-\mathbf{AX})^T\mathbf{W}_1(\mathbf{Y}-\mathbf{AX})+(\mathbf{X}-\mathbf{X}^o)^T\mathbf{W}_2(\mathbf{X}-\mathbf{X}^o) \qquad (2)$$

Methods developed for this purpose differ in choices of weight matrices $\mathbf{W}_1$ and $\mathbf{W}_2$ and zero-approximation $\mathbf{X}^o$ of **X**. The most spread methods are based on thesis of Burrus [1], which utilize generalized inversion of matrix **A**. They are realized by codes FERD, FERDOR, FORIST, COOLC [2] and SLOP [12].

The request for variances and even correlations of resulting neutron spectra leads to the utilization of the maximum likelihood principle or the Bayes' theorem. Both theories have been already used for the few-channel unfolding (e.g. STAY'SL [3], FERRET [4] or BASACF [5]) and recently also for the task we are dealing with: BAYES [6]. In this paper another code named DIFBAS based on the Bayes' theorem is presented. After a short description of the method and the code, the main part is devoted to tests of the code and to the question of input data. Results of the code are compared with those of the FORIST [2].

## 2. The Method and the Code

The used theory of unfolding is based on the Bayes' theorem (e.g. in [7]). Its application to the neutron spectrum unfolding has been thoroughly explained e.g. in [4, 5] and leads to a minimization of the quadratic form (2)[1]. The method used for minimization is called covariance filter [5,7], and can be characterized as a procedure to modify a priori spectrum and its covariance matrix by information contained in the p.h. spectrum (and its covariances) using the response matrix without any matrix inversion.

The presented code DIFBAS uses the same theory and minimiza-

---

1 $W_1=Q_Y^{-1}$ resp. $W_2=Q_X^{o\,-1}$ are the inverses of the covariance matrices of the p.h. resp. the a priori neutron spectrum.

tion method (also some subroutines) as the code BASACF (few-channel unfolding). A detailed description of the theory and method is in [5] and here differences are summarized only: integral detector readings are replaced by the p.h. spectrum and the most important difference is that the response matrix **A** (corresponding to $\sigma$ in [5]) is supposed to be precisely known. This assumptions keeps reasonable size of the code (not requiring large mainframe computers) and in practice there are not response matrices available accompanied by covariances. As a consequence of this supposition the quadratic form has just the form (2) and forms of L-D factors are identical with those mentioned as the special case in [5, part 4] also.

The crucial problem of the used method is *availability of input data* especially the covariance matrices of the p.h. and a priori spectra. In the code DIFBAS, they are calculated from input data for respective variances and correlations. Variances of the p.h. spectrum are supposed to reflect statistics of the measurement. They are read together with p.h. spectrum. The "quality" of the a priori neutron spectrum should be represented by coefficients of variation which must be specified by the user.

There is no source of correlations of both the p.h. and the a priori neutron spectra. To prove the code as a tool and to investigate the influence of non-zero correlations, a Gaussian method for generation of artificial correlations [4,5] has been used. It employs a conception that correlations decrease with mutual distance (expressed in the difference of their average energies) of correlated groups approximately according to Gaussian curve:

$$\mathrm{corr}\, x_i x_j = \theta * e^{-\frac{(E_i - E_j)^2}{2\sigma^2}}, \quad \theta \in \langle 0,1 \rangle, \; i \neq j \tag{3}$$

where values of the parameter $\theta$, called "strength of correlation" [4], will be tested. The parameter "dispersion $\sigma$" in this function, also called "range of correlation", is supposed to be dependent on resolution of the spectrometer: FWHM=2.355*$\sigma$.

For the FWHM of i-th group of the p.h. spectrum the following form used in [8] has been employed:

$$\mathrm{FWHM}_i = \left( \alpha^2 + \frac{\beta^2}{h_i} + \frac{\gamma^2}{h_i^2} \right)^{\frac{1}{2}} * h_i, \tag{4}$$

where $h_i$ is the average pulse height in i-th group of the p.h. spectrum and $\alpha, \beta, \gamma$ are parameters characterizing a detection chain of the spectrometer.

Three different formulas characterizing the dependence of FWHM on energy are available in the code to determine the a priori neutron spectrum correlations:

1.Resolution not dependent on energy:

$$\mathrm{FWHM}=\mathrm{const}. \tag{5}$$

2.Resolution linearly dependent on energy:

$$\mathrm{FWHM}_i=\mathrm{const.}*E_i \tag{6}$$

3.The form published by Tschechanski [9]:

$$\mathrm{FWHM}_i = \left(A^2+ \frac{B^2}{E_i}\right)^{\frac{1}{2}} * E_i . \tag{7}$$

where $E_i$ is the average energy of i-th group and A and B are parameters taken from [9].

## 3. The DIFBAS Program Testing

The test has been focused on

- finding out proper input parameters,
- numerical stability of the method,
- estimation of reliability of a posteriori variance,
- consideration about the influence of non-zero correlations of p.h. and a priori spectra.

The base of the tested data is *a model spectrum* (Fig.1) which is similar to the spectrum of d-t neutrons transmitted through 7.5 cm of Fe. A *"measured" spectrum* is the product of the model spectrum and **A** (from [10]). Its variances have been set to 1%. The first derivative of the "measured" spectrum has been used as an a priori guess of the neutron spectrum (Fig.1). The corresponding coefficients of variation of 64 group fluences has been set to 30, 50 and 10% in the secondary peak (1-6 MeV), valley (6-12 MeV) and primary peak (12-16 MeV) ranges respectively. They express the difference of the first derivative from the model spectrum at the secondary peak caused by parasitic reactions on carbon and unreliability of derivation method in the valley range caused by a relatively small amount of neutrons there.

The following parameters were tested: strength of correlation $\theta$ for both p.h. and a priori spectra (eq.3), variances and forms of FWHM (eq.5-7) for a priori spectra. To reduce the amount of output data for reasoning, the spectra and covariance matrices have been condensed from 59 groups (DIFBAS result) to the supergroup structure with dividing points approximately 1, 6, 12, and 16 MeV following a natural partition of spectrum. The influence of varied parameters has been observed by comparing group fluences to those of true ones *(the model spectrum)* in the supergroup structure. In the case of covariance matrix, influence of parameters will be compared only relatively.

<u>The influence of correlations of the p.h. spectrum:</u> The dependence of the a posteriori supergroup fluences on values of the strength of correlation θ from -1 to 1 is negligible. A similar minor influence has been observed also on smoothness of the a posteriori spectrum and its covariances. From this it can be concluded, that *inclusion of the above type of correlations of the p.h. spectrum does not bring a substantial improvement* and then no correlations can be used.

<u>The influence of variance of the a priori spectrum:</u> An alternative second set of the coefficients of variation has been tested: 5, 10 and 5% in the respective supergroups. For this set the a posteriori spectrum is smoother at the secondary peak but the mean of fluence there is shifted. In this case both a posteriori variances and correlations have been reduced also; roughly to 60% of the original value. *A decrease of the a priori uncertainties can thus cause, besides an expected decrease of the a posteriori ones, a shift of the expectation values.*

<u>The influence of correlations of the a priori spectrum:</u> Two of the three forms for FWHM have been tried: Tschechanski's form (7) and the constant resolution form (5). The influence of non-zero correlations of the a priori neutron spectrum on the a posteriori values can be summarized as follows:

- *Expectation values of supergroup fluxes are not influenced.*
- *An increase of correlations causes an increase of smoothing effect.* This must be carefully used for spectra with a peaked character.
- *The a posteriori covariance matrix is influenced:* In the region of increased a priori correlations a small increase of correlation and decrease of variance is observed (Tab.1).

| a priori Gaussian correlation given by eq. | | | | | |
|---|---|---|---|---|---|
| (3)&(7) A=23.3 B=3.2 | | | | (3)&(5)FWHM=700keV | |
| θ=0.0 | | θ=1.0 | | θ=0.98 | |
| 0.77 | 100 | 0.59 | 100 | 0.43 | 100 |
| 8.68 | -20 100 | 7.95 | -19 100 | 7.91 | -23 100 |
| 0.33 | -2 -73 100 | 0.31 | -3 -75 100 | 0.32 | -4 -78 100 |

Tab.1: A posteriori variance and correlations [%] for two forms of FWHM in definition of a priori Gaussian correlations.

From the above test *the following parameters can be chosen as the most suitable: Diagonal covariance matrix of p.h. spectrum and Gaussian correlation matrix defined by eqs.(3) & (5)* of the a priori neutron spectrum (θ=0.98, FWHM=700 keV) - the a posteriori spectrum is on Fig.1. It is necessary to note that the chosen form for FWHM (5) which gives the smoothest a posteriori spectrum does not correspond to published forms which have their source in a measurement. It casts doubts on the idea of connection of FWHM and "range of correlation".

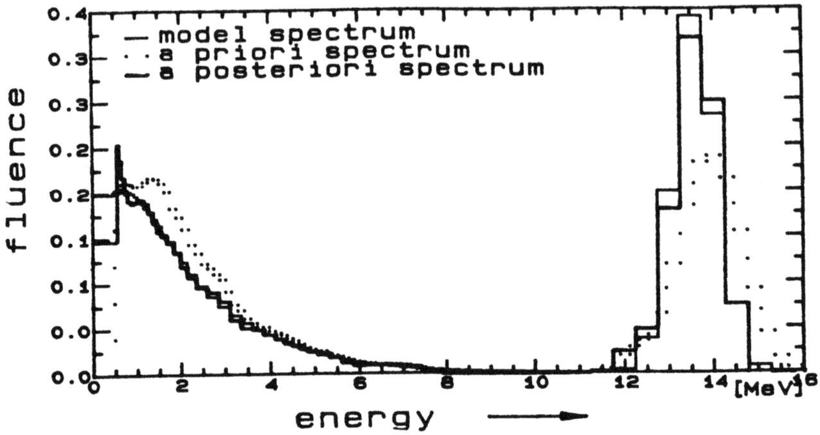


**Fig.1:** Neutron spectra: model spectrum, a priori spectrum (the first derivative of "measured" p.h. spectrum) and a posteriori spectrum with the most suitable parameters.

For these parameters *the difference of the a posteriori and the model spectra is less then one a posteriori standard deviation* in the secondary peak region and in the valley (after corrections of unrealistically low or negative values of the a priori spectrum in this region). For the primary peak the difference is a little bigger. The coefficients of variation of the a posteriori matrix (Tab. 1) seem to be unusually small. This has two reasons:

- The above results do not include an increase of variances due to normalization by a coefficient (e.g. count of a monitor) always known with non-zero variance.
- The negative correlations of group fluences cause a decrease of the variance of integral values; they are in many codes calculated with the unmentioned supposition of diagonal correlation matrix.

By means of randomizing of the p.h. spectrum it was proved that *the described method is numerically stable.* When the a priori spectrum is given in the form of the first derivative of the measured spectrum it must be checked (for negative or unrealistic values).

As it has been mentioned above, the introduction of correlations of the p.h. spectrum does not bring substantial advantages. But even if an a priori spectrum is also supposed non-correlated, the a posteriori covariance matrix has non-zero correlations. In the tested case these correlations were mostly negative and in Tab.2 is shown that omission of them would cause a substantial increase of variances of fluences in the supergroup structure. From it follows that *the a posteriori correlations must be taken into account even if a priori spectrum is considered non-correlated.*

| int. flux range [MeV] | coeff. of variation [%] | |
|---|---|---|
| | correlations not consider. | correlations considered |
| 1-16 | 2.24 | 0.32 |
| 1- 6 | 4.56 | 0.77 |
| 6-12 | 15.29 | 8.68 |
| 12-16 | 1.87 | 0.33 |

Tab.2: The a posteriori variances depending on treatment of the correlations.

The last effect of correlations has been already mentioned: smoothing of the a posteriori spectrum. From this point of view it has a similar role as smoothing procedure in FORIST. Spectra gained by this code are comparable with results of DIFBAS but *estimation of "errors" in FORIST is unrealistic*: one sigma interval is generally narrower and the number of groups where the differences between results and the true spectrum are outside of this interval is greater than in the case of DIFBAS.

## 4.CONCLUSIONS

Utilization of the method based on the Bayes' theorem, already used to few-channel unfolding problem, has been presented. The described and tested code DIFBAS can be used for unfolding of spectrum (of neutrons or gammas) from a pulse-height spectrum for a wide range of types of detectors. It is able to utilize full covariance information on both p.h. and a priori spectrum. Correlations can be omitted when no information is available and described Gaussian correlations are not appropriate. Covariances of response matrix have been ignored to simplify the task and because no source of them is known.

The DIFBAS code is based on the same principle as codes STAY'SL and BAYES, but utilizes a different numerical method for calculation of the a posteriori parameters.

The presented code will be used especially for the processing of p.h. spectra from NE-213 spectrometer (both neutrons and gammas) for benchmarks on materials for fusion technology [11]. The test, covering the influence of input parameters and numerical stability, has shown usefulness of the code in the case of slightly moderated originally monoenergetic fusion spectrum and for the optimized set of input data. Comparison with FORIST demonstrated that while both codes can give similar resulting spectra, DIFBAS produces also correlations and its estimate of a posteriori variances is more realistic.

The possibility to gain correlations of spectra is the main reason to use such methods as the presented one. Their important role in calculations of variances of integral quantities has been also shown.

## 5.REFERENCES


[1] Burrus W.R.: Utilization of A Priori Information by means of Mathematical Programming in the Statistical Interpretaion of Measured Distributions, Thesis, ORNL-3743, Oak Ridge (1965).

[2] Johnson R.H.: A User's Manual for COOLC and FORIST,PNE-75-107, W. Lafayette (1975).

[3] Perey F.G.: Unfolding by the Least-Squares Method, in: Current Status of Unfolding, Oak Ridge 1977
Perey F.G.: Least-Squares Dosimetry Unfolding: The Program STAY'SL, ORNL/TM-6062, Oak Ridge (1977).

[4] Schmittroth F.: A Method for Data Evaluation with Lognormal Distibution, Nucl.Sci.&Eng. 72,1979,19
Schmittroth F.: FERRET Data Analysis Code, HEDL-TME-79-40, Hanford (1979).

[5] Tichy M.: Bayesian Approach to the Neutron Spectrum Adjustment, Nucl.Instr.& Meth. A267(1988)502.

[6] Dickens J.,K. et al: An NE-213 Scintillator - Based Neutron Detection System for Diagnostics Measurements for Neutrons Having Energies > 0.8 MeV..., ORNL/TM-9561, Oak Ridge (1985).

[7] Peterka V.: Bayesian Approach to System Identification, in: Trends and Progress in System Identification, ed. Eykhoff (Pergamon Press, 1981).

[8] Dietze G., Klein H.: NRESP4 and NEFF4, Monte Carlo Codes for Calculation of Neutron Response Functions and Detection Efficiencies for NE 213 Scintillation Detectors, ND-22, Braunschweig (1982).

[9] Tschechanski A., Shani G.: A Deuteron Energy Loss Evaluation in a Solid Ti-T Target Used for the T(d,n)He Reaction, Nucl. Sci.& Eng. 87(1984)189.

[10] Ingersoll D.T. et al:Neutron Response Matrix for Unfolding NE-213 Measurements to 21 MeV, ORNL/RSIC-40, Oak Ridge (1976)

[11] Tichy M., Kralik M., Pulpan J., Novotny T: Benchmark experiments with 14 MeV Neutrons Transmitted through Pb, Fe, Ni and Al, contribution to this Symposium.

[12] Beghian L.E., Wilensky S., Burrus W.R.: A Fast Neutron Spectrometer Capable of Nanosecond Gating, Nucl. Instr. & Meth. 35(1965)34.

# MULTIPLE SCATTERING RESONANCE SELF-SHIELDING FACTORS IN WIRES

MARIA DO CARMO LOPES and JORGE MOLINA ÁVILA

LNETI/ICEN/DEEN - Institute of Nuclear Sciences and Engineering

2685 SACAVÉM - PORTUGAL

ABSTRACT

The new theoretical formalism recently developed by us for the calculation of neutron self-shielding factors, taking into account the effect of the isotropic multiple scattering inside the probes is applied to the infinite cylindrical geometry. Numerical results concern the most important resonances of different materials commonly used in neutron reactor dosimetry both as activation detectors and as emitters of self-powered neutron detectors.

## INTRODUCTION

The calculation of neutron self-shielding factors in the epithermal region of the neutron spectrum, where most of the materials used in neutron metrology exhibit cross-sections with high resonances, has been a permanent problem in reactor dosimetry.

The pure absorber approximation, extensively applied in the thermal energy region, which by definition ignores the multiple scattering effect, is completely inadequate in cases where the scattering cross-section is not negligible. More accurate approximations have already been developed in order to include the scattering effect. Among them, Monte Carlo method [1], a semi-analytical treatment developed by Selander [2] and transport theory [3] have lead to resonance factors in quite good agreement with some of the scarcely reported experimental results.

Recently we developed a new approach to neutron self-shielding factors including the effect of multiple scattering inside the probes [4]. The main important difference of this new method is that it is based on a collision function exclusively determined by the cross-section and the geometry of the probe submitted to the neutron field. This enables one to obtain neutron self-shielding factors for different incident neutron fields quite directly. Another interesting characteristic of the method is that it leads to the definition of the self-shielding factor as a function of the incident neutron energy [5].

In the present work we apply the referred method to the calculation of resonance neutron self-shielding factors in wires. Actually almost all the earlier referred methods were applied to the infinite planar geometry which offers some mathematical simplifications.

The functions developed by us for the calculations of the neutron self-shielding factors in wires, in the second collision approximation [6] are also used in the present application.

## SUMMARY OF THE MODEL

Concerning a probe of arbitrary geometry and volume V, submitted to an external neutron field, we define the absorption probability $\varphi_a(E)$ for incident neutrons of energy E, as the sum of all possible collisional processes ending in absorption.

Analytically we have:

$$\varphi_a(E) = \int \Psi_o(r,E) K_a(r,E) dV \tag{1}$$

where $\Psi_o(r,E)$ is the probability density of first collision and $K_a(r,E)$ is the new collisional function, introduced by the model, which has the physical meaning of being the absorption probability for a neutron of energy E that suffers a collision at the point r. The function $K_a(r,E)$ satisfies an integral equation. To simplify the calculations we used a related function,

$$K(r,E)=(\Sigma(E)/\Sigma_a(E))K_a(r,E)$$

where $\Sigma$ and $\Sigma_a$ are the total and the absorption macroscopic cross-sections.

The integral equation for $K(r,E)$ is:

$$K(r,E) = 1 + \beta(E)\iint Q(r,E',r')(\Sigma_a(E')/\Sigma(E'))\,K(r,E')\,dr'dE' \quad (2)$$

where $\beta(E) = \Sigma_s(E)/\left(\Sigma_a(E)E(1-\alpha)\right)$.

The function $Q(r,E',r')$ is the probability density for a neutron to interact in $r'$ coming from $r$ with energy $E'$. The energy integral is over all possible neutron energies after the isotropic scattering ($\alpha = \left[(A-1)/(A+1)\right]^2$).

The function $K(r,E)$ does not depend on the external field (which is determined exclusively by $\Psi_o(r,E)$) and it reflects the symmetry of the probe.

The equation (2) can be solved, in foils, by sucessive iterations starting with the first-collision approximation, $K=1$. However we have introduced a procedure which accelerates the convergence [4,7], defining a different starting function, which we have called the local interaction approximation, $K_1$.

## APPLICATION TO WIRES

The important geometrical parameter is the radius $R$ of the infinite cylinder. The independent variable is the depth parameter $\rho$ which defines a cylindrical surface over which the collisional probability is constant in an isotropic neutron field.

The probability density of first-collision, $\Psi_o(\rho,E)$, and the integral over $\rho'$ of $Q(\rho,E,\rho')$, which is related to the neutron escape probability after the first collision, are enough to define the $K_1$ approximation in the coordinate in which the integral equation for $K(\rho,E)$ will be solved.

The weak dependence on the energy allows the use of the linear approximation $K(\rho,E+\Delta E) = K(\rho,E) + \Delta K(\rho,E)$, with $\Delta E$ much smaller than the resonance energy width, leading to:

$$K(\rho,E+\Delta E) = \left\{ \frac{(K_1(\rho,E)-1)}{\beta(E)}\, d\beta(E)/dE + \beta(E)\left\{ \frac{\Sigma_a(E)}{\Sigma(E)} K_1(\rho,E) \times \right.\right.$$

$$\left.\left. \times \int Q(\rho,\rho',E)\,d\rho' - \frac{\Sigma_a(\alpha E)}{\Sigma(\alpha E)} K_1(\rho,\alpha E)\,\alpha \int Q(\rho,\rho',\alpha E)\,d\rho' \right\}\right\} \Delta E \quad (3)$$

The solution in $E + \Delta E$ is defined through the knowledge of the function in $E$ and $\alpha E$.

The incident neutron energy dependent self-shielding is:

$$G(E) = \varphi_a(E)/2\,\Sigma_a R \tag{4}$$

and the integral resonance self-shielding is:

$$G_{res} = \frac{\int \Phi(E)\,\Sigma_a(E) G(E) dE}{\int \Phi(E)\,\Sigma_a(E)\,dE} \tag{5}$$

where $\Phi(E)$ is the neutron flux, per unit energy, at $E$.

## RESULTS AND DISCUSSION

The code CYLRES calculates the energy dependent self-shielding factor $G(E)$ over a defined energy region containing a resonance of the neutron interaction cross-section, solving the integral equation (2) in the local interaction approximation. It also calculates the corresponding integral self-shielding, $G_{res}$ (eq.(5)).

The functions that characterize the interaction processes in wires, namely the probability density of first-collision, $\Psi_o(\rho,E)$ (Fig.1a) and the direct escape probability density $W(\rho,E)$ from an inner surface $S(\rho)$ (Fig.1b) involve double angular integrals that are extremely time consuming. To avoid this difficulty, we have constructed tables of these two functions according to some spatial requirements suggested by an exhaustive study of the referred functions, for a wide range of energies. The code CYLRES uses those matrices performing the required energy and spatial interpolations in order to obtain the values of the functions for the given pairs $(\Sigma R, \rho/R)$ of the dimensionaless parameters. That tabulation is the basis of the algorithm created for the increased speed.

The functions $G(E)$, $G_o(E)$ (the single collision self-shielding factor), $K(\rho/R{=}1,E)$ and $\Sigma(E)$ for an infinite wire of cobalt ($R$ = 0.01 cm) are plotted in Fig. 2. The collisional function $K(\rho,E)$ for this scattering resonance ($\Gamma_n/\Gamma$ = 0.916) exhibits a maximum, as a function of the energy, which is situated at an energy somewhat greater than $E_r$(the resonance peak) meaning that the neutrons which enter the probe with that energy are those which suffer the largest number of collisions inside it.

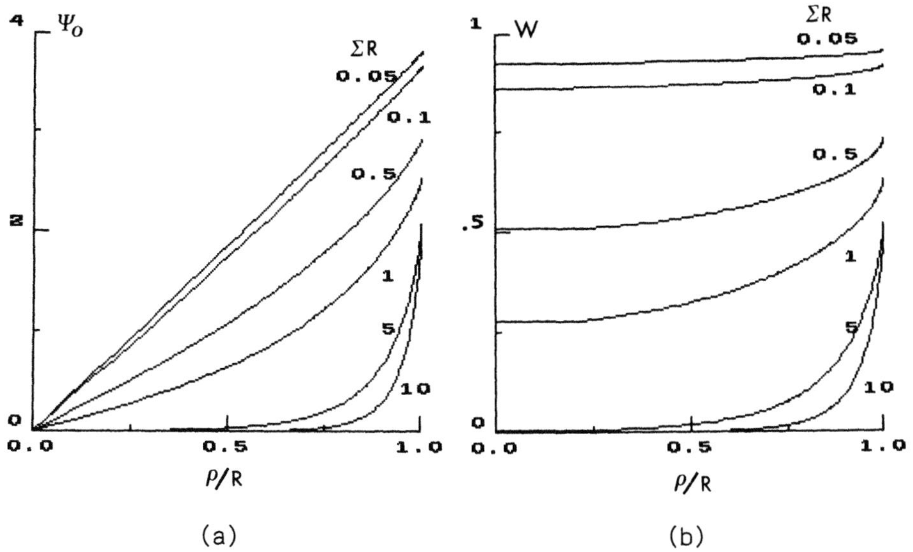


Fig. 1 - Probability density of first-collision (a) and direct escape probability density (b) for wires as a function of the depth parameter $\rho/R$ for different values of $\Sigma R$.

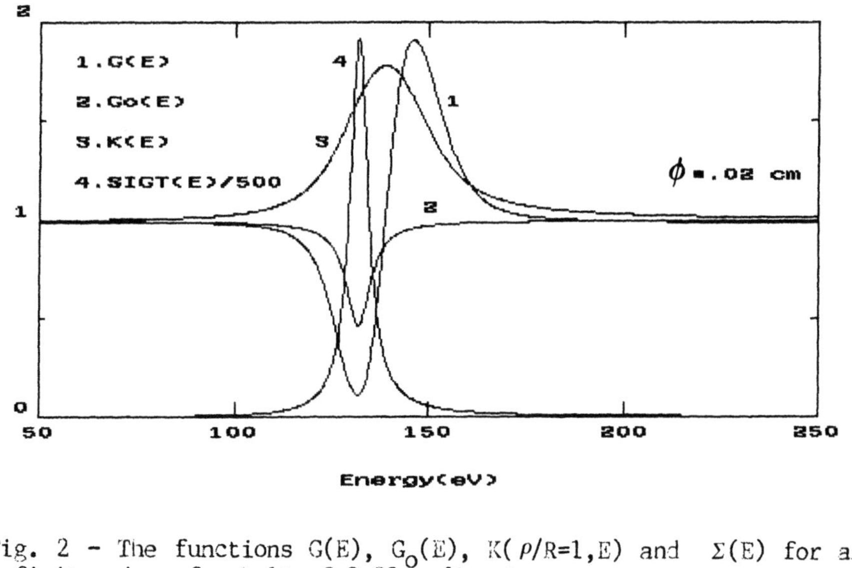


Fig. 2 - The functions $G(E)$, $G_o(E)$, $K(\rho/R=1,E)$ and $\Sigma(E)$ for an infinite wire of cobalt of 0.02cm diameter.

The consideration of multiple scattering leads therefore to values of the factor G(E) greater than unity, for those energies. This accounts for the possibility of those neutrons being absorbed after several collisions, when their energy has reached a value close to $E_r$, where the absorption cross-section is maximum. In a single-collision treatment (curve $G_o(E)$) neutrons which enter the probe with $E>E_r$ cannot be absorbed with an energy other than E with a probability determined by $\Sigma_a(E)$ and the self-shielding effect just reflects the behaviour of the curve $\Sigma_a(E)$, meaning that the activation actually induced in the probe is smaller than that corresponding to linear absorption.

In Table I we present the results of our calculations of the integral self-shielding factors, $G_{res}$ for infinite wires of different radia, for some materials commonly used in neutron reactor dosimetry either as activation detectors, such as Co, Au and Ag or as emitters in self-powered neutron detectors like, Co, Rh, Ag, Hf, Ho and Er.

Erbium has two resonances in the thermal energy region and we have used our treatment to calculate the self-shielding correction in this material despite the fact that multiple scattering is not important there. Nevertheless this kind of energy averaging can cause the integral self-shielding factor to decrease more than 20% (depending on the wire diameter) when compared with a monoenergetic calculation corresponding to $E_o$=0.0253 eV [7].

Very few results are found in the literature concerning the theoretical treatment or the measurement of resonance self-shielding factors in wires. Nevertheless, our results are in quite good agreement with those measured by Eastwood et al. [8] in cobalt wires (Fig. 3).

Concerning the case of gold we can see, in the same figure that our calculations are systematically below the experimental results obtained by McGarry [9] , which can be due to some temperature effect as we are not considering Doppler broadening. Erben [10] has calculated epithermal self-shielding factors for materials used as emitters in self-powered neutron detectors but his treatment apparently does not take scattering into account as the results concerning rhodium, for instance, are close to ours (Fig.3) while those for cobalt are largely displaced upwards.

| R [cm] | Ag | Au | Co | Er | Hf | Ho | Rh |
|---|---|---|---|---|---|---|---|
| 0.010 | 0.287 | 0.202 | 0.458 | 0.850 | 0.348 | 0.468 | 0.371 |
| 0.015 | 0.242 | 0.170 | 0.400 | 0.790 | 0.290 | 0.401 | 0.302 |
| 0.020 | 0.214 | 0.151 | 0.364 | 0.740 | 0.253 | 0.358 | 0.259 |
| 0.025 | 0.195 | 0.138 | 0.338 | 0.700 | 0.228 | 0.328 | 0.229 |
| 0.030 | 0.181 | 0.128 | 0.318 | 0.660 | 0.208 | 0.306 | 0.206 |
| 0.035 | 0.169 | 0.120 | 0.302 | 0.630 | 0.193 | 0.288 | 0.188 |
| 0.040 | 0.160 | 0.113 | 0.289 | 0.600 | 0.180 | 0.274 | 0.173 |
| 0.045 | 0.153 | 0.108 | 0.279 | 0.570 | 0.170 | 0.262 | 0.161 |
| 0.050 | 0.146 | 0.103 | 0.269 | 0.550 | 0.161 | 0.252 | 0.151 |
| 0.055 | 0.140 | 0.099 | 0.261 | 0.520 | 0.153 | 0.243 | 0.142 |
| 0.060 | 0.135 | 0.095 | 0.254 | 0.500 | 0.146 | 0.235 | 0.134 |
| 0.065 | 0.131 | 0.092 | 0.248 | 0.480 | 0.140 | 0.228 | 0.127 |
| 0.070 | 0.127 | 0.089 | 0.242 | 0.470 | 0.134 | 0.222 | 0.121 |
| 0.075 | 0.124 | 0.087 | 0.237 | 0.450 | 0.129 | 0.216 | 0.116 |
| 0.080 | 0.120 | 0.084 | 0.233 | 0.440 | 0.125 | 0.211 | 0.111 |
| 0.085 | 0.117 | 0.082 | 0.228 | 0.420 | 0.121 | 0.206 | 0.106 |
| 0.090 | 0.115 | 0.080 | 0.224 | 0.410 | 0.117 | 0.202 | 0.102 |
| 0.095 | 0.112 | 0.078 | 0.221 | 0.400 | 0.113 | 0.198 | 0.099 |
| 0.100 | 0.110 | 0.076 | 0.218 | 0.390 | 0.110 | 0.194 | 0.095 |

Table I - Integral resonance self-shielding factors in wires

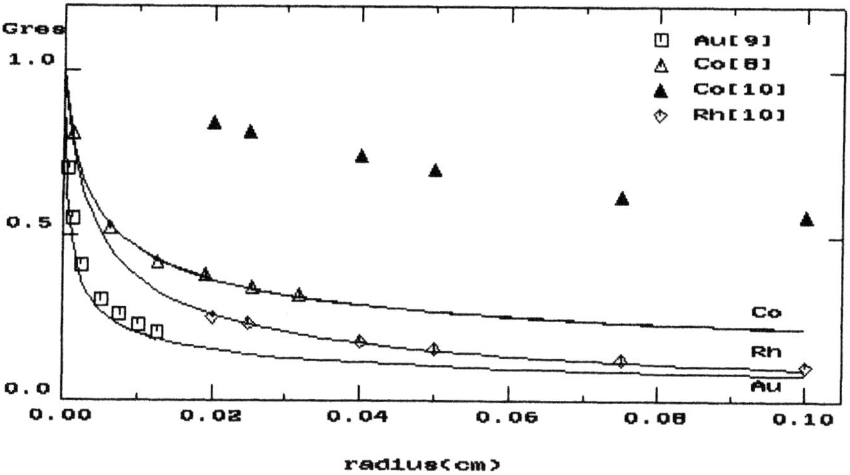


Fig. 3 - Resonance self-shielding factors calculated by the code CYLRES and other experimental and theoretical results.

## CONCLUSIONS

The multiple scattering treatment of resonance self-shielding factors was applied to the infinite cylindrical geometry with the code CYLRES. As expected, all the functions involved in the calculations, namely $K(\rho,E)$ and $G(E)$ exhibit the same behaviour for this geometry as they do for infinite foils [11].

In the highly scattering resonances we can see how drastically the multiple scattering changes the absorption probability inside the probe if we compare it with that calculated in the pure absorber approximation.

We believe that the present work can be a useful contribution to better calculations of epithermal sensitivities of self-powered neutron detectors.

## REFERENCES


1. S.M.Jefferies, T.D.MacMahon, J.G.Williams, A.Ahmad, T.B.Ryves; K.H.Bochoff (ed.) Nucl. Data for Sci. and Techn. (1983) 681
2. W.N.Selander, Rept. AECL - 1077 (1960)
3. S.C.Mo, K.O.Ott, Nucl. Sci. Eng. 95 (1987) 241
4. J.Molina, M.C.Lopes, Kerntechnik 52, 3 (1988) 197
5. M.C.Lopes, J.Molina, Nucl. Sci. Eng. 104, 1 (1990) 40
6. J.Molina, M.C.Lopes, Kerntechnik 51, 3 (1987) 194
7. J.Molina, M.C.Lopes, Proc. of the 4$^{th}$ Int. Conf. on Nucl. Data for Sci. and Techn. , 1988, Mito (Japan) p. 577
8. T.A.Eastwood, R.D.Werner, Nucl. Sci. Eng. 13 (1962) 385
9. E.D.McGarry, Trans. Am. Nucl. Soc. 7 (1964) 86
10. O.Erben, Jaderná Energie, 26, 2 (1980) 47
11. M.C.Lopes,J.Molina, Nucl. Instr. Methods in Physics Research A280 (1989) 304

# PC PROGRAMS FOR THE CONVERSION OF NEUTRON SPECTRA AND CROSS SECTIONS BETWEEN DIFFERENT GROUP AND POINT REPRESENTATIONS

T.O. Serén
Technical Research Centre of Finland, Reactor Laboratory
P. O. Box 200, SF-02151 ESPOO, Finland

## ABSTRACT

The vastly increased speed and mass storage capacity of cheap personal computers have made them a serious alternative to mainframe and mini systems in many applications related to reactor neutron metrology. A frequent task in this context is the conversion of spectrum and cross section data between various pointwise and groupwise representations. A set of PC programs has been developed for flexible and consistent conversion of data between various point and group structures. The programs are written in Turbo Pascal running under MS-DOS. They utilize cubic spline interpolation imposed on either pointwise or integral data. Special attention is paid to the preservation of group integrals when converting from coarse to fine group structures. Output to files for further processing or for export to popular graphics packages is provided. Simpler plots are generated directly for quick checking on screen or hardcopy. Examples of applications and results are presented.

The programs will run on any IBM or compatible personal computer. A graphics adapter (CGA, EGA, Hercules or VGA) is necessary for on-screen inspection, but is not otherwise required.

---

## INTRODUCTION

The need to convert data between different groupwise or pointwise representations arises frequently in connection with several tasks related to reactor neutron metrology. Although the numerically heavier tasks - such as transport calculations and neutron spectrum adjustments - are perhaps still best handled by more powerful computers, personal computers (PCs) can well be used for less demanding tasks such as data conversion. The low cost, flexibility, user-friendliness, and large availability of application programs give them certain advantages over larger computer systems. Since PCs can nowadays be easily connected to most computer nets, the transfer of data files from one system to another is also usually no problem.

The problem of data conversion between different representations has been discussed e.g. by Végh [1], who introduced the application of cubic splines to the cumulative integral function of coarse-group spectra. The introduction of powerful and fast compilers for PCs, such as Turbo Pascal [2], and program packages for numerical analysis (e.g. [3]), has greatly facilitated the application of similar methods on PCs.

Three programs written in Turbo Pascal running under the MS-DOS operating system have been developed: BUGTOFIN, for the conversion of neutron spectra in the upper 35 BUGLE-80 groups to an arbitrary fine-group structure; SIGMACON, for the conversion of pointwise data (e.g. reaction cross sections) to an arbitrary coarse-group structure; and POTOFINE, for the conversion of pointwise data to an arbitrary fine-group structure.

## NEUTRON SPECTRUM CONVERSION FROM COARSE TO FINE GROUPS:THE PROGRAM BUGTOFIN

This program converts a neutron spectrum in the upper 35 groups of the BUGLE-80 group structure (3.3546 keV to 17.333 MeV) to an arbitrary fine-group structure using cubic spline interpolation imposed on the cumulative integral spectrum, thus preserving the original group fluence rate values and ensuring a certain degree of smoothness.

The coarse-group structure is fixed in the program (typed constant). The use of a different coarse-group structure thus requires minor modifications of the source code. Four auxiliary energy groups are added above 17.333 MeV (upper boundaries 18, 19, 20, and 21 MeV). The input spectrum is read from a text file in the same form as it is output e.g. from ANISN, i.e. in descending energy order and in the form of integrated group fluence rate values. The input format is free: spaces and line feeds serve as delimiters (this holds for the input of all the programs discussed here). The fine-group boundaries are read in ascending order from a file. The energy range spanned by the fine-energy groups must be within the range 3.36 keV...21 MeV. This should be sufficient for most calculations in fast-neutron dosimetry (the recommended lower energy limit for calculating the effective dpa cross section is 10 keV). There are no restrictions on the number of fine-energy groups apart from the limitations set by available computer memory.

After the input data has been read, the upper tail (above 17.333 MeV) group values are determined by scaling a Watt-type fission spectrum so that the group fluence rate in the uppermost BUGLE-80 group is preserved. A table of integral fluence rates above the lower coarse-group boundaries is then created. A cubic spline is constructed from the logarithm of the tabulated integral coarse-group fluence rates using a slightly modified version of the subroutine (procedure) CubicSplineFree from the Turbo Pascal Numerical Methods Toolbox [3]. The spline is "free" in the sense that the second derivative is assumed to be zero at the endpoints. The procedure returns an array of interpolated values at the specified lower fine-group boundaries. The integral fine-group fluence rates are calculated from the logarithms, and the fine-group neutron spectrum is finally formed by taking the differences between successive integral values and dividing by the group widths.

The fine-group integral and differential spectra are stored in user-specified files in ascending energy order. A file SHOW.GIA for quick examination using Graph-in-the-Box Analytic (GBA) [4] is created. Another file PLOT.DAT with data for a combined step and line chart is created for ASCII import to some graphics program capable of high-quality plotting (e.g. SigmaPlot [5]).

Fig. 1 shows the original and converted (upper 280 SAND-II groups) neutron spectrum based on ANISN calculations for the surveillance chain (SC) position in the Loviisa VVER-440 reactors (reduced, low-leakage core) [6].

## POINTWISE CROSS SECTIONS TO COARSE GROUPS: THE PROGRAM SIGMACON

This program was primarily designed for the conversion of pointwise cross section data to an arbitrary coarse-group structure (such as BUGLE-80, ELXIR, FERRET etc.) with optional spectrum weighting. A cubic spline drawn through the data points is binned into the given group structure such that the integral of the spline between the group boundaries is preserved. In the case of spectrum weighting two spline curves are constructed: one passing through the response points (cross section x point fluence rate) and one through the neutron spectrum data points.

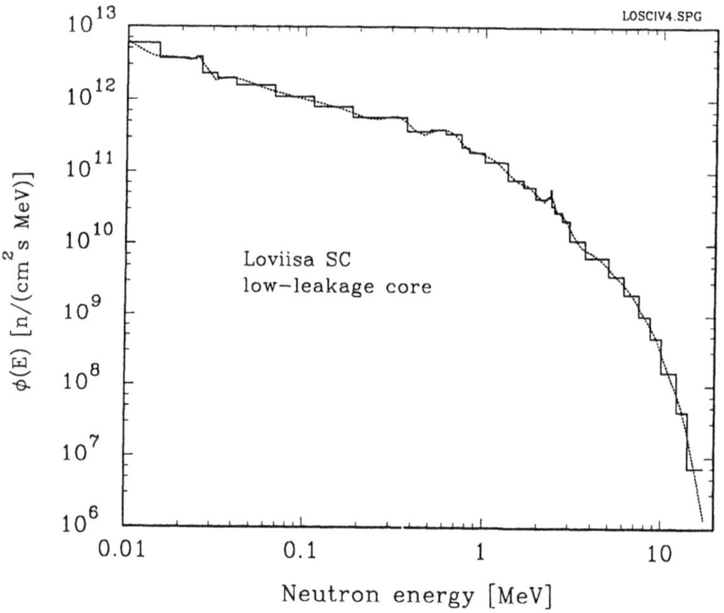


**Figure 1.** Neutron spectrum at the surveillance chain position in the Loviisa VVER-440 reactors (low-leakage core) calculated using the ANISN code in the BUGLE-80 group structure [6] (step chart) and converted to the upper 280 SAND-II energy groups (dashed line) using the program BUGTOFIN.

If spectrum weighting is applied, the spectrum must be given as point values at exactly the same energies as the cross section data. The group boundaries and point data are read from text files and must be in strictly increasing energy order.

Since integration of the spline curve(s) is involved, somewhat larger modifications had to be made to the procedure CubicSplineFree than in the case of BUGTOFIN. Because of the arbitrary point and group structures used it is also necessary to determine the first group with non-zero cross section. For that group the integration is carried out only from the zero-crossing of the extrapolated curve inwards. The number of knots (energy points) within each group (zero, one or more) must also be taken into account and each case treated separately.

Once the group integrals and averages have been determined the procedure is straightforward. In the case of no spectrum weighting, the cross section for each group is simply obtained as the group integral divided by the group width. If spectrum weighting is applied, the group cross section is obtained as the response average divided by the fluence rate average.

Four output files are created. The group boundaries and the calculated group cross sections are stored in a user-specified file. A GBA file SHOW2.GIA is created for quick comparison of the point and group cross sections. Since it is not known *a priori* whether the point or group data will produce longer data columns, two output files for high-quality plotting are produced: XPLOT1.DAT (point values) and XPLOT2.DAT (group values).

Fig. 2 shows the $^{93}$Nb(n,n')$^{93m}$Nb cross section in pointwise and groupwise (BUGLE-80) form with no spectrum weighting applied. The point cross sections are those recommended by Gayther et al. [7].

In the case of coarse group structures the effect of spectrum weighting on the group cross sections may be considerable. In Fig. 3 the $^{93}$Nb(n,n')$^{93m}$Nb cross section in the BUGLE-80 group structure is shown without weighting and weighted with the neutron spectrum in Fig. 1. The differences are appreciable only in the region around 10 MeV and the effect on the spectrum-averaged cross section is only about 0.46 %. However, no generalizations can be made from this single example, and as a rule spectrum weighting should always be applied if possible.

## POINTWISE CROSS SECTIONS TO FINE GROUPS: THE PROGRAM POTOFINE

When finer group structures are applied SIGMACON sometimes produces unjustified oscillations in the group cross sections. In such cases a much simpler method may be used for the conversion of pointwise data to the desired group structure. The group values can be approximated by the values at the group midpoints of a cubic spline curve drawn through the data points. Spectrum weighting is also not necessary.

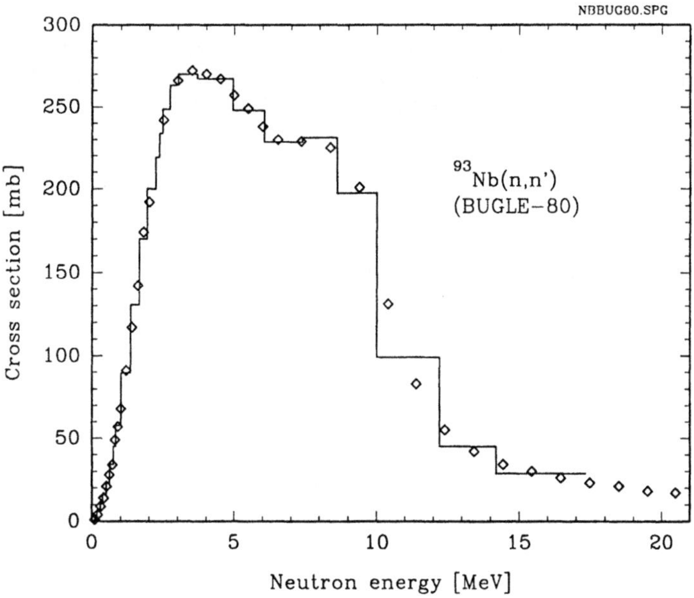


**Figure 2.** Pointwise [7] and converted groupwise (SIGMACON, BUGLE-80) $^{93}$Nb(n,n') cross section (no weighting).

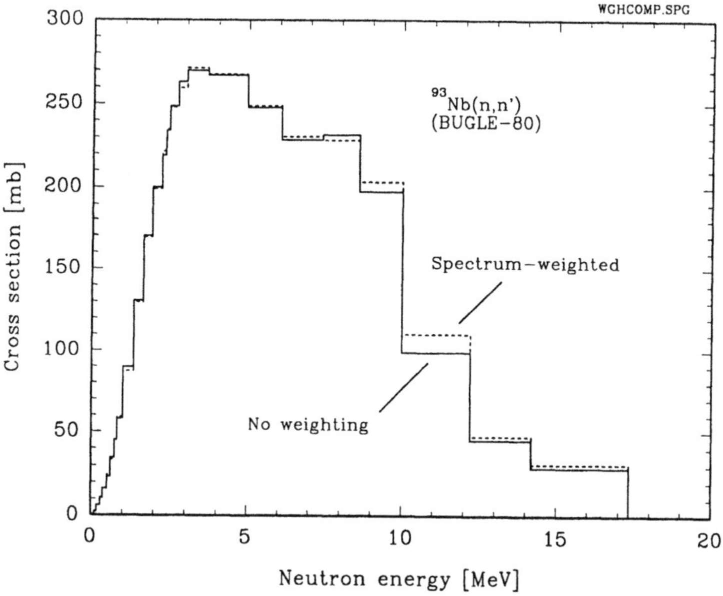


**Figure 3.** $^{93}$Nb(n,n') cross section [7] converted to the BUGLE-80 group structure without spectrum weighting and weighted with the neutron spectrum in Fig. 1.

This simpler method has been implemented in the program POTOFINE. It utilizes the procedure CubicSplineFree with the same modifications as in BUGTOFIN, and the output is identical in form to that of SIGMACON.

Fig. 4 shows the $^{93}$Nb(n,n')$^{93m}$Nb cross section in point form and converted to a 180-group structure with even energy spacing between 0.1 MeV and 18.1 MeV.

## OTHER APPLICATIONS

Even though SIGMACON and POTOFINE were primarily developed for the conversion of cross section data, they can equally well be used for the conversion of any kind of pointwise data to various group structures. An example of this is given in Fig. 5, where a $^{238}$U fission spectrum [8] converted to the BUGLE-80 group structure using SIGMACON is shown together with the original pointwise data.

Fine-group data can easily be converted to pointwise data by simply assigning the group values to the group midpoints. After such a conversion SIGMACON can be used without modifications to perform essentially the same tasks as the program FITOCO [9].

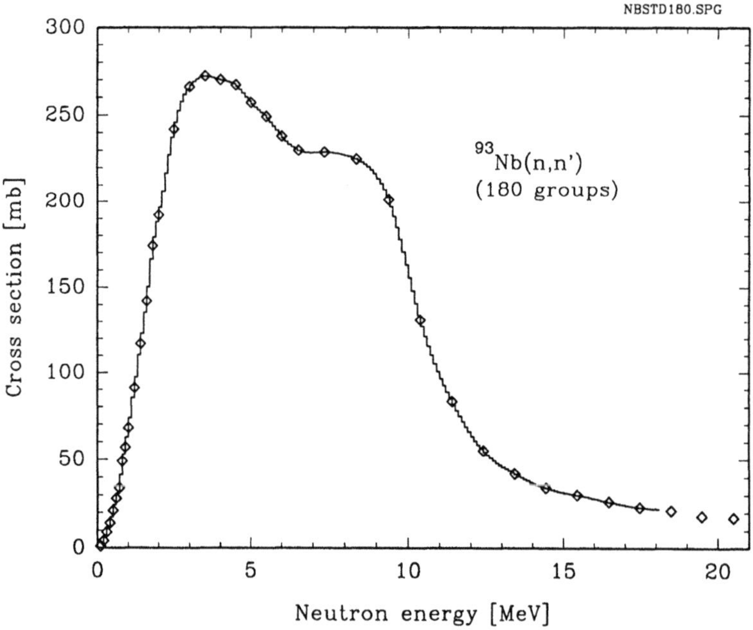


**Figure 4.** Pointwise [7] and converted groupwise (POTOFINE, 180 groups) $^{93}$Nb(n,n') cross section.

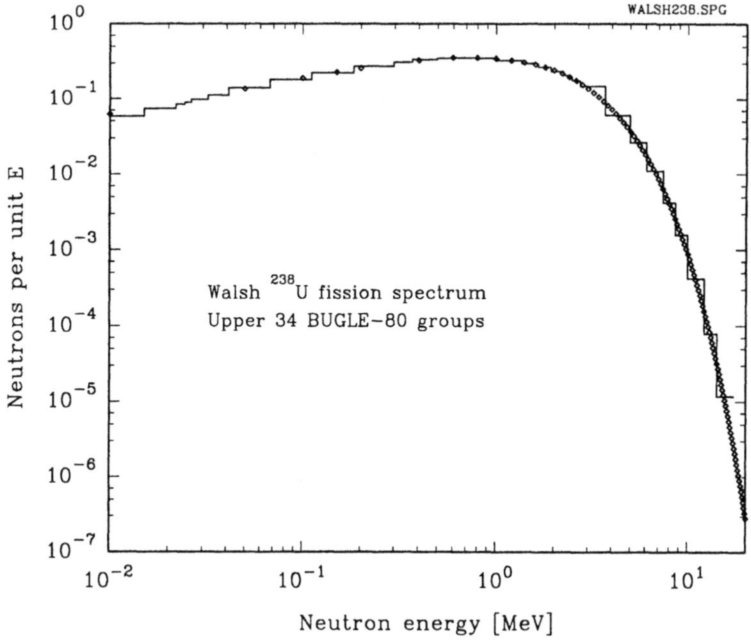


**Figure 5.** The $^{238}$U fission neutron spectrum according to Walsh [8] converted to the BUGLE-80 group structure using SIGMACON.

## SUMMARY AND CONCLUSIONS

A set of easy-to-use PC programs has been written for the conversion of spectral data from coarse to fine group structures and pointwise data to arbitrary group structures using cubic spline interpolation. The programs are written in Turbo Pascal and will run on any IBM or compatible personal computer. A graphics adapter is needed for quick on-screen inspection of the result, but is not otherwise necessary.

With slight modifications of the source code the programs can be used for a large variety of tasks. The programs are available from the author on request, both as source code and in compiled form. Any suggestions for improvement are welcome.

## REFERENCES


1. J. Végh, "On the conversion of coarse group spectra to fine group spectra (using a continuity principle)", Proceedings of the Fifth ASTM-EURATOM Symposium on Reactor Dosimetry, Geesthacht, FRG, 1984; D. Reidel Publ. Co. (1985).
2. Turbo Pascal Reference Guide, Version 5.0, Borland International (1988).

3. Turbo Pascal Numerical Methods Toolbox, Version 4.0, Borland International (1988).

4. Graph-in-the-Box Analytic, Release 1, New England Software, Inc. (1988).

5. SigmaPlot Version 4.0, Jandel Scientific (December 1989).

6. F. Wasastjerna, Technical Research Centre of Finland, Nuclear Engineering Laboratory, Private Communication (November 1989).

7. D. B. Gayther et al., The $^{93}Nb(n,n')^{93m}Nb$ Cross Section, Report AERE R 12612 (May 1987).

8. R. L. Walsh, "Spin-dependent calculation of fission neutron spectra and fission spectrum integrals for six fissioning systems", Nucl. Sci. & Eng. **102** (1989) p. 119-133.

9. H. Ch. Rieffe, FITOCO - A program for the conversion of fine group flux density and cross section data to coarse group values, Report ECN-99 (April 1981).

# A REVIEW OF THE IMPORTANCE OF NEUTRON SELF SHIELDING IN ACTIVATION DETECTORS AND THE INFLUENCE OF COVERS AROUND THE DETECTORS

Ç. ERTEK[x]
Çekmece Nuclear Research Center P.O. Box1
Hava Alanı, Istanbul, Turkey

ABSTRACT

In this work, I would like to present my personal views and findings on the reasons for a need for a separate and detailed study on the effect of neutron self shielding in activation detectors, and on the influence of covers around the detectors(eg Cd, Gd, and/or stainless steel wall of the reactor PV surveillance irradiation capsules). The cross correlations of cover material cross sections vs neutron spectrum developed due to the presence of the cover should be investigated in detail. The work reviews in detail the specific activation rates (activity per nucleus per second) for a variety of Indium target foil thicknesses ranging from $1.10^{-4}$ cm to $23.10^{-4}$ cm around the In(n,$\gamma$) 1.456 eV resonance energy. The calculational model of S. Pearlstein and E. V. Weinstock is used but it is applied to a beam of mono-energetic and mono-directional neutrons perpendicular to the Indium target. Special attention is paid to the energy distribution of neutrons after going into the thin targets.

X Assoc. Prof. Dr. Rer. Nat. Çetin Ertek
Present address: International Atomic Energy Agency
Wagramerstrasse 5 P.O. Box 200, A-1400, Vienna, Austria.

In this work, I would like to present my findings on the reasons for a need for a separate and detailed study on the effect of neutron self-shielding in activation detectors, and on the influence of covers around the detectors (eg Cd, Gd, and/or the stainless steel wall of the reactor PV surveillance irradiation capsules). These type of studies are recommended in various Advisory Group Meetings (1,2) and Consultant Meeting (3) and the 3th ASTM-EURATOM Meeting (4).

In this present meeting , we will hear more results on a series of International calculational exercises like REAL-8o (5), REAL-84 (6) and REAL-38 (7,8,9 ) which were initiated by the distinguished participants at the 3th ASTM-EURATOM Meeting Ispra, Italy in 1979.

These exercises started among the Leading Laboratories and have served as a technology transfer instrument and a learning tool to the newcomers to this field. However neither procedures nor the neutron self shielding corections and influence of covers like Cadmium and Gadalinium were transparent to the participant(1o).

It would be highly desirable to select a Pressure Vessel Surveillance capsule foil package geometry and concentrate on the self shielding and influence of covers, their order in positioning for some capture and fission foils including the burn-up, temperature and other effects. That will eventually lead us to a set of procedure, evaluated values and an additional file-another step towards standardization.

The Ref15 covers over 13o references on the subject matter. Although extensive effort has been spent on the thermal neutron self shielding and perturbation effects, it is still of great interest to investigate the situation in other energy regions. The validation of applied correction factors and their International comparison will certainly lead to fruitful results, as in the past.

These effects have been investigated but in general, the quantitative significance of these effects has been estimated by semi-intuitive calculations grafted on to the infinite probe analysis. Furthermore, the grafting procedures themselves appear to be semi-intuitive in the majority of instances, so that the development of a consistent, successive approximation analysis is needed.

The calculation of the self shielding factor and the absorption probability of probes in a neutron field requires the solution of the space integral transport equation for the flux, if the scattering is to be properly taken into account. But the convergence of Neumann series depends on the ratio of the scattering and total macroscopic cross sections of the probe. A new approach (11) with multiple scattering is proposed but it does not allow to know the flux as it stands now.

In the present work it is briefly indicated that some reactor physics parameter calculations are directly related to this type of perturbation calculations. Further interaction between ideally thin or very thin foils with predicted reaction rates of these foils is very challenging if the activation cross section measurement is rather old. That automatically brings us to the problem of activation of impurities and to the gamma ray component of the reactor environment(1o).(photo fission etc).

It has been recognized that in the fast neutron flux density measurements by activation and fission foils, the existence of cadmium covers introduces some systematic errors due to excitation responses and photo-fission effects. It could be of great benefit to understand the neutron physics(i.e. scattering, absorption and associated transport problems) of boron covers. Boron ratio of e.g. of U-238, could be a very useful reactor physics parameter.(12).

Application of new neutron capture measurement techniques to the activation foil methods will create new dimensions to the physical understanding of the problem in question.

The revalidation of burn-up codes after 33ooo MWd/T is necessary from the reactor safety point of view and this also involves important self-shielding problems. Control rod worths, conversion ratios, neutron flux density spectrum inside the fuel elements and exact determination of Gd quantity as burnable poisson are all neutron self-shielding related phenomena. The sensitivity to the lack of self-shielding treatment in some isotopes is clearly indicated in Ref.(13).

Let us first, discuss some of the conclusions of the mono-energetic mono-directional activation experiment. The detailed explanation of this work can be found in Ref.(14). In this work we have investigated in great detail the specific activity of Indium-115 in its first 1.456 eV resonance energy for a series of covered and uncovered indium foils as a function of neutron energy. The thickness of the indium detector is chosen to be 23 microns because this thickness is in the range which makes the statistical error for the transmission measurements minimum(actual minimum is at 14.7 micron). Primary activities and secondary activities, after scattering inside the detector and covers, have been calculated as a function of energy. For the first five arrangements the detector thickness was kept at 23 micron and the cover thicknesses for the first arrangement were 23 micron, for the second 1o micron, for the third 5 micron, for the fourth 3 micron and the fifth 1 micron. Even for 1 micron cover thickness, considerable influence on the detector activity could be seen around 1.456 eV as could be observed in the fifth arrangement. In the target preparation tenth of a micron thickness difference has a strong influence in the experimental determination of peak cross-section of peak cross-section point. The results (16) clearly indicate that the evaporated foils tested were far more uniform (thickness variations $\leq$1%) than those foils of a similar thickness produced by rolling. The great majority of foils are rolled for transmission measurements. The dimensions of the non-uniformities in the rolled foils are consistent with the suggestion that they arise from surface irregularities in the rollers or sandwich material between which the foils are folded.

If we decrease the detector thickness from 23 micron to 21 micron in the sixth arrangement and further down to 13 micron in the seventh arragement, the specific saturation activity increases by 34 % for this seventh arrangement.

The neutrons coming from the first cover cause secondary activation in the detector foilwith a minimum at 1.456 eV and two maxima. The maxima at the lower energy is higher than that at the higher energy. The reason is that the energy loss between (o-9o) degrees is smaller than the energy loss between (9o-18o) degrees. So, the neutrons coming with resonance or higher energy suffering a small energy loss are captured by the detector foil. The neutrons coming from the first cover and detector without suffering any interaction can be scattered back but their probability of being at the resonance energy is rather small. If the incoming neutron energy is higher than the resonance energy (e.g. 1.55 eV) then they can scatter back from the second cover and make the second maximum greater than the first.

Inside the detector, secondary activation coming from the first cover is higher than the secondary activation from the back cover because the first cover and the detector filters the neutrons.

In Fig. III of Ref(14) we have obtained the activity per nucleus per second as a function of indium thickness. Interestingly, this curve shows a saturation type behavior for ultra-thin detector thicknesses. This curve also shows that for 2o or 3o micron thicknesses specific activity decreases drastically as the thickness increases. The stable region starts below o.2 micron thickness. 2o or 3o micron thickness region is very sensitive to the small thickness variations. On the other hand this is the thickness with which most transmission measurements are performed.

In Fig. IV of Ref(14), the statistical error as a function of sample thickness in transmission measurements is shown. For 1ooo counts per second, in order to make the statistical error minimum(1.439%), the transmission (T) should be equal to o.11. The thickness for this should be 14.7 micron, for the case of indium. This thickness makes the $N\sigma d$ value equal to 2.2 which does not at all obey the $N\sigma d \ll 1$ theoretical criterion. So,there is a statistical reason to take that thickness but it causes a systematic error. How much systematic error is involved? If we take the work performed by C. M. Fleck (17) we see that he used 1.4 micron thick Sn-In alloy. This gives us a value of o.2 for $N\sigma d$. According to Fig. III of Ref.(14), the corresponding error on the activity per nucleus per second is about 1o% (for 1.4 micron thickness). If we agree that the saturation point is achieved below the detector thickness of $10^{-5}$ cm, this estimate can be easily performed. 1o% specific saturation activity error reflects only 1% in the cross-section estimate. In this case the39300 barns value given by Fleck should be 39758 barns. In Fleck's work mono-energetic neutrons are not used, Instead two different temperature (293° and 4.2°K) are applied to the detector. Since he had no possibility of using monó-energetic neutrons, he used ultra-thin foils. If our conclusions are satisfactory, the other transmission cross-section values from the literature should give lower cross-section values which is the case if one takes T. A. Moore's(18) result 385oo barns.

The experimental results as shown in Fig.V (14) indicates clearly the deep penetration of neutrons toward the sample more than foreseen. This result is in agreement with the results obtained by L. Kuipers(19). These results are also in agreement with Ref. (2o,21).

Interestingly,there are some solid state and surface (22) effects that need to be further investigated.

Secondly, as stated by S. Pearlstein and E.V. Weinstock(23) "the results of the calculations show that scattering effects on foil activation are significant. The effects are largest for thin resonance detectors and 1/v detectors of almost any practical thickness in a beam flux, and for thick resonance detectors in an isotropic flux. The major contribution is from scattering in the cadmium covers". The development has been restricted, in the isotropic case, to foils irradiated in a void(23) or in water (24, 25) whereas often the foil or its cadmium cover is embedded in a scattering and absorbing medium. In the latter case, the impinging flux may no longer be isotropic, thereby violating one of the basic assumptions of calculations such as given in Ref.(23). The magnitude of the scattering effect found by Pearlstein does not agree with Tsuchihashi and Iisumi (26). Walker(24) concludes that contrary to earlier predictions(25), that the flux perturbation theories based upon isotropic models cannot be used to predict the perturbations induced in all unisotropic fluxes. Extensive stre ming calculations(e.g.27,28) should be considered for most of the cases. The parametrization of R- Matrix multi-level formula(29,3o) including resolution and Doppler broadening for heavy nuclei is needed. Resonance self shielding calculations using the probability table method(PTM) is promising toward the resolution of the problems stated(31).

Thirdly, the most recent experiments and calculations done by R.N. Hill, K. O. Ott and J. D. Rhodes(32) suggest direction-dependent (flux-component-weighted) group constants use. This investigation indicates that the differences between forward-and backward-directed group constants are large enough to have a significant effect on the calculation. This is in agreement with Ertek's experimental result(14).

Therefore, it is highly desirable to merge these findings and methods in a very specific problem (e.g. one isotope one resonance in a primary reference field) using a well known data set. The absolute values, amplitutes and shapes must be taken into account(including the partial widths and totals simultaneously). The summery of the 114 references can be found in Ref.(15). A comprehensive work on reaction rate uncertainties is presented in Ref.(33)with 2o3 related references.

## REFERENCES


1. N.„ KOCHEROV, edited by IAEA Advisory Group Meeting on Nuclear Data for Radiation Damage Assesment and Related Safety Aspects", Vienna, 12-16 Oct 1981.
2. N. KOCHEROV, edited by, "IAEA Advisory Group Meeting on the status and Requirements of Nuclear Data for Radiation Damage and Related Safety Aspects", Vianna, Sept. 19-22, 1989.
3. E. M. ZSOLNAY and H. J.NOLTHENIUS "Proc. of the IAEA Consultants Meeting on the Assesment of the Results of the REAL-84 Exercise" Edited by V. PIKSAIKIN, March 1987, IAEA INDC(NDS)-19o/G F R.
4. Third ASTM-EURATOM Symposium on Reactor Dosimetry, Ispra, Italy, Oct. 1-5, 1979, EUR 6813 EN-FR.
5. W.L. ZIJP, E. M. ZSOLNAY, H.J. NOLTHENIUS, E.J. SZONDI, G.C.H.M. VERHAAG, D.E. CULLEN, and C. ERTEK "Final Report on the REAL-8o Exercise", ECN-128, Netherlands Energy Research Foundation(1983).
6. W.L. ZIJP, E.M. ZSOLNAY, H.J. NOLTHENIUS, E.J. SZONDI "Final Report On The REAL-84 Exercise", ECN-212 BME-TR-RES-18/88 IAEA-NDS-212 Nov. 1988.
7. H.J. NOLTHENIUS "Results Of The REAL-88 Exercise" ECN-89-o95 Petten, June 1989.
8. W.L. ZIJP, E.M. ZSOLNAY, D.E. CULLEN, "Information Sheet for the REAL-84 Exercise", Report INDC(NDS)-167, IAEA March1, 1985.
9. H.J. NOLTHENIUS, "Information for the REAL-88 Exercise" ECN-88-o23(ECN, Petten, Jan. 1988).
10. N. KOCHEROV, "IAEA Reactor Neutron Dosimetry, 5oth Anniversary of Nuclear Fission Conference", V.G. KHLOPIN Radium Institute, Leningrad, USSR, 16-2o Oct. 1989.
11. J. MOLINA AVILA and M.C. LOPES, "A New Approach to the Neutron Self-Shielding Factor With Multiple Scattering" Kerntechnik 52 (1988) No. 3.
12. C. ERTEK, A. YALCIN and Y. INEL, Nucl. Sci. Eng., 36, 2o9-221, 1969.
13. N.E. PATINO, M.J. ABBATE and M.M. SBAFFONI, "Resonance Shielding in HCLWR", Int. Conf. PHYSOR-9o Marseille, France April 23-27, 199o.
14. C. ERTEK, "On the Penetration of Mono-Energetic Neutrons in Detector Foils", IAEA RL-44, Feb. 1977, Vienna.
15. C. ERTEK Neutron Activation Foil Cover Effects IAEA/RL/57 Feb.1979.
16. C.J. SOFIELD, et. al., "Uniformity of Aluminium Foils", Nuc. Inst. and Meth. 138(1976) 411-413.
17. C.M. FLECK, "Resonance Parameter Determination of In by Activation", Nucleonik, Band 12, Heft 6.
18. T.A. MOORE, Journal of Applied Physics Rev. 1o9,417,1958.

19. L. KUIPERS, "Experimental Model Studies for a Fusion Reactor Blanket", PhD Thesis, Kernforschung Julich, FRG.
20. K. PONNERT, G. MAGNUSSON, I. BERGQVIST, Physica Scipta 10(1974), 35-41.
21. M.C. MOXON, D.B. GAYTHER, M.G. SOWERBY, "Some Problem Areas in Capture Cross-Section Measurements", AERE-R 8082, July 1975, p.12.
22. K.L. WOLF, D.E. LAWSON, J.C. WASS and N.J.C. PACKHAM, Proc. of the EPRI/NSF Conf. on Cold Fusion Phenomena, Washington, D.C., Oct. 1989.
23. S. PEARLSTEIN., E.V. WEINSTOCK, "Scattering and self-Shielding Correction in Cadmium-Filtered Gold, Indium and 1/v Foil Activation Measurements", Nuc. Sci. and Eng. 29(1967) 28-42.
24. J.V. WALKER, "The Effects of Flux Anisotropy on Thermal Neutron Flux Perturbations", Nuc. Sci. and Eng. 22(1965) 94-101.
25. G.R. DALTON, R.K. OSBORN, "The Effect of Flux Anisotropy on Neutron Detector Foils", Nuc. Sci. and Eng. 20(1964) 481-492.
26. K. TSUCHIHASHI, M. Iizumi, "Calculation of Effective Cut-off Energies Taking into Account Scattering and Higher Resonance Filters", J. Nucl. Sci. Tech. 2(1965) 506.
27. G.P. BEER, "Differential Neutron Albedos for Cylindrical Water Surfaces", Nucl. Eng. and Design 33(1975) 422-434.
28. G.P. BEER, "Neutron Streaming Through Conical Ducts in a Water Shield", Nuc. Eng. and Design 33(1975) 435-442.
29. A.M. LANE, R.G. THOMAS, Rev. Mod. Physic. 30(1958) 257.
30. F.H. FRÖHNER, "A Multi-Level Shape Analysis Program for Resonance Parameter Determination by Least-Squares Fitting of Several Sets of Neutron Transmission Data Simultaneously", KFK 2129, Jan. 1976.
31. E.F. PLECHATY, D.E. CULLEN, "Resonance Self Shielding Calculations Using the Probability Table Method", UCID-17230, Aug. 1976.
32. R.N. HILL and K.O. OTT, J.D. RHODES, "Directional Effects in Transitional Resonance Spectra", Nucl. Sci. and Eng. 103, 25-36 (1989).
33. C. ERTEK, "Reaction Rate Measurements, Neutron Spectrum Unfolding, Fluence, Radiation Damage, Embrittlement and Safety for Fission Reactors and Their Uncertainties", IAEA/RL/72, Oct. 1980.

# MEASUREMENT OF NEUTRON SPECTRA AT KUR AND SPR BY MULTI-FOIL ACTIVATION METHODS

Katsuhei Kobayashi·
Research Reactor Institute, Kyoto University
Kumatori-cho, Sennan-gun, Osaka 590-04 Japan

Itsuro Kimura
Department of Nuclear Engineering, Kyoto University
Yoshidahonmachi, Sakyo-ku, Kyoto 606 Japan

Li ZhaoHuan and Wang YongQing
Institute of Atomic Energy, P. O. Box 275(54)
Beijing, People's Republic of China

The neutron spectra at the Kyoto University Reactor, KUR of Research Reactor Institute, Kyoto University, KURRI and the Swimming Pool Reactor, SPR of Institute of Atomic Energy, IAE have been obtained by unfolding multi-foil activation data using both NEUPAC and NEUSPEC codes. The Kyoto University group (KK and IK) made 11 kinds of activation measurements and analyzed the KUR spectrum by the NEUPAC code. The IAE group (LZH and WYQ) selected 26 kinds of activation reactions and unfolded the data by the NEUSPEC code. Both groups exchanged the multi-foil activation data and each group obtained other group's reactor spectrum with its own unfolding code. Two unfolded spectra for both reactors were compared with each other, and the results showed general agreement for both cases.

## INTRODUCTION

An accurate description of the neutron spectrum is of great importance for understanding the reactor characteristics and evaluating the quantity of radiation effect with neutrons in a research reactor. Recently, much interest has been taken in the neutron spectrum for the quantitative comparison with the radiation effects caused by 14 MeV neutrons[1-3].

The multi-foil activation method is one of candidate spectrometries to experimentally characterize the neutron spectrum in and around a reactor. Two of the present authors (KK and IK) who belong to the Kyoto University (KU group) obtained the neutron spectrum at the Hyd facility of the Kyoto University Reactor (KUR) by the multi-foil activation method using the unfolding code NEUPAC[4]. On the other hand, other two (LZH and WYQ) of the present authors who belong to the Institute of Atomic Energy (IAE group) obtained the neutron spectra at the core and the reflector regions of the Swimming Pool Reactor (SPR) of IAE by the multi-foil activation method using the NEUSPEC code[5] which had been prepared by revising the SAND-II code[6]. Both KU and IAE groups exchanged their activation data and unfolded the neutron spectrum of the other group. Intercomparison of the unfolded neutron spectra has been made.

## EXPERIMENTS

### 1) Facilities

The KUR at the Research Reactor Institute, Kyoto University KURRI, is a light-water-moderated research reactor, whose nominal power is 5 MW. The fuel element, 8.1 x 7.7 cm squares and 72 cm long, is of MTR-type, made of 93 % enriched uranium and aluminum.

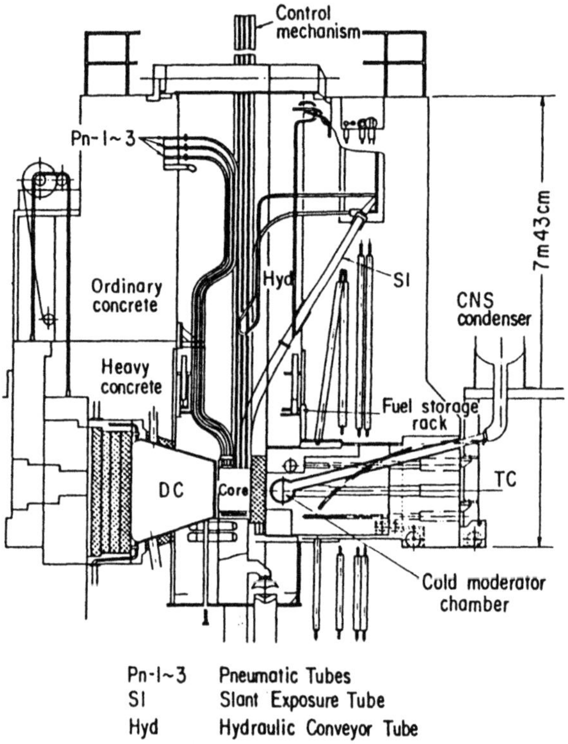


Fig. 1 Experimental facility of KUR (vertical cross section).

Figure 1 shows the vertical cross section of the experimental and irradiation facilities at KUR. A typical core configuration is illustrated in Fig. 2. Neutron fluxes and/or neutron spectra at main experimental facilities have been measured by the KU group using the activation techniques[7]. At the core center of the KUR, where one can obtain the highest neutron flux, there is a hydraulic conveyor, Hyd facility for high fluence irradiation. An aluminum capsule can be conveyed in the Hyd facility even during reactor operation.

The SPR of the Institute of Atomic Energy, IAE is also a light-water-moderated research reactor of 3.5 MW designed for multi-purpose uses. A fuel pin, 10 mm in diameter and 50 cm in active length, is made of 10 % enriched uranium-dispersed oxide in magnesium base and covered with 1.5 mm thick aluminum cladding. Each fuel assembly consists of 16 fuel pins. The reactor is normally loaded by 42 fuel assemblies about 25 % burn-up on average. A typical core loading is shown in Fig. 3. The neutron spectra were measured at the F-5 and D-11 positions of the fuel and reflector regions, respectively[8]. The inner and outer reflectors are made of beryllium and graphite, respectively. The reactor core is made of aluminum lattice frame which supports the fuel assemblies, reflectors and irradiation tubes.

2) Reaction Rate Measurements

Eleven kinds of activation reactions were employed for the neutron spectrum measurement at KUR. After the irradiation for 30 minutes at the Hyd facility, gamma-rays from the induced activities of foils were measured with a Ge(Li) detector of 53 cc, whose detection efficiency had been calibrated with a standard source with mixed radioactive nuclei. The reaction rates measured at KUR are summarized in Table 1, where the errors in percent are due to the statistical ones, detection efficiency and so on, as treated before[9].

For the measurement of neutron spectrum at SPR, 26 kinds of activation reactions were selected. Several sets of the foils were separately irradiated to get appropriate induced activities with different reactor power and operating time. Most of the

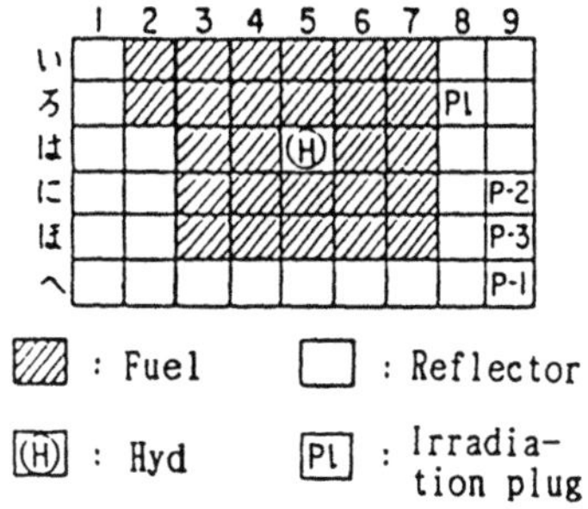


Fig. 2 A typical core configuration of KUR in Japan.

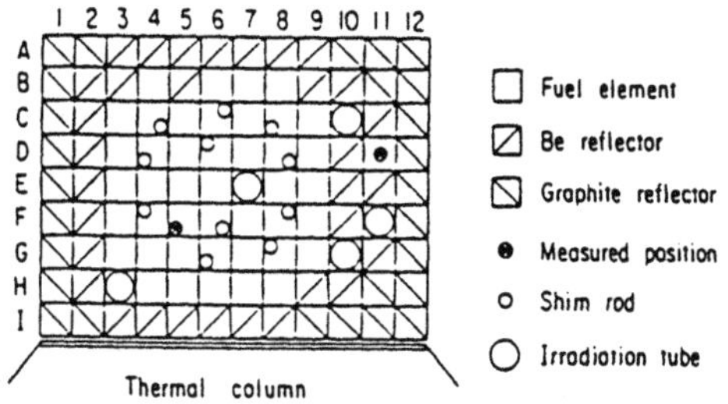


Fig. 3 A typical core loading at SPR in China.

Table 1 Reaction rate data used for spectrum unfolding at KUR.

| Reaction | Reaction rate & error | | Data used for unfolding by NEUPAC | NEUSPEC |
|---|---|---|---|---|
| $^{197}Au(n,\gamma)^{198}Au$ | 1.611E-08 * | 3.73 (%) | o | o |
| $^{58}Fe(n,\gamma)^{59}Fe$ | 1.293E-10 | 3.38 | o | |
| $^{59}Co(n,\gamma)^{60}Co$ | 4.087E-09 | 3.40 | o | o |
| $^{27}Al(n,p)^{24}Na$ | 2.699E-14 | 2.84 | o | o |
| $^{46}Ti(n,p)^{46}Sc$ | 4.175E-13 | 3.12 | o | o |
| $^{47}Ti(n,p)^{47}Sc$ | 6.704E-13 | 3.36 | o | o |
| $^{48}Ti(n,p)^{48}Sc$ | 1.132E-14 | 2.94 | o | o |
| $^{54}Fe(n,p)^{54}Mn$ | 3.139E-12 | 3.02 | o | o |
| $^{58}Ni(n,p)^{58}Co$ | 3.887E-12 | 3.02 | o | o |
| $^{90}Zn(n,2n)^{89}Zr$ | 4.361E-15 | 3.55 | o | |
| $^{93}Nb(n,2n)^{92m}Nb$ | 1.774E-14 | 3.14 | o | |

* Read as $1.611 \times 10^{-8}$.

induced activities were measured by a $4\pi\ \beta\ \gamma$ coincidence counter and/or a Ge(Li) detector. The reaction rates and their uncertainties are tabulated in Table 2, where the positions F-5 and D-11 are in the fuel and reflector regions at SPR, respectively.

## SPECTRUM UNFOLDING

The KU group has successfully obtained neutron spectra[7,10-12] by unfolding multi-foil activation data using the NEUPAC code, which was developed by Nakazawa et al.[4]. The principle of this code is due to the J-1 type unfolding method[4,13,14] which gives an approximate solution based on the Bayes' theorem.

The NEUPAC code contains energy dependent group cross section libraries including the error matrices for main important neutron dosimetry reactions in ENDF/B-V. These constants with 144 groups were processed by the N-JOY code[15] in the energy range from 0.01 eV to 16.4 MeV. In the NEUPAC calculation, variance-covariance errors for every input data and initial neutron spectrum are required. The output gives not only neutron spectrum with its covariance errors but also relating integral quantities and their uncertainties. The chi-squares test in the calculation can be performed to check a physical validity of the resultant spectrum.

Table 2  Reaction rate data used for spectrum unfolding at SPR.

| Reaction | Reaction rate and error at F-5 position | | Reaction rate and error at D-11 position | | Data used for unfolding by NEUPAC | Data used for unfolding by NEUSPEC |
|---|---|---|---|---|---|---|
| $^{55}$Mn(n, γ)$^{56}$Mn | 1.336E-14* | 2.16 (%) | 6.454E-15 | 2.16 (%) | | o |
| $^{164}$Dy(n, γ)$^{165}$Dy | 2.476E-12 | 2.65 | 1.210E-12 | 2.61 | | o |
| $^{176}$Lu(n, γ)$^{177}$Lu | 4.169E-12 | 2.76 | 1.828E-12 | 2.76 | | o |
| $^{151}$Eu(n, γ)$^{152}$Eu | 2.660E-12 | 3.17 | 1.252E-12 | 3.12 | | o |
| $^{191}$Ir(n, γ)$^{192}$Ir | 9.003E-13 | 2.55 | 3.868E-13 | 2.50 | | o |
| $^{115}$In(n, γ)$^{116m}$In | 2.695E-13 | 1.31 | 8.866E-14 | 1.31 | o | o |
| $^{197}$Au(n, γ)$^{198}$Au | 2.003E-13 | 1.82 | 5.635E-14 | 1.82 | o | o |
| $^{98}$Mo(n, γ)$^{99}$Mo | 7.151E-16 | 2.68 | 1.062E-16 | 2.76 | | o |
| $^{186}$W(n, γ)$^{187}$W | 4.899E-14 | 2.94 | 1.843E-14 | 2.79 | | o |
| $^{59}$Co(n, γ)$^{60}$Co | 3.655E-14 | 2.30 | 1.677E-14 | 2.30 | o | o |
| $^{63}$Cu(n, γ)$^{64}$Cu | 4.276E-15 | 2.48 | 1.967E-15 | 2.72 | o | o |
| $^{23}$Na(n, γ)$^{24}$Na | 5.139E-16 | 2.30 | 2.419E-16 | 1.71 | o | o |
| $^{235}$U(n, f) | 5.199E-13 | 2.44 | 2.381E-13 | 3.01 | | o |
| $^{238}$U(n, f) | 5.598E-15 | 3.41 | 1.115E-15 | 3.82 | | o |
| $^{47}$Ti(n, p)$^{47}$Sc | 1.920E-17 | 4.63 | 7.425E-19 | 4.77 | o | o |
| $^{115}$In(n, n')$^{115m}$In | 1.916E-16 | 2.22 | 6.876E-18 | 2.25 | o | o |
| $^{54}$Fe(n, p)$^{54}$Mn | 9.014E-17 | 2.43 | 3.451E-18 | 3.67 | o | o |
| $^{58}$Ni(n, p)$^{58}$Co | 1.151E-16 | 2.83 | 4.157E-18 | 2.46 | o | o |
| $^{64}$Zn(n, p)$^{64}$Cu | 3.551E-17 | 4.68 | 1.363E-18 | 4.38 | o | o |
| $^{32}$S(n, p)$^{32}$P | 6.970E-17 | 2.25 | 2.572E-18 | 2.30 | | o |
| $^{46}$Ti(n, p)$^{46}$Sc | 1.322E-17 | 2.45 | 5.041E-19 | 2.63 | o | o |
| $^{56}$Fe(n, p)$^{56}$Mn | 1.392E-18 | 3.69 | - - - | - - | o | o |
| $^{48}$Ti(n, p)$^{48}$Sc | 2.988E-19 | 2.34 | 1.291E-20 | 2.60 | o | o |
| $^{24}$Mg(n, p)$^{24}$Na | 1.690E-18 | 1.78 | 7.592E-20 | 1.78 | o | o |
| $^{27}$Al(n, α)$^{24}$Na | 7.885E-19 | 1.76 | 3.380E-20 | 1.75 | o | o |
| $^{63}$Cu(n, α)$^{60}$Co | 4.574E-19 | 2.34 | 1.893E-20 | 2.83 | o | o |

* Read as $1.336 \times 10^{-14}$.

The SAND-II code[6] has been widely applied to the spectrum derivation using multi-foil activation techniques in a wide energy range from 1.0 x $10^{-10}$ to 20 MeV. The code named NEUSPEC was prepared by the IAE group who revised the original SAND-II code to deduce the analytical uncertainties using the Monte Carlo method[5]. The cross section libraries for the NEUSPEC code have been produced from either ENDF/B-V or the data which the IAE group evaluated by themselves. The number of energy group is 640, and its structure is almost the same as those in Ref. 8. In addition to the group cross sections, 427 group constants of scattering and total cross sections have been prepared for the calculation correcting neutron self-shielding in the foil. The NEUSPEC code gives the neutron flux spectrum and its uncertainty by a Monte Carlo method.

## RESULTS AND DISCUSSION

### 1) KUR Spectrum

The neutron spectrum at the Hyd facility of KUR was obtained by the NEUPAC code using 11 kinds of activation data, shown in Table 1. For the spectral analysis with the NEUSPEC code, 8 data available in the NEUSPEC cross section libraries were employed. The NEUPAC and NEUSPEC results are compared in Fig 4, where the spectrum calculated by the ANISN code[7] is also shown. The initial spectrum for NEUPAC was referred to the ANISN calculation, while that for NEUSPEC was taken from the Maxwellian and 1/E spectrum below 50 keV[8] and the NEUPAC calculation above the energy. General agreement can be seen between the NEUPAC and NEUSPEC spectra below 10 keV. In the higher energy region, however, the NEUSPEC spectrum is lower for energies between 20 keV and 3 MeV and higher between 3 and 7 MeV than the NEUPAC

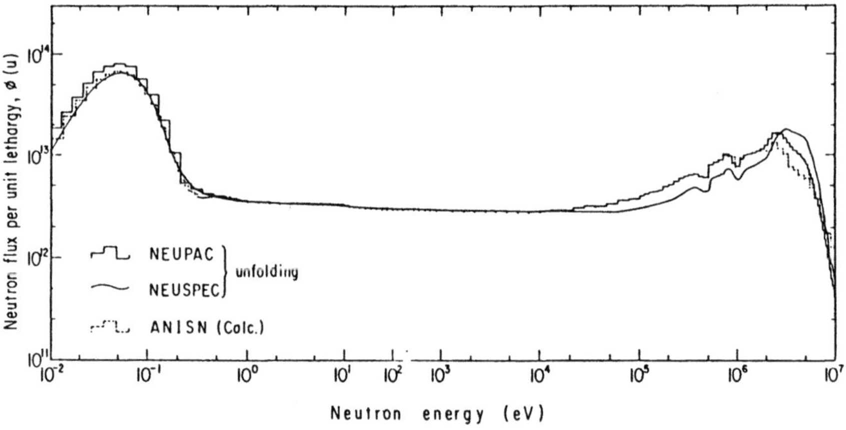


Fig. 4 Neutron spectrum at Hyd facility of KUR in Japan.

spectrum. The discrepancy may be due to the difference of the initial spectrum used in the NEUPAC and NEUSPEC codes. Considering the uncertainties in the neutron spectra by the NEUSPEC and NEUPAC codes, both results in the higher energy region seem to be close to each other in their error range. For the $^{235}U$ fission neutron spectrum unfolding with the $^{235}U$ fission spectrum-averaged cross sections, the NEUPAC and NEUSPEC calculations using the same initial spectrum of Watt showed a general agreement with each other[12].

2) SPR Spectrum

The neutron spectra in the fuel region F-5 and the reflector region D-11 at SPR were obtained by the NEUSPEC code using 26 kinds of activation data. In the NEUPAC cross section libraries, 16 data of the above 26 kinds of activation data were applicable to the spectrum unfolding. The results by the NEUSPEC and the NEUPAC codes are illustrated in Fig. 5. The initial spectrum for the NEUSPEC was deduced from Maxwellian, 1/E and fission neutron spectra[8], in principle, and the NEUSPEC result was used as an initial spectrum for the NEUPAC calculation. It is realized that the flux spectrum in the reflector region is softer than that in the fuel region. The analytical errors are 5 to 9 percent in both calculations. Very good agreement can be seen between the NEUSPEC and the NEUPAC calculations in the whole energy region.

## CONCLUSION

The research group of the Kyoto University measured eleven kinds of activation data at the KUR core and unfolded neutron spectrum by using the NEUPAC code. The research group of the Institute of Atomic Energy measured 26 kinds of activation data in the fuel and reflector regions at SPR and obtained neutron

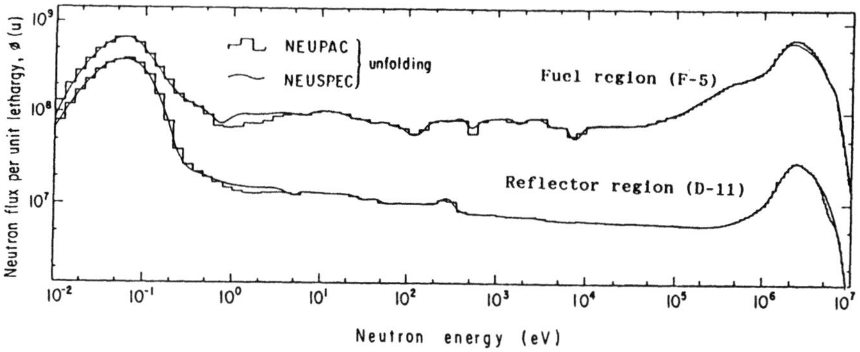


Fig. 5 Neutron spectrum at SPR in China.

spectra there with the NEUSPEC code. The groups have mutually exchanged the activation data and analyzed the other group's data by their own unfolding code. For the unfolding spectrum at KUR, a good agreement was seen below 50 keV in the NEUPAC and NEUSPEC analyses. In the higher energy region, both spectra have shown a difference which is close to maximum value in the analytical error. The discrepancy may be due to the difference with the initial spectrum for NEUPAC and NEUSPEC calculations. For the SPR spectrum, very good agreement was seen between the NEUPAC and NEUSPEC calculations in the whole energy region.

## REFERENCES


1. G. R. Odtte and D. R. Doiron, Nucl. Technol., 29, 346 (1976).
2. M. A. Kirl and L. R. Greenwood, J. Nucl. Materials, 80, 159 (1979).
3. T. Okada et al., J. Nucl. Materials, 133&134, 791 (1985).
4. M. Nakazawa and A. Sekiguchi, Proc. 2nd ASTM-Euratom Symp. on Reactor Dosimetry, NEUREG/CP-0004, Vol.3, p.1423 (1977).
5. Li Zhaoliuan et al., Submitted to this ASTM-Euratom Symp. on Reactor Dosimetry.
6. W. N. McElroy et al., Nucl. Sci. Eng., 27, 533 (1967).
7. K. Kobayashi, I. Kimura et al., Annu. Rep Res. Reactor Inst. Kyoto Univ., Vol.20, 1 (1987).
8. Li Zhaoliuan et al., Japan-China Symp. on Research and Test Reactors, Feb.29-Mar.2, 1988, JAERI, UT-01 (1988).
9. K. Kobayashi, I. Kimura et al., J. Nucl. Sci. Technol., 19, 341 (1982).
10. K. Kobayashi, I. Kimura et al., Annu. Rep. Kinki Univ. Atomic Energy Res. Inst., Vol.25, 21 (1988).
11. I. Kimura and K. Kobayashi, Nucl. Sci. Eng., To be published.
12. K. Kobayashi, I. Kimura, Li Zhaoliuan and Wang Yongqing, Submitted to this ASTM-Euratom Symp. on Reactor Dosimetry.
13. T. Taniguchi et al., NEUT Res. Rep. 83-10 (1983).
14. M. Nakazawa et al., J. Fac. Eng., Univ. Tokyo, A-22, 44 (1984).
15. R. E. MacFarlane et al., LA-9303 M (1982).

# THE $^{235}$U FISSION NEUTRON SPECTRUM ADJUSTED WITH MULTI-FOIL ACTIVATION DATA

Katsuhei Kobayashi
Research Reactor Institute, Kyoto University
Kumatori-cho, Sennan-gun, Osaka 590-04 Japan

Itsuro Kimura
Department of Nuclear Engineering, Kyoto University
Yoshidahonmachi, Sakyo-ku, Kyoto 606 Japan

Li ZhaoHuan and Wang YougQing
Institute of Atomic Energy, P. O. Box 275(54)
Beijing, People's Republic of China

Using the NEUPAC and the NEUSPEC codes, the $^{235}$U fission neutron spectrum was adjusted with experimentally obtained average cross section data for 18 threshold reactions. The reaction rates and the covariance matrix as input data for the calculation have been taken from data-sets measured by Mannhart and Kobayashi et al., respectively. The adjusted neutron spectrum, which was obtained from the reaction rate data combining the both data-sets with the least squares procedure, showed a good agreement with the Watt-type spectrum, except for the energy region above about 12 MeV where the adjusted spectrum was higher by about 30 %. The correlation matrix for the $^{235}$U fission neutron spectrum is given also. General agreement can be seen between the adjusted spectra using the NEUPAC and the NEUSPEC codes.

## INTRODUCTION

The neutron spectrum of the thermal neutron-induced fission of $^{235}$U ($^{235}$U fission neutron spectrum) is of great importance as one of the fundamental data in reactor physics. In order to obtain the fission neutron spectrum accurately, a large number of experiments have been carried out[1], and it was tried to fit the results with semi-empirical formula such as Maxwellian and/or Watt-type spectrum[2,3]. Also theoretical studies have been carried out to predict the $^{235}$U fission neutron spectrum[4-7].

The fission spectrum was recommended as a reference neutron field in addition to the standard neutron spectrum field of the spontaneous fission neutrons of $^{252}$Cf in the fast neutron energy region[8]. Therefore, many research groups have prepared and established the $^{235}U$ fission neutron fields for the measurement of spectrum-averaged cross sections[9-15], which are often used for the integral assessment of energy dependent cross sections. In recent years, much attention has been given to the application of uncertainties in the calculation of spectrum-averaged cross sections. However, except for the work by Petilli[16], few paper gives a covariance matrix of the energy dependent $^{235}U$ fission neutron spectrum which has been derived from the data based on experiments, although some of recent cross section libraries contain the energy dependent covariances of the cross sections.

In the present work, the $^{235}U$ fission neutron spectrum has been obtained by making use of the recent $^{235}U$ fission spectrum-averaged cross sections measured by Mannhart[13] and Kobayashi et al.[15,17]. The Kyoto University group (KK and IK) adjusted the foil activation data with the NEUPAC code[18], and the group of Institute of Atomic Energy, IAE (LZH and WYQ) obtained the neutron spectrum with the NEUSPEC code[19,20]. The ratios in respect to the Watt-type spectrum were given and compared.

## $^{235}U$ SPECTRUM-AVERAGED CROSS SECTIONS

In recent years, experimental data of the $^{235}U$ fission spectrum-averaged cross sections have been carefully measured with detailed experimental uncertainties. Mannhart measured the average cross sections of 17 reactions of importance in reactor metrology by making use of the $^{235}U$ fission neutron field of the BR1 at Mol[13]. He gave a complete uncertainty covariance matrix to his results, and the experimental uncertainties in the cross section values were 4 to 6 %.

Two of the authors (KK and IK) and colleague (Kobayashi et al.) measured very recently the $^{235}U$ fission spectrum-averaged cross sections for 12 threshold reactions, relative to the value of 0.706 $\pm$ 0.028 mb for the $^{27}Al(n,\alpha)^{24}Na$ reaction measured by Mannhart[13], with a large fission plate installed at the heavy water thermal neutron facility of KUR[15,17]. Not only the average cross sections but also the covariance matrix were carefully obtained. The experimental procedure and the data reduction are similar to those as the authors did before[21].

The NEUPAC and the NEUSPEC data libraries contain cross section values for 15 reactions of the Mannhart data and for 13 reactions data measured by Kobayashi et al. In order to apply additional reaction data for neutron spectrum adjustment, we have combined the above two data-sets by the least squares method using the BOLIK code[22]. The resultant data-set of $^{235}U$ fission spectrum-averaged cross sections which were employed for the spectrum adjustment is tabulated in Table 1, as the Mannhart-Kobayashi data.

Table 1 The U-235 fission spectrum-averaged cross sections and the correlation matrix obtained by combining the data-sets measured by Mannhart and Kobayashi et al.

| Reaction | Cross section (mb) | Error (%) | # Ref. | Correlation matrix (x 100) | | | | | | | | | | | | | | | | | |
|---|---|---|---|---|---|---|---|---|---|---|---|---|---|---|---|---|---|---|---|---|---|
| 24Mg(n,p)24Na | 1.495 | 4.20 | M | 100 | | | | | | | | | | | | | | | | | |
| 27Al(n,p)27Mg | 3.858 | 4.91 | M,K | 81 | 100 | | | | | | | | | | | | | | | | |
| 27Al(n,α)27Na | 0.706 | 3.97 | M,K | 95 | 83 | 100 | | | | | | | | | | | | | | | |
| 46Ti(n,p)46Sc | 11.42 | 4.67 | M,K | 86 | 85 | 86 | 100 | | | | | | | | | | | | | | |
| 47Ti(n,p)47Sc | 17.72 | 4.65 | M,K | 85 | 84 | 85 | 91 | 100 | | | | | | | | | | | | | |
| 48Ti(n,p)48Sc | .2965 | 4.68 | M,K | 86 | 84 | 86 | 96 | 89 | 100 | | | | | | | | | | | | |
| 54Fe(n,p)54Mn | 78.72 | 4.69 | M,K | 86 | 86 | 87 | 94 | 92 | 92 | 100 | | | | | | | | | | | |
| 56Fe(n,p)56Mn | 1.067 | 4.83 | M,K | 84 | 86 | 84 | 87 | 87 | 87 | 88 | 100 | | | | | | | | | | |
| 58Ni(n,p)58Co | 106.7 | 4.63 | M,K | 86 | 86 | 87 | 94 | 93 | 92 | 99 | 88 | 100 | | | | | | | | | |
| 59Co(n,α)56Mn | .1574 | 5.26 | M | 77 | 79 | 77 | 80 | 79 | 80 | 80 | 87 | 80 | 100 | | | | | | | | |
| 59Co(n,2n)58Co | .1966 | 4.73 | M | 85 | 84 | 86 | 91 | 88 | 92 | 94 | 87 | 95 | 80 | 100 | | | | | | | |
| 58Ni(n,2n)57Ni | 4.145-3* | 5.11 | M | 56 | 53 | 56 | 58 | 57 | 59 | 56 | 56 | 55 | 52 | 58 | 100 | | | | | | |
| 90Zr(n,2n)89Zr | .09909 | 4.89 | M,K | 84 | 82 | 84 | 91 | 86 | 91 | 91 | 85 | 91 | 78 | 92 | 57 | 100 | | | | | |
| 115In(n,n')115mIn | 189.5 | 4.77 | M,K | 83 | 87 | 83 | 88 | 90 | 86 | 90 | 83 | 90 | 75 | 86 | 55 | 84 | 100 | | | | |
| 197Au(n,2n)196Au | 3.457 | 4.74 | M | 85 | 81 | 85 | 88 | 87 | 88 | 88 | 85 | 88 | 79 | 89 | 60 | 86 | 85 | 100 | | | |
| 64Zn(n,p)64Cu | 31.93 | 5.24 | K | 70 | 70 | 71 | 75 | 73 | 74 | 75 | 71 | 75 | 65 | 73 | 46 | 72 | 72 | 70 | 100 | | |
| 93Nb(n,2n)92mNb | .4838 | 5.59 | K | 70 | 70 | 70 | 74 | 73 | 73 | 75 | 71 | 76 | 65 | 73 | 45 | 72 | 72 | 70 | 64 | 100 | |
| 103Rh(n,n')103mRh | 673.5 | 8.11 | K | 73 | 70 | 75 | 73 | 74 | 73 | 74 | 71 | 74 | 64 | 72 | 46 | 71 | 72 | 70 | 63 | 61 | 100 |

* Read as $4.145 \times 10^{-3}$.

\# M : data which Mannhart measured[13].

K : data which Kobayashi et al. measured[15,17].

## SPECTRUM ADJUSTMENT

Both NEUPAC and NEUSPEC codes were used to adjust the $^{235}U$ fission neutron spectrum. The NEUPAC code was developed by Nakazawa et al.[18] and was frequently applied for neutron spectrum adjustments[23-26]. The principle of this code is based on the J-1 type unfolding method, and is also regarded as equivalent to the STAY'SL code[27]. The NEUPAC code contains energy dependent group cross section libraries which were mainly produced from ENDF/B-V. A 135 groups cross section library including the covariance matrices in the energy range from 0.01 eV to 16.4 MeV was produced by employing the N-JOY code[28]. The obtained reaction rate data and variance-covariance values shown in Table 1 were used as input data for the $^{235}U$ fission spectrum adjustment. For the initial trial spectrum, the following Watt-type $^{235}U$ fission neutron spectrum proposed by CSEWG[29] was employed:

$$\chi(E) \propto \exp(-E/0.988)\ \sinh(\sqrt{2.249E})$$

where, E is neutron energy in MeV. Concerning the relative uncertainties of the initial spectrum for the spectrum adjustment, low weighted value of 200 % was given in the whole energy group.

The output of the adjustment gives not only the adjusted spectrum with its covariance matrix but also the related integral quantities with their uncertainties, and the chi-square value. This value is used to check the physical validity of the resultant spectrum.

The IAE group has newly developed the NEUSPEC code[19,20] for adjustment of a neutron energy spectrum by using activation data. This code has been successfully applied for spectrum adjustment using multi-foil activation data[19,20]. The NEUSPEC code is principally based on the SAND-II code[30]. The main revision is that uncertainties are given in the calculated spectrum by a Monte Carlo method. The number of energy groups is 640 and the group structure is almost the same as that in the SAND-II library. The cross section library was prepared from the evaluated energy dependent cross sections based on the ENDF/B-V data and/or the evaluated ones by the authors (LZH and WYQ). As initial spectrum, the Watt spectrum was used.

## RESULTS AND DISCUSSION

Using 18 kinds of activation data for the $^{235}U$ fission spectrum-averaged cross sections, which were obtained by combining the data-sets measured by Mannhart and Kobayashi et al. with the least squares method, neutron energy spectrum with the covariance matrix for the output fluxes was adjusted with the NEUPAC code. The result of the normalized spectrum is depicted in Fig. 1, as ratios to the Watt-type spectrum. In the figure, supplemental comparison was made with the result adjusted by using 13 activation data measured by Kobayashi et al., in order to investigate the effects appeared in the adjustments by changing the

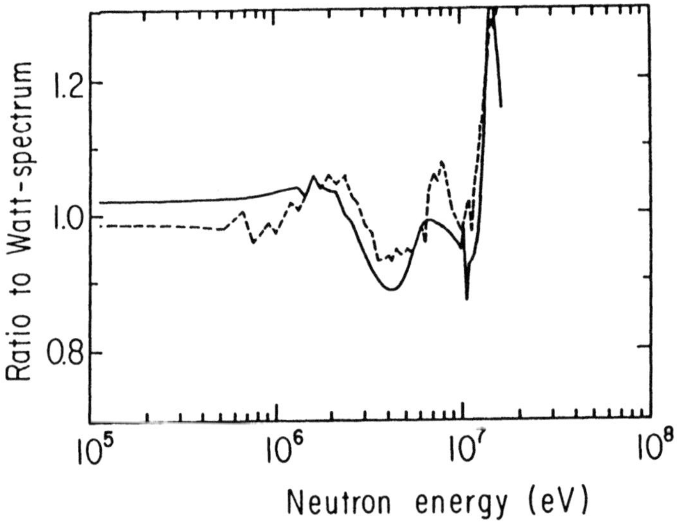


Fig. 1 $^{235}U$ fission neutron spectrum as a ratio to the Watt spectrum. ———— spectrum adjusted with Mannhart-Kobayashi data shown in Table 1. --------- spectrum adjusted with the data measured by Kobayashi et al.[17].

number of activation data and/or the data-set values. Considering that the uncertainties in the adjusted spectrum are several percent, it is found that the two spectra are in good agreement with each other and close to the Watt spectrum except for the energy region above about 12 MeV, although some peak and dip structures are observed. Group fluxes, uncertainties and the correlation matrix adjusted with the Mannhart-Kobayashi data are shown in Table 2, where ten energy group data are given by collapsing the original NEUPAC output results of 135 energy groups. Since the activation data used in the present adjustment have major responses in the MeV energy region, we display the results in the energy region above 0.1 MeV.

Comparing with the spectrum adjusted with the NEUPAC code, Figure 2 shows the neutron spectrum which the IAE group adjusted the Mannhart-Kobayashi data, using the NEUSPEC code. Both ratio spectra to the Watt spectrum show a general agreement within 10 % in the energy region below 12 MeV, although large deviations can be observed at higher energies. Discrepancies between the both spectra are seen from 6 to 7 MeV and from 9 to 12 MeV, and their spectral peaks and dips seem to be shifted a little with energy. Recently, Ikeda et al. measured many dosimetry cross sections at energies between 13 and 15 MeV[31]. They found that the cross section values in ENDF/B-V, which were used to generate the NEUPAC and NEUSPEC code libraries, are lower by 10 to 13 % for the $^{47}$T(n,p)$^{47}$Sc and $^{58}$Ni(n,2n)$^{57}$Ni reactions and higher for the $^{48}$Ti(n,p)$^{48}$Sc and $^{58}$Ni(n,p)$^{58}$Co reactions. One of the reasons that the ratio spectra show some fluctuations may be due to the inappropriate cross section values in the libraries. More investigation is required to understand the discrepancy.

Table 2 The obtained group fluxes, uncertainties and correlation matrix of the adjusted neutron spectrum of U-235.

| No. | Energy ( MeV ) | Flux $\phi$ (E) | Error ( % ) | Correlation matrix ( x 100 ) | | | | | | | | | |
|---|---|---|---|---|---|---|---|---|---|---|---|---|---|
| 1 | 16.40 | 6.307-5* | 3.27 | 100 | 13 | 26 | 28 | 24 | 23 | 25 | 19 | 14 | 17 |
| 2 | 13.72 | 3.396-4 | 3.49 | | 100 | 14 | 17 | 22 | 20 | 22 | 16 | 11 | 13 |
| 3 | 11.48 | 1.802-3 | 3.64 | | | 100 | -2 | -0 | 18 | 16 | 12 | 9 | 11 |
| 4 | 9.31 | 9.001-3 | 3.27 | | | | 100 | -22 | 23 | 27 | 15 | 9 | 10 |
| 5 | 7.33 | 3.530-2 | 3.53 | | | | | 100 | 1 | 5 | 18 | 10 | 11 |
| 6 | 5.53 | 1.012-1 | 3.67 | | | | | | 100 | -32 | 11 | 18 | 13 |
| 7 | 4.00 | 2.666-1 | 3.64 | | | | | | | 100 | -5 | -2 | 6 |
| 8 | 2.50 | 5.761-1 | 5.37 | | | | | | | | 100 | -2 | 7 |
| 9 | 1.16 | 7.727-1 | 5.30 | | | | | | | | | 100 | 26 |
| 10 | 0.504 | 6.659-1 | 4.50 | | | | | | | | | | 100 |
| 11 | 0.100 | | | | | | | | | | | | |

* Read as 6.307 x $10^{-5}$.

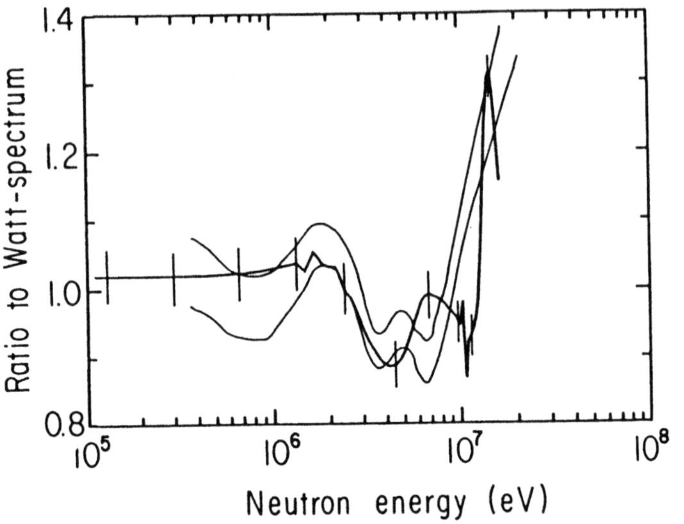


Fig. 2 $^{235}U$ fission neutron spectrum adjusted with Mannhart-Kobayashi data, as a ratio to the Watt spectrum.
—┼┼┼┼— spectrum adjusted with the NEUPAC code.
═══ spectrum obtained with the NEUSPEC code.

## CONCLUSION

The $^{235}U$ fission neutron spectrum was adjusted with the fission spectrum-averaged cross sections of 18 threshold reactions, using the NEUPAC and the NEUSPEC codes. The reaction rates and the covariance matrix for the input data were obtained by combining the data-sets measured by Mannhart and Kobayashi et al. The neutron spectra adjusted with the both codes showed a general agreement and were close to the Watt-type spectrum, although the adjusted spectra were higher by about 30 % than the Watt spectrum at energies above about 12 MeV. The present adjustment using the NEUPAC code gave the energy dependent covariance matrix for the $^{235}U$ fission neutron spectrum, which could be interesting as input data for the next adjustment and/or the spectrum-averaged cross section calculations.

## Acknowledgements

The authors would like to express their thanks to Dr. Tooru Kobayashi of Research Reactor Institute, Kyoto University for his useful discussion and suggestion in carrying out the fission spectrum-averaged cross section measurement. They are much grateful to Prof. M. Nakazawa of University of Tokyo for his guidance to the NEUPAC calculation. They also thank Prof. T. Ohsawa of Kinki University Atomic Energy Research Institute for his useful discussions in carrying out the study.

## REFERENCES

1. B. H. Armitage and M. G. Sowerby (ed.), AERE-R 8636 (1977).
2. P. I. Johansson and B. Holmqvist, Nucl. Sci. Eng., 62, 695 (1977).
3. W. L. Zijp and J. H. Baard, ECN-70 (1979).
4. D. G. Madland and J. R. Nix, Nucl. Sci. Eng., 81, 213 (1982).
5. H. Märten et al., IAEA-TECDOC-335, p.255 (1985).
6. J. C. Browne and F. S. Dietrich, Phys. Rev., C10, 2545 (1974).
7. T. Ohsawa, to be published in KURRI-TR (1991).
8. M. F. Vlasov(ed.) INDC(NDS)-85/L+M (1976).
9. A. Calamand, IAEA Technical Report Series No.156, p.273 (1974).
10. A. Fabry et al., Proc. 4-th ASTM-Euratom Symp. on Reactor Dosimetry, NUREG/CP-0029, Vol.2, 665 (1982).
11. J. A. Grundl, Proc. Consultants Meeting on Prompt Fission Neutron Spectra, STI/PUB/329, p.29 (1972).
12. J. A. Grundl and C. M. Eisenhauer, NBSIR 85-3151 (1986).
13. W. Mannhart, Proc. 5-th ASTM-Euratom Symp. on Reactor Dosimetry, D.Reidel Publ. Co. Dordrecht, Vol.2, p.813(1985).
14. D. M. Gilliam et al. ibid., p.867 (1985).
15. I. Kimura and K. Kobayashi, Nucl. Sci. Eng., 106, 332 (1990).
16. M. Petilli and D. M. Gilliam, Proc. 5-th ASTM-Euratom Symp. on Reactor Dosimetry (1984).
17. K. Kobayashi and T. Kobayashi, NEANDC(J)-155/U, p.52 (1990).
18. M. Nakazawa and A. Sekiguchi, Proc. 2nd ASTM-Euratom Symp. on Reactor Dosimetry, NUREG/CP-0004, Vol.3, 1423 (1977).
19. Li ZhaoHuan et al., Japan-China Symp. on Research and Test Reactors, UT-01, JAERI (1988).
20. Li ZhaoHuan et al., Submitted to this ASTM-Euratom Symp. on Reactor Dosimetry.
21. K. Kobayashi et al., J. Nucl. Sci. Technol., 19, 341 (1982).
22. M. Petilli, ENEA-RT/FI(83)5 (1983).
23. K. Kobayashi et al., Annu. Rep. Res. Reactor Inst. Kyoto Univ., Vol.20, 1 (1987).
24. K. Kobayashi et al., Kinki Univ. Atomic Energy Res. Inst., Vol.25, 21 (1988).
25. I. Kimura and K. Kobayashi, IAEA-TECDOC-410, p.171 (1987).
26. K. Kobayashi and I. Kimura, Proc. of the Int. Conf. on Nuclear Data for Sci. and Technol., JAERI, p.261 (1988).
27. F. G. Percy, ORNL/TM-6062 (1977).
28. R. E. MacFarlane et al., LA-7584 M (1978).
29. B. A. Magurno, Proc. 3rd ASTM-Euratom Symp. on Reactor Dosimetry, EUR 6813 EN-FR, Vol.2, p.903 (1980).
30. W. N. McElroy et al., Nucl. Sci. Eng., 27, 533 (1967).
31. Y. Ikeda et al., JAERI 1312 (1987).

# Adjusted Spectra for Three GODIVA-Type Reactors

W. W. Sallee

Nuclear Effects Laboratory
White Sands Missile Range, New Mexico, USA

## INTRODUCTION

A series of ASTM standards were written in the mid-seventies to characterize the damage to electronic devices from neutrons in terms of 1 MeV mono-energetic neutron exposures (1). Central to these standards was the use of radiometric foil data and the SAND-II adjustment code (2). Prior to acceptance by the community, several validation experiments were performed by Verbinski, et. al., at typical testing facilities (3). Validation consisted of: 1) the 1 MeV damage fluence concept and 2) the SAND-II code to obtain the neutron energy distribution from radiometric monitors.

A powerful new tool with distinct computational and philosophical advantages over the SAND-II method first became widely available with the appearance of STAY'SL (4). However, the advantages of the least squares methodology were long negated by the lack of an evaluated dosimetry cross section set with the required covariance data <u>and</u> a method for converting the cross section, fluence and covariance data from one group structure to another. With the release of the LSL-M2 code package (5) which contains an evaluated cross section set with covariances and the FLXPRO utility for converting between different group structures, the test facility dosimetrist finally had available all the tools necessary to take advantage of the least squares codes.

Recently, the Nuclear Effects Laboratory (NEL) and co-workers at the University of Illinois have published two papers with the purpose of validating, in a fashion similar to that described in ref. (3), the LSL code for use by the hardness testing community. The first (6) concentrated on the adjustment codes by comparing the accepted SAND-II methodology to LSL and STAY'SL. The second paper (7) re-examined the underlying 1 MeV equivalent damage assumptions and methodology described in ASTM Standard E722. Some of the discussion in the latter work is based upon previously undocumented LSL adjusted spectra for the White Sands Missile Range's Fast Burst Reactor (FBR), Aberdeen Proving Ground's Army Pulsed Reactor Facility (APRF) and Sandia National Laboratory's (SNL) Sandia Pulsed Reactor III (SPR-III). All three are super-prompt critical assemblies of highly enriched $^{235}U$ alloyed with molybdenum and are capable of burst or steady state operations. The present work details procedures used to obtain adjusted spectra and damage parameters useful for essentially free field steady state exposures of electronic devices at these reactors.

## ACTIVATION AND CROSS SECTION DATA

While new activation data were available for the SPR-III and FBR facilities, none were available for APRF. The reactions used, the measured reaction probabilities, and the ascribed uncertainties are given in Table 1. The cross sections were the ones supplied with LSL. Foil covers were used to reduce activities from impurities and to establish artificial thresholds for the fission dosimeters. Effective cross sections were calculated using the equation:

$$\sigma_{f,i}^{eff}(E) - \sigma_{f,i}(E)\ e^{-X\,\sigma_{c,i}(E)} \qquad (1)$$

where $i$ refers to the cross section group, $f$ refers to the foil's cross section, $c$ refers to the cover's cross section and $X$ refers to the thickness of cover material. In the case of the fission foils were both cadmium and boron where present, the cross sections are processed sequentially for each cover. Cross correlations were ignored during the adjustment procedure.

The activation data used for APRF were taken from Table 3, Run 11 as described in the Verbinski report (3). The foils were exposed on the APRF Hardness Assurance Table. The SPR-III activation data were provided by J. Kelly (8). The monitors were exposed on Oct 5, 1988, behind a copper control rod at 43.2 cm from the core center. Activation data for the FBR were obtained from a 61 cm (relative to core center) exposure on Feb 21, 1989.

As the Verbinski report contains no uncertainty estimates for the activation data, the uncertainties are those thought typically attainable given good experimental and counting practices. Discussions with J. Kelly indicated that the uncertainties in SPR-III data were typically the same as those for the FBR. Uncertainties for the FBR were assigned based upon positioning, counting geometry, counter

Table 1
Measured Reaction Probabilities[1] With Uncertainty Assignments

| Reaction | APRF | (%) | SPR-III[2,3] | FBR | (%) |
|---|---|---|---|---|---|
| Au197(n,γ) Au198 | | | 4.734E-11 | | 4 |
| Na23 (n,γ) Na24 | | | 5.346E-14 | | 8 |
| Sc45 (n,γ) Sc46 | | | 5.943E-13 | | 4 |
| Au197(n,γ) Au198 | 4.170E-11[4] | 5 | 4.056E-11[4] | 2.151E-11[4] | 4 |
| Na23 (n,γ) Na24 | | | 1.795E-14[4] | | 8 |
| Cu63 (n,γ) Cu64 | | | 6.335E-13[4] | | 6 |
| U235 (n,F) | 1.110E-10 | 8 | 3.029E-11[5] | | 8 |
| Pu239(n,F) | | | 4.171E-11[5] | 1.617E-11[5] | 8 |
| Np237(n,F) | 8.090E-11 | 8 | 2.505E-11[5] | 1.036E-11[5] | 8 |
| U238 (n,F) | 1.370E-11 | 8 | 4.604E-12[5] | 1.922E-12[5] | 8 |
| Th232(n,F) | 3.860E-12 | 8 | | | 8 |
| In115(n,n')In115 | | | 3.287E-12[4] | | 4 |
| Ti47 (n,P) Sc47 | | | 2.702E-13 | | 4 |
| Fe54 (n,P) Mn54 | 3.660E-12 | 5 | 1.242E-12 | 5.381E-13 | 4 |
| Zn64 (n,P) Cu64 | | | 5.683E-13 | 2.492E-13 | 4 |
| Ni58 (n,P) Co58 | 4.650E-12 | 5 | 1.604E-12 | 7.011E-13 | 4 |
| S32 (n,P) P32 | | | 1.025E-12 | | 3 |
| Ti46 (n,P) Sc46 | | | | 7.488E-14 | 4 |
| Fe56 (n,P) Mn56 | 4.380E-14 | 5 | 1.515E-14 | 7.155E-15 | 4 |
| Ti48 (n,P) Sc48 | | | 4.317E-15 | 2.097E-15 | 4 |
| Mg24 (n,P) Na24 | 5.800E-14 | 5 | 2.137E-14 | 9.803E-15 | 4 |
| Al27 (n,α) Na24 | | | 1.020E-14 | 4.660E-15 | 4 |
| Zr90 (n,2n)Zr89 | 4.800E-15 | 5 | | | 4 |
| I127 (n,2n)I126 | | | | 6.349E-15 | 4 |

[1] Only reactions actually used are presented. Foils were discarded only if 1) there was a known problem with the measurement or 2) there was no cross section in the LSL evaluated library.

[2] I gratefully acknowledge the permission of J. Kelly and Sandia National Laboratories to use these unpublished data.

[3] Non-fission threshold foils were actually exposed in cadmium but effective cross sections were not available at the time of this run. Cadmium processing typically has little effect on the reaction cross sections at high energies.

[4] Cadmium covered activity.

[5] Cadmium and Boron-10 covered activity.

calibrations and nuclear data constraints. For the fission reactions an additional uncertainty exists. At APRF, the fact that these foils were not covered with either boron or cadmium raises the possibility that part of the observed activity may come from other fission isotope impurities. For SPR-III and FBR, transmission through the boron cover modifies the exposure environment. This effect is partially compensated for by the use of eq. 1 which handles the absorption process. But, it does not compensate for down scattering processes. A rough scattering factor has been applied to the FBR data but not to the SPR-III data. These additional uncertainties are reflected in the 8% entries in Table 1.

## TRANSPORT CALCULATIONS

For the APRF adjustment, the 37 group leakage spectrum calculation tabulated in Annex 1 of ref. (9) was used. It was modified to account for room return components in the exposure environment by tieing a 1/E + 20-degree Maxwellian shape at 3 keV. During the mid-eighties, Sartor performed a series of MORSE calculations (10) for the SPR-III facility, one of which was a 43.2 cm leakage spectrum complete with room components. FLXPRO was used to convert its 27 group structure into the 37 group DLC-31 (11) structure. The calculation for the FBR was provided by Flanders and Sparks (12). It was a two step MCNP 37 group calculation for the leakage spectrum at 61 cm from the core's centerline. Above 10 MeV, FLXPRO was used to replace groups with poor statistics with a fission shape.

As is typically the case, there exists no covariance data for any of the above calculations. Following the discussion in ref. (6), covariance matrices were constructed in 37 groups for each of the above calculations using the following equation:

$$r_{ij} = e^{-\frac{|\ln(E_i) - \ln(E_j)|}{A}} \tag{2}$$

where $E_i$ and $E_j$ are the geometric mean energies of groups $i$ and $j$ respectively, $r_{ij}$ is the correlation coefficient between these groups and $A$ is a variable parameter which can be viewed as the stiffness of the matrix. It is believed that the correlations between fluences in the groups above 10 keV are driven by production and single scattering processes within the core. This should not result in strong correlations across large energy ranges. A value of A=3 or less should be appropriate. Neutrons below 10 keV are known to arise from room scattering processes and should be strongly correlated across these energies. A value of A=6 or greater should be more applicable.

Uncertainties assigned to the group fluences varied significantly between the various transport calculations. Values of 100% were normally assigned to the data 3.35 keV and less. Values of 15-25% were used between 3.35 keV and 10 MeV. Values of 30-50% were used above 10 MeV where FLXPRO was used to restructure (APRF, SPR-III) or extrapolate a fission spectrum (FBR). Standard A=6 and A=3 matrices were tried for each of the spectra but only yielded satisfactory results for SPR-III. Results from the A=6 matrix with diagonal elements of 15% above and 50% below 3.35 keV are presented for SPR-III. For APRF and FBR, the diagonal elements between 3.35 keV and 10 MeV were 20% and 22%, respectively. Rather than choose A=3 or A=6 for the entire spectrum, the covariance matrix for each of these spectra were disjoint at 3.35 keV with A=3 above, A=6 below and zero correlations across this boundary. Secondary considerations supporting this approach are: 1) the damage function below 10 keV is very small and 2) less than 3% of the neutrons in the spectra have energies below 10 keV.

## RESULTS AND DISCUSSION

The adjusted spectra for each reactor are shown in Figure 1. LSL uses the Chi$^2$ per degree of freedom statistic to indicate the overall consistency of the data. Acceptable values were obtained for all three adjustments: 1.149 for APRF, 1.090 for SPR-III and 0.4255 for FBR. The FBR value indicates that uncertainties are over estimated. This is not surprising considering the 100% uncertainties for fluences below 3.35 keV. However, because of the disjoint nature of the matrix, the shape of the resulting spectrum above 10 keV should not be affected.

All three facilities monitor the magnitude of the $\Phi$ > 3 MeV with sulfur dosimeters. Thus, the 1 MeV equivalent fluence of ASTM standard E722 is calculated from:

$$H.\ P. = \frac{\int_{0.01}^{18} \Phi(E)\cdot K_D(E)\ dE}{K_D(E_{1MeV})\int_{0.01}^{18}\Phi(E)\ dE} \tag{3}$$

$$S.\ I. = \frac{\Phi > 10\ keV}{\Phi > 3\ MeV} \tag{4}$$

where ***H.P.*** is the hardness parameter defined in eq. 2 of E722, $K_D(E)$ is the displacement kerma function, and ***S.I.*** is the spectral index (referred to as $M_r$ in the standard). Notice that when defined this way, both the S.I. and H.P. are only dependent upon the shape of the fluence spectrum. The fluence from the sulfur dosimeters is multiplied by these two factors to give the 1 MeV equivalent fluence. Table 2 provides these damage parameters for each facility for the damage kerma function tabulated in E722 and for a new kerma function (SNL-NJOY) calculated by P. Griffin (13) using NJOY and the soon to be released ENDF-B/VI silicon cross sections[1]. Also provided in Table 2 is an estimate of the tissue dose per unit fluence.

The present LSL-M2 analysis is deficient only in the lack of a validated covariance matrix for the transport calculations. Currently, no information similar to the work of Maerker et. al. (14)

---

[1] Since ENDF-B/VI has not officially been released, these results must be considered preliminary. But, this new kerma calculation or one very similar is likely to replace the table in E722.

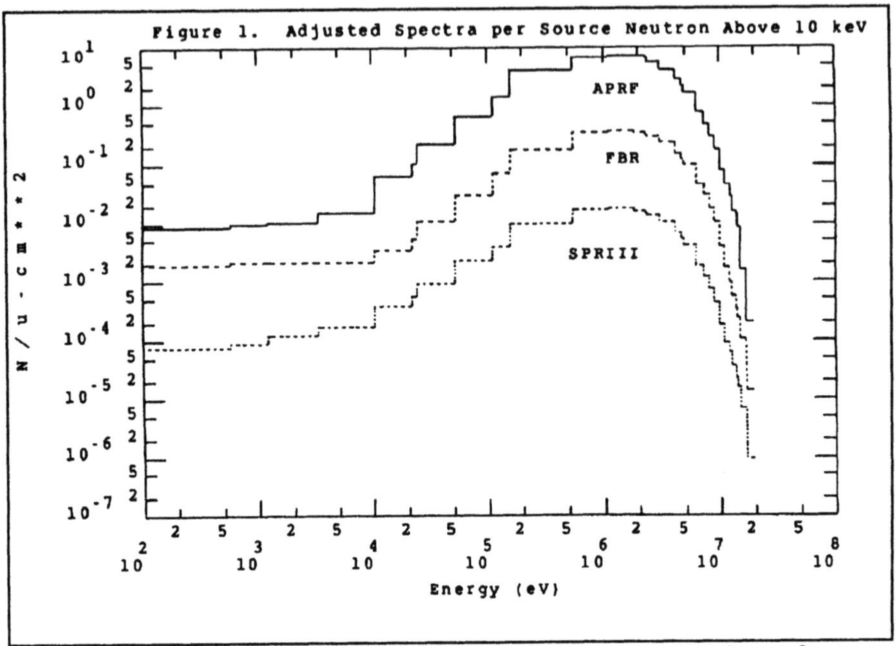


**Figure 1.** APRF is offset by a factor of 15 - SPR-III by a factor of 0.05.

for calculated spectra at the Oak Ridge Reactor Pool Side Facility is available for fast reactor transport calculations. However, it is known that the activation data dominates the adjusted spectrum in those energy regions where there is good cross section coverage and the fluence spectrum's covariance matrix only drives the adjustment in those regions where there is not good coverage (6). Since each of the reported spectra have high quality activation data covering all energies greater than 10 keV, the reported spectral indexes and hardness parameters should be valid even though the fluence covariance matrices are not yet known. Confidence in the output uncertainties suffers most from the lack of a validated fluence covariance matrix since they are mainly dependent upon the input uncertainties and to a lesser extent upon the covariance matrices.

## CONCLUSIONS

When taken in conjunction with references (6) and (7), this work has shown that the LSL-M2 code with it's utilities can be effectively used to obtain damage related parameters required for the testing of electronic devices. However, without an evaluated covariance matrix for the transport calculation, it is not possible to take full advantage of the philosophical advantages of the least

Table 2
Damage Parameters and Uncertainties After Adjustment

| Response Function | APRF Value | ± % | SPR-III Value | ± % | FBR Value | ± % |
|---|---|---|---|---|---|---|
| Spectral Index | | | | | | |
| Φ>0.01 / Φ>3 MeV | 8.29 | 10.1 | 8.40 | 4.5 | 7.30 | 7.7 |
| Uncertainties reflect input and adjustment uncertainties. | | | | | | |
| Hardness Parameter | | | | | | |
| ASTM E-722 | 0.986 | 2.8 | 0.960 | 1.3 | 1.005 | 2.4 |
| SNL-NJOY | 0.899 | 2.4 | 0.879 | 1.2 | 0.917 | 2.1 |
| Uncertainties reflect input, adjustment and normalization uncertainty. Uncertainties in the groupwise response function are not known and are not included in the above estimates. | | | | | | |
| Tissue Dose | | | | | | |
| Henderson | 2.45-9 | 2.4 | 2.44-9 | 1.1 | 2.53-9 | 1.9 |
| DNA Free-in-Air | 2.31-9 | 3.1 | 2.27-9 | 1.4 | 2.38-9 | 2.6 |
| Tissue kerma factors in 37 groups were taken from reference (11). The value cited has units of Rads(tissue)/n-cm$^{-2}$ above 10 keV. Uncertainties reflect input and adjustment uncertainties only. Uncertainties in the tissue response kerma functions are not included in the above analysis. | | | | | | |

squares methodology. Even when the activation data has good energy coverage, the covariance data drives the uncertainty estimates on the output parameters. While SAND-II does not provide uncertainty estimates on the output, it has a long history of effective use. It is the experience of this author who has used both codes for several years, that estimating the uncertainties on the SAND-II output is no more difficult than estimating the uncertainties on the LSL input. It must be conceded that when reasonable uncertainty data for the transport calculation can be obtained, the fact that LSL does not treat the cross section data as absolutely correct (which is an inherent part of the SAND-II method) is a definite advantage and LSL should be the code of preference. Otherwise, results from either code should provide useable results for the hardness testing application.

This work should not be considered as a replacement for the considerable data bases which are maintained by the respective facilities. This work was conducted in support of the work reported in ref. (7) and gives a consistent (LSL) basis for comparing the transistor data reported there.

**ACKNOWLEDGMENTS**

The author wishes to express his appreciation to C. Heimbach (APRF), D. Vehar (SNL), J. Kelly (SNL), P. Griffin (SNL), T. M. Flanders (NEL), and M. Sparks (NEL) for their technical assistance. All code packages for transport, spectrum adjustment, and cross sections were obtained or are obtainable from the Radiation Shielding Information Center, Oak Ridge National Laboratory. This work was supported by the US Army Test and Evaluation Command's Methodology Division and by the US Army Office of Research and Technology.

**REFERENCES**

1. ASTM Standard Methods E720-88, E721-88, E722-88, Annual Book of ASTM Standards, ASTM, Philadelphia, MD, Vol 12.02, 1988.

2. W. N. McElroy, S. Berg, T. B. Crockett and R. G. Hawkins, SAND-II - Neutron Flux Spectra Determination by Multiple Foil Activation - Iterative Method, Vol. I-IV, AFWL-TR-67-41, Air Force Weapons Laboratory, Albuquerque, NM, Aug 1967.

3. V. V. Verbinski, C. Cassapakis, R. L. Pease, and H. L. Scott, Simultaneous Neutron Spectrum and Transistor-Damage Measurements in Diverse Neutron Fields: Validity of $D_{Si}(E_n)$, NRL Memorandum Report 3929, Naval Research Laboratory, Washington, DC, March 1979.

4. F. G. Perey, Least Squares Dosimetry Unfolding: The Program STAY'SL, ORNL/TM-6062, Oak Ridge National Laboratory, TN, 1977.

5. F. W. Stallmann, LSL-M2: A Computer Program For Least-Squares Logarithmic Adjustment of Neutron Spectra, NUREG/CR-4349, Nuclear Regulatory Commission, Washington, DC, 1985.

6. J. G. Williams, R. O. Peach, A. M. Ougouag, W. W. Sallee, T. M. Flanders, and J. L. Meason, "Neutron Field Characterization for Hardness Testing of Electronics by Least Squares Adjustment Methods," Proc. Int. Reactor Physics Conf., Vol. III, p. 219-228, 1988.

7. M. H. Sparks, T. M. Flanders, J. G. Williams, J. G. Kelly, W. W. Sallee, M. Roknizadeh, and J. L. Meason, "Energy Dependence of Neutron Damage in Silicon Bipolar Transistors," IEEE Trans. Nucl. Sci., Vol. NS-36, No. 6, p. 1904-1911, 1989.

8. Private Communication from J. Kelly, Sandia National Laboratories, Spring 1989.

9. A. H. Kazi, C. R. Heimbach, R. C. Harrison, L. Schänzler, and F. W. Buchholz, Measurements of the Free Field Radiation Environment at

the APRD Reactor, Test Agency Report No. APG-MT-5279, Army Harry Diamond Laboratories, ATTN: DELHD-NT, Adelphi, MD, July 1979.

10. R. F. Sartor, Monte Carlo Determination of the Neutron-Gamma Spectrum Behind Cadmium Loaded Polyethylene Slabs Irradiated by the Sandia Pulse Reactor III, Master's Thesis, Texas A&M University, College Station, TX, May 1986.

11. D. E. Bartine, J. R. Knight, J. V. Pace III, and R. W. Roussin, Production and Testing of the DNA Few-Group Coupled Neutron- Gamma Cross -Section Library, ORNL/TM-4840, Oak Ridge National Laboratory, TN, 1977.

12. T. M. Flanders and M. H. Sparks, "Monte Carlo Calculations of the Neutron Environment Produced by the White Sands Missile Range FBR," Proc. Int. Reactor Physics Conf., Vol. II, p. 473-482, 1988.

13. Private Communication from P. Griffin, Sandia National Laboratories, June 1990.

14. R. E. Maerker, M. L. Williams, B. L. Broadhead, J. J. Wagshall and C. V. Fu, Revision and Expansion of the Data Base in the LEPRICON Dosimetry Methodology, EPRI NP-3841, Electric Power Research Institute, Palo Alto, CA, 1985.

# NEUTRON FLUX CALCULATIONS ON FAST BREEDER REACTORS WITH THE PROGRAMME DLS

Kl. Müller, U. Quade

Interatom GmbH

Bergisch Gladbach 1, Federal Republic of Germany

## ABSTRACT

The main objectives of shielding on Liquid Metal Fast Breeder Reactors are discussed.

The two-dimensional shielding code DLS is described. Within material-filled regions diffusion theory is used, while cavities are treated using the line-of-sight-methode.

The calculations of the SNR-300 reactor are described in view of radiation damage on in-vessel components.

Parameter studies of the European Fast Reactor in view of the special problems of pool-type reactors like secondary sodium activity, instrumentation and radiation damage are reported.

With regard to shielding the main differences of fast reactors and thermal reactors are

- the spectrum of leakage neutrons
- primary cooling without moderator effect and
- the power density.

Fig.1 shows the leakage spectrum of a thermal reactor compared to that of a fast breeder reactor in the core and in the blanket. The absence of a moderator in a LMFBR enhances the fast neutron leakage. The energy of the neutrons which determine the shield penetration lies in the intermediate region between 10 and 500 KeV. In Table 1 are power densities of LMFBRs compared with these of thermal reactors. Due to the greater surface area to volume ratio of the smaller fast breeder reactor cores the leakage rate is higher. The leakage neutrons are about 5 % of the fission neutrons for LMFBRs and only about 1 - 2 % for PWRs. As a consequence of this the primary shield of a LMFBR has to give a high attenuation of intermediate and thermal neutrons.

The discrete ordinates and the Monte Carlo methods represent the most rigorous solutions for neutron transport problems. But the application of these methods on the region of a whole reactor system yields different problems, especially in computer capabilities and calculation time. The limitation of diffusion theory arises from the assumption of nearly isotropic angular flux, which is realistic only in the case of weak absorption.

But in fast reactor shielding, where in a large sodium pool the intermediate and low energy fluxes are nearly isotropic, the diffusion theory is a fair approximation. The application of conventional diffusion programs is not possible when important cavities are included in the shield. To solve this kind of shielding problem the programme DLS has been developed at Interatom.

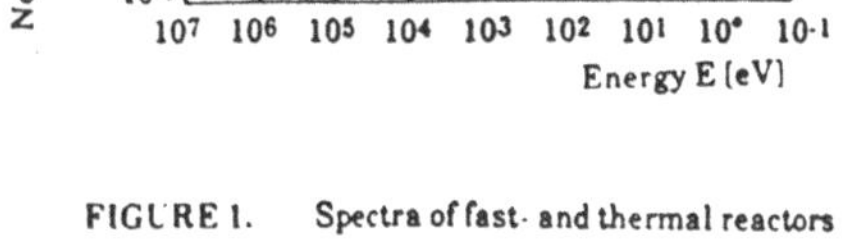


FIGURE 1. Spectra of fast- and thermal reactors

| | HTR | PWR | Fast Reactors | | |
|---|---|---|---|---|---|
| | | | KNK II[1] | SNR-300 | EFR |
| Average Power Density [kW/L] | 5 | 80 | 150/440 | 320 | 220 |

1) Driver-/Test Zone

TABLE 1. Power densities of fast- and thermal reactors

## The Programme DLS

The two-dimensional (r, z) diffusion code DLS (1,2) is a proven economic tool to study neutron flux propagation in LMFBR. An important option of this code is the calculation of neutron transport in cavities by the "line-of-sight" method. As is well known, the numerical solution of the diffusion equation is based upon balance formula for all volume elements containing terms of source, absorption and leakage.
The net current $J_{0,1}$ between the material filled volume elements 0 and 1 (see Fig. 2a) is described by the term

$$J_{o,1} = \frac{\Phi_o - \Phi_1}{\frac{\Delta r_o}{2 D_o} + \frac{\Delta r_1}{2 D_1}}$$

where

| | |
|---|---|
| $\Phi_0, \Phi_1$ | flux densities corresponding to the element 0 and 1 respectively |
| $\Delta r_0, \Delta r_1$ | radial mesh with |
| $D_0, D_1$ | Diffusion coefficient of the material in element 0 and 1 respectively |

At the boundary of a cavity the terms for the currents entering and leaving must include the geometrical properties of the cavity (Fig. 2b). The net current $J_0$ between the material filled element 0 and the cavity is given by the relationship

$$J_o = \lambda_o \cdot \Phi_o - \frac{1}{F_o} \sum_{i=1}^{n} \lambda_i \cdot \Phi_i \cdot F_i \; p_{i \to o}$$

where

| | |
|---|---|
| $\lambda_0, \lambda_i$ | leakage factor, general $\lambda = 0.25$ is used (isotropic flux density) |
| $\Phi_0, \Phi_i$ | flux density |
| $F_0, F_i$ | cavity surface area |
| $p_{i \to o}$ | view factor, i. e. the probability that a neutron leaving $F_i$ enters $F_0$ |

The view factor $p_{i \to o}$ represents the probability that neutrons leaving the cavity surface of an element i enter the element 0. They are identical to the angle factors, which are known from heat radiation calculations (3). There are five different cases for the determination of view factors as shown in Fig. 2c. The general formula for the view factors is

$$p_{i \to o} = \frac{1}{\pi \cdot F_o} \int_{F_o} \int_{F_i} \frac{\cos\theta_o \cdot \cos\theta_i}{\rho^2} \cdot dF_o \cdot dF_i$$

where

| | |
|---|---|
| $\rho$ | distance from a emitting element $dF_0$ to a target element $dF_1$ |
| $\Theta_0, \Theta_i$ | angles formed by the respective normals and the connecting line $\rho$ between the elements. |

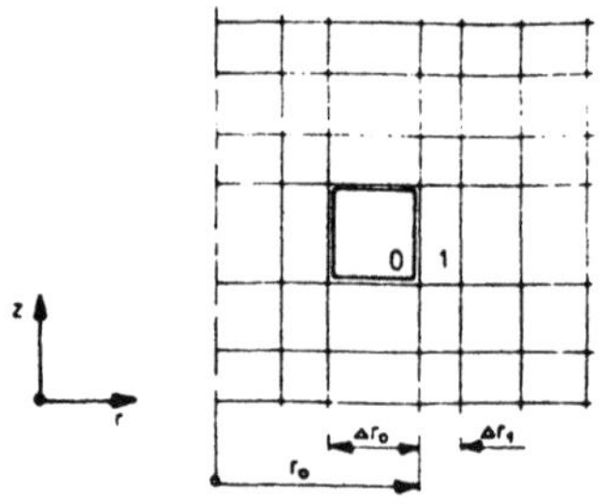

FIGURE 2A.
Treatment of material filled regions in DLS

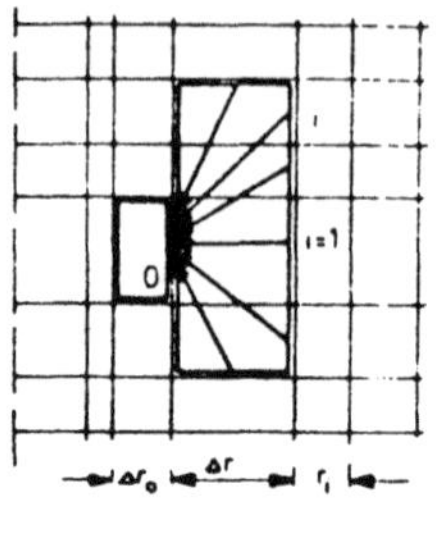

FIGURE 2B.
Treatment of cavity boundaries in DLS

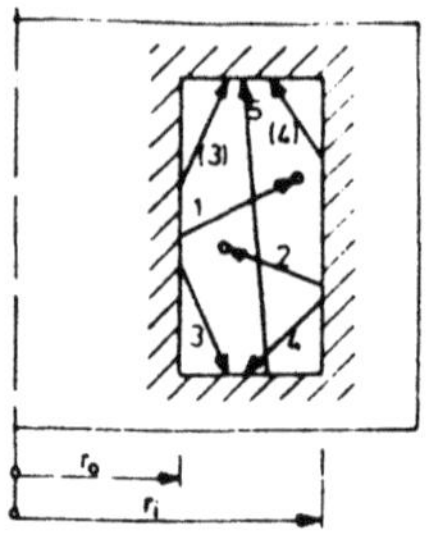

FIGURE 2C. Types of view factors

## Shielding Calculations for the SNR-300

The neutron flux distribution inside the reactor vessel of the SNR-300 (4) has been calculated by the DLS code. The three important objectives of in-vessel shielding are the limitation of

- radiation damage flux on components and structures
- nuclear heating
- activation of coolant and structures.

The radial reflector serves as the first shield between the core and the radial structures. It is made of 96 hexagonal steel subassemblies and located outside the radial blanket zone. The attenuation of the total neutron flux in the blanket is nearly an order of magnitude.
Due to the high amount of sodium around the core no further shielding inside the vessel is required. Fig. 3 shows the real geometry of the reactor vessel with internals, who has to be simplified in r,z-geometry for a DLS calculation. Fig. 4 gives the DLS model which includes in axial direction the region from the bottom of the vessel up to the cover gas plenum and in radial direction up to the guard vessel.
The inner boundary condition is given by the neutron flux densities on the surfaces of the radial- and axial breeding zones. The calculation were performed in a 26-energy group structure. The basic nuclear data originated from ENDF/B-IV data file.
Fig. 5 shows isolines of the total neutron flux density.
Further results are given concerning radiation damage. Radiation damage to a material depends on the fast neutron fluence, the neutron energy spectrum and the probability of a neutron causing an atom to be displaced from the lattice (5).
The neutron flux above 0.1 MeV is widely used in estimates of the lifetime of structural components. To calculate the atomic displacements the well known NRT-Standard model (6) has been used.
Another radiation induced phenomenon causing metal embrittlement is helium production by the (n, α) reaction occurring with the boron impurity in stainless steel and due to the two-step reaction Ni 58 (n, γ) Ni 59 (n, α) Fe 56 (6,7).
The B10 content has been assumed to 10 ppm, while analysis result contents between 1 - 13 ppm B10 in the used stainless steel. Nickel content

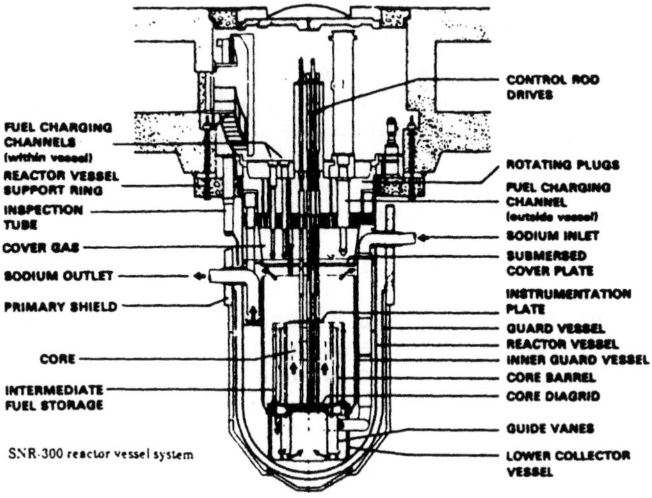


FIGURE 3. SNR-300 reactor vessel system

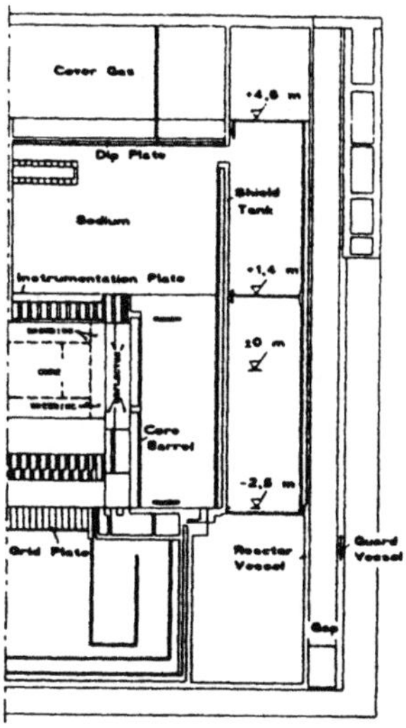


FIGURE 4. SNR-300 DLS-calculation model reactor vessel with internals

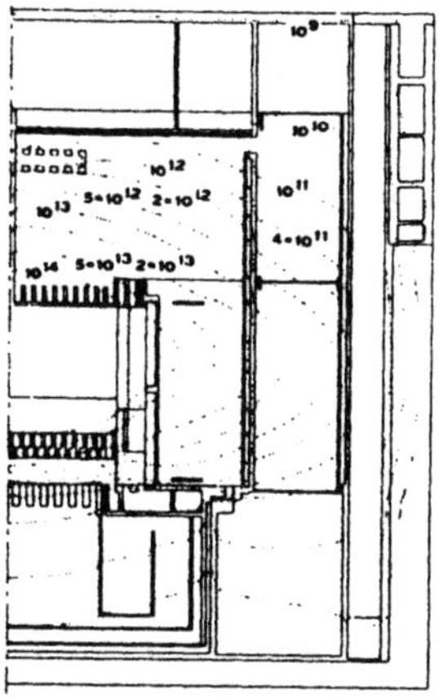


FIGURE 5. SNR-300 total neutron flux isolines inside reactor vessel

| | Fast Fluence E ≥ 0 1 MeV (cm⁻²) | DPA (-) | Helium[1] Concentration (atoms cm⁻³) |
|---|---|---|---|
| Instrumentation Plate | 6 0 E + 21 | 2 10 | 2 0 E + 18 |
| Core Barrel | 1 0 E + 22 | 3 50 | 9 0 E + 17 |
| Shield Tank | 3 0 E + 20 | 0 13 | 8 0 E + 17 |
| Grid Plate | 2 0 E + 19 | 8 5 E -03 | 4 0 E + 16 |
| Reactor Vessel | 2 0 E + 18 | 1 3 E -03 | 6 0 E + 16 |

1) Nickel Content 11 %
B10 Content 10 ppm in stainless steel 1 4948

**TABLE 2. SNR-300 radiation damage data on in-vessel components for exposure time $10^5$ h**

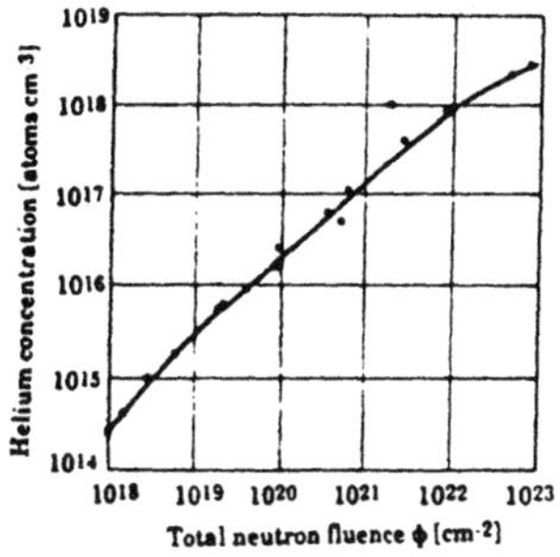


FIGURE 6. Helium concentration as function of total neutron fluence

The B10 content has been assumed to 10 ppm, while analysis result contents between 1 - 13 ppm B10 in the used stainless steel. Nickel content in the used stainless steel is 11 %.
The calculated values of fast fluence, displacements per atom (DPA) and helium content at main components after $10^5$ h exposure time are compared in Table 2.
The produced helium amount shows a good correlation to the total neutron fluence as shown in Fig. 6.

## Calculations for the EFR

The abbreviation EFR means: European Fast Reactor - this is the common project of several European partners and the development of all national projects are included in the layout of the EFR, mainly the projects SPX from France, CFR from Great Britain and SNR from DeBeNe.
The following presentation is based on the "First Consistent Design", with references to improvements which may be practicable and will be realized in the "Consistent Design".
Fig. 7 shows the EFR-Pool. The characteristics are: the compact pool with a diameter of about 17 m and an "In-vessel Fuel Store (IVS)" with 127 spent fuel positions inside the outer diameter of the gridplate of about 8 m. The subassembly lattic pitch is 190 mm.
Fig. 8 shows a cross-section of the core and the subassemblies around the core. Due to the compact pool it is necessary to have a very effective shielding around the core and around the store (to reduce fissions inside the store and to limit the secondary sodium activity in the Intermediate Heat Exchangers (IHX). This can be realized by the use of special shielding elements with a high content of $B_4C$. At the top of Fig. 8 four types of shielding elements with $B_4C$ are shown. There are 19 pin or 7 pin bundle hexagonal elements, 7 pin bundle round elements and round block elements. In all cases granules or solid blocks are in discussion, dependent on this the $B_4C$-amount varies from 41 % to 56 %.
Two variants of store are in discussion: a forced convection cooled store and a natural convection cooled store.
The presented flux calculations have been performed for the forces convection cooled store because this is the preferred option at this time.

Fig. 9 shows the calculation model. The neutron flux calculations were done with the above explained DLS code. The pool geometry is divided in about 16,000 meshes. The boundary fluxes outside the fissile zones (core and blanket and store) are taken from core calculations.

The main tasks of the shielding are:

- to decrease the fission sources in the store,
- to decrease the activation of the secondary sodium in the IHX,
- to balance the shielding in radial and axial direction,
- to protect the sensitive structures against neutrons, so that the material damage is acceptable.

An additional problem is the location of the neutronic instrumentation.

The main results of the performed calculations are:

- The secondary sodium activity is less than 20 Bq x $cm^{-3}$, that means that the steam generator building can be classified as supervised area (D $\leqq$ 7.5 µSv/h).
- The integrated He-production in the lower part of the "Above Core Structure (ACS)" is about .007 atoms ppm, this will require to use reduction factors in stress analysis.
- The DPA-value in the upper plate of the diagrid is about 1 over the lifetime of the reactor - this is acceptable without any reservations.
- The maximum equivalent thermal flux at the location of the neutron chambers in the ACS is about $10^8$ n x $s^{-1}$ x $cm^{-2}$ ($>10^9$ n x $s^{-1}$ x $cm^{-2}$ is required for the neutron chambers - this will be possible by a reduction of the thickness of axial shield in inner core-zone - additional studies are not finished yet).
- The maximum thermal flux at the location of the delayed neutron monitors in the IHX is about 100 n x $s^{-1}$ x $cm^{-2}$, which seems to be acceptable.

Fig. 10 shows isolines of the total flux of the "First Consistent Design" and Fig. 11 the sodium activation rates. It is clear to see the axial streaming and the streaming around the subassemblies. Also it can be seen the increase of the rates due to fission sources in the store. The graph of the fluxes and rates are very similar next to the core and around the core but very different outside the store and inside the IHX due to the different neutron spectra in these zones. The values at the right side of Fig. 11 show the percentage activation rates of each meter height inside the IHX.
The specific rates at the outer side of the IHX at level core-midline and level above subassembly (about 3 m higher) have the same values. But the integral rates (integral across the radial volume of the IHX) are very different at the same levels up to a factor of 3 due to different neutron spectra.
It has to be pointed out that at the bottom of the IHX inside the reservoir there is 100 % of secondary sodium, while in the tube-zones the value is less than 30 %.

An outlook on the "Consistent Design" is:

- The gap in the axial region of the steel reflector will be closed.
- The $B_4C$ amount inside the shield will increase from 40 % up to 56 %.
- The in-vessel store will be located outside the shield.
- The axial thickness of $B_4C$ inside the inner core zone will be reduced.

With these provisions we hope to decrease the absolute values of the sodium activation rates by a factor of 5 and the shield around the core is well balanced in axial and radial direction.

Fig. 7: EFR Core and Primary System, Cross Section

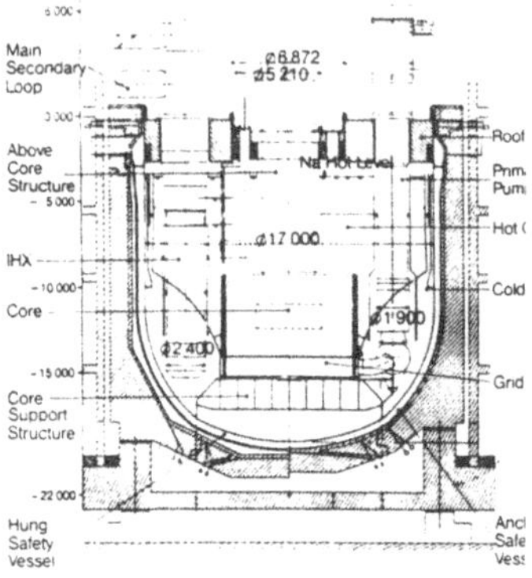


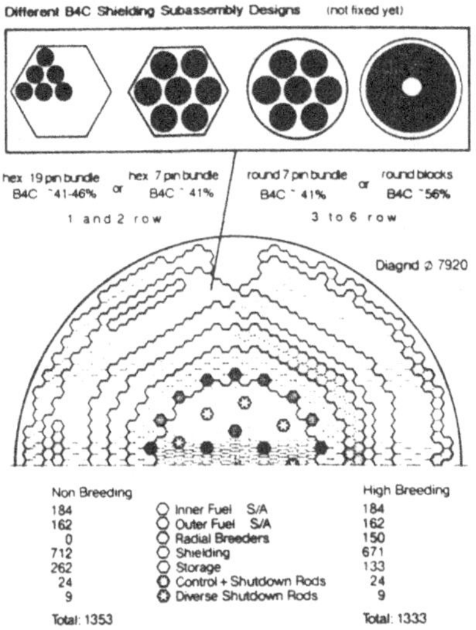


Fig. 8: EFR Core Layout

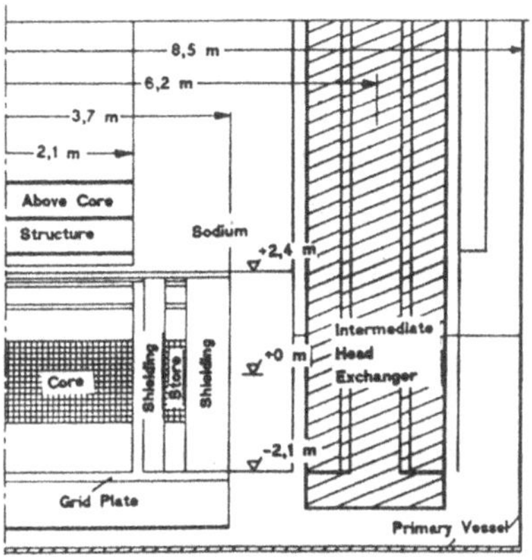


Fig. 9: EFR DLS Calculation Model
(Forced Convection Cooled Core Version)

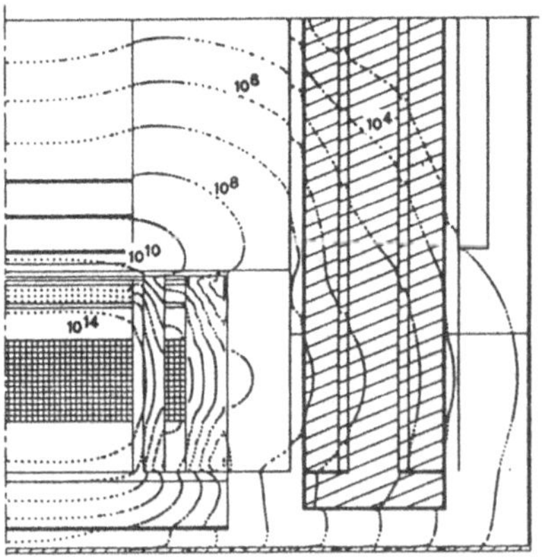


Fig. 10: EFR Total Neutron Flux Isolines
($s^{-1} \times cm^{-2}$)

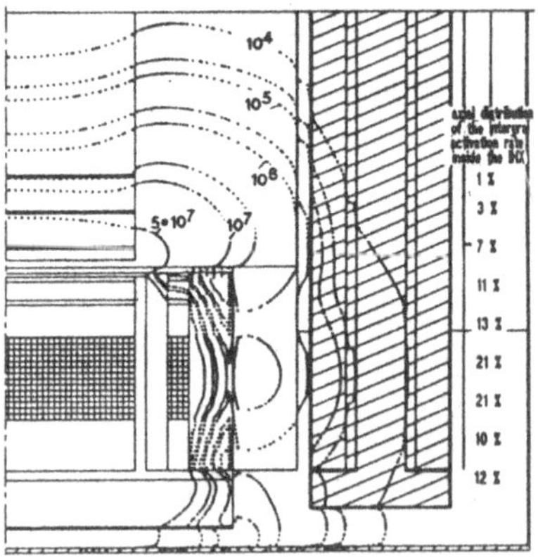


Fig. 11: EFR Isolines of the Na 24
Activation Rate ($s^{-1} \times cm^{-3}$)

## References

[1] H. Vossebrecker
Two-dimensional shielding calculations using a combination of diffusion theory and line-of-sight method"
Atomkernenergie, Bd. 19 (1972). Lfg. 1, S. 14 - 16

[2] Amin, E. et al.
Intercomparison of different methods to calculate neutron transport along sodium ducts
Paper E.5 of the International Conference on Reactor Shielding
Paris 9th - 13th October 1972

[3] Sparrow, E. M., Cess R. D.
Radiation Heat Transfer
Books/Cole Publishing Company Belmont, California
(Revised Edition)

[4] Nuclear Power Plant Kalkar (SNR-300)
Nuclear Technology, Vol. 8, Sep. 1987

[5] Etherington, E. W.
The calculation of atomic displacement dose in fast reactors
TRG-Report 2805(D), April 1976

[6] Norget, M. J. and others
Proposed standard for calculating displacement dose rates
Paper to IAEA specialists' meeting on radiation damage units for ferritic and stainless steels
Seattle, 31 October 1972

[7] Birss, I. R.
Helium production in reactor materials
Journal of Nuclear Materials 34 (1970) 241 - 249

[8] Simons, R. L.
Helium production in fast breeder reactor out-of-core structural components
ASTM STP 683, J. A. Sprague and D. Kramer, Eds., American Society for Testing and Materials, 1979, pp. 365 - 379

# USES AND ABUSES OF ADJUSTMENT

## A CRITICAL REVIEW OF COMMON PRACTICE

J.J. Wagschal and Y. Yeivin

The Hebrew University of Jerusalem, Israel

We begin with an elementary derivation of the "classical" adjustment prescriptions, which is based on geometrical considerations in the parameter space. This leads to the complete prescriptions, and helps to elucidate what Adjustment is all about. The range of validity of linearized adjustment is considered next. The specific combination of the relevant quantities which should be "negligible" in order for such adjustment to be meaningful is derived. We then deal with the mutual consistency of the input data needed in actual applications. A criterion for the consistency of the input responses and given parameter library is established, and a practical method to order the responses by their consistency is discussed. The proposed analysis identifies the less consistent responses and any questionable parameters. Finally, we show why in using mock-ups to improve predicted responses and to reduce uncertainties in them, adjustment is definitely superior to bias-factors.

---

## INTRODUCTION

In the analysis of experimental data in general, and in that of measurements of so-called integral quantities - or responses - in complex experimental facilities in particular, a major concern should always be the utilisation of these data to the greatest possible advantage. In typical analyses the measured responses are compared with their respective calculated values. But once the measurements are judged to be "in agreement" with the corresponding calculations, that, quite often, is where the matter rests. However, the purpose of any response measurement cannot certainly be just to "validate" the nominal values of the parameters which are the arguments of the calculated responses. It should rather be to improve the set of these parameters, i.e. to modify their given values and to reduce the uncertainties in them, and thus to produce a more reliable set of parameters on which future calculations could be based. This exactly is what adjustment purports to accomplish.

Adjustment has been amply vindicated by its extensive application to reactor-core design. It has been also successfully applied to LWR PV surveillance dosimetry.[1] Yet the Dosimetry community seems not to be sufficiently familiar with the subject. We therefore propose to review adjustment briefly and then to warn against some common pitfalls in its application.

## ADJUSTMENT FOR PEDESTRIANS

Consider a given parameter library $p \equiv \{p_n\}$ and a measured response r. The calculated value of the response, $\bar{r}=\bar{r}(p)$, is generally different from its measured value. The problem is how to make use of r to improve the library p. The measured response determines an (N-1)-dimensional manifold in the N-dimensional parameter space. Each point p' on this manifold represents a library which reproduces r. The given library is represented by the point p, which is obviously outside the manifold. Then, by what we may term "the principle of least adjustment," the manifold point closest to the point p represents the optimally adjusted library. Consequently, to find this felicitous library we have to minimize $|p'-p|^2$, subject to the condition that the point p' lie on the manifold defined by $\bar{r}(p')=r$.

A major difficulty in solving this conditional-minimum problem is that there generally is no tractable mathematical expression for the manifold on which p' must lie. To overcome it we may replace the manifold by its tangent hyper-plane at the point p' nearest to point p. In other words, in minimizing $|p'-p|^2$ we could constrain p' to the tangent hyper-plane rather than to the manifold itself.

Further, if the points p and p' are close enough, then

$$\bar{r}(p') \simeq \bar{r}(p) + \frac{\partial \bar{r}}{\partial p}(p'-p) \, , \tag{1}$$

where $\partial \bar{r}/\partial p$ is the gradient of $\bar{r}$ at p, and the last term is a scalar product of the two vectors. By definition $\bar{r}(p')=r$, so that the approximation expresses a linear relationship satisfied by the components of p'. In fact, letting p' vary under this constraint, we obtain a linear manifold which is practically the tangent hyper-plane we were talking about.

Let the adjustments of the individual parameters, the sensitivities of the calculated response to each parameter, and the deviation of the calculated response from its measured value, be respectively denoted by

$$x_n = p'_n - p_n \,, \quad s_n = \partial\bar{r}/\partial p_n \,, \quad d = \bar{r}(p) - r \,. \tag{2}$$

Also let s be a *row* vector. Then Eq.(1) reduces to

$$sx + d = 0 \,. \tag{3}$$

We have transformed the unknown vector p' to x=p'-p. In the x-space the origin represents the given library and Eq.(3) defines the hyper-plane tangent to the manifold $\bar{r}(p+x)=r$ at its point closest to the origin.

Our problem now is to find the vector x which satisfies sx+d=0 and minimizes the sum of squares $|x|^2$. Although this problem is so simple that its solution can be stated right away (x is parallel to s), we still derive the formal solution in order to illustrate the method which should apply in solving the problem of adjusting a library by several responses.

The way to handle conditional-minimum problems is the method of Lagrange multipliers. It reduces any conditional-minimum problem to that of an unconditional minimum. Our particular problem reduces to the unconditional minimization of the form

$$R(x) = x^\dagger x + 2\lambda(sx + d) \,, \tag{4}$$

where $2\lambda$ is the Lagrange multiplier (the factor 2 is intruduced for convenience). We accordingly have to solve the N equations

$$\frac{\partial R}{\partial x} = 2(x^\dagger + \lambda s) = 0 \,, \tag{5}$$

which together with sx+d=0 constitute N+1 equations for our N+1 unknowns x and $\lambda$. Thus

$$x = -\lambda s^\dagger \,, \quad sx + d = -\lambda ss^\dagger + d = 0 \;\longrightarrow\; \lambda = d/ss^\dagger \,,$$

and, finally,

$$x = -(d/ss^\dagger)s^\dagger \,. \tag{6}$$

When several, say I, measured responses are given by which a parameter library is to be adjusted, the generalization of our foregoing result is straightforward. We should then consider the intersection of the I manifolds generated in the x-space by the given responses, and find the point on it which is closest to the origin. Again, we replace the intersection of the manifolds by that of their respective tangent hyper-planes, i.e. by the linear manifold Sx+d=0, the matrix-equation analog of Eq.(3), where S is now the sensitivity *matrix* of the responses, $S_{in}=\partial\bar{r}_i/\partial p_n$, and d the *vector* $d_i=\bar{r}_i(p)-r_i$. The formal solution of the problem at hand is analogous to what we have done for I=1, and Eq.(6) generalizes to

$$x = -S^{\dagger}(SS^{\dagger})^{-1}d \ . \tag{7}$$

This may seem just fine, except that we should by no means treat all parameters equally: a parameter associated with a large uncertainty is less reliable than one in which the uncertainty is small. Therefore, we should rather consider the parameters in terms of their standard deviations, and accordingly the weighted adjustments $\xi_n = x_n/\sigma_n$. Let $\Sigma$ denote the diagonal matrix of which the elements are $\sigma_n$; then $x=\Sigma\xi$. The $\xi$-space manifolds generated by the given responses are $\bar{r}_i(p+\Sigma\xi)=r_i$, and the intersection of their respective tangent hyper-planes is defined by $S\Sigma\xi+d=0$. Without much ado we can right away state that the vector $\xi$ which minimizes $|\xi|^2$ and satisfies $S\Sigma\xi+d=0$ is $\xi=-\Sigma S^{\dagger}(S\Sigma^2S^{\dagger})^{-1}d$. Note that $\Sigma^2$ is the uncertainty matrix C associated with the parameter library. Since we have tacitly assumed that the parameters are uncorrelated, C is indeed diagonal, and its diagonal elements are the variances $\sigma^2$ of the parameters. And if we now left-multiply both sides of the equation by $\Sigma$, we obtain the prescription for the actual adjustments,

$$x = -CS^{\dagger}(SCS^{\dagger})^{-1}d \ , \tag{8}$$

which expression, incidentally, is also the adjustment prescription even if the given parameters are correlated.

The adjusted library $p'=p+x$, with the x of Eq.(8), reproduces all the given responses. This requirement should be definitely relaxed. After all, uncertainties in the measured responses are always finite. We should indeed assign these responses appropriate weights and treat them accordingly. Their "true," adjusted, values r' are not necessarily their given values: $\bar{r}(p')=r'\neq r$. Returning to Eq.(1), subtracting r from both sides, and denoting $y=r'-r$, the linearity assumption becomes $y=Sx+d$. Then, similarly to the transformation from x to $\xi$, we consider the weighted response adjustments $\eta_i=y_i/\sigma_i$ (here the $\sigma_i$ are the standard deviations of the measured responses). With all that, adjustment, in the end, reduces to the minimization of $|\xi|^2+|\eta|^2$ subject to the condition $y=Sx+d$. It is now a straightforward matter to apply the method of Lagrange multipliers, and arrive at the definitive adjustment prescriptions[2]

$$x = -C_pS^{\dagger}(C_r+SC_pS^{\dagger})^{-1}d \ , \quad y = C_r(C_r+SC_pS^{\dagger})^{-1}d \ , \tag{9}$$

where the subscripted C's are the respective uncertainty matrices associated with the given parameters and responses.

## THE LIMITS OF LINEARIZED-RESPONSE ADJUSTMENT

Neat as the adjustment prescription may be, it still is just the solution of the linearized problem. The approximation of Eq.(1), on which the prescription is based, is a rather strong assumption. And yet many actual adjustments have resulted in rather small modifications of the parameters, so that the linearity assumption had been *a posteriori* justified. Even so, at the conclusion of every actual application of linearized adjustment, one should verify that the calculated responses *at p'* are about equal to the adjusted responses r'.

Anyhow, to study the domain of validity of the prescription based on $\bar{r}(p')=\bar{r}(p)+S(p'-p)$, we consider the next approximation,

$$\bar{r}(p') = \bar{r}(p) + S(p'-p) + \tfrac{1}{2}(p'-p)^{\dagger}T(p'-p) \ , \tag{10}$$

where T is the 3-index array $\partial^2\bar{r}_i/\partial p_m\partial p_n$, and the double summation implied in the last term runs over the parameter indices m and n. Of course, conventional adjustment is valid not only if the responses are quasi-linear, but also if their associated uncertainties are relatively large, or the deviations $\bar{r}-r$ small. We will indeed find out which specific combination of these quantities should be negligible, and with respect to what.

The adjustment problem now is, again, to minimize $|\xi|^2+|\eta|^2$, which in terms of x and y is

$$Q(x,y) = x^{\dagger}C_p^{-1}x + y^{\dagger}C_r^{-1}y \ , \tag{11}$$

subject not to y=Sx+d, but rather to

$$y = \tfrac{1}{2}\, x^{\dagger}Tx + Sx + d \ . \tag{12}$$

If we substitute this expression for y in Q(x,y), then the equations $\partial Q/\partial x=0$ lead, after linearization, to[3]

$$x = -\ (C_p^{-1}+S^{\dagger}C_r^{-1}S+d^{\dagger}C_r^{-1}\cdot T)^{-1}S^{\dagger}C_r^{-1}d \ , \tag{13}$$

where the dot multiplication (•) refers to summation over the response index i. If T=0, then indeed the remaining expression for x is identical to the linearized prescription of Eq.(9), which is not too difficult to demonstrate.

Evidently, the last term in the parenthesis is the key to understanding the quality of linearized-response adjustments. These are valid as long as the elements of the last matrix in the parenthesis are significantly smaller than the sums of the corresponding elements of the first and second matrices.

## CONSISTENCY OF PARAMETER AND RESPONSE INPUT DATA

The indiscriminate incorporation of every seemingly relevant experimental-response data in the input file of a data-adjustment project might well result in a practically worthless adjusted library, a library which might not even improve the agreement between the calculated and measured values of some responses by which the library was adjusted.

It is evidently desirable to establish a criterion for the *joint* consistency of all input data in any adjustment problem. A common measure of "the quality of an adjustment" is the minimal value of the form Q(x,y), Eq.(11). This value is the so-called "$\chi^2$ (chi square) of the adjustment." Incidentally, the generalization of the conventional definition of $\chi^2$, relating to vector variates of independent components, to vectors x, of means $\mu$, associated with *non-diagonal* uncertainty matrices C, is

$$\chi^2 = (x-\mu)^{\dagger}C^{-1}(x-\mu) \ . \qquad (14)$$

The minimum of Q(x,y) reduces to an expression which involves only input data, and is the $\chi^2$ corresponding to the vector $d=\bar{r}-r$:

$$\min Q(x,y) = d^{\dagger}C_d{}^{-1}d \ , \quad C_d = SC_pS^{\dagger} + C_r \ . \qquad (15)$$

Thus it actually measures the agreement between the calculated and measured responses, or essentially the consistency of the latter with the given library.

It also measures the mutual consistency of the responses. So, to disqualify an entire input-data set because of a high value of $\chi^2$ might be premature, since even one response might degrade the set's overall consistency. It is then desirable to check whether only few responses are "irregular," while the others are consistent with the given library. Such an inquiry would, in principle, require the examination of all partial sets of the given array of responses with respect to their consistency with the given library. This, however, is out of the question: even for I=10, which is relatively small in this context, it would mean the examination of about 1000 response subsets. An inquiry such as we now discuss would be manageable only if a practical method could be devised to order the input responses by some sort of a "consistency index." One could then assign a $\chi^2$ value to the most consistent response, then to the subset of the first two responses in the consistency sequence, and so on. In this manner it would be possible to determine which responses to reject.

A method which has been successfully applied,[4] begins with the identification of the response which is the least consistent with the given library. To accomplish this, we eliminate the first of the I given responses, and evaluate $\chi^2$ for the remaining I-1 responses, which we denote by $\chi^2(I-1|1)$. The first response is then returned to the response set, the second response is eliminated, and $\chi^2(I-1|2)$ is evaluated. In this manner the remaining $\chi^2(I-1|i)$ are successively evaluated. The response of which the elimination yields the lowest $\chi^2(I-1)$ value is obviously the least consistent. This procedure is applied to the remaining I-1 "more consistent" responses, and the second least consistent response is determined. The procedure is then repeatedly applied to the successive, and diminishing, partial response sets, until the complete consistency sequence is established. Such an analysis should identify less-consistent responses, and would also indicate the level of consistency of all response subsets along the consistency sequence.

We have yet to consider the possibility of the existence of irregular parameters, and discuss their identification. Once we determine $\chi^2$ for any response subset, it is rather trivial to actually adjust the library by these responses. Thus, as we proceed step by step along the consistency sequence, we may also examine the actual individual parameter adjustments induced by the respective response subset, and find out which parameters vary more than the others, and by how much. We expect the adjustments by the more consistent subsets to be marginal, and that only with the addition of each of the less-consistent responses would substantial adjustments occur in certain parameters. The proposed analysis, then, allows the identification of the less consistent responses and that of the questionable parameters.

## BIAS FACTORS MAY BE HARMFUL TO YOUR PREDICTIONS

Mock-up experiments are used to improve the nominal values of, and to reduce the uncerainties in, calculated responses of planned power reactors. The problem is mathematically identical to that of extrapolating surveillance dosimetry information to predict radiation damage in reactor pressure vessels. This, of course, is what adjustment is all about. Reducing uncertainties is just as important as improving nominal values: any investment in setting up critical-assembly experiments is only justified by the substantial cost savings expected in the construction of the planned reactor; these expected savings should derive from reduced design safety margins, which in turn would result from reduced uncertainties in predicted reactor responses. To complement the adjustment prescriptions, we now add the expression for the (positive-definite) uncertainty matrix associated with the adjusted parameters, $p'=p+x$, which is just as easy to derive,[2]

$$C_{p'} = C_p - C_x \ , \quad C_x = C_p S^{\dagger} C_d^{-1} S C_p \ . \qquad (16)$$

Evidently, the variance of any predicted response evaluated with $C_{p'}$ is smaller than the corresponding variance evaluated with $C_p$.

The traditional bias-factor method used to be applied to improve predicted-response values. For whatever reasons it still is, sometimes. To estimate the uncertainties in these values, the following analysis was proposed.[5,6] The prescription for the "improved" predictions is then

$$r_i' = \bar{r}_i \times (r_{oi}/\bar{r}_{oi}) \ , \qquad (17)$$

where the subscript-zero quantities refer to the mock-up experiments, and from which it trivially follows that

$$\frac{\delta r_i'}{r_i'} = \frac{\delta \bar{r}_i}{\bar{r}_i} + \frac{\delta r_{oi}}{r_{oi}} - \frac{\delta \bar{r}_{oi}}{\bar{r}_{oi}} = \frac{\delta r_{oi}}{r_{oi}} + \sum_{n=1}^{N} (S_{in} - S_{oin}) \frac{\delta p_n}{p_n} . \qquad (18)$$

Here S and $S_o$ are *relative* sensitivity matrices. And so, since $\langle(\delta r_i'/r_i')(\delta r_j'/r_j')\rangle$ are the elements of the *relative* uncertainty matrix C', and since the parameters p and the mock-up measurements $r_o$ are mutually independent, we immediately obtain

$$C' = C_o + (S-S_o)C_p(S-S_o)^{\dagger} \simeq \Delta S C_p \Delta S^{\dagger} , \qquad (19)$$

where $\Delta S = S - S_o$ and, for the sake of simplicity and clarity, we assume that the experimental mock-up uncertainties, $C_o$, may be neglected. The difference between this result and the adjustment-method C' is manifest: the bias-factor-method C' employs the uncertainty matrix $C_p$ of the *given* parameter library, whereas (under the assumption $C_o \simeq 0$) the proper $C' = \Delta S C_p' \Delta S^{\dagger}$ derives from the uncertainty matrix of the *adjusted* library, the library adjusted by the very same integral data $r_o$ which were used to generate the bias factors. And further, insofar as the mock-up data are really good, the expected uncertainties in the adjusted parameters would be substantially smaller than those in the unadjusted library.

## REFERENCES


1. J.J. Wagschal, R.E. Maerker and Y. Yeivin, "Extrapolation of Surveillance Dosimetry Information to Predict Pressure Vessel Fluences," Trans.ANS 34, 631 (1980)

2. J.J. Wagschal and Y. Yeivin, "The Significance of Lagrange Multipliers in Cross-Section Adjustment," Trans.ANS 34, 776 (1980)

3. Y. Yeivin, "Parameter Adjustment by Non-Linear-Response Measurements," Unpublished note (1986)

4. Y. Yeivin, J.J. Wagschal, J.H. Marable and C.R. Weisbin, "Relative Consistency of ENDF/B-IV and V with Fast Reactor Benchmarks," p.182 in Nuclear Cross Sections for Technology, J.L. Fowler, C.H. Johnson and C.D. Bowman, Editors, Nat.Bur.Stand.(U.S.), Spec.Publ. 594 (1980)

5. T. Kamei and T. Yoshida, "Error Due to Nuclear Data Uncertainties in the Prediction of Large Liquid-Metal Fast Breeder Reactor Core Performance Parameters." Nucl.Sci.Eng., 84, 83 (1983)

6. J.J. Wagschal and Y. Yeivin, "Comment on 'Error Due to Nuclear Data Uncertainties...'" Nucl.Sci.Eng., 86, 121 (1984)

# REFERENCE DATA FILE FOR NEUTRON SPECTRUM ADJUSTMENT AND RELATED RADIATION DAMAGE CALCULATIONS

É. M. Zsolnay*, H. J. Nolthenius**, L. R. Greenwood***,
E. J. Szondi*

*Technical University of Budapest BME
H-1521 Budapest, Müegyetem rkp. 9, Hungary
**Netherlands Energy Research Foundation ECN
NL-1755 ZG Petten, Westerduinweg 3, The Netherlands
***Argonne National Laboratory ANL
Presently: Pacific Northwest Labs. POB 999,Richland,WA 99352

ABSTRACT

The REAL-88 interlaboratory exercise organized by IAEA resulted in a neutron metrology file (NMF-90) comprising problem dependent data for benchmark neutron fields, furthermore, nuclear data and computer programs for neutron spectrum adjustment and radiation damage parameter calculations for the service life asseessment of nuclear facilities. Calculation results of some experienced laboratories are also present. This paper describes and analyses the content of the neutron metrology file and outlines the most important problems and tasks to be solved in the field of radiation damage parameter calculations.

## 1. INTRODUCTION

IAEA Nuclear Data Section (NDS) has organized the REAL-80, REAL-84, AND REAL-88 interlaboratory exercises (1-3) in order to improve the quality of the results of radiation damage calculations for the service life assessment of nuclear facilities. The emphasis was concentrated on radiation damage (exposure) parameters for reactor pressure vessels and related nuclear technology. Improved damage parameters can be achieved by using good quality input data and proper calculation methods. The combined effort of the participants and evaluators of the exercises has contributed to the solution of numerous basic mathematical and physics problems in the neutron spectrum adjustment procedure for radiation damage purposes (1,2,4-8). The output of the exercises resulted in a neutron metrology file which is available at IAEA for distribution to interested laboratories. The neutron metrology file has been developed by a joint team from the Netherlands Energy Research Foundation, Petten (H. J. Nolthenius, W. L. Zijp), Technical University of Budapest (É. M. Zsolnay, E. J. Szondi), Argonne National Laboratory, Argonne (L. R. Greenwood), Physikalisch-Technische Bundesanstalt, Braunschweig (M. Matzke), and Oak Ridge National Laboratory (F. W. Stallmann).

The content of the neutron metrology file was discussed by a workshop at the IAEA Advisory Group Meeting on the Status and Requirements of Nuclear Data for Radiation Damage and Related Safety Aspects, held in Vienna, 19-22 September, 1989.

## 2. STRUCTURE OF THE NEUTRON METROLOGY FILE (NMF-90)

The structure of NMF-90 is shown by Fig. 1. It contains the following information in the form of four separate modules:

1. User's guide and help module
2. Reference data file for neutron spectrum adjustment and related radiation damage calculations (module BENCHMARK),
3. Relevant cross section libraries (module XSECLIB),
4. Software packages and utility programs.

The users have the freedom to request one or more modules from IAEA depending on their interests and tasks.

### 2.1 Reference data file

The reference data file has been developed in order to create consistent data sets for neutron spectrum adjustment and subsequent radiation damage parameter calculations, which can be considered as benchmarks for interested scientists in testing their relevant calculation procedures. Of course, the comparison is correct only when the complete REAL-88 input set is used in combination with the IRDF-85 cross section library.

On the other hand, certain data of the benchmark fields (eg input spectrum covariance matrices) can be used for estimating missing data in the calculations of similar environments.

The reference data file contains problem dependent input data for 6 different spectrum cases and also the solutions of some experienced laboratories for these spectra.

Table 1. Technical details for the input information of the REAL-88 benchmark spectra

| Data set | Description | Measured reaction rates | Covariances for measured reaction rates | Calculated input spectrum | Covariances for calculated input spectrum |
|---|---|---|---|---|---|
| ANO | Pressure vessel cavity of the Arkansas Power and Light Reactor (Arkansas Nuclear One-1). | 6 reaction rates (no subcadmium response) | available | 55 groups | 16 groups |
| PS1 | Oak Ridge Research Reactor, Pool Side Facility in the metallurgical irradiation exp. (simulated surveillance capsule). | 10 reaction rates | only variances | 37 groups | 37 groups (partly estimated) |
| PS2 | Oak Ridge Research Reactor, Pool Side Facility in the metallurgical irradiation exp. (¼T position in the simulated pressure vessel wall). | 6 reaction rates | only variances | 37 groups | 37 groups (partly estimated) |
| U35 | Fission spectrum of the U-235. | 23 reaction rates | only variances | 24 groups | 24 groups |
| CFR | Neutron spectrum in the centre of the Coupled Fast Reactivity Measurement Facility (CFRMF). | 23 reaction rates | only variances | 26 groups | 26 groups |
| RTN | Fusion simulation spectrum measured at the RTNS-II, a 14 MeV neutron source at the Lawrence Livermore Laboratory. | 12 reaction rates | available | 60 groups | good estimates 60 groups |

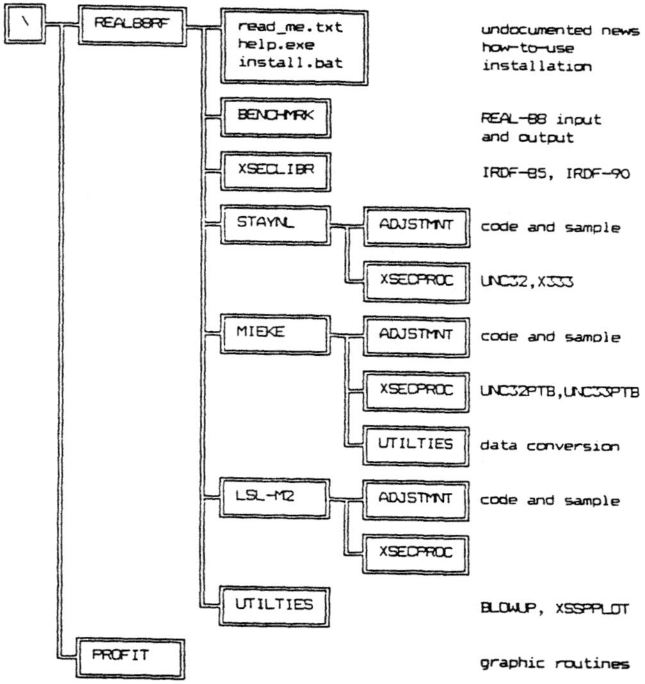


Figure 1.
The structure of the NMF-90

Problem dependent input data. In Table 1 information is given on the problem dependent input data sets (3). The collection comprises all the data (ie reaction rates and input spectra accompanied with their covariances) required for the neutron spectrum adjustment. The reaction rates can be considered to have negligible neutron self-shielding effects.

All detectors of the set PS1 have been irradiated inside a gadolinium cover, therefore the gadolinium total cross section values are also incorporated in the corresponding data set to facilitate the calculation of the cover shielding.

For all the data sets a detailed weighting neutron spectrum in 640 energy group structure has been made available.

Calculation results. A detailed discussion of the results of the REAL-88 exercise with a summary for each spectrum was prepared in the form of progress reports which were then distributed among the participants. The reference data file contains spectrum characteristic integral data (such as integral neutron fluence rate values above some specific energy limits, displacement and gas production rates in iron for the neutron spectra in question, etc), together with uncertainties as calculated by some experienced laboratories using the REAL-88 input data set. The users can then compare their output results with the ones of the reference file.

## 2.2 Cross section libraries

Two cross section libraries, IRDF-85 and IRDF-90, are given in the XSECLIB module of the NMF-90.

1. The International Reactor Dosimetry File IRDF-85, Version 2 (9) contains the cross sections and their uncertainty information to be used in the calculations on the REAL-88 reference spectra. The library has the ENDF-5 format.
   A simplified damage characterization for steel has been introduced in the REAL-88 tasks, ie the relevant damage parameters are calculated for pure iron.

2. The International Reactor Dosimetry File IRDF-90 (10) contains improved cross section and uncertainty information extended with several dosimetry reactions as compared with the IRDF-85 This library is written in the ENDF-6 format, which might cause difficulties in processing the data with the old computer codes interpreting the input only in the ENDF-5 format. The IRDF-90, which became available recently in restricted distribution, has not yet been widely tested by neutron metrology calculations.

## 2.3 Software packages and utility programs

Three software packages (STAYNL, MIEKE, and LSL-M2) recently developed by different laboratories have been made available for the users. Each package provides a complete neutron metrology calculation for radiation damage purposes: they start with cross section data processing, perform neutron spectrum adjustment by the generalized least squares method and terminate in radiation damage (exposure) parameter calculation.

For convenience of the users some utility programs independent of the software packages have also been included in the NMF-90.

The programs are available as standard FORTRAN-77 source files and they can be compiled either for running on PCs or on mainframe computers. (Syntactic extensions over the FORTRAN-77 are not present.) The distribution of executable (load) programs was avoided owing to the copyright restrictions of the runtime subroutine libraries.

The STAYNL module. This program package has been developed by the Institute of Nuclear Technics of the Technical University of Budapest (BME) and the Netherlands Energy Research Foundation (ECN) with the aim of producing "integrated software" for neutron metrology calculations. It means that one cross section processing code, one neutron spectrum adjustment and radiation damage calculation program and one plotting program are concatenated in the module and each program is running only once for one job.

The cross section library is processed by the UNC32 (11) and X333 codes. The latter one is a significantly extended version of the code UNC33 (11). The program X333 calculates weighted, group averaged, self-shielded cross section values and their relative covariance matrix in the user's energy grid, starting from the files No. 3 and No. 33 of the ENDF-5 or ENDF-6 format data sets. It calculates the damage and gas production cross sections as well. The output of the cross section codes can be used directly by the neutron spectrum adjustment code and utility programs.

For neutron spectrum adjustment and damage parameter calculation an updated version of the code STAYNL is used. It is a version of Perey's STAY'SL code (12) extended in its algorithm and modified in the input and output organization. The code STAYNL calculates the normalization factor in the least squares procedure (4), and it takes into consideration the cross-product [nonlinear = NL (5)] terms in derivation of the covariance matrix of the calculated reaction rates. The program calculates the requested damage (exposure) parameters related to the adjusted spectrum, and supplies them in the output together with other spectrum characteristic data. The whole set of the results is available for further computerized processing. As a sample problem, the AND task is evaluated.

The MIEKE module. This module has been developed at the Physikalisch-Technische Bundesanstalt, Braunschweig (13). It is not an integrated software package since it consists of several small programs which have to run sequentially.

The cross section data processing is done by a PTB modified version of the codes UNC32 and UNC33 (UNC33PTB), followed by numerous utility programs. Sometimes manual editing is necessary. The shielding effect of detector covers is taken into consideration by the utility program MFITIB. Radiation damage and gas production cross sections are also calculated. This program collection can handle only ENDF-5 format data at the moment, therefore, parts of the cross section library IRDF-90 containing ENDF-6 format data sections cannot yet be processed by it.

The neutron spectrum adjustment is done by the code MSITER (13). This program is a STAY'SL type code, performing the normalization in frame of a non-linear least squares procedure. Radiation damage parameters together with other spectrum characteristic data are also given in the output.

The programs of the module generate special data files for plotting using compiler specific program elements. Therefore the plotting programs can run only in the case where the compilation of the module has been done by the suitable FORTRAN compiler. As a sample problem, the PS1 task is evaluated.

The LSL-M2 module. This program package has been developed at the Oak Ridge National Laboratory (14). It consists of a cross section processing code, a neutron spectrum adjustment program, and several utility programs.

The cross section processing code requires a cross section library which deviates from the ENDF format and is included in the package.This library was generated from the IRDF-85 data file using the SAND-II 640 group structure for the cross section values and combining the covariance information for each detector reaction into a single upper triangular matrix with the energy group structure as given in the original ENDF covariance files. The processing code converts this information into the energy group structure requested by the user. The neutron spectrum is adjusted by the code LSL-M2 (14). This program performs a least squares adjustment in a logarithmic space. The main advantage of this code is that neutron spectra at several different locations can be processed simultaneously. The output contains spectrum characteristic integral data and damage parameters as well. The sample problem of the module is taken from the ORNL PSF HSST experiments.

The utility programs. The BLOW UP program (8) makes it possible for the user to generate a "physics based" fine group structure weighting spectrum from poor (few groups) spectrum information if the physics characteristics of the neutron field to be investigated are known.

The program XSSPPLOT accomplishes the plotting of cross section data, neutron spectra, the corresponding ratios, and covariance information on CGA, EGA or HERCULES display, or on HP 7475A plotter. The plotting program uses Calcomp or Calcomp-like subroutine libraries. A collection of the necessary subroutines is given in the separate PROFIT file.

The content of the IRDF-85 and IRDF-90 cross section libraries, and the input/output data sets of the BME-ECN program package can directly be plotted by the XSSPPLOT, while the corresponding files of the other (PTB and ORNL) modules require manual editing before plotting by this program.

## 3. CONCLUSIONS

1. The NMF-90 will contribute to the improvement of the accuracy of radiation damage calculations. It will supply good quality, consistent input data and proper calculation methods to the interested laboratories for testing and improving their neutron metrology procedures.
2. Regular updating of the content of NMF-90 will be needed.
3. It is recommended that the data of the IRDF-90 cross section library should be tested by calculations in benchmark neutron fields before distributing them for routine neutron metrology applications.
4. Unfortunatelly, not all the problems of the radiation damage parameter calculations could be solved by developing the reference file NMF-90. A detailed study is still required on a number of topics:
   - generation of input spectrum covariances
   - interpretation and elimination of inconsistencies
   - neutron self-shielding in activation detectors
   - shielding effects of detector covers
   - interpolation and extrapolation procedures for neutron spectra measured at different locations (eg within a surveillance capsule, reactor pressure vessel wall, etc)
   - improved damage characterization

## 4. REFERENCES


1. W. L. Zijp, É. M. Zsolnay, H. J. Nolthenius, E. J. Szondi, G.C. H. M. Verhaag, D. E. Cullen, C. Ertek: Final report on the REAL-80 exercise. Report ECN-128;INDC(NED)-7;BME-TR-RES-6/82. Netherlands Energy Research Foundation, Petten, February 1983

2. W. L. Zijp, É. M. Zsolnay, H. J. Nolthenius, E. J. Szondi: Final report on the REAL-84 exercise. Report ECN-212; BME-TR-RES-18/88; IAEA-NDS-212. Netherlands Energy Research Foundation, Petten, November 1988.
3. H. J. Nolthenius: Information for the REAL-88 exercise. Report ECN-88-023. Netherlands Energy Research Foundation, Petten, January 1988.
4. É. M. Zsolnay, W. L. Zijp, H. J. Nolthenius: Effect of normalization on the neutron spectrum adjustment procedure. Report ECN-139. Netherlands Energy Research Foundation, Petten, Oct. 1983.
5. É. M. Zsolnay, H. J. Nolthenius, E. J. Szondi, W. L. Zijp: "New aspects in least squares adjustment methods." In: J. P. Genthon, H. Röttger (Ed.-s): Reactor dosimetry. Proceedings of the Fifth ASTM-EURATOM Symposium on Reactor Dosimetry, GKSS Research Centre, Geesthacht, F. R. G., September 24-28, 1984. D. Reidel Publishing Company, Dordrecht, 1985.
6. H. J. Nolthenius, É. M. Zsolnay, W. L. Zijp, E. J. Szondi: "Nuclear data aspects encountered in the REAL80 and REAL84 intercomparisons." In: V. Piksaikin (Ed.): Nuclear data for radiation damage estimates for reactor structural materials. Proceedings of an IAEA Consultants' Meeting, Santa Fe, NM, USA, 20-22 May 1985. Report INDC(NDS)-179/G. p. 95-105. International Atomic Energy Agency, Vienna, June 1986.
7. W. L. Zijp, É. M. Zsolnay, H.J. Nolthenius: "Improvement of accuracy assessment in radiation damage predictions." In: Proceedings of the 1986 International Symposium on the Effects of Radiation on Materials, Seattle, June 23-25, 1986.
8. E. J. Szondi, A. P. Bosznay: "Physically based weighting spectrum generation for cross section calculations." Publication in progress as an IAEA Technical Report on the IAEA Advisory Group Meeting on the Status and Requirements of Nuclear Data for Radiation Damage and Related Safety Aspects. Vienna, 19-22 September, 1989.
9. D. E. Cullen, P. K. McLaughlin: The International Reactor Dosimetry File (IRDF-85). Report IAEA-NDS-41, Rev. 1. International Atomic Energy Agency, Vienna, April 1985.
10. The International Reactor Dosimetry File IRDF-90. Publication in progress by the Nuclear Data Section of the International Atomic Energy Agency.
11. H. J. Nolthenius: Computer programs for calculating group cross section covariances from uncertainty parameters in the ENDF/B-V format, UNC 32/33. Report ECN-202. Netherlands Energy Research Foundation, Petten, Oct. 1987.
12. F.G. Perey: Least-squares dosimetry unfolding: the program STAY'SL. Report ORNL/TM-6062, ENDF-254. Oak Ridge National Laboratory, October 1977
13. M. Matzke, publication in progress as a PTB report
14. F. W. Stallmann: LSL-M2: A computer program for least-squares logarithmic adjustment of neutron spectra. Report NUREG/CR-4349, ORNL/TM-9933. Oak Ridge National Laboratory, Oak Ridge, TN, March, 1986.

# UNCERTAINTIES IN DAMAGE PREDICTION - COMPARISON OF TWO METHODS

M. Matzke, J. Koban*,

Physikalisch-Technische Bundesanstalt,
D-3300 Braunschweig
*Siemens AG KWU, D-8520 Erlangen
Fed. Rep. of Germany

ABSTRACT

To calculate damage (exposure) parameters in reactor dosimetry from the reaction rates of a few detectors, the spectral neutron fluence must be known. In a widely used method the spectral fluence is determined by combining the pre-information values resulting from a transport code calculation with the measured data by means of an adjustment procedure.

In the present paper the necessity of this adjustment is investigated. Taking three typical reactor spectra, it is shown that the direct determination of damage parameters, without adjustment, using only the pre-information spectral fluence, leads to nearly the same values as those obtained with the adjustment method.

---

## 1. INTRODUCTION

In radiation damage (exposure) prediction (e.g. at the wall of the pressure vessel of a light water reactor (LWR)) a damage parameter H is to be determined which is derived from the integral over the product of a so-called damage cross section and the spectral neutron fluence. For a precise prediction of neutron exposure the "long-time" value of that parameter is required. For its determination four sources of information are combined :

- measured activities of suitable "long-time" detectors,
- measured activities of detectors for the spectral fluence,

- cross sections for all reactions,
- spectral fluence values determined by a transport code calculation.

In practice, there are often not enough detectors for the spectrum determination available and the damage parameter H considered is evaluated separately for each "long-time" detector without unfolding (more exactly, without adjusting) the spectrum. This is the procedure proposed in the German DIN Standard 25456 (1), and it is often applied (e.g. (2,3)).

On the other hand, in the REAL exercises (4) the emphasis was on spectrum determination. The required damage parameters H were evaluated from the determined spectrum after adjustment. The most important result of the REAL exercises was the strong influence of the pre-information spectrum on the damage parameters H. It was found that this spectrum was not altered significantly by the adjustment procedure. The reason for this was traced to the small rank of the pre-information fluence covariance matrix (5,6). Due to the small rank the manifold of the possible solution spectra was greatly restricted.

It may therefore be concluded that only little information can be gained by fluence unfolding or adjustment. This part of determining a damage parameter is probably only a consistency check for measured and calculated quantities.

In the present paper an attempt is made to determine damage parameters directly and avoid adjusting the neutron spectrum. For this purpose, the damage parameter H considered is the fluence above 1 MeV ( $\Phi( E > 1 \text{ MeV})$) which is proposed in Ref. (1) as a limiting quantity with respect to the damage. Three sample spectra (a typical LWR spectrum and the ANO and the PS1 spectrum of the REAL exercises (4)) are investigated.

## 2. DIRECT DETERMINATION OF DAMAGE PARAMETERS

The direct determination of integral parameters is straightforward and has been described in many publications (e.g. (2,3)). The short overview given here follows ref. (1), where the damage parameter $H = \Phi( E > 1 \text{ MeV})$ is thought to be evaluated separately for each (long-time) detector. H is derived from the equation

$$H = A \cdot F / ( N_0 \cdot h_{eff} ) = a / h_{eff} \qquad (1)$$

(A: sample activity at the end $t_n$ of the irradiation, $N_0$:number of nuclei for activation in the sample). The quantity a may be called the number of reaction events. In F the operational history of the reactor is taken into account:

$$F = \sum_{i=1}^{n} f_i \Delta t_i \ / \ \sum_{i=1}^{n} f_i \ (1-e^{-\lambda \Delta t_i}) \ e^{-\lambda(t_n-t_i)} \qquad (2)$$

F is sometimes written as an effective time multiplied by a so-called history factor (e.g. Ref. (2)). In Eq. (2) $\lambda$ is the decay constant, the quantities $f_i$ are proportional to the neutron flux density and it is required that they are constant in each time interval $\Delta t_i$ at $t_i$. The ratio

$$h_{eff} = \int_0^\infty dE\ \sigma(E)\ g_{n,E}(E) \Big/ \int_0^\infty dE\ h(E)\ g_{n,E}(E) \qquad (3)$$

is assumed as constant during the whole irradiation time. In Eq. (3) $g_{n,E}(E)$ is the relative spectral fluence at the irradiation position obtained by a transport code calculation. In the example considered here, h(E) is set equal to a dimensionless step function ( h(E) = 0 for E < 1 MeV, h(E) = 1 elsewhere), $h_{eff}$ is called the effective cross section, termed $\sigma_{eff}$ (3), and thus H = $\Phi$(E>1MeV).

There are no statements in Ref. (1) for the combination of several values of $H_i$ (i = 1,...,k) obtained with Eq. (1) for several detectors i. However, if a set of measurements is described by the vectors $\underline{a}$, $\underline{h}_{eff}$ and $\underline{H}$ together with covariance matrices $\underline{K}_a$ , $\underline{K}_h$ , $\underline{K}_H$ ($\underline{a}$ and $\underline{h}_{eff}$ assumed to be uncorrelated), a least-squares condition for a final value $H_q$ can be established by minimizing

$$\chi^2 = ( \underline{H} - H_q\ \underline{e} )^T\ \underline{K}_H^{-1}\ ( \underline{H} - H_q\ \underline{e}) \qquad (4)$$

with respect to $H_q$. In Eq. (4), T indicates transposition, $\underline{e}$ is the unit vector. The solution of this equation for $H_q$ and the variance $s^2(H_q)$ yields :

$$H_q = \sum_{i,j=1}^{k} (\underline{K}_H^{-1})_{ij} \cdot H_j \Big/ \sum_{i,j=1}^{k} (\underline{K}_H^{-1})_{ij}, \qquad (5)$$

$$s^2(H_q) = 1 \Big/ \sum_{i,j=1}^{k} (\underline{K}_H^{-1})_{ij} \qquad (6)$$

It can be seen from Eqs. (1) to (6) that no adjustment of the spectral fluence is performed for the determination of $H_q$. In the evaluation, only the fluence pre-information values $\underline{g}_n$ with covariance matrix $\underline{K}_g$ are used for $\underline{h}_{eff}$ in Eq. (3). $\underline{K}_H$ can be expressed by the relative covariance matrices $\underline{K}_a^{rel}$ and $\underline{K}_h^{rel}$:

$$\underline{K}_H = \underline{H}^T\ ( \underline{K}_A^{rel} + \underline{K}_h^{rel} )\ \underline{H} \qquad (7)$$

Although the computer time necessary to obtain the results Eqs. (5) and (6) may be as much as with the adjustment programs (4) (the covariance matrices exhaust most of the storage and computing time) these equations look very simple and attractive. The results obtained with them should be compared with the corresponding adjustment results.

## 3. REVIEW OF ADJUSTMENT - NORMALIZATION FACTOR

In the REAL exercises (4) all participants used adjustment codes based on the least-squares algorithm. The STAY'SL code (7) was the most frequently used program, its essential feature being the consistent application of uncertainty propagation.

In the STAY'SL algorithm it is assumed that the pre-calculated spectral fluence $\underline{g}_n$ is at the same absolute level as that used for the determination of the reaction events $\underline{a}$ . In practical applications, however, this assumption may not be valid. Although absolute transport code calculations have been performed (8), in most cases the integral over the spectral fluence is chosen arbitrarily. The values $\underline{g}_n$ therefore must be scaled with a normalization constant c before the real adjustment is made with the STAY'SL code. c is generally unknown; it is determined within the STAY'SL algorithm by using the measured data on activities in a manner not quite correct from the viewpoint of uncertainty propagation. In some other codes based on the least-squares method, difficulties concerning c may also occur; these have been correctly dealt with in references (11) and (12).

The problem concerning normalization in the STAY'SL code has been discussed in Ref. (9). It was stated that both the determination of c and the fluence adjustment should be part of the same procedure (9). A new adjustment code was developed (10) for that purpose.

When a correct evaluation is included in the STAY'SL code, the question arises whether the results obtained thereafter agree with those obtained with Eqs. (4) to (7). Both methods should in principle agree, because they use the same amount of information. But since the "c-modified" STAY'SL code is a non-linear least-squares approximation (13), the mathematical identity is not easy to prove.

In the following a modified STAY'SL code, MSITER (13), was used for a comparison with the direct evaluation method of Eqs. (4) to (7). In MSITER, the normalization constant c and the fluence are adjusted together in a non-linear least-squares algorithm iterated linearly.

## 4. EXAMPLES: LWR SPECTRUM, ANO, PS1

In the REAL exercises (4) emphasis was placed on the spectrum determination for which fast decaying nuclides can also be used. In practice, only "long-time" irradiations provide information on "long-time" exposure.

In the following, three example spectra are investigated. The first is a typical LWR spectrum (14) for which the quantities entering Eqs. (1) to (3) were simulated numerically using the $\underline{g}_n$ values mentioned in Ref. (14). For the two other spectra ( ANO and

PS1 ) the data were taken from the results of the REAL exercises (4). The LWR spectrum (14) deals with 4 "long-time" detectors.

As regards Eqs. (1) and (2), a considerable contribution to the uncertainties of the reaction events $a_i$ is expected from the irradiation history. To simulate a realistic case, a 10 year-reactor operation was split into 19 intervals with $f_i$ values of between 0.8 and 1.0. For all $f_i$ values an uncorrelated relative standard deviation (RSD) of 20 % and a fully correlated RSD of 10 % were assumed. An assumed uncorrelated RSD of 6 % due to the activity determination was added to obtain the values in table 1. The simulated $a_i$ numbers in table 1 were determined by folding the LWR spectrum mentioned in (14) (100 groups) with IRDF-file (15) cross sections and altering these values randomly within the RSD of 6 %.

Table 1: Uncertainties and Correlation Matrix for LWR

| | History F | | | | | $a_i$ | | | | | |
|---|---|---|---|---|---|---|---|---|---|---|---|
| Reac-tion | RSD % | Corr. Matrix (times 100) | | | | $a_i$ | RSD % | Corr. Matrix (times 100) | | | |
| NB93N | 0.9 | 100 | | | | 4.84E8 | 6.1 | 100 | | | |
| CU63A | 2.8 | 99 | 100 | | | 6.53E8 | 6.6 | 6 | 100 | | |
| FE54P | 13.8 | 85 | 89 | 100 | | 3.64E8 | 15.0 | 12 | 34 | 100 | |
| NI58P | 22.4 | 69 | 74 | 94 | 100 | 4.37E8 | 23.2 | 10 | 30 | 83 | 100 |

For the determination of $\sigma_{eff}$ (see section 2) the greatest problem is the covariance matrix of the pre-information spectral fluence which cannot be provided by most of the transport codes. A well-established method, not quite satisfactory from the viewpoint of statistics, is the computer variation of trial correlation coefficients (14) to study the influence on $\sigma_{eff}$. In the present case a diagonal covariance matrix $\underline{K}_f$ for the 100 group spectrum with an RSD between 30 % and 100 % was assumed.

Table 2: $\sigma_{eff}$ values and relative standard deviations for LWR. RSD : relative standard deviation of $\sigma_{eff}$, RSDF: spectral fluence contribution part to RSD

| Reaction | $\sigma_{eff}$ (barn) | RSD % | RSDF % | Correlation Matrix (times 100) | | | |
|---|---|---|---|---|---|---|---|
| NB93N | 2.32E-1 | 19.9 | 2.1 | 100 | | | |
| CU63A | 3.31E-3 | 11.4 | 10.2 | 1 | 100 | | |
| FE54P | 1.74E-1 | 6.7 | 5.9 | 4 | 45 | 100 | |
| NI58P | 2.22E-1 | 8.1 | 5.6 | 3 | 36 | 61 | 100 |

In Table 2 the results for $\sigma_{eff}$ obtained with IRDF cross sections using Eq. (3) are listed. The RSD of 19.9% for Nb in Table 2 is mainly influenced by the cross section uncertainty taken from Ref. (15). Disregarding the latter, Nb is best for detecting $\Phi$( E>1 MeV ) since the fluence contribution to the RSD of $\sigma_{eff}$ is quite small (14).

For the comparison of the directly determined damage parameter H = $\Phi$( E>1 MeV) using equations (1) to (7) with the adjustment result, the least-squares code MSITER (13) was changed to calculate all relevant quantities.

Table 3: H = $\Phi$(E>1 MeV) for LWR, ANO and PS1 for the two Methods.

$s_{rel}$ : relative standard deviation, $\Phi_i$ : value for each reaction according to Eq. 1, $\Phi_q$ : resulting from Eq. (5), $\Phi_s$ : resulting from STAY'SL (MSITER), $\chi^2/M$: Chi-square per degree of freedom, The exponential numbers of all $\Phi$(E>1 MeV) values are omitted.

| Reaction | $\Phi_i$ | $s_{rel}$(%) | Correlation matrix times 100 | | | | | | | | | |
|---|---|---|---|---|---|---|---|---|---|---|---|---|
| NB93N | 2.09 | 20.8 | 100 | | | | | | | | | LWR |
| CU63A | 1.97 | 13.2 | 1 | 100 | | | | | | | | LWR |
| FE54P | 2.09 | 16.5 | 4 | 31 | 100 | | | | | | | LWR |
| NI58P | 1.97 | 24.6 | 3 | 24 | 79 | 100 | | | | | | LWR |
| $\Phi_q$ | 2.03 | 10.3 | $\chi^2/M$ = 0.082 | | | | | | | | | LWR |
| $\Phi_s$ | 2.04 | 10.4 | $\chi^2/M$ = 0.065 | | | | | | | | | LWR |
| CU63A | 1.92 | 18.6 | 100 | | | | | | | | | ANO |
| TI46P | 2.01 | 19.9 | 53 | 100 | | | | | | | | ANO |
| NI58P | 2.00 | 13.2 | 40 | 51 | 100 | | | | | | | ANO |
| FE54P | 2.12 | 13.7 | 44 | 57 | 71 | 100 | | | | | | ANO |
| U238F | 2.17 | 12.3 | 16 | 16 | 20 | 22 | 100 | | | | | ANO |
| NP237F | 1.82 | 14.1 | 8 | 3 | 5 | 2 | 28 | 100 | | | | ANO |
| $\Phi_q$ | 1.98 | 8.7 | $\chi^2/M$ = 0.285 | | | | | | | | | ANO |
| $\Phi_s$ | 2.00 | 8.6 | $\chi^2/M$ = 0.196 | | | | | | | | | ANO |
| U235F | 2.31 | 35.8 | 100 | | | | | | | | | PS1 |
| CO59G | 2.44 | 33.5 | 90 | 100 | | | | | | | | PS1 |
| SC45G | 2.64 | 40.0 | 95 | 86 | 100 | | | | | | | PS1 |
| FE58G | 1.95 | 37.9 | 78 | 81 | 76 | 100 | | | | | | PS1 |
| NP237F | 2.66 | 12.5 | 7 | 7 | 7 | 6 | 100 | | | | | PS1 |
| NI58P | 2.61 | 8.4 | -1 | -2 | -1 | -1 | -1 | 100 | | | | PS1 |
| FE54P | 2.43 | 6.6 | -2 | -3 | -2 | -2 | -1 | 23 | 100 | | | PS1 |
| TI46P | 2.73 | 14.5 | -1 | -1 | -1 | -1 | 0 | 15 | 22 | 100 | | PS1 |
| CU63A | 2.54 | 9.7 | -3 | -3 | -2 | -2 | 0 | 23 | 33 | 27 | 100 | |
| $\Phi_q$ | 2.49 | 5.0 | $\chi^2/M$ = 0.472 | | | | | | | | | PS1 |
| $\Phi_s$ | 2.51 | 4.4 | $\chi^2/M$ = 0.422 | | | | | | | | | PS1 |

In table 3 the results for the examples discussed are shown. The small $\chi^2/M$ values indicate overestimation of the uncertainties. Some short-lived activities (taken here as reaction events $a_i$) are included for the ANO and the PS1 spectrum of the REAL exercises. Although the uncertainties of the $a_i$ are probably not realistic for a simulated investigation of "long-time" exposure this is of no relevance for the comparison of the direct method (Eqs. (1) to (7)) with the adjustment method. There is good agreement between the values $\Phi_q$ and $\Phi_s$ obtained by the two methods.

## 5. DISCUSSION

From the results in table 3 it may be concluded that it is not necessary to adjust neutron spectra when aiming at integral parameters. The reason for the good agreement between the two methods is probably the strong dependence of the results on the input neutron spectrum. The problem of determining neutron exposure parameters is therefore shifted to the problem of spectrum determination. This was one of the findings of the REAL exercises (4).

Looking at the uncertainties assigned to different reaction rates, we see that the RSD for $^{93}Nb(n,n')^{93m}Nb$ is scarcely influenced by the uncertainties of irradiation history and fluence. This again demonstrates the suitability of Nb detectors already stated in Ref. (14).

For practical applications the formula of Eq.(1) for determining one damage parameter with one detector (1) looks very simple and attractive. However, for the combination of the results from i different reactions the whole covariance matrix is required in Eq. (5). The weighted mean of different $H_i$ values using the reciprocal relative variances as weights and neglecting correlations should in most cases lead to a conservative estimate of $H_q$ if the resulting variance $\sigma^2(H_q)$ is scaled with the $\chi^2/M$ value after adjustment. In the exact evaluation according to Eq. 4-6 the $\chi^2$ per degree of freedom calculated after the adjustment of $H_q$ may serve in the sense of a statistical test.

## REFERENCES


1. DIN 25456, Part 1 to Part 6, "Neutron Fluence Measurement", Beuth Verlag, Berlin 1990

2. E. M. Zsolnay and F. Divós, "Surveillance Neutron Dosimetry at the Hungarian Nuclear Power Plant". Reactor Dosimetry: Methods, Applications, and Standardization (Proceedings of the 6th ASTM-Euratom Symposium) Harry Farrar IV, E.P. Lippincott, Editors, STP 1001, ASTM, Philadelphia 1989, p.105

3. W.G. Alberts, W. Schneider, "Reactor Dosimetry: Standardization of Procedures and Considerations on Further Developement for LWR PV Surveillance in the Fed. Rep. of Germany". Ref. 2, p. 38

4. W.L. Zijp, E.M. Zsolnay, H.J. Nolthenius and E.J. Szondi, "On the problems Encountered in Activation Spectrometry and Subsequent Radiation Damage Predictions", Ref. 2, p. 415

5. M. Matzke, The Covariance of Neutron Spectra Used in the REAL-84 Exercise, PTB-Report, PTB-FMRB-108, Physikalisch-Technische Bundesanstalt, Braunschweig, Germany 1986

6. E.J. Szondi, The Covariance Matrix of Neutron Spectra. Nuclear Reactor Departm., Techn. Univ. Budapest,Report BME-TR-RES-10/86

7. F.G. Perey, Least Squares Dosimetry Unfolding: The Program STAY'SL, Report ORNL/TM-6062, ENDF-254k, Oak Ridge, USA, 1977

8. G. Prillinger, "Neutron Exposure of Gundremmingen-A Pressure Vessel Trepans". Ref. 2, p. 165

9. E.M. Zolnay, H.J. Nolthenius, E.J. Szondi, W.L. Zijp, "New Aspects in Least Squares Adjustment". Proc. of the 5th ASTM-Euratom Symposium on Reactor Dosimetry, Reactor Dosimetry, Vol. 2, D.Reidel Publ. Company, Dordrecht, Netherlands 1985

10. A.P. Bosznay, E.J. Szondi, "Comparison of Neutron Spectrum Adjustment Methods named STAY'SL and EAGLE". Proc. of the IAEA Consultant's Meeting on The Assessment of the Results of the REAL-84 Exercise, Rep. INDC(NDS)-190/G+F+R, IAEA, Wien 1987

11. F.W. Stallmann, LSL-M2: A Computer Program for Least-Squares Logarithmic Adjustment of Neutron Spectra. NUREG/CR-4349, ORNL /TM-9933, U.S. Nuclear Regulatory Commission, Washington 1986

12. R.E. Maerker, B.L. Broadhead, J.J. Wagschal, "Theory of a New Unfolding Procedure in Pressurized Water Reactor Pressure Vessel Dosimetry and Development of an Associated Benchmark Data Base", Nucl. Sc. Eng. 91 (1985) 369

13. M. Matzke, H.J. Nolthenius, E.J. Szondi, E.M. Zsolnay, "Normalization of the Neutron Spectrum in Least-Squares Adjustment". to be published

14. J. Koban, "Bewertung der Unsicherheiten von Fluenzen schneller Neutronen". Tagungsbericht Jahrestagung Kerntechnik, Inforum Verlags- u. Verwaltungsgesellschaft mbH, Bonn, Germany 1989

15. D.E. Cullen and P.K. McLaughlin, The International Reactor Dosimetry File (IRDF-85), IAEA(NDS)-84, Wien 1985

# THE *MACH-1* CODE FOR COUPLED NEUTRON, PHOTON AND ELECTRON TRANSPORT BY THE MONTE CARLO METHOD (*)

Fabrizio CLERI [1] *and* Gueorgui K. VOYKOV [2]

[1] E.N.E.A. - Fast Reactors Department, Project Methods Division
C.R.E. Casaccia, P.O.Box 2400, I-00100 Roma A.D. (Italy)

[2] Institute for Nuclear Research and Nuclear Energy
72, Boulevard Lenin, BG-1784 Sofia (Bulgaria)

ABSTRACT

The MACH-1 code for the coupled neutron, photon and electron transport, resulting from the connection of the two Monte Carlo codes MORSE-CG, for neutron-photon transport, and ACCEPT-E, for photon-electron transport, is presented. The MORSE code covers the neutronic part of the problem, generating batches of photons as source for ACCEPT. Energy deposition by the various species of radiation can be obtained as output, together with other quantities like flux, current, charge deposition and so on.

## 1. INTRODUCTION.

The calculation of energy deposition from radiating sources is currently feasible either by deterministic or by stochastic (Monte Carlo) techniques. In both cases, either multistep calculations, in which the results of a first step of the transport calculation serve as source term for the next one, or fully coupled calculations can be done, in which the generation of secondary radiation showers is globally taken into account. However, the mi-

(*) This work, begun on early 1988, is entitled to the memory of Ernst Mach, in the 150th Anniversary of his birth.

gration of the secondary radiation produced in the photon transport step (*i.e.* electron transport) is generally not accounted for, under the hypothesis that the gamma-ray energy transmitted to the secondary electrons is deposited at the point of interaction ("on the spot"). This procedure can either be acceptable or totally misleading, depending on the dimensions and on the environment of the zone of interest in which gamma-heating is estimated. Large errors can be made at the interface between different materials, or in small zones surrounded by different materials (like detectors), for which the Fano theorem (1) is no longer valid.

The effects of secondary electron migration are treated, where necessary, as perturbations to the local energy deposition, and are corrected by means of "cavity correction" factors, estimated in most cases by the Bragg-Gray theory of cavity ionization (2), or by more refined versions of the same (3). However, it has been ascertained that in many cases the approximation involved by this process can be unacceptable (4), and in general it requires a non-negligible supplementary effort, which can also overwhelm that of basic nuclear heating calculations (5). It is thus felt that setting up a fully coupled calculation including all the main secondary radiations after the primary one, especially in the case of neutronic primary radiation, can be a powerful tool to solve the problem of the nuclear heating calculation. All available codes for electron transport are based on Monte Carlo methods, although discrete ordinates calculations could also, in principle, be suitable (6). We therefore consider the Monte Carlo route as the best candidate for this program.

The MACH-1 code ( Morse and ACcept Hybrid - version 1 ) results from the linking of the Monte Carlo codes MORSE-CG (7) and ACCEPT-E (8), and allows for performing the complete transport of the most important secondary species created by a neutron source, *e.g.* photons and positive or negative electrons. Protons and heavier charged particles can be supposed to deposit their kinetic energy in a very little volume around the point of emission, due to their extremely short ranges, except at very high energies where complete transport should be accounted for. The code has been developed in a basic version at the ENEA Center of Casaccia, paying also attention to its envisaged use in fields other than nuclear reactor technology (*e.g.* medical physics, radiation damage, material science and so on). In the following, the main characteristics of the MACH-1 code will be described.

## 2. BASICS ABOUT THE MACH-1 CODE.

Both MORSE-CG and ACCEPT-E adopt the same kind of geometrical description (9), based on a pre-defined set of elementary bodies to be combined by Boolean operations into the desired geometry zones. Together with the relatively efficient groupwise neutron treatment (considering the very huge amount of computer work required in a complete simulation of this kind), this was one of the main reasons that led us to identify MORSE as the best candidate to be interfaced with ACCEPT. The two codes use the "Monte Carlo" method for the calculation of physical quantities of interest, like flux, current, reaction rates and so on. Basically, the Monte Carlo method is grounded on the homology between the integral formulation of the transport problem and the simulation of the physical path of the particle, with all its events of capture, scattering and so on (6). It can be shown that, with an appropriate definition of the scattering and transport kernels, the integral Boltzmann transport equation can be reduced to the statistical average over a large number of simulated particle tracks (the "histories"), directly giving estimates for the quantities of interest at the sites of collision.

### 2.1 The MORSE-CG Neutron-Photon Code.

MORSE (7) is a "multigroup" Monte Carlo neutron-photon transport code, *i.e.* it treats the energy variable by discretization on a given set of energy boundaries common to all the cross-section data for the various events considered, to the energy slowing-down and to the energy-differential scoring of the quantities of interest. Macroscopic materials (*i.e.* homogeneous mixtures of atomic compounds) are defined on a continuous 3-D geometric description. Nuclear data relative to these materials are used in form of tabulated values, and the interactions take place with the entire material (no individual nuclide collisions are treated); the details of the collision mechanics, like energy conservation or angular distribution of the reaction products, are absorbed in the multigroup cross-section tables, which are developed in Legendre expansion over discrete angular bins to describe angular anisotropy. Such treatment can substantially save computer time, with acceptable precision of the results in most cases. The data are read in ANISN format, so that any library of this kind can easily be adapted to the MORSE runs. All the major neutron-matter interaction events can be modelled by MORSE, like radiative capture, scattering, fission, (n,2n) and so on. Scoring can be made for point flux, surface flux or current, volume averaged flux, reaction rates (via particular reaction cross-section libraries), energy release, secondary neutron and photon production. As the photon transport in MACH-1 is treated by ACCEPT-E, we will give here further detail of the photon treatment in MORSE.

### 2.2 The ACCEPT-E Photon-Electron Code.

ACCEPT (10) is a Monte Carlo transport code devoted to the treatment of both photons and electrons on a continuous energy range. All the cross-section data result from analytic distributions, tabulated at given energy values (more than 2700 photon points on the interval 0.001-1000 MeV for photons, on a variable energy grid for electrons in the same interval), and linearly interpolated between these values. For photons, all the major kinds of interaction are treated, like photoelectric absorption, Compton scattering from free electrons, pair creation, bremsstrahlung and fluorescence radiation, secondary Auger-electron shell relaxation. The ACCEPT-E version (8), developed at ENEA Casaccia by one of the authors, includes also the Rayleigh coherent scattering and binding corrections to Compton scattering. Efficient photon specific variance reduction techniques, such as biased photoelectric capture and photon spatial splitting, further improve the capabilities of ACCEPT-E over the standard version.

The electron transport is treated on the basis of the "condensed history" Monte Carlo method (11,6). In this family of algorithms the physical electron trajectory is replaced by an "average electron" path cumulating the effect of many individual collisions, characterized by energy-loss steps, multiple scattering probability distributions in place of single scattering events, and so on. The low-energy treatment of ACCEPT is believed, in particular, to be the most accurate among the codes devoted to electron transport (8,12). The ACCEPT final scores include both photon and electron energy-angular differential flux, volume-averaged flux, volume-averaged energy and charge deposition, spectra of the absorbed energy, photon and electron escape fractions and spectra.

### 2.3 MACH-1 Operation.

The general operation of the MACH-1 program can be derived by the above general informations about the structure and the operation of the codes MORSE-CG and ACCEPT-E. The two subprograms are linked into a unic executable module, with a very complete input deck. The physical connection lies in a blank common area, into which the photon bank is built. Each batch is performed by tracking some neutrons at first, banking the parameters of each neutron-produced photon. Then, secondary photons are picked from the bank one at a time, and the complete photon-electron shower is followed for each photon, until the bank is exhausted. A new batch of primary particles is now started, and so on. Common statistics are drawn, both at the end of each batch and at the end of the calculation.

The problem of determining the nuclear heating of a structure exposed to a neutron source is, as usual, decomposed into the coupled estimation of the neutron heating and of the gamma heating

$$H = H_n + H_\gamma \qquad [1]$$

The neutron component is written as the sum over each nuclide present in the material with density $N_j$, of the energy integral of the neutron flux times the neutron KERMA $k_n$ for the given nuclide,

$$H_n(\vec{r}) = \sum_j N_j(\vec{r}) \int_0^{E_{max}} \phi_n(\vec{r}\,,\, E_n)\, k_{nj}(E_n)\, dE_n \qquad [2]$$

In MACH-1 calculations the neutron flux is groupwise estimated; then, neutron heating is derived by multiplying and summing group by group the flux times the corresponding response function at the appropriate energies; at this step, other quantities of interest in radiation material testing can be unexpensively estimated, by concurrent use of other appropriate scoring functions (gaseous elements production like tritium or $^4$He, structural materials DPA, and so on). The neutron KERMA for use in MACH-1 must be defined as

$$k_{nj} = \sigma_{tj}\left( E_n + \sum_i \frac{\sigma_{ij}}{\sigma_{tj}} Q_{ij} + \sum_{i'} \frac{\sigma_{i'j}}{\sigma_{tj}} E^D_{i'j} - \sum_m \frac{\sigma_{mj}}{\sigma_{tj}} E^{n'}_{mj} - \frac{1}{\sigma_{tj}} S_{\gamma,j} \right) [3]$$

$E_n$ being the incident neutron energy, $\sigma_{tj}$ the total microscopic collision cross-section, and terms in parentheses being the energies contributed or taken by a particular reaction weighted by the relative probability of that reaction. $Q_{ij}$ is the energy resulting from mass conversion in reaction i, $E^D_{i'j}$ is the average decay energy per reaction i', $E^{n'}_{mj}$ is the average secondary neutron energy per reaction m, and $S_{\gamma,j}$ is an integral of the energy carried away by any secondary photon. The contribution from radioactive decay of the reaction products is clearly time-dependent, and is normally included under saturation hypothesis; if nuclides with very long excited state lifetimes are present, this approximation can induce some error. Moreover, electrons and positrons created in beta-decays cannot be stored by MORSE, to be tracked by ACCEPT, and this approximation too can have some impact.

Neutral secondary particles like neutrons and photons being explicitly followed, by MORSE and ACCEPT respectively, their contribution is subtracted in eq.[3]. The definition [3] assumes that charged particles other than electrons or positrons have very short ranges in bulk matter, so that their energy can be considered to be absorbed "on the spot". On the contrary, electrons just at energies of a fraction of MeV are relativistic, and can travel over quite long distances, so that this motivates the inclusion in MACH-1 of the electron transport step to evaluate the gamma-heating. Gamma-heating is thus estimated on a direct basis, without use of photon KERMA functions. The energy deposited by secondary electrons set in motion by photons is properly scored in the ap-

7. H.B.Emmett, The MORSE Monte Carlo Radiation Transport Code System, Report ORNL-4972/R2, Oak Ridge (USA), 1984.

8. F.Cleri, Nucl.Instr.Meth., **A295** (1990), p.231.

9. W.Guber *et al.*, A Geometric Description Technique suitable for Computer Analysis of both the Nuclear and Conventional Vulnerability of Armored Military Vehicles, Report MAGI-6701, Math.Appl.Group Inc. 1967.

10. J.A.Halbleib, T.A.Melhorn, Nucl.Sci.Eng., **92** (1985), p.338.

11. M.J.Berger, "Monte Carlo Calculation of the Penetration and Diffusion of Fast Charged Particles", in Methods of Computational Physics vol.1, p.135, B.Alder and S. Fernbach eds., Academic Press, New York 1963.

12. D.W.Rogers, A.F.Bielajew, "A comparison of EGS and ETRAN", in Monte Carlo Transport of Electron and Photons below 50 MeV, p.323, T.M.Jenkins, W.R.Nelson and A.Rindi eds., Plenum Press, New York 1989.

13. F.Cleri, Gamma-ray Dosimetry Calculations for the "TAPIRO" Fast Source Reactor at Casaccia: a Primer, ENEA Report FT/WHC-42, CRE Casaccia (Italy), 1988.

14. V.Rado *et al.*, BABEL: a New Multigroup Neutron and Gamma Library for Fast Reactor Shielding Design, OECD-NEACPL SECU Bulletin n.4, NEA Saclay (France), Nov. 1979.

# PARAMETER AND FLUENCE-RATE COVARIANCES IN LEPRICON

R. E. Maerker

Oak Ridge National Laboratory

P.O. Box 2008, Oak Ridge, TN 37831-6364

The LEPRICON code system is now available from the Radiation Shielding Information Center at ORNL as PSR-277. The system consists of modules that involve both the calculation of neutron fluence rates through PWR pressure vessels and the adjustment of these fluence rates with reduced uncertainties based on surveillance dosimetry. This paper describes in detail the manner in which important parameter uncertainties are partitioned and quantified as part of the input to the adjustment procedure, and to what degree they are applicable to all reactor designs.

## INTRODUCTION

A derivation of the solution to the generalized linear least-squares equations to obtain the best estimate of the group fluence rates in the pressure vessel of a power reactor and their uncertainties is given elsewhere.(1) It consists essentially of two steps. First, the simultaneous combination of all the measured and calculated reactor dosimeter activities with similar benchmark data leads to adjustments in the calculational input data which minimize the discrepancies between all the calculated and measured activities subject to estimates of all the uncertainties involved. Second, adjustments in the calculated pressure vessel group fluence rates (i.e., "unfolding") are determined from the adjustments in the differential data to which they are sensitive. The uncertainties in these adjusted group fluence rates are in general reduced in magnitude from their corresponding values before the adjustment. These best estimates of the group fluence rates are then converted to accumulated group fluences.

## THE BENCHMARK DATABASE

The foundation of the LEPRICON system is the "built-in" benchmark database. This database consists of the results of 37 integral measurements in the Intermediate Standard Neutron Field (ISNF), $^{252}$Cf, Pool Critical Assembly (PCA), and the Poolside Facility (PSF) fields, in which detailed analyses using the same calculational method and input nuclear data as in the power reactor analysis have previously been performed. A necessary adjunct to this database is a benchmark covariance library which provides values of the estimated group-wise uncertainties of those parameters found to be the most significant in the analyses. These parameters consist of the following nuclear data: the fission spectra from $^{252}$Cf, $^{235}$U, and a modified $^{235}$U spectrum in the ISNF, ten dosimetry cross sections for those reactions analyzed, and the total inelastic cross sections for iron, chromium, and nickel. The remaining parameters are non-nuclear in origin and assume the form of group-dependent fluence-rate bias factors. These bias factors provide a means for correcting the calculations for known deficiencies or approximations in the transport method employed (DOT-IV), and uncertainties in both the absolute source strength and its spatial distribution. Two types of bias factors are incorporated into this library: the first, applicable to both the PCA and PSF experiments, represents a correction for the 3-D effects of finite source height; the second, applicable to the PSF only, represents a correction to the absolute magnitude of the leakage from the Oak Ridge Research Reactor determined solely from a VENTURE calculation. The symmetric covariance matrix is shown in Fig. 1, where each shaded square represents the existence of a 15x15 submatrix corresponding to a 15-energy group structure above 0.11 MeV, and which is a subset to the one used in the transport analysis. Blanks indicate that the submatrices are not included in the library. (The first partition, denoted IR in Fig. 1, refers to the covariances of the measured integral responses, and is a 37x37 submatrix.)

## EXTENSION TO POWER REACTORS

The extension of the benchmark covariance matrix in Fig. 1 to include additional data for a power reactor first requires an expansion of the integral response submatrix to contain uncertainties in the dosimetry measurements for the reactor. More bias factors must also be appended to the list of parameters in order to introduce a number of possibly significant uncertainties present in the reactor calculations because of the lack of information on the "as-built" physical properties. The bias factors in LEPRICON, whether they apply to the benchmarks or the power reactor, usually have a value of unity, as will be described below. Their uncertainties, however, influence the adjustment procedure, and the adjusted values of the bias factors are no longer unity.

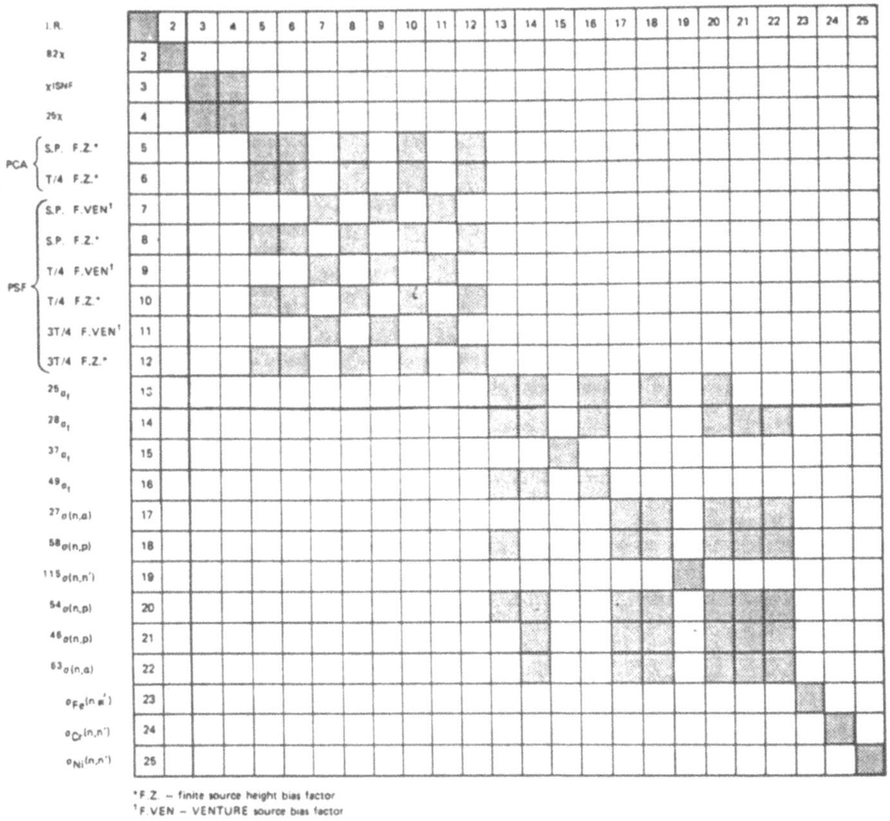


Fig. 1. Partition extended benchmark covariance matrix

## Description of the Bias Factors

Eight types of bias factors applicable to a power reactor are considered in LEPRICON. The first arises as a result of the uncertainty in the exact location of the dosimetry. Although the most probable location (i.e., according to the specifications) is used in the transport calculation, leading to a bias factor of unity, there is an estimated uncertainty in the as-built location of ±1.27cm (see Ref. 2), leading to the considerable calculated group fluence-rate uncertainties at the in-vessel surveillance location shown in the column headed "Bias 1" in Table 1. Note that this bias is only valid for a downcomer surveillance position (SP) since any cavity dosimetry (CAV) is not sensitive to uncertainties in its exact location.

A second type of bias arises as a result of an out-of-roundness of the pressure vessel in the vicinity of the location in the presexist between fluence rates in the same group at the different locations, and these are also assigned high values independent of energy. Thus, to within the degree of detail warranted, the complete correlation between a group fluence rate at one location and another group fluence rate at another location has been approximated as the product of the cross-correlation and autocorrelation coefficients.

Thus far the only power reactors included in the supporting analysis of the LEPRICON system have been of B&W design (AND-1) and Westinghouse design (HBR-2). However, the coding is sufficiently general to allow additional reactor designs to be included. This generality must not be misconstrued to imply that other designs are automatically valid using the current set of defaults shown in Table 1 and without the possible need to introduce new types of bias factors. Additional work might have to be performed outside of the LEPRICON system to arrive at credible estimates of uncertainties for a new type of bias factor peculiar to that reactor. The generality in the code is such that when these new bias factor uncertainties are determined the code will allow them to be substituted as part of the standard input. This redefinition and/or overriding of the default uncertainties of the bias factors combined with the parameterization of two important sensitivities to be described next allows virtually any reactor to be treated by LEPRICON.

## SENSITIVITIES AND FLUENCE-RATE UNCERTAINTIES

Propagation of the uncertainties in the calculated fluence rates are determined in a straightforward manner from the uncertainties in the input differential data. They are obtained as quadratic products of the data covariances and the fluence-rate sensitivities, where the sensitivities are defined as fractional changes in the group fluence rates per unit fractional change in each group datum:

$$S_\phi = \frac{\partial \phi/\phi}{\partial p/p} \quad , \tag{1}$$

and

$$C_{\phi\phi} = S_\phi C_{pp} S_\phi^T \quad . \tag{2}$$

In Eq. (1) $\phi$ has dimensions Gx1 and p has dimensions PGx1, where G is the number of groups and P is the total number of parameter types, including both nuclear and non-nuclear data. In Eq. (2) $C_{\phi\phi}$ is the fluence-rate fractional covariance matrix propagated from the total parameter covariances $C_{pp}$, also expressed in fractional values, by the sensitivities $S_\phi$ ("T" represents the transpose). $S_\phi$ has dimensions GxPG ("sensitivity profiles"), $C_{pp}$ has dimensions PGxPG, and $C_{\phi\phi}$ has dimensions GxG. For the bias factors, since they represent multiplicative correction factors to the calculated fluence rates,

$$\phi_{g_c} = \phi_g \prod_j (BF_j)_g \quad , \tag{3}$$

and the sensitivities are either zero or unity.

The nuclear data used in a power reactor transport calculation that result in the greatest ex-core fluence rate uncertainties are $^{25}\chi$, the spectrum from thermal fission of $^{235}U$, and $\sigma_{STEEL}$ (n,n'), the total inelastic "steel" cross section. In order that they may be used with any type of reactor design, the sensitivities of the group fluence rates in a power reactor geometry to these two types of differential data have been approximated by empirical expressions. From analysis of rigorous sensitivity calculations for the PCA and PSF benchmarks as well as the experiments in ANO-1, the fluence-rate sensitivities to $\sigma_{STEEL}$ (n,n') may be represented to acceptable accuracy as

$$S_\phi = K_\phi t_{STEEL} \quad , \tag{4}$$

where $K_\phi$, the sensitivity of group fluence rate g to the total inelastic steel cross section in group k per cm of steel penetrated, is represented by a spatially invariant GxG matrix. This approximation implies the effect of any intervening coolant is small and the effect to be linear in the intervening steel thickness. The latter effect indicates that uncertainties in the cross sections are propagated to all ex-core flux locations. The resulting fluence rate covariances thus vary with position in the reactor, increasing with increasing thickness. A very important consequence of this is that these cross sections affect significantly the overall correlation between the fluence rates at any two locations - e.g., dosimetry and in the pressure vessel - thus causing the parameter adjustments determined at the dosimetry location to result in fluence rate adjustments of different amounts at the two locations. This readily follows from the expression for the covariances between the fluence rates at the two locations:

$$C_{\phi_A \phi_B} = S_{\phi_A} C_{pp} S^T_{\phi_B} \quad , \tag{5}$$

in which $S_{\phi A}$ and $S_{\phi B}$ represent sensitivity profiles to p at locations A and B respectively.

A similar analysis of sensitivities to changes in the fission spectrum was performed and led to an even simpler expression than Eq. (4):

$$S_\phi = K \quad , \tag{6}$$

which implies that for a wide range of ex-core locations the sensitivities rapidly approach asymptotic values independent of the specific nature of any intervening material media or the distance from the core. These "universal" sensitivities are represented by another invariant matrix, this time leading to virtually complete correlations to this parameter over all ex-core fluence-rate points enclosed within the concrete shield.

## CONCLUDING REMARKS

In concluding, it can be stated that the fluence rate uncertainties at a given pressure vessel location tend to be similar for a given class of reactors since the bias factors and sensitivities to the nuclear data tend to be similar. While the same conclusion is expected to remain generally valid for any location - in particular that of a dosimeter - the use of different in-vessel dosimetry locations as well as the possible introduction of ex-vessel dosimetry will result in different fluence-rate uncertainties at these new dosimeter locations, but these should not affect to any significant degree the resulting reduced fluence-rate uncertainties and adjustments at locations within the pressure vessel.(7)

## REFERENCES

1. R. E. Maerker, B. L. Broadhead, and J. J. Wagschal, "Theory of a New Unfolding Procedure in Pressurized Water Reactor Pressure Vessel Dosimetry and Development of an Associated Benchmark Data Base," Nucl. Sci. & Eng. 91, 4, 369 (1985).

2. Minutes of a 1979 Meeting in San Francisco to Quantify Some of the Uncertainties in Reactor Environments.

3. P. Silvennoinen, Reactor Core Fuel Management, p. 133, Pergamon Press, Elmsford, N.Y. (1976).

4. D. P. Bozarth and H. D. Warren, IEEE Trans. Nucl. Sci. NS-26, 1, 924 (1979).

5. R. E. Maerker, B. L. Broadhead, B. A. Worley, M. L. Williams, and J. J. Wagschal, "Application of the LEPRICON Unfolding Procedure to the Arkansas Nuclear One-Unit 1 Reactor," Nucl. Sci. & Eng. 93, 2, 137 (1986).

6. N. Snidow, F. Walters, and D. Nitti, private communication (1987).

7. R. E. Maerker, "LEPRICON Analysis of Pressure Vessel Surveillance Dosimetry Inserted into H. B. Robinson-2 During Cycle 9," Nucl. Sci. & Eng. 96 4, 263 (1987).

# EXPERIENCES WITH Nb DOSIMETERS FOR NEUTRON FLUX MEASUREMENTS

B. Bärs and K. Uusheimo

Technical Research Centre of Finland, Reactor Laboratory

P.O.Box 200 (Otakaari 3A), SF-02151 Espoo, Finland

ABSTRACT

The aim was to improve the quality of neutron flux estimates based on the reaction $^{93}Nb(n,n')^{93m}Nb$ and to develop and to select proper methods for routine measurements of the specific $^{93m}Nb$ activities.

Experiences from or references to the following methods are given: X-ray energy spectroscopy ($^{93m}Nb$ activity and mass), liquid scintillation counting (LSC, $^{93m}Nb$ and $^{182}Ta$ activities), gamma spectroscopy ($^{94}Nb$, $^{95}Nb$, $^{95}Nb$, $^{182}Ta$ etc and relative mass), spectrophotometry (Nb conc.), atomic emission and absorption spectrometry (Nb mass), mass spectrometry (Nb mass), laser Resonance Ionization Spectroscopy (isomer ratio $^{93m}Nb/Nb^{93}$), polarography (Nb conc.) and chromatography (Nb separation).

---

## 1. INTRODUCTION

The main objective of the work has been to develop and to experimentally validate methods for determination of the population ratio $^{93m}Nb/^{93}Nb$. This isomer ratio is needed at neutron flux estimations based on the reaction $^{93}Nb(n,n')^{93m}Nb$. An additional objective is to develop routine methods in Nb dosimetry.

We have used two main sources of irradiated Nb material for dosimetric purposes a) prefabricated foil and wire dosimeters, which contain a few hundred ppm of Ta and b) Nb from the inner cladding of the reactor pressure (RPV) of several PWR units.

Nb was applied for neutron spectrum checks at the surveil-

lance chains (SC) and at ex-vessel (cavity) positions in conne-
xion with other prefabricated dosimeters and for flux and dose
estimations in the RPV and at ex-vessel locations.

The 30 keV gammas from the $^{93m}Nb$ are internally completel
converted into electrons and and X-rays (main energies 16.6
18.6 keV). The experimental estimation of the $^{93m}Nb$ population i
presently based on X-ray detection or on electron and X-ra
detection (by Liquid Scintillation Counting, LSC).

The RPV cladding contains small (<1%) and varying amounts o
Nb. The $^{93m}Nb$ induced X-rays and internal conversion electrons i
nonpurified RPV samples cannot be resolved. The dominating $^{60}C$
activity ($^{94}Nb$ : $^{93m}Nb$ : $^{60}Co$ ≈ 0.03 : 1 : 10000 Bq/mg) als
prevented accurate measurements of the $^{94}Nb$ activity in the non
purified RPV samples from the Loviisa NPP (3,4). RPV samples nee
purification before $^{93m}Nb$ and accurate $^{94}Nb$ activity measurement
and preferably also before Nb mass determination. Anion exchang
(liquid) chromatography was applied to separate and purify Nb
Due to losses the Nb mass has to be measured after the extrac
tion. Several mass determination methods have been applied.

Reference is also made to published literature on som
alternative and potential mass estimation methods and to researc
work on laser resonance ionization spectroscopy (RIS) in Finlan
aiming at direct measurement of the isomer ratio $^{93m}Nb/^{93}Nb$.

## 2. MASS DETERMINATION OF Nb

Nb wire and foil dosimeters can be weighed and the mas
flow further followed e.g. by measuring the $^{94}Nb$, $^{95}Nb$ or $^{182}T$
activities. Weighing of Nb cannot be applied on RPV samples.

### Spectrophotometry

We have used spectrophotometers to measure the Nb concentra
tion in liquid samples (3,4). The spectrophotometers were cali
brated with nonactive standard Nb solutions. The transmissic
(T=0...100) of 385 µm wavelength light through the samples wa
measured. The limited applicable concentration range (≈2 to 30 µ
Nb/ml) constitutes no problem since proper dilutions can be made.

The electrodeposited Nb masses on Cu disks (3,4) we
determined by first measuring the Nb concentrations in the sto
solutions. In addition the a) relative decrease in the $^{93m}Nb$ a
tivities in the electrolytes by LSC or b) the $^{94}Nb$ or $^{95}Nb$ acti
vities of the stock solutions and of the deposits were measure
c) The Nb masses have also been independently measured with XRF.

### Atomic emission spectrometry

The Nb content in nonactive RPV cladding samples has also been remeasured with an optical emission spectrometer (VTT, Metallurgy laboratory). The Nb amount varies largely between different samples and sample specific mass determinations are required. The sensitivity of an inductively coupled plasma optical emission spectrometer (ICP-OES) is good, 20 µg Nb/l) (6).

### Atomic absorption spectrometry, AAS.

In an atomic absorption spectrophotometer (AAS) the sample is vaporized (atomized) and the atoms brought into a light beam and the absorption of the characteristic spectrum lines is measured. Standards are used. The literature on AAS is extensive (7). Flame atomic absorption spectrometry (FAAS), where the sample is atomized in an acethylen flame, is relatively insensitive for Nb analysis (≈20 000 µgNb/l), (8). The amount of Nb material required is also large (≈ mg), radioactive contamination problems exist and the equipment are expensive.

It was not possible to use the (electrothermal) graphite furnace atomic absorption spectrophotometer (GFAA) in our laboratory to measure the mass of purified and active Nb samples of RPV origin many years ago due to the expected radioactive contamination of the equipment. Hegedüs has reported (5) AAS measurements of Nb, Fe and Ni contents in nonpurified RPV material.

### Mass spectrometers

We have also preliminarily investigated (1984-85) Nb mass determination by using a mass spectrometer with a W atomizer in a three wire surface ionization. The results indicate that the method is applicable for Nb masses around 1 µg or more. Standards and careful calibration of the ionization circumstances are needed. The primary goal to determine the $^{94}Nb/^{93}Nb$ (expected $10^{-3}...10^{-4}$) ratio of an active sample was not successful.

Inductively coupled plasma mass spectrometers (ICP-MS) are very sensitive (.02 µg Nb/l) but the equipment are expensive (6).

### Radioactive contamination problems

Atomic absorption, emission and mass spectrometers require special measures to prevent or to reduce the radioactive contamination of the equipment at the atomization of the samples in flames or in graphite furnaces.

We have also tried to determine the mass of Nb and other elements by reactivating parts of purified and nonpurified RPV samples together with known standards based on the induced shortlived $^{94m}Nb$ (6.3 min, 871 keV) or the additional longlived $^{94}Nb$ (20 000 y, 702 & 871 keV) activity (4). Nondesired activities degraded the measurements and the method needs further development.

## Polarography or voltametry

Based on the literature (8) polarographic or voltametric determination of transition metals e.g. with an aqueous-organic background electrolyte is possible. A linear dependence between the peak currents (max. 6 µA) and ion concentrations of tested solutions was observed in the concentration range $10^{-4}$ to $10^{-3}$M (≈10 to 100 mg Nb/l) for Nb (V) (8).

## Primary and secondary (atomic fluorescence) X-radiation, XRF

The intensity of primary X-radiation induced in Nb targets by accelerated electrons (e.g. in electron microscopes) and also the intensity of secondary or fluorescent X-radiation (XRF) produced by photoexcitation of Nb with X- or gamma rays are proportional to the mass of the Nb target.

The variations in the Nb/Fe and Ni/Fe weight fractions of RPV material was measured by Hegedüs (5) with the XRF method.

Kondurov et al. (9) have used X-ray energy spectrometry for both activity and mass determination of Nb depositions (Cu, Ni disks etc) with the same equipment and in the same geometry. The X-ray emission rate ($C_X$) induced by $^{93m}Nb$ activity was measured with a Si detector. The additional X-ray emission rate ($C_M$) arising from subsequent excitation by an external radionuclide or a superior X-ray (tube) source was proportional to the Nb mass (total emission rate = $C_X + C_M$).

Szaloki et al (10) have reported the use of XRF with an external $^{125}I$ source to determine the weight of thin Nb samples.

We have used Si(Li) and Ge detectors to measure the X-ray emission rates from high and medium active $^{93m}Nb$ samples deposited on Cu and Ni disks and on low background polycarbonate (millipore) filter papers (3,4). A 5 mCi $^{109}Cd$ source was used for additional excitation of the Nb atoms in the same geometry to obtain mass estimates of the Nb deposits. The detection limit with 3 h counting time is less than 1 µg Nb on polycarbonate filter paper (Fig. 1).

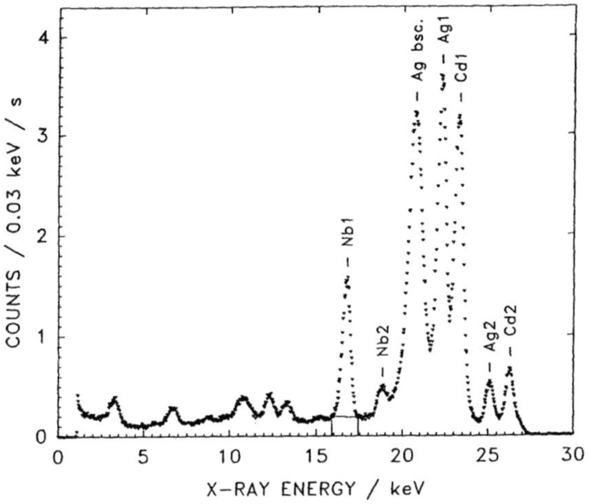


FIG. 1.
XRF measurement of Nb mass (58 µg) on polycarbonate filter paper. Excitation with a 5 mCi $^{109}Cd$ source. ≈ 1 % unc. in Nb1 (16.6 keV) net peak area in 50 channels, with background from 2*5 channels and 1200 s counting time. Ge det. eff. ≈ 0.01 cps/X-ray. The Ag and Cd peaks are from the source and shielding (M3NB58B3).

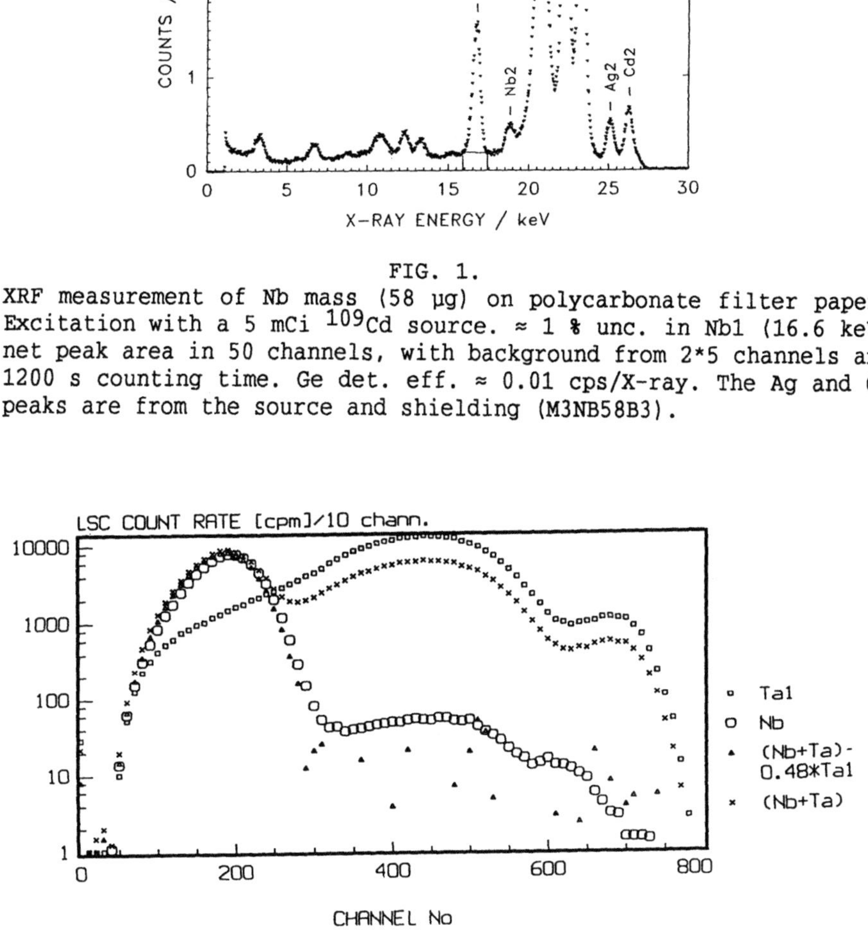


FIG. 2.
LSC counting rates in 10 channel windows from $^{93m}Nb$ and $^{182}Ta$ in a) nonpurified (Nb+Ta) sample, b) purified Nb sample c) Ta1 sample and d) the calculated net difference (Nb+Ta) - 0.48*Ta1 corresponding to pure Nb. (File: 19NBGW10.GIA 30.8.88).

## 3. $^{93m}$Nb ACTIVITY MEASUREMENTS

Our main method for routine $^{93m}$Nb activity measurements is based on LSC with automatic sample changing and using LUMAGEL scintillator in glass bottles. The efficiency is high (≈1 cps(LSC)/Bq) both for $^{93m}$Nb and $^{182}$Ta activities, which can be accurately resolved in LSC samples from ordinary Nb wires despite the low energy resolution.

The X-ray emission rates of medium (≈300 Bq $^{93m}$Nb) and highly active samples have been measured with calibrated Si(Li) and Ge detectors and then converted to absolute $^{93m}$Nb activities. The equipment were calibrated with a standard $^{93m}$Nb X-ray source and by using a known ($^{93m}$Nb gamma to X-ray) internal conversion ratio (≈0.11). The sensitivity in our X-ray measurements is about three magnitudes smaller than in LSC and e.g. Nb samples from irradiations in the cavity give poor accuracy. The same X-ray equipment are also used for mass determination.

Different methods (A, B, C) have been applied to subtract or reduce $^{182}$Ta and other nondesired activity contributions at LSC.

A) After purification of RPV scraping and of Nb wire samples with anion exchange chromatography the contributions from other than $^{93m}$Nb activities are insignificant (Fig. 2), (3,4).

B) In the graphical-mathematical fitting method (Fig. 2) the LSC spectra for the sample (Nb+Ta) and for a pure $^{182}$Ta sample (Ta) were measured. The $^{182}$Ta spectrum (Ta) was subtracted from the sample spectrum (Nb+Ta), based on fitted intensities in the $^{182}$Ta peak region and so that no main parts of the net spectrum ((Nb+Ta) - K*Ta) are negative. No $^{182}$Ta calibration is needed.

C) In present LSC equipment, with automatic sample changing and data storing, the LSC rates CPSw in several spectrum windows (m) can be measured quickly and accurately. The counting rate CPS in window w is a sum of responses to n types of (nuclide) activities Ak multiplied by the corresponding efficiencies EFkw or

CPSw = A1*EF1w +..+ Ak*EFkw +...+ An*EFnw (w=1..m,k=1..n) (1)

For m windows we obtain m equations from which the activities Ak are solved (m≥n). Due to the poor resolution many activities and activities with similar responses cannot be resolved. The LSC sensitivities EFkw in all windows and to all involved activities are obtained by measuring pure standards with known activities.

Fig. 2 displays measured LSC spectra from a nonpurified Nb+Ta, a purified Nb and a Ta sample. The $^{93m}$Nb and $^{182}$Ta activities can be accurately determined from the counting rates in two windows (w1=1-300, w2=301-1024 chann.) as also reported by McCrohen et al (11). The counting rates were corrected for changes in the attenuation (quenching) of light in the LSC samples by applying the Spectral Quenching Parameter SQP(E) for an exteral $^{226}Ra$ source as an amplification factor. In methods B and C the activities Ak of gamma emitting nuclides (e.g. $^{182}Ta$) were also obtained with gamma spectrometers and the above methods may be slightly modified.

## 4. LASER RESONANCE IONIZATION SPECTROSCOPY (RIS) FOR DIRECT MEASUREMENT OF THE ISOMER RATIO $^{93m}Nb/^{93}Nb$

Additional possibilities to determine the $^{93m}Nb$ (and $^{93}Nb$) populations or the isomer ratio $N[^{93m}Nb]/[^{93}Nb]$ directly arise from the fact that "the hyperfine splittings (hfs) of the electronic levels of the stable isotope and the isomer differ considerably due to the large difference in the nuclear spin: I = 9/2 for $^{93}Nb$ and I= 1/2 for $^{93m}Nb$. Moreover, the differences in the nuclear magnetic dipole and electric quadrupole moments of the two nuclei cause additional changes and shift in the energy levels. Due to these spectral changes, it may be possible to determine the isomer ratio by laser resonance ionization spectroscopy (RIS), which provides good enough spectral selectivity and sensitivity" (12,13).

The feasibility of several two- and three-step schemes for resonance ionization of Nb with field ionization of Rydberg levels for determination of the $^{93m}Nb/^{93}Nb$ isomer ratio has been investigated. Experiments to measure Rydberg level spectra and hyperfine structure of the intermediate levels in the excitation schemes with stable $^{93}Nb$ were carried out on our RIS spectrometer using two or three excimer pumped dye lasers (12, 13). Due to needed basic research, needed improvements in the equipment and due to the expected radioactive contamination of the vacuum oven equipment radioactive Nb samples have so far not been measured.


## Acknowledgements

The present investigations have been financially supported by the Ministry for Trade and Industry in Finland and by the Imatran Voima Power Company Ltd. We are indepted to Dr. E. Häsänen for valuable help with sample preparation, to MSc. I. Auterinen for assistance in the XRF measurements and to MSc. T. Serén for cross section and other calculations. The basic laser RIS investigations were carried out in a research collective including MSc. I. Auterinen, Dr. H. Lauranto, Prof. R. Salomaa and others.

REFERENCES


1. Fifth ASTM-EURATOM Symposium on Reactor Dosimetry, Geesthacht, FRG, Sept. 24-28, 1984. Genthon J.P., Röttger H. (ed), Vol. I and II, 1985.

2. Sixth ASTM-EURATOM Symposium on Reactor Dosimetry, Jackson Hole, Wyoming, USA, 1987. H. Farrar and E.P. Lippicott (ed), STP 1001, ASTM, PA, USA.

3. Bärs B., Karnani H., In ref. (2), p. 319-325.

4. Bärs N., Karnani H., Kernenergie 29, 10, 1986, p. 369-376.

5. F. Hegedüs, Fast neutron dosimetry by means of the scraping method. in ref (2) p. 381-390.

6. W. Slavin, "Flames, furnaces, plasmas. How do we choose ?" Analytical Chemistry, Vol. 58, No 4, April 1986, p. 589-597.

7. J.A. Holcombe and D.A. Bass, Atomic absorption, atomic fluorescence, and flame emission spectrometry. Anal. Chem. Vol. 60., No 12, June 15, 1989, 226-252R.

8. D.I.Kurbatov, A.V.Trubachev, Polarographic determination of transition metals with an aqueous-organic background electrolyte. International laboratory, Nov. 1989, p. 16-23.

9. I. Kondurov et al. (Private communication 1988). Leningrad Nuclear Research Institute, USSR.

10. I. Szaloki, S. Szegedi, M. Varadi, Cs. M. Buczko, A. Pazsit, F. Divos, An improved method for the determination of the neutron fluence in fission power reactors based on the $^{93}$Nb(n,n')$^{93m}$Nb reaction. Ann. nucl Energy, Vol. 15, No 10/11, pp. 523-525, 1988

11. R. McCrohen, M. E. Plested, and A. J. Fudge, Measurement of $^{93m}$Nb for fast neutron metrology. in (2) p. 245-252.

12. I. Auterinen, B. Bärs, H. Lauranto, K. Nyholm, R. Salomaa, Resonance ionization spectroscopy of niobium: determination of $^{93m}$Nb/$^{93}$Nb isomer ratio for fast neutron fluence measurements. Fourth Int. Symp. on Resonance Ionization Spectroscopy & Its Applications 1988. Inst.Phys.Conf Ser. No 94: Section 8.

13. H.M. Lauranto, I.H. Auterinen, T.T. Kajava, K.M. Nyholm, R.R.E. Salomaa, Determination of hyperfine structures and Rydberg convergence limits of selected optical transitions in $^{93}$Nb using resonance ionization spectroscopy. Appl. Phys. B50, 323-329, 1990

# NUCLEAR DATA

# NUCLEAR MODEL CALCULATION OF DOSIMETRY REACTIONS ON $^{93}$Nb

Naoteru Odano, Shin Iwasaki and Kazusuke Sugiyama

Department of Nuclear Engineering, Tohoku University

Aramaki-Aza-Aoba, Aoba-Ku, Sendai 980, Japan

## ABSTRACT

Cross sections for the dosimetry reactions, the $^{93}$Nb$(n,n')^{93m}$Nb and $^{93}$Nb$(n,2n)^{92m}$Nb reactions, have been calculated by the statistical model code with preequilibrium correction, the ELIESE-GNASH joint program. The calculations have been performed for not only the cross sections of the dosimetry reactions but also the other reaction cross sections and particle emission spectra. The model calculation have also been carried out for the secondary reactions on the residual nucleus of the $^{93}$Nb$(n,n')^{93m}$Nb reaction for the purpose of estimating the burn-up in $^{93m}$Nb. The burn-up rate would be negligible under assumed irradiation condition in a power reactor. The $^{93}$Nb$(n,n')^{93m}$Nb reaction cross section has been evaluated from the result of the nuclear model calculation and available experimental data by the generalized least squares program, GMA, for the use of the cross section in the reactor dosimetry.

## INTRODUCTION

The $^{93}$Nb$(n,n')^{93m}$Nb and $^{93}$Nb$(n,2n)^{92m}$Nb reactions are of importance for dosimetry applications. Specifically, the $^{93}$Nb$(n,n')^{93m}$Nb reaction is considered as a reaction for the measurement of fast neutron fluence by the activation in the vicinity of the pressure vessel of light water reactor. The methodology of the use of the reaction for fast neutron dosimetry was discussed in the previous ASTM-EURATOM Symposium (1).

In spite of the importance of the $^{93}$Nb$(n,n')^{93m}$Nb reaction, the experimental data of the cross section is still sparse and the uncertainties of the data are large due to the difficulty of the activity measurement. Recently, a few experimental efforts have been performed to obtain more accurate data (2,3). A new theoretical calculations have also been presented by Strohmaier (4). In the present study, our aim was focused on obtaining a reliable excitation function of the $^{93}$Nb$(n,n')^{93m}$Nb reaction cross section. The nuclear model calculations were performed not only for the $^{93}$Nb$(n,n')^{93m}$Nb reaction cross section but also the other reactions and particle emission spectra for neutrons, charged particles and $\gamma$-rays to obtain a consistent data set of the Nb cross sections. Adjustment procedures of several parameters in nuclear model codes are also presented. Evaluation of the covariance matrix of the $^{93}$Nb$(n,n')^{93m}$Nb reaction will be described. The burn-up in produced $^{93m}$Nb state by long term irradiation in a power reactor was estimated.

## NUCLEAR MODEL CALCULATION

### Nuclear Model Codes

In this study, a joint program EGNASH (5), composed of the codes ELIESE (6) and GNASH (7), and the code DWUCK (8) were used for the calculation of the neutron induced reaction cross sections. The EGNASH code is the main code in this study. The code is based on the statistical nuclear model with the preequilibrium correction by the exciton model. The code provides reaction cross sections, level production cross sections and spectra (neutron, $\gamma$-ray and charged particles) resulting from particle induced reactions. The transmission coefficients for the GNASH code were prepared by the ELIESE code which is implemented in the EGNASH. A feature of the EGNASH code is that, it enables us to calculate the cross sections for isomeric state as target nucleus. The DWUCK code, distorted wave Born approximation code, was used to calculate the direct inelastic scattering cross sections. The contribution of the direct inelastic scattering cross sections were taken into account in the calculation of the EGNASH code.

### Optical Model Potential Parameters

The modified Walter-Guss neutron potential parameters (7), which are the default ones implemented in the EGNASH code, were used as an initial guess parameter set. The calculated total cross section by this set could not reproduce the experimental data below a few MeV. To improve the degree of agreement, an optical model potential parameter set by Smith et. al. (Table 1) (9) was adopted, among several sets. According to the SPRT criterion (10) for choosing the potential, the s- and p- wave neutron strength functions, scattering radius and total cross section (SPRT data) were checked. The calculated SPR data are listed in Table 2 with the experimental data and ones by other surveyed potential sets.

Table 1. Neutron optical model potential parameter set derived from the volume integral values given by Smith et. al. (9). The geometrical parameters are expressed in fm, while the potential strength and incident neutron energy $E$ are in MeV.

| real | surface imaginary | spin orbit |
|---|---|---|
| $V = 47.34 - 0.25E$ | $W = 5.841 + 0.366E$ | $V_{so} = 6.0$ |
| $r_0 = 1.25$ | $r_I = 1.3$ | $r_{so} = 1.25$ |
| $a_r = 0.698$ | $a_I = 0.465$ | $a_{so} = 0.698$ |

Table 2. The s-, p- wave neutron strength functions (s0, s1) and scattering radius R′ which calculated using Smiths' potential parameter set were compared with ones using other parameter sets.

| | s0($\times 10^{-5}$) | s1($\times 10^{-5}$) | R′(fm) |
|---|---|---|---|
| Experiment (11) | $0.6 \pm 0.07$ | $5.8 \pm 0.6$ | $6.9 \pm 0.1$ |
| Smith (9) | 0.71 | 5.3 | 6.8 |
| mod. Walter-Guss (7) | 0.40 | 6.7 | 6.6 |
| Delaroche (10) | 0.43 | 4.8 | 6.6 |
| Lagrange (12) | 0.44 | 4.3 | 6.9 |

## Level Density Parameters and Discrete Level Data

In the original EGNASH code, the Gilbert-Cameron level density formula was adopted. In this study, the energy dependence of the level density parameter developed by Ignatyuk (13), which is effective for the nuclei in the neighborhood of closed shell, e.g. Nb, was implemented. The energy dependence of the level density parameter can be expressed by the following formula,

$$a(U) = a(1 + f(U)dW/U), \tag{1}$$

where $U$ is the excitation energy and "$a$" is the asymptotic value for the Fermi gas parameter occurring at higher energies. The term $dW$ denotes the shell correction term and is evaluated by the mass formula of Cameron and Elkin (14). The term $f(U)$ describes the energy dependence of the parameter by $f(U) = 1 - exp(-0.05U)$.

The discrete level data (spin and parity of discrete levels, and $\gamma$-ray decay scheme) are important input data in the present calculation. The data for $^{93}$Nb were taken from the evaluation by Demanins et. al. (15). The data of the Table of Isotopes (16) were used for the other nuclei.

## Summary of Calculated Reaction Cross Sections

The calculated excitation functions of the $^{93}$Nb$(n,n')^{93m}$Nb, $^{93}$Nb$(n,2n)^{92m}$Nb, $^{93}$Nb$(n,2n)^{92m+g}$Nb, $^{93}$Nb$(n,\alpha)^{90m}$Y, and $^{93}$Nb$(n,\alpha)^{90m+g}$Y were compared with their experimental and existing evaluated data. The excitation function of the $^{93}$Nb$(n,n')^{93m}$Nb reaction is slightly lower than the experimental data in the energy range from 2 MeV to 4 MeV as shown in Fig.1. The excitation function of the isomeric state production cross section via the $(n,2n)$ reaction agreed well with the recent accurate experimental data (Fig.2). The sum of production of the isomeric state and the ground state is somewhat higher than experimental data from threshold energy to 12 MeV. For the $(n,\alpha)$ reaction, two excitation functions of the isomeric state production and the sum of production of the ground state and the isomeric state agreed with experimental data. The calculated $(n,\gamma)$ reaction cross section reproduced the experimental data in acceptable level. The particle emission spectra for neutrons, charged particles and $\gamma$-rays, were also compared with experimental data or evaluated data at 14 MeV. Agreement was rather good.

## Discussion on $^{93}$Nb$(n,n')^{93m}$Nb Reaction Cross Section

The degree of agreement with experimental data was improved by adopting the energy dependence of the level density parameter compared with the previous work by the authors (17). As mentioned in the previous sub-section, however, the calculated excitation function of the $^{93}$Nb$(n,n')^{93m}$Nb reaction was slightly lower than the recent experimental data and the calculated excitation function by Strohmaier (4). At 14 MeV, there exists only one experimental data by Ryves et. al. (18). The calculated cross section was out range of the error. Because very few experimental data exist in the energy region from 6 MeV to 20 MeV, it is difficult to deduce the reliability of the calculated excitation function at the present stage.

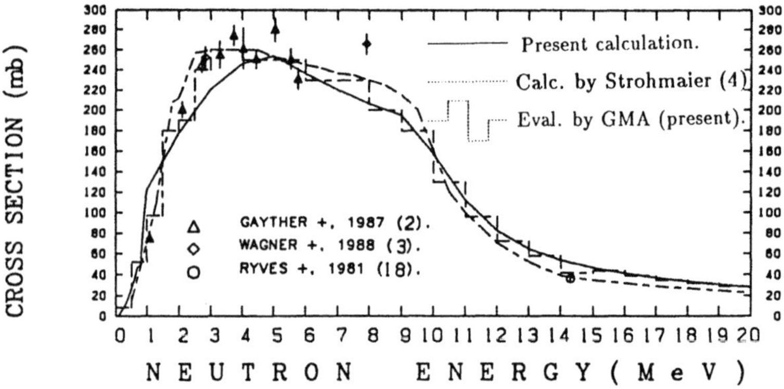


Fig.1. The $^{93}$Nb$(n,n')^{93m}$Nb cross sections; the present calculation by EGNASH, the Strohmaier's and the present GMA evaluation (see text), as well as the three experimental data by Gayther et. al., Wagner et. al. and Ryves et. al.

$^{93}Nb(n,2n)^{92m+g}Nb$

- # PAULSEN+, Z. Physik, Vol.238, p.23 (1970).
- ⧗ MATHER+, Report EANDC(UK)-142-AL (1972).
- ✳ LYCHAGIN+, Report FEI-1385 (1983).
- ★ HOLUB+, J. Phys. G2, p.405 (1976).
- □ HERMSDORF+, Reports ZfK-262 and ZfK-271 (1974).
- △ HARING+, Z. Physik, Vol.244, p.352 (1978).
- ◇ VESSER+, Phys. Rev. C16, p.1792 (1977).
- ○ FREHAUT+, Proc. Symp. Neutron Cross Sections, Upton, Vol.1,p.399 (1980).

$^{93}Nb(n,2n)^{92m}Nb$

- ✱ WEN+, Phys. Rev. C1, p.358 (1970).
- X TEWES+, Report UCRL-6028-T (1960).
- \+ PAULSEN+, Z. Physik, Vol.238, p.23 (1979).
- # NETHAWAY+, J. Inorg. Nucl. Chem., Vol.40, p.1285 (1978).
- ⧗ MANNAN+, Phys. Rev. C39, p.630 (1988).
- ✳ HUDSON+, Ann. Nucl. Energy, Vol.5, p.589 (1978).
- ★ HOLUB+, J. Phys. G2, p.405 (1976).
- □ CSIKAI, Int. Conf. Nucl. Data for Sci. and Tech., Antwerp, p.414 (1983).
- △ WOELFE+, Appl. Radiat. Isot., Vol.39, p.407 (1988).
- ◇ LU HANLIN+, IAEA-NDS-30637 (1982).
- ○ IKEDA+, JAERI-1312 (1988).

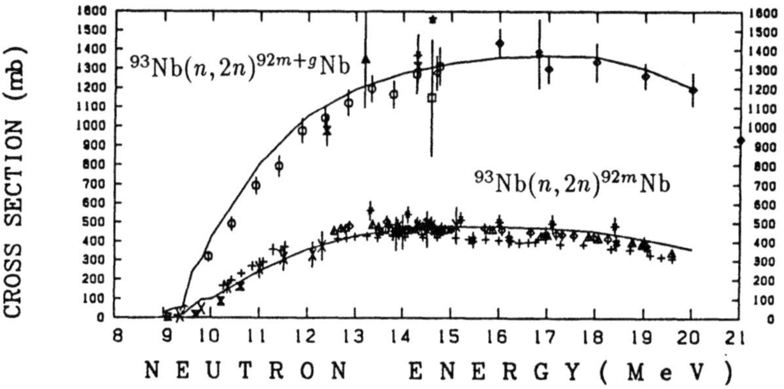


Fig.2 The calculated $^{93}Nb(n,2n)^{92m}Nb$ and $^{93}Nb(n,2n)^{92m+g}Nb$ reaction cross sections with experimental data. The references are shown in this figure by abbreviation.

## EVALUATION OF $^{93}Nb(n,n')^{93m}Nb$ CROSS SECTION

A Pascal version of GMA code (19) was used for evaluating the cross section of the $(n,n')$ reaction with its covariance matrix based on the above theoretical calculation taking into account the recent experimental data. The result of the present calculation was considered to be one of the experimental data with the appropriately assumed error band and their correlation. Below 6 MeV, experimental data by

Gayther (2) and Wagner (3) in the vicinity of 2 MeV were mainly taken into account, while the weight of the experimental data by Wagner in the vicinity of 8 MeV and by Ryves (18) at 14 MeV was reduced in order to eliminate unnatural structure in the excitation function. The result of the evaluation reasonably follows the experimental data below 6 MeV, as shown in Fig.1.

The spectrum averaged cross section for the $^{93}$Nb$(n,n')^{93m}$Nb in the $^{252}$Cf fission spectrum (20) was calculated. The cross section value was 145 mb and in good agreement with the recent experimental results by Alberts et.al. (145 $\pm$ 5 mb) (21), Williamson et.al. (144 $\pm$ 5 mb) (22) and Williasms et.al. (146.4 $\pm$ 3.8 mb) (23).

## ESTIMATION OF BURN-UP OF $^{93m}$Nb

At long term irradiation of Nb foils in a power reactor, burn-up in the isomeric state, $^{93m}$Nb, caused by secondary reaction on the $^{93m}$Nb must be considered. The cross sections of the reactions $^{93m}$Nb$(n,x)$Y, where $x$ means $\gamma$, $n'$, $2n$, $p$, $\alpha$,..., and Y means all residual nuclei except for the $^{93m}$Nb, are necessary for estimation of the effect. The nuclear model parameters of the calculation were the same as those in the case of the stable target. The contribution of the resonance cross section was not included in the calculation. The sum of the cross sections of all of the $^{93m}$Nb$(n,x)$Y reactions was 3.7 b at the thermal neutron energy. The amount of the loss effect was estimated under the following typical conditions : Average neutron spectrum in the core of a 110MW PWR with the flux of $3.2 \times 10^{14} n/cm^2/sec$ for twenty years of the irradiation period. In this case, there would be only 0.6% decrease in the $^{93m}$Nb produced via the $^{93}$Nb$(n,n')^{93m}$Nb reaction.

## CONCLUSIONS

The reaction cross sections for $^{93}$Nb were calculated with statistical model code EGNASH and a consistent data set has been obtained. The calculated excitation function of the $^{93}$Nb$(n,n')^{93m}$Nb reaction was slightly lower than recent experimental data in the energy range from 2 MeV to 4 MeV. For the purpose of use as dosimeter, a reasonable cross section and covariance matrix of the $^{93}$Nb$(n,n')^{93m}$Nb reaction have been evaluated by the GMA code taking into account the present calculated data in addition to the experimental data. The evaluated data reasonably followed the recent experimental data and can be used for dosimetry applications. The spectrum averaged cross section was calculated and agreed with recent integral experimental data. Further experimental work on the cross section especially above 4 MeV is required for this reaction. Numerical data of the evaluated data and their covariance matrix are available from the authors.

## REFERENCES


1. H. Farrar IV and E. P. Lippincott, Editors, Reactor Dosimetry: Methods, Applications, and Standardization, ASTM Special Technical Publication; 1001 (1989).

2. D. B. Gayther, C. A. Uttley, M. F. Murphy, W. H. Taylor and K. Randle, "A Measurement of the $^{93}$Nb$(n,n')^{93m}$Nb Cross-Section in the Energy range 1 to 6 MeV", Proc. Int. Conf. Nucl. Data for Science and Technology, Mito, Japan, May 30 - June 3, 1988, p.1057 (1988).

3. M. Wagner, G. Winkler, H. Vonach and H. Liskien, "Measurement of the $^{93}$Nb $(n,n')$ $^{93m}$Nb Cross Section at 7.9 MeV", ibid., p.1049 (1988).

4. B. Strohmaier, "New Issue of Nuclear Model Calculations of Cross Sections for Neutron-Induced Reactions on $^{93}$Nb to 20 MeV", Ann. Nucl. Energy, Vol. 16, (1989), p. 461.

5. S. Igarashi, "Modified Method of Neutron Cross-Section Calculations", J. Nucl. Sci. Tech., Vol.12, (1975), p.67.

6. P. G. Young and E. D. Arthur, "GNASH : A Preequilibrium, Statistical Nuclear-Model Code for Calculation of Cross Section and Emission Spectra", LA-6947 (1977).

7. N. Yamamuro, "A Nuclear Cross Section Calculation System with Simplified Input-Format Version I, (SINCROS-I)", JAERI-M 88-140 (1988) (in Japanese).

8. P. D. Kunz, "DWUCK", University of Colorado Report (1974).

9. A. B. Smith, P. T. Guenther and R. D. Lawson, "On the Energy Dependence of the Optical Model of Neutron Scattering from Niobium", ANL/NDM-91 (1985).

10. J. P. Delaroche, Ch. Lagrange and J. Salvy, "The Optical Model with Particular Consideration of the Coupled-Channel Optical Model", Proc. of a Consultant Meeting on the Use of Nuclear Theory in Neutron Nuclear Data Evaluation organized by the IAEA Nuclear Section in Co-Operation with the International Center for Theoretical Physics, Trieste, IAEA-190 (1975).

11. S. F. Mughabghab, M. Divadeenam and N. E. Holden, Neutron Cross Sections, Volume 1, Neutron Resonance Parameters and Thermal Cross Section, Part A, Z=1-60, Academic Press (1981).

12. Ch. Lagrange and A. Lejeune, "Phenomenological and Microscopic Optical Model Analysis of the Interaction of Low-Energy Nucleons with $^{93}$Nb", Phys. Rev. C, Vol.25 (1982) p.2278.

13. A. V. Ignatyuk, "Statistical Properties of Excited Atomic Nuclei", INDC (CCP)-233/L (1985).

14. A. G. W. Cameron and R. M. Elkin, "Role of the Symmetry Energy in Atomic Mass Formulas", Can. J. Phys., Vol.43 (1965) p.1288.

15. F. Demanins, U. Abbondano and F. Raicich, "Level and Decay Scheme of $^{97}$Mo and $^{93}$Nb", Report INFN(REP)-014/88, Instituto Nazionale de Legnaro, Annual Report 1987 (1988).

16. C. M. Lederer and V. S. Shirley, Table of Isotopes, Seventh Edition, John Wiley & Sons, Inc., 1978.

17. N. Odano, S. Iwasaki and K. Sugiyama, "A Preliminary Nuclear Model Calculation for $^{93}$Nb$(n,n')^{93m}$Nb Reaction Cross Section", NETU-52, Department of Nuclear Engineering, Tohoku University, Research Report (1989).

18. T. B. Ryves and P. Kolkowski, "The Nb$(n,2n)$, Nb$(n,\alpha)$ and $(n,n')$ Cross Sections At 14 MeV", J. Phys. G, Vol.7 (1981) p.529.

19. W. P. Poenitz, "Data Interpretation, Objective Evaluation Procedures and Mathematical Techniques for the Evaluation of Energy - Dependent Ratio, Shape and Cross Section Data", Proc. of the Conference on Nuclear Data Evaluation Methods and Procedures, BNL-NCS-51363, (1981) p.249. Modified by M. Sugimoto for PC (Pascal) version.

20. W. Mannhart, "Evaluation of the Cf-252 Fission Neutron Spectrum Between 0 MeV to 20 MeV", IAEA-TECDOC-410, (1987) p.158.

21. W. G. Alberts, U. Schotzig and H. Siergert, "Nuclear Data for the Use of Niobium as Neutron Fluence Monitor", Proc. of the 6th ASTM-EURATOM Symposium on Reactor Dosimetry, ASTM STP 1001, p.223 (1989).

22. T.G. Williamson, T.A. Ayers, W.L. Hammersten, G.P. Lamaze and F.J.Schima, "Measurement of the $^{93}$Nb$(n,n')$ Fission Spectrum Cross Section", ibid., p.229 (1989).

23. J. G. Williams, C. O. Cogburn, L. M. Hodgson, S. C. Apple, E. D. McGarry, G. P. Lamaze, F. J. Schima, JW Rogers, R. J. Gehrke, J. D. Baker and F. J. Wheerer, "Measurements of Fission Spectrum Averaged Cross Section for the $^{93}$Nb$(n,n')^{93m}$Nb Reaction", ibid., p.235 (1989).

# IMPROVEMENTS IN ENDF/B-VI IRON AND POSSIBLE IMPACTS ON PRESSURE VESSEL SURVEILLANCE DOSIMETRY

C. Y. Fu, D. M. Hetrick, C. M. Perey, F. G. Perey,
N. M. Larson, D. C. Larson

Oak Ridge National Laboratory

Oak Ridge, TN 37831

## ABSTRACT

The ENDF/B-VI cross-section evaluations for the four iron isotopes are summarized, emphasizing the major improvements over ENDF/B-V. The evaluations were mostly based on a preliminary file generated in 1986 for natural iron that has been used for re-calculating several neutron-transport experiments, all of which showed improved agreement. These re-analyses, including those for pressure-vessel surveillance dosimetry, are also discussed.

## INTRODUCTION

The results (Ref. 1 and references therein) from combining integral, differential, and calculated dosimetry data in the pressure vessels of several existing power reactors, using ENDF/B-V nuclear data, indicated that the iron inelastic scattering cross section in ENDF/B-V is about 8% too large from 3 to 8 MeV. This indication was confirmed by newly available $(n, n'\gamma)$ data[2] and a new model calculation for $^{56}$Fe using an improved nuclear model code[3] to fit all of the available $(n, n')$ data and other relevant cross sections. All neutron-producing reaction cross sections above 3 MeV in the ENDF/B-V iron evaluation were revised using the new model fits, resulting in a preliminary file[4] for ENDF/B-VI. In this preliminary file, the energy-angle correlations for the inelastically scattered neutrons in the continuum were also introduced. The processed cross-section set[5] in the Vitamin-E group structure[6] has been collapsed to various broad-group structures for the re-analyses of several neutron-transport experiments,[7−11] including the surveillance dosimetry mentioned above, all with improved agreement. The re-analysis[7] of the neutron leakage spectrum of a thick iron-sphere experiment[12] with a central 14-MeV source showed that about half of the 100% increase in the leakage spectrum between 3 and 10 MeV is due to the lowered inelastic scattering cross section in the new evaluation and the other half is due to the forward-peaked angular distributions introduced for the inelastically scattered neutrons in the continuum.

The test results from using the preliminary ENDF/B-VI evaluation for iron are summarized first. Other major improvements to be discussed are: (1) the separation of the evaluation into four isotopic files, (2) the use of the new File-6 format for better representation of correlated energy-angle distributions of the outgoing neutrons and for introducing the energy distributions of the charged-particle and recoil spectra, and (3) the extension of the resonance parameter region of each isotope to higher neutron energy and the application of measured angular distributions for elastically scattered neutrons to determine the spins and parities of the resonances.

## TEST OF THE PRELIMINARY EVALUATION

The preliminary ENDF/B-VI file for iron was intended to remove several outstanding discrepancies between integral data and calculations which occurred with the ENDF/B-V iron evaluation.

For a 14-MeV source, the $(n, n')$ cross section for the continuum (threshold = 4.5 MeV) is important for any neutron transport problem. It is well known[13] that inelastically scattered neutrons in the continuum are forward-peaked. Forward-peaked neutrons are more penetrating, hence the isotropic assumption for the inelastically scattered neutrons in the continuum in ENDF/B-V iron is no doubt a source of error for thick-shield analysis involving fusion sources.[12]

In the cases of fission problems, analyses of three integral measurements consistently indicated that the Fe$(n, n')$ cross section in ENDF/B-V above 3 MeV was too large. The three problems are: (1) the pressure vessel surveillance dosimetry for several existing power reactors,[1,8] (2) the source terms for the Hiroshima and Nagasaki dosimetry reevaluation program,[10] and (3) the Tower Shielding Facility deep penetration measurements.[11]

New Fe$(n, n'\gamma)$ data for the 846-keV level[2] and the new TNG code[3] capable of calculating angular distributions for continuum neutrons were used for addressing the above two types (fission and fusion) of problems and generating the preliminary file. The corresponding multigroup cross-section set[6] was applied to the four experiments mentioned above, all showing improved agreement with the data.[7–11] Smaller discrepancies remain, but they are rather difficult to resolve. For example, Maerker[8,9] suggested that the Fe$(n, n')$ cross section below 3 MeV is still a few percent too large but we don't have any new experimental data or theoretical means (because the cross section below 3 MeV is structured) to answer this question at present. Additional high-resolution (5 keV) and/or high-accuracy (2%) measurement in this energy region are needed.

The ENDF/B-VI iron evaluation retains the basic features of the preliminary file. Figure 1 compares the Fe$(n, n')$ cross sections between ENDF/B-VI and -V. The newly available data of Larson[2] are also shown. The ENDF/B-VI results shown are abundance-averages of the isotopic evaluations and closely resemble those shown for the preliminary file,[4] where older data were also plotted.

Figure 2 illustrates the calculated and experimental[13–15] double differential Fe($n,xn$) cross sections for 14.5-MeV incident neutrons and several secondary energy ranges. Equally good agreement between calculation and experiment has been shown[4] for the incident energy of 26 MeV, suggesting reliability of the model used for the calculation. The ENDF/B-V data, if plotted in Fig. 2, would appear as horizontal lines.

## INTRODUCTION OF ISOTOPIC EVALUATIONS

The isotopic evaluations enabled us to make use of additional data measured for the individual isotopes. For example, careful analyses of the inelastic scattering cross sections of the minor isotopes make the values of the inelastic cross section for natural iron between 1.4 and 3.4 MeV about 2% smaller, a change that would not have drawn our attention without evaluating the cross sections of the minor isotopes separately. This reduction in the inelastic scattering cross section of natural iron may result in improved agreement between calculated and measured $^{237}$Np($n,f$) responses in the pressure vessels of existing power reactors. However, confirmation of it will await processing of the new evaluation and re-analysis of the surveillance dosimetry data.

The separation of the natural iron evaluation into four isotopic evaluations allows the representation of activation and dosimetry reaction cross sections in the general purpose files. For example, the dosimetry reactions $^{54}$Fe($n,p$) and $^{56}$Fe($n,p$) are now contained in the respective general purpose files. These two cross sections were updated for ENDF/B-VI primarily by using the ENDF/B-VI $^{235}$U($n,f$) and $^{238}$U($n,f$) standards in the GLUCS package containing 14 correlated reactions.[16] The evaluation of 37 high-priority dosimetry and activation cross sections available in the general purpose files for which we are responsible has been summarized.[17]

In an evaluated cross-section file for an element containing several isotopes, it is not possible to represent the energies of the reaction products correctly, hence the energy balance cannot be guaranteed. For example, the Q-values for the ($n,\gamma$) events for $^{54}$Fe and $^{56}$Fe are, respectively, 9.3 and 7.8 MeV. This difference in the capture gamma-ray energies can be correctly accounted for only in the new isotopic evaluations, ensuring energy balance and more reliable estimates of derived quantities such as Kerma factors. In the area of radiation damage studies, the new File-6 format plays an important role, which is explained below.

## USE OF FILE-6 FORMAT

Correlated energy-angle distributions of outgoing neutrons can now be given efficiently in the new File-6 format in ENDF/B-VI. In the case of the iron isotopes, correlated energy-angle distributions are given for the outgoing neutrons for each neutron-producing reaction, and energy distributions are available for the emitted protons, alpha-particles, heavy recoils, and gamma rays. The incident neutron energy and the sum of the average energies of all reaction products have been checked to be consistent within 1%. The fact that energies are balanced and the fact that energy distributions of heavy recoils are given should have a positive impact on derived quantities such as displacement cross sections for radiation damage studies.

## RESONANCE PARAMETERS

SAMMY,[18] an R-matrix code based on the Reich-Moore formalism and Bayes' Equations, was used for the resonance region of the iron isotopes. For $^{56}$Fe, the analysis was based on measured cross sections for total[19] and capture,[20] and the angular distributions for neutron elastic scattering.[19] The resolved resonance region for all isotopes were extended to higher energy than available in ENDF/B-V. In the case of $^{56}$Fe, the resonance region was increased from 400 to 850 keV. An example is shown in Fig. 3 where the measured transmission data[19] and elastic angular distributions[19] jointly determine the spin and parity assignments of several resonances. The new ENDF/B-VI formats allow the calculation of elastic angular distributions in the resonance region directly from the resonance parameters. It has been shown[19] that resonance parameters evaluated without considering the scattering data often resulted in incorrect spin and parity assignments and hence poor results for the calculated angular distributions. The ENDF/B-V data in the energy region shown in Fig. 3 agree well with the ENDF/B-VI results shown; therefore they are omitted. However, angular distributions for the elastically scattered neutrons, such as shown in the three lower graphs of Fig. 3, cannot be obtained from ENDF/B-V.

## CONCLUDING REMARKS

Five major improvements in the ENDF/B-VI evaluation for iron over Version V have been summarized, namely, (1) the consideration of integral data in a preliminary ENDF/B-VI evaluation, (2) the introduction of energy-angle correlations for the nonelastically scattered neutrons in the continuum, (3) the separation of the evaluation into four isotopic files, (4) the use of the new File-6 format for charged-particle energy distributions and energy balance assurance, and (5) the extension of the resolved resonance parameters to higher energies and the ability to calculate angular distributions for the elastically scattered neutrons in the resonance energy range. The first two have been positively tested, particularly in the surveillance dosimetry area. The other three improvements are solidly based on advanced model analyses of a rather large data base.

## ACKNOWLEDGMENTS

The authors wish to thank Drs. J. K. Dickens and R. E. Maerker for helpful comments and discussions. This research was sponsored by the Office of Energy Research, U. S. Department of Energy, under contract DE-AC05-84OR21400 with Martin Marietta Energy Systems, Inc.

## REFERENCES

1. R. E. Maerker et al.,"Combining Integral and Differential Dosimetry Data in an Unfolding Procedure with Application to the Arkansas Nuclear One-Unit 1 Reactor," p. 477, *Proc. Int. Conf. Nucl. Data for Basic and Applied Sciences*, Santa Fe, May 13-17, 1985.
2. D. C. Larson, "High-Resolution Structural Material $(n, x\gamma)$ Production Cross Sections for $E_n$ from 0.2 to 40 MeV," p. 71, *Proc. Int. Conf. Nucl. Data for Basic and Applied Sciences*, Santa Fe, May 13-17, 1985.
3. C. Y. Fu, *Nucl. Sci. Eng.* **100**, 61 (1988).
4. C. Y. Fu and D. M. Hetrick, *Update of ENDF/B-V Mod-3 Iron: Neutron-Producing Reaction Cross Sections and Energy-Angle Correlations*, ORNL/TM-9964 (ENDF-341), 1986.
5. R. W. Roussin and J. I. Bartley, Oak Ridge National Laboratory, private communication (1986).
6. C. R. Weisbin et al., *Vitamin-E: An ENDF/B-V Multigroup Cross-Section Library for LMFBR Core and Shield, LWR Shield, Dosimetry and Fusion Blanket Technology*, ORNL-5505 (ENDF-274), Oak Ridge National Laboratory (1979).
7. C. Y. Fu and D. M. Hetrick, "Evaluation and Testing of Double Differential Fe$(n, n')$ Cross Sections," *Trans. Am. Nucl. Soc.* **53**, 409 (1986).
8. R. E. Maerker, *Nucl. Sci. Eng.* **96**, 288 (1987).
9. R. E. Maerker, *Analysis of the NESDIP2 and NESDIP3 Radial Shield and Cavity Experiments*, ORNL/TM-10389, Oak Ridge National Laboratory (1987).
10. J. V. Pace, Oak Ridge National Laboratory, private communication (1986).
11. C. O. Slater, JASPER Workshop, Oak Ridge National Laboratory, April 1987.
12. R. H. Johnson, J. J. Dorning, and B. W. Wehring, *Trans. Am. Nucl. Soc.* **22**, 799 (1975).
13. J. L. Kammerdiener, *Neutron Spectra Emitted by Fe Irradiated by 14 MeV Neutrons*, UCRL-51232, Lawrence Livermore Laboratory, 1972.
14. D. Hermsdorf et al., *Differential Neutron Emission Cross Sections from Iron by 14.6 MeV Neutrons*, ZfK-277, Institute of Nuclear Physics, University of Dresden, 1975.
15. A. Takahashi et al., "Double Differential Neutron Emission Cross Sections, Numerical Tables and Figures," A-83-01, Intense 14-MeV Neutron Source Facility, Osaka University, 1983.
16. C. Y. Fu and D. M. Hetrick, "Experience in Using the Covariances of Some ENDF/B-V Dosimetry Cross Sections: Proposed Improvements and Addition of Cross-Reaction Covariances", p. 877, *Proc. 4th*

*ASTM-EURATOM Symposium on Reactor Dosimetry*, NUREG/CP-0029, Gaithersburg, Maryland, March 22-26, 1982.
17. C. Y. Fu, "A Review of Activation Cross Sections in the ENDF/B-VI General Purpose Files for Cr, Fe, Ni, Cu, and Pb," p. 275 , *Proc. NEANDC Specialists' Meeting on Neutron Activation Cross Sections for Fission and Fusion Energy Applications*, NEANDC-259 'U', Argonne National Laboratory, Sept. 13-15, 1989.
18. N. M. Larson and F. G. Perey, *Users' Guide for SAMMY: A Computer Model for Multi-Level R-Matrix Fits to Neutron Data Using Bayes' Equations*, ORNL/TM-7485 (1980); Updates ORNL/TM-9179 (1984), ORNL/TM-9179/R1 (1985) and /R2 (1989).
19. C. M. Perey et al., Oak Ridge National Laboratory, to be published (1990).
20. F. Corvi et al., *Proc. Consultants Meeting on Nuclear Data for Structural Materials*, Vienna, Austria, 1983, International Nuclear Data Committee Report INDC(NDS)-152 L, 1984 and private communication to C. M. Perey, 1984.

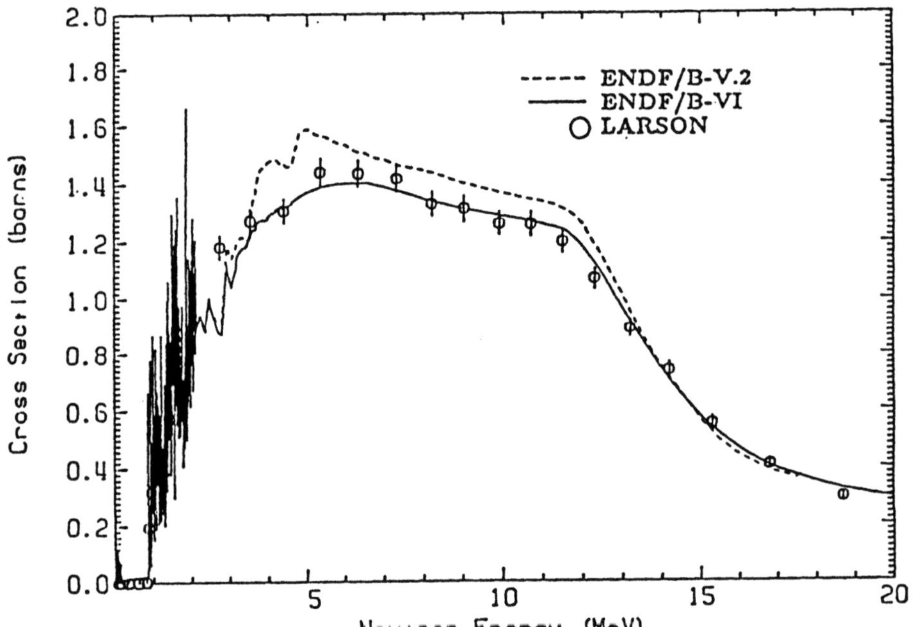


Fig. 1. Comparison of the Fe inelastic cross sections of ENDF/B-VI with ENDF/B-V and with experimental data of Larson.[2]

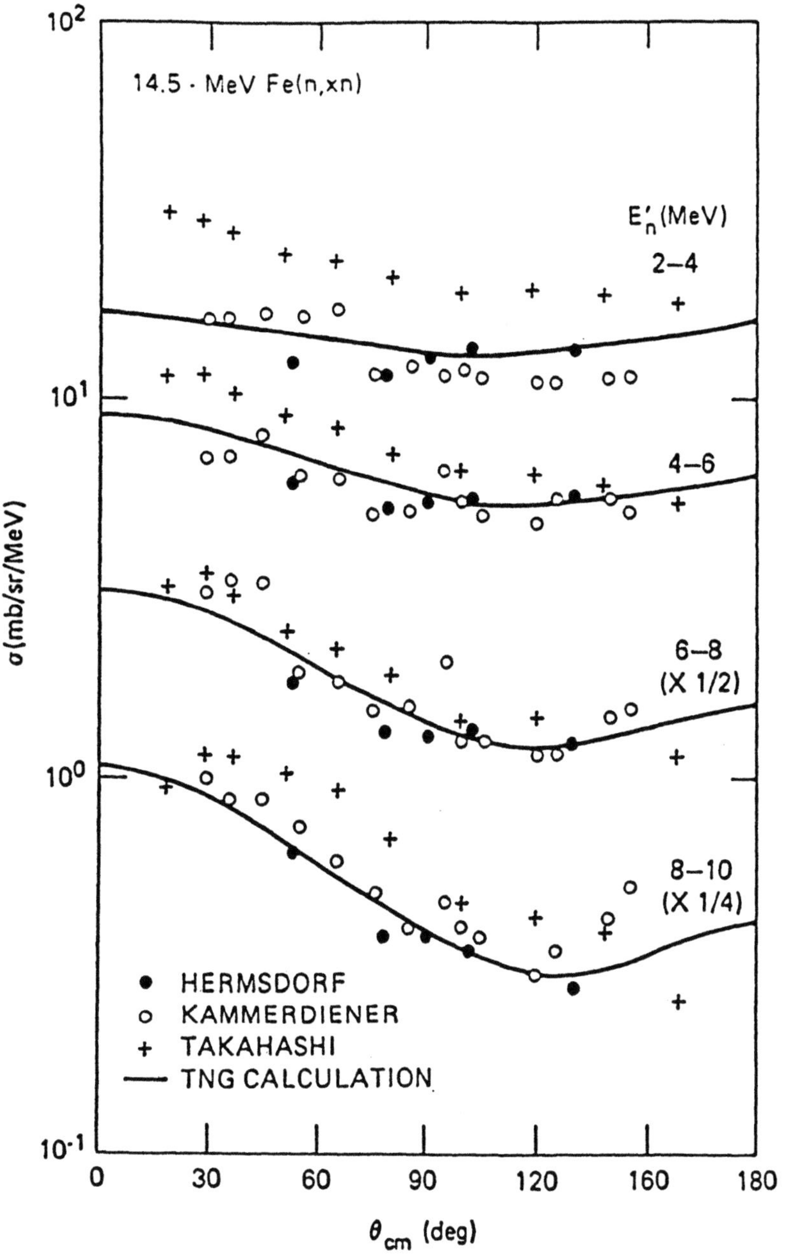


Fig. 2. Calculated and experimental[13,14,15] double differential Fe$(n, xn)$ cross sections for 14.5-MeV incident neutrons and several secondary energy ranges.

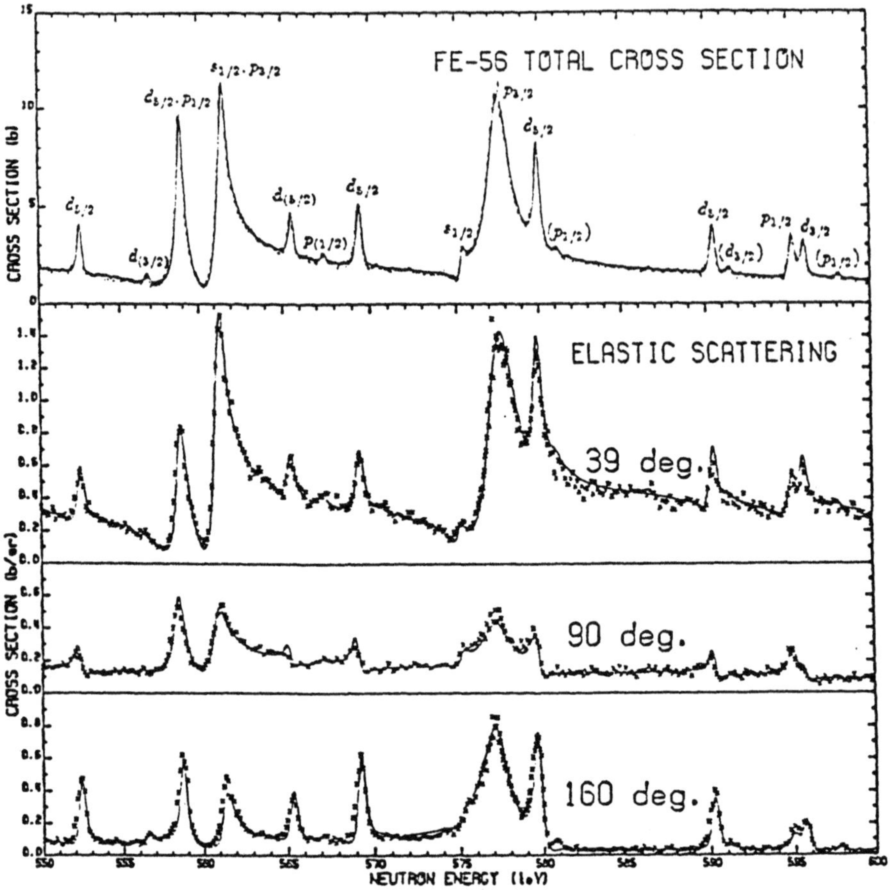


Fig. 3. In the upper graph the theoretical total cross sections, calculated with the parameters of ENDF/B-VI $^{56}Fe$, are compared with the data.[19] Parentheses are used to indicate uncertain spin and parity assignments. The assignments were made using the differential elastic scattering data[19] shown in the three lower graphs.

# INTERNATIONAL REACTOR DOSIMETRY FILE (IRDF-90)

## STATUS AND TESTING

N.P. Kocherov
International Atomic Energy Agency
P.O. Box 100, A-1400 Vienna, Austria

H.K. Vonach
Institut für Radiumforschung und Kernphysik
A-1090 Vienna, Austria

ABSTRACT

The new version of the International Reactor Dosimetry File is put together by combining the ENDF/B-VI Dosimetry File and evaluations made at IRK in Vienna, Austria. The inclusion of some evaluations made at the Chinese Nuclear Data Centre by Prof. Cai Dunjiu and his group is also under consideration. The original evaluations were processed to 640 group format. Comparisons with the previous evaluations and existing experimental data base were made.

---

The majority of the evaluations in the previous version of the International Reactor File (IRDF-85) which was issued in 1985 have been superseded by new evaluations and this explains the necessity and timeliness of its revision.

The 1990 version 1 of the International Reactor Dosimetry File (IRDF-90, V.1) contains 31 neutron dosimetry reaction cross-sections from the ENDF/B-6 library and evaluations of 14 reactions at IRK Vienna by Prof. H. Vonach and co-workers documented in [1]. The data in the original ENDF/B-VI format were processed to 640 group extended SANDII format in the Nuclear Data Section of the IAEA using the processing codes LINEAR, RECENT and GROUPIE by D.E. Cullen [2]. The covariance information is not processed by these codes and it is contained in IRDF-90 in the original ENDF/B-6 format.

In particular the users of the file should pay attention to the fact that some of the US evaluations use the new LB=8 format of presentation of covariance information and should check whether their codes handle them.

The list of reactions and the origins of evaluations are given in Table 1. As we did not have any new sets of standard damage cross-sections the ones from IRDF-85 were kept here with the same special notations.

Table 1. Contents of the IRDF-90

| Nuclide | MAT No. | Reactions and* Uncertainties | Author & Lab | Date |
|---|---|---|---|---|
| 3-Li-6 | 325 | 3 105; 33 105 | G. Hale et al., LANL | 1989 |
| 5-B-10 | 525 | 3 1; 3 107; 33 107 | G. Hale et al., LANL | 1989 |
| 9-F-19 | 925 | 3 16; 33 16 | H. Vonach et al., IRK | 1990 |
| 12-Mg-24 | 1225 | 3 103; 33 103 | H. Vonach et al., IRK | 1990 |
| 13-Al-27 | 1325 | 3 103; 33 103 | US evaluation, ORNL | 1990 |
| 13-Al-27 | 1325 | 3 107; 33 107 | H. Vonach et al., IRK | 1990 |
| 15-P-31 | 1525 | 3 103; 33 103 | H. Vonach et al., IRK | 1990 |
| 16-S-32 | 1625 | 3 103; 33 103 | US evaluations, ORNL | 1989 |
| 21-Sc-45 | 2125 | 3 102; 33 102 | B. Magurno et al., BNL | 1979 |
| 22-Ti-46 | 2225 | 3 103; 33 103 | US evaluation, ORNL | 1989 |
| 22-Ti-47 | 2228 | 3 103; 33 103 | US evaluation, ORNL | 1990 |
| 22-Ti-48 | 2231 | 3 28; 33 28 | C. Philis et al., ANL | 1977 |
| 22-Ti-48 | 2231 | 3 103; 33 103 | US evaluation, ORNL | 1990 |
| 24-Cr-52 | 2431 | 3 16; 33 16 | H. Vonach et al., IRK | 1990 |
| 25-Mn-55 | 2525 | 3 16; 33 16 | K. Shibata et al., JAERI, ANL | 1988 |
| 26-Fe-54 | 2625 | 3 103; 33 103 | D. Hetrick et al., ORNL | 1989 |
| 26-Fe-56 | 2631 | 3 103; 33 103 | C. Fu et al., ORNL | 1989 |
| 26-Fe-58 | 2637 | 3 102 | N. Larson et al., ORNL | 1989 |
| 27-Co-59 | 2725 | 3 16; 33 16 | H. Vonach et al., IRK | 1990 |
| 27-Co-59 | 2725 | 3 102; 3 107; 33 107 | A. Smith et al., ANL | 1989 |
| 28-Ni-58 | 2825 | 3 103; 33 103 | N. Larson et al., ORNL | 1989 |
| 28-Ni-58 | 2825 | 3 16; 33 16 | H. Vonach et al., IRK | 1990 |
| 28-Ni-60 | 2831 | 3 103; 33 103 | N. Larson et al, ORNL | 1989 |
| 29-Cu-63 | 2925 | 3 16; 33 16 | H. Vonach et al., IRK | 1990 |
| 29-Cu-63 | 2925 | 3 16; 3 102; 3 107; 33 16; 33 107 | C. Fu et al., ORNL | 1989 |
| 29-Cu-65 | 2931 | 3 16; 33 16 | C. Fu et al., ORNL | 1989 |
| 30-Zn-64 | 3025 | 3 103; 33 103 | H. Vonach et al., IRK | 1990 |
| 39-Y-89 | 3925 | 3 16; 33 16 | R. Howerton, A. Smith, D. Smith, LLNL, ANL | 1990 |
| 40-Zr-90 | 4025 | 3 16; 33 16 | H. Vonach et al., IRK | 1990 |

Table 1. (cont.)

| Nuclide | MAT No. | Reactions and* Uncertainties | Author & Lab | Date |
|---|---|---|---|---|
| 41-Nb-93 | 4125 | 3 16; 3 51; 33 16; 33 51 | H. Vonach et al., IRK | 1990 |
| 45-Rh-103 | 4525 | 3 51; 33 51 | H. Vonach et al., IRK | 1990 |
| 48-Cd-0 | 4800 | 3 1; 3 102 | S. Pearlstein, BNL (translated from UK) | 1990 |
| 49-In-115 | 4931 | 3 51; 33 51 | S. Chiba et al., ANL | 1990 |
| 64-Gd-00 | 6400 | 3 1; 3 102 | Mixed from EN-6 isotope data by N. Kocherov, IAEA | 1990 |
| 79-Au-197 | 7925 | 3 102; 33 102 | P. Young, LANL | 1989 |
| 79-Au-197 | 7925 | 3 16; 33 16 | H. Vonach et al., IRK | 1990 |
| 90-Th-232 | 9040 | 3 18; 3 102; | M. Bhat et al., BNL,ANL | 1977 |
| 92-U-235 | 9228 | 3 18; 33 18 | L. Weston et al., ORNL, LANL | 1989 |
| 92-U-238 | 9237 | 3 18; 33 18 | L. Weston et al., ORNL LANL | 1989 |
| 93-Np-237 | 9346 | 3 18 | P. Young et al., LANL | 1989 |
| 94-Pu-239 | 9437 | 3 18; 33 18 | P. Young et al., LANL | 1989 |
| 26-Fe-00 | 8000 | ASTM Damage Cross Sections | Priv. Comm. W. Zijp | 1979 |
| 26-Fe-00 | 8001 | Eur. Damage Cross Sections | Priv. Comm. W. Zijp | 1979 |
| 24-Cr-00 | 8002 | Eur. Damage Cross Sections | W. Zijp, Petten | 1985 |
| 28-Ni-00 | 8003 | Eur. Damage Cross Sections | W. Zijp, Petten | 1985 |

* Note: The following ENDF notations for reactions are used 1-total, 16-n,2n, 18-fission, 28-n,np, 102-nγ, 103-np, 107-nα. 51 means total population of the 1st level from all channels (not an ENDF notation)

The benchmark spectra were not included in IRDF-90 V.1 because only scarce data are available to us now. We shall include these data in a later version when a more complete set becomes available.

Some of the evaluations show considerable changes in limited neutron energy regions. To illustrate this Fig. 1 gives a comparison of $^{93}$Nb(n,n') evaluations for IRDF-85 and IRDF-90. After discussions it was decided to take this item from the NEANDC/INDC discrepancy list because the experimental and evaluated data sets have now become consistent. There are other cases when the new evaluations show very little changes in comparison with the previous version. A good example of this situation will be

27-Al(n,α)24-Mg reaction. Fig. 2 shows the comparison between the new measurement data and the evaluated curve from IRDF-85. It can be seen that the new measurements confirm the old evaluation and even make it more accurate. Comparison between the 1985 and the 1990 evaluations of 27-Al(n,α) reaction is shown in Fig. 3. The evaluated values remained practically the same, but the errors became smaller.The plots comparing evaluations for all reactions from IRDF-85 and IRDF-90 together with experimental data base can be found in [3].

The preliminary version of IRDF-90 was distributed among the participants of the International Intercomparison exercise REAL-88 for detailed testing. Some of the necessary corrections have already been introduced but much more testing will be made in the future.


REFERENCES

1. M. Wagner, H. Vonach. A. Pavlik, B. Strohmaier, S. Tagesen, J. Martinez-Rico, "Evaluation of Cross-Sections for 14 Important Neutron Dosimetry Reactions," Physics Data, Karlsruhe (in press)

2. D.E. Cullen, P.K. McLaughlin, "The 1989 ENDF/B Preprocessing Codes", Report IAEA-NDS-39 Rev. 4.

3. N.P. Kocherov, "The International Reactor Dosimetry File (IRDF-90)", Report IAEA-NDS (to be published).


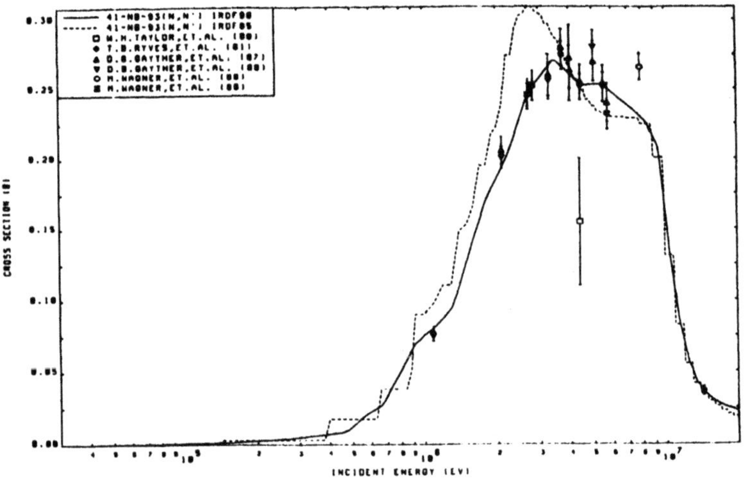


Fig. 1 Reaction 41-Nb-93(n,n'). Reactor Dosimetry Files IRDF-85 & IRDF-90 compared with experiment

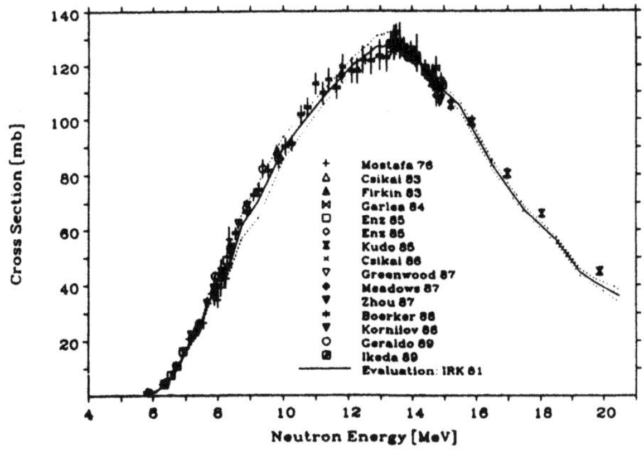


Fig. 2 Recent experimental cross section data for the reaction $^{27}Al(n,\alpha)^{24}Na$ compared to the results of the previous evaluation carried out at the TRK [1].

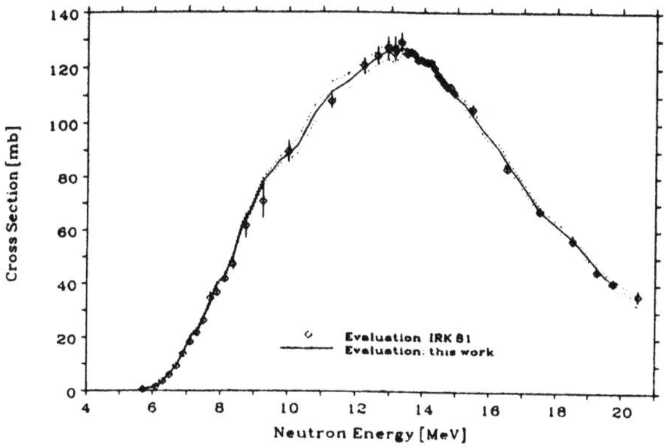


Fig. 3 The evaluated cross section for the reaction $^{27}Al(n,\alpha)^{24}Mg$: Results of this work compared to the previous evaluation carried out at the IRK [1].

## MEASUREMENT OF THE ISOTOPIC ABUNDANCE OF $^{54}Fe$ IN NATURAL IRON BY MEANS OF NEUTRON ACTIVATION

B. Bärs and T. Serén

Technical Research Centre of Finland, Reactor Laboratory

P.O.Box 200 (Otakaari 3A), SF-02151 Espoo, Finland

### ABSTRACT

The reaction $^{54}Fe(n,p)^{54}Mn$ is probably the most commonly used activation reaction in fast neutron metrology based on prefabricated dosimeters or on reactor structure materials. An accurate knowledge of the isotopic abundance of $^{54}Fe$ is therefore of utmost importance. The commonly used value for the isotopic abundance of $^{54}Fe$ in natural iron is given with fairly large uncertainty (0.058 ± 0.001), [1].

The isotopic abundance of $^{54}Fe$ was experimentally determined based on simultaneous irradiation of natural iron and enriched iron foils of Soviet origin with accurately known enrichments 93.8 %, 99.2 % and 99.68 % in $^{54}Fe$. 13 dosimeter foils were packed into a quartz ampoule and irradiated for a period of about two months in the TRIGA Mark II reactor in Finland.

The $^{54}Mn$ activities of the irradiated samples were carefully measured using two separate gamma spectrometers with HPGe detectors. An estimate (1.0013 ± 0.24 %) was obtained for the ratio of the counting rate per $^{54}Fe$ atom in natural and in enriched iron, assuming the recommended abundance value 0.058.

The obtained experimental estimate (0.058 ± 1.0013) for the isotopic abundance of $^{54}Fe$ in natural iron is indeed close the recommended value 0.058. The stated 1 $\sigma$ uncertainty ± 0.001 (± 1.7 %) could be significantly reduced to ± 0.00014 (± 0.24 %).

## 1. INTRODUCTION

The reaction $^{54}Fe(n,p)^{54}Mn$ is probably the most commonly used activation reaction in fast neutron metrology in nuclear fission reactors. The objective of the work was to experimentally check the isotopic abundance of $^{54}Fe$ in natural iron. The recommended and commonly used value for the isotopic abundance of $^{54}Fe$ in natural iron is given with fairly large uncertainty (X(Fe54,nat)= 0.058 ±0.001), [1]. This recommended value is based on measurements by G.E. Valley and H.H. Anderson (1947), who found no variation in 7 terrestrial and 12 meteoritic samples of iron. We have noticed a small but significant discrepancy between fast-neutron fluxes obtained from natural iron and nickel activation. This indicates discrepancies in the reaction cross sections or/and in the isotopic abundance of $^{54}Fe$.

The isotopic abundance of $^{54}Fe$ was experimentally determined based on simultaneous irradiation of 7 natural iron and 6 enriched iron foils of Soviet origin with accurately known enrichments 93.8 %, 99.2 % and 99.68 % in $^{54}Fe$. These 13 dosimeter foils were tightly packed into a closed quartz glass tube and irradiated for a period of about two months in the dry activation tube (F3) in our TRIGA Mark II reactor.

The relative $^{54}Mn$ activity of the irradiated samples was carefully measured using two separate gamma spectrometers with HPGe detectors. The quantities compared are "CPA" = count rate/$^{54}Fe$ atom for natural and for enriched iron in a fixed measuring configuration.

## 2. DOSIMETER MATERIALS

Seven natural iron dosimeters and six highly enriched dosimeters, with three different enrichments (Low: 93.8 %, Medium: 99.2 %, High: 99.68 % $^{54}Fe$) were irradiated. The properties of the irradiated dosimeter materials are given in Table 1. Natural iron dosimeters were cut from a 0.25 mm thick iron plate (Goodfellow Metals, batch 1). Two independent analyses at the Metallurgy Laboratory (optical emission spectrophotometer) and at the Chemical Laboratory (atomic absorption spectrometer) of the Technical Research Centre of Finland showed that the total amount of other elements is 0.322 % and 0.325 % respectively, Mn being the main extra element (0.26 %). Thus we have used a mass fraction of Fe of 0.9967 ± 0.0005 (± 0.05 %).

The enriched dosimeters (H1, H2, M2, M2, L1, L2) and their properties, were provided by the Soviet suppliers (Table 1). The content of other elements were given (L: 0.00028, M: 0.00126). The

## TABLE 1

Basic data for natural and enriched (High, Med., Low) Fe foils irradiated in TRIGA Mk II.

**A: Masses and mass uncertainties of Fe foils**

| | Foil | Mass [mg] | Foil | Mass [mg] |
|---|---|---|---|---|
| Nat.: | N1 | 20.56 ± 0.01 | N2 | 20.565 ± 0.01 |
| | N3 | 20.565 ± 0.01 | N4 | 20.607 ± 0.01 |
| | N5 | 20.63 ± 0.01 | N6 | 18.335 ± 0.01 |
| | N7 | 19.57 ± 0.01 | | |
| High: | H1 | 4.280 ± 0.005 | H2 | 5.040 ± 0.005 |
| Med.: | M1 | 5.35 ± 0.01 | M2 | 5.60 ± 0.01 |
| Low: | L1 | 7.67 ± 0.01 | L2 | 7.525 ± 0.01 |

**B: Atomic and material data**

Avogadro's constant $N_A = 6.0221367 \times 10^{23}$ $mol^{-1}$ ± 0.59 ppm [3].

Relative atomic masses of Fe isotopes [2]:

| $^{54}Fe$ | $^{56}Fe$ | $^{57}Fe$ | $^{58}Fe$ |
|---|---|---|---|
| 53.9396 | 55.9349 | 56.9354 | 57.9333 |

Isotopic mole fractions, relative atomic masses $A_r$ and atoms/mg $N_m$:

| | Mole fractions | | | | | |
|---|---|---|---|---|---|---|
| | $^{54}Fe$ | $^{56}Fe$ | $^{57}Fe$ | $^{58}Fe$ | $A_r$ | $N_m/10^{19}$ |
| Nat. | **5.8**[*)] | 91.72 | 2.2 | 0.28 | 55.8470 | 1.07833 |
| High | 99.68 ± 0.2 | - | - | - | 53.9460 | 1.11633 |
| Med. | 99.2 ± 0.3 | 0.76 | 0.016 | 0.005 | 53.9555 | 1.11613 |
| Low | 93.8 ± 0.5 | 5.9 | 0.3 | - | 54.0663 | 1.11384 |

Mass fractions (%) of Fe in foils:

| Nat. | High | Medium | Low |
|---|---|---|---|
| 99.67 ± 0.05 | 100 ± 0.01 | 99.972 ± 0.02 | 99.874 ± 0.02 |

---

*) Quantity to be adjusted.

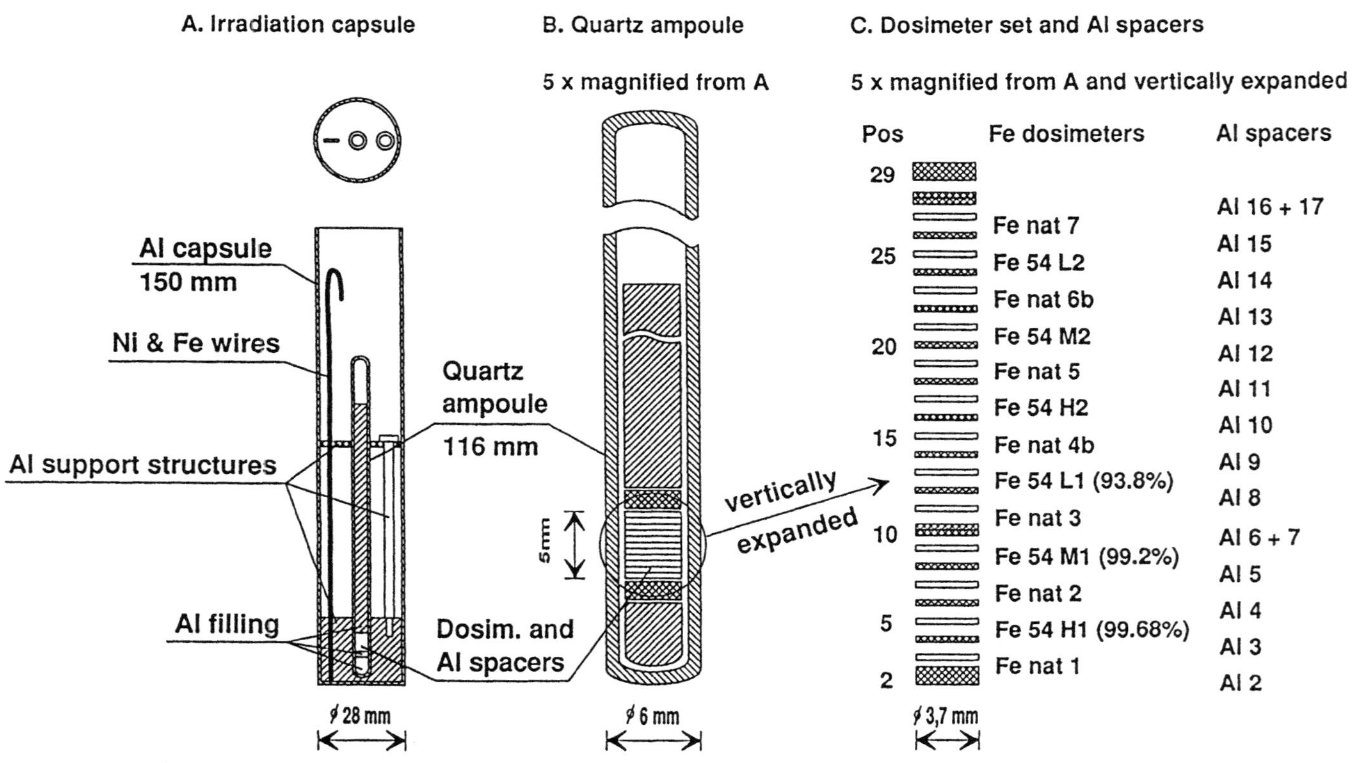


Figure 1. Arrangement of natural and enriched Fe foils in quartz glass tube and irradiation capsule.

uncertainties in the enrichments (L: 0.938 ± 0.005, M: 0.992 ± 0.003, H: 0.9968 ±0.002) were not supplied but were pessimistically taken as 0.2 to 0.5 per cent.

The atomic masses of the iron isotopes are from ref. [2] and Avogadro's constant 6.0221367x$10^{23}$/mole ± 0.59 ppm from ref. [3] The atomic masses and Fe-atoms per unit mass of the enriched foils (with some other isotopes present as well) were calculated from this data and are given in Table 1.

The notations, masses and enrichments of the dosimeter foils are also given in Table 1.

## 3. IRRADIATION

Figure 1 shows the irradiation capsule and the arrangement of the dosimeters. The dosimeter foils (diam. 3.7 mm) and Al spacers were packed into a quartz glass tube (diam. 4.3 mm) so that every second dosimeter consisted of natural iron (Fenat). All Fe dosimeters were separeted by thin aluminium foils. The total length of the dosimeter and aluminium foils was about 6 mm with about 5 mm long Al bars at the bottom and a 80 mm long Al bar on the top. At least to start with all foils were located at one side of the quartz tube. The quartz tube was located in a rack in the centre of a larger Al capsule (28 mm diam.).

The dosimeter capsule was irradiated in the dry irradiation tube (F ring position 3) in the 250 kW TRIGA Mark II research reactor in Finland from Dec. 12, 1988 to Febr. 6, 1989 in a neutron flux $\phi$(>1 MeV) of about $10^{12}$ n/$cm^2$s. The normal operation time was about 6 hours per day at most five days a week. The relative flux was monitored with Ni-wires. The capsule was in the axially flat flux region and was once temporarily taken out and reloaded. The radial flux gradient was large, about 2% /mm.

## 4. RELATIVE GAMMA ACTIVITY MEASUREMENTS

The counting rates from the $^{54}$Mn nuclide (835 keV) in the irradiated natural and enriched dosimeters were measured with two different gamma spectrometers including HPGe detectors, amplifiers, PCA-8000 analysers and PCs (1: Olivetti M24 and 2: AST 286). In spectrometer 1 two measuring distances (131.1/131 mm sample changer, and 99.1/99 mm; nat/enr) and in spectrometer 2 one fixed distance (99.1/99 mm) were used. The dead-time and pile-up correction was carried out with the pulser method. A 50 Hz pulser fed pulses to a parallel input of the preamplifier and the amplitude was adjusted to the upper almost empty part of the gamma spectra. The net counting rates in the 835 keV $^{54}$Mn peak were multiplied by a correction factor (50/measured count rate in pulser peak).

From Febr. 13 to 24, 1989 the following nine series of measurements were carried out in three geometries:

| SPECTROMETER | CONFIGUR. nat/enr. | No of MEAS. | CODE |
|---|---|---|---|
| 1 | 131.1/131 | 3 | OLI131 |
| 1 | 99.1/99 | 2 | OLI99 |
| 2 | 99.1/99 | 3 | AST99 |

In configuration OLI131 the foils were in plastic capsules. In the other configurations the foils were put in thin mylar foils directly on an aluminium rack on the detector assembly.

About 200 gamma spectra were measured and stored on floppy disks. The measuring times were in the range 1000 to 100000 s and the net counting rates of the order 20...200 cps. One additional measurement series with spectrometer 2 (foils in plastic capsules, 100.1/100 mm) and short counting times was rejected due to poorer statistical accuracy.

## 5. DATA PROCESSING

GWBASIC programs were used to retrieve the spectra from the floppy disks and calculate several estimates such as peak areas, statistical uncertainties and the experimental counting rates per Fe54 atom (CPA) in the samples on Febr. 13, 1989

$$CPA = (COUNT\ RATE\ in\ 835\ keV\ peak)/[N_m*MASS*X(Fe)*X(Fe54)] \quad (1).$$

In all geometries the greater thickness of the natural Fe foils led to 0.1 mm longer measuring distance and + .15 or 0.2 % corrections of the counting rates.

The quantities $R=CPA(nat)/CPA(enr)$ (2),

including statistical counting uncertainties were estimated with statistical weighting for all adjacent foil pairs from the measured results in all three configurations and from the data in Table 1. The results are given in Table 2.

The statistical smoothing of the counting rates from many measurements reduced the uncertainties from counting distance effects (± 0.15 mm/99 mm, ± 0.3 mm/131 mm) to < 0.1 %. Due to the random displacements (max. ± 0.3 mm) and orientations of the 12 foil pairs within the quartz tube the estimated uncertainty from the radial flux gradient was reduced to about 0.06 %.

## 6. RESULTS

The statistically averaged experimental values obtained for the ratio R = CPA(nat)/CPA(enr) are given in Table 2 for each enrichment type as well as the average of all results based on the material and nuclear data in Table 1 and assuming X(Fe54,nat) = 5.8 %. A summary of the results is given below. The total uncertainties include standard deviations from counting and quadratically added other errors.

| $^{54}Fe$ ENR. (%) X(Fe54) in foil 2 | RATIO R | UNCERTAINTY (%) Count.+geom. | TOT 1 | DIST.+ FL.GRAD. | TOT 2 |
|---|---|---|---|---|---|
| 99.68 | 0.9996 | ± 0.095 | ± 0.29 | | |
| 99.2 | 1.00569 | ± 0.056 | ± 0.38 | | |
| 93.8 | 0.9994 | ± 0.049 | ± 0.55 | | |
| stat. weigh. aver. | 1.0013 | | ± 0.21 | ± 0.12 | ± 0.24 % |

Since the average ratio R = 1.0013 (2) is greater than 1 the assumed X(Fe54,nat) = 0.058 is slightly too low and the $^{54}Fe$ abundance estimate based on our results is

$$X(Fe54,nat) = 0.058*1.0013 = 0.05808 \pm 0.00014 \ (1\sigma)$$

The assumptions on the enrichment uncertainties (0.2...0.5 %) are crucial. On the other hand the above results agree within the error limits. *)

The recommended [1] and present estimates for the abundance of $^{54}Fe$ in natural iron are consistent and within the old (± 1.7%) and also within the present (± 0.24 %) tighter error limits. The previous error limits could be reduced by a factor of seven.

*) The present results also agree with the results (0.05810 ± 0.00005) obtained with the Sputtered Neutral Mass-Spectrometry method as reported by P. Wille and R. Jede at this conference.

## REFERENCES


1. N.E. Holden, Isotopic composition of the elements and their variation in nature. BNL-NCS-50605, March 1977

2. CRC Handbook of Chemistry and Physics. 67th Edition, 1986-87.

3. E.R. Cohen, B.N. Taylor, "The Fundamental Physical Constants", Physics Today, Aug. 1990.

## TABLE 2

Statistically weighted averages of the ratios R=CPA(nat.)/CPA(enr.) from 2-3 pairs of measurements of adjacent Fe foils in three measurement configurations for high (H), medium (M), and low (L) enriched Fe foils. The weighted averages for each type of enriched foils (<R>) and the final results (<<R>> and X(Fe54)) are also given. (CPA = cps at ref. time / (Nm*mass*X(Fe)*X(Fe54)))

**Fe-nat./high enrichment:**

| | AST, 99 mm | | Oliv. 131 mm | | Oliv. 99 mm | |
|---|---|---|---|---|---|---|
| Samples | R | +- % | R | +- % | R | +- % |
| N1/H1: | 0.9995 | 0.190 | 0.9916 | 0.189 | 1.0013 | 0.145 |
| N2/H1: | 0.9973 | 0.195 | 0.9998 | 0.930 | 1.0016 | 0.062 |
| N4/H2: | 0.9941 | 0.125 | 1.0032 | 0.304 | 0.9965 | 0.157 |
| N5/H2: | 0.9993 | 0.198 | 1.0029 | 0.218 | 1.0008 | 0.153 |

Weighted average <R(N/H)>: 0.9996

Uncertainties (%): 0.095 (count. stat. + geom.) 0.29 (total)

**Fe-nat./medium enrichment:**

| | AST, 99 mm | | Oliv. 131 mm | | Oliv. 99 mm | |
|---|---|---|---|---|---|---|
| Samples | R | +- % | R | +- % | R | +- % |
| N2/M1: | 1.0028 | 0.195 | 1.0078 | 0.933 | 1.0062 | 0.062 |
| N3/M1: | 1.0072 | 0.095 | 1.0108 | 0.359 | 1.0109 | 0.194 |
| N5/M2: | 1.0027 | 0.263 | 1.0055 | 0.142 | 1.0060 | 0.122 |
| N6/M2: | 1.0004 | 0.222 | 1.0015 | 0.159 | 1.0053 | 0.067 |

Weighted average <R(N/M)>: 1.00569

Uncertainties (%): 0.056 (count. stat. + geom.) 0.382 (total)

**Fe-nat./low enrichment:**

| | AST, 99 mm | | Oliv. 131 mm | | Oliv. 99 mm | |
|---|---|---|---|---|---|---|
| Samples | R | +- % | R | +- % | R | +- % |
| N3/L1: | 0.9986 | 0.103 | 1.0005 | 0.354 | 1.0032 | 0.215 |
| N4/L1: | 0.9975 | 0.122 | 1.0027 | 0.288 | 1.0000 | 0.159 |
| N6/L2: | 0.9975 | 0.194 | 0.9968 | 0.541 | 0.9982 | 0.069 |
| N7/L2: | 0.9944 | 0.254 | 1.0015 | 0.565 | 0.9957 | 0.204 |

Weighted average <R(N/L)>: 0.9984

Uncertainties (%): 0.049 (count. stat. + geom.) 0.553 (total)

| Final values: | | | | |
|---|---|---|---|---|
| <<R>> = | 1.0013 | +- | 0.243 | % |
| X(Fe54) = | 0.05808 | +- | 0.00014 | |

Chi-sqr = 0.

# MEASUREMENT OF THE $^{93}$Nb(n,2n)$^{92m}$Nb CROSS SECTION IN A $^{235}$U FISSION SPECTRUM

T.G. Williamson*
Department of Nuclear Engineering
and Engineering Physics
University of Virginia
Charlottesville, VA 22901

G.P. Lamaze
National Institute of Standards and Technology
Gaithersburg, MD 20899

ABSTRACT

When niobium is used as a fast neutron monitor the isotopes $^{93m}$Nb and $^{92m}$Nb are produced. Much interest has been placed on the $^{93}$Nb(n,n')$^{93m}$Nb reaction because of the long half life of the product isotope and the low threshold energy of the reaction. On the other hand, the $^{93}$Nb(n,2n)$^{92m}$Nb reaction product has a half life of 10.1 days and the reaction has a threshold energy of about 9 MeV. It is useful for relatively short irradiations and for cases in which information is desired about high energy neutrons, such as in fusion applications. The spectrum-averaged cross section for the reaction $^{93}$Nb(n,2n)$^{92m}$Nb in a $^{235}$U fission spectrum has been measured. The measured value is 0.433 ± 0.017 mb. Comparisons are made with calculated values.

## INTRODUCTION

When niobium is used as a fast neutron monitor the isotopes $^{93m}$Nb and $^{92m}$Nb are produced. Much light water reactor dosimetry interest has been placed on the

*Present address: Westinghouse Savannah River Laboratory, Aiken, SC.

$^{93}Nb(n,n')^{93m}Nb$ reaction because of the long half life of the product isotope and the low threshold energy of the reaction. On the other hand, the reaction $^{93}Nb(n,2n)^{92m}Nb$ produces an isotope with a 10.15 day half life, and has a threshold energy of about 9 MeV so is less useful for pressure vessel monitoring than the other reaction. This reaction is of interest though as a monitor for fusion studies. In the course of measurements of the $^{93}Nb(n,n')^{93m}Nb$ reaction rate, spectrum-averaged cross sections for the $^{93}Nb(n,2n)^{92m}Nb$ reaction in a $^{235}U$ fission spectrum have been determined.

An early measurement of the cross section for the production of $^{92m}Nb$ in a fission spectrum was made by Hegedus (1) when he reported a value of approximately 0.4 mb. Nasyrov (2) reported a value of 0.37 ± 0.03 mb and Fabry (3) measured a cross section of 0.47 ± 0.03 mb. Kimura et al (4) report cross-section measurements for several threshold reactions including the $^{93}Nb(n,2n)^{92m}Nb$ reaction. These measurements were made in a water cooled reactor and in the field of an enriched uranium fission plate. The measured values are 0.432 ± 0.033 mb in the reactor spectrum and 0.402 ± 0.034 mb with the fission plate. Calamand (5) in 1974 reviewed all of the fission spectrum data up to that time, renormalized the data, and recommended a value of 0.48 ± .04 mb for this reaction in the $^{235}U$ fission spectrum.

## EXPERIMENT

Five niobium foils (1.27 mm diam. and ranging in thickness from .004 to .125 mm thick) from two commercial suppliers were irradiated in the cavity fission source(6) of the National Institute of Standards and Technology Reactor. Calibration of this facility has been described by McGarry et al (7). In this facility, two $^{235}U$ metal fission source disks are placed outside a cadmium box coaxial with the detector stack. This assembly is placed in the center of a 0.3-m-diameter cavity in the graphite thermal column. Three of the foils were irradiated at one time and the other two included in separate irradiations. Each foil was exposed for approximately 100 hours. The neutron fluences were determined by fluence transfer with nickel foil irradiations (8). In the fluence transfer method, the fluence is, by a series of ratio measurements, traceable to NBS-1, the national standard Ra-Be neutron source, which has been absolutely calibrated with the manganous sulfate bath technique. No cross sections enter directly into the fluence determination. Corrections have been made for deviations from a pure fission field due to scattering. These corrections were made with a Monte Carlo technique and the correction varies from 1.7 ± 1.0 to 2.9 ± 1.0 % depending on the number and thickness of other foils irradiated with the niobium. The one standard deviation uncertainty in the fluence measurement is 2.1%.

The foils were counted with germanium detectors and the analysis done with the 934-keV gamma ray. The detectors were calibrated with standard point sources and have an uncertainty in the detector response efficiency of 3% at this energy. A gamma-ray abundance of 0.9915 and half life of 10.15 days was used in the analysis. The activity of each foil was measured several times over a two week decay period. All uncertainties were combined in quadrature to give an uncertainty of 4% for each individual determination of the cross section.

The results of the individual measurements are shown in Table I. The average determination of the spectrum-averaged cross section from the five foils is 0.433 mb with a one standard deviation from the average of 2.6%. Combining all uncertainties gives a final result of 0.433 ± 0.017

TABLE I

FOIL CONDITIONS

| Foil | Mass mg | Fluence Rate $n/cm^2$-s | Cross Section mb |
|---|---|---|---|
| RE-3 | 131 | $1.75 \times 10^{10}$ | 0.431 ± 0.017 |
| RE-A | 100 | $4.13 \times 10^{10}$ | 0.431 ± 0.017 |
| RE-B | 130 | $4.00 \times 10^{10}$ | 0.434 ± 0.017 |
| RE-C | 3.6 | $4.19 \times 10^{10}$ | 0.417 ± 0.016 |
| TS-A | 95.9 | $3.85 \times 10^{10}$ | 0.453 ± 0.018 |
| | | AVERAGE | 0.433 ± 0.011 (2.6%) |

## DISCUSSION

Many measurements of the differential cross section for the $^{93}$Nb(n,2n)$^{92m}$Nb reaction have been reported since Vonnach (9) in 1959. These data sets are reviewed by Strohmaier (10) and compared with her new calculations based on the compound nucleus and exciton model. Using the calculated cross section of Strohmaier (10) (values were interpolated from Figure 9 of that work) and ENDF/B-V (11) and NIST (6) evaluated fission spectra, a

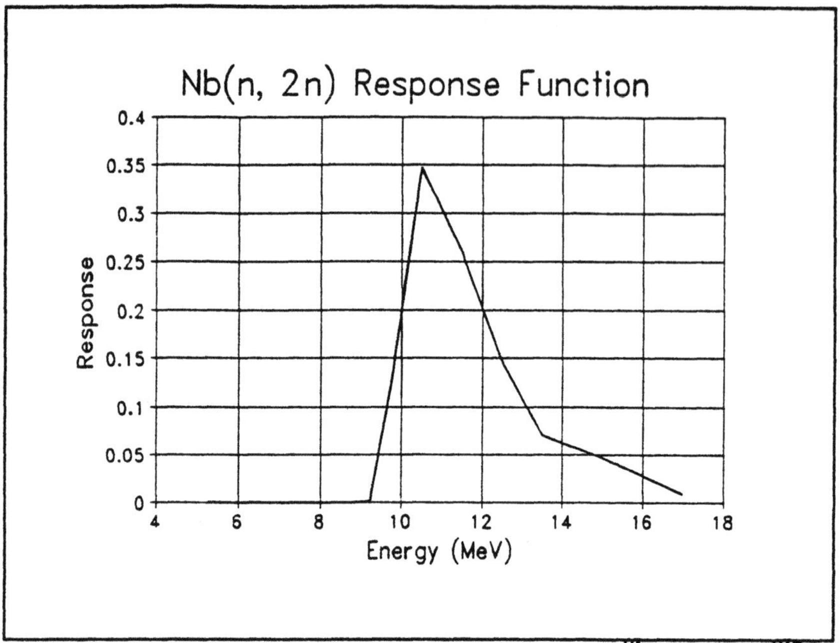


**Figure 1.** Response function of the $^{93}$Nb(n,2n)$^{93m}$Nb reaction using the differential cross section of Strohmaier (10) and the ENDF-B/V (11) $^{235}$U fission neutron spectrum.

calculated value of the $^{93}$Nb(n,2n)$^{92m}$Nb spectrum-averaged cross section has been obtained. Figure I presents the response function for the $^{93}$Nb(n,2n)$^{92m}$Nb reaction obtained using the Strohmaier differential cross section and the ENDF/B-V $^{235}$U fission neutron spectrum. Table II lists the present results compared with some previous results and the calculated cross section.

Table II

$^{93}$Nb(n,2n)$^{92m}$Nb Spectrum Averaged Cross Section

| Value (mb) | Reference | Comments |
|---|---|---|
| 0.433 ± 0.017 | This work | $^{235}$U fiss. spec. |
| 0.432 ± 0.033 | Ref. 4 | Reactor Spec. |
| 0.402 ± 0.034 | Ref. 4 | Fission Plate |
| 0.47 ± 0.03 | Ref. 3 | $^{235}$U fiss. spec. |
| 0.44 | Ref. 10 and 11 | Calculated |
| 0.43 | Ref. 10 and 6 | Calculated |

## SUMMARY

A spectrum-averaged cross section of 0.433 ± 0.017 mb has been measured for the reaction $^{93}Nb(n,2n)^{92m}Nb$ in a $^{235}U$ fission spectrum. The measurement compares favorably with a calculated value using the Strohmaier calculation and either ENDF/B-V or the NIST evaluated $^{235}U$ fission neutron spectrum. The uncertainty is about half that of previously reported measurements. This reaction product has a half life of 10.1 days and the reaction has a threshold energy of about 9 MeV. It is useful for relatively short irradiations and cases in which information is desired about the high energy end of reactor spectrum neutrons.

## ACKNOWLEDGEMENTS

The authors would like to thank C.M. Eisenhauer for the calculations of the $^{93}Nb(n,2n)^{92m}Nb$ response function and the calculated spectrum-averaged cross sections.

## REFERENCES

1) Hegedus, F., Helv. Phys. Acta., 39 pp. 179 (1966).
2) Nasyrov, F., Sov. J. At. Energy, 25, pp. 59 (1968).
3) Fabry, A. Rep. BLG-465 (1972).
4) Kimura, H., Kobayashi, K., and Shibata, T., J. of Nucl. Sc. and Tech. 8(2) p. 59-69 (1971).
5) Calamand, A., "Cross-sections for Fission Neutron Spectrum Induced Reactions", Handbook on Nuclear Activation Cross-Sections, IAEA, Technical Report Series No. 156 pp 273-309 (Vienna)(1974)
6) Grundl, J.A., and Eisenhauer, C.M., "Compendium of Benchmark Neutron Fields for Reactor Dosimetry," NBSIR 85-3151, Jan. 1986.
7) McGarry, E.D., Eisenhauer, C.M., Gilliam, D.M., Grundl, J., and Lamaze, G.P., "The U.S. U-235 Fission Spectrum Standard Neutron Field Revisited," in Proceedings, Fifth ASTM-Euratom Symposium on Reactor Dosimetry, Geesthact, West Germany, Sept. 1984.
8) Lamaze, G.P., McGarry, E.D., and Schima, F.J., in Proceedings, Int. Conf. on Nuclear Data for Science and Technology, Antwerp, Belgium, Sept. 1982.
9) Vonach, H., and Munzer, H., Anz. Osterr. Akad. Wiss., math.-naturwiss. Klasse 96 (1959)p.120.
10) Strohmaier, B., Ann. Nucl. Energy, 16 (1989)p. 461.
11) Magurno, B.A., Kinsey, R.R., Scheffel, F.M., Guidebook for the ENDF/B-V Nuclear Data Files, EPRI NP-2510 (1982)

# TRANSMISSION OF NEUTRONS OF 20 keV TO 1 MeV THROUGH IRON

*M. Matzke, K. Knauf, E. Dietz, A. Plewnia, W.G. Alberts*

Physikalisch-Technische Bundesanstalt,

D-3300 Braunschweig, F.R. of Germany

## ABSTRACT

The transmission of reactor neutrons through thick blocks of iron has been investigated at a beam tube of the PTB Research and Measurement Reactor (FMRB). The iron filter provides a collimated neutron beam with a stable, multi-line spectrum which is very suitable for neutron spectrometer and dosimeter investigation and development. Spectrometry with proton recoil and $^{3}He$ proportional counters was performed at this beam.

Measurements of the area of the peaks between 24 keV and 1 MeV are compared with calculations using JEF-1 neutron cross section data.

Lines with a width of a few keV could be resolved by the proton recoil proportional counter method. Investigation of the transmission through different slabs of iron allowed conclusions to be drawn on how sensitively the transmission depends on the cross sections around the minima between the resonances.

## 1. INTRODUCTION

Neutron filters are a valuable tool for various purposes: Owing to their ability to preferentially transmit neutrons in the low-keV energy range, they provide special beams for neutron therapy of cancer (1,2) or quasi-monoenergetic beams for dosimetric investigations (3). Some of them can also serve to produce a neutron spectrum with very stable and well separated narrow lines which can be used for the energy calibration of high-resolution neutron spectrometers. Neutron transmission, in particular through thick layers of iron, has also been of interest to reactor neutron dosimetrists for a long time (4,5) because of the neutron leakage from iron shieldings such as the pressure vessel and other structural components in a reactor.

Straight-forward methods of determining the energy-dependent transmission of neutrons through an iron filter are either the calculation using energy-dependent total cross sections from libraries like JEF (6) or the direct measurement using a white neutron source and a time-of-flight method (7). If the spectrum on the detector side of the iron filter is known, e. g. from calculations with the transmission data, the beam is suitable for studying the characteristics of neutron spectrometers.

Calculation and measurement of cross sections and transmission and high-resolution spectrometry behind the iron-filtered beam at the PTB's Research and Measurement Reactor (FMRB) furnish a system of interdependent information. In initial studies, discrepancies were detected between the measured and calculated transmission data for filters of different length and of different material composition. Additional measurements were therefore done. The results are discussed with respect to their consistency with the data base used for calculations.

## 2. CALCULATION OF TRANSMISSION FUNCTIONS

The spectral fluence of the filtered neutrons is the product of the transmission function T(E) and the spectral fluence $\Phi_E(E)$ at the input end of the filter. For a well collimated beam, when scattered neutrons which reach the detector are neglected, the transmission of neutrons through a material of length L can be described by

$$T(E) = \exp(-\Sigma_t(E) \cdot L), \qquad (1)$$

where $\Sigma_t(E)$ is the total cross section density (macroscopic cross section) at the neutron energy E. A correction factor to T(E) for scattering can be estimated at $K = 1 + \Sigma_s(E) \cdot L \cdot \Delta\Omega / 4\pi$, with the scattering cross section density $\Sigma_s(E)$ and the solid angle $\Delta\Omega$ being defined by the detector-filter distance and the filter outlet face.

Measured proportional counter spectra show separated narrow neutron lines in energy intervals $\Delta E$ (Fig. 1). Integrals over these lines can be represented by

$$\Phi_K = \int_{\Delta E} dE\ \Phi_E(E) \cdot T(E). \qquad (2)$$

If the input spectral fluence in $\Delta E$ is assumed to be constant, i.e. $\Phi_E(E) = C$, $\Phi_K$ reduces to

$$\Phi_K = C \int_{\Delta E} dE \cdot T(E) = C \cdot I_{TR}. \qquad (3)$$

The applicability of Eq. (1) was checked by comparing transmission measurements with a filter composed of 352 mm Fe + 230 mm Al + 75 mm S (the original PTB filter set-up (8)) by time-of-flight spectrometry at the GELINA accelerator (7) with calculations according to Eq. (1) using JEF-1 cross section data (6). For the 24 keV line there was good agreement between measured and calculated T(E)-values. The transmission integrals $I_{TR}$ between 18.5 keV and 29 keV were determined as 0.196 keV (measured) and 0.184 keV (calculated). However, there were discrepancies at higher energies, e.g. up to 50% for the 128 keV line (7), which were ascribed to JEF-1 cross-section errors, mainly those of S and Al.

In the discussion of such discrepancies, it should be borne in mind that for small T(E) (i.e. large $|\ln T(E)|$) a small variation of $\Sigma_t(E)$ in Eq. (1) entails a large variation of $\delta T/T$, according to

$$\delta T/T = \delta\Sigma_t/\Sigma_t \cdot \ln T \qquad (4)$$

As there was good agreement between calculated and measured values at 24 keV, Eqs. (1) to (3) were used in consistency checks in which the spectrum of filters of different lengths L was measured in order to explain the discrepancies.

## 2. EXPERIMENTS

A radial beam tube of the reactor was used for the experiments. Three cylinders of pure natural iron (armco) with L=352 mm, L=482 mm, and L=672 mm were investigated in succession. The cylinders were positioned at the outer end of the beam tube, tightly fitted into a hollow polyethylene cylinder (50 mm inside diameter) enriched with 30 % of $B_4C$ (8), producing a collimated beam with a plateau diameter of at least 50 mm in the irradiation position. A cylindrical proton recoil proportional counter was irradiated perpendicular, and a $^3$He-counter parallel to its axis at a distance of 2.8 m from the end of the iron filter. The available solid angle (see section 2) was $\Delta\Omega / 4\pi = 2.0 \cdot 10^{-5}$ leading to a scattering correction for T(E) of less than 0.5 %. The fluence rate for 24 keV neutrons was expected to be $5 \cdot 10^3$ $cm^{-2}s^{-1}$ with L = 352 mm. A cycle of measurements with cylinder lengths of 672 mm, 482 mm, 352 mm and again 672 mm (for stability checks) was performed.

The detector and the spectrometer arrangement for the proton recoil proportional counter measurement have been described in Ref. (9). Due to the geometrical conditions (neutrons impinging perpendicular to the counter axis), the response function of the spectrometer could be computed precisely, idealized by disregarding blurring due to the counter resolution (10). In this case the response matrix is triangular, and unscrambling of the neutron spectrum from the measured pulse height distribution is reduced to a simple recursion. The resulting spectrum, however, is still broadened by the counter resolution. On the assumption of a Gaussian broadening and by a comparison with the calculated transmission spectrum (line by line), the relative resolution vs. energy was determined (e.g. 10 % at 100 keV and 5 % at 400 keV). Using this information a 'resolution-free' spectrum was unfolded by a modified MIEKE code (11). This two-step unfolding process yields lines with a relative FWHM of a few percent only. The results for L = 672 mm are shown in Fig.1. For a single unfolded line the shape of the corresponding transmission line is obviously reproduced (compare e.g. the small satellite line at 0.128 MeV). When comparing the transmission function with the measured and unfolded specta in Fig. 1, it should be taken into account that the latter is the product $\Phi_E(E)$ T(E), where $\Phi_E(E)$ is the neutron fluence at the reactor side of the filter (see Eq. (2)).

## 3. RESULTS FOR DIFFERENT LENGTHS OF IRON

In order to check the consistency of the data involved, the C values (Eq. (3)) were determined for the three filter lengths using measured integrals between 10 keV and 1 MeV after the first unfolding step and using the JEF-1 (6) cross-section file (MAT=4260).

The results are listed in Table 1. The second column contains the transmission integrals $I_{TR}$ obtained with the JEF-1 data. The C values in the third column are normalized to those obtained for the first measurement (L = 672 mm). The uncertainty of $C = \Phi_K/I_{TR}$ should combine the uncertainties of both components. The uncertainty of $\Phi_K$ includes beam monitoring uncertainties, possible fluctuations of the beam profile for different L values and the uncertainties from the first unfolding procedure. The relative standard deviation of each $\Phi_K$ was estimated from additional measurements to be less than 8 % (15 % for the $^3$He measurement, because of the more complex line evaluation). The uncertainty of $\Sigma_t(E)$ and consequently that of $I_{TR}$ was not available from the data base.

The increase of the C values (up to 85 %) for a single energy interval with decreasing L suggests that there may be either hidden errors in the determination of C or that Eq. (3) is not valid.

For transmission functions T(E) of about 0.001 to 0.01 - as they are found here - errors in $\Sigma_t \cdot L$ lead to considerably amplified errors in T(E) and thus $I_{TR}$ (Eq. 4). Therefore, to explain the discrepancies, scaling of the JEF-1 cross section of Fe (in the minima) was tried. The term $\Sigma_t(E) \cdot L$ in the exponent of Eqs. (1) and (3) was modified to $\Sigma_t(E)\ L\ (1+\alpha)$ so that the adjusted values $C(\alpha)$ fluctuate in the expected range of uncertainties. For each energy interval the respective parameters $\alpha$ are given in Table 1.

This scaling of the $\Sigma_t \cdot L$ values means a scaling of the evaluated cross sections, since the density and the length are well known. However, this is only one possible explanation of the measured inconsistencies. Another reason may be the shape of the reactor spectrum, but it is unlikely that the reactor spectrum has similarly narrow lines at all the peaks investigated, all the more as the material of the PTB reactor is mainly aluminium and not iron.

It should be mentioned that similar discrepancies have been detected in the interpretation of the EURACOS IRON BENCHMARK, where gold foil activation and calculations were compared (12).

From the present results it can be concluded that the cross section minima of natural iron of the JEF-1 file between 10 keV and 500 keV are confirmed within a range quantified by the $\alpha$-values of Table 1, i.e. 3 % to 9 %.

## 4. CONCLUSION

An attempt has been made to explain some discrepancies between measured and calculated neutron spectra for filters of different lengths by a relatively small variation of the cross-section in the minima between 10 keV and 1 MeV. As the spectrum at the reactor side of the filter was unknown, questions are still open concerning the amount of variation. The experiments have shown in addition that lines with a relative FWHM of a few percent can be unfolded from proportional counter measurements. This may be of general importance for the determination of disturbing high energy lines for reactor filters, when they are used for calibration purposes in metrology and dosimetry.

In further discussions, it remains to be seen what will be the result of future calculations with the newly evaluated JEF-2 or ENDF B/VI cross section files.

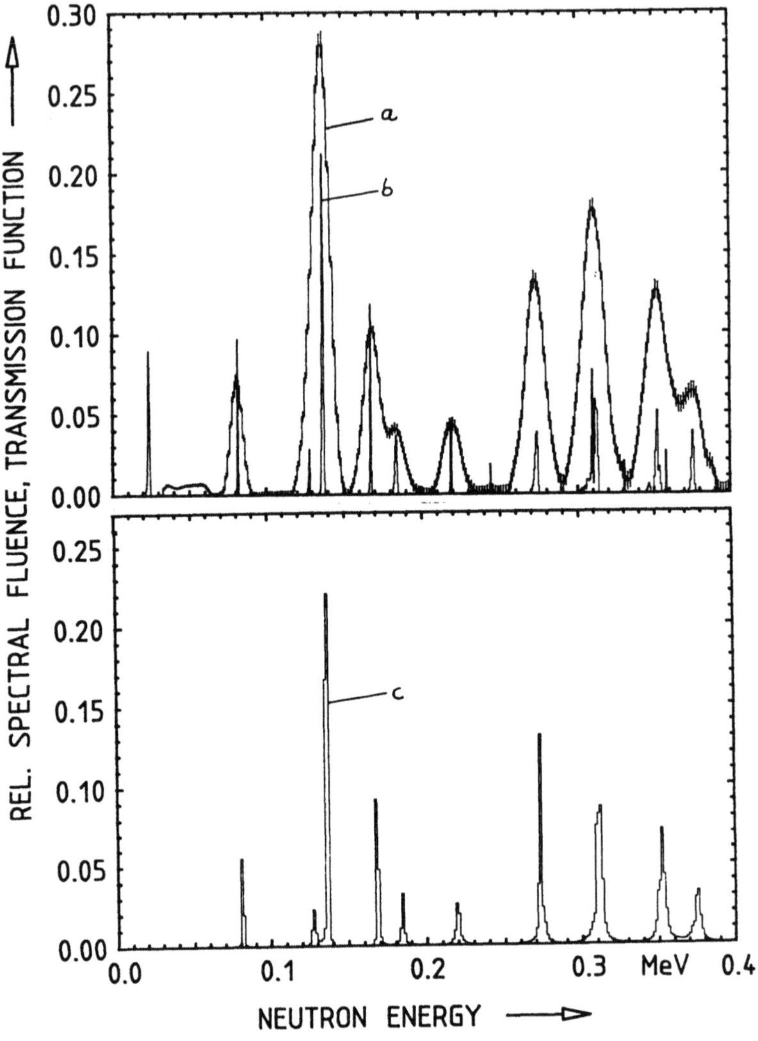


Fig. 1: Relative spectral neutron fluences (Curves (a) and (c)) and transmission function (Curve (b)) vs. neutron energy for the Fe filter with L=672 mm.

(a) : Measured spectrum between 0.03 MeV and 0.4 MeV after the first unfolding step, arbitrarily scaled.

(b) : Transmission function (see Eq.1), calculated from JEF-1 data

(c) : Unfolding results (second step) from curve (a) using the MIEKE code with Gaussian resolution functions.

Curve (c) is normalized to curve (b) so that maximum values agree.

Table 1: Transmission integrals $I_{TR}$, C- and $\alpha$-values

In each energy group the first two lines are for $L_1$ = 672 mm (first and last measurement of the cycle) followed by lines for $L_2$ = 482 mm and $L_3$ = 352 mm. C($\alpha$) : adjusted values using the $\alpha$-value indicated. All C- and C($\alpha$)-values are normalized to the first one.

| Energy interval in keV and $\alpha$ | $I_{TR}$ in keV | C | C($\alpha$) | | Energy interval in keV and $\alpha$ | $I_{TR}$ in keV | C | C($\alpha$) |
|---|---|---|---|---|---|---|---|---|
| 18.5 - 29 | 0.116 | 1.00 | 1.00 | $L_1$ | 53 - 89 | 0.056 | 1.00 | 1.00 |
| | 0.116 | 0.99 | 0.99 | $L_1$ | | 0.056 | 0.88 | 0.88 |
| | 0.276 | 1.32 | 0.87 | $L_2$ | | 0.130 | 1.34 | 0.94 |
| | 0.530 | 1.85 | 1.02 | $L_3$ | | 0.249 | 1.53 | 0.94 |
| $\alpha$ = 0.070 | | | | | $\alpha$ = 0.076 | | | |
| 114 - 151 | 0.289 | 1.00 | 1.00 | $L_1$ | 155 - 198 | 0.140 | 1.00 | 1.00 |
| | 0.289 | 1.01 | 1.01 | $L_1$ | | 0.140 | 1.00 | 1.00 |
| | 0.570 | 1.33 | 0.97 | $L_2$ | | 0.366 | 1.08 | 0.96 |
| | 0.998 | 1.60 | 1.01 | $L_3$ | | 0.772 | 1.22 | 1.01 |
| $\alpha$ = 0.088 | | | | | $\alpha$ = 0.061 | | | |
| 206 - 232 | 0.040 | 1.00 | 1.00 | $L_1$ | 256 - 288 | 0.089 | 1.00 | 1.00 |
| | 0.040 | 1.05 | 1.05 | $L_1$ | | 0.089 | 1.08 | 1.08 |
| | 0.105 | 0.99 | 0.95 | $L_2$ | | 0.263 | 1.07 | 1.04 |
| | 0.216 | 1.13 | 1.05 | $L_3$ | | 0.589 | 1.09 | 1.04 |
| $\alpha$ = 0.033 | | | | | $\alpha$ = 0.025 | | | |
| 288 - 329 | 0.240 | 1.00 | 1.00 | $L_1$ | 329 - 396 | 0.233 | 1.00 | 1.00 |
| | 0.240 | 1.07 | 1.07 | $L_1$ | | 0.233 | 1.08 | 1.08 |
| | 0.641 | 1.15 | 1.05 | $L_2$ | | 0.677 | 1.10 | 1.05 |
| | 1.349 | 1.19 | 1.03 | $L_3$ | | 1.500 | 1.12 | 1.04 |
| $\alpha$ = 0.054 | | | | | $\alpha$ = 0.036 | | | |
| 680 - 720*) | 0.063 | 1.00 | 1.00 | $L_1$ | 893 -1003*) | 0.074 | 1.00 | 1.00 |
| | | -- | -- | $L_1$ | | | -- | -- |
| | 0.258 | 1.11 | 0.95 | $L_2$ | | 0.360 | 1.03 | 1.03 |
| | 0.716 | 1.28 | 1.00 | $L_3$ | | 1.189 | 0.99 | 0.99 |
| $\alpha$ = 0.052 | | | | | $\alpha$ < 0.01 | | | |

*) Results from $^3$He-proportional counter measurement.

Acknowledgement: The authors gratefully acknowledge the support given by Dr. M. Mattes, IKE Stuttgart, for the JEF-1 data preparation.

REFERENCES


1. G. Constantine, J.A.B. Gibson, K.G. Harrison, P. Schofield: "Harwell research on beams for neutron capture therapy", Strahlentherapie und Onkologie, 165 (1989) p. 92.

2. G. Constantine, P.R.D. Watkins, C.A. Perks, H.J. Delafield, D. Ross, W.P. Voorbraak, A. Paardekooper, W. Freudenreich, F. Stecher-Rasmussen, R.L. Moss: "Progress in neutron beam development at the HFR-Petten", this conference.

3. H. Schuhmacher, W.G. Alberts, H.G. Menzel and G. Bühler, "Dosimetry of Low-Energy Neutrons Using Low-Pressure Proportional Counters", Radiation Research 111 (1987) p. 1.

4. R. Nicks, G. Perlini, H. Rief: "Fission Neutron Penetration in Iron and Sodium", Ann. Nucl. Energy, 15 (1988), p. 457 and 16 (1988), p. 189.

5. G. Perlini, M. Carter, S. Acerbis: Neutron Spectrometry Measurements in Iron, Report EUR 11767 EN, Brussels-Luxembourg 1988.

6. Index to the JEF-1 Nuclear Data Library, JEF Report 1, NEA Data Bank, Paris 1985.

7. H.G. Priesmeyer, A. Brusegan, M. Matzke: "Investigations of an optimized 24 keV Neutron Beam Filter". to be published.

8. W.G. Alberts, E. Dietz: Filtered Neutron Beams at the FMRB. Phys.-Techn. Bundesanstalt, Braunschweig, Report PTB-FMRB-112 (1987).

9. K. Knauf, J. Wittstock, Neutron Spectrometry with Recoil Proton Proportional Counters at the Research and Measurement Reactor Braunschweig, Phys.-Techn. Bundesanstalt Braunschweig, Report PTB-FMRB-114 (1987).

10. K. Knauf, W. Vorbrugg, to be published.

11. M. Matzke, K. Weise, "Neutron Spectrum Unfolding by Monte Carlo Methods", Nucl. Instr. and Methods A 234, (1985), p. 324.

12. G. Perlini, H. Rief, E. Vittone, K. Burn, "Interpretation of the EURACOS Iron and Sodium Benchmark", Proc. of the 7th Intern. Conf. on Radiation Shielding, Bournemouth, England, 1988

# Radiometric measurement of cumulative mass chain fission yields for uranium-235 and plutonium-239 from hard fast reactor core spectrum to reflector/blanket degraded neutron spectra

M. Angelone
ENEA C.R.E. Frascati, Italy

P. D'hondt, A. Fabry
SCK/CEN, Mol, Belgium

ABSTRACT: Radiometric measurements of cumulative mass chain fission yields for Uranium-235 and Plutonium-239 were performed at the Tapiro fast source copper-reflected reactor of ENEA-CASACCIA and at the MASURCA facility (CEA) Cadarache.

The measurements were carried out over the whole neutron energy range relevant to fast breeder and fast test reactors, using the radiometric technique. The spectral index F8/F5 was used as spectral parameter, as well as the mean neutron spectrum energy. An interesting stability was observed for Ba-La-140 and Zr-95 fission yields versus neutron spectra energy both for U-235 and Pu-239, while the other fission yields investigated showed subtle variations.

This work agrees well with the results of the US Interlaboratory Reaction Rate (ILRR) programme, whose applicability, limited to rather hard spectra, is thus considerably extended.

KEY WORDS: fast neutron dosimetry, fission yields vs neutron energy

## INTRODUCTION

Fission yield data files currently available for U-235 and Pu-239 in general discern only between thermal, fast and 14 MeV fission yields. With respect to the measured fast fission yields, it is often difficult to conclude whether or not there is any dependence of the fission yields versus neutron energy since observed differences are in general well within the claimed experimental uncertainties.

The neutron energy dependence of the fission yields has however clearly been observed for monoenergetic neutron induced fission of U-235 [1], Pu-239 [2], U-238 [3] and Th-232 [4].

In the frame of the TAPIRO characterization programme [5], an attempt was made to measure fast fission yields for U-235 and Pu-239 from hard fast reactor core spectrum to reflector/blanket degraded neutron spectrum and to see how major conclusions of this study can be brought into line with the above-mentioned monoenergetic neutron study. Our work was concentrated on the Zr-95, Ru-103, I-131 and Ba-La-140 mass chain yields, most widely used for reactor dosimetry.

## EXPERIMENTAL METHOD

The fission yield can be defined as :

$$Y_j = R_j/F \quad (1)$$

where $Y_j$ and $R_j$ are respectively the fission yield and the production rate of the j-th fission product. F is the absolute fission rate.

The absolute fission rates and the spectral index F8/F5, used as spectral parameter, were measured by means of microfission chambers calibrated at SCK/CEN-MOL, against NBS-type double fission chambers [6]. In the case of U-235 fission yields measured at Masurca (CEN Cadarache, France) during the IRMA campaign [7], the absolute fission rates and the spectral index F8/F5 were measured using a NBS-type double fission chamber.

Fission product generation rates were determined measuring the gamma-ray activity of the selected fission products. The used radiometric technique has been reported in [8]. The measured activities were corrected for self-absorption in the foils, self-shielding, minor isotope contribution to the measured fission rate, fission fragment escape from the fission foils [9] and for counting geometry. The nuclear data were taken from [10].

With respect to the experimental uncertainties, a quadrature propagation law was applied. The uncertainties are about 2.5 to 3.0 % at the 1 $\sigma$ level. The main contributions to the uncertainty arise from the fission chamber measurements (2.0 to 2.5 %), the Ge-detector efficiency (1.5 to 2.0 %), the gamma-ray transition probability (0.1 to 1.5 %) and the counting statistics (< 1 %). However, since all measurements are made under identical conditions, relative uncertainties on fission yields versus neutron energy are much less (< 1 %).

## RESULTS

The U-235 fission yields measured at TAPIRO and MASURCA are reported in table 1, where they are compared to the ILRR-results [11]. All results are depicted in figure 1.

TABLE 1 : U-235 Fission Yields measured at TAPIRO, MASURCA and at the ILRR-programme

| F8/F5 | $^{140}Ba-^{140}La$ | $^{95}Zr$ | $^{103}Ru$ | $^{131}I$ |
|---|---|---|---|---|
| .0495 + | .0609 ±2.0% | .0644 ±2.0% | .0330 ±2.0% | .0328 ±4.3% |
| .0375 ++ | .0610 ±1.9% | .0646 ±1.9% | .0337 ±1.9% | .0334 ±4.1% |
| .0362 * | .0611 ±2.6% | .0651 ±2.9% | .0331 ±2.8% | .0346 ±2.7% |
| .0342 ** | .0612 ±2.9% | .0661 ±3.2% | .0332 ±3.0% | .0346 ±3.3% |
| .0142 * | .0605 ±2.6% | .0646 ±3.0% | .0325 ±2.8% | .0330 ±2.8% |
| .00633 * | .0613 ±2.7% | .0656 ±3.0% | .0327 ±2.9% | .0336 ±2.7% |
| .00302 * | .0607 ±2.7% | .0652 ±2.9% | .0322 ±2.9% | .0327 ±2.7% |
| .00153 * | .0611 ±2.7% | .0651 ±2.9% | .0323 ±2.8% | .0325 ±2.7% |

\+ ILRR-CFRMF [11]
++ ILRR-BIGTEN [11]
** MASURCA
* TAPIRO

From figure 1, one observes the remarkably constant behaviour of the Ba-La-140 and the Zr-95 yield in contrast to the slight increase of the I-131 and Ru-103 yield with increasing F8/F5 spectral index. This interesting observation with respect to the Ba-La-140 and Zr-95 yield, makes these two fission products very useful for fission rate determination in all regions of fast reactors without introducing relevant uncertainties due to the neutron spectrum variation. According to the definition reported in [12] :

$$\bar{E} = \frac{\int_0^{14MeV} \Phi(E)EdE}{\Phi_{tot}} \qquad (2)$$

where $\Phi(E)$ is the differential neutron flux and $\Phi_{tot}$ is the total flux, the calculated mean neutron energy corresponding to the measured spectral index in TAPIRO is gathered in table 2. The neutron spectra used, were obtained from transport calculations [13].

TABLE 2 : Comparison between Mean Neutron Spectrum Energy and F8/F5

| Position | F8/F5 | Mean Energy |
|---|---|---|
| Core Centre | 1.40E-1 | 1.650 MeV |
| 1 | 3.62E-2 | 0.568 MeV |
| 2 | 1.42E-2 | 0.356 MeV |
| 3 | 6.33E-3 | 0.261 MeV |
| 4 | 3.03E-3 | 0.205 MeV |
| 5 | 1.53E-3 | 0.169 MeV |

Very similar results are obtained for Pu-239. The Pu-239 fission yields are reported in table 3 where they are compared to the ILRR-results [11]. All results are depicted in figure 2.

TABLE 3 : Pu-239 Fission Yields measured at TAPIRO and at ILRR program

| F8/F5 | $^{140}Ba-^{140}La$ | $^{95}Zr$ | $^{103}Ru$ | $^{131}I$ |
|---|---|---|---|---|
| .0495 + | .0533 ±2.1% | .0480 ±2.1% | .0705 ±2.1% | ----- |
| .0375 ++ | .0530 ±2.0% | .0479 ±2.0% | .0709 ±2.0% | ----- |
| .0362 * | .0537 ±2.6% | .0487 ±2.9% | .0709 ±2.8% | .0431 ±2.7% |
| .00633 * | .0538 ±2.6% | .0489 ±2.9% | .0712 ±2.8% | .0427 ±2.7% |
| .00302 * | .0535 ±2.6% | .0485 ±2.9% | .0698 ±2.8% | .0416 ±2.8% |

+ ILRR-CFRMF [11]
++ ILRR-BIGTEN [11]
* TAPIRO

The experimental results obtained at TAPIRO clearly show that observed variations of the fission yields are small compared to the corresponding variations observed for the already mentioned monoenergetic neutron fission yield data [1,2]. Indeed, in the case of U-235, variations of 20 % are observed for the Ba-La-140 yield in the neutron energy range 170 keV to 8.1 MeV, while for Zr-95 yield a variation of 12 % is observed in the same energy range.

In order to check how both observations can be brought into line, weighted yield values were calculated for the TAPIRO measuring positions, the weighting functions being the corresponding neutron spectra. The fission yields were calculated using following relation :

$$\langle Y \rangle = \frac{\int_0^{14MeV} Y(E)\ \sigma_f(E)\ \Phi(E)\ dE}{\int_0^{14MeV} \sigma_f(E)\ \Phi(E)\ dE} \tag{3}$$

where : $\Phi(E)$ : differential neutron flux

$\sigma_f(E)$ : differential fission cross-section

$Y(E)$ : function representing the monoenergetic yield versus neutron energy.

The neutron spectra used, were obtained from transport calculations [13], while Y(E) was calculated by least square fitting through the data reported in [1,2]. It should be stressed that the fission yields reported in [1,2] were measured in the energy range from 170 keV to 8.1 MeV. Below 170 keV down to thermal energy, a linear fit through the corresponding yield values was adopted. Less important is the lack of data above 8.1 MeV due to the little fraction of neutrons in this energy range. The 14 MeV yield data were however taken into account. The results obtained for U-235 are represented in figure 3, where they are compared to the fission yields measured at TAPIRO. The calculations seem to be able to reproduce both the observed trends and the absolute fission yield values. For Zr-95 the calculated fission yields are however slightly higher than the measured ones ($\sim$ 2 %), while for I-131 the opposite observation is made. For Pu-239 similar calculations were performed for Ba-La-140 and Ru-103. Similar observations as in the case of U-235 are made.

## DISCUSSION AND CONCLUSION

The data obtained within the ILRR-programme were compared to results obtained at the TAPIRO reactor in fast reactor spectra with varying hardness. Through these results, the ILRR-data, with limited applicability to rather hard spectra, are extended to all fast spectra of interest in fast reactor experimental physics.

The results obtained give an answer to the currently open problem concerning the correlation of fission yields versus neutron spectra energy. Variations were pointed out from hard fast reactor core spectrum to reflector/blanket degraded neutron spectrum, for some fission yields of interest in reactor dosimetry. In the case of Zr-95 and Ba-La-140 we found, both for U-235 as for Pu-239, a stability of these fission yields over the whole energy spectrum range investigated.

The possibility to evaluate the absolute fission yields and their trends versus neutron spectra energy was demonstrated using monoenergetic neutron yield data. The lack of data below 170 keV represents however the main limitation of the method. This method for fission yield calculation is rather straightforward and might be applied to calculate the absolute fission yield in media for which there is a lack of experimental data.

## REFERENCES


1. L.E.GLENDENIN, J.E.GINDLER, D.J.HENDERSON and J.W.MEADOWS, Phys.Rev.C. Vol.24, n°6, (1981), p 2600.

2. J.E.GINDLER, L.E.GLENDENIN, D.J.HENDERSON and J.W.MEADOWS, Phys.Rev.C. Vol.27, n°5, (1983), p 2058.

3. S. NAGY, K.F. FLYNN, J.E. GINDLER, J.W. MEADOWS and L.E. GLENDENIN, Phys.Rev.C. Vol.17, n°1, (1978), p 163.

4. L.E. GLENDENIN, J.E. GINDLER, I. AHMAD, D.J. HENDERSON and J.W. MEADOWS, Phys.Rev.C. Vol.22, n°1, (1980), p 152.

5. A.FABRY, P.MOIOLI, M.ANGELONE, H.AIT ABDERRAHIM, P.D'HONDT, G. and S. DE LEEUW, M. CARTA, G. MINSART, This symposium, paper 166.

6. J.A. GRUNDL, D.M. GILLIAM, N.D. DUDEY and R.J. POPEK, Nuclear Technology, Vol.25, (1975), p 237.

7. W. SCHOLTYSSEK, M. ANGELONE, B.L.H. BURBRIDGE, P. D'HONDT, K. GMUR, G. GRANGET, P. MARIMBEAU and D. MADISSON, Proceedings of the 1988 International Reactor Physics Conference, 18-21 September 1988, Jackson Hole (Wyoming, USA), Vol.II, p 271.

8. P. D'HONDT and A. FABRY, "Reactor Dosimetry : Methods, Applications and Standardization," ASTM STP 1001, (1989), p 634.

9. M. ANGELONE and P. D'HONDT, Nucl. Instr. and Methods in Phys. Research, A286 (1990) p 301.

10. J.H. BAARD, W.L. ZIJP and H.J. NOLTHENIUS, "Nuclear Data Guide for Reactor Neutron Metrology," Kluwer Academic Publishers, ISBN 0-7923-0486-1.

11. D.M. GILLIAM, R.G. HELMER, R.C. GREENWOOD, J.W. ROGERS, R.R. HEINRICH, R.J. POPEK, L.S. KELLOG, E.P. LIPPINCOT, G.E. HANSEN and W.H. ZIMMER, Proceedings of the 2nd ASTM-Euratom Symposium on Reactor Dosimetry, Palo Alto, Calif., NUREG/CP-0004, (1977), p 1289.

12. G. LAMMER and M. LAMMER, IAEA-report 208, Vol.I, p 301, Vienna (1978).

13. M. CARTA, A. D'ANGELO and M. ANGELONE, "Transport theory analysis of the TAPIRO reactor," private communication.

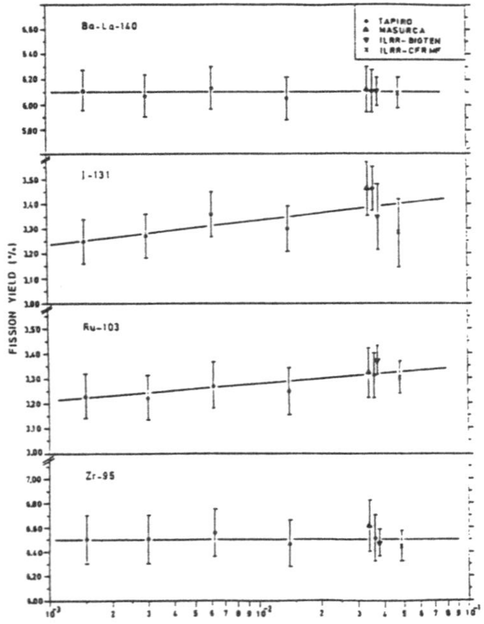


FIGURE 1 : U-235 FISSION YIELDS AS A FUNCTION OF F8/F5

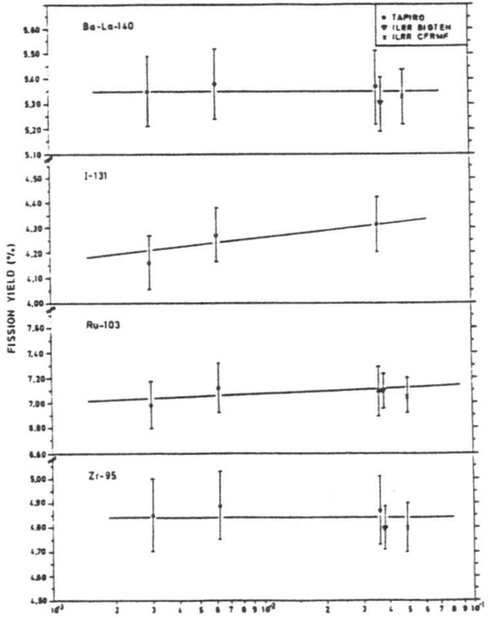


FIGURE 2 : Pu-239 FISSION YIELDS AS A FUNCTION OF F8/F5

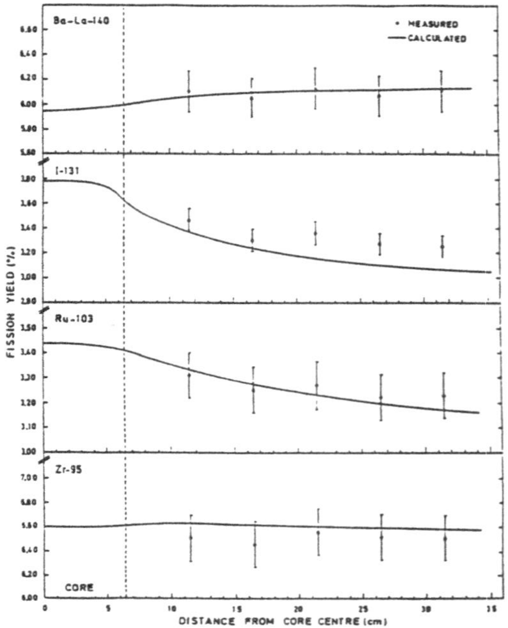


FIGURE 3 : COMPARISON BETWEEN MEASURED AND CALCULATED FISSION YIELDS FOR U-235

# ACCURATE CALCULATIONS OF NEUTRON KERMA AND DAMAGE FROM ENDF/B-VI EVALUATIONS FOR SILICON, CHROMIUM, IRON AND NICKEL, AND COMPARISON WITH ENDF/B-V RESULTS

D. C. Larson, D. M. Hetrick, C. Y. Fu and S. J. Epperson*

Oak Ridge National Laboratory
Oak Ridge, TN 37831-6356

and

R. E. MacFarlane

Los Alamos National Laboratory
Los Alamos, NM 87545

## I. INTRODUCTION

Accurate calculations of kerma (Kinetic Energy Released in MAterials) factors and displacement cross sections are fundamental to studies of neutron heating and neutron radiation damage. Damage and heating studies are important elements in both fission and fusion reactor design. For fusion reactor studies, heating in the blanket helps determine the efficiency, and heating studies in the superconducting magnet regions determine properties of the blanket and shield, thereby affecting the economics of the reactor. Radiation damage is important for reactor core lifetime studies and is related directly to the economy and safety of both fusion and fission reactors.

Calculation of kerma factors and displacement cross sections require nuclear data for neutron-induced reactions. These data are normally obtained from evaluated nuclear data libraries, such as ENDF/B (US), JENDL (Japan), JEF/EFF (Europe) and BROND (USSR). Unfortunately, most of these evaluated libraries do not contain sufficient information for a direct calculation of these quantities, and various approximations must be employed. The accuracy of the kerma factors and displacement cross sections, and thus the heating and damage functions, are directly related to the available information from the data libraries. In this paper we describe improvements

* Summer Employee, University of Florida, Nuclear Engineering Department, 1989

incorporated in the latest version of the US library, ENDF/B-VI, which allow exact calculations of these quantities for many of the structural materials included in the library.

In Section II we define kerma in terms of the nuclear data required, and review the various approximations used to estimate kerma. In Section III we describe the new information available in ENDF/B-VI which allows exact kerma calculations, and list the materials for which this information is available. Section IV compares heating and damage functions calculated from ENDF/B-V and ENDF/B-VI for the materials 28Si, chromium, iron and nickel. A summary and conclusion are given in Section V.

## II. CALCULATION OF KERMA

Discussions of the concept of kerma are available (AB73, FA89, MA82). The neutron kerma factor is the energy released locally due to neutron reaction products, and is defined as

$$k_{ij} = \sigma_{ij}(E) * E_{ij}(E) \ , \tag{1}$$

where

$k_{ij}(E)$ = neutron kerma factor for reaction $i$ in material $j$ at neutron energy $E$

$\sigma_{ij}(E)$ = neutron cross section for reaction i in material $j$ at energy $E$

$E_{ij}(E)$ = energy released locally for reaction $i$ in material $j$ induced by a neutron of energy $E$

The cross sections $\sigma_{ij}$ are given as a function of neutron energy in the evaluated libraries. The fundamental quantity required for kerma studies is $E_{ij}$. Since we are considering only energy deposited locally, $E_{ij}$ is the sum of the kinetic energies of the recoil nucleus and emitted charged particles (and radioactive decay contributions to energy deposition which will not be discussed here). For two-body reactions where the kinematics are well defined, $E_{ij}$ can be directly calculated from the evaluated libraries. However, in the continuum energy region (typically beginning at 3-5 MeV excitation energy for the structural materials) the two-body kinematics utilize energy distributions to define the average outgoing particle energy. As an example, consider $E_{ij}$ for the $(n, n\alpha)$ reaction on $^{56}$Fe at 14 MeV. We need the average energy for the recoil nucleus $^{52}$Cr and the average energy of the outgoing alpha particle. For this case, the average energy is obtained as

$$E_{ij} = \int E' P_{ij}(E') dE' \ . \tag{2}$$

Herein lies the problem. In the past, evaluated neutron cross section libraries have not provided the outgoing energy spectrum $P(E')$ for charged particles. Without this information, the average energy cannot be calculated in a direct manner.

For kerma calculations, two approximations have been commonly used to circumvent this problem. In the first one, known as the "energy-balance method," the energy carried away by neutrons and photons is subtracted from the available energy $(E + Q)$ to obtain the energy carried away by the charged particles. This approach utilizes energy distributions for the outgoing neutrons and photons from the libraries. There are several problems with this approach. The first is that often the kerma depends on the difference of two large numbers and must be carefully calculated. The method also assumes that the evaluation conserves energy which has not always been the case as discussed in (MA79). A problem important at low neutron energies (below 10 keV) is the neglect in evaluations of recoil energy due to photon emission which can be important especially for high-energy capture gamma-ray emission. Finally, since most structural materials have had elemental and not isotopic evaluations, average reaction $Q$-values had to be used to obtain the average energy, thereby introducing added inaccuracies.

The second method for estimating kerma, known as the "kinematic" method, uses conservation of momentum and energy to estimate average energies of recoil and reaction product charged particles. For binary reactions, such as elastic scattering, this is exact, but for the more complicated reactions simplifying assumptions are employed which affect the accuracy of the calculation.

To meet applied needs several computer codes have been developed to create libraries of kerma factors utilizing the above approximations. Two of the more well known of these are the MACK-IV code (AB78), and its associated library MACKLIB-IV (GO78), based on ENDF/B-IV, and KAOS-V (FA89) and its library KAOS/LIB-V (FA89), based on ENDF/B-V. A nice summary of these (and other) libraries is given in (FA89). Kerma factors in these libraries all suffer from one or more of the problems noted above, even though much attention has been given to reducing the magnitude of the problems.

To summarize: (1) previous methods to calculate kerma rely upon approximations to estimate the average energy in Eq. 1, and (2) the accuracy of a neutron kerma calculation depends upon the accuracy and availability of nuclear data.

We also note that calculation of the displacement cross sections and damage functions are more sensitive than kerma to the details of the recoil and reaction product charged particle spectra (MA82). While kerma requires the spectral shape only to obtain the average energy, the radiation damage function results from folding the Lindhard partition function, (which is a function of the recoil energy) with the spectral shape. Previous studies of radiation damage have used various approximations for the spectral shape, based primarily on kinematics.

## III. IMPROVEMENTS TO ENDF/B-VI FOR KERMA AND DAMAGE CALCULATIONS

As a result of serious problems with energy balance for many evaluations in ENDF/B-V (MA79), much thought was given to expanding the nuclear data available in evaluations for ENDF/B-VI to reduce these problems. Principal improvements included (1) use of isotopic evaluations for the important structural materials to eliminate the need for average $Q$ values, (2) use of sophisticated nuclear model codes benchmarked against an improved experimental data base to provide energy distributions for all outgoing charged particles, and (3) use of the File 6 formats which allowed energy-angle correlated data to be stored, as well as recoil spectra.

Advances in nuclear model theory, including work in precompound emission theory and level densities, were implemented in model codes. The code TNG (FU80) was used in this work to simultaneously reproduce experimental data in a global manner for all reactions and incident neutron energies from 1 to 20 MeV. Details of the calculations can be found in (HE87) for chromium and (HE87a) for nickel. This method takes advantage of constraints imposed by the measured total cross section and results in smaller uncertainties in the final results than evaluation methods which use several unrelated codes to calculate the various partial cross sections and related quantities.

Examples of data pertinent to the calculation of average energy released (Eq. 1) but not available in earlier versions of ENDF are shown in Figs. 1 and 2. Figure 1 shows an example of the measured proton emission spectrum (GR79) for $^{60}$Ni at an incident energy of 14.8 MeV, compared with the calculated spectrum from TNG. Figure 2 shows an example of the recoil spectrum of $^{24}$Mg, resulting from the $(n, n\alpha)$ reaction on $^{28}$Si. Utilizing such $P(E')$ information in Eq. 2, the average outgoing charged particle energies can be calculated from the evaluated data, and combined with the cross sections to provide the neutron kerma factors for each of these reactions at a particular incident neutron energy. Summing the kerma factors for an isotope over all the reactions gives the total neutron kerma for that energy, from which the heating can be calculated.

## IV. COMPARISON OF KERMA AND DAMAGE CALCULATED FROM ENDF/B-V AND VI

As noted earlier, with the new information for outgoing charged- particle spectra now included in ENDF/B-VI, the kerma factor can be calculated directly from Eq. 1, without approximation. The quality of the kerma factor now depends only upon the quality of the nuclear data, and not upon the severity of the approximation required to get the average energy. Thus, the kerma factors obtained from ENDF/B-VI should be the standard against which approximation schemes for obtaining kerma are compared.

The HEATR module of the NJOY processing code (MA82, MA90) was used to calculate the kerma factors and heating for isotopes of silicon, chromium, iron and nickel. In each case the heating calculated from ENDF/B-V used the energy balance approximation, while the ENDF/B-VI results used kerma factors calculated from Eq. 1. Figure 3 compares the heating in chromium calculated from ENDF/B-V and -VI. The ENDF/B-VI evaluations for the chromium isotopes contain all the necessary information to calculate kerma directly for each isotope, and the results are combined to calculate heating for the natural material. Heating calculated from ENDF/B-V is almost meaningless and conclusions regarding heating drawn at any energy other than near 14 MeV should be viewed with skepticism.

The basic problem in this evaluation is lack of energy balance in ENDF/B-V; the negative heating factors (especially from 5-12 MeV) indicate that more energy is emitted in the form of neutrons and gamma rays than is brought into the system. Such problems often arise when experimental photon production data are used directly in the evaluation. As pointed out in (MA79), negative heating factors calculated from ENDF/B-V are not uncommon.

Figure 4 shows a similar comparison for heating in iron calculated from ENDF/B-V and -VI. We note the agreement is much better than for chromium (differing by a maximum of 20%), reflecting increased attention to energy balance for the ENDF/B-V iron evaluation. Finally, Fig. 5 displays heating calculated for nickel. As for chromium, large differences are seen and are attributable to the direct use of experimental photon production data in the evaluation for ENDF/B-V.

Figure 6 compares the damage function (MA82) for iron calculated from ENDF/B-V and -VI. The results agree within about 10%, with the ENDF/B-VI results generally lower below 15 MeV. Similar results are available for the damage function in silicon, chromium, and nickel.

## V. SUMMARY AND CONCLUSIONS

Neutron radiation damage studies, heating and kerma calculations require cross section data and associated spectral distributions for outgoing charged particles, including recoil nuclei. Prior to ENDF/B-VI, evaluations did not contain this information, so only approximate calculations (based on energy balance or kinematics) could be made for the above quantities of interest. The energy balance method did not yield accurate heating results because energy conservation was only approximately achieved in ENDF/B-V.

With advances in nuclear theory being implemented in nuclear model codes, an improved experimental data base, more flexible ENDF formats, and use of isotopic evaluations, the present silicon, chromium, iron, and nickel evaluations for ENDF/B-VI are significantly improved over previous versions for radiation damage and heating studies. Separate complete evaluations are given for all isotopes of these materials. Cross sections for all important reactions, with pre-equilibrium effects included, are given. Experimental results, including new data, are used to benchmark the calculations. Charged particle spectra (including recoil spectra), are provided. Energy conservation is achieved for all reactions at all energies to better than 1%. For the first time, all necessary nuclear data are given to allow kerma, displacement cross sections, etc., to be calculated directly from information available in the evaluation, without approximation and dependent only upon the quality of the evaluated data. Significant differences are found between ENDF/B-V and -VI for heating in chromium and nickel, with smaller differences for iron. ENDF/B-VI evaluations for carbon, silicon, chromium, iron, nickel, copper, and lead contain detailed information to allow direct calculation of heating, kerma and damage functions. Based on the present findings, we suggest that earlier results may have large errors and should be recalculated with ENDF/B-VI.

## ACKNOWLEDGEMENTS

Research sponsored by the Office of Energy Research, Division of Nuclear Physics, U.S. Department of Energy, under contract DE-AC05-84OR21400 with Martin Marietta Energy Systems, Inc.

## REFERENCES

AB73 M. A. Abdou, C. W. Maynard and R. Q. Wright, *MACK: A Computer Program to Calculate Neutron Energy Release Parameters (Fluence-to-Kerma Factors) and Multigroup Neutron Reaction Cross Sections From Nuclear Data in ENDF Format*, Report ORNL-TM-3994, Oak Ridge National Laboratory (July 1973).

AB78 M. Abdou, Y. Gohar and R. Q. Wright, *MACK-IV, A New Version of MACK: A Program to Calculate Nuclear Response Functions from Data in ENDF/B Format*, Report ANL/FPP-77-5, Argonne National Laboratory (July 1978).

FA89 Y. Farawila, Y. Gohar and C. Maynard, *KAOS-V Code: An Evaluation Tool for Neutron Kerma Factors and Other Nuclear Responses*, Report ANL/FPP/TM-240, Argonne National Laboratory (September 1989).

FA89a Y. Farawila, Y. Gohar and C. Maynard, *KAOS/LIB-V: A Library of Nuclear Response Functions Generated by KAOS-V Code From ENDF/B-V and Other Data Files*, Report ANL/FPP/TM-241, Argonne National Laboratory (April 1989).

FU88 C. Y. Fu, *Nucl. Sci. and Eng.* **100**, 61 (1988), and K. Shibata and C. Y. Fu, *Recent Improvements of the TNG Statistical Model Code*, Report ORNL/TM-10093, Oak Ridge National Laboratory (August 1986).

GO78 Y. Gohar and M. Abdou, *MACKLIB-IV: A Library of Nuclear Response Functions Generated with the MACK-IV Computer Program from ENDF/B-IV*, Report ANL/FPP/TM-106, Argonne National Laboratory (March 1978).

GR79 S. M. Grimes, R. C. Haight, K. R. Alvar, H. H. Barschall and R. R. Borchers, *Phys. Rev.* **C19**, 2127 (1979).

HE87 D. M. Hetrick, C. Y. Fu and D. C. Larson, *Calculated Neutron-Induced Cross Sections for 52Cr from 1 to 20 MeV* and *Comparisons with Experiments*, Report ORNL/TM-10417, Oak Ridge National Laboratory (September 1987).

HE87a D. M. Hetrick, C. Y. Fu and D. C. Larson, *Calculated Cross Sections for 58,60Ni from 1 to 20 MeV and Comparisons with Experiments*, Report ORNL/TM-10219, Oak Ridge National Laboratory (June 1987).

MA79 R. E. MacFarlane, "Energy Balance of ENDF/B-V," *Trans. Am. Nucl. Soc.* **33**, 681, (1979).

MA82 R. E. MacFarlane, D. W. Muir and R. M. Boicourt, *The NJOY Nuclear Data Processing System, Volume II: The NJOY, RECONR, BROADR, HEATR and THERMR Modules*, Report LA-9303-M, Vol. II, Los Alamos National Laboratory (May 1982).

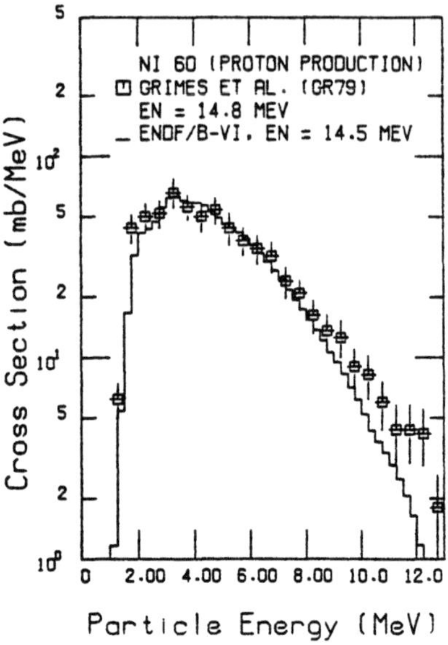


Fig. 1. Measured spectrum of outgoing protons from the $^{60}$Ni($n$,x$p$) reaction at 14.8 MeV, compared with ENDF/B-VI.

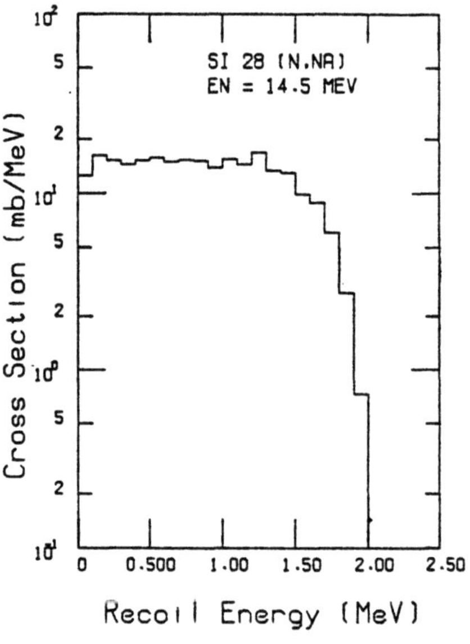


Fig. 2. ENDF/B-VI recoil spectrum for $^{24}$Mg from the $^{28}$Si$(n,n\alpha)$ reaction.

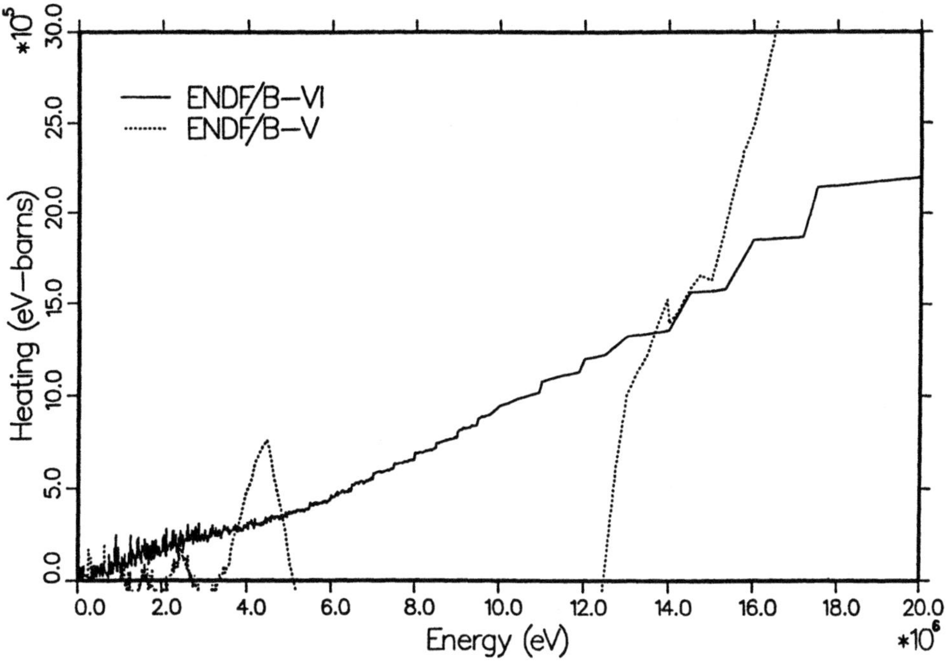


Fig. 3. Comparison of heating in chromium calculated with ENDF/B-V and -VI.

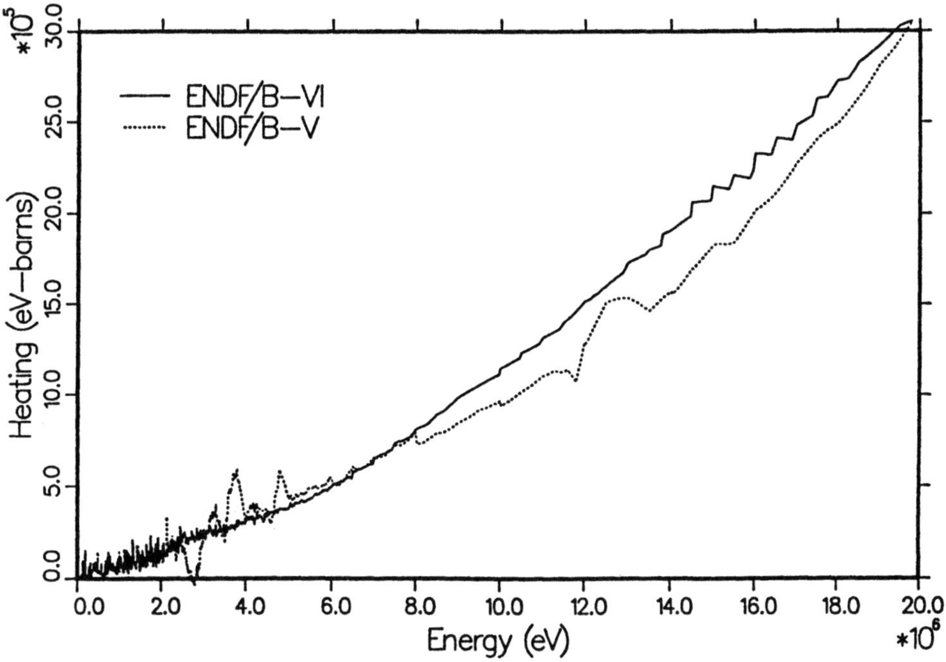


Fig. 4. Comparison of heating in iron calculated with ENDF/B-V and -VI.

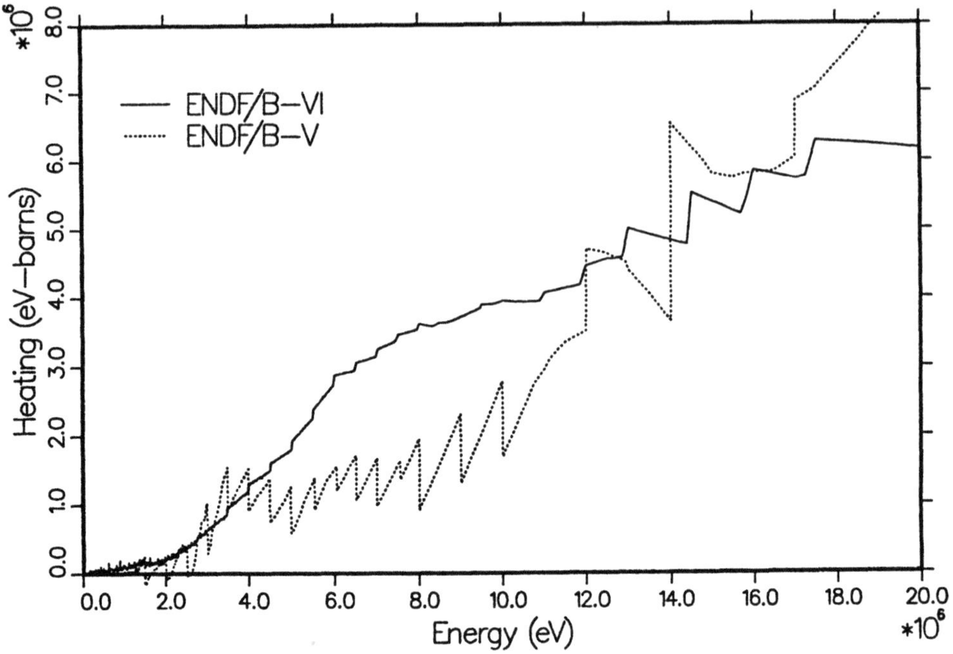


Fig. 5. Comparison of heating in nickel calculated with ENDF/B-V and -VI.

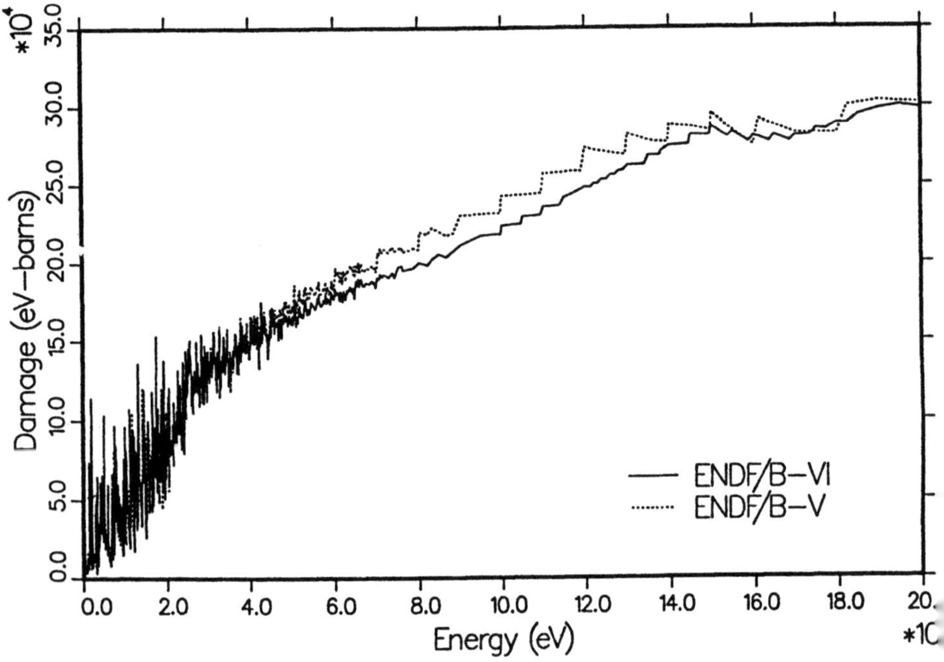


Fig. 6. Comparison of the damage function for iron calculated with ENDF/B-V and -VI.

# ISOTOPIC ABUNDANCE IN NATURAL IRON

P. Wille and R. Jede*

GKSS Research Centre, Institute for Materials Research,
*Leybold AG, Köln, Germany

It has been suggested [1] that the accepted value for the abundance of $^{54}Fe$ in natural iron might be wrong. Neutron fluence measurements based on the activation of nickel, iron, and iron enriched in $^{54}Fe$ yielded different results, which were explained as stemming partly from a wrong value for the isotopic abundance of $^{54}Fe$, for which 5,8% is commonly accepted [2].

Using Sputtered Neutral Mass Spectrometry (SNMS), the isotopic abundance of the stable isotopes in natural iron was determined in a preliminary measurement. The value of 0.28% for the isotopic abundance of $^{58}Fe$ thus obtained, confirmed the reliability of the method. Earlier investigations [3], [4], [5] gave similar results. These investigations were performed carefully, because previously recommended values for $^{58}Fe$ ranged from 0.34% to 0.28% according to different reference tabulations [4], [5].

The value for $^{54}Fe$, however, did not yield such differences. The commonly accepted value of 5.8% was confirmed in our preliminary measurements. The iron used was CBNM-certified-nuclear-reference-material. The values obtained for the abundance of the stable iron isotopes $^{54}Fe$, $^{56}Fe$, $^{57}Fe$ and $^{58}Fe$ were

5.810 ± 0.005
91.695 ± 0.02
2.211 ± 0.07
0.2842 ± 0.0004

The SNMS method allows an analysis of all elements and isotopes with quantitative detection down to ppm concentrations. The method does not suffer from artifacts caused by preferential sputtering and matrix effects.

---

## INTRODUCTION

Iron is a commonly used monitor element for fast neutron fluence measurements. The activation of the iron isotope $^{54}Fe$ via the (n,p)-reaction yields $^{54}Mn$. Its activity can be measured by detecting $^{54}Mn$'s characteristic gamma radiation of 834 keV. Its intensity, the mass of the inserted $^{54}Fe$, and a neutron-spectrum-averaged cross section for the $^{54}Fe$ (n, p)-reaction are required if the fluence of the fast neutrons which activate the iron monitor is to be determined. Thermal neutrons activate $^{58}Fe$ producing radioactive $^{59}Fe$.

For practical reasons pure iron with its natural isotopic composition is used as monitor material. Thus the isotopic abundance of $^{54}Fe$ must be known. Its accuracy is the lower limit of the accuracies for the fluence values which one obtains with iron activation detectors.

In the past numerous measurements of the isotopic abundances in natural iron have been made (1-10), for example on $^{58}Fe$.

A critical evaluation of some isotopes including the abundance of $^{58}Fe$ was published by Simonits et. al [5]. From 48 original papers and compilations, only 11 give a value between 0.28 and 0.29%, which is recommended by [4, 5, 6]. Commonly used was a value of 0.31% [9]. The adopted value for $^{58}Fe$ was taken as a measure of the reliability of our method [11]. It yields the abundances of all stable iron isotopes if the sum of the measured data for the four iron isotopes is set to 1 (=100%).

## EXPERIMENTAL METHOD

The need for quantification in surface and thin-film analysis has increased the interest in Sputtered Neutral Mass Spectrometry (SNMS) in recent years, since this method is more sensitive and does not suffer from preferential sputtering effects like the classical techniques, Auger Electron Spectroscopy (AES) and Photoelectron Spectroscopy (XPS). In addition, SNMS is far less affected by matrix effects than Secondary Ion Mass Spectrometry (SIMS). It shows a smaller amount of signal interferences by molecular ions and it yields the isotopic abundances.

The key feature of the SNMS technique is that the sputtering and ionization processes are completely separated.

In sputter equilibrium, the flux of sputtered neutrals is directly proportional to the isotopic concentrations at the sample surface, thus enabling both thin film depth profiling and bulk trace analysis. Post-ionization of the neutral particles by electron impact in the RF discharge region results in a linear relationship between measured signal intensities and true concentrations, which deviates from linearity only by the scatter of the ionization probabilities for each element.

Relative sensitivity factors can thus be established easily. For isotopes of one element sensitivity factors are not relevant.

## RESULTS

Pure iron (CBNM Certified Nuclear Reference Material No. 524) was taken as the sample. Thirty-one independent measurements [Fig.1] were made each showing the analysis of a deeper surface. The countrates of the four iron isotopes: $^{54}Fe^+$, $^{56}Fe^+$, $^{57}Fe^+$ and $^{58}Fe^+$ were measured and their sum was normalized to 100% [Fig.2].

Calculation gives the averages and the standard deviation of the averages.

| isotope | abundance at[%] aver. of 31 runs | standard deviation |
|---|---|---|
| $^{54}Fe$: | 5.8097 | ± 0.005 |
| $^{56}Fe$: | 91.6954 | ± 0.013 |
| $^{57}Fe$: | 2.2107 | ± 0.07 |
| $^{58}Fe$: | 0.2842 | ± 0.0004 |

## DISCUSSION

By Sputtered Neutral Mass Spectrometry (SNMS) the isotopic abundance of all stable iron isotopes in pure natural iron was determined.

Our measured value for the abundance of $^{58}Fe$ by SNMS are in good agreement with data measured by neutron activation analysis (NAA) [3, 4] and with data measured by Secondary Ion Mass Spectrometry (SIMS) [6, 7]. The interference from FeH-complexes in SNMS is only about 1% of the interference known from SIMS, which was estimated to be less than $2 \cdot 10^{-4}$ [7]. This gives a production rate of FeH between $10^{-6}$ and $10^{-5}$ for SNMS, which is negligible. Remark: the low-energy tail seen on the $^{56}Fe$-peak [Fig.2] is due to effects in the ion optics. Three evaluations [5, 8, 10] recommend 0.28% for $^{58}Fe$. Our value for $^{54}Fe$ shows a difference of 0.1% and 1% compared with the values found by the two SIMS-measurements. Our value for $^{57}Fe$ shows a somewhat larger difference of 3.5% compared with the two SIMS-measurements, but it is consistent concerning the error bars. This shows the degree of reliability (systematic uncertainties) of our data. For comparison the following table is instructive:

| Method | Abundances [%] in natural iron for | Authors |
|---|---|---|
| NAA | $^{58}Fe$ 0.2832 ±0.01 | P.F. Schmidt; J.E. Riley [3] |
| NAA | $^{58}Fe$ 0.280 ±0.003 | W.D. James; J.J. Carni [4] |
| Evaluation | $^{58}Fe$ 0.280 ±0.003 | A. Simonits et al. [5] |
| Evaluation | $^{58}Fe$ 0.280 ±0.01 | P. de Bièvre et al. [8] |
| Compilation | $^{54}Fe$ 5.8<br>$^{56}Fe$ 91.7<br>$^{57}Fe$ 2.19<br>$^{58}Fe$ 0.3 ±0.1 | J. H. Baard, W. L. Zijp [9] |
| Compilation | $^{54}Fe$ 5.8 ±0.1<br>$^{56}Fe$ 91.72 ±0.3<br>$^{57}Fe$ 2.2 ±0.1<br>$^{58}Fe$ 0.28 ±0.01 | J.H. Baard; W.L.Zijp; H.J. Nolthenius [10] |
| SIMS | $^{54}Fe$ 5.819<br>$^{56}Fe$ 91.779<br>$^{57}Fe$ 2.130<br>$^{58}Fe$ 0.271 | H. Nishimura; H. Takeshi; J. Okano [7] |
| SIMS | $^{54}Fe/^{56}Fe$ 0.062669 (normalization)<br>$^{57}Fe/^{56}Fe$ $0.023261 \pm 2 \cdot 10^{-6}$<br>$^{58}Fe/^{56}Fe$ $0.0031132 \pm 1.1 \cdot 10^{-6}$ | J. Volkening; D.A.Papanastassiou [6] |
| | $^{54}Fe$ 5.7545 norm.<br>$^{56}Fe$ 91.8237<br>$^{57}Fe$ 2.13591±0.00018<br>$^{58}Fe$ 0.28587±0.0001 | was calculated from ratios given in [6] |
| SNMS | $^{54}Fe$ 5.810 ±0.005<br>$^{56}Fe$ 91.695 ±0.02<br>$^{57}Fe$ 2.211 ±0.07<br>$^{58}Fe$ 0.2842 ±0.0004 | this work |

## REFERENCES


[1] L.B. Bärs, T.O. Serén, F. Wasastjerna: Reactor Dosimetry: Methods, Applications, and Standardization, ASTM STP 1001, H. Farrar, E.P. Lippincott, Eds., Philadelphia (1989) 127

[2] W.L. Zijp, J.H. Baard: Nuclear Data Guide for Reactor Neutron Metrology, ECN-70, August (1979)

[3] P.F. Schmidt, T.E. Riley: Analytical Chemistry 51, Nr. 2 (1979) 306

[4] W.D. James, J.T. Carni: Journal Radioanalyt. Chem. 57 (1980) 223

[5] A. Simonits et al: Journal Radioanalyt. Chem. Articles, 81/2 (1984) 369

[6] J. Volkening, D.A. Papanastassiou: Astrophys. J., 347 (1989) (1, Pt.2), L43-L46

[7] H. Nishimura, H. Takeshi, J. Okano: Proc. Secondary Mass Spectrometry, 6th Int. Conf., 1987, p.113; Edited by: A. Benninghoven, A.M. Huber, H.W. Werner Wiley: Chichester UK

[8] P. de Bièvre, M. Gallet, N. E. Holden, I.L. Barnes: J. Phys. Chem. Ref. Data 13 (1984) 809

[9] J. H. Baard, W.L. Zijp: Nuclear Data Guide for Reactor Neutron Metrology (1979); EUR-7164-EN

[10] J. H. Baard, W.L. Zijp, H.J. Nolthenius: Nuclear Data Guide for Reactor Neutron Metrology (1989); ECN-89-027

[11] R. Jede, K. Seifert, G. Dünnebier: Combined electron gas SNMS and SIMS instrument for trace and depth profile analysis with high dynamic range, Fresenius Z. Anal. Chem. 329 (1987) 116

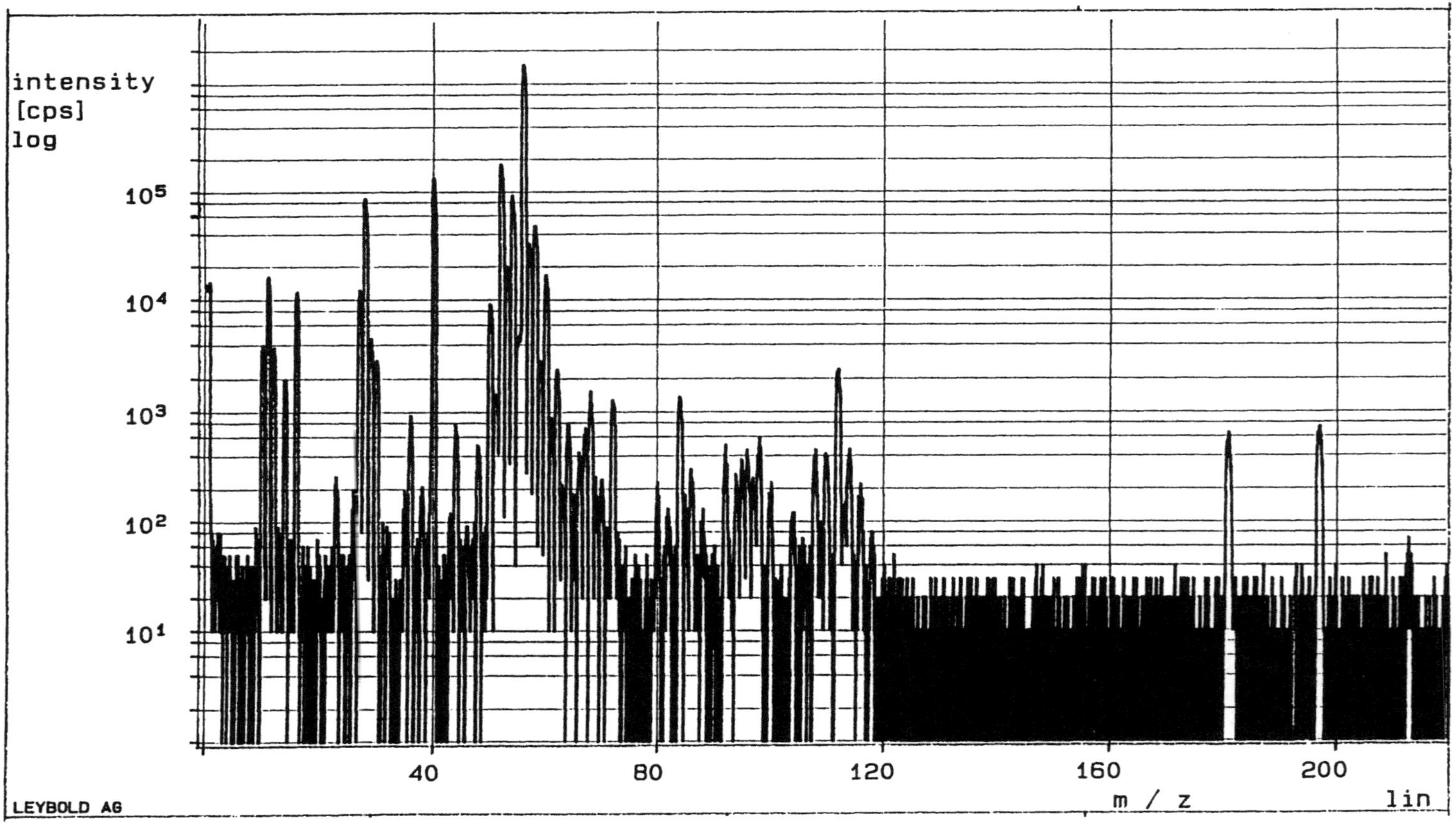


Fig. 1: Mass spectrum of sputtered neutral mass spectometry on pure iron during one run

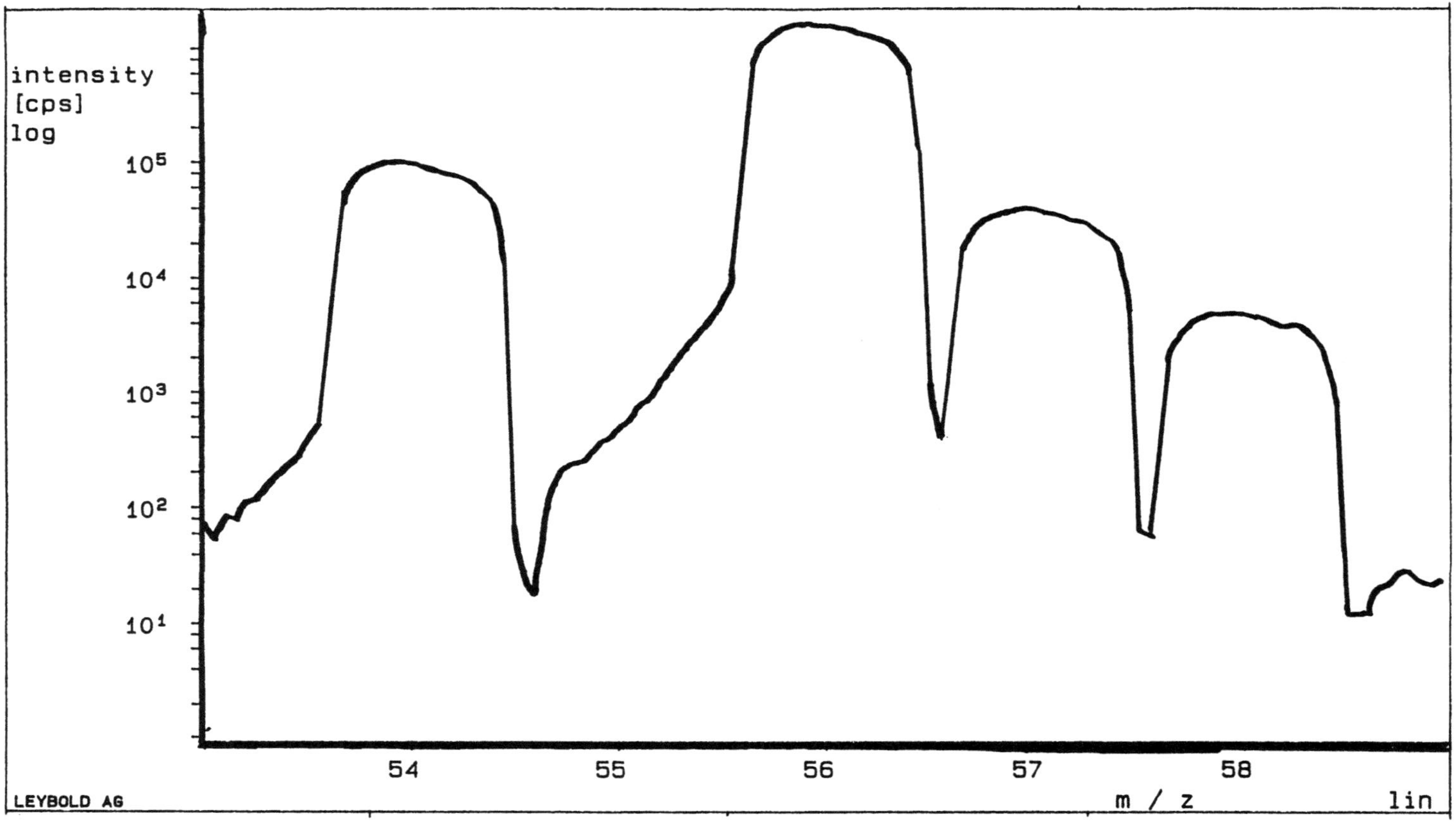


Fig. 2: Mass spectrum of sputtered neutral mass spectometry on pure iron. The countrates of the four ironisotopes

# FUSION

# NUCLEAR DATA NEEDS FOR FUSION PROGRAMS

E.T. Cheng
TSI Research
225 Stevens Avenue, Suite 110
Solana Beach, California USA 92075

## Abstract

A review is given in areas relevant to nuclear data needs for fusion programs. These areas address the motivation and direction of fusion nuclear data development activities and include fusion fuel cycles, nuclear data needs, low-activation fusion motivations, and progress on fusion nuclear data development activities. A list of important activation cross sections is presented whose measured data are needed to help assess some fusion reactor design issues.

## Introduction

The development of fusion as a long-term energy source for mankind has been focusing on more promising, energetically viable, fusion fuel cycles: deuterium (D) and tritium (T), deuterium and deuterium (DD), and deuterium and helium-3 ($^3$He). The DT fuel cycle is the most promising one and is expected to be the fuel cycle employed in the first generation fusion reactors. Neutrons are generated as a primary reaction product for the DT and DD fuel cycles, and are produced as secondary products for the D$^3$He fuel cycle.

Tritium breeding is one of the vital functions in self-sustaining DT-fueled fusion reactors. Radiation shielding to protect the reactor components such as the superconducting magnet and materials activation are important areas that need to be considered in the development of fusion reactors. Accurate assessments of the required blanket and shield compositions and dimensions, and of the activation, after-heat, and waste characteristics of fusion reactors are critical to the development of fusion concepts. Nuclear data of potential reactor materials are required to perform the design evaluations of potential reactor concepts. Neutron and gamma-ray transport data necessary for the determination of the neutron and gamma-ray flux level and spectrum in various reactor components are fundamental to the development of blanket and shield concepts for fusion reactors.

In this paper, the following areas relevant to the motivation and direction of fusion nuclear data development activities are briefly reviewed: fusion fuel cycles, nuclear data needs, low-activation fusion motivations, and progress on fusion nuclear data development activities. In conclusion, a list of activation cross sections whose measured data are needed to help assess some important fusion reactor design issues is presented.

## Fusion Fuel Cycles

There are presently three energetically viable fuel cycles for fusion energy research to investigate: DT, DD, and $D^3He$.

The DT fuel cycle is the most promising fuel cycle because of its high reaction cross sections at relatively low plasma temperatures. DT fusion produces an alpha particle (3.5 MeV) and a neutron (14.1 MeV), and releases 17.6 MeV after each reaction. However, DT fusion reactors require tritium breeding for fuel self-sustaining because tritium, which is radioactive with a half-life of 12.3 years, does not exist in nature.

The DD fuel cycle is the ultimate, unlimited fuel cycle because deuterium is abundant. DD fusion has two reaction channels and produces $^3He$, T, a proton, and a neutron (2.5 MeV) as reaction products. It releases an average energy of 3.6 MeV per reaction, or 12.4 MeV per reaction if T is consumed in the plasma during the reaction.

The $D^3He$ fuel cycle is perhaps the most attractive fuel cycle because $D^3He$ fusion produces only charged particles, an alpha and a proton, after each reaction.

Neutrons are produced only in the secondary DD reactions. A total of 18.3 MeV energy is released per reaction. However, $D^3He$ fusion reactors will need to overcome the $^3He$ availability issue, as $^3He$ is rare.

## Nuclear Data Needs

Since a majority of fusion energy appears in the form of neutrons (35% for simple DD, and 80% for DT fuel cycles), a blanket that consists of a material medium will be needed to extract the nuclear energy deposited by the interaction of neutrons with blanket materials.

Shielding of neutrons and induced gamma-ray radiation is also needed for fusion reactors with all energetically viable fuel cycles to protect the reactor components such as the superconducting magnet. Neutron-induced radioactivity in blanket and shield materials also needs to be understood in detail in order to assess the safety and environmental aspects of various reactor concepts, and to help with the selection and development of fusion materials.

Nuclear data needed for fusion energy development depend primarily on fuel cycles and reactor concepts [1]. The following fusion functional needs for nuclear data were identified in a 1986 IAEA Advisory Group Meeting [2]:

a. Neutron flux determination - total cross sections; neutron emission cross sections (double differential data); and dosimetry cross sections.

b. Tritium production - $^6Li(n,\alpha)$ and $^7Li(n,n'\alpha)$ cross sections.

c. Neutron multiplication - $^9Be(n,2n)$ and Pb(n,2n) cross sections.

d. Radiation hazards - activation cross sections and decay data.

e. Radiation damage - displacement per atom (DPA) data (recoil spectra and charged-particle production cross sections); and gas (hydrogen and helium) production cross sections.

f. Power (heat) generation - kinetic energy release per material atom (KERMA) data (recoil spectra and charged-particle production cross sections); and gamma-ray production cross sections.

g. Fusion reactions - DD, DT, and $D^3He$ charged-particle nuclear and physics data.

Potential fusion reactor materials also were identified for each reactor component under investigation. These materials are occasionally updated based on various on-going reactor study programs aimed at identifying promising reactor blanket and shield concepts. Listed below are such potential materials, identification based primarily on DT-fueled reactor concepts.

a. Structure - Fe, Cr, Ni, Mn, V, Ti, W, Ta, Al, Si, and C.

b. Tritium breeder/coolant - Li, H, O, Pb, F, He, Be, Al, Si, and Zr.

c. Neutron multiplier - Be and Pb (nonfissioning).

d. Magnet shield - C, B, W, Pb, H, O, and structure.

e. Magnet - Cu, Sn, Ti, Nb, Al, N, He, and structure.

f. First wall/divertor/limiter - C, Cu, W, Be, and structure.

g. Biological shield - Ca, Si, Ba, O, C, H, concrete and structure.

h. Hybrid blanket - Th, U, and Pu.

Note that some of the materials, such as niobium and molybdenum, have been seriously considered as structural materials for DT-fueled reactors because of their high temperature capabilities and other attractive features [3,4]. However, these materials are determined to produce large quantities of long-lived radionuclides such as $^{94}Nb$ (half-life 20,300 years), and therefore, are recently being avoided as promising structural materials. It is also noted that materials such as Th, U, and Pu, for hybrid blankets are drawing only regional interests, particularly in China and the U.S.S.R.

The status of nuclear data for fusion reactor technology has been a subject of constant review [1,2]. At present, major fusion nuclear data review and data testing activities are being conducted due to the recent release of all major nuclear data evaluations.

## Low-Activation Fusion Motivations

The goal of fusion energy development is to achieve a low-activation, inexhaustible energy source for future generations of mankind. This goal can be reached through fusion processes because no energetically viable fusion fuel cycles produce radioactive products. Although tritium (half-life 12.3 y) is required for the DT-fuel cycle, it will be carefully handled in the reactor systems. The tritium inventory in the reactor fueling system and blanket tritium extraction system can be minimized such that no excess quantity of tritium will be produced other than that needed for reactor operation.

Radioactivity in fusion is induced by neutrons produced in fusion reactions. Areas of concern for fusion reactors due to induced radioactivity are: maintenance (contact or remote), safety (after-heat, radioactive hazards released to public), and waste management (reuse or disposal; shallow-land or deep geological burial disposal).

Reuse or low-level nuclear waste disposal of fusion reactor materials is an attractive goal for fusion energy development. This goal can be achieved by the use of lower activation materials. Specifically, this implies (a) use of elements that do not produce large quantities of long-lived radionuclides; and (b) control of alloying, compounding, and impurity elements in fusion reactor materials [5]. This would involve selections of low-activity structural materials, such as modified ferritic steels in control of alloying elements, and very low activity materials, such as SiC, in the use of elements. The impurity elements, such as Nb and Mo, must be controlled below approximately 1 ppm for DT-fueled reactors.

## Nuclear Data Needs Coordination Program

The U.S. Department of Energy, Office of Fusion Energy (OFE), in 1982 established a coordination program to promote the development of nuclear data for fusion energy research. The objectives of this program are: (a) to improve communication between fusion nuclear data users and nuclear data developers; and (b) to help assure that data developers are focusing limited funding on the most critical data needs of fusion energy development programs. In addition to participation in the review of nuclear data status and the conceptual design of fusion reactors, the OFE program also has been conducting data analysis and testing to identify data deficiencies and to implement courses of action for reducing data deficiencies in the fusion nuclear database. This program has recently been promoting international collaboration in order to focus worldwide efforts on critical nuclear data needs for fusion reactor technology.

Highlights of progress on recent fusion nuclear data development activities are: (a) a major review on nuclear data measurements was conducted during the Sixth Coordination Meeting of the U.S. DOE Office of Nuclear Physics Program to Meet the High-Priority Data Needs of the Office of Fusion Energy, September 1989, at Ohio University, Athens, Ohio [6]; (b) the status of activation cross sections was reviewed at both the NEANDC Specialists' Meeting on Activation Cross Sections for Fission and Fusion Energy [7], and the IAEA Consultants' Meeting on the Measurement of Activation Cross Sections Leading to the Production of Long-Lived Radionuclides [8], September 1989, at Argonne National Laboratory; (c) the IAEA Nuclear Data Section has been sponsoring a nuclear data library development activity to establish a current, unrestricted, international fusion nuclear data library, the fusion evaluated nuclear data library (FENDL). The first version of this library, FENDL-1, is expected to be available in early 1991 for fusion users to employ in fusion reactor design studies [9]; and (d) a coordinated integral 14 MeV neutronics experiment on beryllium is being conducted under the U.S./China, and SWINPC (China)/OKTAVIAN (Japan) cooperation programs [10].

## Some Important Activation Data Needs

A list of activation cross sections whose measured data are recently identified as needed to help assess some important fusion reactor design issues follows.

a. Radiological hazard verification. Lead is a promising neutron multiplier for DT-fueled fusion blankets. However, due to the production of a large quantity of $^{210}$Po (half-life 138 days; alpha emitter) in Pb, it has been considered, particularly in the United States, to avoid the use of lead as a blanket material. The reaction cross sections to be verified regarding this issue include $^{208}$Pb(n,g)$^{209}$Pb, and $^{209}$Bi(n,g)$^{210}$Bi at energies from thermal to about 1 keV.

b. Waste disposal and material recycle concerns. Production of long-lived radionuclides is always a concern in structural and divertor materials. The following single or multi-step reactions need to be investigated in detail: $^{58}Fe(n,g)^{59}Fe(n,g)^{60}Fe$(half-life 1.5e+06 years); $^{182}W(n,n'\alpha)^{178m}Hf$ (half-life 31 years); $^{61}Ni(n,2p)^{60}Fe$; $^{63}Cu(n,\alpha)^{60}Co(n,p)^{60}Fe$; $^{63}Cu(n,p)^{63}Ni(n,\alpha)^{60}Fe$. Calculations using the presently available activation data library indicate that $^{60}Fe$ activity in the exposured (20 MW-y/m$^2$) ferritic steel first wall is about 5 orders of magnitude lower than the limiting concentration for near-surface waste disposal.

c. Low activation materials. Very-low-activation materials, such as Al and SiC, are especially attractive for fusion applications. Studies have confirmed the possible production in these materials of medium- and long-lived radionuclides that are affecting the assessments of shut-down radiation dose rate and waste disposal ratings. These cross sections include the following: $^{27}Al(n,n'\alpha)^{23}Na(n,2n)^{22}Na$ (half life 2.6 years), and $^{28}Si(n,n'p)^{27}Al(n,2n)^{26}Al$ (half-life 7.2e+05 years).

d. Data testing needs. The most urgent need in this area is to obtain a better than 5% accuracy for $^{16}O(n,\alpha)$ cross sections at 14 MeV energy. This is very important for the recent integral experiments on beryllium, using neutron moderating water tanks, to achieve a better than 3% accuracy [11,12]. The presently available accuracy for $^{16}O(n,\alpha)$ cross section is estimated to be 30% [13].

Acknowledgment

This work was supported by the U.S. Department of Energy, Office of Fusion Energy.

References

1. E.T. Cheng, "Review of the Nuclear Data Status and Requirements for Fusion Reactors," Nuclear Data for Science and Technology (1988 Mito), Japan Atomic Energy Research Institute, 1988; see also references cited in this paper.

2. V. Goulo, ed., "Report on the IAEA Advisory Group Meeting on Nuclear Data for Fusion Reactor Technology, Gaussig, German Democratic Republic, 1-5 Dec. 1986," IAEA Memorandum INDC/P(87)-3.

3. D. Steiner and M. Tobias, Nucl. Fus., 14 (1974) 153.

4. B. Badger, et al., UWMAK-III, High-Performance,

Noncircular Tokamak Power Reactor Design, UWFDM-150, University of Wisconsin, Madison, 1975.

5. S. Fetter, E.T. Cheng, F.M. Mann, "Long-Term Radioactivity in Fusion Reactors," Fus. Eng.Dsg., 6, 123 (1988).

6. S. Grimes, E.T. Cheng, ed., "Sixth Coordination Meeting of the Division of Nuclear Physics Program to Meet High-Priority Nuclear Data Needs of the Office of Fusion Energy, September 19-21, 1989, Ohio University, Athens, Ohio," CONF-8909333, U.S. Department of Energy, June 1990.

7. M. Wagner, H. Vonach, ed., "Proc. Specialists' Meeting on Neutron Activation Cross Sections for Fission and Fusion Energy Applications, Argonne National Laboratory, USA, 13-15 September 1989," NEANDC-259 'U', Organization for Economic Co-Operation and Development, 1990.

8. Wang Dahai ed., "Activation Cross Sections for the Generation of Long-lived Radionuclides for Importance in Fusion Reactor Technology, Proc. IAEA Consultants' Meeting, Argonne National Laboratory, 11-12 September 1989," INDC(NDS)-232-L, IAEA memorandum, January 1990.

9. V. Goulo, ed., "Fusion Evaluated Nuclear Data Library (FENDL), Proc. IAEA Specialists' Meeting on FENDL, Vienna, 8-11 May 1989," IAEA memorandum, INDC(NDS)-223/GF, August 1989.

10. Liu Lian-yan, E.T. Cheng, "Preliminary Analyses of SWINPC Beryllium Neutron Multiplication Integral Experiments," Fusion Technology, to be published, May 1991.

11. J. Smith, "Neutron Multiplication in Bulk Beryllium," and J.W. Davidson, "Calculational Analysis of the INEL Beryllium Multiplication Experiment," Proc. Ninth Topical Meeting on the Technology of Fusion Energy, Oct. 1990, Oak Brook, IL, to be published in Fusion Technology.

12. Yuan Chen, "Measurements of Neutron Multiplication in Beryllium Assemblies," ibid.

13. P.G. Young, Los Alamos National Laboratory, private communication.

# MEASUREMENT OF AVERAGE CROSS SECTIONS FOR SEVERAL DOSIMETRY REACTIONS IN A THICK TARGET Li(d,n) NEUTRON FIELD

J.R. Dumais, S. Iwasaki, S. Tanaka, N. Odano, and K. Sugiyama

Department of Nuclear Engineering, Tohoku University

Aramaki-Aoba, Aoba-Ku, Sendai 980, Japan

## ABSTRACT

An accelerator-based neutron source employing Li(d,n) reaction has been proposed for the validation of the dosimetry cross sections. The neutron field which had continuous energy spectrum from a few hundred keV to 17 MeV was provided by bombarding a lithium target of 1-mm thickness with 2-MeV deuterons accelerated by the 4.5-MV Dynamitron at the Fast Neutron Laboratory in Tohoku University. Average cross sections for several dosimetry reactions have been measured in this neutron field by means of the activation method. The present measurement was performed to make the Li(d,n) reaction usable as neutron source for the integral test, and to validate cross sections of the recently evaluated JENDL-3 dosimetry file and the IRDF-85.

---

## INTRODUCTION

Validation of the dosimetry cross section data can be performed with the integral experiment. The fission neutrons from reactors and $^{252}$Cf sources are very often used in such kind of measurement. In the energy spectra of these neutrons, however the neutron fluence of high energy region is relatively weak with large uncertainties, and such integral test is not suitable for reliable verification of reactions having high threshold energies which are especially important for the fusion energy development. Complementary neutron sources which have continuous energy spectrum with relatively high intensity above 10 MeV, can be obtained using accelerators. For example, Greenwood (1) applied the $^9$Be(d,n) reaction for the integral experiment at Argonne National Laboratory (ANL), and recently, a deuteron beam of 7 MeV from the Tandem accelerator at ANL was bombarded on a beryllium target; the source provided neutrons having continuous energy spectrum up to 11 MeV with relatively high intensity (2).

In the present study, we propose a neutron source employing $^7$Li(d,n) reaction to obtain high energy neutrons with relatively strong intensity using a small accelerator. The feasibility of this reaction was studied using the 4.5-MV Dynamitron accelerator at Tohoku University. In the Li(d,n) neutron field it was attempted to test the cross section data of several reactions of the recently evaluated JENDL-3 dosimetry file (3) and the IRDF-85 (4) by measuring average cross sections (5).

## CHARACTERISTICS OF THE NEUTRON SOURCE

### Lithium Metal Target

The neutron target was made from natural metal lithium. A small piece of lithium metal was melted on a copper backing plate in a vacuum and shaped into a disk with 0.8-cm diameter and about 1-mm thickness in a nitrogen gas environment. The lithium target was mounted in a chamber as shown in Fig.1, which was installed at the end of the beam duct of the Dynamitron accelerator. An air cooling system was used for this target during irradiation.

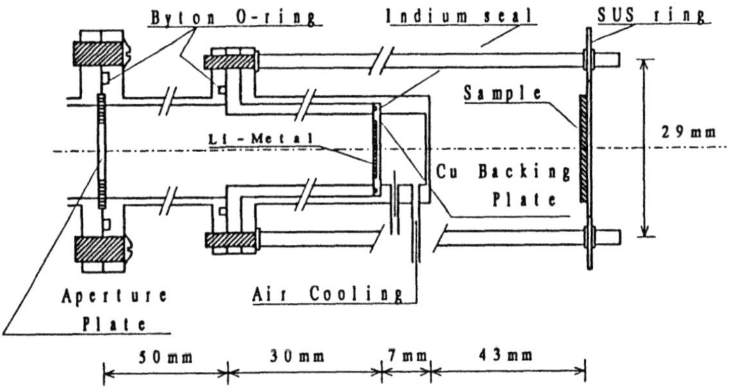


Fig.1. Target assembly with a light mass support of samples for the irradiation experiment.

In the integral experiment, the target was bombarded by 2-MeV deuterons at a direct beam current of 10 $\mu$A. Since the Q-value for the $^{7}$Li(d,n) reaction is very high (+15.027 MeV), neutrons with energies from a few hundred keV to about 17 MeV can be produced.

Determination of the neutron energy spectrum was accomplished by the neutron time of flight method. In this case, a 1.0-MHz pulsed beam with 0.2 $\mu$A of deuteron was used. The neutrons were detected by an NE-213 scintillation counter (2"×2"), which was placed at a distance of 5.3 m from the target and collimated by a shield of paraffin in the direction of zero degrees with respect to the deuteron beams. Two detectors of 1"×1" NE-213 scintillation counter and a $BF_3$ counter were used to monitor the neutron fluence, placed at 30 degrees and 60 degrees, respectively.

The relative efficiencies of the neutron detector in the energy range up to 6 MeV were obtained by the measurement of $^{252}$Cf spectrum (6), while those above 6 MeV were given by a Monte Carlo calculation code, O5S (7). The absolute values of the neutron flux at the irradiation position (see Fig.1 and next section) were obtained by measuring the reaction rate for the reference reaction $^{27}$Al(n,$\alpha$)$^{24}$Na (8) at 15.45 MeV, using the T+d neutron source in the same experimental arrangement. The energy group energy spectrum ($\phi$(E)) at 0 degrees is shown in Fig.2(a) together with cross section curves ($\sigma$(E)) taken from the JENDL-3 file for the several reactions to be studied in the integral experiment. In the spectrum, three main broad peaks in the vicinities of 16 MeV, 13 MeV and 3 MeV correspond to the ground states, and the first and fourth broad excited states of the residual nucleus $^{8}$Be, respectively. The degradation of the incident deuteron energy in the target made the shape of the each peak broaden. A peak around 2 MeV may correspond to $^{12}$C(d,n) reaction due to the contaminated carbon. We have some experiences of observing the distinctive carbon peak in the neutron spectrum, which has changed in the intensity from a target to other target.

In Fig.2(b), are shown the calculated sensitivity profile functions (S(E)=$\phi$(E)$\sigma$(E)) of four reactions in the Li(d,n) neutron spectrum. The functions for the $^{48}$Ti(n,p)$^{48}$Sc and $^{56}$Fe(n,p)$^{56}$Mn reactions were almost the same with that of the $^{27}$Al(n,$\alpha$)$^{24}$Na reaction. From this figure, we can emphasize two points: first, these functions present how the cross section curve of each reaction contributes to the average cross section (discussed later); the regions correspond to the peaks of the profile would be tested in the integral experiment; secondly, their maximum values are rather large even for the (n,2n) reactions with high threshold energies; this situation is contrast to the case of the fission sources.

Angular spectra were also measured from 0 to 90 degrees using a goniometer equipped with a heavy shield-collimator system. The angular distributions of the neutrons in some energy intervals are nearly flat over this angular range, except for the high energy interval from 15 to 18 MeV as shown in Fig.3.

## INTEGRAL EXPERIMENT

The following six nuclear reactions have been tested in the present integral experiment, i.e., $^{27}$Al(n,$\alpha$)$^{24}$Na, $^{48}$Ti(n,p)$^{48}$Sc, $^{56}$Fe(n,p)$^{56}$Mn, $^{58}$Ni(n,p)$^{58}$Co, $^{58}$Ni(n,2n)$^{57}$Ni and $^{59}$Co(n,2n)$^{58}$Co. Two packages of high purity metal foils of Al, Ti, Fe, Ni and Co with 19-mm dia. and 0.1~0.2-mm thickness for each foil were prepared. Each package was fixed on a thin SUS ring at a distance of 50 mm (see Fig.1) and irradiated along 10 and 24 hours, respectively, depending on the half lives of the residual nuclei. Activities of the irradiated samples were measured using the gamma ray counting method with a calibrated 80-cc HP-Germanium detector.

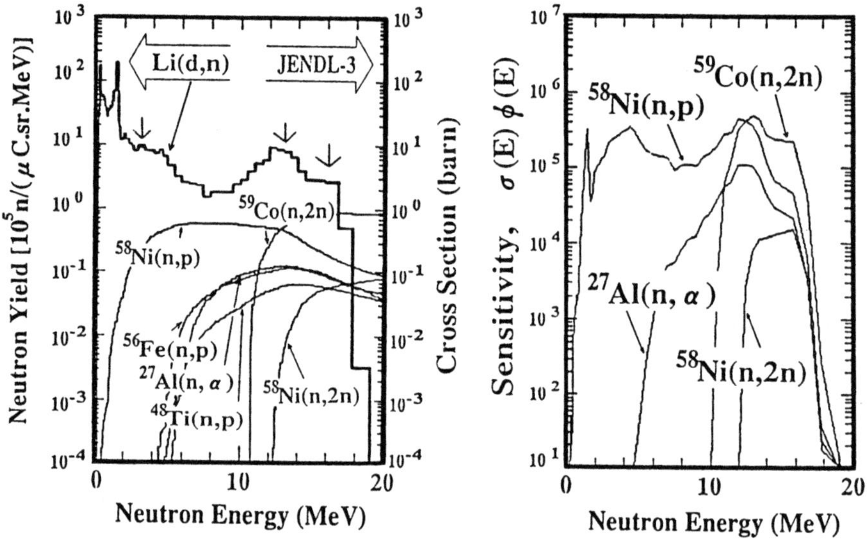


(a) (b)

Fig.2.(a) Neutron energy spectrum of the Li(d,n) reaction at the irradiation position with six cross section curves taken from JENDL-3 (arrows indicate three broad peaks correspond to the states of the residual nucleus $^{8}$Be). (b) Sensitivity profile functions of four reactions in the spectrum.

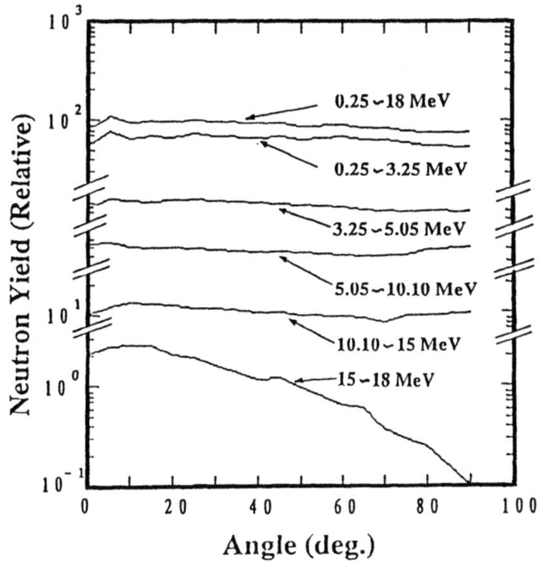


Fig.3. Angular distributions of emitted neutrons, summed over the indicated neutron energy intervals.

## AVERAGE CROSS SECTIONS

The average cross section ($<\sigma>$) can be derived from the following equations,
- Calculated value (C-Value):

$$<\sigma>_C = \frac{\sum \phi_i \sigma_i \Delta E_i}{\sum \phi_i \Delta E_i} \quad (1)$$

- Experimental value (E-value):

$$<\sigma>_E = \frac{R}{\sum \phi_i \Delta E_i} \quad (2)$$

where, $\phi$ : energy group neutron flux, $\sigma$ : energy group cross section, $\Delta E$ : energy interval, $R$ : measured reaction rate, $i$ : index of the energy group (group structure is the same as that given in Ref.6).

Responses of the calculated average cross sections from three reactions on the angular spectra were evaluated, in order to estimate the effect of the neutron anisotropy. Corrections for this effect were estimated to be less than 1%.

Errors of the average cross sections were evaluated using variance-covariance analysis based on the method by Smith (9). Covariance data of JENDL-3 dosimetry file tentatively taken (4) from IRDF-85 were used for the analysis. The uncertainties of the reaction rates which were derived from the measurement of the activities are 1.0 ~ 3.7%. The uncertainties of the source neutron spectrum were estimated about 15.0 ~ 20.0% for energy range of 0.25 ~ 0.35 MeV, 1.0 ~ 3.0% for 0.35 ~ 5.5 MeV, 5.0 ~ 6.0% for 5.5 ~ 17.0 MeV and 15.0 ~ 20.0% for 17.0 ~ 18.0 MeV.

We have compared the average cross sections of the calculated values with those of the experiments. The C/E values and their uncertainties are given in Table 1 and Table 2 for JENDL-3 dosimetry file and IRDF-85, respectively, and compared in Fig.4. The result of integral experiment can be summarized as follows:

- With respect to the low (0.3 MeV) and intermediate (around 5 MeV) threshold energy reactions, i.e., $^{58}$Ni(n,p)$^{58}$Co and $^{27}$Al(n,$\alpha$)$^{24}$Na, $^{48}$Ti(n,p)$^{48}$Sc and $^{56}$Fe(n,p)$^{56}$Mn, show that their calculated average cross sections using JENDL-3 and IRDF-85 are consistent with those of the experiments. The low threshold reactions possess the 90% of the total sensitivity, i.e., reaction rate, in the wide energy range from 2 to 16 MeV, while the intermediate threshold reactions do in rather high energy region from 10 to 16 MeV.
- As for the two (n,2n) reactions with high threshold energies, the result for the $^{59}$Co(n,2n)$^{58}$Co reaction (about 100% sensitivity from 11 to 17 MeV) is satisfactory for both the data of JENDL-3 and IRDF-85. Whereas, the $^{58}$Ni(n,2n)$^{57}$Ni reaction having about 100% sensitivity at the energy of 13 ~ 17 MeV, shows that the result by JENDL-3 is consistent, but inconsistent by IRDF-85. It is suggested that further evaluation of the cross section data of this reaction in IRDF-85 is needed for energy range from 13 to 17 MeV.

Table 1. Results for average cross sections from the measured and the calculated values using JENDL-3 dosimetry file.

| Reaction | Average Cross Section (b) | | C/E (% Error) |
|---|---|---|---|
| | Calc. (% Error) | Exp. (% Error) | |
| $^{27}$Al(n,$\alpha$)$^{24}$Na | $2.24\times10^{-2}$ ( 6.3) | $2.11\times10^{-2}$ (7.2) | 1.06 ( 9.6) |
| $^{48}$Ti(n,p)$^{48}$Sc | $1.07\times10^{-2}$ (13.1) | $1.04\times10^{-2}$ (7.1) | 1.03 (14.9) |
| $^{56}$Fe(n,p)$^{56}$Mn | $2.05\times10^{-2}$ ( 5.0) | $2.14\times10^{-2}$ (7.3) | 0.96 ( 8.8) |
| $^{58}$Ni(n,p)$^{58}$Co | $1.55\times10^{-1}$ ( 6.5) | $1.62\times10^{-1}$ (7.1) | 0.96 ( 9.6) |
| $^{58}$Ni(n,2n)$^{57}$Ni | $2.95\times10^{-3}$ (19.0) | $3.37\times10^{-3}$ (7.9) | 0.88 (20.6) |
| $^{59}$Co(n,2n)$^{58}$Co | $8.99\times10^{-2}$ (11.0) | $9.30\times10^{-2}$ (7.8) | 0.97 (13.5) |

Table 2. Results for average cross sections from the measured and the calculated values using IRDF-85.

| Reaction | Average Cross Section (b) | | C/E (% Error) |
|---|---|---|---|
| | Calc. (% Error) | Exp. (% Error) | |
| $^{27}Al(n,\alpha)^{24}Na$ | $2.29\times10^{-2}$ ( 5.2) | $2.11\times10^{-2}$ (7.2) | 1.09 ( 9.0) |
| $^{48}Ti(n,p)^{48}Sc$ | $1.07\times10^{-2}$ (13.1) | $1.04\times10^{-2}$ (7.1) | 1.03 (14.9) |
| $^{56}Fe(n,p)^{56}Mn$ | $2.00\times10^{-2}$ ( 5.0) | $2.14\times10^{-2}$ (7.3) | 0.93 ( 8.8) |
| $^{58}Ni(n,p)^{58}Co$ | $1.60\times10^{-1}$ ( 6.9) | $1.62\times10^{-1}$ (7.1) | 0.99 ( 9.9) |
| $^{58}Ni(n,2n)^{57}Ni$ | $2.62\times10^{-3}$ (18.3) | $3.37\times10^{-3}$ (7.9) | 0.78 (19.9) |
| $^{59}Co(n,2n)^{58}Co$ | $9.11\times10^{-2}$ (11.0) | $9.30\times10^{-2}$ (7.8) | 0.98 (13.5) |

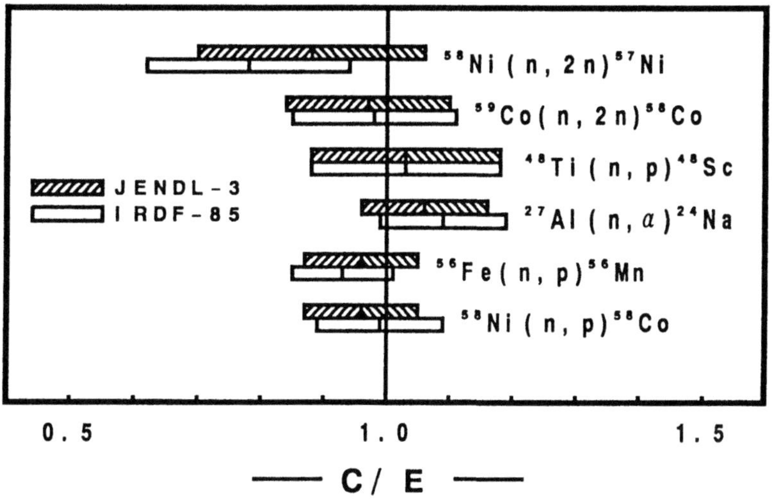


Fig.4. Schematic comparison for the average cross sections of JENDL-3 dosimetry file and IRDF-85.

## CONCLUSION

An accelerator-based neutron source employing the thick target Li(d,n) reaction has been proposed for the purpose of the validation of the dosimetry cross sections and its good characteristics have been presented. Feasibility of the source was studied by benchmarking the six representative dosimetry reactions in the field. The field was proved to be appropriate for the integral test, because the C/E values showed the consistent results for the five established reactions of both the dosimetry files, and clearly different results for the $^{58}Ni(n,2n)^{57}Ni$ reaction – the cross sections by both files are quite different with each other above 14 MeV. According to the result for the $^{58}Ni(n,2n)^{57}Ni$ reaction, the source has advantage on producing large sensitivities even for the high threshold reactions with the small cross sections. The neutron source was adopted as one of the benchmark fields for the test of the JENDL-3 dosimetry file, and the result for 14 reactions will be presented soon.

## REFERENCES


1. L.R. Greenwood, R.R. Heinrich, R.J. Kennerly and R. Medrzychowski, "Development and Testing of Neutron Dosimetry Techniques for Accelerator-Based Irradiation Facilities", Nuclear Technology 41, p.109 (1978).
2. D.L. Smith, J.W. Meadows and M.M. Bretscher, "Integral Cross Section Measurements for $^{7}Li(n,n't)^{4}He$, $^{27}Al(n,p)^{27}Mg$, $^{58}Ni(n,p)^{58}Co$ and $^{60}Ni(n,p)^{60}Co$ Relative to $^{238}U$ Neutron Fission in the Thick-Target $^{9}Be(d,n)^{10}B$ Spectrum at Ed=7MeV", Report ANL/NDM-93, Argonne National Laboratory (1985).
3. M. Nakazawa and Dosimetry Sub-WG, "Benchmark Test of JENDL-3 Dosimetry File", JAERI-M 90-025, Japan Atomic Energy Research Institute, p.103 (1990).
4. D.E. Cullen and P.K. McLaughlin, "The International Reactor Dosimetry File (IRDF-85)", IAEA-NDS-41 Rev.1 (1985).
5. J.R. Dumais, S. Tanaka, N. Odano, S. Iwasaki and K. Sugiyama, "Average Cross Section Measurement for Several Dosimetry Reactions in Thick Target Li(d,n) Neutron Field", Proc. of the 1989 Seminar on Nuclear Data, JAERI-M 90-025, Japan Atomic Energy Research Institute, p.277 (1990).
6. W. Mannhart, IAEA-TECDOC-410, p.158 (1987).
7. R.E. Textor and V.V. Verbinski, "O5S Monte Carlo Code", ORNL-4160, Oak Ridge National Laboratory (1968).
8. S. Tagesen and H. Vonach, INDC-36/LN, Nuclear Standards FileVersion '80 (1981).
9. D.L. Smith, "Covariance Matrices and Applications to the Field of Nuclear Data", Report ANL/NDM-62, Argonne National Laboratory (1981).

# STUDY OF NEUTRON DIAGNOSTICS FOR THE COMPACT TOKAMAK IGNITOR

J Källne[1], P Batistoni[2], B Coppi[3], R Dierckx[4], G Gorini[5],
M Martone[2], L Perasso[6] and S Rollet[2].

[1]Dept. of Neutron Research, Uppsala University, S-751 21 Uppsala, Sweden; [2]Assoc. EURATOM-ENEA,C.R.E. Frascati, I-00044 Frascati, Italy; [3]Dept. of Physics, MIT, Cambridge, Ma 02136, USA; [4]JRC, I-21020 Ispra, Italy; [5]Ist. Fisica Plasma, Assoc. EURATOM-ENEA-CNR, via Bassini 15, I-20133 Milano,Italy; [6]Ansaldo Ricerche, I-16161 Genova,Italy.

## ABSTRACT

This paper discusses diagnostic measurements of neutrons produced in high power plasmas of tokamaks, especially the next step devices dedicated to fusion burn studies. Examples are given of detection methods based on classical principles (for neutron yield determinations) and on new developments (for neutron emissivity profile and spectral measurements) with illustrations of system studies for the compact tokamak Ignitor. The plasma information to be provided is demonstrated in comparison with similar studies for other planned (large) tokamaks and in perspective of what has been achieved on operating ones.

## INTRODUCTION

Diagnostic measurements using the neutron emission from thermo-nuclear plasmas play an important role in the present fusion research. The demands and opportunities for neutron diagnostics (**ND**) are expected to increase, as one now embarks on tokamak studies for plasmas with significant fusion power. Not only is the fusion power a gauge of progress in fusion research but it will also become an important dynamic factor for the plasma equilibrium that needs to be studied, understood and eventually, controlled. Since the fusion reactivity is manifested in the neutron emission ($d+t \longrightarrow \alpha+n$), the fusion power is obtained through the measurement of the neutron yield ($Y_n$) for which an absolute accuracy of <10 % is required. Moreover, collimated neutron flux measurements of the neutron production are especially challenging to perform with good differential resolution (in time and space) since they determine directly the $\alpha$-particle birth distribution. The $\alpha$-particle heating (delayed by the time $\tau_\alpha$ it takes for the 3.6-MeV $\alpha$ particles to thermalise in the plasma) will affect the magnitude of neutron yield so that ND could be used to study predicted fusion burn phenomena (e.g., collective $\alpha$-particle driven instabilities representing plasma regimes to avoid) and to explore the unknown physics of burning plasmas.

The present main alternatives to attain plasma burn conditions are the next step machines of large dimensions[1], ITER and NET, and of compact size[2], Ignitor and CIT. All have as a goal, first of all, to attain **b**urning DT **p**lasmas (**BP** tokamaks). The large machines will, in addition, address the technical question of nuclear furnace designs which entails operation with long burn times ($10^2$ s) and use of substantial radiation protection (or breeding) blankets around the plasma. The compact machines are planned without a blanket, have shorter burn times ('flash ignition' of a few seconds) and their objective is mainly to study the physics of burning plasmas. The two approaches present certain differences to ND, especially with regard to the determination of $Y_n$ based on global neutron flux measurements and plasma (burn) control aspects. A common feature is the intense collimated neutron fluxes they will produce, typically up to $\leq 10^{11}$ n/s·cm$^2$, and this would set future ND apart from present. As a consequence, collimated neutron measurements (time and space differential) may be expected to gain in importance. This calls for neutron detectors of high performance, with regard to count rate capability (>1 MHz), calibration accuracy (≈1%) and energy resolution, to form the basis for new designs of neutron spectrometers and collimator arrays (cameras). Such ND system studies have been given early attention as pilot exercises for diagnostics planning on the BP tokamaks under study as reported already both for Ignitor[3] and the large tokamak NET[4] and, recently, also for ITER[5,6]. Experience on the operation of ND systems has been gained on present tokamaks JET and TFTR and to this would be added experience of DT plasma ND on JET in the near future. Here we shall discuss the salient features of ND on BP tokamaks illustrated with examples taken from the Ignitor study.

## 2. NEUTRON DIAGNOSTICS ON TOKAMAKS

There are several technical reasons to use ND on BP tokamaks in addition to other methods: i) The high radiation level, accompanying high fusion power, is of concern to diagnostics. For ND, this is ameliorated by the facts that the signal to background ratio is not deteriorated and the intense neutron signal would improve the counting statistics. ii) The interface between diagnostic and tokamak is generally a difficult problem curtailing the information output and restricting the use of methods; ND have fewer interface problems than most other installations. iii) Access to the plasma (port space) is limited and the amount of diagnostic information that can be provided per unit port space is important (the benefit/cost factor); ND can provide a broad range of data. Thus, ND can be expected to be widely used to cover many important plasma diagnostic functions.

Taking the presently used ND systems as a reference, we note that **fission chambers** are in regular use[7] on tokamaks to determine $Y_n$ and its time dependence $Y_n(t)$; these would find use on future BP machines, too. The same is true for the **activation technique** used to determine $Y_n$. For the crucial calibration of these two systems, one can draw on previous experience including a recent bench mark experiment[8]. This applies to 2.5-MeV neutrons of D plasmas and extrapolations are needed for determining $Y_n$ of DT plasmas. For 14-MeV neutrons, one would in part rely on camera measurements aiming for an accuracy of ≤10 % absolute and ≤5 % relative. **Neutron cameras** will be used[9,10] to measure the azimuthal and radial neutron intensity distributions, I(Z,t) and I(R,t). The objective, though, would be more ambitious than has been experimentally possible before with regard to detail, accuracy, absolute calibration and scope of the data but could, in the future, be attainable due to the expected data rates (>>1 MHz) for DT plasmas. **Neutron spectrometers** have been used earlier [11] but on the BP tokamaks they would have a wider use. For instance, on Ignitor one would depend on neutron spectrometry to provide the ion temperature of DT plasmas as matter of routine, besides the neutron energy distribution in general, as well as to measure the absolute neutron flux[12]. The neutron cameras and spectrometers for DT plasma diagnostics should also have an ancillary capability to detect 2.5-MeV neutrons.

The time scale for ND measurements on BP tokamaks is the thermalisation time of α-particles, $\tau_\alpha$, which is of the order $\tau_\alpha$=$10^{-2}$ s in compact machines compared to $\tau_\alpha$=$10^1$ s in the large ones; this is also the delay time between an α-particle burst and its subsequent fusion heating pulse. The minimum time resolution requirement is thus $\Delta t \leq \tau_\alpha$ but $\Delta t \ll \tau_\alpha$, is desirable to detect the much faster α-particle driven transients that can occur.

Lacking a fitting plasma scale length for the spatial resolution (Δx), the collimated ND measurements should aim for a lateral resolution of the order 1/5 of the width (FWHM) of the plasma neutron emission profile (typically, corresponding to ≈1/30 of the full plasma height/width)[3]. Usually, Δx>5-10 cm must be maintained in order not to sacrifice count rate and hence Δt, especially, in plasma regions where the local neutron emissivity, $y_n(r,\theta)$, is low. This applies to cameras measuring $I_\lambda=\iint_a y_n(r,\theta)\cdot da \cdot dA_\lambda$, i.e., the line integration of $y_n(r,\theta)$ along a sight line chord λ (with cross section area $a_\Lambda$) in the poloidal plane. Good 3-D (volume) resolution for the derived $y_n(r,\theta)$ depends on good spatial resolution of the 2-D camera data on $I_\lambda(Z)$ and $I_\lambda(R)$, and on the accuracy of the rather complicated (tomographic or other reconstruction) analysis[9].

For the energy resolution (ΔE/E) there are two standards. Spectrometers need sufficient resolution (e.g., ΔE/E≤2.5 %) to measure the typical base ion temperature for producing burning plasmas (typically, $T_i$>3 keV). For the camera counters, a resolution of ΔE/E≈15 % is needed in order to discriminate against energy degraded background neutrons, especially in the peripheral channels; without energy resolution, camera measurements lose precision and become impossible outside the plasma core region.

The energy of fusion neutrons is characteristic of the reaction producing them viz., $\acute{E}_n$= 2.45 and 14.0 MeV, respectively, for d+d–>$^3$He+n (or d+d–>t+p) and d+t–>α+n; the charged fusion products are born at $E^o{}_{^3He}$≈$E^o{}_t$=1.0 MeV and $E^o{}_\alpha$=3.6 MeV and they form populations in the plasma with energies in the range $T_i \leq E \leq E^o$. Normally, the fusion neutrons arise from **thermal** reactions thus having a Gaussian energy distribution centered at $\acute{E}_n$; the widths, determined by the Doppler broadening, are $\Delta=82\cdot\sqrt{T_i}$ and $\Delta E=177\cdot\sqrt{T_i}$, respectively, (FWHM, in keV). If there are ions with **non-thermal** (and anisotropic) velocity components, the $E_n$-distribution will be a distorted Gaussian while the anisotropy would cause Doppler shifts (and hence asymmetry relative to $\acute{E}_n$) as well. For the examples considered here, it is tacitly assumed that thermal plasma conditions prevail which would be the case when little auxiliary power is used to heat the plasma. Thus the direct plasma neutrons are identified on the basis of the characteristic energy distribution tion while those of lower energies are attributable to (multiple) scattering inter loci birth and detection.

The (omnidirectional) global neutron radiation from the plasma is detectable with fission chambers at positions **external** of the tokamak. It consists of direct ($\Phi_o$) and backscattered ($\Phi_1$, from the plasma facing 'first' wall) neutron fluxes as well as a flux scattered from external structures ($\Phi_2$); all scattering sources are also γ-sources. The radiation flux at the detector is $\Phi=\Phi_\gamma+\Phi_n$ where $\Phi_n=\Phi_o+\Phi_1+\Phi_2$. The last term is important since proportionality between Φ and $Y_n$ exists only so long as $\Phi_2$ does not change relative to the other contributions. Clearly, insensitivity to

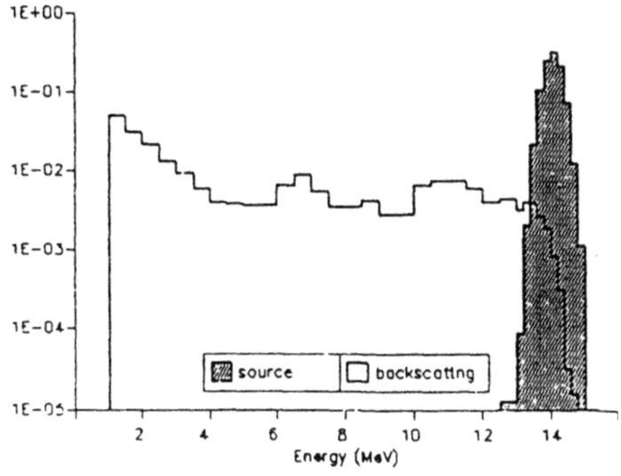


Fig.1 Calculated spectrum of direct source and backscattered neutron flux ($F_1$+$F_2$) from the plasma core of Ignitor ($T_i$=10 keV).

$\gamma$-radiation is essential for determining the **relationship between detector response and $Y_n$ ($Y_n$ calibration)** and crucial for maintaining it during varying operation conditions. The initial calibration can be done with radiological sources in the plasma vessel for 2.5-MeV neutrons; for 14-MeV, a neutron generator would have to be used in situ which is not a realistic proposition, especially, at the accuracy of <10%. Repeat calibrations are needed to validate the fission chamber response which is a complicated task. The large BP tokamaks with blankets present a special problem as the $Y_n$ measurements are subject to using the complex flux of streaming neutrons.

For **internal** positions (near the plasma), the multifoil activation technique can be used (as could possibly also micro fission cambers). The sample is then exposed to the flux $\Phi=\Phi_\gamma+\Phi_n$ where $\Phi_n=\Phi_o+\Phi_1$. Both $\Phi_o$ and $\Phi_1$ can be calculated[8] with neutron transport codes so that the relationship between the (measurable) sample activation and $Y_n$ can be determined for both 2.5 and 14-MeV neutrons. Again, $\gamma$:s are present in proportion to the $\Phi_1$ flux so that $\gamma$-insensitive activation reactions are to be used.

Detectors in collimated sight lines would see a flux $F=F_n+F_\gamma$ where $F_n=F_o+F_1+F_2$ and $F_3$ is due to (collimator) in-scattering and the $\gamma$:s would come mostly from the opposite first wall so that $F_\gamma \propto F_1$. While $F_o$ dominates for central channels (see Fig. 1) it would be reduced by a factor of $10^2$ to $10^3$ for the peripheral ones of the camera. For these, $F_1$ (and $F_\gamma$) would remain high and $F_3$ can be important resulting in $F_o/F \ll 1$. The camera counters must, therefore, be insensitive to $\gamma$:s and have good energy resolution (e.g., $\Delta E/E<15\%$) in order to maintain a high signal/ noise ratio $F_o/F_1>10$ in the recorded data so as to obtain essentially background free data after soft ware correction. In addition, the detector must be of a type highly selective to the 14-MeV neutron component of the direct radiation flux in order that the count rate is not down-loaded by extraneous background.

## 3. NEUTRON CAMERAS AND SPECTROMETERS ON IGNITOR

Characteristic of the Ignitor design[1] is high magnetic field (≤12 T) and current density (≈1 kA/cm$^2$) which permit reaching high plasma density ($n_d \approx n_t \approx n_e/2 \approx 10^{15}$ cm$^{-3}$). As a consequence, the ohmic heating alone is projected to bring the temperature to about $T_i$=5 keV from where the plasma burn would commence with auxiliary heating (from RF generators) available to boost $T_i$ as needed. The fusion power would raise $T_i$ further and bringing on 'full' plasma burn conditions (viz, 150 MW or $Y_n=5\cdot10^{19}$ n/s at $T_i\geq10$ keV) corresponding to 'flash ignition' within about 2 s. Since auxiliary power can produce strong supra-thermal neutron production, the Ignitor scenario would mean prevailing thermal plasma conditions favourable to interpretation of ND data. With regard to dimensions, the machine is compact having a major radius of $A_R$=127 cm, and minor radii of $a_R$=46 and $a_Z$=83 cm; the plasma volume is <10 m$^3$ compared to >100 m$^3$ for the large BP machines. A sketch of Ignitor can be seen in Fig 2 also showing the diagnostic access possibilities through the large horizontal (17x80 cm$^2$) and limited vertical ports. These would be used by the cameras and spectrometers to be discussed below. There would be several fission chambers outside the machine and transfer lines to positions near the plasma for activation samples ; these would be similar to set-ups on the existing machines[7].

The (main) **horizontal neutron camera** would be a vertical array of collimators and detectors viewing the plasma through one of the large ports (see Fig. 3); about 2/3 of the plasma height (azimuthally) can be covered as desired. The spatial resolution of each camera channel is chosen to be $\Delta x \approx 10$ cm (FWHM) which is attained with collimators 4.5 cm wide and 2.5 m long at 6 m distance from the plasma. The channel separation at the plasma is chosen to match $\Delta x$ implying 13 channels in total; the maximum permitted chan nel separation at the detectors is ≈35 cm. The viewing access from above is more limited,

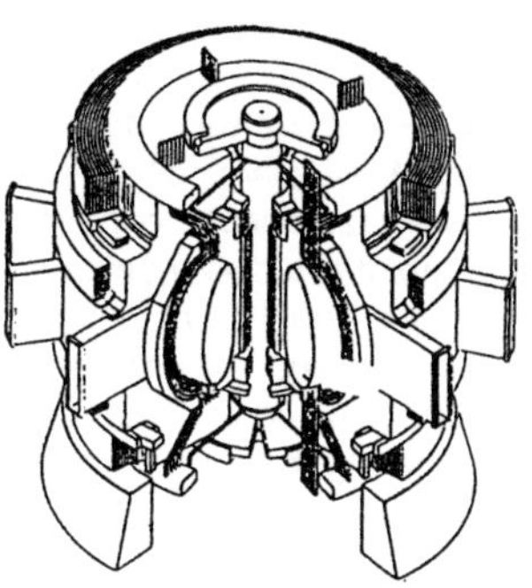

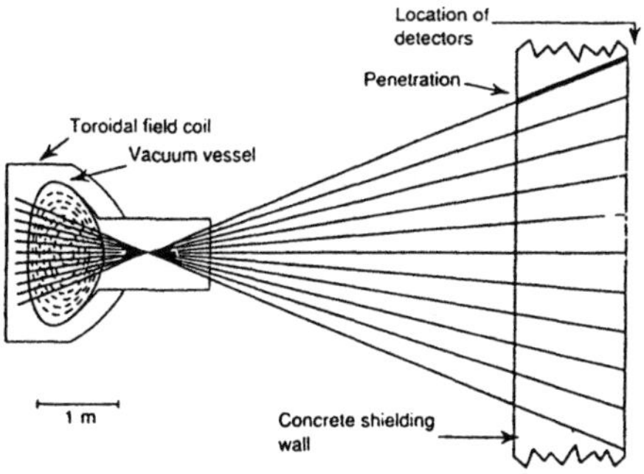


Fig.2 Schematics of the Ignitor tokamak. Fig.3 Schematics of the neutron camera

both with regard to aperture size and number. However, a second **vertical camera** can probably be accommodated (making use of the space between two adjacent toroidal coils) covering the plasma core region up to about $2/3 \cdot a_R$ These, two orthogonal cameras would make an adequate **2-D neutron camera** system for the determination of $y_n(\rho)$ if complementing plasma shape information can be obtained from magnetic field measurements[9]. Finally, there would be up to three **high resolution neutron spectrometers** viewing the plasma through collimators (area 10 $cm^2$ and length 2 m, and with $\Delta x \approx 12$ cm); the sight lines would be aimed towards Z=0 and Z≈15 cm.

Regarding detection methods for 14-MeV neutrons with energy resolution, the choice is rather limited. All are based on nuclear reactions producing charged energetic particles that are detectable through their ionising power, for instance, $n+^3He \longrightarrow p+t$ and $n+^6Li \longrightarrow t+^3He$ or elastic scattering such as $n+p \longrightarrow n'+p'$. If the neutron flux is collimated, it is sufficient to measure energy and angle of one charged particle to determine $E_n$. Likewise, n' can be used for this purpose, but this then takes another nuclear reaction to detect the scattered neutron (besides the recoil) as time-of-flight (TOF) is the only viable method for determining $E_{n'}$.The differential elastic scattering has an advantage in large cross sections ($d\sigma/d\Omega \approx \approx 10^{-25}$ $cm^2$), especially, in the forward angle dir-ection where $d\sigma/d\Omega(\theta)$ peaks and $\Delta E_R/d\theta_R=0$. This can be utilised with the **m**agnetic **p**roton **r**ecoil (**MPR**) technique[3]. For MPR, the recoil protons are produced in a $CH_2$ foil, focused, deflected and the momentum analysed in mag-nets, and detected in a scintillator array (Fig 4). Its principal advantages are the high count rate capability ($C_R^{MAX}$=100 MHz), high precision and predictability in the response to neutrons, and stability in operation.

The **MPRc (counter)** would have an efficiency of $\varepsilon \approx 10^{-3}$ $cm^2$ at $\Delta E/E$=15 %. Used in the Ignitor camera, it would afford determination of the neutron **profile** with an accuracy of ±10% and the time resolution would be the required $\Delta t \lesssim 20$ ms over the full DT burn phase (dynamic range D= $=10^2$ for $F_n^{MAX}=5 \cdot 10^{10}$ n/s·$cm^2$ at $Y_n^{MAX}= =5 \cdot 10^{19}$); the spatial

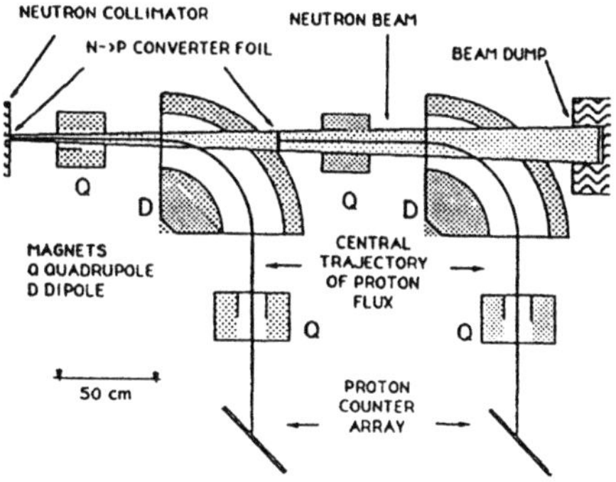


Fig.4 Sketch of neutron detection based on the magnetic proton recoil (MPR) method; shown is a tandem spectrometer for both 14- and 2.5-MeV neutrons.

dynamic range would be $D_S \gtrsim 10^2$. The **MPRs spectrometers** would have $\Delta E/E=2.6$ % (by choice) and $\varepsilon=2\cdot10^{-5}$ $cm^2$ and would be able to determine the (central) $T_i$ with an accuracy of$\pm$10 % and $\Delta t<20$ ms with D=60; this would satisfy the requirements to cover the Ignitor burn phase. Although, the MPR technique is most effective for 14-MeV neutrons it could be used also for 2.4-MeV. A tandem MPRs system (as illustrated in Fig 4) could simultaneously measure the 2.4-MeV neutron **minority** flux (from d+d reactions in DT plasmas) and the 14-MeV flux[3,5]. Similarly, MPRc in the camera could be modified for use in the D phase operation which is especially attractive for determining $Y_n$ of D and DT plasma on the same footing[8]. It may be noted that having the sight ending in recesses on the opposite plasma chamber wall could help lower the backscattering flux thus improving the conditions for measuring minority neutrons in particular.

## 4. PLASMA INFORMATION FROM NEUTRON DATA

The horizontal and vertical cameras on Ignitor would provide data on $I_\lambda(Z)$ and $I_\lambda(R)$ from which the neutron brightness distributions $B_\lambda(Z)$ and $B_\lambda(R)$ can be determined (viz

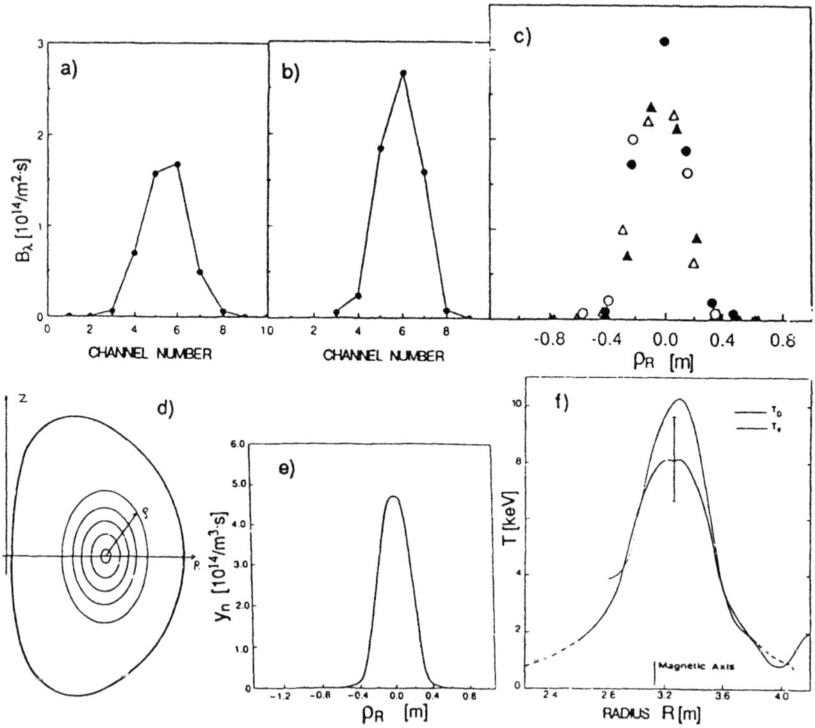


Fig.5 Illustration of neutron camera data, analysis and deduced results from [9]: (a,b) data on azimuthal and radial brightness profiles; (c) combined brightness profile; (d) contours (given by parameter $\rho$) of magnetic flux surfaces; (e) result on the neutron emissivity distribution; (f) deduced result on the ion temperature distribution compared with $T_e(\rho)$.

$B_\lambda = \int y_n(r,\theta) \cdot d\Lambda \propto I_\lambda / a_\Lambda$ where $a_\Lambda$ is the effective intersected cross section area of the plasma[9]. From such data, using a suitable algorithm, the neutron emissivity profile can be determined which in turn can be used to deduce the profile $T_i(r,\theta)$.

An example of neutron brightness data on deuterium plasma from JET and their analysis is shown in Fig. 5[9]. Data on $B_\lambda(Z)$ and $B_\lambda(R)$ (Fig 5a,b) can be combined (Fig 5c) to $B_\lambda(\rho)$ using the assumption that the neutron emissivity, $y_n(\rho)$, is constant on a magnetic flux surface, $\mu(\rho)$, of the plasma (Fig. 5d); $\rho$ is the radial parameter of the contours with a small off-set in the centers of $y_n(\rho)$ and $\mu(\rho)$. From the $B_\lambda(\rho)$ data one can deduce $y_n(\rho)$ which is the most fundamental neutron parameter of the plasma (Fig 5e); this also shows $\Delta\rho$ off-set. More over, for a thermal plasma, $y_n$ is a known (strong) function of $T_i$ so that $T_i(\rho)$ can be extracted from the $I_\lambda$ results using (for D plasmas) $y_n(\rho)=n_d(\rho)^2 \cdot f(T_i(\rho))$ where the $n_d(\rho)$ data come from other sources; it is interesting to compare $T_i(\rho)$ and $T_e(\rho)$ distributions as shown in Fig. 5f. Similar results could be obtained for DT plasmas on Ignitor and with expected high statistical and absolute accuracy of the $B_\lambda$ data, one could expect to cover a spatial dynamic range of $D_S>10^2$ with $\Delta t<20$ ms. This would make it possible to provide $T_i(r)$ information over the value range from say a typical peak of $T_i(\rho=0)=10$ keV to <2.5 keV in the periphery with statistics of $\Delta T_i/T_i<10$ %. This accuracy would be comparable to that of other data (e.g., from electrons) so the time evolution of the difference $T_i(\rho)-T_e(\rho)$ could be studied which would be of special interest, for instance, during the flash ignition periods. A primary task of the neutron camera on Ignitor would be studies of transient plasma phenomena at the time scale $\Delta t \leq 20$ ms over a full $B_\lambda$ profile and $\Delta t \geq 10$ $\mu$s in the plasma core.

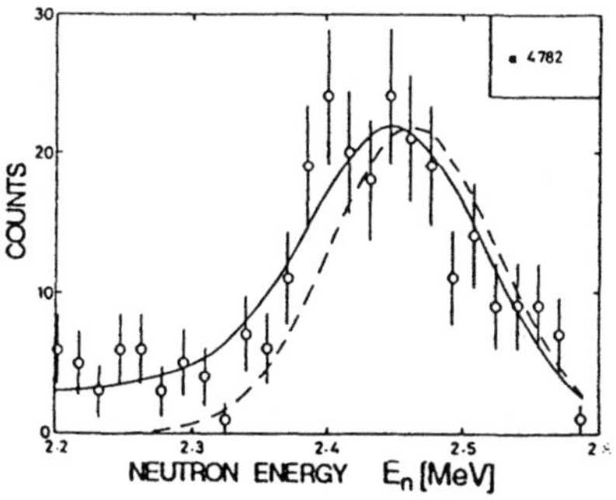


Fig.6 Neutron spectrum recorded by a $^3He$ detector with $\Delta t=7$ s; the dashed curve is a fitted Gausian due to $T_i \approx 3$ keV[9].

Another task for the neutron camera is to determine the absolute value of $Y_n$; if this would be limited to only certain plasma discharges or conditions, it would suffice for calibrating the fission chambers. It is done by integrating the $B_\lambda(R)$ data over the full brightness profile.[4]. The accuracy attainable is set by the absolute calibration of the detector efficiency (which is the n+p cross section for MPR detectors) and the geometrical solid angle of the camera channels, i.e., an absolute accuracy of 5 to 10 % is possible. These $Y_n$ data are of great importance since the calibration is essentially experimental with quantifiable systematic uncertainties in contrast to the $Y_n$ determinations from global flux measurements. When this is done also for 2.5-MeV neutrons of D plasmas, the ratio of $Y_n$ of D and DT plasma could be determined with good precision (≤5%). For D and DT plasma discharges of the same basic characteristics, these $Y_n$ measurements would afford information on the effectiveness of the plasma burn for various base conditions constituted by the D plasma as a reference; furthermore,the camera would be instrumental for the characterisation of the reference conditions of the D plasma through the $I_\lambda$ data on the profile.

The spectrometers would measure the neutron energy distribution $I_\lambda(E_n,t)$ from which $T_i(t)$ can be determined directly (Fig 6). Most important are data on $T_i(Z=0)$ but another point at $T_i(Z\neq0)$ from a second spectrometer would be needed to provide $T_i(\rho)$ constraints that can be used in model calculations of the plasma equilibrium. These $T_i(Z)$ data would also afford valuable normalisation for the $T_i(\rho)$ distribution deduced (conditionally) from the camera data. Since $y_n$ is a known function of $T_i$, $y_n=n_d \cdot n_t \cdot f(T_i)$ for 14-MeV neutrons in DT plasmas, one can, in principle, determine the (fuel) density product, $n_d \cdot n_t$ from $I_\lambda(E_n,t)$; the accuracy depends on the absolute $T_i$ and $I_\lambda$ with the estimate for the contributing plasma volume being a large error source. With a spectrometer system one could also measure the 2.5-MeV (minority) neutrons and thus determine $n_d$ and $n_t$ individually, and, especially, $n_d/n_t$. The density data are important since the deuterium and tritium fuelling can occur in different ways and the correct balance is both a diagnostic tool and a control issue.

## 5. CONCLUSION.

We have discussed the diagnosis of burning fusion plasmas of tokamaks and its application for Ignitor using their intense neutron emission, especially in spectrometry and profile measurements. These depend critically on development of dedicated neutron detection methods. The diagnostic capabilities of neutron spectrometers and cameras can be greatly enhanced by use of the magnetic proton recoil (MPR) technique which were illustrated in view of the expected diagnostics requirements. Systems for the compact tokamak Ignitor were discussed in some detail demonstrating that neutron diagnostics will be an indispensable tool in the studies of burning fusion plasmas on this and other tokamaks now under design study.

ACKNOWLEDGEMENT.

Some of the authors (PB, GG and JK) are pleased to acknowledge the privileged experience of work on neutron diagnostics at JET that has been of great value for carrying out studies for the next generation plasma studies part of which has been reported here.

References

1 J Gilleland,et al, Nucl.Fusion 29 (1989), 1191

2 B Coppi et al, Com. on Plasma Physics and Cont. Fusion 11 (1987) 47.

3 J Källne, Comments on Plasma Physics and Cont. Fusion 12(1989)235,and"Neutron measurements in the study of plasma fusion physics with tokamaks", MIT Report PTP 88/11 (1988), to be published.

4 M Martone,et al "Analysis and design of neutron diagnostic for NET; Part I: Plasma diagnostics", NET EUR-FU7XII-80/88-86, July 1988.

5 J Källne, "Neutron diagnostics for burning plasmas of NET/ITER tokamak",Prog. rep. for EURATOM/NET (Feb -90), to be published.

6 J Källne, G Gorini and P Batistoni, in 'Proc. of '5th Nat. topical meeting on high temperature plasma diagnostics', Minsk (June 1990), to appear.

7 H Hendel, et al, Rev Sci. Instr 61(1990)1900, and G Sadler, et al., Nucl. Fusion (1990), to appear.

8 M Angelone, et al, Rev. Sci. Instr. **61**(1990)3157.

9 G Gorini, J Källne, and S Rollet, in 'Proc. of 17th Eur Conf on fusion and plasma heating', Amsterdam 1990, EPS 14B/IV (1990)1652, G Gorini and N Gottardi, 'Inversion methods for analysis of neutron brightness measurements in tokamaks', JET report JET-R(90)04, G Gorini and N Gottardi, Il Nuovo Cimento xx(1990)xx, to appear, and G Gorini, "Measurements and Interpretation of Fusion Neutrons for Plasma Diagnostics", PhD Thesis, Scuola Normale Superiore, Pisa, Italy, (1990).

10 JM Adams et al, in 'Proc. of 14th EPS Conf on Contr. Fusion and Plasma Physics', Madrid 1987 and 16th EPS Conf. Venice, 1989, 13BI(1989)63

11 G Gorini, et al, in 'Proc. of workshop on basic physics processes in fusion plasmas', CEC Brussels, 2 (1985), and ON Jarvis et al, Rev. Sci. Instr.**57**(1986)654.

12 J Källne, Rev. of Sci. Inst. **61**(1990)3160.

# MEASUREMENTS OF ACTIVATION CROSS SECTIONS FOR FUSION REACTOR APPLICATIONS

L.R. Greenwood[a] and D.L. Bowers

Argonne National Laboratory

Argonne, IL 60439 USA

## ABSTRACT

Neutron activation cross section data are needed for fusion reactor materials to facilitate neutron dosimetry and to determine activity levels for waste disposal, safety, and reactor maintenance applications. Measurements are reviewed for 36 different reactions, many leading to long-lived nuclides. Irradiations were performed at 14 MeV at the RTNS II to fluences of $10^{18}$ n/cm$^2$. Activities were determined by either gamma and x-ray analysis or liquid scintillation counting following radiochemical separations to remove interfering activities. Results are compared to calculations with the ALICE computer code. Activity levels are predicted for the STARFIRE fusion reactor design and compared to previous estimates. Other work is summarized including rare earth irradiations in Be(d,n), H(t,n), and T(d,n) accelerator-based neutron sources with average energies of 3, 10, and 14 MeV, respectively.

---

[a]Pacific Northwest Laboratory, P.O. Box 999, Richland, WA 99352 USA. PNL is operated for the U.S. Department of Energy by Battelle Memorial Institute under Contract DE-AC06-76RLO 1830.

## INTRODUCTION

Neutron activation cross sections are needed for fusion reactor materials to facilitate neutron dosimetry and to determine activity levels for waste disposal, safety, and reactor maintenance appli-cations.[1] Although such data are widely available for fission reactor applications, the high fluences of higher energy (14 MeV) neutrons found in fusion reactors favor different reaction channels leading to different radioisotopes. In particular, (n,2n) and charged particle reactions are favored, thus tending to produce lower mass isotopes than the target material, whereas fission reactor irradiations tend to produce higher mass products through successive (n,gamma) capture reactions.

Unfortunately, we do not have any sources of high-energy neutrons at the fusion reactor levels of $10^{15}$ $n/cm^2$-s. Hence, we cannot experimentally verify theoretical predictions[2,3] of the production of radioisotopes that are produced by two or more successive reactions. To illustrate the problem, suppose that all of the (n,2n) and charged particle reactions have a cross section of about 1 barn ($10^{-24}$ $cm^2$) and that the 14 MeV fusion reactor flux is about $10^{15}$ $n/cm^2$. If all the products are long-lived, then in 1 year we will produce an atomic ratio of 0.01 for the primary product, $10^{-4}$ for the reaction product from that isotope, and so on. On the experimental side, the most intense source of 14-MeV neutrons was the Rotating Target Neutron Source II (no longer in operation) with a flux of about $10^{12}$ $n/cm^2$. This means that in a 2-week irradiation we can produce a neutron fluence of about $10^{18}$ $n/cm^2$, which is only capable of producing atom ratios of $10^{-6}$ for the primary reaction product and only $10^{-12}$ for the secondary, and so on. Although, in some cases we can measure each sequential reaction rate separately, this is only rarely possible since many of the intermediate isotopes are radioactive. Hence, we have little hope of directly measuring all of the predicted activation products barring the operation of a high-flux, high-energy neutron source.

Consequently, we have attempted to measure the primary reaction products at 14 MeV on a variety of fusion reactor materials. Comparisons between measured and calculated reaction rates can then help to refine our predictions of activities.

## OVERVIEW OF MEASUREMENTS

Measurements have been performed at a wide range of neutron energies from thermal up to 450 MeV. Results presented in this paper will focus on measurements of the production of long-lived isotopes near 14 MeV and attempts to measure long-lived isotopes in rare earth materials. Previously, we have reported measurements of activation cross sections of 24 other reactions at 14 MeV,[4] including the $^{27}Al$ and $^{54}Fe$(n,2n) reactions[5,6] that are useful for fusion plasma diagnostics. Above 14 MeV, neutron dosimetry cross sections have been measured and tested at energies up to 40 MeV in Be(d,n) fields,[7,8] including new work presented separately in these proceedings.[9] At higher energies, spallation cross sections have been measured for Al and Cu[10,11] for irradiations at spallation neutron sources.

## MEASUREMENTS NEAR 14 MEV

In the absence of any high-flux source of 14-MeV neutrons, we have irradiated samples in a variety of available facilities. Prior to decommissioning in 1987, many different samples were irradiated at the Rotating Target Neutron Source II at Lawrence Livermore National Laboratory. Fluences up to $10^{18}$ n/cm$^2$ were obtained during 2-week irradiations in close geometry, as well as during prolonged irradiations at more remote locations. The source and neutron spectra have been described previously.[12,4] Data have been published[13,14] for 12 different reactions leading to long-lived isotopes. The results are summarized in Table I. Results have been compared to previous measurements and theoretical calculations with the ALICE computer code, which calculates energy-dependent nuclear cross sections,[15] and the ratio of calculated to experimental values (C/E) are listed in Table I along with the data and half-lives.

Examination of the data in Table I reveals rather large uncertainties in some of the measurements and rather large deviations from unity in the C/E ratios. The uncertainties in the nuclear decay data[16] are described in detail in Table II. As can be seen, the half-life and x-ray intensity uncertainties are substantial, especially for $^{59}Ni$, $^{91}Nb$, and $^{93}Mo$. Comparison with Table I shows that the decay data uncertainties are usually the largest part of the total error, the remainder being counting statistics and efficiency calibrations. In general, this is a serious problem in attempts to measure any of the longer-lived isotopes, and we note a need to review and improve the nuclear decay data.

The C/E ratios, which compare ALICE calculations with the experimental data, may indicate problems with the nuclear data. Although the ALICE results may have a large uncertainty (100% or more), relative comparisons may indicate systematic trends. In particular, the case of $^{91}Nb$ is quite unusual with a C/E value of only 0.33, whereas more generally ALICE appears to over predict the (n,2n) cross sections. We note that the half-life is only known to about 20%. If, in fact, the half-life were closer to 300

Table I Production of Long-Lived Isotopes Near 14.8 MeV (Measured Cross Sections)

| Reaction | $T_{1/2}$, y | σ, mb | C/E[a] |
|---|---|---|---|
| $^{50}V(n,2n)^{49}V$[b] | 0.90 | 258±39 | 2.8 |
| $^{56}Fe(n,2n)^{55}Fe$ | 2.7 | 454±35 | 1.4 |
| $^{63}Cu(n,p)^{63}Ni$ | 100. | 54±4 | 0.91 |
| $^{64}Ni(n,2n)^{63}Ni$ | 100. | 958±64 | 1.2 |
| $^{60}Ni(n,2n)^{59}Ni$ | $7.5 \times 10^4$ | 104±25 | 6.0 |
| $^{94}Mo(n,p)^{94}Nb$ | $2.0 \times 10^4$ | 55±6 | 0.82 |
| $^{Nat}Mo(n,x)^{94}Nb$ | $2.0 \times 10^4$ | 7.8±0.8 | - |
| $^{92}Mo(n,2n\beta)^{91}Nb$ | 680.(?) | 603±119 | 0.33 |
| $^{94}Mo(n,2n)^{93}Mo$ | 3500. | 550±136 | 2.5 |
| $^{94}Mo(n,d+np)^{93m}Nb$ | 16.1 | 5.7±0.9 | 2.0 |
| $^{95}Mo(n,t+nd)^{93m}Nb$ | 16.1 | 1.36±0.27 | - |
| $^{Nat}Mo(n,x)^{93m}Nb$ | 16.1 | 0.75±0.11 | - |

[a]Ratio of ALICE calculation to experimental measurement
[b]Vanadium data measured at 14.3 MeV

Table II Nuclear Decay Data for Long-Lived Isotopes

| Isotope | Half-life, y | Radiation, keV | Branching, % | Net, %[a] |
|---|---|---|---|---|
| $^{49}V$ | 0.90±0.04 | $X_K$ 4.5-4.9 | 0.196±0.017 | 9.8 |
| $^{59}Ni$ | $7.5 \pm 1.3 \times 10^4$ | $X_K$ 6.9 | 29.5±1.5 | 18.2 |
| $^{63}Ni$ | 100.1±2.0 | $\beta^-$ 66. | 100. | 2.0 |
| $^{91}Nb$ | 680±130 | $X_K$ 15.7 | 54.2±1.8 | 19.4 |
| $^{93m}Nb$ | 16.13±0.15 | $X_K$ 16.6 | 8.98±0.19 | 2.3 |
| $^{93}Mo$ | 3500±700 | $X_K$ 16.6 | 62.0±2.3 | 20.3 |

[a]Net uncertainty (sum of half-life and branching)

years rather that 680 years, our results would be in better agreement with the theoretical predictions. A similar situation is found for $^{59}$Ni, where the C/E value is 6.0 and the half-life is only known to about 20%. If the half-life were 3x10$^{5}$ or longer, the results would agree with calculations and systematics of other nickel isotopes. Obviously, more measurements of the cross sections and half-lives are needed to resolve these questions.

## PREDICTED FUSION REACTOR ACTIVITIES

The measured activation cross sections in Table I can be used to predict the production of these long-lived isotopes in operational fusion reactors. We have performed such calculations using the first wall spectrum of the STARFIRE reactor design,[16] and the results are listed in Table III. In most cases the 14 MeV cross section is all that is needed for this calculation due to the high thresholds of the (n,2n) reactions. In the case of $^{63}$Cu(n,p)$^{63}$Ni and $^{94}$Mo(n,p)$^{94}$Nb, lower energy neutrons do contribute and we have used the ALICE calculation to estimate these contributions. We have not made any allowance for burnup of the isotopes, as has been done in previous calculations;[2,3] however, we estimate that such effects would be small for these isotopes. The calculations assumed six full-power years operation followed by a 3000-day cooling period. These results can be compared to previous calculations as well as to landfill disposal requirements.

## OTHER WORK IN PROGRESS

Argonne is participating in a Coordinated Research Program sponsored by the International Atomic Energy Agency to measure the production of long-lived activities in rare earth materials. The laboratories and facilities include Argonne, using a Be(d,n) neutron spectrum produced by 7-MeV deuterons stopped in a thick beryllium target; Los Alamos National Laboratory, using 10-MeV neutrons from the H(t,n) reaction; and the Japan Atomic Energy Research Establishment, using the 14-MeV T(d,n) Fast Neutron Source. Foils and pressed powders of copper, silver, europium, terbium, and hafnium, as well as titanium, iron, and nickel dosimeters, were irradiated in each facility. The reactions being studied are listed in Table IV. Preliminary results have been obtained for each irradiation; however, data have not been corrected for interferences from low-energy neutrons nor for multiple scattering effects. Estimates show that such effects are less than 7% for all reactions except for $^{153}$Eu(n,2n)$^{152}$Eu where the $^{151}$Eu(n,g) reaction may produce up to half of the observed activity due to the very high thermal cross section (5900 barns). In the case of $^{179}$Hf(n,2n)$^{178m}$Hf, the activity has not

Table III Calculated Fusion Reactor Activities
(STARFIRE First Wall, 21.6 MW-y/$m^2$, 3000 d decay)

| Material | Isotopes | Half-Life,y | Activity, mCi/cc |
|---|---|---|---|
| Vanadium | $^{49}V$ | 0.9 | 2.8 |
| Molybdenum | $^{91}Nb$ | 680. | 243. |
| | $^{93m}Nb$ | 16.1 | 106. |
| | $^{94}Nb$ | 20,000. | 0.8 |
| | $^{93}Mo$ | 3500. | 28. |
| Iron | $^{55}Fe$ | 2.7 | 25,000. |
| Copper | $^{63}Ni$ | 100. | 1,795. |
| Nickel | $^{63}Ni$ | 100. | 227. |
| | $^{59}Ni$ | 75,000. | 1. |

Table IV Long-Lived Rare Earth Isotopes
[Irradiations at Be(d,n), 10, and 14 MeV]

| Reaction | Half-life | Comments (1-2 year decay) |
|---|---|---|
| $^{107}Ag(n,2n)^{106m}Ag$ | 8.5 d | Measurements complete |
| $^{109}Ag(n,2n)^{108m}Ag$ | 127 y | Only seen at 14 MeV |
| $^{151}Eu(n,2n)^{150g}Eu$ | 35.8 y | Measurements complete |
| $^{153}Eu(n,2n)^{152}Eu$ | 13.3 y | Strong interference $^{151}Eu(n,g)$ |
| $^{159}Tb(n,2n)^{158}Tb$ | 150 y | Measurements complete |
| $^{176}Hf(n,2n)^{175}Hf$ | 70 d | Measurements complete |
| $^{179}Hf(n,2n)^{178m2}Hf$ | 31 y | Not seen - very weak |
| $^{180}Hf(n,2n)^{179m2}Hf$ | 25 d | Measurements complete |

been observed after nearly 2 years decay due to the higher activities of other hafnium isotopes.

## FUTURE WORK

Other measurements have been made at the Be(d,n) facility at Argonne and work is in progress to determine cross sections for the reactions $^{93}$Nb(n,n')$^{93m}$Nb (16.1 y), $^{63}$Cu(n,p)$^{63}$Ni (100 y), and $^{60}$Ni(n,p)$^{60}$Co (5.27 y). Other materials were also irradiated to high fluences at RTNS II at 14 MeV and analyses are planned for the reactions $^{94}$Zr(n,2n)$^{93}$Zr ($1.5 \times 10^6$ y) and $^{93}$Nb(n,2n)$^{92}$Nb ($3.6 \times 10^7$ y). As discussed previously, measurements are needed for many other reactions to cover the gamut of fusion reactor materials.

## ACKNOWLEDGEMENTS

This work was supported by the Office of Fusion Energy, U.S. Department of Energy.

## REFERENCES

1. E. T. Cheng, Nuclear Data Needs for Fusion Programs, these proceedings.

2. G. J. Butterworth and L. Giancarli, J. Nucl. Mater. 155-157 (1988) 575-580.

3. F. M. Mann, Fusion Technol. 6 (1984) 273.

4. L. R. Greenwood, ASTM STP 956 (1987) 743-749.

5. R. K. Smither and L. R. Greenwood, J. Nucl. Mater. 122, (1984) 1071-1077.

6. R. K. Smither and L. R. Greenwood, Damage Analysis and Fundamental Studies Quarterly Progress Report, DOE/ER-0046/17 (1984) 11-13.

7. L. R. Greenwood, R. R. Heinrich, M. J. Saltmarsh, and C. B. Fulmer, Nucl. Sci. Eng. 72 (1979) 175.

8. D. L. Smith, J. W. Meadows, P. T. Guenther, and L. R. Greenwood, ASTM STP 1001 (1989) 253-260.

9. D. L. Smith, J. W. Meadows, and L. R. Greenwood, Microscopic Integral Cross Section Measurements in the Be(d,n) Neutron Spectrum, these proceedings.

10. L. R. Greenwood and R. K. Smither, Reactor Dosimetry (1985) 251-261.

11. L. R. Greenwood and R. K. Smither, Damage Analysis and Fundamental Studies Quarterly Progress Report, DOE-ER/0046/18 (1984) 11-16.

12. L. R. Greenwood, D. G. Doran, and H. L. Heinisch, Phys. Rev. 34C (1987) 76-80.

13. L. R. Greenwood and D. L. Bowers, Proc. Specialists' Meeting on Neutron Activation Cross Sections for Fission and Fusion Applications, NEANDC-259 'U' (1989) 85-94.

14. L. R. Greenwood and D. L. Bowers, ASTM STP 1001 (1989) 508-514.

15. M. Blann, Lawrence Livermore National Laboratory Report UCID 19614 (1982).

16. E. Browne and R. B. Firestone, Table of Radioactive Isotopes, Ed. V. S. Shirley, John Wiley, New York (1986).

16. STARFIRE: Commercial Tokamak Fusion Power Plant Study, ANL/FPP-80-1 (1980).

# MICROSCOPIC INTEGRAL CROSS SECTION MEASUREMENTS IN THE Be(d,n) NEUTRON SPECTRUM FOR APPLICATIONS IN NEUTRON DOSIMETRY, RADIATION DAMAGE AND THE PRODUCTION OF LONG-LIVED RADIONUCLIDES

D.L. Smith, J.W. Meadows and L.R. Greenwood[a]

Argonne National Laboratory[b]
Argonne, Illinois 60439, U.S.A.

## ABSTRACT

Integral neutron-reaction cross sections have been measured, relative to U-238(n,fission), for 27 reactions of interest in nuclear applications (e.g., neutron dosimetry, radiation damage and nuclear waste disposal). The neutron field was produced by bombarding thick Be-metal accelerator targets with 7-MeV deuterons. The experimental results and their associated uncertainties are reported. These data are also compared with corresponding calculated values, based on the known differential cross sections and the neutron spectrum. Some conclusions are offered concerning this method for neutron-reaction data testing.

## INTRODUCTION

Integral measurements in continuum neutron fields provide valuable information for the testing of differential neutron cross section data. When the yield of activity from a particular reaction is low, integral measurements may offer the only hope for acquiring experimental information. Such measurements are often made in reactors which produce fission neutron spectra with fairly low average energies. Thermal and epithermal components

---

[a] Present address: Battelle Pacific Northwest Laboratory, Richland, Washington 99352, U.S.A.

[b] This work was supported by the U.S. Department of Energy.

in reactor spectra can generate strong interfering activities which complicate the study of weaker processes. A variety of fast-neutron fields (with small low-energy components) can be conveniently produced using accelerators. We have explored applications for the neutron field produced by bombarding thick Be-metal targets with 7-MeV deuterons. This work was reported at the Sixth ASTM-Euratom Symposium (hereafter referred to as Paper I) [1]. The present paper reports on the experimental results which have been obtained from this project.

## EXPERIMENTAL PROCEDURE, DATA ANALYSIS AND EXPERIMENTAL ERRORS

The facility and methodology employed for these measurements is described more fully in Paper I. A brief summary is provided here for completeness. Neutrons are produced by directing a collimated beam of 7-MeV deuterons onto a Be-metal target which is sufficiently thick to stop the deuterons. The maximum neutron energy is $\approx$ 11.5 MeV. Recent neutron time-of-flight measurements with a fission chamber have improved our knowledge of this spectrum at low energies [2]. Relatively few neutrons are observed with energies < 200 keV, when a low-mass Be-target assembly is used and extraneous scatterers are avoided in the irradiation environment [1,3-5]. Activation samples are affixed to a fission chamber which employs U-238(n,fission) as the standard. Samples of fissionable material (assayed by low-geometry alpha counting [8]) are placed either singly or back-to-back in the fission chamber. Fission yields are measured in real time during the irradiation [4,8]. The neutron-induced activities are determined using conventional techniques [4,6,7]. The measurements involved three separate experiments. Thus, the 27 accumulated data points are grouped into distinct sets: Set I (Nos. 1-5), Set II (Nos. 6-12) and Set III (Nos. 13-27).

Various corrections were applied to the experimental data to convert them to integral cross section ratios. Relevant details are provided in Refs. 4 and 8. Table 1 lists the reactions which were studied in this work. It also gives the decay half lives and detected radiations [9,10]. The major data corrections treat: low-energy fission events masked by noise, alpha particles and gas-recoil pulses; loss of fission fragments in the deposit; decay of the induced sample activities; radiation branching parameters; detector calibrations; neutron absorption and multiple scattering; geometric effects; anisotropic emission of the Be(d,n) neutrons. Experimental errors were estimated as described in Refs. 4 and 11. The principle random errors (magnitudes) are: fission counting (0.3 - 2.4%); activity counting (0.3 - 3.0%); random geometric effects ($\lesssim$ 1.4%); miscellaneous random influences on the observed experimental reproducibility ($\lesssim$ 10.3%). The principle systematic errors (magnitudes) are: fission deposit atoms (0.5 - 2.0%); fission measurement effects (0.3 - 1.5%); activity decay half life ($\lesssim$ 0.4%); activity radiation detection efficiency (0.6 - 3.1%); sample atoms (0.1 - 4.7%); activity decay branching ($\lesssim$ 5.9%); neutron absorption (0.1 - 0.5%); neutron multiple scattering ($\lesssim$ 1.5%); systematic geometric effects ($\lesssim$ 1.4%); neutron source properties ($\lesssim$ 10.1%).

## EXPERIMENTAL RESULTS AND NEUTRON SPECTRUM ADJUSTMENT

The measured integral cross section ratios and their errors appear in Table 1. These measured ratios can be converted to integral cross sections by employing the absolute integral cross-section of U-238(n,fission). The value 0.46097 barn was used for this purpose. It was calculated using the ENDF/B-V [13] U-238(n,fission) cross section and a representation for the Be(d,n) spectrum obtained as described below.

Prior to the present work, knowledge of the thick-target Be(d,n) neutron spectrum for 7-MeV incident deuterons was based entirely on data reported in Refs. 3,5 and 14, and on unpublished results from this laboratory [2]. The first step taken to improve this situation was to generate a hybrid zero-degree spectrum representation which incorporated the best features of earlier time-of-flight measurements (involving a scintillation detector for neutron energies above 4 MeV [14] and a fission chamber for lower energies [2]). This spectrum served as the a priori for a spectrum adjustment procedure (utilizing code STAYS'L [15]) which incorporated all 27 reactions from Table 1. This procedure required uncertainty information for the trial spectrum and for the differential cross sections. Uncertainties for the spectrum were estimated from statistics and systematic errors associated with the time-of-flight measurements. Uncertainties for the differential cross sections were derived from ENDF/B-V covariance files (where available) or were estimated. The adjusted spectrum is given in Table 2. This analysis led to a modest increase in the neutron fluence at energies below 4.5 MeV (< 5%), a modest decrease in the range 4.5-5.5 MeV (< 5%), and a significant decrease at higher energies (5-25%). As a result, the calculated integral cross sections for low-threshold reactions changed by only small amounts while significant reductions were seen for the higher threshold reactions. The adjustments produced a chi-square of 1.36 per degree of freedom, indicating pretty good overall consistency.

## COMPARISON OF MEASURED AND CALCULATED CROSS SECTIONS

A detailed comparison has been made between the experimental (E) integral cross sections (see Table 1) and those calculated

(C) using the neutron spectrum described above. ENDF/B-V [13] values (generally obtained from the Dosimetry File) were employed for this analysis, except as noted for the following reactions: Ti-47(n,p)Sc-47 [16]; Co-59(n,p)Fe-59, Co-59(n,2n)Co-58 and Co-59(n,$\alpha$)Mn-56 [17]; Zn-64(n,p)Cu-64 [18]; Cu-65(n,p)Ni-65 [19]; In-113(n,n')In-113m [19]; and In-115(n,n')In-115m [20]. Errors (not listed) for the calculated results were estimated from available information. Results from this analysis are shown in Fig. 1, excluding Co-59(n,2n)Co-58. Eighteen reactions out of 27 (67%) exhibit quite good C/E consistency (near unity). The C/E comparisons were marginally consistent with unity (just slightly beyond the combined errors) for four reactions (15%): Al-27(n,$\alpha$)Na-24, U-234(n,fission), In-113(n,n')In-113m and In-115(n,n')In-115m. Only five reactions (18%) were indisputably discrepant based on C/E: Li-7(n,n't)He-4, Al-27(n,p)Mg-27, Ni-60(n,p)Co-60, Co-59(n,$\alpha$)Mn-56 and Co-59(n,2n)Co-58. For each discrepant reaction, most of the integral response comes from neutron energies above 5 MeV (where the neutron spectrum uncertainty is the greatest). The differential cross section for Li-7(n,n't)He-4 is poorly known near threshold while that for Ni-60(n,p)Co-60 is not very well established at any energy. The agreement is especially poor for Co-59(n,2n)Co-58 (C/E $\approx$ 2.38). This is not surprising since its reaction threshold is quite high (10.633 MeV). There is no explanation for the poor agreement observed for Al-27(n,p)Mg-27 and Co-59(n,$\alpha$)Mn-56, since their differential cross sections are supposed to be rather well known in the maximum response range for this neutron spectrum.

## CONCLUSIONS

We conclude that the neutron field produced by 7-MeV deuteron bombardment of thick Be-metal targets is fairly well established, and that it is quite useful for the integral testing of differential neutron cross-section data as long as the effective reaction threshold is of the order of a few MeV or less. The relatively high yield of neutrons (about 3 x 10(9) n/sr/$\mu$C at zero degree [21]) adds to its utility for investigating certain low-yield reactions at a small accelerator facility. The Be(d,n) source (for incident deuterons $\lesssim$ 7 MeV) loses much of its appeal in testing reactions involving higher thresholds, due to anisotropic neutron emission (necessitating large corrections) and to the relatively low yield of neutrons above 6 MeV. This is not a basic limitation to the method described here. The Be(d,n) neutron source can be used very effectively to study higher-threshold reactions when higher deuteron bombarding energies are available (e.g., see Ref. 22). The present experiment offers sufficient confidence in the method to encourage applications to the study of reactions which are not at all well known, e.g., reactions which yield long-lived activities implicated in low-level nuclear waste management.

## REFERENCES


1. D. Smith, J. Meadows, P. Guenther and L. Greenwood, Reactor Dosimetry: Methods, Applications, and Standardization, ASTM STP 1001, H. Farrar, IV and E. Lippincott, eds., Amer. Soc. for Testing and Materials, Philadelphia, p. 253 (1989).

2. J. Meadows, Argonne National Laboratory, unpublished.

3. D. Smith, J. Meadows and P. Guenther, ANL/NDM-90, Argonne National Laboratory (1985).

4. D. Smith, J. Meadows and M. Bretscher, ANL/NDM-93, Argonne National Laboratory (1985).

5. D. Smith and J. Meadows, ANL/NDM-95, Argonne National Laboratory (1986).

6. J. Meadows and D. Smith, ANL/NDM-60, Argonne National Laboratory (1984).

7. D. Smith, J. Meadows, M. Bretscher and S. Cox, ANL/NDM-87, Argonne National Laboratory (1984).

8. J. Meadows, ANL/NDM-97, Argonne National Laboratory (1986).

9. J. Tuli, Nuclear Wallet Cards, National Nuclear Data Center, Brookhaven National Laboratory (1985).

10. C. Lederer et al., Table of the Isotopes, 7th Edition, John Wiley and Sons, New York (1978).

11. J. Meadows, ANL/NDM-98, Argonne National Laboratory (1987).

12. D. Smith, Nuclear Instruments and Methods in Physics Research A257, 365 (1987).

13. "Evaluated Nuclear Data File, ENDF/B-V", ENDF/B Summary Documentation, compiled by R. Kinsey, ENDF-201, 3rd Edition, Brookhaven National Laboratory (1979).

14. A. Crametz, H. Knitter and D. Smith, Nuclear Data for Science and Technology, ed. K. Boeckhoff, D. Reidel Publishing Company, Dordrecht, Holland, p. 902 (1983).

15. F. Perey, ORNL/TM-6062, Oak Ridge National Laboratory (1977).

16. W. Mannhart, D. Smith and J. Meadows, Proceedings of a Specialists' Meeting on Neutron Activation Cross Sections for Fission and Fusion Energy Applications, NEANDC-259 'U',

eds. M. Wagner and H. Vonach, OECD Nuclear Energy Agency, Paris, France, p. 121 (1990).

17. P.Guenther, R. Lawson, J. Meadows, M. Sugimoto, A. Smith, D. Smith and R. Howerton, ANL/NDM-107 (1988).

18. S. Tagesen, H. Vonach and B. Strohmaier, Physics Data, No. 13-1, Fachinformationszentrum Karlsruhe, Karlsruhe, Fed. Republic of Germany (1979).

19. V. McLane, C. Dunford and P. Rose, Neutron Cross Sections: Vol. 2, Neutron Cross Section Curves, Academic Press, Inc., Boston (1988).

20. A. Smith, S. Chiba, D. Smith, J. Meadows, P. Guenther, R. Lawson and R. Howerton, ANL/NDM-115, Argonne National Laboratory (1990).

21. K. Weaver, J. Anderson, H. Barschall and J. Davis, Nuclear Science and Engineering 52, 35 (1973).

22. L. Greenwood, R. Heinrich, R. Kennerly and R. Medrzychowski, Nuclear Technology 41, 109 (1978).

Table 1: Reaction parameters and experimental results

| Data Point | Reaction | Half Life[a] | Detected Radiation[b] | Measured Ratio[c] | Error (%)[d] |
|---|---|---|---|---|---|
| Set I: | | | | | |
| 1 | Li-7(n,n't)He-4 | 12.33 y | $\beta$- [t] | 0.1217 | 4.0 |
| 2 | Al-27(n,p)Mg-27 | 9.462 m | $\gamma$ | 0.02380 | 3.9 |
| 3 | Al-27(n,$\alpha$)Na-24 | 15.020 h | $\gamma$ | 0.006155 | 5.9 |
| 4 | Ni-58(n,p)Co-58 | 70.916 d | $\gamma$ | 0.5298 | 4.1 |
| 5 | Ni-60(n,p)Co-60 | 5.271 y | $\gamma$ | 0.01750 | 3.7 |
| Set II: | | | | | |
| 6 | Th-232(n,fission) | Various | FF | 0.2579 | 2.2 |
| 7 | U-233(n,fission) | Various | FF | 4.053 | 1.7 |
| 8 | U-234(n,fission) | Various | FF | 2.894 | 1.6 |
| 9 | U-235(n,fission) | Various | FF | 2.646 | 1.4 |
| 10 | U-236(n,fission) | Various | FF | 1.646 | 1.7 |
| 11 | Np-237(n,fission) | Various | FF | 3.296 | 1.5 |
| 12 | Pu-239(n,fission) | Various | FF | 3.995 | 2.0 |
| Set III: | | | | | |
| 13 | Ti-46(n,p)Sc-46 | 83.83 d | $\gamma$ | 0.07371 | 4.6 |
| 14 | Ti-47(n,p)Sc-47 | 3.345 d | $\gamma$ | 0.08769 | 6.7 |
| 15 | Ti-48(n,p)Sc-48 | 43.7 h | $\gamma$ | 0.002526 | 6.5 |

Table 1: (continued)

| | | | | | |
|---|---|---|---|---|---|
| 16 | Fe-54(n,p)Mn-54 | 312.5 d | $\gamma$ | 0.4205 | 4.8 |
| 17 | Fe-54(n,$\alpha$)Cr-51 | 27.704 d | $\gamma$ | 0.007011 | 10.5 |
| 18 | Fe-56(n,p)Mn-56 | 2.5785 h | $\gamma$ | 0.007849 | 5.6 |
| 19 | Co-59(n,p)Fe-59 | 44.496 d | $\gamma$ | 0.009159 | 5.6 |
| 20 | Co-59(n,2n)Co-58 | 70.916 d | $\gamma$ | 0.0003028 | 14.8 |
| 21 | Co-59(n,$\alpha$)Mn-56 | 2.5785 h | $\gamma$ | 0.001192 | 6.2 |
| 22 | Cu-63(n,$\alpha$)Co-60 | 5.271 y | $\gamma$ | 0.004122 | 7.3 |
| 23 | Cu-65(n,p)Ni-65 | 2.520 h | $\gamma$ | 0.003770 | 11.6 |
| 24 | Zn-64(n,p)Cu-64 | 12.701 h | $\gamma$ | 0.1745 | 4.0 |
| 25 | Nb-93(n,2n)Nb-92m | 10.15 d | $\gamma$ [IT] | 0.004780 | 9.8 |
| 26 | In-113(n,n')In-113m | 1.658 h | $\gamma$ [IT] | 0.5248 | 6.3 |
| 27 | In-115(n,n')In-115m | 4.486 h | $\gamma$ [IT] | 0.6173 | 3.7 |

[a] Reaction product half lives from Ref. 9. Fission produces many half lives. Units: m (minute), h (hour), d (day), y (year).

[b] Detected radiations: $\beta$- [t], tritium beta rays; $\gamma$, gamma rays; $\gamma$ [IT], isomeric-transition gamma rays; FF, fission fragments.

[c] Integral cross-section ratios in the Be(d,n) neutron field for the indicated reaction relative to U-238 neutron fission. Multiply these ratios by 0.46097 barn to obtain absolute integral cross sections (see text).

[d] Total error (random and systematic) in the measured ratio.

Table 2: Energy spectrum for Be(d,n) neutrons near zero degrees[a]

| E[b] | $\phi$[b] | E | $\phi$ | E | $\phi$ | E | $\phi$ |
|---|---|---|---|---|---|---|---|
| $\leq$ 0.1826 | $\approx$0.0 | 4.5600 | .13930 | 6.4000 | .02161 | 8.7400 | .01341 |
| 0.2238 | .09214 | 4.6500 | .13519 | 6.4700 | .01996 | 8.8500 | .01362 |
| 0.2693 | .10542 | 4.7300 | .13315 | 6.5500 | .01848 | 8.9700 | .01316 |
| 0.3178 | .10773 | 4.8500 | .13168 | 6.6200 | .01721 | 9.0900 | .01279 |
| 0.3662 | .11597 | 4.9700 | .13088 | 6.7000 | .01664 | 9.2100 | .01231 |
| 0.4373 | .11234 | 5.0400 | .12816 | 6.8100 | .01604 | 9.3300 | .01096 |
| 0.5328 | .11993 | 5.0900 | .12347 | 6.9700 | .01610 | 9.4600 | .01024 |
| 0.6302 | .12479 | 5.1400 | .11960 | 7.0900 | .01558 | 9.5900 | .00939 |
| 0.7262 | .13649 | 5.1900 | .11802 | 7.1700 | .01629 | 9.7200 | .00848 |
| 0.8220 | .18599 | 5.2400 | .11233 | 7.2600 | .01642 | 9.8500 | .00776 |
| 0.91856 | .18047 | 5.3000 | .10482 | 7.3500 | .01634 | 9.9900 | .00756 |
| 1.05474 | .17289 | 5.3500 | .10209 | 7.4300 | .01659 | 10.130 | .00728 |
| 1.24055 | .18021 | 5.4300 | .09985 | 7.5200 | .01726 | 10.270 | .00720 |
| 1.4313 | .18697 | 5.5500 | .09753 | 7.6100 | .01647 | 10.420 | .00662 |
| 1.61385 | .19302 | 5.6300 | .09538 | 7.7100 | .01583 | 10.560 | .00603 |
| 1.7949 | .20674 | 5.7200 | .09522 | 7.8000 | .01469 | 10.720 | .00536 |
| 2.1208 | .19354 | 5.8400 | .09355 | 7.9000 | .01454 | 10.870 | .00462 |
| 2.5803 | .19934 | 5.9400 | .09020 | 8.0000 | .01424 | 11.030 | .00416 |

Table 2: (continued)

| | | | | | | | |
|---|---|---|---|---|---|---|---|
| 3.03595 | .19832 | 6.0000 | .08644 | 8.1000 | .01448 | 11.190 | .00356 |
| 3.25723 | .18923 | 6.0600 | .07440 | 8.2000 | .01371 | 11.350 | .00313 |
| 3.4785 | .18034 | 6.1300 | .05065 | 8.3000 | .01386 | 11.370 | .00302 |
| 3.9009 | .17493 | 6.2000 | .03200 | 8.4100 | .01407 | 11.450 | .00137 |
| 4.3500 | .14546 | 6.2600 | .02596 | 8.5100 | .01380 | 11.550 | .00012 |
| 4.4600 | .14340 | 6.3300 | .02333 | 8.6200 | .01372 | ≥11.650 | ≈0.0 |

[a] Neutron spectrum from 7-MeV deuterons on a thick Be-metal target. Energy (E) in MeV. Energy-differential flux ($\phi$) in neutron/MeV. Flux is normalized to unit total fluence.

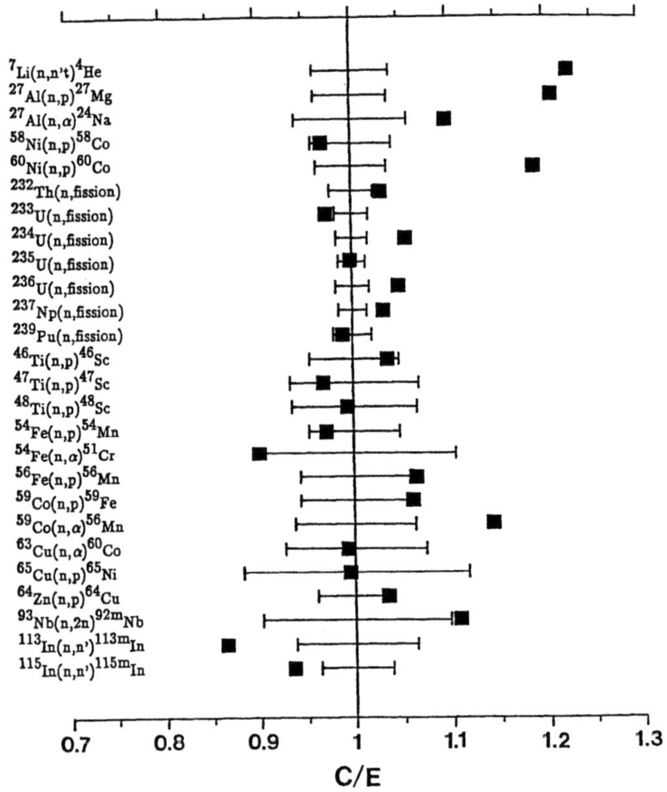


Figure 1: Comparison of experimental (E) and calculated (C) integral cross sections. Experimental values are centered on unity (horizontal bars indicate errors). Calculated quantities (solid squares) show C/E deviations from unity.

# STATUS OF RADIATION DAMAGE EVALUATION IN FUSION MATERIALS IRRADIATION EXPERIMENTS

A. Alberman *, H. Nolthenius **, W.P. Voorbraak **

* Centre d'Etudes Nucléaires (CEA), Saclay, 91191 Gif-sur-Yvette, France
** Netherlands Energy Research Foundation ECN, P.O. Box 1, 1755 ZG Petten, The Netherlands

## ABSTRACT

The EURATOM Working Group on Reactor Dosimetry (EWGRD) has issued in the past several documents in order to establish standardized procedures and recommendations on neutron spectrum information, fluence measurements, damage cross-section data, etc.

The main goal of this paper is to review the suitable material irradiation characterization parameters, in such a way that experimental results, obtained in a research reactor environment, can be applied in the design of fusion power plants. Recent developments in fusion reactor technology programs and mainly the large European component qualification tests undertaken for NET (Next European Torus) have led the EWGRD to consider new requirements.

Particularly the application to ceramics (tritium breeding blankets, insulators) addresses new requirements: damage to sublattices, relevance of "dpa" as irradiation parameter, etc.

This paper presents the status of available dosimetry methods, recommended cross-sections and damage assessment, relevant to fusion technology material irradiations.

## 1. INTRODUCTION

Technology development programmes require tests for materials preselection. These tests have to be performed in relatively short irradiation times. Current design of fusion reactors, based on D-T reactions, imply high neutron fluence levels on structural materials. The source neutron energy is 14 MeV in the plasma region, slowing down to lower energies beyond the first wall of the vessel, that is to say the blanket region.

The lack of high-intensity 14 MeV neutron sources brings up the necessity to carry out irradiations in existing Materials Testing Reactors or Fast Breeder Reactors, which are more appropriate to reach high fluence values. The neutron spectrum in reactor experimental facilities is significantly different from the fusion environment; therefore applicability of the irradiation results may be difficult without proper dosimetry information.

The EURATOM Working Group on Reactor Dosimetry (EWGRD) has issued these past years several documents [1] and [2] in order to establish standardized procedures and recommendations on neutron spectrum information, fluence measurements, damage cross-sections etc.

In the framework of the European Fusion Technology Programme, material irradiations are performed in different reactors: SILOE, OSIRIS and PHENIX in France, DIDO and KNK in Germany and the HFR in the Netherlands. The purpose of this paper is to review some recommendations, after describing briefly the damage processes and available dosimetry methods in test reactors relevant to fusion technology materials irradiations.

## 2. RADIATION EFFECTS AND DAMAGE CROSS SECTIONS

A review of radiation damage effects to metals and ceramics can be found in the proceedings of international conferences on fusion reactor materials (ICFRM) [3]. Main studies refer to plasma facing components such as first wall, limiters, divertors, etc. The first wall is bombarded with high energy neutrons (14 MeV) and charged particles leaking from the plasma (e.g. helium from D-T reactions) creating atomic displacements. Transmutations and gas production occur from a series of different neutron reactions at high energy, and may lead to further damage.

If a material is tested in a research reactor, these radiation effects cannot easily be achieved, and simulation techniques must be implemented [4].

The number of displacements per target atom (dpa) is generally recommended as a suitable parameter for characterization of the damage in fission reactors, because it comprises neutron spectrum effects [5]. In most cases damage in the test material is better correlated with this irradiation parameter, than with a neutron fluence value.

Here we shall briefly describe the radiation damage dosimetry data applicable to these materials.

### 2.1. Metals

Displacement cross-sections up to 14 MeV are available for most metals. They are usually based on the Lindhard's energy partition model [6]. The relationship between energy available for atomic displacements E, and the number of displaced atoms was further developed by Norgett, Robinson and Torrens (NRT) [7], leading to the following equation:

$$N_d = k\, E / 2\, E_d\,, \qquad (1)$$

where $E_d$ is the displacement threshold energy, and k the displacement efficiency (usually 0.8 for metals). Recent work suggests that the physical processes in a fusion environment could be different from those in fission reactors [8], [9]. Therefore the displacement efficiency could differ from 0.8. Nevertheless, eq. 1 is still used.

The EURATOM Working Group on Reactor Dosimetry (EWGRD) has proposed a recommended set of cross-sections [2], as well as metrology procedures for irradiation dosimetry [1].

## 2.2. Ceramics and Alloys

Radiation damage to ceramics in a fusion environment is a very complex problem. Most displacements are created by fast neutrons, but a variety of crystal structures and the magnitude of local charge may be the cause that not all displacements have an equal weight in the balance of stable defects. The effect of gamma-ray exposure may be not negligible. Of course, fundamental and experimental studies are needed to improve the knowledge on basic defects responsible for the primary damage.

In the particular case of alloys, such as austenitic steels, where atoms have atomic and mass numbers which differ not much, averaging the displacement cross-section with the atomic abundance is acceptable. But more generally it is not correct to do so, and a more detailed description is needed.

General codes for the derivation of damage cross-sections for alloys with atomic constituents which differ much in mass number have been issued recently (COMPOSI [10], SPECOMP [11]). They are based on the same binary collision model using Lindhard's energy partition. They are able to compute separately the number of displaced atoms for each atom type (sub-lattice), from Primary Knock-on Atoms (PKA) spectra already computed for single elements.

The model requires a displacement threshold energy for each type of atom. Of course, the total number of displacements depends very much on these threshold values. A general discussion on this matter can be found in ref. [12].

Solid breeder materials development work for fusion blanket applications has raised the question on the contribution of extra displacements created by energetic recoils of charged particles. This indirect damage process involves thermal-neutron induced reactions:

$$^{6}Li + n \rightarrow T + \alpha + 4.8 \text{ MeV}$$

and may induce, depending on the neutron spectrum at the test position, a relatively large amount of displacements. Methods for computation of these displacements in lithium ceramics are given in [10].

## 3. IRRADIATION PARAMETERS DERIVATION

The purpose of neutron metrology is to derive parameters for the neutron field during the irradiation of the experiment, which facilitates the comparison of the irradiation condition with planned or already performed irradiations which may be done in other neutron fields.

The parameters of general interest can be split in two groups, i.e.

1) Characterization of the neutron spectrum in terms of fluence rate above a given energy, like :
   $\phi(E>1.0MeV)$, $\phi(E>0.1MeV)$, etc.
   Also thermal and intermediate neutrons are taken into account, if necessary.
2) Characterization by data derived from the neutron spectrum like :
   - the number of displacements per atom per second and the total number of displacements for the test material of interest,
   - the gas production (rate) in the test material of interest,
   - the number of atom mutations in the test material,
   - the fission rate and number of fissions.

The above-given irradiation parameters can be derived from the neutron spectrum for the position of interest and special metrology libraries with energy dependent cross-sections [13],[14]. The neutron spectrum can be calculated, but it is a normal practice to validate the calculated information with the help of experimental data. The experimental data can be obtained from irradiations of special sets of activation or fission monitors in the irradiation experiment.

In a number of cases the material experiment of interest is irradiated in a neutron spectrum which is thoroughly measured or which can be supposed to be the same as a reference spectrum. Then only a rather simple normalization with experimental activation monitor information is needed to determine the irradiation parameters.

In other cases the damage rate for a particular irradiation field can be determined directly in a short pre-irradiation with special damage rate monitors [15],[16].

If experimental information is available for an irradiated experiment, then a fit of these data to the neutron spectrum information which was calculated or available for the reference field is needed. Ideally the fit should be done with the generalized least-squares technique, but in practice the uncertainty data needed for this method are often not available, so that a more simple technique has to be applied [17],[18]. The disadvantage is that in this situation the output values will be somewhat dependent on the procedure which is followed actually.

## 4. STATUS OF AVAILABLE DATA

### 4.1. Displacement Cross-Sections in Metals

Data are derived from several codes using Lindhard's model (see 2.1). The DAMSIG library [19] comprises evaluations of aluminium, carbon, chromium, copper, iron, molybdenum, nickel, niobium, silicon, steel, vanadium and zirconium. In the near future an update of the DAMSIG library is foreseen, then also beryllium will be included.

### 4.2. Gas Production Cross-Sections

Cross-section data for a number of reactions producing hydrogen, helium, deuterium and tritium are compiled in the DAMSIG library [19]. Data are given for aluminium, beryllium, boron, carbon, chromium, cobalt, copper, iron, lithium, manganese, nickel, silicon, titanium and vanadium.

### 4.3. Displacement Cross-Sections in Ceramics and Alloys.

The DAMSIG library does not comprise displacement cross-sections in ceramics yet. Its update is now under consideration. Damage cross-sections for a few ceramic materials, in particular lithium compounds, have been computed with the code COMPOSI at CEA. Table 1 gives the damage cross-sections averaged over a fission neutron spectrum and a specification of the damage due to α and T per capture in $^{6}Li$. The differential cross-section curve for $Li_2ZrO_3$ is depicted in figure 1.

## 5. RECOMMENDATIONS

### 5.0. Neutron Spectrum Information

The neutron spectrum in which the irradiation was carried out should be specified in a sufficiently detailed group structure and the method by which the spectrum was obtained should be described.

### 5.1. Metals

For fusion applications undertaken in research reactors, it is proposed to apply the general EURATOM dpa recommendations [1],[2], which are very consistent with ASTM and IAEA cross-section sets (Lindhard + NRT). For the calculation of the gas production use should be made of special energy-dependent data. A selected set is available in the DAMSIG library [19].

### 5.2. Ceramics and Alloys

In the absence of physically more acceptable damage models, it is recommended to quote the number of displacements for each type of atom (sub-lattice), as well as for the "average dpa". The "average dpa" is calculated as the total number of displacements in the constituent atoms divided by the number of atoms present in the lattice of the alloy.

In the case of breeder ceramics, it is recommended to quote separately the dpa contribution of fast neutrons (direct damage process) and the dpa contribution of charged particles (T + α).

- Even though it is not fully satisfactory, nevertheless it is proposed to keep for the moment a constant displacement threshold energy of $E_d$=40 eV for all types of atoms in ceramic material, in order to avoid inconsistencies when values from different evaluators are used in one calculation. This working method presents the advantage to focus more on the damage sensitivity to different neutron spectra rather than on the absolute dpa determinations.

- It is also proposed to quote -in parallel- graphite and iron dpa values in order to provide spectrum information applicable to other metal damage studies.

- The total absorbed gamma dose in ceramics should be given.

## 6. REFERENCES


1. J.P. Genthon, B.W. Hasenclever, W. Schneider, P. Mas, S.B. Wright, and W.L. Zijp, "Recommendations on the measurements of irradiations received by reactor structural materials", Report EUR 5274 (1975).
2. A. Alberman, J.P. Genthon, W. Schneider, P. Mas, L. Weise, S.B. Wright, and W.L. Zijp, "Introduction of neutron metrology for reactor radiation damage", Report EUR 6182 (1978).
3. Proceedings ICFRM-3 in Journal of Nuclear Materials, vol. 155-157 (1988).
4. R. Dierckx and C.R. Emigh, "Characterization for fusion first wall damage studies of using tailored D-T neutron fields"; Proceedings 3rd ASTM-EURATOM Symposium on Reactor Dosimetry, Ispra, October 1979, EUR 6813, vol. I, p. 230.
5. A.L. Lowe Jr and A. Alberman, "Report of workshop on radiation damage correlations", ASTM STP 1001 Reactor Dosimetry methods, applications and standardization (1989), pp. 778,779.
6. J.L. Lindhard, V. Nielsen, et al., "Matematisk-fysiske Meddelelser Kongelige Danske Videnskabernes Selskab, Vol. 33 (1963), p. 10.
7. M. Norgett, M. Robinson and I.Torrens, "Nuclear Engineering and Design 33 (1975).
8. R. Dierckx, "The importance of the PKA-energy spectrum for radiation damage simulation", Journal of Nuclear Materials 144, pp. 214-227 (1987).
9. H.L. Heinisch, "Computer simulation of displacement cascades in copper", HEDL-TME 83-17 (1983).
10. A. Alberman and D. Lesueur, "Determination of damage function for polyatomic materials irradiated in research reactors", ASTM STP 1001, Reactor Dosimetry methods applications and

standardization (1989), pp. 576-591.
11. L.R. Greenwood, "Radiation damage calculations for compound materials", Advisory Group Meeting on Nuclear Data for Radiation Damage Assessment, IAEA Vienna, September 1989.
12. G.P. Pells, "Radiation effects and damage mechanisms in ceramic insulators and window material", Journal of Nuclear Materials, (1988), pp. 155-157.
13. D.E. Cullen, P.K. McLaughlin, "The International Reactor Dosimetry File (IRDF85)", Report IAEA-NDS-41, REV. 1 (1985).
14. W.L. Zijp, H.J. Nolthenius, and G.C. Verhaag, "Cross-section library DOSCROS84", Report ECN-160 (1984).
15. M. Cance, J.P. Genthon, "Le detecteur neutronique au graphite GAMIN", Report CEA.N.1823 (1975).
16. A. Alberman, J.P. Genthon, "Le detecteur miniaturise au tungstene", Dosimetry methods for fuels, cladding and structural materials, EUR-6813, VOL. 2, (1980) pp. 1104-1118.
17. F.G. Perey, "Least-squares dosimetry unfolding: the program STAY'SL", Report ORNL/TM-6062, ENDF-254 (1986).
18. W.N. McElroy, et al. :" A computer automated iteration method for neutron flux spectra determination by foil activation, SAND II", Report AFWL-TR-67-41, VOL. I-IV (1967).
19. W.L. Zijp, H.J. Nolthenius, and G.C. Verhaag, "Damage cross-section library DAMSIG84", Report ECN-159 (1984).

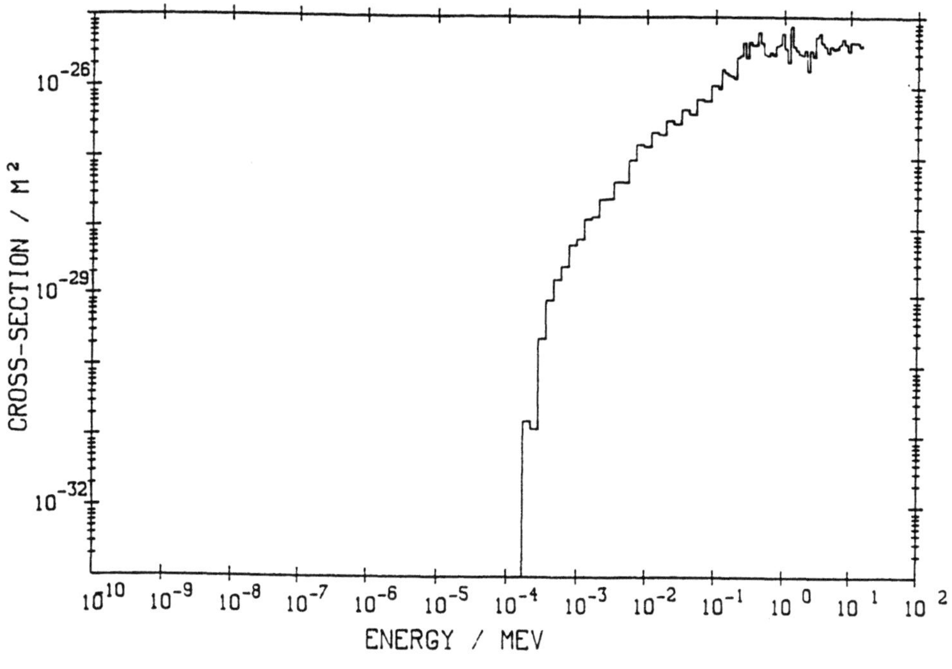


Figure 1. Damage cross-sections for $Li_2ZrO_3$.

Table 1. Damage cross-section for ceramics averaged over a fission neutron spectrum (also some values for other materials are included)

$E_d$ threshold energy 40 eV for each atom. The cross-sections are given in dpa.barn/n.

| Alloy | atom | Direct damage (averaged U-235 fiss. spectrum) | | Indirect damage (α + T) Number of displacements/ capture in Li-6 |
|---|---|---|---|---|
| $LiAlO_2$ | Li | 333 | | 12 |
| | Al | 713 | | 26 |
| | O | 679 | | 47 |
| | avg. | 601 | total: | 85 |
| $Li_2SiO_3$ | Li | 320 | | 17 |
| | Si | 660 | | 19 |
| | O | 619 | | 47 |
| | avg. | 526 | total | 83 |
| $Li_4SiO_4$ | Li | 300 | | 23 |
| | Si | 573 | | 12 |
| | O | 549 | | 42 |
| | avg. | 441 | total | 77 |
| $Li_2ZrO_3$ | Li | 235 | | 15 |
| | Zr | 384 | | 18 |
| | O | 448 | | 43 |
| | avg. | 366 | total | 76 |
| $Li_2O$ | Li | 248 | | 35 |
| | O | 396 | | 31 |
| | avg. | 297 | total | 66 |
| $Al_2O_3$ | Al | 896 | | |
| | O | 813 | | |
| | avg. | 846 | | |
| Graphite (Thompson-Wright) | | 720 | | |
| Iron (dpa EURATOM) | | 835 | | |

# RADIATION FIELDS CHARACTERISTICS

DETERMINATION OF HIGH ENERGY NEUTRON

FLUX (E>10 MeV) IN TWO LIGHT WATER REACTORS

Jean-Jacques BONNIN

CEA-DRE

C.E.N.G. GRENOBLE 85 X 38041 GRENOBLE CEDEX FRANCE

ABSTRACT

In this experiment, the fast neutron flux has been measured considering 3 components i.e. the "usual" fast neutron flux (E>1 MeV) and the fast neutron flux for E > 6 MeV and E > 10 MeV.

The experiments were performed in the research reactors MELUSINE (8 MW) and SILOE (35 MW). The measurement positions are located in the reactor core and in the light water reflector up to 30 cm from the core.

Three sets of activation foils were selected for the measurement of reaction rates. They cover the spectrum from 1 MeV up to 14 MeV approximatively, using 10 threshold reactions.

After irradiation, reaction rates were deduced from gamma-ray counts using a germanium detector.

The calculations were performed using 3 codes i.e. 1-D ANISN, 2-D DOT and the French 3-D code TRIPOLI using the Monte-Carlo technique. The average cross-sections of each dosimeter at each position of measurement were derived from the calculated neutron spectra.

Different types of the U235 neutron fission spectrum (such as MADLAND's) were used.

Spectral indexes, values of the fast flux, penetration in water were compared to measurements. They are generally in good agreement. The results are discussed.

## INTRODUCTION

Over the last 40 years intensive research has been devoted to the study of the neutron spectrum released by the fission of Uranium 235. Although neutrons with an energy of up to 18 MeV have been detected, few investigations have concerned the energy range over 10 MeV, even in pure spectra, not to mention in the complex radiation field of research reactors.

To characterize this high energy component, activation dosimetry techniques were chosen since they can be readily and economically implemented as a routine practice.

### Description of the MELUSINE (8 MW) and SILOE (35 MW) reactors

They are both open core pool type research reactors. The core is composed of plate elements in an aluminium-93 %-enriched uranium alloy. The coolant is light water which also serves as the reflector.

Several types of locations are arranged to accommodate the experiments, such as core locations within the special fuel elements, and locations on the periphery of the core in the water reflector.

The measurement positions chosen are one core position and a number of positions in the light water reflector up to within 30 centimeters from the core surface.

## CHARACTERIZATION OF THE FAST NEUTRON FIELD

In this experimental work, the fast neutron field was studied with respect to three components :

- En > 1 MeV, the "usual" fast flux,
- En > 6 MeV,
- En > 10 MeV.

For this purpose, activation dosimeters with the following reaction characteristics were selected :

Neutron fast flux En > 1 MeV

| | | |
|---|---|---|
| 58 Ni (n,p) 58 Co | T = 70.78d | Threshold : 1 MeV |
| 54 Fe (n,p) 54 Mn | T = 312.3d | Threshold : 1 MeV |

Neutron fast flux En > 6 MeV

| | | |
|---|---|---|
| 63 Cu (n,$\alpha$) 60 Co | T = 5.272y | Threshold : 3 MeV |
| 56 Fe (n,p) 56 Mn | T = 2.578h | Threshold : 4 MeV |
| 27 Al (n,$\alpha$) 24 Na | T = 15 h | Threshold : 5 MeV |

Neutron fast flux, En > 10 MeV

| | | | |
|---|---|---|---|
| 51 V (n,α) 48 Sc | T = 43.7h | Threshold : | 8 MeV |
| 55 Mn (n,2n) 54 Mn | T = 312.3d | Threshold : | 10.5 MeV |
| 59 Co (n,2n) 58 Co | T = 70.78d | Threshold : | 10.5 MeV |
| 90 Zr (n,2n) 89 Zr | T = 78.43h | Threshold : | 12 MeV |
| 58 Ni (n,2n) 57 Ni | T = 36.08h | Threshold : | 12.5 MeV |

At the same position, 2 sets of dosimeters are displayed in 2 boxes :

- one box contains the short life, dosimeters that must be counted very shortly after irradiation,
- the other box contains the long life dosimeters or those presenting spurious activities for which counting can be deferred.

In order to keep spurious thermal and epithermal reactions to a minimum, the dosimeters are put under a boron nitride filter.

The dosimeters used are gamma emitters and their activity is measured by means of a germanium detector.

Figure 1 shows the energy range where each reaction used in a fission spectrum (WATT) gives a 90 % response rate.

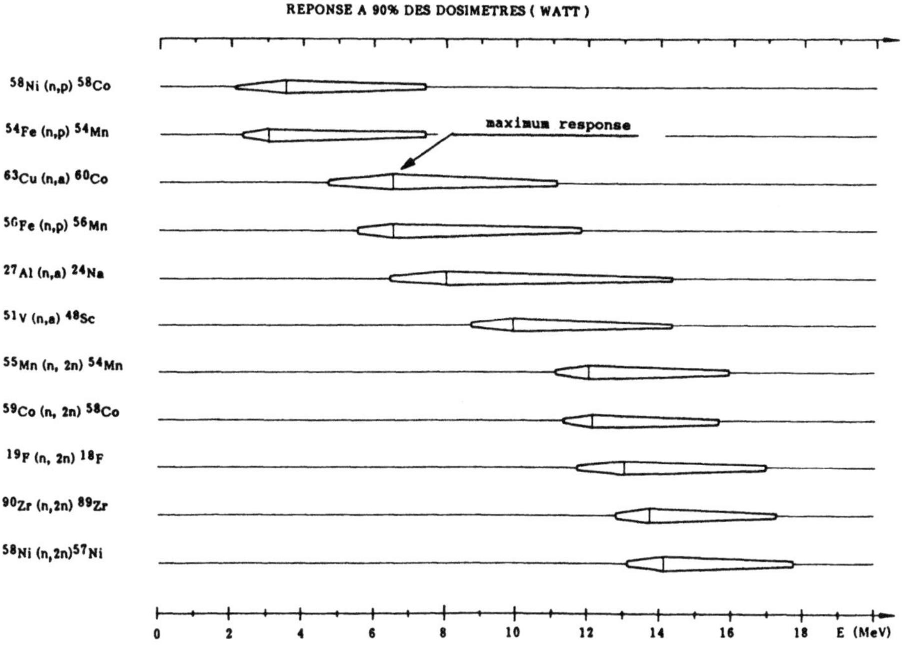


Fig.1 - Energy range where the response of reactions in a fission spectrum is 90 %

## CALCULATION METHOD

The neutron energy spectrum was calculated using 1-D and 2-D transport codes. The 1-D calculations were performed using the ANISN code. The 2-D calculations in x-y geometry used the DOT code, version 5.1 (1).

The french 3-dimensional code TRIPOLI (2) (using Monte Carlo methods) was also applied to obtain the neutron spectra. In this way any vertical geometrical discontinuities, such as a control rod not fully inserted or the aluminium grid supporting the core, could be taken into account.

From the multigroup neutron spectra issued from the calculations, 2 kinds of approximation (polygonal and exponential) were used to restore the continuity of the spectra. These continuous spectra were then used to calculate the mean cross-sections for each dosimeter at each measurement position.

A recent form of the fission spectrum of Uranium 235 (MADLAND) was also used to calculate the mean cross-sections of the 5 high energy reactions.

The basic cross-section data for this work was taken from ENDF/B-IV in the form of the IRDF-82 library (3) except for the data for the 51 V (n,α) 48 Sc reaction that was taken from a separate publication (4).

It is interesting to note here that as the dosimeters were chosen in order to characterize the energy field where their rate of response is maximum, the value of the mean cross-section seems to be practically independent of the spectrum.

Example : Reaction 63 Cu (n,α) 60 Co

| (σ) calculated ANISN | <σ>En>1 MeV | <σ>En>6 MeV |
|---|---|---|
| Core measurement position | 0.57 mb | 27.7 mb |
| 10.1 cm in water reflector | 1.19 mb | 22 mb |
| 30.3 cm in water reflector | 2.46 mb | 22.2 mb |

The values of mean cross-sections calculated with MADLAND's spectrum (5) for high energy reactions, are as follows :

| Reaction | <σ>En>10 MeV |
|---|---|
| 51 V (n,α) 48 Sc : | 9.61 mb |
| 55 Mn (n,2n) 54 Mn : | 171 mb |
| 59 Co (n,2n) 58 Co : | 156 mb |
| 90 Zr (n,2n) 89 Zr : | 66.8 mb |
| 58 Ni (n,2n) 57 Ni : | 2.51 mb |

## RESULTS

NEUTRON FLUX for En > 6 MeV :

The results are given in Figure 2. The measured value is the mean value obtained for the 3 selected reactions. Scatter with respect to this mean value is not very important (3.6 % ± 1.9 %). There is a good agreement between calculated and measured values.

NEUTRON FLUX for En > 10 MeV :

The results are shown in Figure 3. The measured value is the mean value obtained from the reactions chosen. Two categories of results are obtained systematically :

- for manganese and cobalt, the results are below the mean value,
- for zirconium and nickel, the results are above the mean value (see example Figure 4).

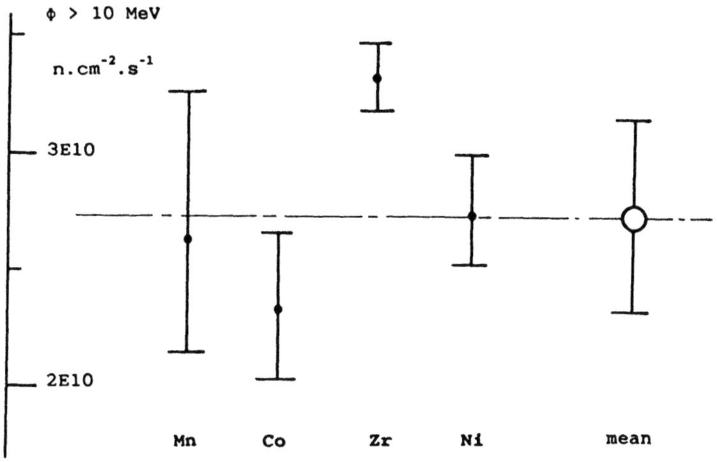


Fig. 4 - Values measured at position 2 cm below the reflector water

SPECTRAL INDEX $\Phi > 6/\Phi > 1$ MeV

This spectral index is given in Figure 5. The experimental spectral index lies between the curves calculated by ANISN-DOT and the curve calculated by TRIPOLI.

SPECTRAL INDEX $\Phi > 10/\Phi > 1$ MeV (Figure 6)

Experimental results show that the calculations underestimate the proportion of high energy neutrons. Note that the fission spectrum used in the 3 codes is that of CRANBERG. MADLAND's spectrum has a high energy (En > 10 MeV) portion that is 30 % higher than that of CRANBERG's spectrum. If MADLAND's spectrum was used in the codes, the calculated index would undoubtly be very close to the measured index (this calculation has not been performed).

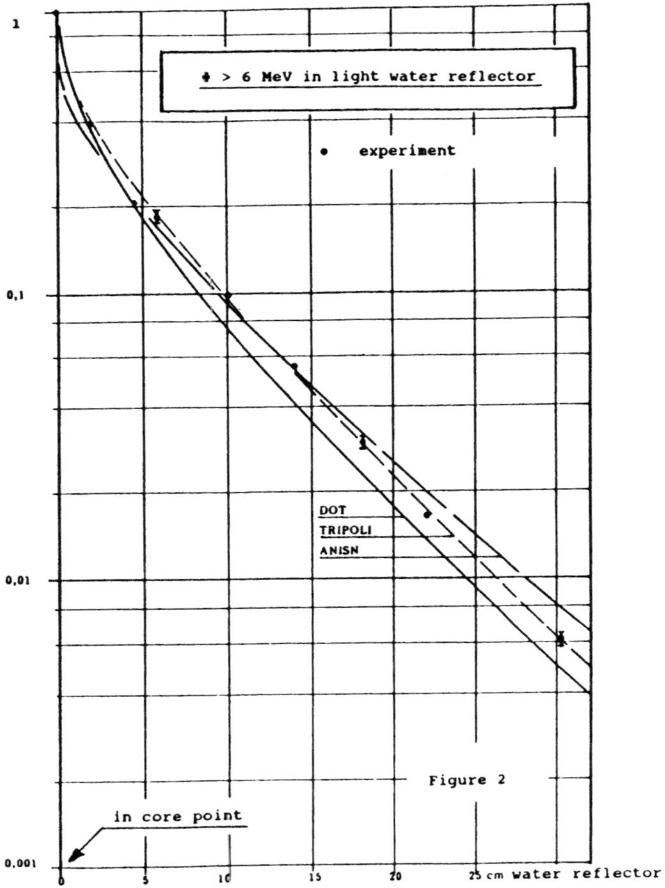


Figure 2

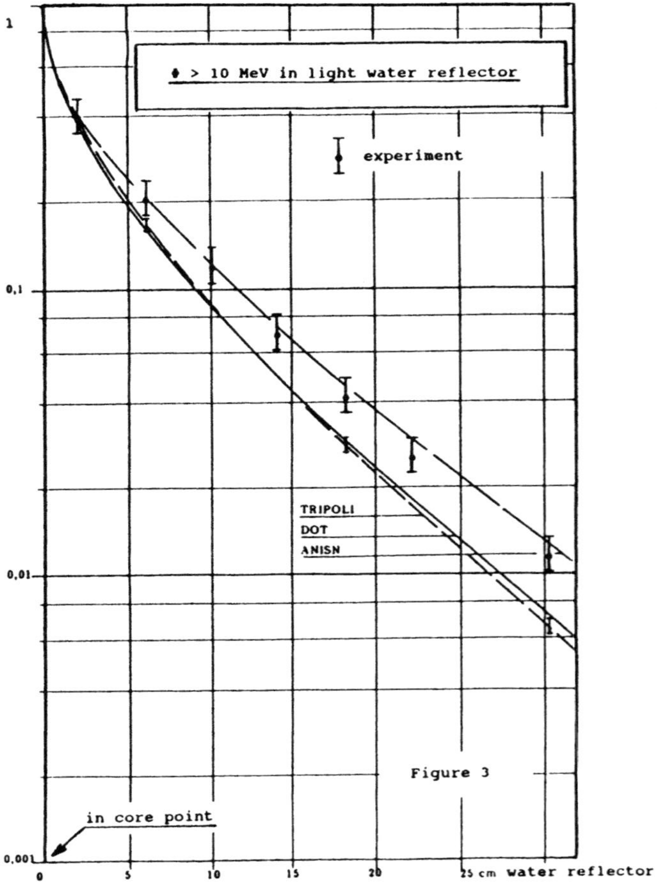


Figure 3

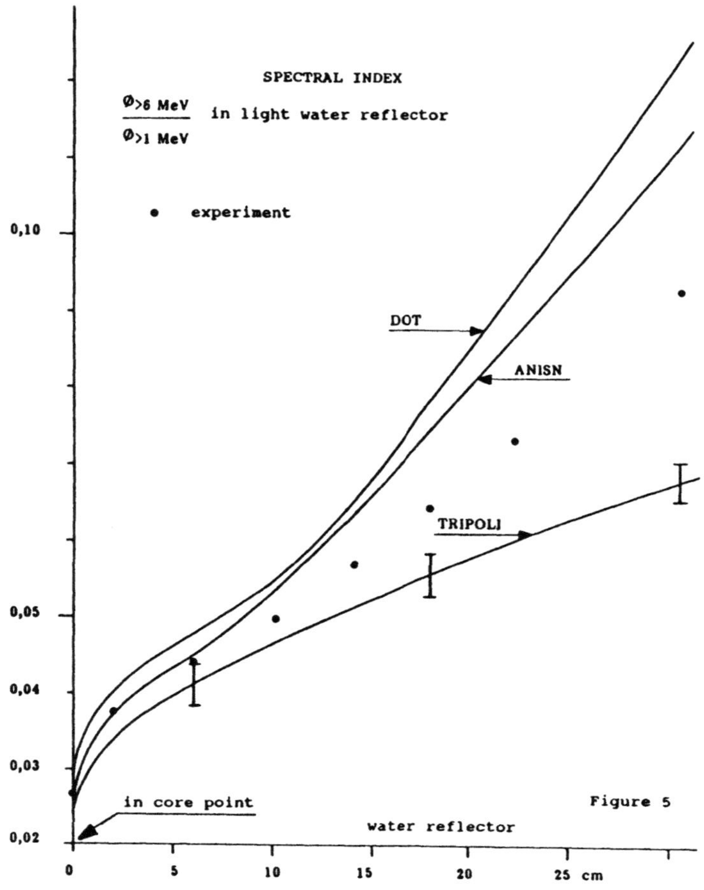


Figure 5

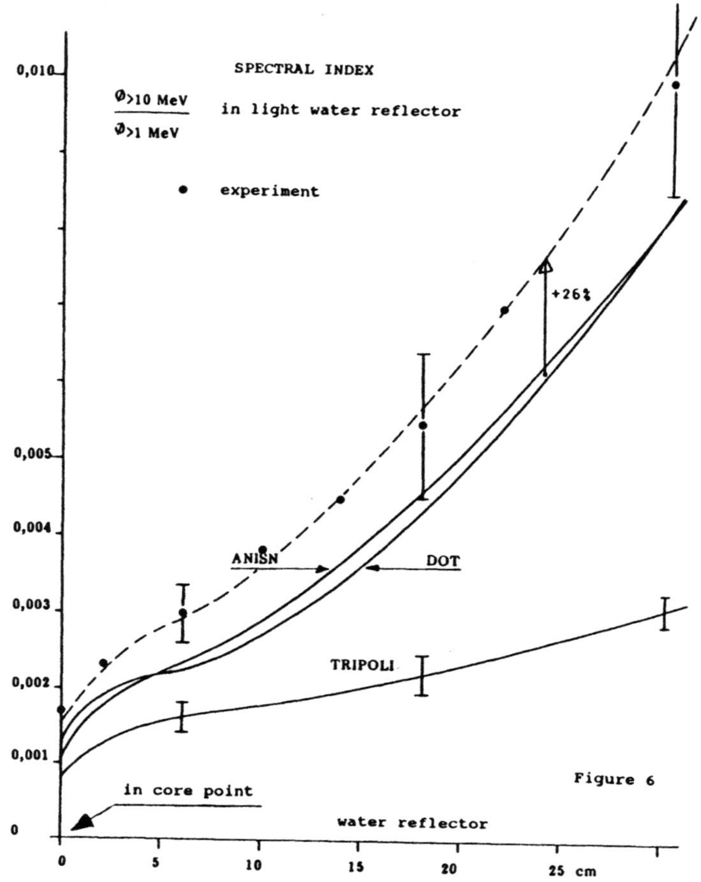


Figure 6

## CONCLUSIONS

By activation dosimetry methods it was possible to measure the high energy neutron fluxes (En > 10 MeV) near the core of the MELUSINE and SILOE research reactors, despite the spurious thermal and epithermal components in the neutron spectrum.

The precision of these high energy neutron flux measurements depends on the cross-section data ; in this respect further study of the reaction 51 V (n,$\alpha$) 48 Sc would be beneficial, and the most consistent measurements were obtained using MADLAND's spectral form.

In the SILOE reactor at a nominal power of 35 MW, measurements in a core location provided the following values :

| | | |
|---|---|---|
| $\Phi$ > 1 MeV | : | 2.0 E14 n.cm-2.s-1 |
| $\Phi$ > 6 MeV | : | 5.6 E12 n.cm-2.s-1 |
| $\Phi$ > 10 MeV | : | 3.0 E11 n.cm-2.s-1 |

## REFERENCES


1. W.A. Rhoades, R.L. Childs, DOT 5-1. An updated version of the DOT one and two dimensional neutron-photon transport code. ORNL - 5851 Report (1982).

2. A. Baur, L. Bourdet, G. Dejonghe, J. Gonnord, A. Monnier and J.C. Nimal, Programme de Monte Carlo polycinétique à trois dimensions TRIPOLI 02. Tomes 1, 2, 3.

3. D.E. Cullen, N. Kocherov, The International Reactor Dosimetry File (IRDF-82), AIEA Nuclear data section (1982).

4. H. Lu, W. Zhao, P. Fan, Cross-sections between neutrons and matter. Nuclear Instruments and Methods A255, pp 103-106 (1987).

5. D.G. Madland, J.R. Nix, R.J. Labauve, Uranium 235 thermal fission spectrum. Bande LANL - Communication personnelle (Mai 1988)

# PRESENT STATE OF NEUTRON METROLOGY IN THE HIGH FLUX REACTOR

A.N. Polle

Netherlands Energy Research Foundation ECN,
P.O. Box 1, 1755 ZG Petten, The Netherlands

ABSTRACT

This work gives a summary of applied hardware, software and quality control procedures at the Netherlands Energy Research Foundation to arrive at detailed neutron fluence,neutron spectrum information and characteristic spectrum parameters derived from activation monitors.

Most of the neutron metrology work is related to the material and fuel testing programs carried out in the High Flux Reactor (HFR) at Petten.

An exchange of information on the practical aspects of the neutron metrology work can serve to establish improved metrology procedures.

## 1. INTRODUCTION

Material testing reactors as the High Flux Reactor (HFR) [1] are particularly suitable to investigate the radiation damage effects in fuel materials (such as uranium oxides and uranium carbides), moderator materials (such as graphite and beryllium) or reactor structural materials (such as steel, tungsten, carbon composites). Usually the materials are irradiated in capsules or loops during long periods, receiving neutron fluences in the range $2 \cdot 10^{23}$ - $1 \cdot 10^{26}$ neutrons per $m^2$.

As the neutron fluence rate is not constant during these long periods, due to variations of the power level, to changes in the fuel, to fuel burn-up, to movement of control rods, to changes in neighbouring experimental assemblies, etc. the neutron fluence can not be predicted with sufficient accuracy from calculations or from fluence rate measurements at low power during short runs.

The fluence and related irradiation parameters received by an experimental assembly (and possibly also its spectral distribution) have therefore often been determined with monitors installed in or on the assembly during the whole irradiation period.

## 2. MONITOR SETS APPLIED FOR NEUTRON METROLOGY

In our metrology practice we are interested in a $\gamma$- or X-ray active reaction product. In most of the neutron metrology tasks we use a uniform type of monitor set. A description of the applied materials and the composition is given in figure 1. For the measurement of the neutron spectrum we use a set as given in figure 2. The main characteristics of the uniform monitor set are given in the figures 3 and 4.

## 3. COUNTING PROCEDURE

In our laboratory we use $\gamma$-ray spectrometers, containing the Nuclear Data multichannel analyser of the type ND66, coupled to a Ge-$\gamma$-ray detector. The analysers are operated with a channel width of 0.4 keV and the data storage is usually performed in 4096 channels.

For each $\gamma$-ray spectrometer efficiency curves for several source detector distances are calculated.

The analyses of the measured $\gamma$-ray spectra will be performed on a host (CDC CYBER 855 mainframe) computer system. Programs of the NDDT package [2] running on the CDC computer read the spectrum data header blocks and channel content blocks and generate instructions for the data-processing to obtain the activity results for the different nuclides which are present in the counted monitor material. Nuclear data [3] as well as the peak efficiency data sets are available as libraries.

## 4. DETERMINATION IRRADIATION CHARACTERISTICS

### 4.1 Determination of Saturation Activity

The program VILLA [4] calculates the neutron fluence, fluence rate and saturation activity from measured activity values. For small changes in the local fluence rate, for example, due to a change in irradiation position of the irradiation rig, a correction will be superimposed on the actual power irradiation history. Information for the changes can be obtained from previous measurements or, as usually used in ECN practice, from reactor physics calculations.

The thermal and the fast fluence rate values in the HFR are of the same order of magnitude. However, the cross-sections for the thermal range are larger than the averaged cross-sections for the fast range. That means that the burn-up effects are predominately caused by thermal neutrons [5]. Therefore the calculations for the thermal reactions are done without burn-up caused by fast neutrons. The calculations for the fast reactions are corrected for burn-up caused by thermal neutrons.

When in an irradiation experiment two or more monitors are applied at the same position, we always derive the spectrum indices, defined by the relation:

$$SI_{ij} = \left(\frac{A_{sat}}{N}\right)_i \Big/ \left(\frac{A_{sat}}{N}\right)_j = \frac{\int_0^\infty \sigma_i(E)\ \phi_i(E)\ dE}{\int_0^\infty \sigma_j(E)\ \phi_j(E)\ dE} .$$

The saturation activities can be derived directly from the measured activities. Its accuracy is related to the accuracy of the activity determination, the accuracy of the irradiation history, the accuracy of the monitor material composition and the accuracy of the mass determination of the monitor.

### 4.2 Spectrum Characteristic Results

The computer program which is used to determine neutron spectra based on activity measurements of irradiated monitors gives a solution which has a larger number of energy groups (640) than the number of activation reactions induced in the monitors. The program requires a-priori knowledge of the fluence rate spectrum in the form of an input spectrum which represents the available information from physics experiments or reactor physics calculations. The adjustment by the CHARDAT procedure [6] used in our laboratory is based on a modified SAND II spectrum adjustment method [7].

The modifications of the input spectrum are limited to:
- normalization of the input spectrum to the metrology results;
- adjustment of the thermal and intermediate region;
- adjustment of the intermediate and fast region.

In energy regions not covered by the monitor response the adjusted spectrum is solely determined by the input spectrum. This does not necessarily mean that within these regions the output spectrum equals the input spectrum. However, the modifications must of course be consistent with the activity values for the given monitor set. The number of different reactions is important to the extent that more coverage and resolution is possible with more monitors. A good quality input spectrum is a spectrum for which the normalized activity ratios do not deviate by more than 10% from unity. For the solution, deviations of this ratio larger than 5% are not accepted.

In principle many mathematical solutions are possible, but not all of these are physically acceptable. It is rather difficult to give quantitative relations for the influence of uncertainties in the cross section values, of uncertainties in the activities, or of the choice of the input spectrum.

The quality of the solution spectrum is influenced by the following points:
- set of monitor materials;
- measurement uncertainties in the input activities;
- uncertainties and inconsistencies in the input cross section values;
- input spectrum;
- mathematical model;
- number of energy points or energy groups used in the analysis.

It is emphasized that the choice of the input spectrum is the more important as the monitor response is poorer.

### 4.3 Neutron Damage

Damage cross section data have been obtained from the ENDF/B-V cross section file. A damage cross section library DAMSIG84 [8] is used for the neutron metrology work at ECN. This library is an updated and extended version of a previous ECN damage cross section library and comprises also the EWGRD recommendations and some ASTM data for steel. This library contains about 24 energy dependent group cross section data sets for a number of materials, to facilitate the calculations of damage production and about 37 cross section data sets for the gas production, e.g. due to $(n,\alpha)$, $(n,t)$ and $(n,p)$ reactions.

For composite materials the damage will be calculated as the sum of all contributions to the damage for the number of atoms of the elements of interest in the composite.

## 5. SYSTEM RELIABILITY AND QUALITY CONTROL

The uniform monitor set applied in a considerable number of irradiation experiments enables us to obtain a large set of equivalent data originating from the same activation reactions and processed with a uniform data treatment. Especially spectrum index values serve as an important indicator of reliability and quality. Usually the spread within the set of spectrum index values for one irradiation experiment is comparable with the measurement uncertainty. Large deviations indicate discrepancies. Table 1 gives spectrum index data for different regions in the reactor core of the HFR.
The 'redundant' number of monitors within one set yields the monitor set as a system for which the consistency can be tested.

As part of our quality control program once a week we measure under controlled conditions a $^{152}$Eu reference source to verify the repeatability of the general performance of the analysis of gamma ray spectra. This includes the manual action, the stability of the electronic instrumentation of the counting arrangement and the peak area determination. The FWHM parameter is used to examine the time behavior of the semi-conductor detector.

The scheme of the main uncertainty contributions in the neutron metrology is given in figure 5.

## 6. REFERENCES


1. Röttger, H. et al. "High Flux Materials Testing Reactor HFR Petten; Characteristics of facilities and standard irradiation devices"; Report Eur 5700 EN, Revised Edition 1986/1987 (CEC, JRC, Petten, 1986).
2. Polle, A.N. "Diagrammatic Summary of the Program Package NDDT for the Processing of a Measured γ-ray Spectrum", Unpublished Laboratory note FYS/RASA 87/11 (ECN, Petten, April 1987).
3. Baard, J.H., Zijp, W.L., Nolthenius, H.J. "Nuclear Data Guide for Reactor Neutron Metrology", ISBN 0-7923-0486-1 (Kluwer Academic Publishers, 1989); Report EUR 12354 EN (CEC, JRC, Petten, 1989).
4. Baard, J.H., Nolthenius, H.J. "Progam Description of VILLA", Unpublished Laboratory note FYS/RASA 89/05 (ECN, Petten, March 1989).
5. Freudenreich, W.E. "Burn-up Effects in Neutron Activity Metrology", Unpublished Laboratory note FYS/RASA 89/19 (ECN, Petten, September 1989).
6. Freudenreich, W.E. "CHARDAT, a Program Package for the Calculation of Neutron Spectrum Characteristics, Radiation Damage and Gas Production in Irradiation Experiments", Unpublished Laboratory note FYS/RASA 89/07 (ECN, Petten, April 1989).
7. Freudenreich, W.E., Nolthenius, H.J. "Neutron Spectrum Unfolding Code SANDP01", Unpublished Laboratory note FYS/RASA 87/17 (ECN, Petten, June 1989).
8. Zijp, W.L. "Damage cross-section library DAMSIG84 (in a 640 Group Structure of the SAND II type)", Report ECN-159 (ECN, Petten, October 1984).

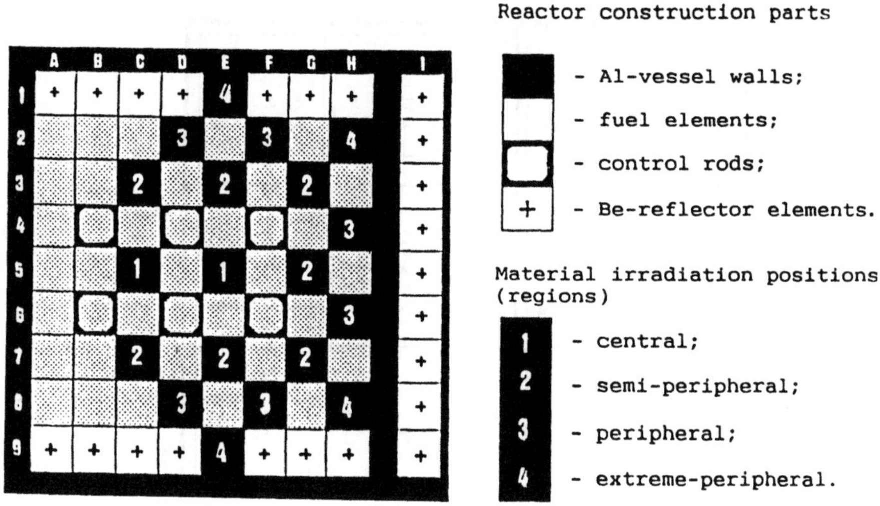


| core region | spectrum index $\frac{\langle\sigma(^{54}Fe)\rangle}{\langle\sigma(^{46}Ti)\rangle}$ | $\frac{\langle\sigma(^{59}Co)\rangle}{\langle\sigma(^{58}Fe)\rangle}$ | $\frac{\langle\sigma(^{94}Nb)\rangle}{\langle\sigma(^{58}Fe)\rangle}$ | $\frac{\langle\sigma(^{59}Co)\rangle}{\langle\sigma(^{94}Nb)\rangle}$ | $\frac{\langle\sigma(^{93}Nb)\rangle}{\langle\sigma(^{54}Fe)\rangle}$ |
|---|---|---|---|---|---|
| 1 | 7.94- 8.11 | 37.39- 37.38 | 2.34- 2.47 | 14.59- 16.22 | 2.34- 2.59 |
| 2 | 7.96- 8.22 | 38.00- 38.71 | 2.43- 3.15 | 12.62- 15.43 | 2.48- 2.64 |
| 3 | 7.79- 7.95 | 35.75- 36.47 | 1.93- 2.23 | 16.57- 18.48 | 2.64- 2.73 |
| 4 | 7.83- 8.03 | 36.82- 38.44 | 2.384- 2.643 | 14.51- 15.49 | 2.73- 2.89 |

Table 1. Spectrum index values for different core regions. Intervals for a confidence level of 95%.

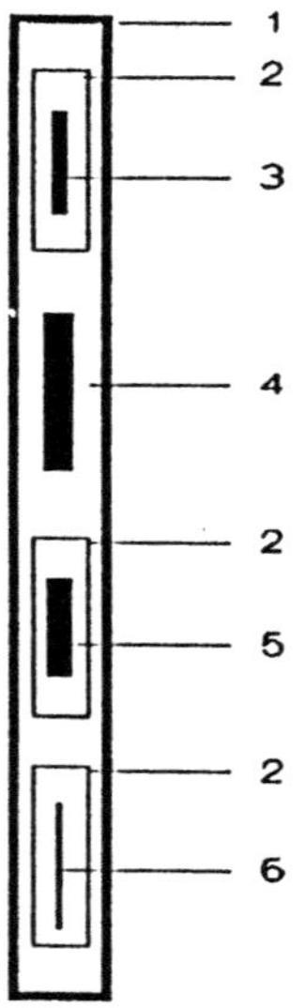


1. Quartz minitube, l=17 mm, ø 1.5/0.7 mm.
2. Quartz minitube, l= 5 mm, ø 0.6/0.2 mm.
3. Nb monitor.
4. Ti monitor.
5. Fe monitor.
6. NiCo monitor.

Fig. 1. Monitor set: composition and encapulation.

top

| Foil material | Dimensions (in mm) | Composition |
|---|---|---|
| $^{238}U$-Al | ø 5 x 0.1 | 22.83% $^{238}U$ |
| W-Al | ø 2 x 0.2 | 1.00% W |
| Ni | ø 7 x 0.2 | 99.99998% Ni |
| Th-Al | ø 5 x 0.2 | 10% Th |
| Sc | ø 2 x 0.1 | 99.99% Sc |
| Ag-Al | ø 7 x 0.2 | 1% Ag |
| $^{235}U$-Al | ø 5 x 0.1 | 4.49% $^{235}U$ |
| NaF | 2 x 2 x 1 | 54.8% Na |
| Cu-Al | ø 7 x 0.2 | 10% Cu |
| Al | ø 7 x 1 | 100% Al |
| Ti | ø 7 x 0.4 | 100% Ti |
| Au-Al | ø 5 x 0.2 | 0.1% Au |
| La-Al | ø 5 x 0.2 | 5.00% La |
| Fe | ø 7 x 0.52 | 100% Fe |
| In-Al | ø 7 x 0.2 | 0.2% In |
| Co-Al | ø 7 x 0.2 | 1% Co |
| Mn-Al | ø 7 x 0.2 | 1% Mn |

bottom

Fig. 2. Monitor set for neutron spectrum measurement.

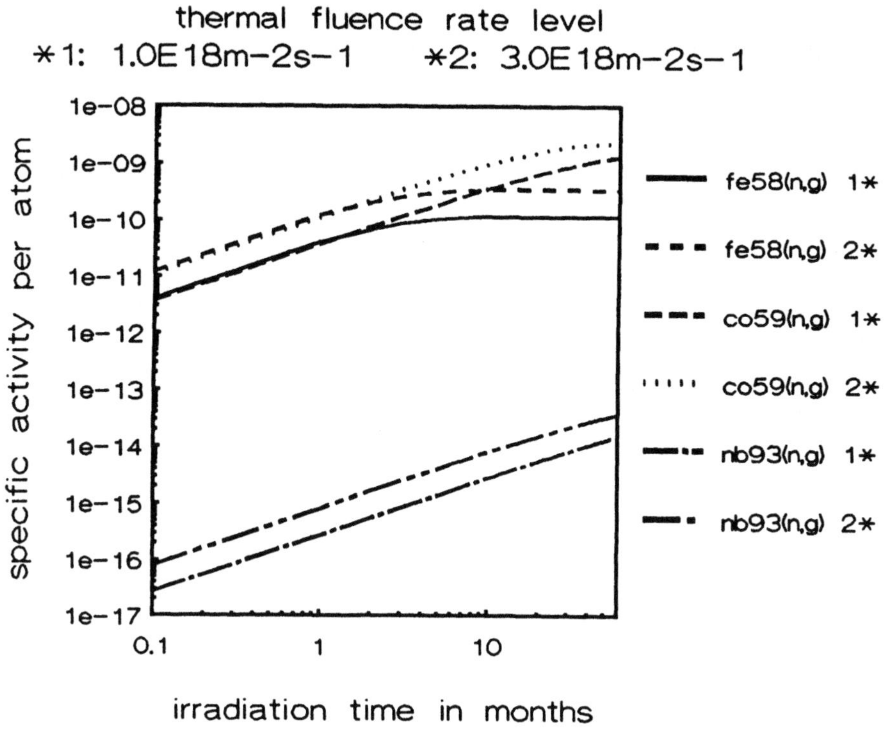


Fig. 3. Activity increase for thermal neutrons.

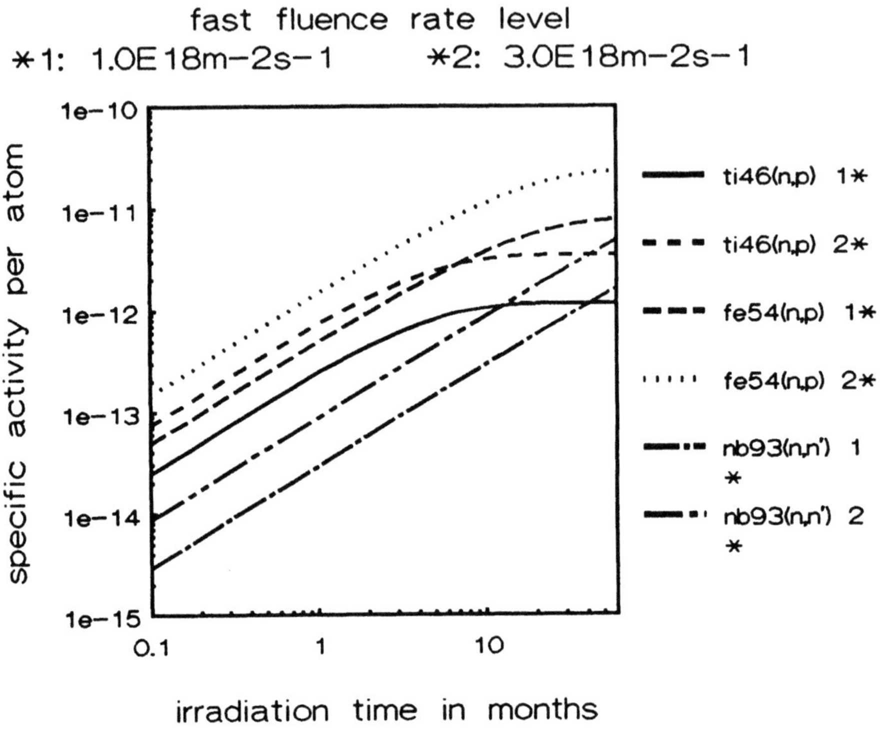


Fig. 4. Activity increase for fast neutrons.

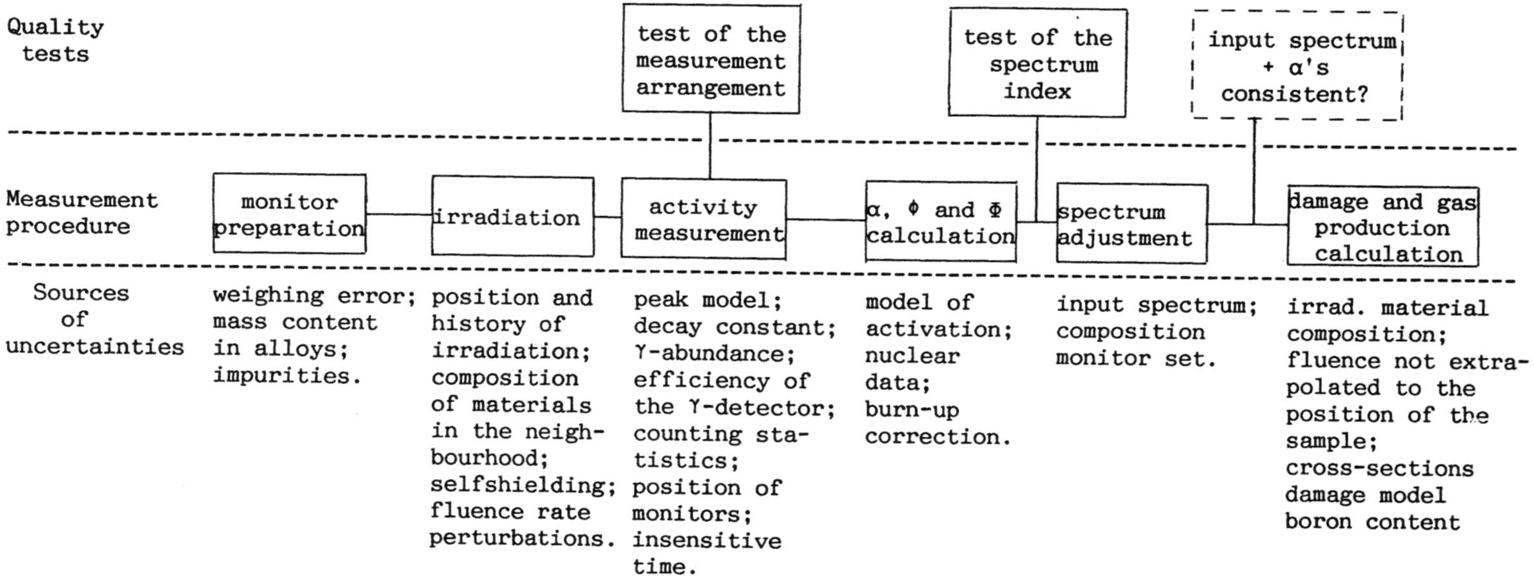


$\alpha$: saturation activity per target atom
$\phi$: neutron fluence rate
$\Phi$: neutron fluence

Fig. 5. Scheme of the measurement procedure

# HIGH DOSE HIGH TEMPERATURE DOSIMETRY USING RADIOTHERMOLUMINESCENT MATERIALS

C. HICKMAN, J. BARTHE and G. PORTAL

Commissariat à l'énergie atomique, IPSN/DPHD/SIDR

B.P. 6, 92265 Fontenay-aux-Roses Cedex, FRANCE

## ABSTRACT

A means for performing in-situ dosimetry in intense radiation fields at high temperature (e.g. inside a nuclear reactor at temperatures up to 400°C or in the vicinity of highly active nuclear wastes, at temperatures up to 300°C) using thermoluminescent (RTL) dosimetry is proposed. Glow curves corresponding to deep traps in Alumina ($Al_2O_3$) and Calcium Sulfate ($CaSO_4$ : Dy) have been studied. Irradiations performed at above ambient temperature can lead to signal enhancement. Such characteristics have been determined and accurate dose determinations performed.

## 1. INTRODUCTION

It is necessary for safety reasons to be able to measure radiation dose rates and integrated doses in intense radiation fields at high temperatures. Examples are :

- 1) The safe handling, transport and final disposal of highly active radioactive wastes.

- 2) Nuclear reactors for which it is necessary to know the behavior of structural and confining materials in order to establish guidelines for their safe operating lives.

Today, many countries, and in particular, the USA, Belgium, Germany, UK and France are studying, in the framework of their radioactive waste disposal programs, the behaviors of disposal materials and media under real conditions.

As the DPHD/SIDR has extensively studied deep traps suitable for high temperature/high dose dosimetry (1) and also has had successful practical experience in carrying out dosimetry in hot reactor environments (Phenix, St. Laurent des Eaux etc,) we were asked to participate in the HAW experiment (Highly Active Waste) in the ASSE salt mine near Brunswick in Germany. In this experiment, the effects of the high dose rates and elevated temperatures generated by highly active radioactive wastes on salt and confining materials are studied. Steel canisters containing controlled amounts of homogeneously distributed cesium and strontium are used to simulate real wastes. The DPHD/SIDR has been given the responsibility of accurately evaluating the corresponding dose rates in intense radiation fields (up to $10^3$ Gy $h^{-1}$) at temperatures up to 250°C.

## 2. THERMOLUMINESCENCE (TL) METHODS EMPLOYED BY THE DPHD/SIDR TO MEASURE HIGH DOSES

Although attempts have been made by our laboratory to use Electron Spin Resonance (quartz and $SiO_2$) and Optical Absorption (single crystals of lithium fluoride) for high dose/high temperature dosimetry, it was found difficult to take temperature effects into consideration and such studies have been postponed, at least for the time being, in favor of thermoluminescence. Such studies have however been continued in Germany (Regulla and coworkers GSF Munich (2,3)).

### 2.1. Ambient temperatures

Various different dosimetric products can be used for high dose dosimetry. We restrict our present discussions to $Al_2O_3$ (dosimetry peak : 270°C) [see Figure 1] and Ca $SO_4$ : Dy (dosimetry peak : 230°C) to be used in the ASSE experiment.

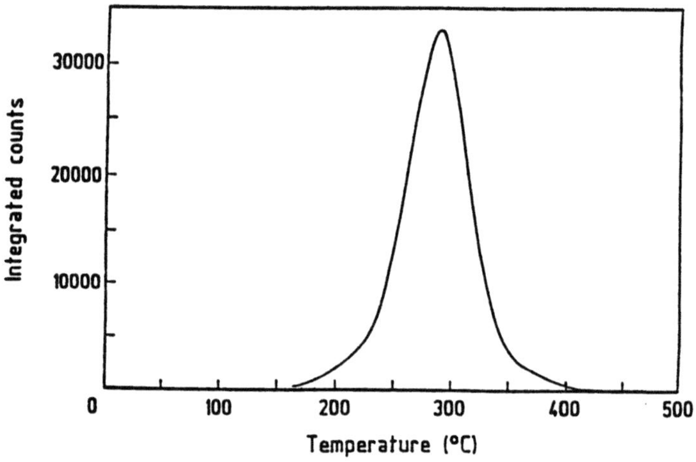


**Figure 1 : Alumina (classical dosimetry peak).**

### 2.2. High temperatures

Two competing phenomena intervene when integrated doses at temperatures higher than about 50°C are to be measured :

- THERMAL FADING,
- TISE (Temperature Induced Signal Enhancement).

a) THERMAL FADING is the term given to the disapperance of the integrated signal due to recombination effects.

From the Randall and Wilkins theory : the frequency factor corresponding to temperature induced de-trapping is given by the equation :

$$S(T) = So \exp(- E/kT)$$

b) TISE is the increased radiation sensitivity (enhanced signal) observed when certain dosimetry products are irradiated at high temperatures.

Alumina : $Al_2O_3$

Figure 2 shows the high temperature peak of $Al_2O_3$. This peak located at a relatively high temperature undergoes little thermal fading until quite high temperatures are reached (300°C) : the calculated theoretical lifetime of this peak at 300°C is 9.1 x $10^2$ years (4).

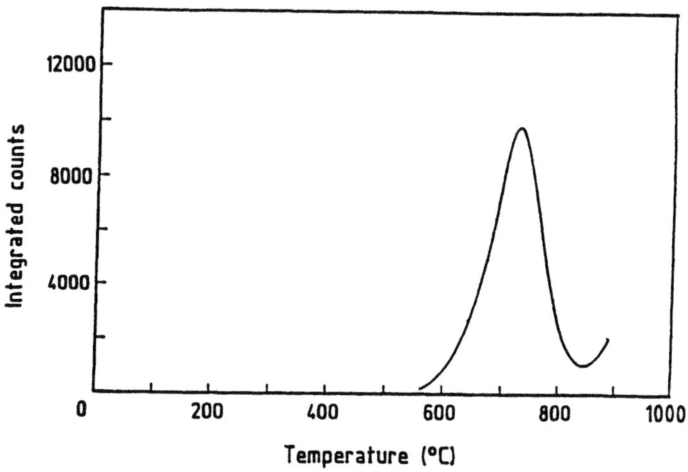


Figure 2 : Alumina (high temperature peak).

The spectral emission of $Al_2O_3$ is a very broad band extending from below 300 nm up to, beyond 700 nm. In order to measure the high temperature glow curve of this product, it is best to observe emission in the ultraviolet and to filter out as much as possible of the black body emission.

Figure 3 shows how the integrated glow curve signal varies with temperature for different integrated doses.

Figure 4 shows this phenomenon ; as the percentage signal enhancement has been measured at different temperatures up to 300°C, doses can be established within this temperature range, provided that the irradiation temperature is known.

Figure 3 : $Al_2O_3$ irradiated at different temperatures (high temperature peak : ≃ 700°C).

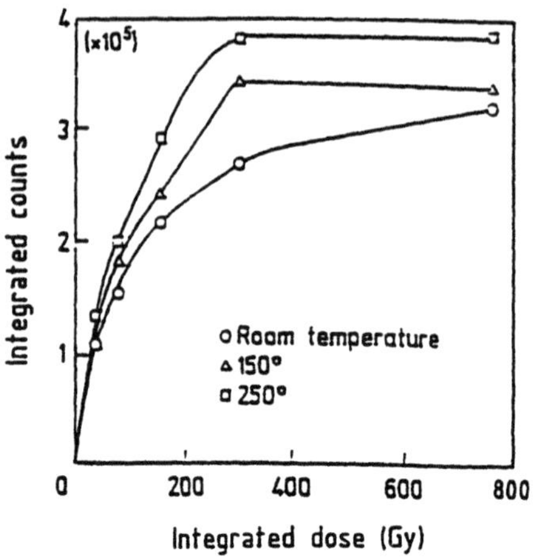


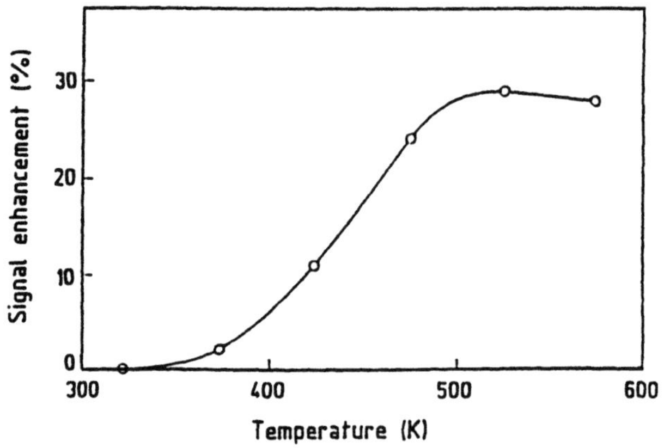


Figure 4 : $Al_2O_3$ percentage increase in signal in comparison with the signal (integrated counts) measured at 50°C. (Integrated dose ≃ 200 Gy).

## Dysprosium doped calcium sulfate $CaSO_4$ : Dy

Figure 5 shows the high temperature peak of $CaSO_4$ : Dy. This peak exhibits similar thermal fading (4) and TISE effects to those of the high temperature peak of $Al_2O_3$. Unlike the broad band spectral emission of $Al_2O_3$, this product, exhibits an emission in the form of relatively narrow bands. Bands centered about 482, 572 and 654 nm (Fig. 6) are normally used for classical dosimetry glow curve measurements. The theory and properties of emissions at these wavelengths are well understood (5).

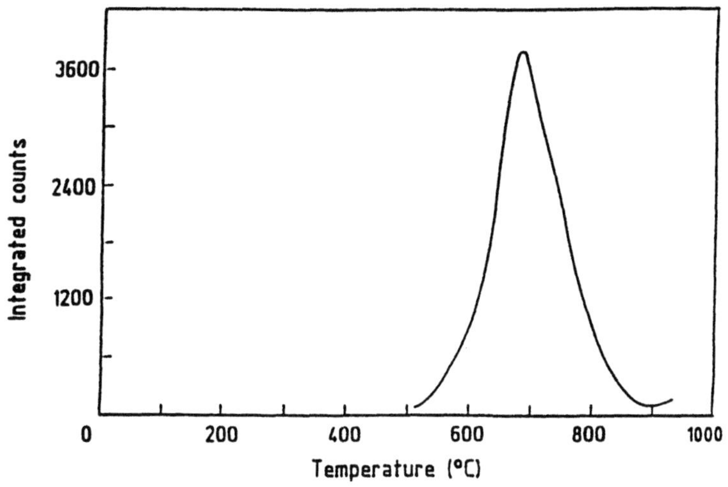


Figure 5 : Calcium sulfate $CaSO_4$ : Dy (high temperature peak).

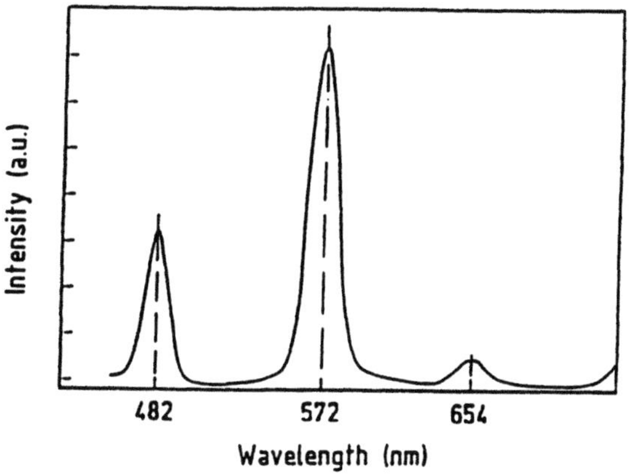


Figure 6 : $CaSO_4$ : Dy, wavelengths used for classical dosimetry peak. (Measured with a multichannel optical analyzer "DILOR").

Emissions at these wavelengths cannot however be used for the high temperature peak because of the copious black body emission camouflaging the TL signal. It is for this reason that studies were undertaken by the DPHD/SIDR to discover whether any emission peaks could be found in the ultraviolet. A "Dilor" multichannel optical analyzer was used for these measurements. Figure 7 shows that emission peaks do in fact exist in the ultraviolet. Accurate high temperature, high dose measurements can be made by isolating these peaks from the black body emission. Figure 8 shows how the TL signal (integrated counts) varies with integrated dose.

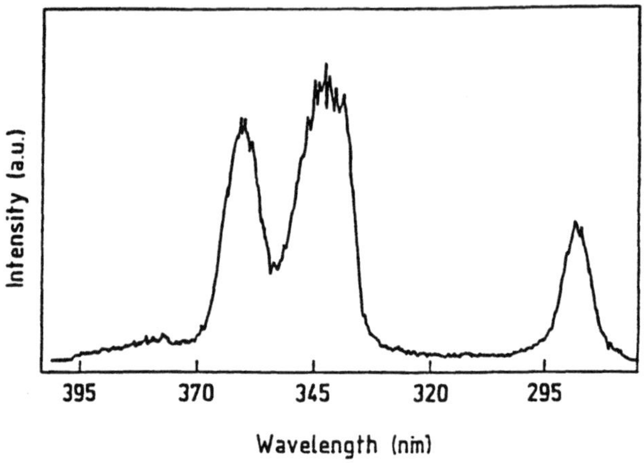


Figure 7 : $CaSO_4$ : Dy, wavelengths used for high temperature peak. (Measured with a multichannel optical analyzer "DILOR").

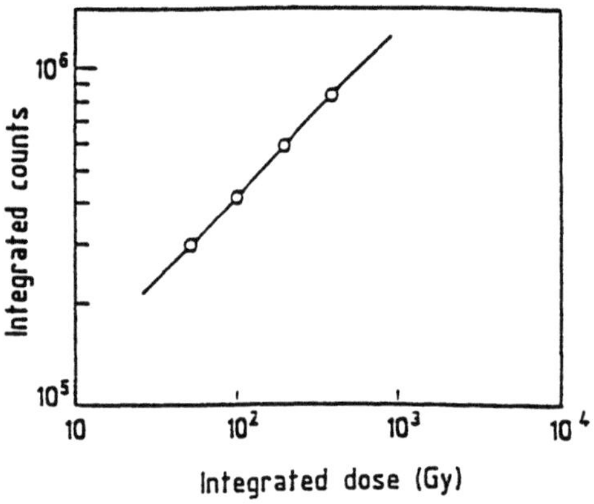


Figure 8 : $CaSO_4$ : Dy, signal/dose relation (high temperature peak ≃ 700°C).

## 3. CONCLUSIONS

Both $Al_2O_3$ and $CaSO_4$ : Dy are useful materials for high dose/high temperature dosimetry.

Although $CaSO_4$ is one of the earliest (1951 Watanabe) and most widely studied TL materials, it still possesses useful dosimetric properties that have not yet been exploited.

Acknowledgments

The authors would like to express their gratitude to C. Moulin and P. Mauchien of the "Division d'Etudes de Retraitement et de Déchets Radioactifs, Section d'Etudes Analytique du Cycle Combustible", CEN. Fontenay-aux-Roses and to E. Da Silva and J.P. Caron of the DILOR Company (Lille) for their help and advice on optical measurements.

They are also wish to thank M. Raynal and J.M. Palut of ANDRA (Agence Nationale pour la Gestion des Déchets Radioactifs) for facilitating and encouraging the work undertaken in the framework of the HAW project.

## REFERENCES

1. Very deep traps in $Al_2O_3$ and $CaSO_4$ : Dy. Guy Portal, Serge Lorrain, G. Valladas Nuclear. Instruments and Methods 175 (1980) 12-14.

2. Ultra-High Level Dosimetry by ESR Spectroscopy of Crystalline and Glassy Quartz, A. Wieser, D.F. Regulla. GSF Munich, $9^{th}$ International Conference on Solid State Dosimetry, Vienna, 1989.

3. The use of LIF for Dosimetry in the Megagray Range and at Elevated Irradiation Temperatures. A. Waibel, H.Y. Göksu, D.F. Regulla. GSF Munich, $9^{th}$ International Conference on Solid State Dosimetry, Vienna 1989.

4. Etude et Développement de la Radiothermoluminescence. G. Portal Doctorial thesis, Toulouse 1978 and CEA report : CEA-R-4843.

5. Thermoluminescent Mechanisms of Gamma-Irradiated Calcium Sulfate Doped with Dysprosium. M. Morgan, Doctorial Thesis, University of Washington 1986.

# THERMAL DYNAMICS OF CALORIMETER SYSTEMS

J. A . Mason[1], N. Bainbridge[2] and P. J. H. Heffer[3]

1. Imperial College Reactor Centre, Silwood Park, Ascot, Berkshire, England.*

2. National Nuclear Corporation, Booths Hall, Knutsford, Cheshire, England.

3. Nuclear Electric Plc, Berkeley Nuclear Laboratories, Berkeley, Gloucestershire, England.

## ABSTRACT

In the measurement of radiation absorbed dose, indirect techniques such as activation detectors for neutron fields, and ionisation chambers and TLD's for gamma-ray fields, have the disadvantage that they require post-measurement analysis. In contrast, calorimetry which is a direct technique, utilizes the thermal energy generated within a sample material to obtain a measurement of the absolute radiation absorbed dose rate which the sample material has received. For absorbed dose measurement the successful application of calorimetric techniques in their various modes of operation depends crucially on a thorough understanding of the thermal dynamics of the measuring system. The present paper outlines the main results which have been obtained as a consequence of the use of computer modelling techniques in the analysis of calorimeter thermal dynamics. References are made to specific calorimeters operating under experimental and commercial plant conditions. Through a comparison of experimental results and computer simulations, a number of conclusions are drawn concerning the assessment and control of thermal dynamics in calorimeter systems.

## INTRODUCTION

The basis of calorimeter operation is the measurement and control of both heat-transfer and temperature as a means of determining energy deposition. The successful operation of a calorimeter system depends on an understanding of the operating mode of the device, the heat-transfer and nuclear characteristics of the constituent materials and the mechanisms of temperature control. These three aspects determine the calorimeter thermal dynamics.

* Presently on leave of absence seconded to A. N. Technology Ltd.

Calorimetry methods may be broadly divided into three types. In adiabatic calorimeters a baffle which surrounds the sample or measurement region is maintained at (and required to follow) the temperature of the sample. In this way no heat-transfer takes place between the sample and the baffle. As a result any change in sample temperature, which can be measured, is due solely to energy deposition in the sample or measurement region. In isothermal calorimetry the sample or measurement region is maintained at constant temperature. If additional thermal energy is delivered to the sample the controlling source of heat will need to be reduced if isothermal conditions are to be maintained. The required change in the controlling heat source, which can be determined, provides a measure of the unknown additional thermal energy.

Midway between these two extremes is the heat-flow or steady state calorimeter method. If additional thermal energy is delivered to a sample or measurement region, its temperature will rise until a temperature is reached at which the rate of energy delivery matches the rate of energy loss from the sample. A calibration can be established between the rate of thermal energy delivery in the sample and the sample temperature rise, allowing the unknown additional thermal energy to be determined. As the sample is often surrounded by a constant temperature region, this method of calorimeter operation is sometimes erroneously referred to as isothermal operation. The three different calorimeter operating modes lead to different forms of calorimeter thermal dynamic behaviour.

## HEAT-TRANSFER AND TEMPERATURE CONTROL

The heat-transfer characteristics of calorimeter components are governed in part by three heat-transfer mechanisms. The most common and most important of these is conduction which operates in liquids and gases as well as solids, although it is most important in solids where thermal conductivity is usually higher. Convection is only important in liquids and gases. Convective heat-transfer is essentially a function of Rayleigh Number (Ra) which describes the interaction that occurs between buoyancy and sheer forces in a fluid. It can be shown that for Ra $< 1000$ the temperature field across an air gap is identical to that where conduction alone applies. This is the situation in most cases of interest in calorimetry.

The third heat-transfer mechanism is radiative heat-transfer. It is generally less important in calorimetry than conduction although it is strongly temperature dependent. Also it is often difficult to accurately determine material emissivities. Theoretical values are notoriously unreliable and an empirical determination for a given calorimeter geometry and material surface finish is often impossible. All three of the mechanisms of heat-transfer are dependent on the existence of temperature differences and temperature gradients.

The thermal and nuclear properties of the constituent materials of a calorimeter also have a major influence on the thermal dynamic behaviour of the device. Perhaps the most significant of the thermal properties is the thermal defusivity of each of the calorimeter materials. This parameter is defined as the ratio of the thermal conductivity to the product of the density and the specific heat and it gives a measure of the rate of rise of temperature in a material with constant applied thermal energy or the rate of transmission of heat through a material with a constant temperature gradient.

The main significance of the nuclear properties of the constituent materials on calorimeter thermal dynamic behaviour lies in the differing thermal energy deposition rates (differing rates of kinetic energy released in materials or KERMA) for different materials exposed to the same radiation field. A further characteristic of radiation energy deposition in materials is the production of secondary radiation which may lead to the deposition of thermal energy at a distance from the site of the primary radiation interaction. This raises the issue of secondary radiation and charged particle equilibrium between different components in a calorimeter system.

The mechanisms and efficiency of temperature control in calorimeter operation have an important influence on calorimeter thermal dynamics. Temperature control is required in order to maintain calorimeter operating conditions. For example it is necessary for the maintenance of a 'zero' temperature difference between the sample and baffle (adiabatic conditions at the sample surface) in an adiabatic calorimeter. Similarly temperature control is required in sustaining constant temperature conditions (in the sample or measurement region) in isothermal calorimeters and (in the surrounding environment) in some steady state calorimeters. Temperature control is usually achieved by electrical heaters which pass through components of calorimeters or are wound around cylindrical component surfaces. In other cases it may be accomplished by passing cooling air over a heated surface or by the use of the Peltier effect.

## MODELLING OF CALORIMETER THERMAL DYNAMIC BEHAVIOR

In order to understand the heat-transfer characteristics and thermal dynamics of calorimeter systems, a variety of computer-aided modelling techniques have been developed and employed. Initially a simple electrical analogue model was used to study calorimeter dynamics including the operation of the temperature control algorithms in the case of adiabatic microcalorimeters. While providing considerable insight the model had serious limitations. First of all it was a two dimensional model which assumed that the calorimeter was of infinite length. End effects and especially end heat losses were ignored. The model also ignored the thermal resistance of calorimeter materials and the heat capacity of air gaps.

In order to overcome these modelling limitations a two dimensional finite difference heat conduction computer code "TEACH-C", (1) was implemented and calculations were performed in one-and two-dimensional (R-Z) geometry. Versions of the code specific to the calorimeter problem were devised and the temperature-control algorithms which are employed in the calorimeter control systems were incorporated into the model. Cases to simulate both radiation experiments and electrical calibrations were constructed.

Although in the adiabatic microcalorimeter and isothermal plutonium assay calorimeter problems convective heat-transfer is not important and can be ignored, radiative heat-transfer is significant, for example in cases where adiabatic jacket control is not maintained. In order to account for potential heat-losses through thermal radiation, this loss mechanism was included in the model as a correction to the conduction term. Before the code was put to extensive use the effects of varying both the mesh spacing and the convergence criterion were studied as part of the model validation process.

The model was also employed to investigate the behaviour of the temperature control algorithms and to establish optimal values of the gain parameters. In the case of adiabatic microcalorimeters a three term (proportional, integral,derivative) PID control model was found to be inadequate in correcting steady state temperature errors due to a lag in the response of the algorithm during periods of radiation heating or electrical calibration. The algorithm was replaced by one based on a (proportional, integral, integral) $PII^2$ control model. This new control algorithm was implemented in the heat-transfer computer code and new gain parameters were established.

## MEASUREMENTS AND MODEL RESULTS

The benefits of modelling calorimeter thermal dynamic behaviour can be illustrated with reference to three calorimeters types where the performance of each instrument type has been understood and improved as a result of what may be called thermal dynamics analysis. The first and second examples are prototype adiabatic microcalorimeters (2) and commercial steady state power reactor calorimeters (3) designed for the measurement of absorbed dose in reactor mixed radiation fields. The third type are isothermal plutonium assay calorimeters (4). Although these latter calorimeters are not used for absorbed dose measurements, the design concept and measurement principles are applicable to isothermal absorbed dose calorimeters.

The original Mk 2 Imperial College adiabatic microcalorimeters, which had samples made from graphite, aluminium, iron and high density polyethylene, suffered from systematic errors of between 5% and 10% in the measurement of radiation absorbed dose in a calibrated gamma ray facility. However, even more serious errors arose when attempts were made to calibrate the device electrically. Systematic errors of between 10% and 20% resulted from the electrical calibration procedure. The size of the errors varied from one calorimeter sample material to another .

In order to understand the problems the heat-transfer code was used to simulate calorimeter measurements and calibrations to study both heat-transfer effects and the behaviour of the temperature control algorithm. Early in the work the importance was recognised of the differences in heat-transfer that arise due to thermal energy generation from electrical calibration, which is localised, and volumetric thermal energy generation, which occurs with radiation heating. Related to this effect the modelling revealed the two main sources of measurement error. The first was found to be heat loss from the ends of the cylindrical sample and baffles where the temperature control heaters were shown to be relatively ineffective. The second source of error was the significant heat capacity of the PTFE sample supports which acted as a heat loss path from the sample. These two sources of error largely explain the deviation in the radiation measurement and electrical calibration results compared to the expected behaviour of a perfect adiabatic microcalorimeter.

The heat-transfer and control dynamic calorimeter model was then employed in the design of the improved Mk5 microcalorimeters. The code was used to identify inappropriate cases of heat-transfer from the sample in different proposed calorimeter designs and to examine the efficiency of the temperature control algorithms in maintaining adiabatic conditions around the surface of the sample.

Both the Mk2 and Mk5 calorimeter designs are shown in Fig. 1. In the Mk5 design the baffles which are used to maintain adiabatic conditions now completely surround the cylindrical sample. Also the sample is attached by a fine filament rather than by a fixed support. The results of an intercomparison for the worst case of electrical calibration for the graphite microcalorimeters are shown in fig. 2. A significant reduction in the measurement error for the Mk5 device can be seen. Some results of measurements for other microcalorimeters for both electrical and radiation heating are displayed in Table 1.

Table 1

| Material | Electrical Heating (μW/g) | Error Mk5 % | Error Mk2 % |
|---|---|---|---|
| | | | |
| Iron | 84.65 | -0.4 | -8.5 |
| | 20.15 | -1.1 | -10.9 |
| Aluminium | 84.47 | -1.1 | -14.7 |
| | 20.31 | -0.5 | -12.3 |
| Graphite | 88.87 | -2.5 | -21.3 |
| | 21.51 | -2.0 | -19.7 |
| | | | |
| Material | Radiation Heating (μW/g) | | |
| | | | |
| Iron* | 59.58 | -4.1 | -6.2 |
| | 15.77 | -5.2 | -6.2 |
| Aluminium | 64.54 | 0.9 | -3.1 |
| | 17.24 | 0.6 | -3.8 |
| Graphite | 65.83 | -0.9 | -4.5 |
| | 17.57 | -1.2 | -5.1 |

* Errors for the iron microcalorimeters are larger due to attenuation.

$$\text{Error} = \frac{\text{Experiment - Ideal}}{\text{Ideal}}$$

Shortcomings in the design of the early steady state calorimeters (3) employed in British Commercial Advanced Gas-Cooled Reactors (CAGRs) produced results that were inconsistent with the results of other dosimetry measurement techniques. Re-analysis of calorimeter performance by modelling the thermal dynamic behaviour revealed systematic errors in calibration of about 20 % relative to data obtained from other methods.

Three sources of error have come to light as a result of the re-analysis of the thermal and nuclear properties of the calorimeter components. The first source of error concerned the electrical calibration procedure where heat loss from the electrical calibration heater resulted in an over-estimation of the gamma ray dose measured by installed devices. The second source of error was associated with the steel baffle surrounding the graphite sample. In this case electron transport calculations (3) have shown that a dose enhancement of about 6% results from electron disequilibrium. This is in addition to the dose enhancement due to secondary radiation sources in the baffle (3). A third source of error is attributed to the mismatch of materials in particular the structural components in contact with the sample. The presence of materials in thermal contact with relatively high density and high KERMA rate will transfer thermal energy to the sample resulting in an overestimate of the dose rate.

A modified version of the same heat-transfer code has been used to model the thermal dynamic behaviour of an isothermal plutonium assay calorimeter. The model includes the temperature control algorithm which is based on a (proportional,integral) PI controller. The model has been used to study both the effect of design on heat-transfer and the efficiency of the control algorithm. It has been used to establish optimal values of the control gain parameters.

Two areas of calorimeter performance have been improved as a result of the modelling. In an early design the wall thickness of the aluminium measurement chamber (the sample region) was found to be too thin resulting in a significant axial

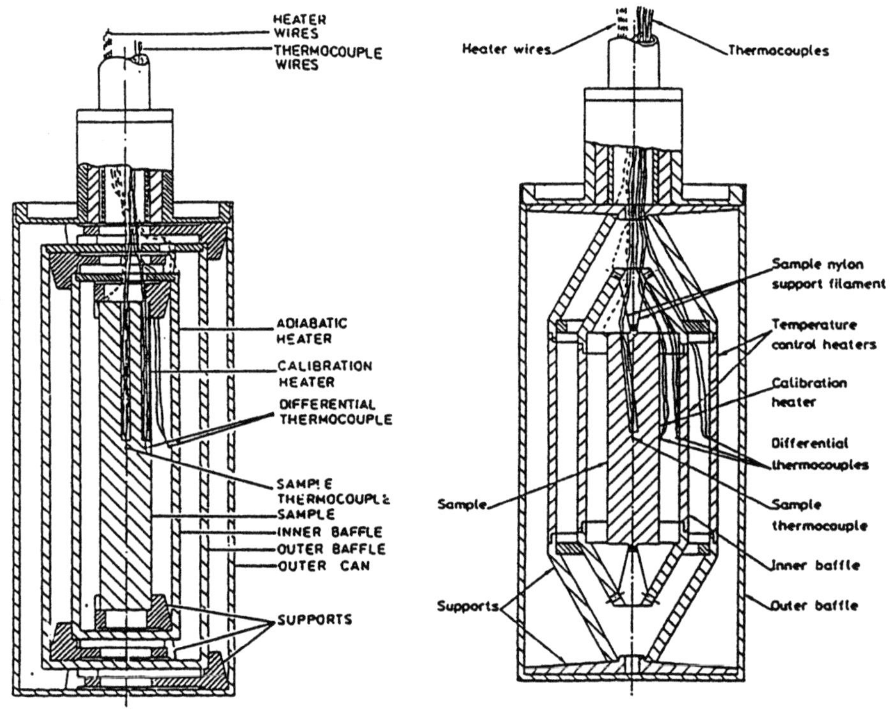


Mk.2 Adiabatic Microcalorimeter

Mk.5 Adiabatic Microcalorimeter

Fig.1

thermal resistance. A second problem was associated with high conductivity through support structures resulting in a higher than desired axial heat loss. Both problem areas have been addressed as a result of the information made available from the thermal dynamic analysis.

## CONCLUSIONS

The use of computer modelling techniques to study the thermal dynamics of calorimeter systems has been shown to be a useful tool both in the analysis of calorimeter measurement and calibration errors and in the design of calorimeters of all types. This paper has described the types of calorimeters used for absorbed dose measurement and their characteristics. The application of modelling of both the heat-transfer and temperature control has been illustrated using three types of calorimeters.

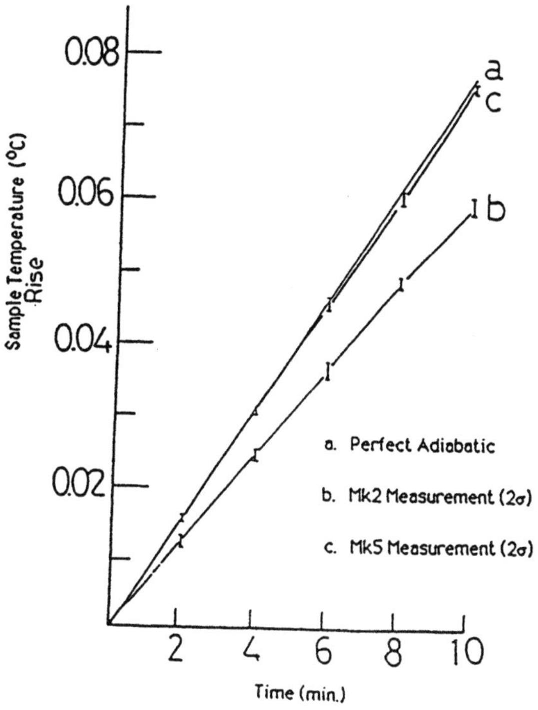


Fig. 2

Results from electrical and radiation heating measurements have been presented for the Mk2 and Mk5 adiabatic microcalorimeters. The results illustrate the effectiveness of the modelling approach in in exposing calorimeter measurement errors.

## REFERENCES


1. A.D. Gosman, B.E. Launder and G.J. Reece, *Computer Aided Engineering Heat-Transfer and Fluid Flow*, Ellis Horwood Ltd., Chichester, 1985.

2. N. Bainbridge, J.A. Mason and S.J. Peerless, Measurements and Model Predictions for Improved Microcalorimeter Design. Proc. 6th ASTM-EURATOM Symp. on Reactor Dosimetry, Jackson Hole, 31 May - 5 June 1987.

3. P.J.H. Heffer, Neutron and Gamma Ray Dose Studies in CAGR Instrumentation and Fuel Components. Proc. 5th ASTM-EURATOM Symp. on Reactor Dosimetry, Geesthacht, 24 -28 September 1984.

4. J.A. Mason, R.W. Wilde, J.C. Vickery, B.W. Hooton and G.M. Wells, Development and Evaluation of a Plutonium Assay Calorimeter Test-Bed, Proc. 29th INMM Annual Meeting, Las Vegas, 26-29 June 1988, 699-704.

# Intercomparison of gamma-ray dose rate measurement techniques and calculation results for several benchmark radiation fields

H. Aït Abderrahim, R. Menil, H. Geens
SCK/CEN, Mol, Belgium

ABSTRACT: In the framework of our activites in the LWR VENUS programme and the characterization of the Italian fast reactor TAPIRO, several spherical fast-neutron and gammaray benchmark fields were investigated in the large cavity located in the vertical thermal column of the BR1 reactor. The gamma-ray field is generated by a spherical shell of variable nature and thickness, surrounded or not surrounded by a natural uranium shell. The gamma dose rate was measured at the centre of each system by means of a graphite ionization chamber and $Al_2O_3$ and BeO thermoluminescent dosemeters (TLDs) under teflon.

Coupled neutron-gamma transport calculations were performed for five different systems; the theoretical results so-obtained were compared with experimental data. The agreement between the different techniques is within $\pm$ 10% whereas the calculation underestimates the gamma dose rate by 10 to 30% depending on the configuration considered.

KEY WORDS: gamma-ray dose rate, TLDs measurements, gamma calculation

## INTRODUCTION

In the framework of the pressure vessel surveillance programme the characterization of the ex-core gamma-ray field is of interest mainly for the following reasons :

- influence of the gamma heating on the temperature profile across the pressure vessel wall
- influence of the gamma heating on the temperature of metallurgical specimens in the PV surveillance capsules during irradiation

- contribution of the photoreaction to the response of some of the neutron dosemeters in the surveillance capsules.

The evaluation of these different contributions is mainly based on calculated gamma-ray fields. It is necessary to qualify the calculational methods used to establish these gamma-ray fields. Such qualification work was completed during the VENUS Engineering LWR Mock-up programme [1]. In support of this work five gamma-ray benchmark fields, with less complex geometry than the VENUS one, were analysed in the large cavity located in the vertical thermal column of the BR1 reactor. The gamma-ray field is generated by a spherical shell of variable nature and thickness, surrounded or not surrounded by a natural uranium shell. The characteristics of the analysed configurations are summarized in table 1.

TABLE 1 : Characteristics of the five configurations analysed

| N° | Configuration | Internal Shell | | $U_{nat}$ Shell | |
|---|---|---|---|---|---|
| | | Inner Radius (cm) | Outer Radius (cm) | Inner Radius (cm) | Outer Radius (cm) |
| 1 | 1 cm Unat | - | - | 16.25 | 17.25 |
| 2 | 6 cm Unat | - | - | 14.25 | 20.25 |
| 3 | 1 cm Unat + 7 cm Fe | 9.25 | 16.25 | 16.25 | 17.25 |
| 4 | 1 cm Unat + 9 cm Cu | 9.25 | 18.25 | 18.25 | 19.25 |
| 5 | 1.3 cm Fe | 9.25 | 10.55 | - | - |

The system is loaded into the centre of the large BR1 (1 m diameter) cavity, as shown in figure 1.

## EXPERIMENTAL RESULTS

The gamma dose rate was measured at the centre of each device loaded into the BR1 cavity using two different types of thermoluminescent dosemeters ($Al_2O_3$ and BeO) and a graphite ionization chamber. The dimensions of the TLD's used are 6 mm Ø x 0.9 mm for the $Al_2O_3$ dosemeters and 6 mm Ø x 0.7 mm for BeO ones. Five TLD's of each type were irradiated per configuration. The ten dosemeters were stacked in a cylindrical teflon pot (I.D. : 7mm and O.D. : 14 mm).

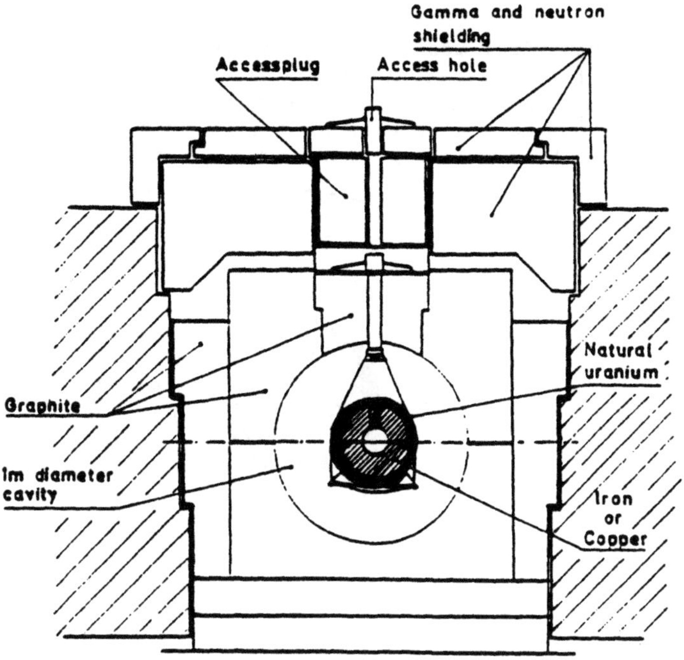


VERTICAL THERMAL NEUTRON GRAPHITE COLUMN

Fig.1. Experimental device for measurement of gamma-ray dose rate in the BR1 large cavity.

The teflon pot, hanging on a nylon wire, was introduced at the centre of the device through the access hole at the top of the BR1 cavity (see fig. 1) after first having stabilized the BR1 reactor power. Ten detectors of each type were simultaneously irradiated in a different but identical teflon pot positioned behind a calibration Co-60 gamma source. The TLD's used for the calibration were taken from the same preselected batch of dosemeters used for the measurements in the BR1 cavity. The graphite carbone dioxide- filled ionization chamber was also calibrated using the reference Co-60 gamma source before being used in the different configurations. This ionization chamber was provided by T.A. LEWIS [2] of CEGB (Central Electricity Generating Board). It was chosen for its very low thermal neutron sensitivity. The so-obtained experimental results were normalized to a reference BR1 power of 600 KW on the basis of the run-to-run monitoring results. A thermal neutron flux of $6.10^{8}$ $n/cm^{2}.s$ at the centre of the large empty BR1 cavity corresponds to the reference power of 600 KW. The normalized experimental results are given in table 3. The background due to decay gamma-ray intensity (at zero power) is negligible.

## CALCULATIONAL RESULTS

The 1-D spherical model, shown on figure 2, was adopted for the analysis of the five configurations. A graphite shell 30 cm thick around a 1 m diameter air cavity was considered together with successive spherical shells of : 1 cm Unat, 6 cm Unat, 1 cm Unat + 7 cm Fe, 1 cm Unat + 9 cm Cu and finally 1.3 cm Fe.

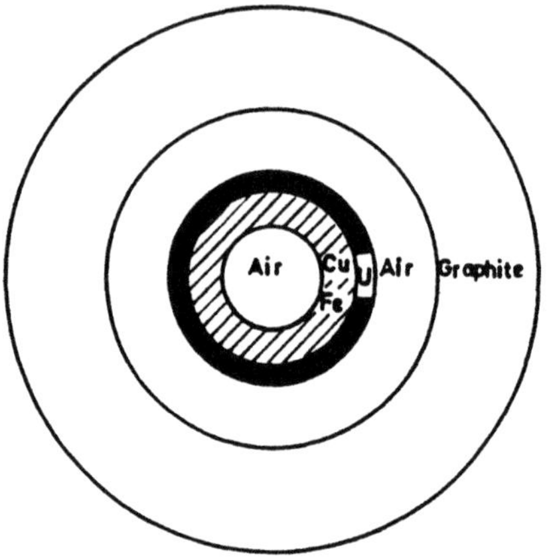


**Fig.2. The 1-D spherical model adopted for the theoretical analysis**

Firstly we calculated the neutron field distribution with the one-dimensional (1-D) discrete ordinate transport code DTF-4 [3], of the CEN/SCK MULCOS [4] code system, in a fixed source mode with S8P0 (transport diagonal corrected) approximations. The fixed source (of 1 n/s.cm$^3$) was distributed on a thermal maxwellian spectrum; it is introduced into the last mesh of the model (the most external mesh in the graphite). The MOL-BR2-40GR [5] 40 group (11 groups beneath the thermal cut-off) cross-section library was used for these calculations. Using the home-developed GAMPREP [6] programme with a 40 neutron group - 20 gamma group (n-$\gamma$) production (including inelastic gamma production) library (derived from the EURLIB-4 coupled n-$\gamma$ library [7]) and the 40 groups neutron flux map, calculated as described above, we prepared the capture (and fission) gamma source (space and energy dependent) as well as the macroscopic gamma interaction cross-sections. These 20 groups gamma source and cross-section sets were used as an input to a one-dimensional S8P5 gamma transport DTF-4 calculation with a fixed source mode to obtain the gamma flux distribution.

The gamma-ray dose rate at the centre of each configuration was calculated by weighting the computed gamma spectrum at these position with the group energy absorption mass attenuation coefficients, taken from reference [8], for the different TLD materials and air, multiplied by the mean energy of the considered group.

As the $Al_2O_3$ and BeO TLD's and the graphite ionization chamber, used for the gamma measurements have non-negligible fast and/or thermal neutron sensitivities, the calculated gamma dose rates (Cγ) must take the fast ($R_n^f$) and thermal neutron ($R_n^{th}$) contributions into consideration. The $R_n^f$ is calculated by multiplying the calculated fast neutron flux at the centre of the configuration ($\phi_n$(E>0.1 MeV)) by the fast neutron sensitivity of the detector considered. In the same way the $R_n^{th}$ contribution is obtained by multiplying the calculated thermal flux ($\phi_n$(E<0.414 eV)) by the thermal neutron sensitivity. Table 2 gives the values of the fast and thermal neutron sensitivities used for the $Al_2O_3$ and BeO TLD's and the graphite ionization chamber.

TABLE 2 : Fast and Thermal neutron sensitivities for $Al_2O_3$ and BeO TLDs and the graphite ionization chamber

| Detector | Fast neut. sensit. $S_n^f$ (R/10E+10 $n.cm^{-2}$) | Th. neut. sensit. $S_n^{th}$ (R/10E+10 $n.cm^{-2}$) |
|---|---|---|
| $Al_2O_3$ TLD | 0.36 [9] | 0.23 [10] |
| BeO TLD | 0.56 [9] | 0. [10] |
| Graphite Ion.Ch. | * [11] | 0.019 [11] |

* An integral fast neutron sensitivity is obtained by integrating the sensitivity function given in reference [11] weighted by the neutron spectrum computed at the centre of each configuration.

As the experimental results are given for a BR1 power level of 600 KW (corresponding to a thermal flux level of $6.10^8 n/cm^2s$ at the centre of the empty cavity), we decided to normalize our calculated results to this thermal flux level. As the thermal flux in the cavity is perturbed by the system introduced into it, we have to chose a normalization position where the perturbation of the thermal flux due to this system is known. We call this perturbation "Shadow Effect". The shadow effect was measured at such a position for a natural uranium shell of 1 cm thickness and for a 1 cm natural uranium shell filled with Cu. This position is situated in the graphite 72.5 cm from the centre of the cavity. The shadow effect factor, defined as the ratio of the non-perturbed to the perturbed thermal neutron flux, measured at the midplane level of the cavity is 1.040 for 1 cm Unat and 1.253 for 1 cm Unat filled with Cu. As the calculated gamma-ray dose rate Cγ, the fast $R_n^f$ and the thermal $R_n^{th}$ neutron contributions are all on the same scale (scale of the DTF-4 neutron run) the normalization adopted is :

$$C_{norm} = C \times \frac{\emptyset^{th}_{n,600KW}\ (72.5\ cm)}{F_{shadow} \times \emptyset^{th}_{n,DTF-4}\ (72.5\ cm)}$$

where $C_{norm}$ = is the calculated Cγ, $R_n^f$ or $R_n^{th}$ normalized to 600 KW power of BR1

$C$ = is the calculated Cγ, $R_n^f$ or $R_n^{th}$ on the scale of the DTF-4 run

$F_{shadow}$ = is the "Shadow Effect" factor (we took 1.04±0.03 for configurations 1 and 5, and 1.253±0.038 for configurations 2, 3 and 4)

$\emptyset^{th}_{n,DTF-4}$ (72.5 cm) = is the flux corresponding to the DTF-4 neutron calculations of the configuration considered (72.5 cm from the centre of the model) ($\Delta\emptyset/\emptyset$=7.%)

$\emptyset^{th}_{n,600KW}$ (72.5 cm) = is the thermal neutron flux corresponding to a DTF-4 neutron run of the empty cavity configuration at 72.5 cm from the centre of model, normalized in such a way to have a calculated thermal neutron flux of 6 $10^8$ n/cm².s at the centre of the model ($\Delta\emptyset/\emptyset$=3.%).

Table 3 summarizes the Cγ, $R_n^f$, $R_n^{th}$ values normalized to 600 KW, as well as the experimental values (E). It also gives the ratios of the energy absorption mass attenuation coefficient, of air and $Al_2O_3$ or BeO at the mean energy (1.25 MeV) of the Co-60 gammas. These ratios are given because the Cγ components have to be multiplied by this ratio, since the E values are expressed in Co-60 equivalent $R_{air}$.

## C/E RESULTS AND DISCUSSION

Examination of the C*/E values given in table 3 shows the consistency of these ratios between the different detectors for a given configuration; better than 3% except for configuration 2 (6 cm Unat), where the dispersion of the C*/E reach 9%. Now considering the ability of calculations to predict the gamma-ray dose rate, we can say that it is gratifying to evaluate the gamma dose rate with an underestimation less or equal to 20% for the three first configurations.

The 0.73 average C*/E value of configuration 4 is not surprising when we consider the incertainties in the neutron and gamma production copper cross-sections whereas it is surprising to find such a low C*/E value for configuration 5.

TABLE 3 : Comparison of the gamma-ray dose rates measured and computed in different gamma-ray fields generated in the BR1 cavity

| Detector | Config. N° | $C\gamma$ (R/h) | $\frac{(\mu/\rho)_{Air}}{(\mu/\rho)_{Det}}$ | $R_n^f$ (R/h) | $R_n^{th}$ (R/h) | E (R/h) | C*/E |
|---|---|---|---|---|---|---|---|
| $Al_2O_3$ | 1 | 939.08 | 1.0210 | 24.32 | 27.36 | 1059.7 | 0.95 |
| | 2 | 212.20 | 0.0210 | 42.57 | 1.96 | 319.4 | 0.82 |
| | 3 | 277.66 | 0.0210 | 16.74 | 0.52 | 323.8 | 0.93 |
| | 4 | 138.46 | 0.0210 | 16.47 | 0.12 | 214.1 | 0.74 |
| | 5 | 623.27 | 0.0210 | 0.00 | 26.67 | 781.4 | 0.85 |
| | 5 Rev. | 664.12 | 0.0210 | 0.00 | 26.67 | 781.4 | 0.90 |
| BeO | 1 | 914.74 | 1.0408 | 37.82 | 0.00 | 1062.8 | 0.93 |
| | 2 | 206.58 | 1.0408 | 66.23 | 0.00 | 309.8 | 0.91 |
| | 3 | 265.04 | 1.0408 | 26.04 | 0.00 | 335.5 | 0.90 |
| | 4 | 131.30 | 1.0408 | 25.62 | 0.00 | 211.8 | 0.77 |
| | 5 | 581.93 | 1.0408 | 0.00 | 0.00 | 774.1 | 0.78 |
| | 5 Rev. | 652.71 | 1.0408 | 0.00 | 0.00 | 774.1 | 0.88 |
| Graphite Ionizat. Chamber | 1 | 952.76 | 1.0000 | 92.02 | 2.38 | 1127.4 | 0.93 |
| | 2 | 215.16 | 1.0000 | 99.57 | 0.17 | 429.8 | 0.73 |
| | 3 | 276.71 | 1.0000 | 44.55 | 0.04 | 378.7 | 0.85 |
| | 4 | 137.21 | 1.0000 | 38.60 | 0.01 | 247.4 | 0.71 |
| | 5 | 609.30 | 1.0000 | 0.00 | 2.32 | 784.6 | 0.78 |
| | 5 Rev. | 650.15 | 1.0000 | 0.00 | 2.32 | 784.6 | 0.83 |

$$C^* = C\gamma \cdot \frac{(\mu/\rho)_{air}}{(\mu/\rho)_{Det}} + R_n^f + R_n^{th}$$

Hence we decided to revise the analysis of configuration 5. To check the validity of our modelisation, we calculated the gamma-ray dose rate in the empty cavity by using the same calculation scheme described above. We obtained a dose rate of 21.75 R/h for the ionization chamber, whereas it measures 80.9 R/h. This discrepancy is due to the insufficiency of the graphite thickness taken in our model. But if we increase this thickness we will increase dramatically the CPU time to converge our neutron problems. As the contribution of the graphite gamma-source is negligible for the other configurations, we revised only configuration 5. We proceeded as follows : we added to the gamma-ray dose rate calculated for configuration 5 the contribution of the graphite gamma source to the dose rate at the centre of the iron shell attenuated by a 1.3 cm thickness of iron. This contribution was calculated as follows :

$$C_{graphite} = (E_{empty\ cavity} - C^{*}_{empty\ cavity})\ K_{atten.}$$

where $K_{atten.}$ is the attenuation factor by calculating the ratio of the computed gamma dose rate at the centre of the empty cavity and the computed gamma dose rate at the centre of the cavity containing a 1.3 cm thickness iron shell sphere and where the gamma source was made zero. The results of this revision are given in table 3 and labelled configuration 5REV.

## REFERENCES


1. A. FABRY et al., "LWR-PV-SDIP : VENUS PWR Core Source and Azimuthal Lead Factor Experiments and Calculational Tests," BLG 586/NUREG CR-3323 report, Under Publication.
2. S.J. KITCHINGS and T.A. LEWIS, Nucl. Inst. and Meth. 227 (1984), 598-600.
3. K.D. LATHROP, "DTF-IV, a Fortran-IV Program for Solving the Multigroup Transport Equation with Anisotropic Scattering," LA-3373, July 1965.
4. G. MINSART, "MULCOS (MULtipurpose COde System) - General Organization of the Code System," CEN/SCK Internal Report, REACTOR STUDY GM/md 63-176/71-20, February 1971.
5. P. VANDEPLAS and J.-C. SCHEPERS, "A forty Group Cross-Section Library for the Calculation of Fast-Thermal Systems," CEN/SCK Internal Report, REACTOR STUDY PV/md 63-206/71-180, December 1971.
6. G. MINSART and L. FOUARGE, "GAMPREP, a Programme to prepare Input for Gamma Calculations," CEN/SCK Internal Report, REACTOR PHYSICS GM/LF/sa, 604/77)01/C, September 1977.
7. E. CAGLIOTI, "EURLIB-4 : Group Coupled Neutron and Gamma Data Library," Euratom, CCR Ispra, April 1978.
8. H. JOFFRE and L. PAGES, "Coefficients d'atténuation Massique et d'Absorption Massique en Energie pour les Photons de 10 keV à 10 MeV," Rapport CEA-R-3655 (1968).
9. H. AIT ABDERRAHIM, E.D. McGARRY and V. SPIEGEL, "Assessment of Fast Neutron Sensitivity of Thermoluminescent Gamma Dosimeters," This Conference.
10. T.A. LEWIS, "The thermal Neutron Response of Beryllium oxide and other Thermoluminescence Dosimeters," TPRD/B/0781/R86, May 1986.
11. S.J. KITCHING and T.A. LEWIS, "An Ionization Chamber for Measuring Gamma Exposure Rates in Mixed Radiation Fields," TPRD/B/0348/N83, October 1983.

# Neutron and Gamma Calculations for Future Irradiation Devices in the BR2 Materials Testing Reactor

Ch. De Raedt
SCK/CEN, Mol, Belgium

ABSTRACT: The present paper describes the analytical investigation of the performances to be expected from new BR2 irradiation devices for research in various fields :

- safety tests on single LMFBR fuel pins, in the framework of the study of melting phenomena during a protected control rod withdrawal incident;
- source term measurements performed on the MOL7C safety experiments for LMFBR fuel pin bundles with local blockage of the sodium coolant flow;
- further irradiations for the post-accident heat removal (PAHR) programme;
- irradiations of single PWR fuel rods, some with high initial burn-ups, with transients and partial fuel melting;
- long-term irradiations of PWR fuel rod bundles, up to very high burn-ups;
- irradiations of large samples of silicon in the outer H channels of BR2 and/or in the tangential beam tubes in the pool outside the reactor;
- isotope production in the H1 channel of the reactor, with a new beryllium plug.

KEYWORDS : BR2, test reactor irradiations for fast and thermal reactors, neutron and gamma calculations, fast breeder safety, post-accident heat removal, silicon doping.

## INTRODUCTION

The high flux materials testing reactor BR2 has been in operation for more than 25 years. Fig.1 gives a cross-section at the reactor mid-plane with a typical loading. Special features of BR2 are the very high fast and thermal neutron fluxes attainable, the possibility of spectrum tailoring, the versatility of the irradiation programmes and the large irradiation space available.

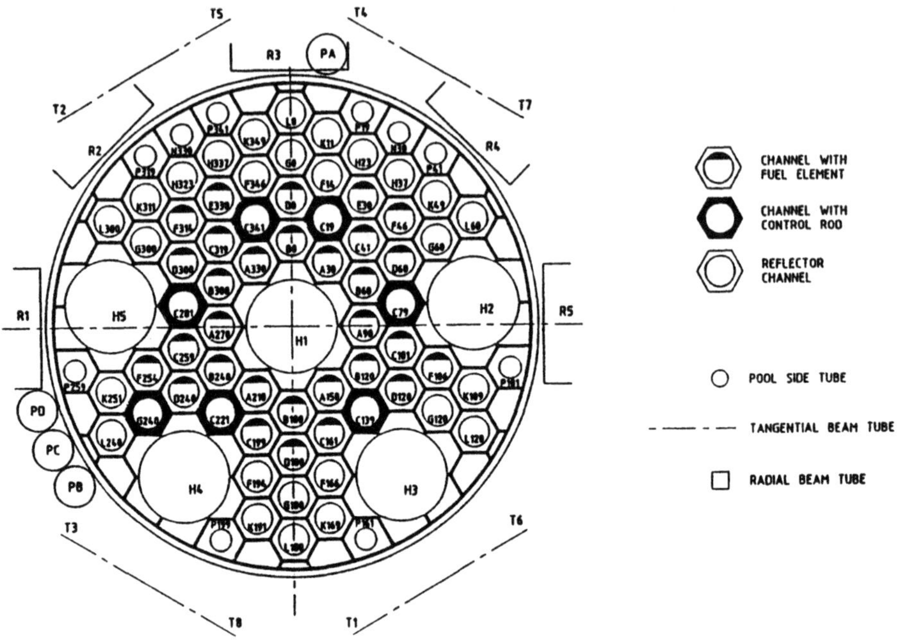


FIG. 1. Typical BR2 core configuration

In the present paper the analytical investigation of the performances obtained in, or to be expected for, some of the new irradiation devices, conceived for tests in a variety of fields, will be presented. Most calculations were performed with the calculation methods, codes and libraries already described in previous papers (see e.g. [1] to [5]).

## SAFETY TESTS ON SINGLE LMFBR FUEL PINS

Irradiations in BR2 have been contemplated in the framework of the study of fuel melting phenomena during a protected control rod withdrawal incident. These transient experiments would take place in the central H1 channel of BR2 : the irradiation programme concerned is called SOLISTE (SOdium Loop for Incidental and Safety Transients Experiments). Extremely high power ratings have to be attained in order to induce fuel pin cladding rupture, i.e. two or three times those of nominal operation according to the burn-up, the type of pellet (full or hollow), the cladding material, the power ramp rate, etc. In the same time a limited radial depression of the fission density in the fuel pin tested must be kept. Various device constructions were examined : some are shown on Fig.2.
In a first series of calculations, variants I, II and III were examined, containing a driver fuel element with 13 (type 213), 10

(type 210) and 6 (type 206) fuel plates, respectively, which enhances the epithermal neutron flux. In principle, such a configuration leads to a flatter radial distribution of the fission power throughout the fuel pin. With these variants, and specially when a cadmium screen is included in order to cut off the thermal component of the neutron flux spectrum, the 1500W/cm required could not be reached in a so-called VESPER1 fuel pin with fission products (for 12.1% burn-up : ID/OD 0.93mm/5.72mm; fuel linear density : 2.43g/cm; Pu/(U+Pu) : 20.3%), i.e. the least favourable fuel pin considered as to the achievement of the very high power requests mentioned above. However, 1500W/cm could be reached in some of the other types of test fuel pins.
In variant III the linear power increases (but the radial distribution becomes less flat) when one rethermalizes the neutron flux spectrum inside the cadmium screen, i.e. when one puts a larger percentage of water in the "Al plug + H2O" zone of Fig.2. Because of the strong rethermalization needed to enhance the linear power, the driver fuel element and Cd screen were no longer considered in the further studies.
A second series of calculations was then performed : in variant V the loop is surrounded by a massive beryllium plug, whereas in variant VI the free space around the loop is occupied by a mixture of aluminium and water (e.g. Al tubes with intermediate H2O layer). The characteristics calculated for variants III, V

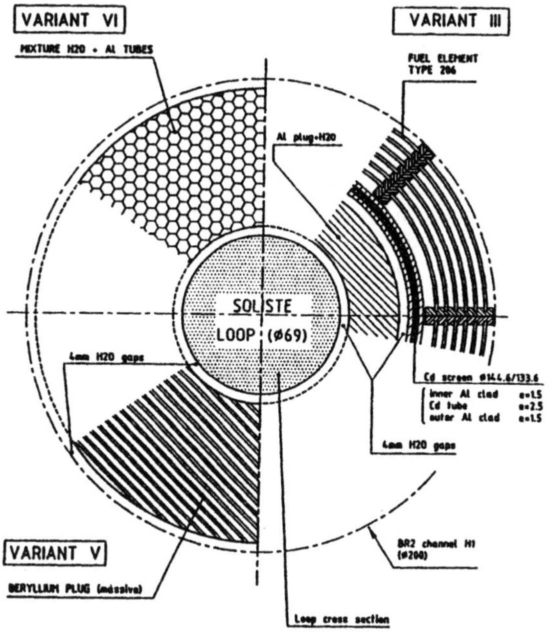


| Variant | Linear power (1)<br>W/cm | Power density depression in the fuel pin<br>outer/inner mesh |
|---|---|---|
| III | 1080 | 2.72 |
| V (2) | 1770 | 2.81 |
| VI (3) | 1530 | 2.85 |

(1) for max. allowable heat flux of 600W/cm2 in the BR2 fuel element plates
(2) subvariant G : 90 vol% H2O
10 vol% Al
(3) subvariant C : 40 vol% H2O
60 vol% Al

FIG. 2. SOLISTE in-pile section in H1 and calculated nuclear characteristics for VESPER1 and VI are summarized in Fig.2 for the irradiation of the VESPER1 fuel pin with fission products. The best results are achieved with variant V (Be plug), where almost 1800W/cm is predicted. When the fission products are omitted in the calculation (which physically corresponds to the situation obtained after a pre-irradiation in a thermal flux), still better characteristics can be obtained : the linear powers increase by about 40% and reach e.g. 2530W/cm for variant V; the power depressions decrease to about 1.78.

## SAFETY TESTS ON LMFBR FUEL PIN BUNDLES

Up to now, the MOL7C series of safety tests, in which the consequences of local cooling disturbances leading to severe fuel pin damage in fuel subassemblies are investigated, has consisted of seven experiments. MOL7C/6 and MOL7C/7, concerning the irradiation in 1988 of a bundle of 30 highly irradiated fuel pins and the irradiation in 1989 of a bundle of 19 thicker fresh fuel pins, respectively, are described e.g. in [6]. Source term measurements (STMs) were added to these tests for the additional investigation of the phenomena described hereunder.

In a hypothetical core disruptive accident, a fission product release to the atomosphere of the containment building of a LMFBR could occur by two different accident modes :

- a core melt-down followed by transfer of fission products and fuel to sodium and by thermal vessel failure;
- an energetic disruption followed by vessel head failure.

The study of the core melt-down accident requires the knowledge of the fuel-to-sodium transfer coefficients for several fission products. The whole fission product inventory is generally subdivided into seven groups, ranging from the noble gases to the lantanides. Transfer coefficients have been recommended by the Nuclear Energy Agency (NEA) [7], taking into account the degree of volatility and the chemical reactivity with sodium of the various fission products. Because the data base of these transfer coefficients is rather restricted, more experimental results are desirable.

The availability of under-water high resolution gamma spectrometry at SCK/CEN prompted the implementation of the last two MOL7C experiments carried out in BR2 with STMs. Details of the experiment are given in [8], together with the results obtained. A view of the experimental set-up is shown in Fig.3. In case of fuel pin failure in MOL7C, some of the fission products are circulated with the sodium coolant. As the upper part of the sodium circuit of the MOL7C in-pile section extends above the reactor top cover, fission product activity measurements can be made in front of the expansion tank, which forms the upper part of the loop. Activity measurements are made with a germanium

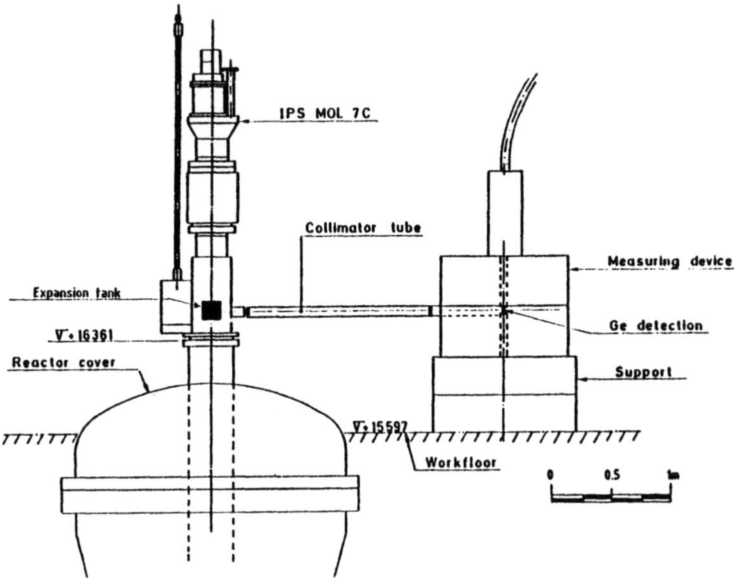


FIG. 3. Experimental set-up for STMs on MOL7C

detector, incorporated in an under-water measuring device which has a weight of about eight tons. This device has been conceived and used in the past for the gamma scanning of LWR fuel elements in fuel storage pools [9]. Between the detector and the MOL7C loop a collimator tube is installed.

Among others, there were experimental problems with obtaining acceptable counting rates, alignment of the collimators, and calibration of the whole detector system. Calculations were performed with the MERCURE4 code [10] to help establish the counting rates. The alignment was difficult because of the weight of the detector system and was effected by temperature changes and reactor vibrations. Due to the short time available during the two experiments, an intensive preparation was necessary each time.

The calculation part of the experiment consisted in the determination, with the aid of the ORIGEN2 code [11], of the fission product inventory in the fuel pins, taking into account their previous irradiation history and cooling time in the case of MOL7C/6 and the actual irradiation in BR2 (in the case of both MOL7C/6 and MOL7C/7) and the determination of the transport of the fission products from the failed fuel to the expansion tank.

From the comparison of the experimental with the calculated results it could be concluded that [8] :

- the STMs are reliable for MOL7C/6 and MOL7C/7;
- although not directly checking the transfer coefficients recommended by NEA, no fundamental disagreements were detected, at least for an instanteous source term release as considered in [7]. More information will be obtained when the actual

amounts of defect fuel will be determined by post-test examinations, planned for 1990-1991;
- more accurate results would need a more advanced analysis of the recorded spectra and a more advanced calculation model (amongst others, ORIGEN2 with updated nuclear data);
- better results would have been obtained if other measurement positions had been chosen (outlet pipe + gas plenum instead of expansion tank).

## SAFETY TESTS IN THE FRAMEWORK OF THE POST-ACCIDENT HEAT REMOVAL (PAHR) PROGRAMME

The second test in BR2 for this programme (PIRAMID2) (PAHR IRradiation According to a Mol Integrated Device) has been partly prepared. It differs from the first test (PIRAMID1) mainly by the presence of steel particles in the debris bed, the fuel particle granulometry, the bed porosity, the temperatures to be reached by, and hence the material used for, the crucible, the dimensions of the in-core tubes constituting the loop. The main analytical work performed was similar to that described in [5] for PIRAMID1.

## SINGLE LWR ROD IRRADIATIONS

A new series of single LWR rod irradiations is being carried out in the high burn-up chemistry (HBC) programme. Some of these irradiations comprise power transients with central fuel melting. On some of the irradiated rods, central temperature measurements will be carried out. Both fresh and pre-irradiated fuel rods are concerned in this programme. For these irradiations, the accurate knowledge of the neutron-induced fission heating and of the gamma heating is required, as one of the purposes of the programme consists in establishing the correlation among the thermal conductivity, the burn-up and the irradiation temperature. Calibration work among various measuring methods and between measurements and calculations is being pursued.

## LONG-TERM IRRADIATIONS OF PWR FUEL ROD BUNDLES UP TO VERY HIGH BURN-UPS

Several bundles of nine PWR fuel rods, in 3x3 arrays, are to be irradiated in pressurized water-cooled loops called CALLISTO (CApabiLity for Light water Irradiation in Steady-state and Transient Operation) (Fig.4), in standard channels in the BR2 reflector, up to very high burn-up values. To successfully achieve these irradiations, the following problems had to be solved :
- the adaptation, based on neutron calculations, of the neutron

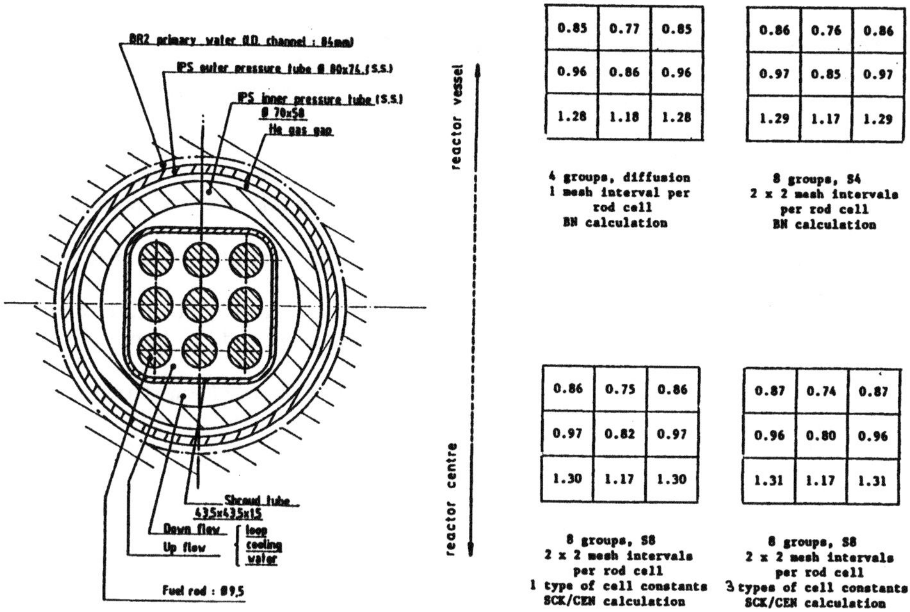


FIG. 4. CALLISTO in-pile section in a reflector channel and preliminary calculated relative power distributions

environment of the irradiation device in order to obtain test conditions, more specifically neutron spectra, as close as possible to the equilibrium conditions in actual LWRs;
- the adaptation of the flux and fission power levels;
- the obtention of the flattest possible radial fission density distribution throughout the fuel rod bundle, taking into account the initial enrichments of the fuel rods.

Because of the geometry of the loop cross-section and, still more, because of the excentered position each loop occupies in BR2, leading to an overall flux gradient, two-dimensional calculations have to be carried out to study the performances of the irradiation devices. As reported in [12], preliminary calculations, performed with the SCK/CEN system MULCOS (DTF-IV with proper library) - DOT3.5 [13...16] and with the Belgonucléaire (BN) system LWR-WIMS - TWOTRAN [17,18] led to practically the same distributions for a simplified two-dimensional model, except for the central fuel rod : see Fig.4.

A measurement campaign in BR2 at low flux was carried out in June 1990 to calibrate the calculation methods. The results are being analysed.

## SILICON IRRADIATIONS IN THE OUTER H CHANNELS OF BR2 AND/OR IN THE TANGENTIAL BEAM TUBES

The neutron transmutation doping of silicon in BR2 is under study. Both the large excentered H channels and the tangential beam tubes in the reactor pool (e.g. T2-T5) are possible irradiation positions (see Fig.1).

The main conditions to be fulfilled are :

- the presence of a sufficiently high thermal neutron flux inside the silicon target, to induce adequate doping within acceptable time limits;
- a flat doping rate throughout the silicon target, both radially and axially;
- a sufficiently low fast to thermal flux ratio, to limit the radiation damage, which after irradiation will have to be annealed;
- a limited temperature level throughout the silicon target and thus a limited heat production (caused mainly by the photons issuing from BR2 and from the device structural components), taking into account the cooling;
- adequate neutron flux and temperature monitoring systems;
- sufficient irradiation space.

Calculations were made to investigate these points for the H channel irradiations. For the structural parts of the irradiation device, aluminium seems to be the most adequate material for the in-pile section. Rotation of the device around its axis and up-and-down translation during the irradiation will ensure a proper distribution of the neutron flux throughout the silicon.

Targets of silicon with diameters up to 127mm (5") could be irradiated in the H channels while targets with still larger diameters could be irradiated in the tangential beam tubes, which each have an inner diameter of 305mm.

## ISOTOPE IRRADIATION IN THE CENTRAL H1 CHANNEL

In order to increase the performances of BR2 as to isotope production at very high thermal flux levels, a new beryllium plug was designed and manufactured for the central H1 channel : Fig.5. The introduction of a fuel element in the central hole of this plug has a very favourable effect on the reactivity and hence on the fuel economy of the reactor loadings. The adoption of this solution was preceded by neutron core calculations (for Be, $H_2O$ and fuel with various degrees of burn-up in the central hole). Measurements in BR2 confirmed the calculations : e.g. for {a Be plug / $H_2O$ / a SVIG fuel element with 13.7% mean burn-up} in the central hole,following reactivities($) were obtained : measured :

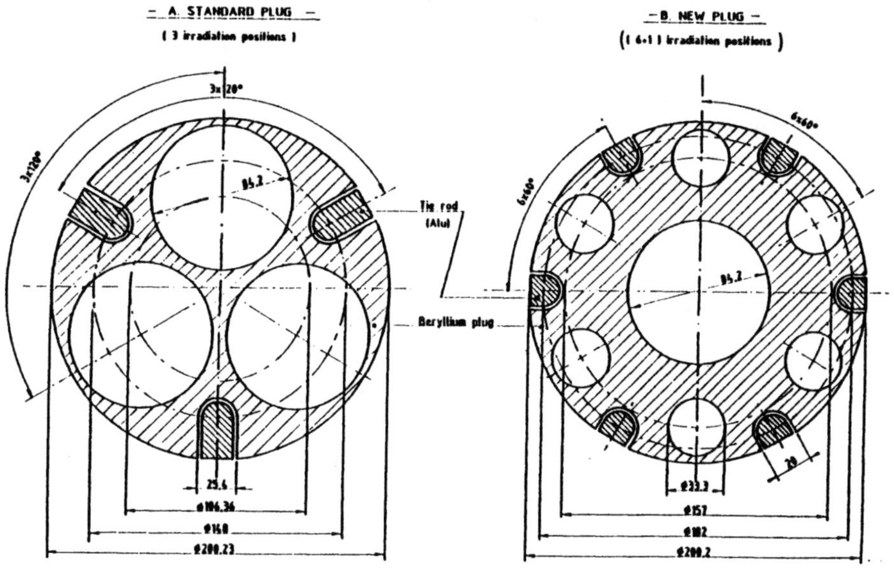


FIG. 5. Beryllium plugs for isotope production in H1

{0.0 (reference) / -2.05 / +5.24}; and calculated : {0.0 (reference) / -2.04...-2.06 / +4.8...+5.7}. Hot spot calculations were also carried out.

## ACKNOWLEDGEMENTS

Useful discussions with J.-M.Baugnet, S.Bodart, J.Dekeyser, L.Leenders and A.Verwimp are acknowledged.

## REFERENCES

1. J. DEBRUE, CH. DE RAEDT, N. MAENE. "Prediction of the Irradiation Characteristics of Fast Reactor Fuel Pins Tested in Cd-screened Loops in BR2." BLG 488, Feb. 1974.
2. J. DEBRUE, G. DE LEEUW-GIERTS, S. DE LEEUW, CH. DE RAEDT, A. FABRY, L. LEENDERS, N. MAENE, R. MENIL. "Dosimetry Work in Connection with Irradiations in the High Flux Materials Testing Reactor BR2." First ASTM-EURATOM symposium on reactor dosimetry. Petten, Sept.22-26, 1975.
3. CH. DE RAEDT. "Methods Used for Calculating Radial Power Distributions in Large Fuel Pin Bundles Irradiated in BR2. Comparison with Experimental Results." IAEA technical commitee meeting on homogenization methods in reactor physics. Lugano, Nov.13-15, 1978. IAEA-TECDOC-231, 1980.

4. CH. DE RAEDT, L. FOUARGE. "Calculation Methods Applied to BR2 Irradiations." Third ASTM-EURATOM symposium on reactor dosimetry. Ispra, Oct.1-5, 1979.
5. L. FOUARGE, P. D'HONDT, CH. DE RAEDT, J. DEBRUE, CH. JOLY. "Neutronic Study of a PAHR Irradiation Experiment in BR2." Sixth ASTM-EURATOM symposium on reactor dosimetry. Jackson Hole, Wyoming, May 31 - June 5, 1987.
6. H. BECKERS, G. GEBOERS, G. VANMASSENHOVE, A. VERWIMP, J. ABERLE, S. JACOBI, A. RAHN, K. SCHLEISIEK, L. SCHMIDT. "Experiments MOL7C 6 & 7. Design, Performance, Results and Preliminary Interpretation." Proceedings of the seminar on the commission contribution to reactor safety research. Varese, Nov.20-24, 1989.
7. "Nuclear Aerosols in Reactor Safety. A State-of-the-Art Report by a Group of Experts of the NEA Committee on the Safety of Nuclear Installations." Nuclear Energy Agency (OECD), June 1979.
8. L. LEENDERS, L. BORMS, Y. YANG. "Fission Product Transfer from Fuel to Sodium in MOL7C 6 & 7 Experiments." Proceedings of the seminar on the commission contribution to reactor safety research. Varese, Nov.20-24, 1989.
9. CH. DE RAEDT, J.-L. GENICOT, L. LEENDERS. "Nuclear Characterization of Spent Fuel by Non-Destructive Examinations at SCK/CEN." Third ASTM-EURATOM symposium on reactor dosimetry. Ispra, Oct.1-5, 1979.
10. C. DEVILLERS, C. DUPONT. "MERCURE4, un Programme de Monte Carlo à Trois Dimensions pour l'Intégration de Noyaux Ponctuels d'Atténuation en Ligne Droite." Note CEA-N-1726 (1974) + updatings.
11. A.G. CROFF. "ORIGEN2. A Revised an Updated Version of the Oak Ridge Isotope Generation and Depletion Code." ORNL-5621, July 1980.
12. CH. DE RAEDT, A. CHARLIER. "Problems in Relation to the Irradiation of LWR Fuel Rod Bundles in Materials Testing Reactors." IAEA technical committee meeting on in-core fuel management. Vienna, Dec.4-7, 1989. IAEA-TECDOC-567, 1990.
13. G. MINSART, G. PEPERSTRAETE, G. VAN ROOSBROECK, J. DANIELS, D. CHRISTYN de RIBEAUCOURT, F. BOSMANS. Internal Reports SCK/CEN, Mol. 1971 ... 1976.
14. P. VANDEPLAS, J.-C. SCHEPERS. "A Forty Group Cross-Section Library for the Calculation of Fast-Thermal Systems." Internal Report SCK/CEN, Mol, Dec. 1971.
15. K.D. LATHROP. "DTF-IV, a FORTRAN-IV Program for Solving the Multigroup Transport Equation with Anisotropic Scattering." LA-3373, Nov. 1965.
16. W.A. RHOADES, F.R. MAYNATT. "DOT III, a Two-Dimensional Discrete Ordinates Transport Code." RSIC Computer Code Collection CCC-276. ORNL-TM-4280, Sept. 1973.
W.A. RHOADES. "Revision of DOT III. Version 3.5." Nov. 1975.
17. M.J. HALSALL. "LWR-WIMS, a Computer Code for LWR Lattice Calculations." AEEW-R 1498, June 1982.
18. K.D. LATHROP. "TWOTRAN - a FORTRAN Program for Two-Dimensional Transport." GA-8747, 1968.

# NEUTRON SPECTRUM STUDIES IN THE ATR†

JW Rogers, R. A. Anderl and M. H. Putnam

Idaho National Engineering Laboratory
EG&G Idaho, Inc., P. O. Box 1625
Idaho Falls, Idaho USA 84115-7111

## ABSTRACT

Neutron spectrum studies have been conducted in the fast and epithermal neutron energy ranges for the various irradiation test facilities and the routine neutron monitoring locations of the Advanced Test Reactor (ATR). These studies consisted of the measurements of neutron monitor reaction rates, calculations of neutron spectra, least-squares data adjustment analyses to obtain optimal characterization of the neutron energy spectra with specified uncertainties, determination of spectrum-averaged cross sections and calculations of neutron-induced displacement damage cross sections for these spectra. The results show a high degree of consistency between the measured reaction rates and the reaction rates calculated using the adjusted spectra. Significant reductions in the uncertainty specifications were obtained for the group and cumulative fluence rates. The spectral-averaged cross sections with reduced uncertainties provide information that allows "fast" neutron monitoring to be done at a higher accuracy.

## INTRODUCTION

The ATR at the Idaho National Engineering Laboratory (INEL) provides intense neutron irradiation fields for radiation on samples of reactor materials (fuel, cladding, control and structural materials) and for production of radionuclides used in industrial and medical applications as well as for scientific research. Routine monitoring of the "thermal" and "fast" neutron levels have been conducted during

†Work performed under U. S. Department of Energy Contract No. DE-AC07-76ID01570.

every operational cycle since its startup in 1970. This routine neutron dosimetry has been primarily accomplished using the $^{59}Co(n,\gamma)^{60}C$ reaction for "thermal" neutrons and the $^{58}Ni(n,p)^{58}Co$ reaction for "fas

neutrons as described in ASTM standard methods E261, E262 and E264.[1] Recently more detailed neutron spectrum studies were conducted for various irradiation locations in the ATR. After a brief description of the ATR, the results of these studies are presented here.

## DESCRIPTION OF THE ATR

Figure 1 provides a cross-sectional view of the uniquely designed ATR core that consists of a serpentine fuel assembly arrangement (four lobes) to allow instrumented in-pile loop experiments, irradiation holes for capsule experiments and safety and control device location. Fueled with forty plate type assemblies (enriched uranium in an aluminum cladding and matrix), the core has an active length of 1.22 meters and it is cooled with light water (75°C at a normal power of ~150 MW). The core is surrounded by a beryllium reflector and by hafnium control cylinders that are used primarily to adjust individual core lobe power. Hafnium fixed (neck) shims, located inside the fuel arrangement (in H-holes or in A-holes), are also used primarily to compensate for fuel burnup. The safety rod followers have holes (SR) in them where the neutron monitors are positioned during routine operation.

Capsule irradiation holes (A, H, B, SR) located in the core and reflector regions provide for sample irradiations at high neutron levels. Lower neutron level irradiation are provided by the I-holes and capsule irradiation tanks located outside the core-reflector tank on the north (ON) and south (OS) sides. Nominal peak fluence rates ($n/cm^2/sec$) per lobe MW in these capsule positions are as follows: A- and H-holes (thermal - 7.5E12, fast - 8.0E12), B-holes (thermal - 1.5E13, fast 5.0E12), SR-holes (thermal - 9.0E12, fast - 7.5E12), I-holes (thermal - 6.5E11, fast - 9.5E10) and OS-5 (thermal - 3.0E10, fast 8.0E09). Here thermal corresponds to equivalent 2200 m/s neutrons and fast corresponds to neutrons with energies >1 MeV. The nominal thermal neutron reaction rate fractions as measured for the bare and cadmium covered Co in the ATR locations, are as follows:

| | | |
|---|---|---|
| H-holes - 0.75 ± 0.01 | Inner B-holes - 0.84 ± 0.02 | I-holes - 0.98 ± 0.01 |
| A-holes - 0.75 ± 0.01 | Outer B-holes - 0.91 ± 0.03 | OS-holes - 0.98 ± 0.01 |
| SR-holes - 0.76 ± 0.01 | | |

## DOSIMETER ACTIVATION

Because the ATR normal operational cycles are too high in power and too long in duration to allow optimum exposure and measurement of the neutron monitors desirable for a detailed spectral analyses, a 10 MW - 8 hour irradiation was performed. The lobe powers were divided as follows: LNW = 1.81 MW, LNE = 1.82 MW, LC = 2.17 MW, LSW = 2.15 MW, LSE = 2.05 MW based on the $^{16}N$ lobe power monitors. Neutron monitors were centered at the reactor midplane elevation in the following locations (see Figure 1): [H-2, H-10, H-14], [A-1, A-10, A-12], [SE(SR)-1, SE(SR)-3], [B-4, B-8, B-11], [I-6, I-8, I-22], OS-5.

The neutron monitor packages consisted of sensor materials encapsulated in cadmium tubes with 7.2-cm outside length, 0.24-cm outside diameter, 0.14-cm inside diameter, 6.0-cm inside length and 0.6-cm crimped ends. The selection, design, irradiation and handling of the sensor materials and encapsulations were based on the standard guidelines[1] as described in ASTM E844. The sensor materials were $^{235}U$, Sc, Au, Co, $^{237}Np$, $^{238}U$, Ni, Fe, Ti, Cu and Al with the fission monitors encapsulated in vanadium.

The neutron-induced radioactivities for each sensor were measured by gamma-ray spectrometry using germanium detector based spectrometry systems standardized and calibrated as described in ASTM Standard E181. Spectrometer detection efficiencies were calibrated with an estimated uncertainty of $\pm 2$ percent ($1\sigma$) throughout the energy range of the measurements. Reaction rates were derived from the measured activations using standard methods as described in ASTM E261 and ASTM E1005. Where they exist, ASTM standard methods were followed for specific types of monitors and reactions. The fission reaction rates were determined from the $^{95}Zr$, $^{103}Ru$, $^{140}Ba$ and $^{140}La$ fission products observed in each fission monitor. The total estimated uncertainties for the measured reaction rates were a combination of the detection efficiency, counting statistics, gamma-ray emission probability, fission yield and target atom uncertainty components. Uncertainties ranged from about 2 percent to 6 percent.

## MULTIGROUP SPECTRUM ANALYSIS

Multigroup characterization of the neutron spectra in the ATR was obtained by least-squares adjustment analyses using the FERRET[2,3] code as per ASTM Standard E944. In these analyses, adjustment of a priori (or input) estimates of the group fluence rates were made to achieve consistency between the measured reaction rates and the reaction rates calculated using the equation:

$$_jR_i = \sum_g {}_jC_g \; {}_i\sigma_g \; {}_j\phi_g$$

where

$_jR_i$ = the reaction rate for reaction i, spectrum j,

$_jC_g$ = group correction factor to account for neutron flux depression and shielding effects due to the use of cadmium covered dosimeters, spectrum j,

$_i\sigma_g$ = infinite dilute group cross section for reaction, i,

$_j\phi_g$ = neutron fluence rate group values for the spectrum identified as spectrum j.

The analyses were made in a group structure employing 51 groups with group energy boundaries chosen to cover in some detail the major structure in the dosimeter cross sections and in the neutron spectra. Input estimates of the 51-group fluence rates, $_j\phi_g$, for each core location were based on a transport calculation of the neutron spectra in the H-hole region of the ATR. Group correction factors, $_jC_g$, were obtained from calculated spectra corresponding to "cadmium-covered" and "no-cadmium" conditions. Infinite-dilute group cross sections, $_j\sigma_g$, were obtained by processing 620-group representations of the dosimeter cross sections, as specified in the ENDF/B-V Dosimetry file,[4] into 51-group cross sections using reactor-spectrum weighting. Least-squares-adjustment analyses of the type performed here also required a detailed prescription of the covariances for the input fluence rates and cross sections. For all of the dosimeter reactions, except for $^{45}$Sc(n,$\gamma$) and $^{59}$Co(n,$\gamma$), the PUFF code[5] was used to process ENDF/B-V covariance files and to generate the appropriate 51-group covariance matrices. Because the ENDF/B-V covariance files for $^{45}$Sc(n,$\gamma$) and $^{59}$Co(n,$\gamma$) were incomplete, a covariance prescription provided by Schmittroth[6] was used. The covariance prescription for the input multigroup/neutron spectra consisted of a fractional normalization uncertainty of 10% and a short-range component in which groupwise uncertainties of 20% were correlated by a Gaussian distribution.[2,3] This prescription assured that the final group fluence rate adjustments were largely determined by the relatively accurate measured dosimeter reaction rates and cross sections.

The calculated neutron spectra for the H-hole region were obtained using one-dimensional $S_6$ transport theory and the SCAMP computer code.[7] To obtain these spectra, various parameters in the SCAMP model were adjusted in an attempt to match the 73-group SCAMP spectral results to the 4-group spectra generated from a PDQ-7 two-dimensional diffusion code[8] calculation. A spline interpolation scheme was used to generate the 51-group input fluence rate spectra from the 73-group SCAMP calculation. The 51-group spectra were modified further by adjusting the spectrum normalization with scale factors based on the $^{235}$U fission rates in the various measurement locations.

Knowledge of the various cross-section group responses are important to identifying those energy regions of the neutron spectra that are most sensitive to the specific dosimeter reactions, i.e., to identifying those regions of the neutron spectra that will be adjusted and better defined as a result of the FERRET analysis. A comparison of the fractional group responses is shown in Figure 2 for the low-energy, resonance dosimeters and in Figure 3 for several high-energy threshold dosimeters. It is important to note from Figure 3 that the threshold dosimeters provide good overlapping responses over the energy range from ~0.5 MeV to greater than 10 MeV. This means that the least-squares-adjustment analysis should provide an improved characterization of this region of the input neutron spectrum, a region of particular significance to accurate prediction of displacement damage production due to fast neutrons. The information shown in Figure 2 illustrates significant narrow response due to resonances in the $^{197}Au(n,\gamma)$, $^{59}Co(n,\gamma)$ and $^{58}Fe(n,\gamma)$ reaction and a generally broad response for the $^{45}Sc(n,\gamma)$ and $^{235}U(n,f)$ reactions. These reactions should provide improved characterization of the input neutron spectrum from approximately 0.4 eV to $10^{-3}$ MeV. The region between $10^{-3}$ MeV and 0.5 MeV will not, in general, undergo significant improvement because no dosimeters have response in this region.

## RESULTS

Individual FERRET analyses were made for the reaction-rate data corresponding to the various reactor regions, namely, A, H, SR, B, I and OS-5. For each region, this meant a simultaneous adjustment of group-fluence rate values corresponding to the input spectra and the respective reaction-rate data for each location in that region, e.g., A-1, A-10 and A-12 for region A.

Complete detailed results of these analyses are not presented here, because of limited text space, and they will be compiled in a future publication. However representative FERRET results are illustrated in Figure 4 where the adjusted and unadjusted (input) group fluence rates are shown for position H-2. In the upper part of the figure, the group values are $n/cm^2$-s per unit lethargy. This comparison indicates that to achieve consistency between the measured and calculated reaction rates, the input group fluence rate values were adjusted upward by 15 to 30% in the low energy range from 0.4 eV to 0.1 MeV and by 3 to 5% in the high energy range above 0.1 MeV. For the H-hole case, initial (before FERRET adjustment) calculated-to-measured (C/M) reaction rate ratios ranged from 0.7 to 0.8 for the low energy dosimeters and 0.8 to 1.1 for the high energy dosimeters. After the FERRET adjustment, these C/M ratios were essentially 1.0 for the low energy dosimeters and between 0.97 and 1.03 for the high energy dosimeters.

The lower part of Figure 4 indicates a significant reduction in the uncertainty specification for the adjusted group fluence rates, especially in the energy ranges of dominant dosimeter response, e.g., by a factor of 2 to 6 in the energy range between 0.4 eV and $10^{-3}$ MeV and by as much as a factor of 3 in the fast neutron energy range 0.1 to 5.0 MeV. Similarly, the FERRET results indicate significant uncertainty reductions in the cumulative fluence rates, by factors of 6, 3 and 3 for the >1.0 MeV, >0.1 MeV and >0.4 eV energy ranges respectively.

A comparison of representative adjusted multigroup spectra from various reactor locations is shown in Figure 5. These spectra show how the high-energy and low-energy spectral regions change as going from an inner position (A-10) to an outer location (OS-5). It should be noted that the relative shape of the input spectrum was identical for all the FERRET analysis. Hence, the initial C/M reaction-rate ratios for the high-energy dosimeters were quite large for the B-11, I-holes and OS-5 location (ranging from a 3 to 6, typically). To achieve consistency between the measured and the calculated reaction rates, the FERRET adjustment resulted in a significant "softening" of the input neutron spectra for these locations as expected. The FERRET-fit C/M reaction-rate ratios were typically between 0.97 and 1.01 for the low-energy dosimeters and between 0.95 and 1.08 for the high-energy dosimeters for the B, I and OS locations. C/M comparisons and FERRET results for the A, SE, H and B-4, B-8 locations were very similar to those stated above for the H-2 case.

ATR reactor-relevant spectrum-averaged cross sections were computed using the adjusted spectra from the FERRET analysis. Representative values for H, B, I and OS locations are presented in Table 1. Cross sections for the low-energy reactions were computed by dividing the respective measured reaction rates by the >0.4 eV, cumulative adjusted fluence rate values and high-energy reaction cross sections were computed by dividing the respective measured reaction rates by the >0.1 MeV, cumulative adjusted fluence rates. Estimated uncertainties in these derived spectrum-averaged cross sections range from ~6% for the low-energy reactions to 7-10% for the high-energy reactions. Spectrum-averaged cross sections corresponding to >1 MeV can be derived for the high-energy reactions listed in Table 1 by multiplying the tabulated >0.1 MeV values by the factors 2.0, 2.7, 3.0 and 3.6 for locations H-2, B-11, I-6 and OS-5, respectively.

Finally, displacement damage integral cross sections were derived for Fe, Ni and Cr in the representative spectra of the ATR. These values, also listed in Table 1, were derived using ASTM damage cross section data obtained from the International Reactor Dosimetry File[9] and adjusted multigroup fluence rates from the FERRET analyses. Uncertainties estimated for these cross sections range from 7-10%.

## CONCLUSION

In this paper it is demonstrated how measured reaction-rate data for low-energy and high-energy threshold reactions have been used in least- squares-adjustment analyses with the FERRET code to obtain a significant improvement in the neutron spectral characterization for various irradiation locations in the ATR. Spectrum-averaged reaction cross sections were derived with uncertainties in the 6 to 10% range for the high response reactions. The displacement damage (dpa) spectrum-averaged cross sections derived in this work provide information not previously available for the prediction of damage to materials in the ATR. The results from these studies have produced information which allows neutron monitoring to be done at a higher accuracy than has previously been possible.

## REFERENCES


1. 1988 Annual Book of ASTM Standards, Volume 12.02, "Nuclear (11), Solar, and Geothermal Energy," American Society for Testing and Materials, Philadelphia, PA (1988).

2. F. Schmittroth, "FERRET Data Analysis Code," US Doe Report HEDL-TME 79-40, Hanford Engineering Development Laboratory (September 1979).

3. F. Schmittroth, Nucl. Sci. Eng., 72, (1979) 19-34.

4. R. Kinsey, "ENDF-102 Data Formats and Procedures for the Evaluated Nuclear Data File, ENDF," BNL-NCS-50496, (ENDF-102) 2nd Edition (ENDF/B-V), UC-80, (General Reactor Technology - TID-4500), Brookhaven National Laboratory, Upton, NY (October 1979).

5. "PUFF2 Determination of Multigroup Covariance Matrices from ENDF/B-V Uncertainty Files," RSIC Computer Code Collection, PSR-157 (1980).

6. Private Communication from F. Schmittroth, Hanford Engineering Development Laboratory (February 1982).

7. G. E. Putnam, TOPIC - A Fortran Program Calculating Transport of Particles in Cylinders, IDO-16968, April 1964. SCAMP is an undocumented internal $S_n$ transport code with one-dimensional slab, cylindrical, and spherical capabilities. In cylindrical form, it is a multigroup version of TOPIC.

8. C. J. Pfeifer, PDQ-7 Reference Manual II, WAPD-TM-947(2), 1971.

9. D. E. Cullen and P. K. McLaughlin, "The International Reactor Dosimetry File (IRDF-85)," IAEA-NDS-41, Rev. 1, IAEA Nuclear Data Section, Vienna, Austria (April 1985).

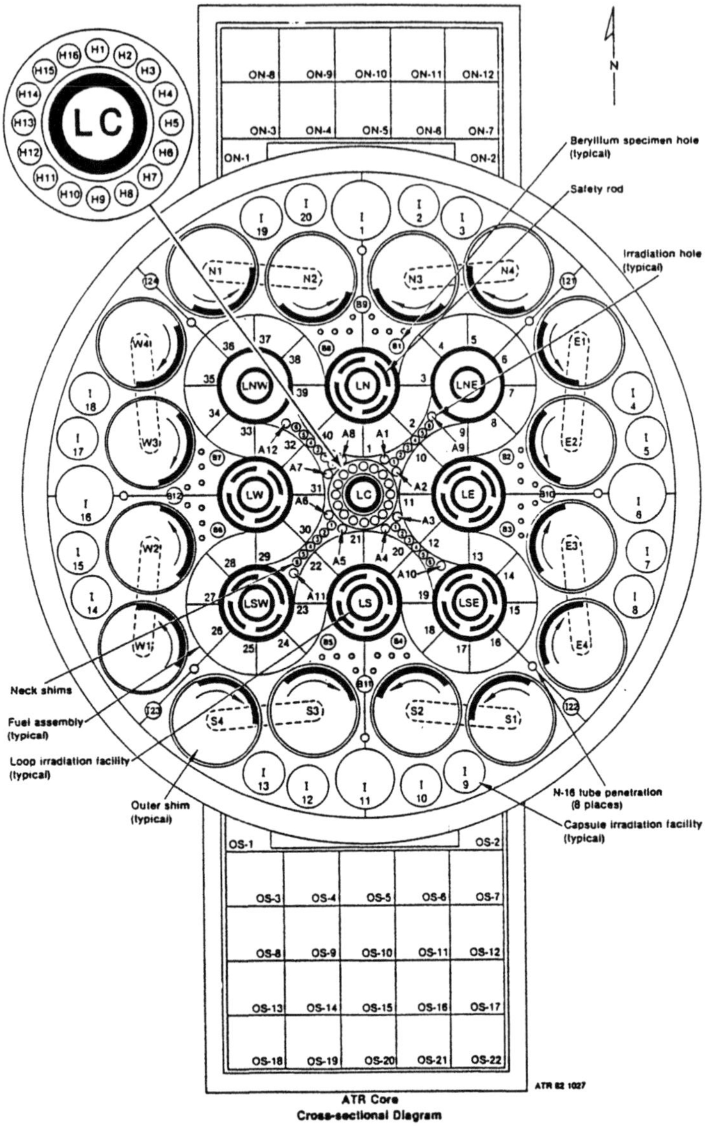


Figure 1. ATR core cross-section diagram.

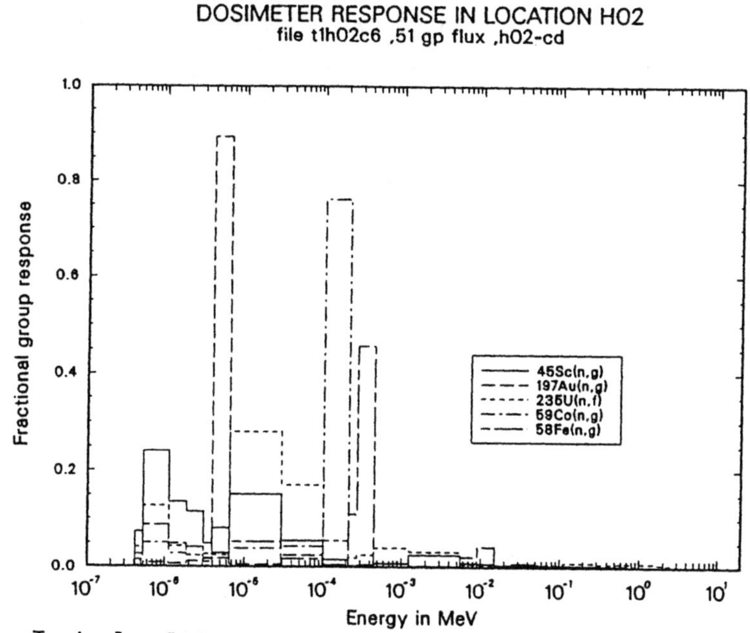


Figure 2. Typical multigroup response for low-energy dosimeter reactions.

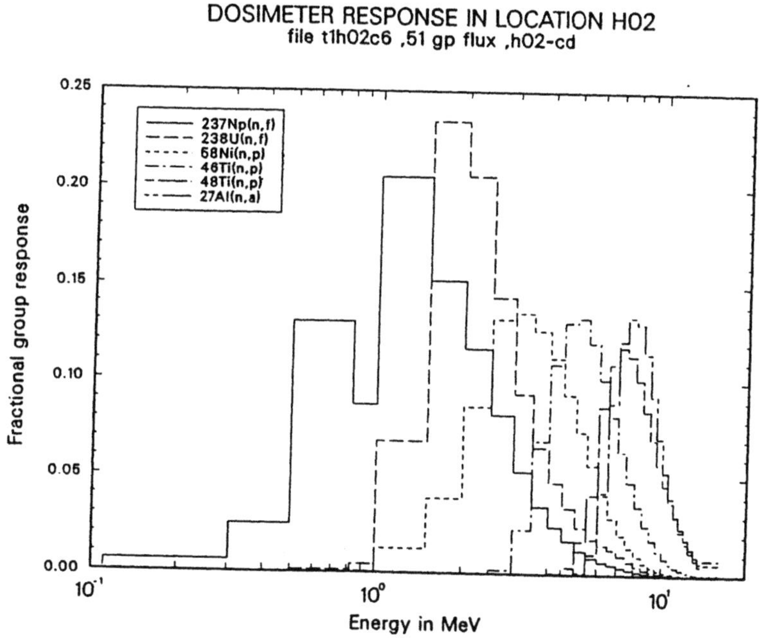


Figure 3. Typical multigroup response for high-energy threshold dosimeter reactions.

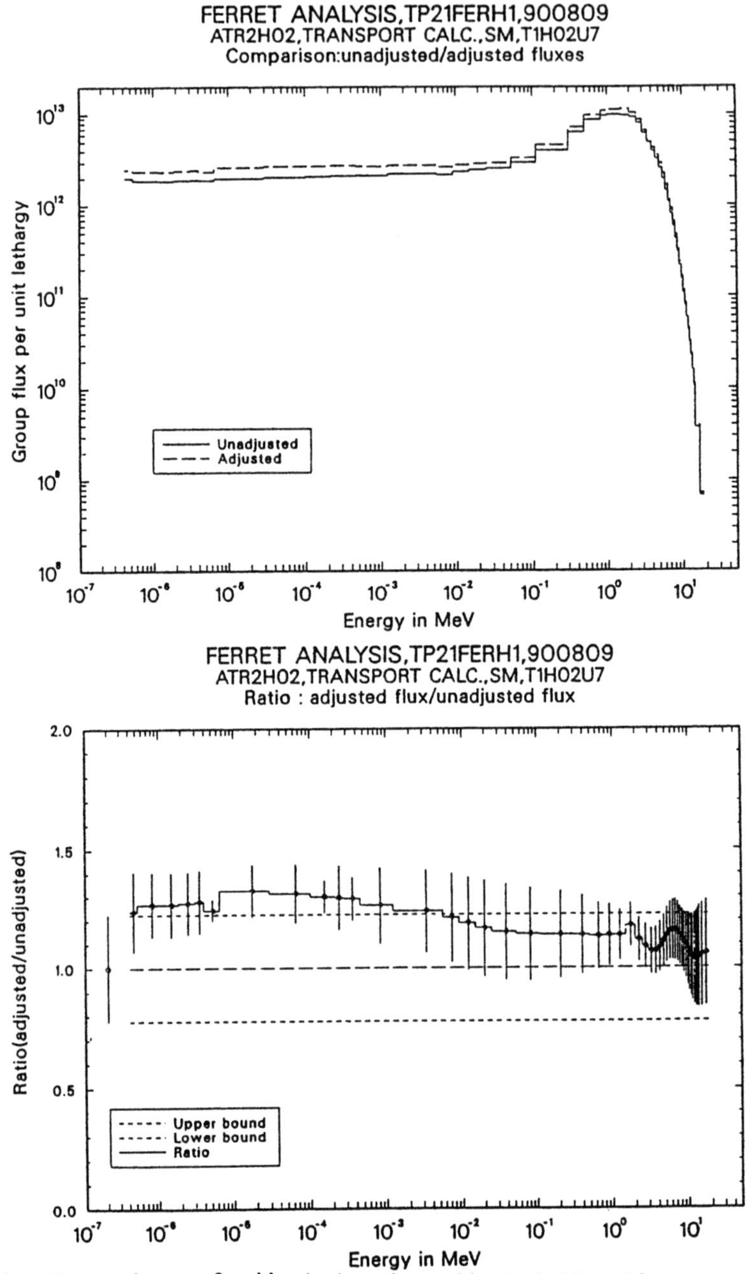


Figure 4. Comparison of adjusted and unadjusted (input) group fluence rates for H-02 location. Upper part of figure shows multigroup fluenc rates. Lower part shows ratio of adjusted-to-unadjusted group values input uncertainty bounds and final uncertainty error bars.

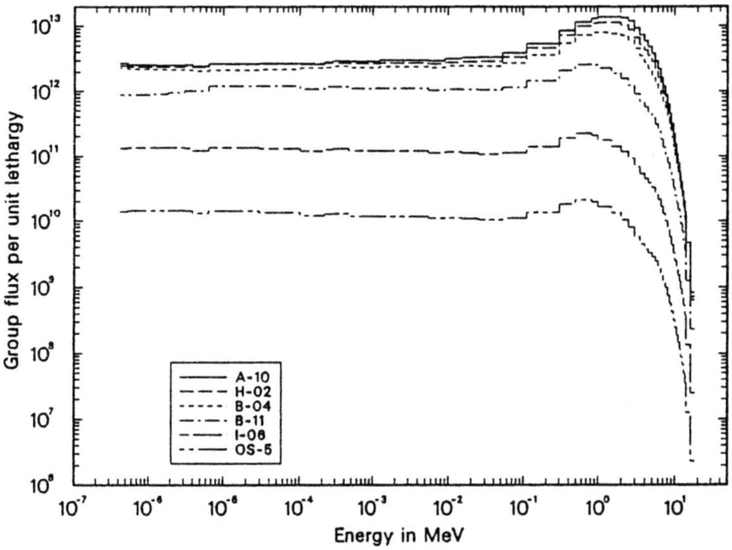


Figure 5. Comparison of final adjusted spectra for ATR locations associated with this study.

Table 1. Spectrum-averaged Cross Sections (barns)

| Reaction | H-2 | B-11 | I-6 | OS-5 |
|---|---|---|---|---|
| $^{45}$Sc(n,$\gamma$)* | 4.66E-1 | 5.85E-1 | 8.40E-1 | 8.40E-1 |
| $^{197}$Au(n,$\gamma$)* | 5.71E+1 | 7.66E+1 | 8.73E+1 | 9.58E+1 |
| $^{235}$U(n,f)* | 1.16E+1 | 1.51E+1 | 1.74E+1 | 1.77E+1 |
| $^{59}$Co(n,$\gamma$)* | 2.90E+0 | 3.56E+0 | 3.89E+0 | 3.97E+0 |
| $^{58}$Fe(n,$\gamma$)* | 5.35E-2 | 7.02E-2 | 8.03E-2 | 8.22E-2 |
| $^{237}$Np(n,f)† | 1.12E+0 | 9.61E-1 | 9.25E-1 | 7.94E-1 |
| $^{238}$U(n,f)† | 2.04E-1 | 1.38E-1 | 1.24E-1 | 1.05E-1 |
| $^{58}$Ni(n,p)† | 6.10E-2 | 3.64E-2 | 3.37E-2 | 2.46E-2 |
| $^{54}$Fe(n,p)† | 4.46E-2 | 2.68E-2 | 2.37E-2 | 2.23E-2 |
| $^{46}$Ti(n,f)† | 5.97E-3 | 3.84E-3 | 4.07E-3 | 3.18E-3 |
| $^{48}$Ti(n,p)† | 1.45E-4 | 1.06E-4 | 1.29E-4 | 9.91E-5 |
| $^{27}$Al(n,$\alpha$)† | 3.47E-4 | 2.53E-4 | 2.62E-4 | 1.93E-4 |
| Fe (dpa)† | 689 | 588 | 578 | 566 |
| Ni (dpa)† | 725 | 648 | 645 | 635 |
| Cr (dpa)† | 769 | 653 | 636 | 622 |

*For neutrons with E > 0.4 eV.
†For neutrons with E > 0.1 MeV.

# Assessment of the Fast Neutron Sensitivity of Thermoluminescent Gamma Dosimeters

H. Aït Abderrahim
Studiecentrum voor Kernenergie/Centre d'Etude de l'Energie Nucléaire (SCK/CEN), Mol, Belgium

E.D. McGarry and V. Spiegel
National Institute of Standards and Technology (NIST), Gaithersburg, Maryland, USA

ABSTRACT: Thermoluminescent gamma dosimeters (TLDs) are primarily sensitive to gamma rays and electrons; however, they are also sensitive to neutrons. The neutron interaction contribution (in a mixed radiation field) must be taken into account before comparison with gamma transport calculations can be made. In principle TLD energy-dependent neutron sensitivities can be calculated. In practice, the sensitivity is significantly dependent upon impurities, mainly in the raw materials but also from contaminants during manufacturing process. This paper concerns the measurement of the fast neutron sensitivities of $^{7}LiF$, $Al_2O_3$ and BeO TLDs to the Cf-252 fission spectrum. The exposures were carried out at the National Institute of Standards and Technology (NIST) who supplied a Cf-252 source with a calibrated neutron emission rate. Integral neutron sensitivities are given with an uncertainty of 5% to 7%, for the various types of TLDs investigated.

KEY WORDS: TLDs neutron sensitivity, mixed radiation field, gamma measurements

## INTRODUCTION

Several studies of gamma-ray energy deposition using TLDs have been carried out in different programmes such as ZEBRA [1], BALZAC [2] or VENUS [3]. One of the objectives of these programmes was the validation of the calculational methods used to evaluate the gamma-ray energy deposition in the reactor stainless steel structures. For such measurements in a mixed gamma and neutron environment, it is necessary to know the neutron sensitivity of the dosimeters and correct the raw results by the neutron contribution before comparing the computed results to the measured one. In principle, TLDs energy-dependent neutron sensitivities can be calculated. In practice, this sensitivity is significantly dependent upon impurities, mainly in the raw materials but also from contaminants during the manufacturing process.

The VENUS programme [3] assessed the integral fast neutron sensitivities of TLD materials, used for the VENUS gamma-ray field characterization, by irradiating dosimeters from the same TLDs batchs as those used during the VENUS campaign. Specimens of $^7LiF$ (HARSHAW cylindrical micro-rod 1 mm Ø x 6 mm), $Al_2O_3$ (CEC (France) chips 6 mm Ø x 0.9 mm) and BeO (6 mm Ø x 0.7 mm) TLDs were irradiated with a Cf-252 fission neutron source surrounded with a 14 cm thick spherical shielding shell. The exposures were carried out at the National Institute of Standards and Technology (NIST, formerly NBS) who supplied the calibrated Cf-252 source.

The material and thickness of the spherical shielding shell were chosen to attenuate the gamma-ray flux without modifying significantly the neutron spectrum. Scoping calculations [4], [5], reveal that a bare Cf-252 source does indeed introduce a dominant gamma-ray contribution in the TLDs response (∿ 98%), but this contribution is reduced to about 20% if the neutron source is shielded by the spherical lead shell of 14 cm thickness (O.D. 16.25 cm) shown on figure 1.

## EVALUATED INTEGRAL FAST NEUTRON SENSITIVITIES

As the thermoluminescence yield of the TLDs can vary significantly from one detector to another (for a same TLD material), it is important to use dosimeters taken from the same preselected batches of each TLD type used in the VENUS programme. Twenty five detectors of each type were sent to NIST. Ten were provided for the irradiation with the shielded Cf-252 source. Ten were for irradiation with the bare source. Five were used to measure background during the travel of the detectors from SCK/CEN to NIST and back. BeO TLDs were supplied by T.A. LEWIS* from Central Electricity Generating Board (CEGB, U.K.).

The TLDs fast neutron sensitivity is generally expressed in Co-60 equivalent Roentgen**-per-neutron fluence unit. This is accomplished by comparing the count rates of TLDs irradiated with a known neutron source to TLDs irradiated with a Co-60 gamma source whose intensity is known.

The measured dose of a TLD in a mixed neutron - gamma-ray

---

*The gamma-ray measurement in VENUS were performed by T.A. LEWIS with the same detector type.

**All along the document Roentgen will be used for comparison with old publications (1 R = 0.876 $10^{-2}$ Gy).

field, such as generated by a Cf-252 source, can be expressed as follows :

$$D_{meas.} = D_\gamma + D_n = D_\gamma + S_n \emptyset_n t \qquad (1)$$

where :

$D_{meas.}$ = is the total measured dose (in R)
$D_\gamma$ = is the partial dose due to the gamma-rays (in R)
$D_n$ = $S_n \emptyset_n t$ = is the partial dose due to the neutrons (in R)
$\emptyset_n$ = is the fast neutron flux (in $n/cm^2.s$)
$S_n$ = is the fast neutron sensitivity of the TLD (in $R/\ n.cm^{-2}$)
$t$ = is the irradiation time (in s).

From expression (1), we can easily derive the fast neutron sensitivity ($S_n$) if we know the total dose measured by the TLD, the gamma dose due to the residual gamma-rays not attenuated by the spherical shielding, the fast neutron flux level at the measurement position, and the time of the irradiation. The following describes the manner in which the values of the above mentioned parameters were obtained.

## Total Dose Measured ($D_{meas.}$)

Ten detectors of each type were wrapped in aluminium foils which were folded as indicated on Figure 2. These aluminium strips were mounted 24.85 cm from the Cf-252 as shown in Figure 3. The $Al_2O_3$ and BeO strips were positioned 1.9 cm (3/4 inch) on both sides of the $^7LiF$ strip which was located at the equator level of the lead sphere. The TLDs were irradiated with a Cf-252 source whose intensity was 3.44 $10^9$ n/s (on July 19, 1986). The duration of the irradiation was 849540 s. The BeO dosimeters were returned to T.A. LEWIS from CEGB who carried out the readout of these detectors. The $^7LiF$ were read at Mol with a CON-RAD 7100 TLD reader, and the $Al_2O_3$ with a TNO reader. Before averaging the 10 results for each TLD type, a distance correction was applied to each result, to obtain values corresponding to a unique distance from the source (24.85 cm). The results are expressed in Co-60 equivalent $R_{air}$ in Table 1. The error ($\Delta D_{meas.}$) quoted in Table 1 is based on the dispersion of the ten measurements for each material. It doesn't take into account the error due to the calibration procedure. When the latter is included, the relative errors become 9.36% (instead of 2.3%) for the $Al_2O_3$ and 4.99% (0.44%) for the $^7LiF$. For the BeO, the error on the calibration is not available but we estimate it as the same order as the other materials. To be consistent, our analysis took the errors given in Table 1.

## Fast Neutron Flux ($\emptyset_n$)

A 40-group one-dimensional S8P0 (diagonal transport

corrected) calculation, using the DTF-4 code [6],[7], was performed to obtain the space dependent neutron flux distribution. The geometrical model was chosen to have a mesh centered on the measurement location (i.e. 24.85 cm). The calculation was performed with a cross-section set prepared from the MOL-BR2-40GR library [8]. It was completed in a fixed source mode. The fixed source of one n/cm³.s was placed in the first mesh of the model (0.182293 cm³) and distributed on the fission spectrum of the Cf-252 recommended by J.A. GRUNDL [9]. The neutron spectrum at the measurement location was selected from this computed flux map and the first eight groups were summed to obtain the neutron flux above 100 keV. This value ($\emptyset_n$ = 2.5915 $10^{-5}$n/cm².s) was normalized to the absolute Cf-252 source intensity level by multiplying it by the ratio of the absolute intensity to the intensity of the source used in the DTF-4 run (3.44 $10^9$/0.182293) and the value of 4.8903 $10^5$n/cm².s was obtained for the fast neutron flux $\emptyset_n$ ($E_n$ >0.1 MeV). To check the validity of the calculation, measured U-238, Np-237 and U-235 fission rates were compared with those computed at 24.85 cm from the Cf-252 source. The flux spectrum at this position was from the above calculation and the fission cross-sections from the IRDF-81 dosimetry-file. The C/E ratios were 0.96 (U-238), 0.96 (Np-237) and 0.72 (U-235). On the basis of these results, the uncertainty on the flux above 0.1 MeV, from the DTF-4 calculation is 4% because it represents nearly 100% of the Np-237 and U-238 responses.

## Residual Gamma Dose (Dγ)

The evaluation of the residual gamma dose is based on gamma transport calculation. The GAMPREP [10] Code was used with a 40 neutron-group and 20 gamma-group production library (derived from the coupled (n-γ) EURLIB4 library) and the already mentioned 40-group neutron flux map to calculate a gamma interaction cross-section set and the capture gamma space and energy dependent source. The Cf-252 gamma source in the first mesh was used to obtain the total gamma source. This 20-group total gamma source was introduced in a DTF-4 run as a fixed source. The calculation was performed with S8P5 approximations to obtain the gamma flux distribution. The gamma spectrum at 24.85 cm was selected and converted into gamma dose by weighting it with group-wise energy absorption mass attenuation coefficients for each TLDs type taken from reference [11]. To convert to Co-60 equivalent $R_{air}$ these results were multiplied by the ratio of the energy absorption mass-attenuation of air to those of the TLD material at the mean energy of Co-60 (1.25 MeV). The results are given in Table 1. The relative error on the Dγ values was estimated to be 6%.

## Fast Neutron Sensitivity of Each TLD Type

From expression (1) the fast neutron sensitivity may be expressed as :

$$S_n = (D_{meas} - D_\gamma)/\phi_n t \qquad (2)$$

The values obtained for the different parameters in this expression and the calculated $S_n$ value for each TLD type are given in Table 1.

TABLE 1 : Fast Neutron Sensitivities for $^7LiF$, BeO and $Al_2O_3$

| TLD Type | $D_{meas} \pm \Delta D_{meas}$ ($R_{air}$) | $D_\gamma \pm \Delta D_\gamma$ ($R_{air}$) | $S_n$ ($R_{air}/10^{10} n.cm^{-2}$) |
|---|---|---|---|
| $^7LiF$ | 25.65 ± 0.11 | 2.78 ± 0.17 | 0.55 ± 0.03 |
| BeO | 26.61 ± 0.40 | 2.77 ± 0.17 | 0.57 ± 0.04 |
| $Al_2O_3$ | 18.06 ± 0.41 | 2.79 ± 0.17 | 0.37 ± 0.03 |

## VALIDATION OF THE RESIDUAL GAMMA DOSE CALCULATION

To check the validity of the calculated residual gamma dose, the procedure was used to evaluate the dominant gamma dose from an irradiation with a bare Cf-252 source. The computed dose was then compared to measurements of TLDs exposed to a bare Cf-252 source for three hours. The calculated fast neutron fluence ($E_n$>0.1 MeV) was 5.1204 $10^9 n/cm^2$. Using this fluence and the sensitivities determined above, one can estimate the neutron contribution. The computed and measured gamma dose are compared in Table 2.

TABLE 2 : Comparison of the computed and measured gamma dose due to the Cf-252 bare source

| TLD Type | $D_{meas} \pm \Delta D^*_{meas}$ ($R_{air}$) | $D_n^{calc} \pm \Delta D_n^c$ ($R_{air}$) | $D_\gamma^{calc} + \Delta D_\gamma^{calc}$ ** ($R_{air}$) | $\frac{D_n^{calc} + D_\gamma^{calc}}{D_{meas}}$ |
|---|---|---|---|---|
| $^7LiF$ | 11.35 ± 0.10 | 0.28 ± 0.03 | 10.70 ± 0.64 | 0.97 |
| BeO | 12.13 ± 0.05 | 0.29 ± 0.03 | 10.65 ± 0.64 | 0.90 |
| $Al_2O_3$ | 10.38 ± 0.36 | 0.19 ± 0.02 | 10.64 ± 0.64 | 1.04 |

* same remark as for Table 1 concerning the errors quoted in this table.

** the computed gamma doses were respectively multiplied by the ratio of the energy absorption mass attenuation coefficient of air to this of the considered TLD material at the energy of 1.25 MeV; (1.0764 for $^7LiF$, 1.0408 for BeO and 1.0210 for $Al_2O_3$).

The irradiated TLDs (10 of each type) with the bare source were measured using the same procedure. The TLDs were loaded in the plexiglass holder of 8 mm thickness (figure 4) to realize the electronic equilibrium. The results show agreement to within 10%, which is in reasonable agreement with the estimated error of 6 %.

## DISCUSSION AND CONCLUSIONS

Comparison with the literature shows the present results for $Al_2O_3$ confirms the result of 0.38 $R/10^{10}\ n.cm^{-2}$ of PORTAL [12]. However, the same author gives a fast neutron sensitivity of 0.73 $R/10^{10}\ n.cm^{-2}$ for a $^7LiF$ (powder). P.H. McGINLEY [13] gives 0.89 $R/10^{10}\ n.cm^{-2}$ for $^7LiF$ (HARSHAW powder). For comparison calculated integral fast neutron sensitivity for $^7LiF$ starting from three differential response functions : 0.54 $R/10^{10}\ n.cm^{-2}$ for the WINGATE [14] evaluation (HARSHAW powder), 0.74 for FURUTA and TANAKA [15] response function (MATSUSHITA powder) and finally 0.40 for A.D. KNIPE [16] response (HARSHAW, square micro-rod). These comparison for $^7LiF$ shows a great dispersion in results and indicate that the sensitivity depends on the nature of the detector (powder, cylindrical micro-rod, square micro-rod ...). Nevertheless our value is contained between the extreme values. For BeO TLDs, the differential response function established by T.A. LEWIS [17] provides a calculated integral sensitivity of 0.50 $R/10^{10}\ n.\ cm^{-2}$ which is in agreement with the 0.57 $\pm$ 0.04 experimental evaluation.
This work presents a practical and accurate approach to the problem of having to make adjustments for the neutron sensitivity of TLDs. For materials corresponding to those employed in the VENUS programme, experimental results with substantiated uncertainties are available for further use. Calculations demonstrate the need for quality assurance of materials and knowledge of the radiation environment before attempting to predict neutron sensitivity of thermoluminescent gamma dosimeters.

## ACKNOWLEDGEMENT

The authors acknowledge Dr. T.A. LEWIS from CEGB who supplied the BeO TLDs and accomplished their calibration and read-out. They also thank Messrs H. GEENS, D. LANGELA, L. VERSTREPEN, J. OEYEN and Mrs S. LENAERTS from the Reactor Physics Department, SCK/CEN for their help with this paper and the posters for the presentation.

## REFERENCES

[1] A.D. KNIPE and R. de WOUTERS, "Gamma-Ray Energy Deposition Measurements in a Heterogeneous Core and their Analysis," IAEA-SM-244/35 (Sept. 1979).

[2] R. de WOUTERS et al., "Measurements and Analysis of Gamma-Ray Energy Deposition in a Critical Assembly containing a central simulated Diluent," Proceedings (Vol.1) of the International Conference on the Physics of Reactors : Operation, Design and Computation, Marseille, April 23-27, 1990.

[3] A. FABRY et al., "LWR-PV-SDIP : VENUS PWR Core Source and Azimuthal Lead Factor Experiments and Calculational Tests," BLG 586/NUREG CR-3323 Report, Under Publication.

[4] H. AIT ABDERRAHIM, "Mesure Intégrale de la Sensibilité aux Neutrons de Dosimètres Thermoluminescents $Al_2O_3$ au Moyen d'une Source Cf-252 (NBS) - Etude de Faisabilité," SCK/CEN Internal Report 620/85-01.

[5] H. AIT ABDERRAHIM, "Study of Fast Neutron Sensitivity of $Al_2O_3$ Thermoluminescent Dosimeters : Cf-252 experiment," SCK/CEN Internal Report 620/86-01.

[6] K.D. LATHROP, "DTF-IV, a Fortan-IV Program for Solving the Multigroup Transport Equation with Anisotropic Scattering," LA-3373, July 1965.

[7] G. MINSART, "MULCOS (MULtipurpose COde System) - General Organization of the Code System," CEN/SCK Internal Report, REACTOR STUDY GM/md 63-176/71-20, February 1971.

[8] P. VANDEPLAS and J.-C. SCHEPERS, "A forty Group Cross-Section Library for the Calculation of Fast-Thermal Systems," CEN/SCK Internal Report, REACTOR STUDY PV/md 63-206/71-180, December 1971.

[9] J.A. GRUNDL et al. - "A Californium-252 Fission Spectrum Irradiation Facility for Neutron Reaction Rate Measurements," Nuclear Technology, Vol. 32, Mars 1977.

[10] G. MINSART and L. FOUARGE, "GAMPREP, a Programme to prepare Input for Gamma Calculations," CEN/SCK Internal Report, REACTOR PHYSICS GM/LF/sa, 604/77-01/C, September 1977.

[11] H. JOFFRE and L. PAGES, "Coefficients d'Atténuation Massique et d'Absorption Massique en Energie pour les Photons de 10 keV à 10 MeV," Rapport CEA-R-3655 (1968).

[12] G. PORTAL, "Contribution à l'Etude et au Développement de la Dosimétrie par Radiothermoluminescence," CEA-R-4697 (Thèse 1975).

[13] P.H. McGINLEY, "Response of LiF to Fast Neutrons," Health Physics, Vol.23 - 105,106 (1972).

[14] G.L. WINGATE, E. TOCHILIN, N. GOLDSTEIN, "Response of Lithium Fluoride to neutrons and Charged particles," Proc. Intern. Conf. Luminescence Dosimetry (Stanford 1965), p.421.

[15] Y. FURUTA et S. TANAKA, "Response of $^6$LiF and $^7$LiF Thermoluminescence Dosimeters to Fast Neutrons," Nucl. Inst. and Meth. 104 (1972), p.365-374.
"Revised Energy Responses of $^6$LiF and $^7$LiF Thermoluminescence Dosimeters to Neutrons," Nucl. Inst. and Meth. 140 (1977), p.395-396.

[16] A.D. KNIPE, "The Fast Neutron Response of $^7$LiF Thermoluminescent Dosimeters," RDD/ADK/769 (Dec. 1982).

[17] L.T. JONES et al., "A Measurement of the Response to Fast Neutrons of Several Materials Dosimeters," TPRD/B/0827/R86 (July 1986).

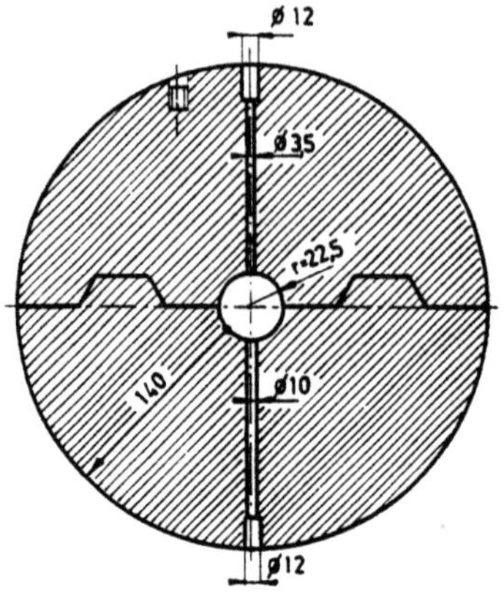


Fig.1: Lead sphere for the Cf-252 experiment

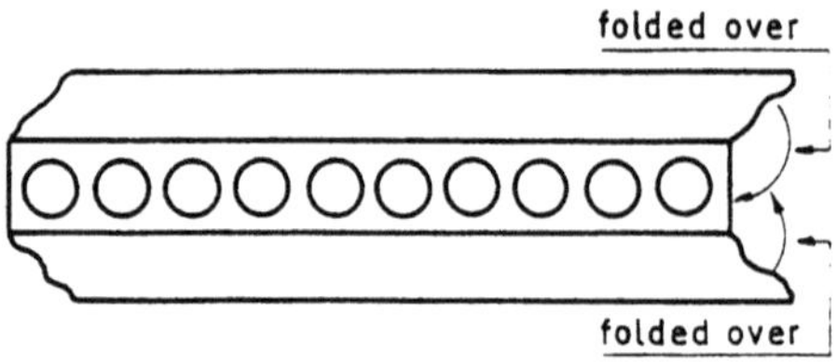


Fig.2: Aluminium wrapping for TLDs (Sketch for disks TLDs)

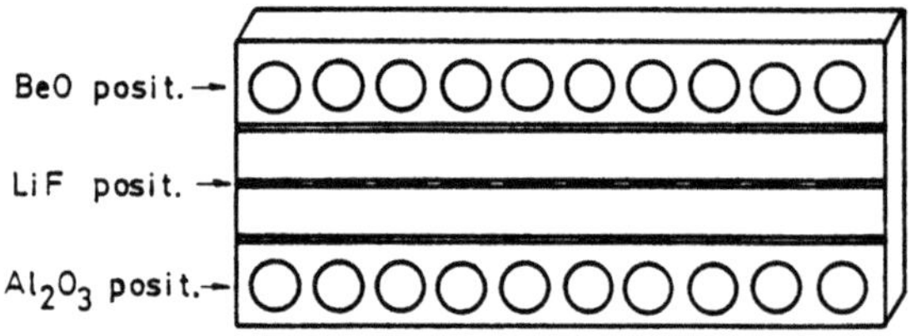


Fig.4: TLD holder for the irradiation with bare source.

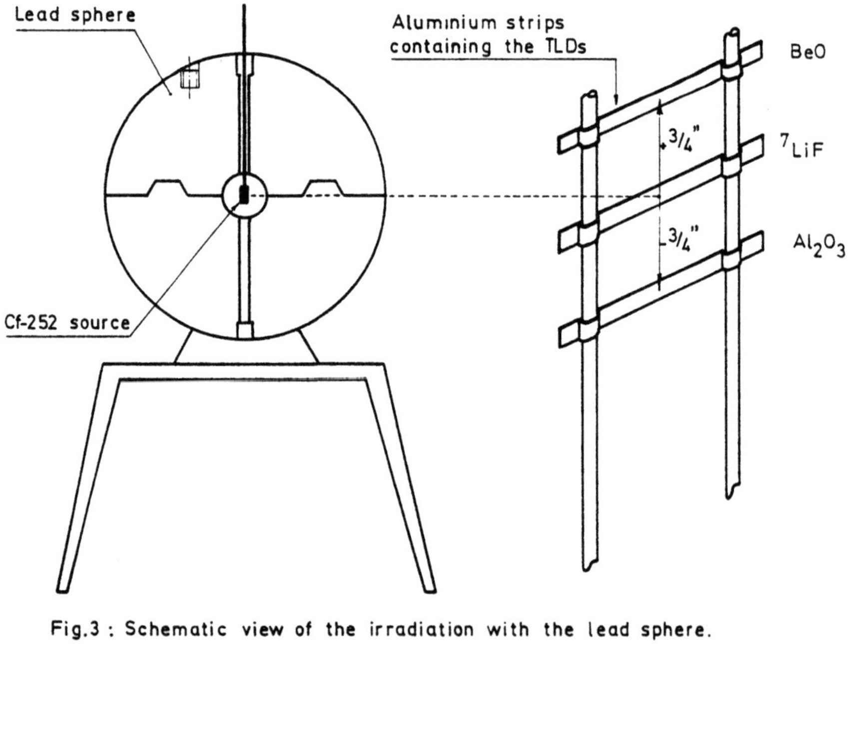


Fig.3 : Schematic view of the irradiation with the lead sphere.

# PHOTOFISSION EFFECTS IN B&W 177-FA REACTOR VESSEL SURVEILLANCE CAPSULE DOSIMETERS

L. Petrusha

B&W Nuclear Service Company

Lynchburg, Virginia, U.S.A.

## ABSTRACT

Neptunium ($^{237}Np$) and uranium ($^{238}U$) dosimeters, used in reactor vessel surveillance capsules, have historically had an error introduced by photofission events associated with them. Specifically, gamma-induced photofission events in the target material result in $^{137}Cs$ product activity that cannot be distinguished from neutron-induced $^{137}Cs$ fission product activity. This condition leads to significantly higher measured dosimeter activities than those predicted by calculational methods, resulting in discrepancies in surveillance capsule fluence analyses. Thus, using a full-scale reactor plant model, photofission reaction rates were calculated for both the U and Np dosimeters irradiated in an actual Babcock & Wilcox (B&W) reactor vessel surveillance capsule. The photofission effect was then calculated using these photofission reaction rates and deriving the fraction of the product isotope specific activity produced by photofissions. This paper presents the photofission fractions developed for both the U and Np dosimeters and used as correction factors in B&W capsule fluence analyses.

---

## BACKGROUND

The Babcock & Wilcox Owners Group (BWOG) Reactor Vessel Surveillance Program was established to monitor the combined effects of irradiation and temperature on the mechanical properties of the beltline region materials in B&W-designed

reactors. The program was designed to irradiate the material specimens in reactor vessel surveillance capsules located in the coolant downcomer region near the inside surface of the reactor vessel wall. These capsules contain samples of weld metal, plate or forging material, heat-affected zone material from the vessel beltline, temperature monitors, and neutron dosimeters for monitoring flux exposure.

Since implementation of the program, the neptunium [$^{237}Np(n,f)^{137}Cs$] and uranium [$^{238}U(n,f)^{137}Cs$] dosimeters have provided fairly accurate results of the integrated activity accumulated during the irradiation period. These dosimeters have a wide effective energy range and a long half-life product. However, both the $^{237}Np$ and $^{238}U$ dosimeters can experience significant photofission events in the target material induced by high-energy gamma rays. Even though it has long been known that an error is introduced by photofission events in these fission dosimeters, and measurements of this error have been made in test reactors, its magnitude in the actual operating B&W-designed reactor environment has been a subject of conjecture.

As discussed by Bowman et al.(1) and Martin and Cogburn(2), the region near the inside surface of the pressure vessel experiences the highest ratio of gamma to neutron flux, implying that this is where photofission is most significant. The major source of the gamma rays in the downcomer region is from thermal neutron capture in the core barrel and thermal shield, due to the relatively high neutron flux in this region(1,3). Neutron capture in iron, chromium, and nickel, elements found in the composition of the core barrel and thermal shield, leads to considerable emission of capture gamma rays in the energy range from 5-10 Mev(4). Thus it follows, and has been shown by Verbinski et al.(5) and Bowman et al.(1), that the photofission ($\gamma$,f) reactions in fission detectors located in the downcomer region can be significant with respect to the threshold neutron-induced (n,f) reactions, causing misleading dosimeter activity measurement data. This, in turn, results in erroneous normalization factors (measured-to-calculated dosimeter activity ratios) during surveillance capsule fluence analyses. Therefore, photofission correction factors have been developed to reduce the absolute magnitude of the normalization factors and achieve more accurate representation of surveillance capsule fluence results.

## CALCULATION OF THE PHOTOFISSION EFFECT

The first step in establishing the photofission effect in B&W reactor vessel surveillance capsule dosimeters was to create a full-scale B&W 177 fuel assembly (FA) plant model for use in the DOT4.3 computer code(6). The Oconee 2-E capsule(7) was chosen as a representative standard B&W surveillance capsule for the photofission calculation. This capsule was initially irradiated in the Oconee Unit 2 reactor for Cycle 1, and then transferred to the Crystal River Unit 3 reactor for irradiation in Cycles 1B-7. The calculational model developed was an R-θ geometric representation of a plan view through the reactor core midplane, as shown in Figure 1. The geometry modelled the core region, through the thermal shield and pressure vessel, and into the concrete shield. In accordance with the results of the B&W Owners Group Flux Perturbation Experiment(8), which determined perturbations in the fast flux due to the capsule to be as high as 47% at the capsule center, explicit modelling of the surveillance capsule and capsule holder tube was also included.

The two-dimensional discrete ordinates neutron transport code DOT4.3 was then employed to generate the neutron and gamma flux distributions in the region of interest. Flux generation in the core was calculated using a fixed distributed source input based on a $^{235}U$ fission spectrum, the input relative power distributions, and a normalization factor to adjust flux level to the desired power density. The calculation also used a $P_3$ order scattering, $S_8$ quadrature, and the DLC-23E/CASK cross sections and 40 group energy structure(9).

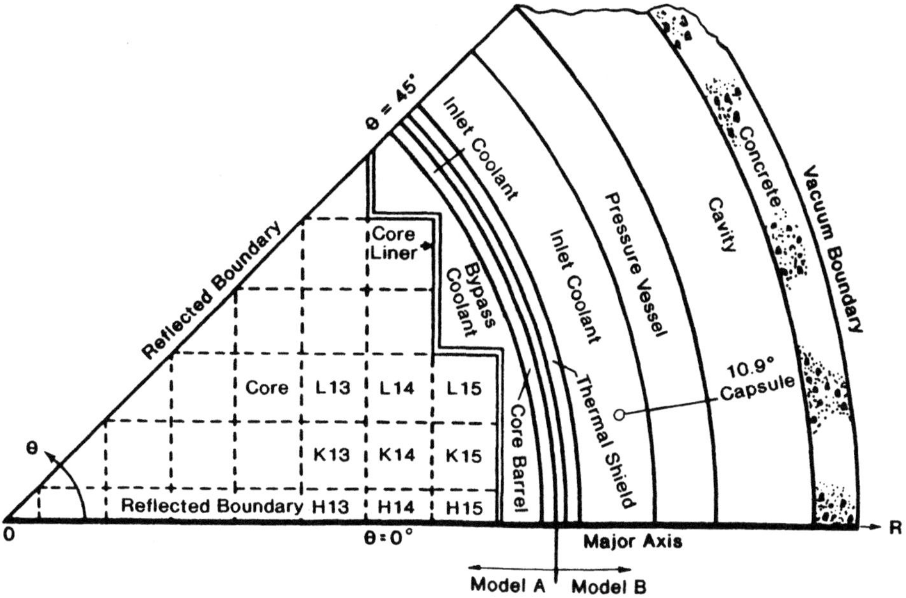


Figure 1: *Plan View Through Reactor Core Midplane (R-θ Calculational Model)*

Using the calculated gamma flux spectrum (i.e., flux weighting function) as generated by the DOT4.3 computer code, the group-dependent flux-weighted average photofission cross sections were calculated. The threshold energy for photofission in the $^{237}Np$ and $^{238}U$ dosimeters is approximately 6 Mev (1,4). The DLC-23E/CASK 40 group energy boundaries are structured with 3 gamma-ray groups in the energy range of interest: Group 23 (10.0-8.0 Mev), Group 24 (8.0-6.5 Mev), and Group 25 (6.5-5.0 Mev). Thus, it was only necessary to acquire average photofission cross sections for the three CASK energy groups above the photofission threshold energy.

More specifically, the flux-weighted group-averaged photofission cross sections were calculated using the following formula:

$$\overline{\sigma_g} = \frac{\sum_{j=E_1}^{E_2} \sigma_{vj}\phi_j}{\sum_{j=E_1}^{E_2} \phi_j} \qquad \text{(Equation 1)}$$

where

$\overline{\sigma}_g$ = average photofission cross section for each energy group g, mb,

$\sigma_{vj}$ = value of photofission cross section for photon energy j, as published by Verbinski et al.(5), mb,

$\phi_j$ = gamma flux generated by DOT4.3 code for photon energy j, neutrons/cm$^2$-sec,

$E_1$ = lower bound of energy group g, Mev,

$E_2$ = upper bound of energy group g, Mev.

The referenced data(5) used energy groups of 0.25 Mev. The corresponding values of the photofission cross sections are listed in Table 1. They are based on the measurements by Caldwell et al.(10), as listed by Verbinski et al.(5). T.G. Williamson has also confirmed that the Caldwell et al. data has been used most by investigators to calculate the photofission effect in the pressure vessel environment and appears to be the best set of cross sections to use for calculating photofractions for U, Th, and Np(4). Table 2 shows the resulting average photofission cross sections calculated for each energy group, using Equation 1, for two dosimeter locations in the surveillance capsule.

TABLE 1

*Photoreaction Cross Sections of Detector Foils $^{238}U$ and $^{237}Np$(5)*

| Photon Energy j (Mev) | Photoreaction Cross Section (mb) $^{238}U$ (γ,f) | $^{237}Np$ (γ,f) |
|---|---|---|
| 10.00 | 47.5 | 167 |
| 9.75 | 43.0 | 135 |
| 9.50 | 39.0 | 110 |
| 9.25 | 34.0 | 88 |
| 9.00 | 30.0 | 72 |
| 8.75 | 25.5 | 57 |
| 8.50 | 22.0 | 47 |
| 8.25 | 18.5 | 40 |
| 8.00 | 16.5 | 37 |
| 7.75 | 14.0 | 37 |
| 7.50 | 12.0 | 36 |
| 7.25 | 11.5 | 35 |
| 7.00 | 11.0 | 33 |
| 6.75 | 9.5 | 31 |
| 6.50 | 7.0 | 28 |
| 6.25 | 7.5 | 23 |
| 6.00 | 7.0 | 17 |
| 5.75 | 3.0 | 10 |
| 5.50 | 1.0 | 3 |
| 5.25 | 0.0 | 0 |
| 5.00 | 0.0 | 0 |

TABLE 2

*Flux-Weighted Group-Averaged Photofission Cross Sections*

| CASK Energy Group | Energy Range (Mev) | Average Photofission Cross Section $\bar{\sigma}_g$(mb) | | | |
|---|---|---|---|---|---|
| | | Dosimeter ED1 | | Dosimeter ED3 | |
| | | $^{238}U$ | $^{237}Np$ | $^{238}U$ | $^{237}Np$ |
| 23 | 8.0-10.0 | 27.40 | 71.35 | 27.55 | 72.31 |
| 24 | 6.5-8.0 | 11.59 | 33.89 | 11.64 | 33.94 |
| 25 | 5.0-6.5 | 4.29 | 13.86 | 4.27 | 13.77 |

These average cross sections, along with the DOT-generated gamma fluxes, were then used to calculate the total photofission reaction rate for each type of dosimeter ($^{238}U$ and $^{237}Np$) at the two locations of interest. The calculation of the total photofission reaction rate was based on the following formula:

$$R_{ph} = \sum_g \phi_g \overline{\sigma_g} n \qquad \text{(Equation 2)}$$

where

- $R_{ph}$ = photofission reaction rate,
- $\phi_g$ = gamma flux generated by the DOT4.3 code for energy group g, photons/cm$^2$-sec,
- $\overline{\sigma}_g$ = average photofission cross section for each energy group g, cm$^2$,
- n = total number of target atoms in the dosimeter, atoms/dosimeter.

Note that the term $R_{ph}/n$ is equivalent to the photofission reaction rate per target atom.

The DOT4.3 computer code calculates the $^{137}Cs$ production from neutron-induced fissions in the U and Np dosimeters at saturation in units of ($\mu$Ci of $^{137}Cs$ activity/g of target). For comparative purposes, the $^{137}Cs$ response function from DOT had to be manipulated into a form similar to the photofission reaction rate shown above. Therefore, the desired form of the $^{137}Cs$ response function calculated by DOT was in units of (neutron-induced fission rate per target atom).

The response function used to represent the production of $^{137}Cs$ in the DOT code was based on the following rate equation:

$$\frac{dN}{dt} = -\lambda N(t) + nf_i \sum_g \phi_g \sigma_g^f . \qquad \text{(Equation 3)}$$

Here $\lambda$ referred to the radioactive decay constant and N(t) denoted the number of $^{137}Cs$ nuclei at time t. The term $\sigma_g^f$ referred to the target fission cross section for energy group g, $\phi_g$ was the neutron flux generated by DOT for energy group g, n was the total number of target atoms in the dosimeter, and $f_i$ denoted the $^{137}Cs$ fission yield for isotope i. The $^{137}Cs$ (n,f) fission yield for $^{238}U$ is 0.0600 and 0.06265 for $^{237}Np$, as listed in the ENDF/B5 library.

Since the DOT4.3 computer code calculates only the time-independent solution, i.e., the saturated activity values, it was necessary to solve Equation 3 for an infinite time period. At saturation, dN/dt=0. Thus, expressing the solution to Equation 3 in terms of the total number of disintegrations occurring per second $\lambda N(t)$, i.e., the $^{137}Cs$ activity, yielded

$$A(t) = \lambda N(t) = nf_i \sum_g \phi_g \sigma_g^f. \qquad \text{(Equation 4)}$$

Deriving the saturated activity of $^{137}Cs$ due to neutron-induced fission events in terms of $\mu Ci/g$ of target led to the following equation for calculating the DOT response:

$$A_D = \frac{nf_i \sum_g \phi_g \sigma_g^f}{m_i(3.7e{+}04)} = \frac{N_{Av} f_i}{M_i(3.7e{+}04)} \sum_g \phi_g \sigma_g^f. \qquad \text{(Equation 5)}$$

$N_{Av}$ represented Avogadro's number, $M_i$ was the atomic weight of the $i^{th}$ isotope, and $m_i$ denoted the target mass of isotope i in grams.

Lastly, manipulating the DOT response into the final form of an equation representing the neutron-induced fission rate per target atom resulted in

$$\frac{R_D}{n} = \sum_g \phi_g \sigma_g^f = \frac{M_i(3.7e{+}04)}{N_{Av} f_i} A_D. \qquad \text{(Equation 6)}$$

Table 3 shows the calculated reaction rates due to photofission and neutron-induced fissions, acquired using Equations 2 and 6, for the $^{238}U$ and $^{237}Np$ dosimeters at the two locations of interest.

TABLE 3

*Calculated Photofission Fractions for $^{238}U$ and $^{237}Np$ Dosimeters*

| Dosimeter | Dosimeter Type | Reaction Rates (per target atom) | | Photofission Fraction (%) |
|---|---|---|---|---|
| | | $R_{ph}/n$ | $R_D/n$ | |
| ED1 | $^{238}U$ | 1.799E-15 | 3.388E-14 | 5.04 |
| | $^{237}Np$ | 5.057E-15 | 2.084E-13 | 2.37 |
| ED3 | $^{238}U$ | 9.597E-16 | 1.848E-14 | 4.94 |
| | $^{237}Np$ | 2.705E-15 | 1.150E-13 | 2.30 |

To quantitatively determine the photofission effect for the U and Np dosimeters, a comparison was made between the calculated photofission and neutron-induced fission reaction rates.

Explicitly, the photofission fraction may be expressed as the ratio of the photofission rate per target atom to the sum of the photofission and neutron-induced fission rates per target atom, or,

$$\text{Photofission Fraction} = \frac{\frac{R_{ph}}{n}}{\frac{R_D}{n} + \frac{R_{ph}}{n}} , \qquad \text{(Equation 7)}$$

where $R_{ph}$ referred to the calculated photofission reaction rate and $R_D$ referred to the calculated neutron-induced fission rate. The photofission fraction values for each type of dosimeter at each location can also be found in Table 3. Thus, averaging the dosimeters, it was concluded that the photofission contribution to the $^{238}U(n,f)^{137}Cs$ measured dosimeter activity is equal to 5.0%, while the photofission contribution to the $^{237}Np(n,f)^{137}Cs$ dosimeter activity is 2.3%.

## CONCLUSIONS

The photofission correction factors that are applied to the normalization factors, measured-to-calculated dosimeter activity ratios, in B&W 177 fuel assembly (FA) surveillance capsule fluence analyses are, therefore, 5.0% and 2.3% for the $^{238}U$ and $^{237}Np$ dosimeters, respectively. The use of these correction factors has enabled us to achieve better agreement between the calculated and measured dosimeter results than has been encountered in past analyses. The most recent five B&W capsule analyses have all yielded measured-to-calculated values that agree within 5% or less. Thus, these photofission corrections and the resulting good agreement between calculations and measurements lead to more accurate predictions of reactor vessel fluence for reactor vessel surveillance and life extension.

Note that as the present phase of the B&W Reactor Vessel Surveillance Program nears completion and all initial reactor vessel surveillance capsules in the downcomer region are removed from their host reactors, cavity dosimetry will be relied upon to monitor vessel fluence. At that time, actual gamma flux measurement data from lithium fluoride (LiF) detector chips will be available and photofission correction factors will be calculated for cavity applications to examine differences that might exist.

## REFERENCES


1. C.D. Bowman, C.M. Eisenhauer, and D.M. Gilliam, "Photofission Effects in Reactor Pressure Vessel Dosimetry," NUREG/CP-004 (1977).

2. G.C. Martin, Jr. and C.O. Cogburn, "Special Considerations for LWR Dosimetry Experiments," p.399-406 in Proceedings of the Fifth ASTM-Euratom Symposium on Reactor Dosimetry, GKSS Research Centre, Geesthacht, F.R.G., September 24-28, 1984, Volume 1, D. Reidel Publishing Company, Dordrecht, Holland, 1985.

3. G.T. Chapman and W.R. Burrus, Nuclear Science and Engineering, Volume 34, p.169 (1968).

4. T.G. Williamson, Evaluation of the Photofission Effect in Pressure Vessel Dosimetry, Department of Nuclear Engineering and Engineering Physics, University of Virginia Report No. UVA/532886/NEEP88/101CN (June 1988).

5. V.V. Verbinski, G.G. Cassapakis, W.K. Hagen, and G.L. Simmons, "Photointerference Corrections in Neutron Dosimetry for Reactor Pressure Vessel Lifetime Studies," Nuclear Science and Engineering, Volume 75, p.159-166 (1980).

6. "DOTIV Version 4.3, One- and Two-Dimensional Transport Code System," CCC-429, Radiation Shielding Information Center, Oak Ridge National Laboratory.

7. A.L. Lowe, Jr., J.D. Aadland, A.D. Nana, L. Petrusha, and W.R. Stagg, Analysis of Capsule OC2-E, Duke Power Company, Oconee Nuclear Station Unit 2, Babcock & Wilcox Report No. BAW-2051 (October 1988).

8. N.L. Snidow, B&WOG Flux Perturbation Experiment at ORNL, Measured and Calculated Dosimeter Results, Babcock & Wilcox Report No. BAW-1886 (September 1985).

9. "CASK, 40 Group Coupled Neutron and Gamma-Ray Cross Section Data," DLC-23E, Radiation Shielding Information Center, Oak Ridge National Laboratory.

10. J.T. Caldwell, E.J. Dowdy, B.L. Berman, R.A. Alvarez, and P. Meyer, Photonuclear Measurements on Fissionable Isotopes Using Monoenergetic Photons, Report LA-UR--76-1615, Los Alamos Scientific Laboratory (1976).

# COMPARISON OF THE CALCULATED TO MEASURED VALUES OF THE INTEGRATED GAMMA DOSE IN THE CAVITY OF A FULL-SCALE OPERATING POWER REACTOR

BY

S. Q. King[1] and D. M. Gilliam[2]

1. B&W Nuclear Service Company, P. O. Box 10935, Lynchburg, Virginia, 24506-0935 USA

2. National Institute of Standards and Technology, Bldg A155, Gaithersburg, Maryland, 20899, USA

**ABSTRACT**

A lithium fluoride gamma dosimeter (LiF chip) has been irradiated for a single cycle in the cavity of one of the B&W Owners Group reactors. The LiF chip was exposed to a high level mixed gamma and neutron field for long enough to ensure a megagray-range absorbed dose. The LiF chip was analyzed at the National Institute of Standards and Technology (NIST) to ascertain its total integrated gamma dose. These measured results were then compared with calculated results that were generated by neutron-gamma transport analysis at B&W. This paper presents the results of that comparison.

A single LiF chip dosimeter was placed in the cavity in a specially designed holder mechanism which also contained a number of neutron dosimeters of various types. The dosimetry was irradiated for one full cycle, experiencing 408.9 full power days of exposure. Detailed analyses, based both on calculations and measurements (dosimeter responses), were performed to ascertain the neutron fluence spectrum in the LiF chip. The resulting fluence spectrum information was then transmited to NIST and used to calculate the fraction of the dosimeter's response attributable to neutron interactions. Independent simultaneous analyses were carried out to determine the best-estimate calculated response and the most accurate measured response of the LiF chip.

## ANALYTICAL PROCEDURE

The two-dimensional transport code, DOT4.3, was used to calculate the energy- and space-dependent neutron and gamma ray fluence rate at all points of interest in the reactor vessel and in the cavity dosimetry.

The neutron population in the core during power operation is a function of energy, space, and time. The time dependence was accounted for in the analysis by calculating the time-weighted average neutron source, i.e., the neutron source corresponding to the time-weighted average pin-by-pin power distribution. The effects of the other two independent variables, energy and space, were accounted for by using an appropriately large number of discrete intervals in energy and space - in each interval the source is assumed to be invarient and independent of all other variables.

The fission spectrum used was an average U + Pu spectrum in the BUGLE-80 group structure, corresponding to the U-Pu mix in the peripheral assemblies.
The horizontal plane calculational model is an R-θ geometric representation of a plane view through the reactor core midplane and includes the core, core liner, core barrel, thermal shield, downcomer, pressure vessel, cavity, cavity liner, and concrete biological shield, as shown in Figure 1. Eighth-core symmetry was used with a third-order LeGendre polynomial scattering approximation and an angular quadrature corresponding to 48 scattering angles ($P_3S_8$). The model contains an explicit representation of the LiF chip dosimeter, its glass shield, and holder mechanism.

The vertical plane (R-Z) calculational model extends in elevation from the lower grid plate to the upper grid plate and in radius from the core centerline through the cavity, ending at a point 60.96 cm into the concrete biological shield. Radial intervals in the R-Z DOT analysis are identical to those in the R-θ analysis. This facilitates the calculation of the fluuence rate at the dosimeter location which lies outside both calculational planes. The flux at the dosimeter location was calculated by the synthesis procedure described in Equation 1.

Equation (1)

$$\Phi(R,\theta,Z) = \Phi(R,\theta) * F_Z$$

$$\text{where } F_Z = \Phi(R,Z) / \bar{\Phi}(R)$$

FIGURE 1
DOSIMETRY LOCATION

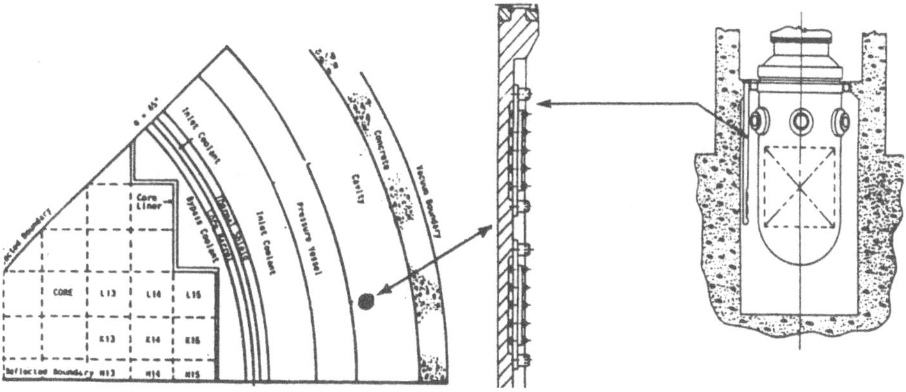


The 67 group BUGLE-80 coupled neutron-gamma cross section set was used. The homogeneous macroscopic material cross sections were generated in the BUGLE-80 group structure using the GIP cross section mixing code. Group cross sections were generated for all materials in the DOT models and a LiF dose rate response function was generated using the MACKLIB lithium and fluorine kerma factors. The dosimeter reaction response functions used were the standard B&W set in the BUGLE-80 structure.

**MEASUREMENT PROCEDURE**

The welded aluminum capsule was found to have very little activation, about 1400 Bq (37nCi) due to trace impurities. The capsule was opened on a lathe in a "warm" shop and the LiF chip was recovered. The optical condition of the LiF chip was found to be essentially perfect, without scratches or any sort of fogging. Over a period of about three weeks, the absorbance of the R and N color centers was measured three times using a different spectrophotometer for the third reading and for the subsequent calibration with $^{60}Co$.

<u>TABLE 1</u>
<u>SPECTROPHOTOMETER READINGS of OPTICAL ABSORBANCE</u>

| Inst ID | $N_2$ (547 nm) | $N_1$ (517 nm) | $R_2$ (374 nm) | $R_1$ (315 nm) |
|---|---|---|---|---|
| C-219 | 0.2908 | 0.3223 | 0.6754 | 0.8856 |
| C-219 | 0.2902 | 0.3230 | 0.6745 | 0.8850 |
| PE-L4C | 0.2971 | 0.3308 | 0.6857 | 0.8978 |

Then the LiF chip was annealed and calibration curves for these color centers were determined using a well-calibrated $^{60}Co$ radiation facility at NIST. The calibration curves are shown in Figure 2. The reproducibility of the calibration was checked by repeating the anneal, radiate, and read steps for the 150 kGy points three more times. In this check, the optical absorbance was found to reproduce with a standard deviation of 1.8% or better for these color centers. The two different spectrophotometers were consistent within about 2.6%.

**MEASUREMENT RESULTS**

The crosses on the calibration curves in Figure 2 are the readings of the LiF chip for the last line of Table 1. The gamma dose values corresponding to these points are listed in Table 2.

<u>FIGURE 2</u>
<u>LiF CHIP CALIBRATION CURVES</u>

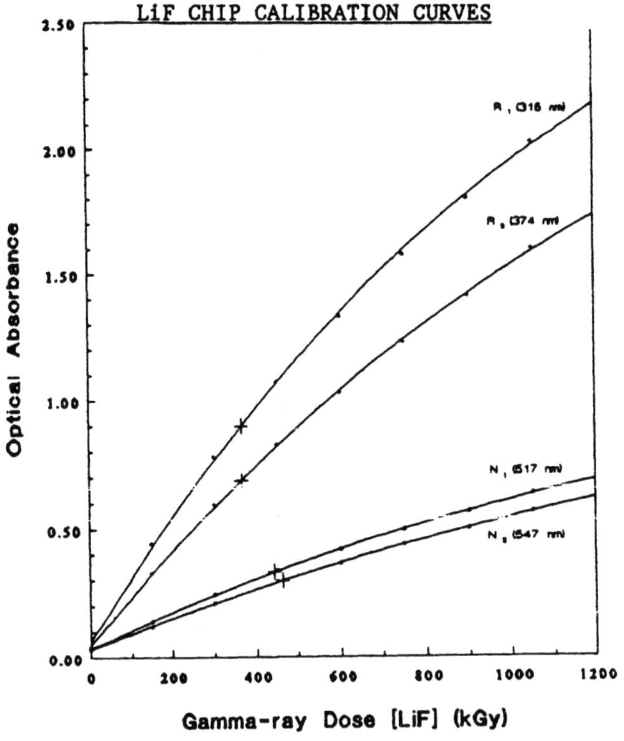

TABLE 2
UNCORRECTED GAMMA DOSE VALUES FROM FIGURE 4
(MGy)

| $N_2$ | $N_1$ | $R_2$ | $R_1$ | Mean | Spread |
|---|---|---|---|---|---|
| 0.463 | 0.443 | 0.366 | 0.367 | 0.410 | 24.0% |

A correction for neutron sensitivity should now be applied to find the portion of the dose indicated in Table 2 which is due to gamma rays only. However, this step turns out to be problematic. The calculated neutron fluence for the 408.9 effective full power days of the exposure was $1.79 \times 10^{17}$ n/cm$^2$. This total fluence number is summed over all groups from thermal to 17 MeV. Since the neutron sensitivity measurement that was previously reported by Gilliam et. al [1] was stated in terms of equivalent gamma dose per fission spectrum fluence, the fluence of the moderated spectrum of this particular cavity must be scaled in some way to represent an effective fission spectrum fluence. Two schemes were tried for this scaling, one on the basis of displaced atoms (DPA) and the other on the basis of kinetic energy deposition (kerma). The scaling factor based on DPA was found to be 0.205, giving an effective fission spectrum fluence of $3.67 \times 10^{16}$ n/cm$^2$. The scaling factor based on kerma was found to be 0.161, giving an effective fission neutron spectrum fluence of $2.88 \times 10^{16}$ n/cm$^2$. The agreement of these two values is acceptable, given the large ($\pm$50%) uncertainties on the neutron sensitivity data of Reference 6. However, when one then calculates the neutron correction based on the neutron sensitivity results of Reference 1, the corrected gamma doses come out unbelievably small or even negative. Since the gamma to neutron ratio is at least five times lower in the present case than had been anticipated in the "typical application" discussed in Reference 1, this difficulty is not completely surprising.

The most probable explanation for this unsatisfactory situation is that the neutron sensitivity values of Reference 6 are significantly overestimated. After all, the difficulties in correcting for the X-rays and other complications were pointed out at some length in Reference 1. Another possible explanation is that some fading (annealing) of the radiation damage took place when the cavity temperature was at its highest (52 °C during the warmest part of the year). However, the experience of Baeyens et al. [2] at CERN makes the possibility of fading seem unlikely. At least at room temperature, Baeyens et al. showed that no fading took place over a period of two years. Furthermore, Baeyens et al. found that a one hour heat treatment at 180 °C produced fading in the R centers but enhancement of the N centers. This response to heating would tend to argue against any uniform fading of both the R and N centers at the 21 °C - 52 °C range in the reactor cavity.

Until better and more complete data are in hand with regard to neutron sensitivity and fading at temperatures in the range of 50 °C, one may at least make a reasonable estimate of the gamma dose rate by assuming that the fading is negligible and using a calculated value for neutron sensitivity based on neutron KERMA. On this basis, the calculated neutron sensitivity correction factor is .045 MGy/($10^{16}$ n/cm$^2$) for fission spectrum neutrons, the neutron correction is 0.13 MGy for this particular irradiation, and the corrected gamma dose inferred from the average value in table 2 is 0.28 MGy. This value is expected to be good within a factor of two.

**CALCULATIONAL RESULTS**

The neutron spectrum in the LiF chip was determined by normalization of the DOT-calculated spectrum to measurements. First the DOT-calculated dosimeter activities were corrected for non-saturation and then combined with the measured activities to obtain measured-to-calculated (M/C) ratios. These M/C ratios were then used to correct the calculated flux spectrum to measurement.

TABLE 3
M/C RATIOS FOR THE NEUTRON DOSIMETERS

| Dosimeter Type | Number by Type | Reaction | M/C |
|---|---|---|---|
| Iron | 4 | Fe(n,p) $^{54}$Mn | 1.05 |
| Nickel | 4 | Ni(n,p) $^{58}$Co | 1.05 |
| Copper | 4 | Cu(n,$\alpha$) $^{60}$Co | 1.05 |
| Neptunium | 1 | Np(n,f)$^{137}$Cs | 0.95 |
| Cobalt | 4 | Co(n,$\gamma$) $^{60}$Co | 0.55 |

The calculational results show that the gamma fluence rate is essentially unperturbed by the LiF chip detector. The gamma energy spectrum is not significantly affected by the dosimeter except in the uppermost energies, where the fluence rate is negligibly small. Figure 3 demonstrates, in more detail, the gamma dose rate profile close to, and through, the dosimeter. The upper curve plots the gamma dose rate for a case that had aluminum only (no Li glass). The lower curve plots the gamma dose rate through the LiF dosimeter package. The maximum difference is about 3.5%. Figure 3 also shows that the lithium glass does not produce significant numbers of radiative capture gamma rays (which would have the effect of flattening out the curve).

## FIGURE 3
## GAMMA DOSE RATE PROFILE IN LiF DOSIMETER

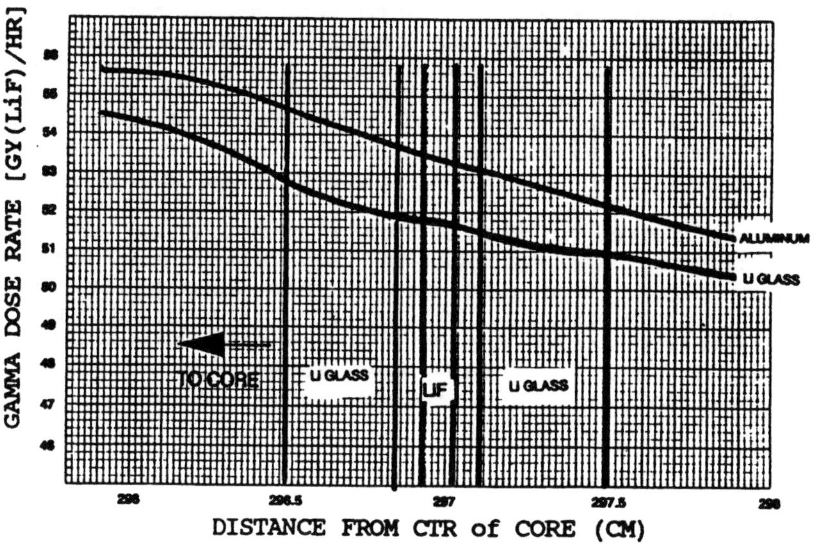


The calculated gamma dose was .50 MGy, which corresponds to a dose rate of 50.9 Gy/hr for an irradiation time of 408.9 full power days.

The gamma fluence rate in the cavity is probably related very closely to the thermal neutron fluence rate in the vessel. If so, then the large disparity between the measured and calculated thermal fluence rate (M/C=.55) is probably applicable to the gamma dose rate as well. The rationale for this hypothesis is that:

(a) most of the gamma ray fluence rate in the cavity is the result of radiative capture events in the reactor vessel;

(b) most of the radiative capture events in the reactor vessel are the result of thermal neutron captures;

then if statements (a) and (b) are true, and it is assumed that the thermal neutron fluence M/C ratio (which was derived from cavity measurements) is valid in the reactor vessel, it follows that the accuracy of the gamma fluence in the cavity is approximately the same as that for the thermal fluence in the cavity. In that case, the gamma dose would be adjusted by the thermal neutron correction to measurement factor of 0.55.

Considerable evidence exists to support statement (a). I. Kodeli and colleagues [4] reported that the gamma fluence produced by secondary gamma rays in the downcomer of a similar reactor was about ten times that from primary gamma ray contributions. This effect would be even more pronounced in the cavity because the secondary-to-primary ratio would increase with increasing penetration. C. D. Bowman and co-workers at NIST concur with this finding (5) as does T. G. Williamson in a recent photo-fission study [6]. Statement (b) is supported by an examination of the balance tables in the DOT code output which shows that the thermal absorptions in the vessel are about six times higher than the sum of the absorptions in all other energy groups. Estimates of the measurement-to-calculation ratio for the integrated gamma dose are given below. These estimates account for the uncertainty in the measured value (page 6) and the calculated value (above).

Using uncorrected calculational results: .28 < M/C < 1.12 ....nominal value = .56

Using corrected results (factor of .55): .51 < M/C < 2.04 ....nominal value = 1.02

**CONCLUSIONS**

The feasibility of LiF chip gamma dosimetry in the environment of a power reactor has been demonstrated, although the slight possibility of response fading at higher cavity temperatures needs to be conclusively ruled out.

The neutron sensitivity of the LiF chips is probably significantly lower than reported in Reference 1, so the suitability of LiF chips as gamma dosimeters in mixed neutron and gamma ray fields would be better than previously expected. However, improved neutron sensitivity data is needed before quantitative results can be obtained when gamma/neutron ratios are as low as they were in this particular cavity.

Finally, this study has shown that LiF chip gamma dosimeters:

(1) do not significantly perturb the field they are measuring

(2) can operate successfully for long periods of time in a cavity-type environment

(3) can be reused

(4) do not become activated.

**ACKNOWLEDGMENTS**

The authors wish to acknowledge the efforts of Mr. Charles M. Eisenhauer of NIST for his invaluable assistance in determining the KERMA- and DPA-based scaling factors used in the neutron sensitivity measurements.

**REFERENCES**

1. D. M. Gilliam, D. G. Hocken, J. C. Humphries, and G. P. Lamaze, "Neutron Sensitivity of LiF Chip Gamma Detectors at Megagray Doses," ASTM/STP1001, p.206

2. B. Baeyens, F. Coninckx, P. Maier, and H. Schonbacher, "Investigations of the use of LiF Crystals for Routine High Level Dosimetry at CERN," High Dose Dosimetry, IAEA, 1985, STI/PUB/671, p.275.

3. W. L. McLaughlin, A. C. Locas, B. M. Kapsar, and A. Miller "Electron and Gamma Ray Dosimetry using Radiation-Induced Color Centers in LiF," Radiation Physics and Chemistry-1979

4. I. Kodeli, B. Glumac, and N. Najzer,"Transport Calculations of the Gamma Field in the Ex-Core Region of the Krsko PWR," ASTM/1001, p.189

5. C. D. Bowman, C. M. Eisenhaur, and D. M. Gilliam, Photo-fission Effect in Reactor Pressure Vessel Dosimetry,"NUREG/CP-004, 1977

6. T. G. Williamson, "A Review of Photofission Measurements,"ASTM/STP1001, p.269

# REACTION RATES OBSERVED IN A MIXED PLUTONIUM-URANIUM SOLUTION CRITICAL ASSEMBLY

Raymond Gold, Christopher C. Preston and James H. Roberts,
Metrology Control Corporation, Richland, Washington, USA

Nubuo Fukumura, Power Reactor and Nuclear Fuel Development Corporation, Oarai, Ibaraki, Japan
and
Tomozo Koyama, Power Reactor and Nuclear Fuel Development Corporation, Tokyo, Japan

## Abstract

Critical assembly experiments have been conducted in an annular water-reflected tank filled with a Pu-U nitrate solution. Absolute neutron reaction rate measurements were carried out using solid state track recorders (SSTR) consisting of mica track recorders and fissile deposits of $^{232}$Th, $^{235}$U, $^{238}$U, $^{237}$Np and $^{239}$Pu. Three sets of these five SSTR fission dosimeters were assembled in aluminum holders and irradiated simultaneously in the critical assembly. These three sets were positioned near mid-plane, i.e. near the half-height of the critical solution, at radial locations of 12.541, 19.685, and 26.788 cm. Absolute SSTR fission rates were determined with an overall uncertainty in the range from approximately 2 to 5 percent($1\sigma$).

Calculations were carried out with Monte Carlo code MCNP-3A, which uses continuous energy range cross section libraries from ENDF/B-IV and ENDL-85. These Monte Carlo calculations are compared with experiment by using relative reaction rates, i.e. using fission rate ratios. On this basis, the calculated-to-experimental (C/E) ratio is shown generally to lie in the 10 - 15 percent ($1\sigma$) range.

## 1.0 INTRODUCTION

Critical experiments have been conducted in the Battelle Pacific Northwest Laboratory's (PNL) Critical Mass Laboratory (CML) using a variety of fast test reactor (FTR) fuel pin configurations and fissile solution concentrations. These experiments are part of a broad collaboration between the U.S. Department of Energy (DOE) and the Power Reactor and Nuclear Fuel Development Corporation, Japan (PNC) concerning criticality data development.[1] In the task 140 series of experiments, absolute neutron reaction rate measurements were carried out with solid state track recorders (SSTR). This particular SSTR neutron dosimetry experiment has been assigned the identification number, CFRP-PNC 2(21/10)-2-087S. Consequently, we will refer to the task 140 SSTR measurements as experiment 087S.

## 2.0 DESCRIPTION OF THE CRITICAL ASSEMBLY

The critical assembly for experiment 087S was an annular water-reflected tank filled with a Pu-U nitrate solution. Figure 1 provides a cross sectional view of this critical assembly. The more relevant characteristics of the critical assembly are summarized in Table 1. A cadmium control blade, located in the water reflector adjacent to the tank, was used to achieve criticality. A period of 67.7 seconds was used to reach a constant power level and the irradiation was conducted for a duration of one hour at this constant power level. The position of the control blade for this constant power level irradiation was 33.496 cm above the bottom of the Pu-U nitrate solution. The actual solution height for the irradiation, which must be slightly higher than the extrapolated critical height, was 49.24 cm.

## 3.0 DESCRIPTION OF SSTR NEUTRON DOSIMETERS

SSTR neutron dosimetry measurements have already been conducted within the framework of this DOE-PNC collaboration, i.e., for the task 220 series of experiments[2] as well as the task 110 series of experiments.[3] A description of the experimental method of SSTR neutron dosimetry can be found in the ASTM Standard E 854.[4] In experiment 087S, three sets of SSTR neutron dosimeters were deployed. Each set contained mica track recorders and fissionable deposits of Th-232, U-235, U-238, Np-237, and Pu-239. Figure 2 provides the axial locations of these three SSTR sets in the critical assembly. Note that the three sets were located at $H_c/2$, where $H_c$ is the critical height of 48.99 cm.

SSTR sets 1 and 3 were deployed on the outer surfaces of the concrete insert and annular tank, respectively. Aluminum holders, shown in Figure 3, were used to insure firm contact between fission deposits and the mica track recorders. The set 1 SSTR were recessed into the surface of the concrete so that the aluminum holders were flush with the outer surface of the concrete insert, with a center-to-center spacing of approximately 3.5 cm between adjacent SSTR. Consequently, the SSTR in set 1 were oriented vertically with the outer surface of the aluminum holders at a radial location of 12.541 cm. For set 3, the SSTR were spaced on the thin aluminum strap with a center-to-center separation of approximately 3.5 cm. This thin aluminum strap was mounted on the outer surface of the annular tank, with the aluminum SSTR holders directly against the tank so that the SSTR of set 3 were oriented vertically. All SSTR of set 3 had a radial location of 26.788 cm.

SSTR for set 2 of 1.91 cm diameter were loaded into a single aluminum capsule of 2.0 cm diameter, 0.63 cm height and 0.05 cm wall thickness. This capsule was inserted in the re-entrant

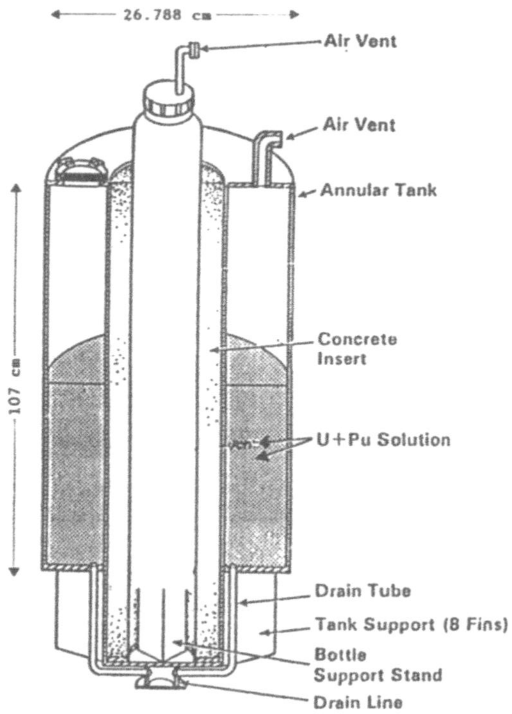


FIGURE 1: Schematic view of the Annular Water-Reflected Critical Assembly. In Experiment 087S, no bottle was placed in the concrete insert.

TABLE 1
CRITICAL ASSEMBLY CHARACTERISTICS FOR EXPERIMENT 087S

| Irradiation Date | Concrete Insert[a] | Pu Density | U Density | Water Density[b] | Free Acid | Critical Height[c] ($H_c$) |
|---|---|---|---|---|---|---|
| | | (g/l) | (g/l) | (g/cm$^3$) | M | (cm) |
| 06/27/86 | 2 wt% $B_4C$ | 102.69 | 364.88 | 1.6780 | 0.77 | 48.99 |

ISOTOPIC DISTRIBUTION OF Pu

| Mass No. | 238 | 239 | 240 | 241 | 242 |
|---|---|---|---|---|---|
| wt % | 0.029 ± 0.001 | 91.12 ± 0.04 | 8.32 ± 0.04 | 0.442 ± 0.004 | 0.091 ± 0.001 |

ISOTOPIC DISTRIBUTION OF U

| Mass No. | 238 | 236 | 235 | 234 |
|---|---|---|---|---|
| wt % | 99.229 ± 0.007 | 0.053 ± 0.002 | 0.710 ± 0.006 | 0.007 ± 0.001 |

a No bottle was placed in the concrete insert for experiment 087S (see Figure 1).

b Water Density measured at 23° C; the level of the water reflector was 24.1 cm below the top of the annular tank.

c Extrapolated Value

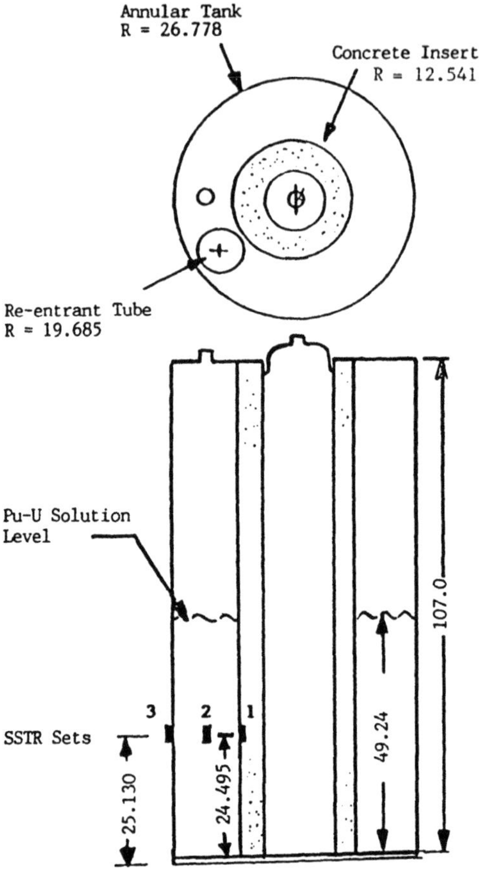


Figure 2: Axial locations of SSTR sets 1, 2, and 3 in experiment 087S. All dimensions are given in centimeters.

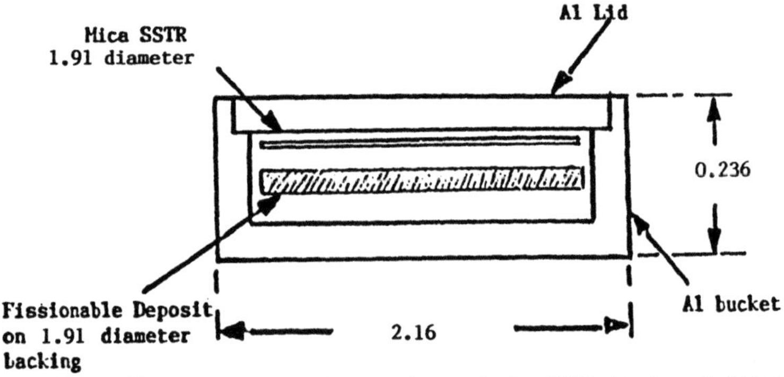


**Figure 3: Geometry and dimensions of the SSTR aluminum holders for SSTR sets 1 and 3.** All dimensions are given in centimeters.

tube, which possesses holes and is therefore filled with the U-Pu nitrate solution. Prior to insertion into the capsule, each mica SSTR and fission deposit was separately wrapped in thin ( ≈ 0.005 cm) aluminum foil. The set 2 capsule was positioned so that all SSTR were oriented horizontally and centered in the re-entrant tube, whose axis lies at a radial location of 19.685 cm.

## 4.0 EXPERIMENTAL RESULTS

SSTR dosimeters have been retrieved and etched. Scanning results are given in Tables 2, 3 and 4 for sets 1, 2 and 3, respectively. As explained in Tables 2 and 4, data was not obtained for the Pu-239 dosimeter of set 1 ( SSTR ID 140-16 ), as well as the U-235 dosimeter of set 3 ( SSTR ID 140-11 ). In addition, some U-Pu solution seeped into the capsule which contained set 2. Solution seepage was revealed by the much higher track density on the bottom edge of the Th-232 SSTR in set 2 (i.e. ID 140-6).

Each SSTR was manually scanned by at least two independent observers. Each observer counted approximately 10,000 tracks, so that the statistical precision of each scan was 1 percent at the 1$\sigma$ level. Consequently, the expected agreement between independent observers should range from 1.4 percent at the 1$\sigma$ level up to 2.8 percent at the 2$\sigma$ level. In most instances the two observers agreed within the 2.8 percent at the 2$\sigma$ statistical limit. This is shown in Tables 2, 3 and 4, where the 1$\sigma$ uncertainty cited for the track density is simply the standard deviation from the mean for the number of scans performed. Hence when this standard deviation was less than 2.8 percent, only two scans were performed.

TABLE 2

FISSION RATE DATA FROM SET 1[a]

| SSTR ID | ISOTOPE | DEPOSIT ID[b] | DEPOSIT MASS DENSITY ($\mu g/cm^2$) | NUMBER OF SCANS | TRACK DENSITY ($Tracks/cm^2$) | FISSION RATE[c] Fissions/(atom • sec) |
|---|---|---|---|---|---|---|
| 140-4 | U-238 | $s_\infty$ | 4.521 E+03 (2.0) | 4 | 1.997 E+06 (3.7) | 4.760 E-17 (5.1) |
| 140-5 | Np-237 | 37K-3 | 9.812 E+01 (1.0) | 2 | 2.381 E+05 (1.9) | 2.638 E-16 (2.3) |
| 140-8 | Th-232 | $s_\infty$ | 4.350 E+03 (2.0) | 2 | 4.264 E+05 (0.7) | 1.029 E-17 (2.1) |
| 140-14 | U-235 | 25A-6-2 | 1.146 E+00 (2.1) | 2 | 7.845 E+04 (2.4) | 7.377 E-15 (3.3) |
| 140-16 | Pu-239 | 49K-4 | 0.786 E+00 (1.0) | d | d | d |

a Value in parenthesis is the 1$\sigma$ uncertainty in percent.

b $s_\infty$ denotes an asymptotically thick metal foil was used. For Th-232 $s_\infty = 4350\ \mu g/cm^2$ and for U-238 $s_\infty = 4521\ \mu g/cm^2$. (See references 5 and 6).

c Exposure duration was (3600+67.7) = 3667.7 sec.

d SSTR 140-16 fell off the concrete insert and consequently representative results could not be obtained for this SSTR.

TABLE 3
FISSION RATE DATA FROM SET 2[a]

| SSTR ID | ISOTOPE | DEPOSIT | | NUMBER OF SCANS | TRACK DENSITY | FISSION RATE[c] |
|---|---|---|---|---|---|---|
| | | ID [b] | MASS DENSITY ($\mu g/cm^2$) | | ($Tracks/cm^2$) | Fissions/(atom sec) |
| 140-1 | Np-237 | 37K-4 | 9.418 E+01 (1.0) | 5 | 2.928 E+05 (3.2) | 3.379 E-16 (3.5) |
| 140-6 | Th-232 | $s_\infty$ | 4.350 E+03 (2.0) | 2 | 4.971 E+05 (2.2) | 1.199 E-17 (3.0) |
| 140-12 | U-238 | $s_\infty$ | 4.521 E+03 (2.0) | 2 | 2.368 E+06 (2.4) | 5.643 E-17 (3.1) |
| 140-13 | U-235 | 25A-6-3 | 1.111 E+00 (2.2) | 2 | 2.135 E+05 (0.7) | 2.071 E-14 (2.5) |
| 140-15 | Pu-239 | 49K-5 | 0.771k E+00 (1.0) | 2 | 2.870 E+05 (1.9) | 4.078 E-14 (2.3) |

[a] Value in parenthesis is the 1σ uncertainty in percent.

[b] $s_\infty$ denotes an asymptotically thick metal foil was used. For Th-232 $s_\infty$ = 4350 $\mu g/cm^2$ and for U-238 $s_\infty$ = 4521 $\mu g/cm^2$. (See references 5 and 6).

[c] Exposure duration was (3600+67.7) = 3667.7 sec.

TABLE 4
FISSION TRACK AND FISSION RATE DATA FROM SET 3[a]

| SSTR ID | ISOTOPE | DEPOSIT | | NUMBER OF SCANS | TRACK DENSITY | FISSION RATE[b] |
|---|---|---|---|---|---|---|
| | | ID[b] | MASS DENSITY ($\mu g/cm^2$) | | ($Tracks/cm^2$) | Fissions/(atom•s |
| 140-2 | Np-237 | 37K-47 | 1.048 E+02(1.1) | 3 | 1.902 E+05(1.4) | 1.973 E-16(2.0) |
| 140-3 | Pu-239 | 49K-6 | 0.788 E+00(1.0) | 5 | 6.437 E+05(3.0) | 8.951 E-14(3.3) |
| 140-7 | Th-232 | $s_\infty$ | 4.350 E+03(2.0) | 3 | 3.178 E+05(2.3) | 7.668 E-18(3.1) |
| 140-10 | U-238 | $s_\infty$ | 4.521 E+03(2.0) | 5 | 1.983 E+06(4.9) | 4.726 E-17(5.3) |
| 140-11 | U-235 | 25A-6-4 | 1.100 E+00(2.2) | d | d | d |

[a] Value in parenthesis is the 1σ uncertainty in percent.

[b] $s_\infty$ denotes an asymptotically thick metal foil was used. For Th-232 $s_\infty$ = 4350 $\mu g/cm^2$ and for U-238 $s_\infty$ = 4521 $\mu g/cm^2$. (See references 5 and 6).

[c] Exposure duration was (3600+67.7) = 3667.7 sec.

[d] For some unknown reason, mica SSTR 140-11 could not be properly etched. Hence track scanning could not be performed on SSTR 140-11 and therefore no results could be obtained from this SSTR.

For four SSTR, namely ID 140-1, ID 140-3, ID 140-4 and ID 140-10, the agreement between the two initial observers was outside the 2.8 percent $2\sigma$ limit. Hence further scans were performed for these four SSTR. The results from more scans provided standard deviations from the mean of 3.2, 3.0, 3.7 and 4.9 percent for SSTR ID 140-1, ID 140-3, ID 140-4 and ID 140-10, respectively. Since these standard deviations still exceed the expected statistical $2\sigma$ limit of 2.8 percent, additional factors exist that are contributing to the uncertainty of these SSTR measurements.

Note that the track density for these four SSTR ranges from 3 E + 05 to nearly 2 E + 06 tracks/cm². In this high track density region, track pile-up exists that can detract from observer objectivity. In manual scanning at high track density, it is difficult to resolve many-fold pile-up, i.e. track overlap where many tracks are involved. In such cases, different observers may conclude different numbers of tracks are involved, so that a systematic bias between independent observers can arise at high track density. Hence it was concluded that this pile-up effect was responsible for the larger than statistical uncertainty observed at higher track density.

As a consequence of this effect, the standard deviation from the mean has been used to define the $1\sigma$ uncertainty in the SSTR track density. With this more conservative uncertainty estimate, the overall uncertainty in absolute fission rate ranges from approximately 2 to 5.3 percent($1\sigma$). This uncertainty range is somewhat larger than the uncertainty that can customarily be attained in SSTR neutron dosimetry experiments.

## 5.0 Comparison of SSTR Results with Calculations

Neutron transport calculations were carried out with the Monte Carlo code MCNP-3A. This code utilizes continuous energy range neutron cross sections from the nuclear libraries ENDF/B-IV and ENDL-85. The calculational model included detailed descriptions of the SSTR aluminum holders and capsules. A total of 27,300 case histories were run.

In order to compare calculation and experiment, reaction rate ratios were formed. For SSTR sets 1 and 2, reaction rate ratios were formed with respect to the U-235 fission rate. Since the SSTR result for U-235 was not available for set 3, reaction rate ratios were formed with respect to the Pu-239 fission rate for set 3. Table 5 summarizes the calculational (C) and experimental(E) reaction rate ratios as well as providing the resulting C/E ratios.

It can be seen from table 5 that the C/E agreement is comparable for all three sets. A measure of the adequacy of the Monte Carlo calculations can therefore be obtained by averaging the C/E results from Table 5. One finds <|C/E-1|> = 0.14, or an average C/E deviation from unity of approximately fourteen percent. Given the uncertainty of calculations (approximately 11 percent 1σ) and experiment (ranging from 2 to 5 percent 1σ), it is concluded that the overall agreement is reasonable.

TABLE 5

COMPARISON OF MCNP CALCULATIONS WITH EXPERIMENT

| SSTR SET | REACTION RATE RAITO | Calculation[a] (C) | Experiment[a] (E) | C/E |
|---|---|---|---|---|
| 1 | U238/U235 | 6.531 E-03 (12.0) | 6.452 E-03 (6.0) | 1.012 |
| | Th232/U235 | 1.556 E-03 (12.0) | 1.395 E-03 (3.9) | 1.115 |
| | Np237/U235 | 3.400 E-02 (11.4) | 3.576 E-02 (4.0) | 0.951 |
| | Pu239/U235 | 2.472 E+00 (21.0) | —— | —— |
| 2 | U238/U235 | 3.225 E-03 (9.7) | 2.725 E-03 (4.0) | 1.183 |
| | Th232/U235 | 7.837 E-04 (9.7) | 5.789 E-04 (3.9) | 1.354 |
| | Np237/U235 | 1.517 E-02 (9.2) | 1.632 E-02 (4.3) | 0.930 |
| | Pu239/U235 | 1.772 E+00 (11.8) | 1.969 E+00 (3.4) | 0.900 |
| 3 | U238/Pu239 | 4.763 E-04 (7.6) | 5.280 E-04 (6.3) | 0.902 |
| | Th232/Pu239 | 1.153 E-04 (7.7) | 8.567 E-05 (4.5) | 1.346 |
| | Np237/Pu239 | 2.379 E-03 (7.0) | 2.204 E-03 (3.9) | 1.079 |

[a] Value in parenthesis is the relative uncertainty in percent at the 1σ level.

## REFERENCES

1. Memorandum of Agreement between the United State Department of Energy and the Power Reactor and Nuclear Fuel Development Corporation, Japan, in the Area of Criticality Data Development, signed August 12, 1983.

2. R. Gold, J. H. Roberts, L. S. Kellog, W. Y. Matsumoto, W. N. McElroy, C. C. Preston, R. L. Simons, and S. R. Bierman, "Reaction Rate Measurements for Nuclear Reactor Analyses and Criticality Data", Nucl. Tracks Radiat. Meas. 14, 387-408 (1988).

3. S. R. Bierman and N. Fukumura, "Fission Rate Measurements at the Vessel/Reflector Interface of a Dilute Plutonium-Uranium Nitrate Solution in Cylindrical Geometry", Trans. Am. Nucl. Soc. 55, 378 (1987).

4. R. Gold, J. H. Roberts, and F. H. Ruddy, Standard

4. R. Gold, J. H. Roberts, and F. H. Ruddy, Standard Method for Application and Analysis of Solid-State Recorder (SSTR) Monitors for Reactor Vessel Surveillance, 1982 Annual Book of ASTM Standards, American Society for Testing and Materials, ASTM E854-81, Philadelphia, Part 45, 1982.

5. R. Gold, R. J. Armani and J. H. Roberts, "Absolute Fission Rate Measurements with Solid State Track Recorders, Nuclear Science Engineering 34, 13, 1968.

6. R. Gold, "Mass and Charge Dependence of SSTR Asymptotic Sensitivity", Proceedings of the First ASTM-EURATOM Symposium on Reactor Dosimetry, Petten, Netherlands, September 22-26, 1975, EUR 5667, Commission of the European Communities, Vol. II, 189-196, 1977.

# CHARACTERIZATION OF GAMMA AND NEUTRON RADIATION FIELDS AT POWER REACTORS IN THE U.S.

GWR Endres, RI Scherpelz, LW Brackenbush
JE Tanner, KL Soldat, DL Haggard

Pacific Northwest Laboratories
Battelle Memorial Institute

Richland, Washington, USA

## ABSTRACT

Personnel from the Pacific Northwest Laboratories have conducted measurements of gamma and neutron dose and energy spectra at about 30 of the operating nuclear power plants in the United States. These measurements have been made in most cases to assess the exposures of the operating plant personnel who must enter high radiation areas while the plant is operating or during maintenance. Many of the measurements were conducted inside reactor containment with the reactor at 100 percent power.

The instruments used for gamma radiation characterization were ion chambers, thermoluminescent dosimeters, and intrinsic germanium spectrometers. For neutron characterizations the tissue equivalent proportional counter (TEPC), $^3$He spectrometers, and lithium iodide detectors used with polyethlene spheres (multisphere spectrometers) were used. Although the intent of these measurements was to assess personnel exposure, the techniques and results would be quite valid for determining the environments to which various system components such as coolant pumps, pressurizers, and electrical equipment are subjected for projections of component lifetimes and for waste assay.

Radiation levels generally were low (1 mR/hr to 100 mR/hr gamma and 1 mrem/hr to 500 mrem/hr neutrons) in PWRs and relatively high in BWRs (100 mR/hr to 2 R/hr gamma and 100 mrem/hr to 20 rem/hr neutron). The neutron spectra were significantly harder (higher energy) in the BWR containments than in the PWR containments. Average neutron energies were in the range of 10 keV to 100 keV in the PWRs and 100 keV to 300 keV in the BWR containments. Gamma radiation spectra were also harder in the BWRs due to a higher component of $^{16}N$ gamma radiation with energies of 6.1 MeV and 7.1 MeV.

Measurements of the activity of activated components have also been conducted to provide the information needed for disposal of the components. Particularly, activated resins and control rods were studied to determine curie content. These measurements were conducted using intrinsic germanium detectors and thermoluminescent dosimeters.

## NEUTRON MEASUREMENTS

Measurements of neutron dose, dose equivalent and energy spectra were taken at nuclear power plants to more accurately characterize the neutron radiation fields and to help define problems with interpretation of neutron monitoring instruments and personnel neutron dosimeters (1). Most of the measurements were taken inside containment with the reactor at or near 100% power, except for boiling water reactors (BWRs) where power levels were usually 50% or less to minimize radiation exposure to personnel conducting the measurements. A diagram showing a typical PWR containment is given in Figure 1 with measurement locations numbered from 1 to 13.

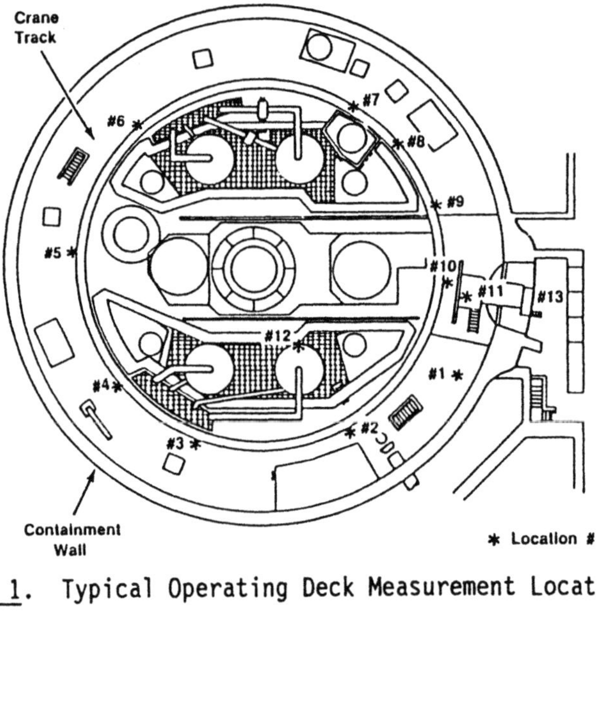


Figure 1. Typical Operating Deck Measurement Locations

The tissue equivalent proportional counter (TEPC) was considered as the standard for neutron dose and dose equivalent determination. All other methods were compared to the TEPC results. The typical results are shown in Table 1.

TABLE 1. TEPC Data for 3 Sites

| Location | | Quality Factor | Absorbed Dose Rate (mrad/hr) | Dose Equivalent Rate (mrem/hr) |
|---|---|---|---|---|
| Site 1 (PWR) | 1 | 10 | 0.12 | 1.2 |
| | 10 | 10 | 0.36 | 3.6 |
| | 11 | 10 | 0.17 | 1.7 |
| | 13 | 10 | 0.03 | 0.3 |
| Site 2 (BWR) | 2 | 11 | 540 | 6000 |
| | 3 | 11 | 18 | 200 |
| | 4 | 11 | 17 | 185 |
| | 5 | 11 | 45 | 495 |
| Site 3 (BWR) | 1 | 11 | 5 | 55 |
| | 3 | 11 | 0.9 | 10 |
| | 4 | 11 | 51 | 560 |
| | 5 | 11 | 26 | 290 |
| | 12 | 11 | 280 | 3080 |

The TEPC is considered to be most accurate because linear energy transfer (LET) information is available directly from the lineal energy spectra produced in the counter. Thus, the quality factor, Q, can be determined directly from this data.

Multisphere spectrometer systems were also used extensively for these measurements. Typical results are given in Table 2 for the multispheres.

TABLE 2. Multisphere Data for Several Sites

| Location | | Average Neutron Energy (keV) | Dose-Equivalent Rate (mrem/hr) |
|---|---|---|---|
| Site 1 (PWR) | 5 | 52 | 5 |
| | 10 | 10 | 2.4 |
| | 11 | 1 | 0.9 |

TABLE 2. Multisphere Data for Several Sites (Cont')

| Location | | Average Neutron Energy (keV) | Dose-Equivalent Rate (mrem/hr) |
|---|---|---|---|
| Site 5 (PWR) | 2 | 50 | 16 |
| | 3 | 62 | 19 |
| | 9 | 65 | 98 |
| Site 6 (PWR) | 12 | 51 | 32 |
| | 14 | 87 | 30 |
| Site 7 (PWR) | 1 | 77 | 45 |
| | 2 | 90 | 24 |
| | 3 | 53 | 8.6 |
| | Interlock | 63 | 1.5 |

Plots of neutron flux versus energy are shown in Figure 2.

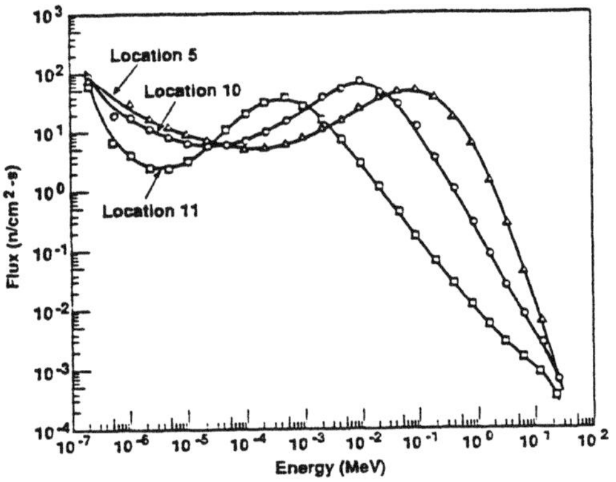


Figure 2. Site 1, Neutron Flux Versus Energy Measured by Multisphere Spectrometer

## GAMMA MEASUREMENTS

High purity germanium, HpGe, and lithium drifted germanium, Ge(Li), detectors were used to obtain photon spectra at various work locations at commercial nuclear power plants (2). Raw data provide estimates of incident spectral distributions. In most cases spectra from reactors that were shut down were dominated by $^{137}Cs$ and/or $^{60}Co$ with a continuum of scattered photons below about 300 keV. Spectra taken in containment at PWRs and near the turbines and heaters of BWRs show the presence of $^{16}N$. Typical spectra are shown in Figures 3 and 4.

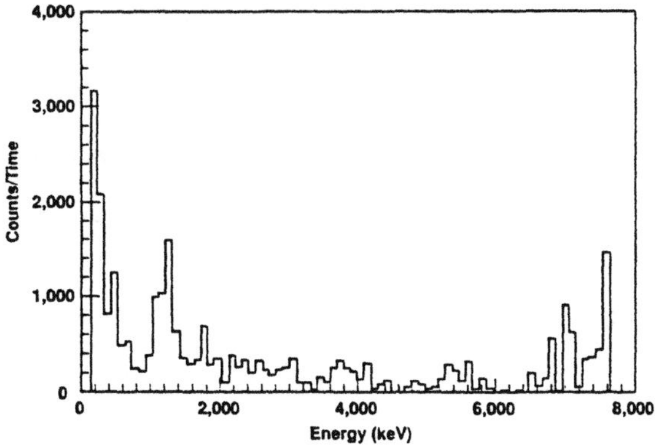


Figure 3. Corrected Field Spectrum (Operating PWR, Near Elevator, 401-ft Elevation)

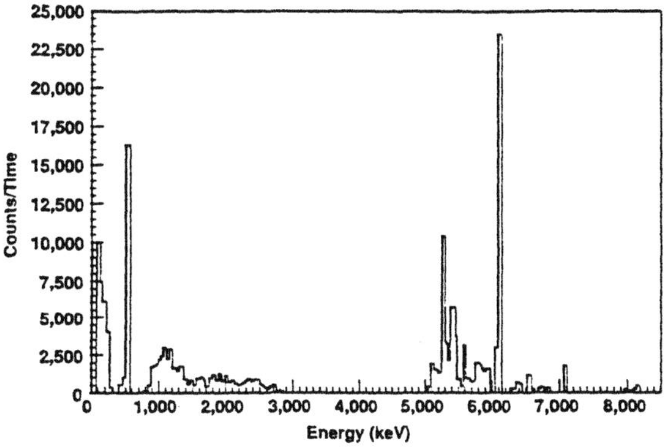


Figure 4. Corrected Field Spectrum, Operating BWR, Entrance Hallway (collimated)

## WASTE ASSAY TECHNIQUES

Direct assay techniques involving gamma-ray spectrometry for qualitative analyses, TLD dosimetry followed by dose-to-curie conversion, and correlation analysis for non-gamma emitters have been developed into a powerful analytical tool for the measurements of radionuclides in nuclear plant radwastes (3). These methods are particularly advantageous for assaying waste packages having high dose rates (up to 22,000 R/hr), and can accommodate a wide variety of waste container geometries (4). This direct assay technology can save utilities time, money, and radiation exposure to personnel by minimizing the need for extensive sampling and complicated radiochemical analyses. This technology has been successfully demonstrated at several nuclear power plants (5).

The recent enactment of 10 Code of Federal Regulation (CFR) 61, "Licensing Requirements for Land Disposal of Radioactive Waste", has necessitated a more complete and accurate analysis of the radionuclide contents of radwaste packages produced at nuclear power plants for disposal at low-level waste burial facilities. The radionuclides specified in the 10 CFR 61, whose concentrations must be reported for each radwaste package, include the following: $^{3}H$, $^{14}C$, $^{59}Ni$, $^{60}Co$, $^{63}Ni$, $^{90}Sr$, $^{94}Nb$, $^{99}Tc$, $^{129}I$, $^{137}Cs$, $^{241}Pu$, $^{242}Cm$, and alpha-emitting transuranic radionuclides (TRU) having half-lives greater than five years (6). The TRU radionuclides of significance in nuclear power station radwastes include $^{238}Pu$, $^{239}Pu$, $^{240}Pu$, $^{242}Pu$, $^{241}Am$, $^{244}Cm$, and $^{246}Cm$ (7).

With the exception of $^{60}Co$ and $^{137}Cs$, the above group of radionuclides are difficult to measure in radwastes since they decay by either beta, alpha, or low-energy photon emissions, which are not directly detectable. The decay of $^{94}Nb$ is accompanied by two high energy gamma rays, but the concentration of $^{94}Nb$ is usually too low for measurement by direct gamma-ray spectrometry. Therefore, present methods for the measurement of these radionuclides in radwastes require sampling and destructive radiochemical analysis which are time consuming, technically difficult, expensive, and contribute to increased radiation exposure to plant personnel.

A typical system for conducting underwater gamma-ray spectrometry of highly radioactive waste stored in spent fuel pools is schematically shown in Figure 5. A portable intrinsic germanium detector and liquid nitrogen cryostat are mounted on the above-pool end of a 3.8-cm (1-1/2 in.) diameter by 12 to 14-m (40 to 45 ft) long steel collimator pipe which is capped on the underwater end. The underwater end of the pipe is placed on top of the waste canister to form a collimated beam of gamma-rays from the canister to detector via the 3.8-cm diameter air-filled pipe.

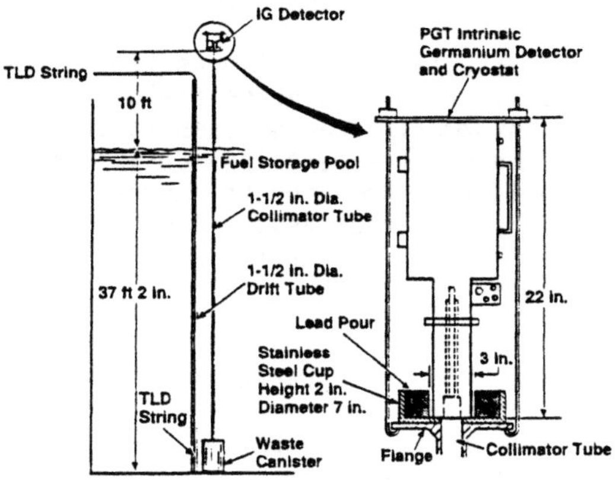


Figure 5. Collimator Tube and Drift Tube Positions for Gamma Spectrum and TLD Measurements on Canisters of In-Core Waste Components

Thermoluminescent dosimeters (TLDs) are used to determine the gamma dose profiles of radwaste packages, and provide input for computer codes such as CYLIN, SLAB, or ISOSHLD which calculate the quantities of radionuclides from gamma exposure rates from the various radwaste package configuration. This technique was used for direct assay of spent BWR control rods and compared to calculations of curie content as shown in Table 3.

Table 3. $^{60}Co$ Curie Content of Each Control Rod

| Control Rod Number | Curie Content Measured | Curie Content Calculated* | % Difference |
|---|---|---|---|
| 1 | 691 | 626 | -9.4 |
| 15 | 606 | 622 | +2.7 |
| 47 | 1210 | 1220 | +0.7 |
| 55 | 801 | 982 | +22.6 |
| 64 | 748 | 713 | -4.6 |

* Torrey Pines Group using code "Path".

## CONCLUSION

The neutron dose and spectral measurements provide much improved characterization of neutron fields around operating reactors. This information can be used to develop better calibrations for both instruments and dosimeters. Gamma spectral measurements provided information to improve calibration of personnel dosimeters. Direct assay of waste packages shows promise as an accurate and versatile method for determining curie contents of rad wastes.

## REFERENCES


1. J. M. Aldrich, L. W. Brackenbush, G. W. R. Endres, L. G. Faust, R. V. Griffith, D. E. Hankins, Neutron Dosimetry at Commercial Nuclear Plants, NUREG/CR-1769, U.S. Nuclear Regulatory Commission, Washington, D.C., May, 1981.

2. G. W. R. Endres, R. A. Fox, D. L. Haggard, K. L. Holbrook, L. A. Rathbun, P. L. Roberson, Spectral and Dosimetric Measurements of Photon Fields at Commercial Nuclear Sites, NUREG/CR-3569, U.S. Nuclear Regulatory Commission, Washington, D.C.

3. U.S. Nuclear Regulatory Commission, Licensing Requirements for Land Disposal of Radioactive Wastes, 10 Code of Federal Regulation 61. Federal Register, Vol. 26, No. 142, Friday, July 24, 1981.

4. R. L. Brodzinski, D. P. Brown, H. G. Rieck, Jr., L. A. Rogers, Design and Operation of a Passive Neutron Monitor for Assaying the TRU Content of Solid Wastes, PNL-4966, Pacific Northwest Laboratory, Richland, Washington, February, 1984.

5. J. E. Cline, et al., Assay of Long-Lived Radionuclides in Low-Level Wastes from Power Reactors, NUREG/CR-4101, U.S. Nuclear Regulatory Commission, Washington, D.C., April, 1985.

6. W. T. Best and A. D. Miller, Radionuclide Correlations in Low-Level Radwastes, Final Report, EPRI NP-4037, Electric Power Research Institute, Palo Alto, California, June, 1985.

7. J. C. Evans, et al., Long-Lived Activations Products in Reactor Materials, NUREG/CR-3473, U.S. Nuclear Regulatory Commission, Washington, D.C., August, 1984.

# CALCULATED NEUTRON SPECTRA OF FAST PULSED REACTORS

T.M. Flanders and M.H. Sparks

Nuclear Effects Laboratory
White Sands New Mexico,
USA

## ABSTRACT

We have calculated the neutron spectrum produced by the White Sands Missile Range Fast Burst Reactor, including the contributions by the major scattering surfaces of the exposure cell. The model of the core and its environment were improved to obtain more accurate information on neutrons leaving the core and their relationship to the neutrons arriving at the detection point.

## INTRODUCTION

The White Sands Missile Range (WSMR) Fast Burst Reactor (FBR) is used for materials testing. The FBR is an unmoderated, unreflected, bare critical assembly. The reactor can operate in either the steady state mode at the several kilowatt level or can be made super-prompt critical, producing a pulse with a full-width at half-maximum time of 45 microseconds or longer. The neutron distribution leaking from the core is a slightly degraded fission spectrum.

Godiva-type assemblies have been studied for both their time-dependent behavior and for leakage spectrum determinations[1-5]. An important parameter that characterizes the significant portion of the spectrum is the spectral index:

$$S.I. = \frac{\Phi > 10\,kev}{\Phi > 3\,MeV} \qquad (1)$$

The early measurements gave spectral indices which ranged from 6.7 to as high as 8.3. Harvey[4] calculated the FBR spectrum using the DOT-III code[6] and found the spectral index to be about 8.6. More recent determinations[7,8] using SAND-II[9], STAY'SL[10] and LSL-M2[11] narrow the range of spectral indices to 6.7 with a high of 7.4. The estimated uncertainties on the spectral index was on the order of 5%.

In 1989 Flanders and Sparks[12] reported new Monte Carlo calculations for the WSMR FBR. They first calculated the leakage spectrum of the FBR using MCNP. The leakage flux was found to have a spectral index of 7.59. The leakage flux was then used as the starting source for a second calculation which treated the FBR as a point source in the exposure cell. The spectral index at 61 cm.(24") from core centerline was 7.96, or about 19% softer than the historical spectrum reported in 1975. The difference between the calculations and the more recent experimental determinations was about 8%

New activation data became available in 1989 for the FBR. The experiment was similar to earlier activation foil experiments, but a wider selection of foils was available. Sparks et. al.[13] reported on several new evaluations of the WSMR spectrum using the latest available activation data. Both LSL-M2 and SAND-II adjustments to the activation data gave spectral indices of 7.3.

In this work, we report the progress of our calculations simulating the measured parameters for the WSMR FBR. We also calculated the spectral distribution at points farther from the core assembly in order to investigate the degradation of the spectrum with increasing distance.

## DISCUSSION

### CORE MODEL AND LEAKAGE CALCULATION

The FBR core is a cylindrical assembly consisting of six annular fuel rings made of highly enriched uranium-molybdenum alloy. The core has an outer radius of 10.16 cm.(4") and it is 27.2 cm.(7.7") high. The fuel rings have an inside radius of 5.08 cm(2"). The safety block is a large fuel element that is inserted into the central cavity of the core. When the safety block is removed from the core, the assembly cannot be made critical. The fuel rings are bolted to a stainless steel support plate by three Inconel metal bolts. There is a stainless steel retaining plate at the top of the core. The power level of the core is adjusted by two control rods of the same fuel alloy, which are inserted into voids in the body of the rings. A third control rod can be quickly driven into the core for burst operations. The core is covered by a cylindrical decoupling shroud, made of aluminum and a B-10 loaded silastic. The decoupling shroud minimizes the effect of reactivity changes due to the experiments and the external environment.

The core geometry is shown in Figure 1. Six materials are used in the core model: air, type 303 stainless steel, aluminum with a B-10 loaded silastic material, fuel, Inconel X-750, and aluminum.

The core calculation was performed using the MCNP code[14]. No external environment was included for this model. The continuous-energy cross section representation was used. For the core calculation, about 300,000 source particles were followed. The core leakage flux and current was calculated at the cylinder that defined the outer surface of the core decoupling shroud. The scoring particles were sorted into discrete energy bins corresponding to the DLC-130 cross-section library[15].

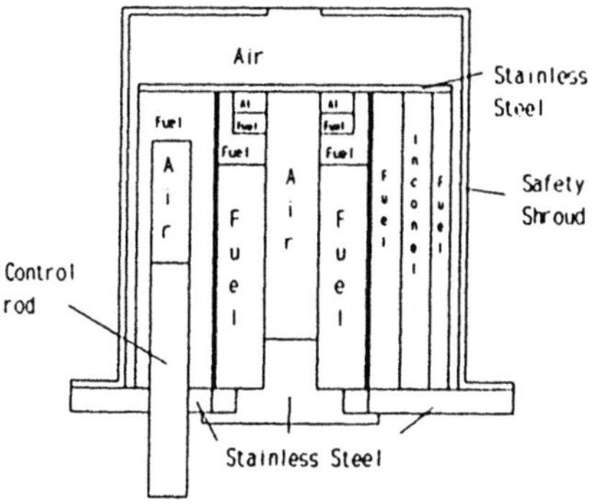


**Figure 1**

**Vertical Cross Section of the Monte Carlo Model of the WSMR Fast Burst Reactor.**

The calculated spectral index using the flux estimator is 7.76. For energy above 3 MeV, the neutron current estimator shows a higher fraction of neutrons than the flux estimator for the same neutron population. In Table I, we examine the dependence of the spectral index on the angular distribution of the neutron field. The angle is relative to the normal direction vector at the scoring surface. As the exit angle relative to the surface normal increases, the spectrum gets progressively softer as demonstrated by the increasing spectral index.

Neutrons leaving the core at high angles are unlikely to make a significant contribution to a detector placed at more distant radial positions at the mid-height of the core. These neutrons are following trajectories which will not intersect the small volume of detectors at remote positions unless they undergo further scattering interactions which direct them toward the detector position. The situation is somewhat different for the case where the detector surfaces are large or relatively close to the core. In this case, the neutron track is more likely to intersect the region of interest regardless of the exit angle from the core.

Table I

Current Tally as a Function of Angle Relative to the Surface Normal for the Godiva System.
About 300,000 Histories.

| Angular Bin | 0° - 30° | 30° - 60° | 60° - 90° | Total |
|---|---|---|---|---|
| Fluence | | | | |
| $\Phi$ > 10 keV | 0.1318 | 0.1966 | 0.04539 | 0.3738 |
| $\Phi$ > 3 MeV | 0.01934 | 0.02764 | 0.00479 | 0.0518 |
| S.I. | 6.81 | 7.11 | 9.48 | 7.22 |
| % of Total | | | | |
| $\Phi$ > 10 keV | 35.3 | 52.6 | 12.1 | |
| $\Phi$ > 3 MeV | 37.4 | 53.4 | 9.2 | |

**SPECTRUM AT 61 CM (24 in.) FROM THE CORE CENTERLINE**

The geometric description of the model used to perform the $k_{eff}$ calculation described only the FBR core. The environment surrounding the core was not included in the model. The FBR is operated in the center of an exposure cell, which is 15.25 meters(50') square by 6.1 meters(20') tall. The walls are 61 cm.(24") of concrete and lined by 20.3 cm.(8") of gypsum wallboard. The 10.2 cm.(4") of wallboard closest to the concrete and the floor is boron loaded. The cell is designed to minimize wall-return neutrons. However, there is a certain number of neutrons which will be reflected from the walls, floor, and ceiling which will eventually reach the experimental position.

In the earlier calculations, the source was calculated as neutrons left the core without considering the exit angle. The resulting neutron energy distribution was used as the starting distribution for a calculation in which the reactor was treated as a point source in the exposure cell. All neutrons were started isotropically. A better starting distribution would have been a joint probability distribution function based on angle and energy.

An alternative method for improving the calculation fidelity is to construct a model which describes the reactor and the exposure cell in a single model. This approach increases the times compared to the point source model, however, the geometry of this model more nearly simulates the real world situation.

Figure 2 shows the calculated spectra (fluence/lethargy-cm$^2$ per fission neutron for several positions using the model which included the exposure cell. The calculated spectral index at 61 cm.(24") is 7.74. It has been reported earlier that the least squares method (using the LSL-M2 spectral adjustment code) gives a spectral index between 7.3 and 7.4. This represents a difference of only 5% between calculation and experiment. The experimental uncertainty of the LSL-M2 determined spectrum is approximately 8%.

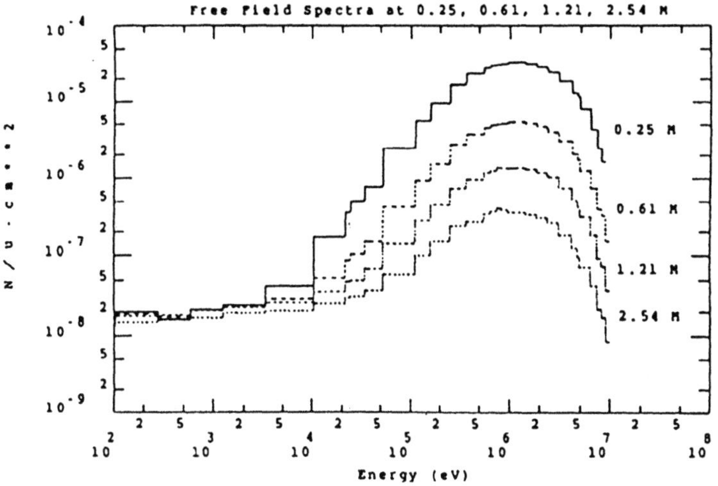


**Figure 2**

Fast Burst Reactor Spectra at 4 Locations.
Distance is Measured From the Core Center.

Table II shows the spectral index for several exposure positions. Note that the change in spectral index is not significant until the detector position exceeds about 61 cm.(24") from the core. However, at 254 cm.(100") the spectrum is 28% softer relative to the 61 cm.(24") position as measured by the spectral index.

Table II

Neutron Spectral Parameters for Several Exposure Positions in the Exposure Cell.

| Distance from Core Centerline cm. | Fluence > 3 MeV $n/cm^2$ | Fluence > 10 keV $n/cm^2$ | Spectral Index | Hardness Parameter |
|---|---|---|---|---|
| Core Surface* | 4.71 E-5 | 3.62 E-4 | 7.69 | 1.01 |
| 25.4 (10") | 1.13 -5 | 8.75 -5 | 7.76 | 1.01 |
| 40.64(16") | 4.21 -6 | 3.24 -5 | 7.69 | 1.01 |
| 60.96(24") | 1.84 -6 | 1.43 -5 | 7.74 | 1.00 |
| 121.9 (48") | 4.52 -7 | 3.76 -6 | 8.33 | 0.971 |
| 254.0(100") | 1.08 -7 | 1.08 -6 | 9.94 | 0.900 |

* Surface Crossing Estimator
Fluences are normed per fission neutron.

In Table II, we have also tabulated the hardness parameter (HP)[16]. The damage fluence to a silicon device is measured by multiplying the fluence greater than 10 keV by the hardness parameter. For these free field conditions, as the spectral index goes up, the hardness parameter goes down. At 61 cm.(24") the 1 MeV equivalent multiplier is 7.74 if the measurable parameter is fluence greater than 3 MeV as determined by sulfur activation measurements[1]. At 254 cm.(100"), the multiplier would be 8.95. This represents an underestimate of about 14% for the 1 MeV equivalent fluence in contrast to the underestimate by 22% of the fluence greater than 10 keV.

$$H.P. = \frac{\int_{10keV}^{E_{max}} \Phi(E)\, K_D(E)\, dE}{K_D(E_0) \int_{10keV}^{E_{max}} \Phi(E)\, dE} \qquad (2)$$

$K_D(E)$ is the silicon neutron displacement kerma as a function of energy

$\Phi(E)$ neutron fluence

$K_D(E_0)$ average value of the neutron displacement kerma near 1 MeV.

---

1 The 1 MeV Equivalent multiplier is the spectral index times the hardness parameter. The equivalent fluence is obtained by multiplying the fluence > 3 MeV by the multiplier.

## CONCLUSIONS

Using the new calculation of the Spectral Index at 61 cm.(24"), the calculated fluence greater than 10 keV will differ from the experimental determination by about 5%. However, if the calculated spectral index at 61 cm. is used for all experimental locations (as is presently the case) the fluence greater than 10 keV is underestimated for the more distant positions. The calculation indicates the fluence at 254 cm.(100") will be underestimated by 22%. This condition will be mitigated somewhat when silicon displacement dose is calculated since the spectrum grows progressively softer as indicated by the hardness parameter. The 1 MeV Equivalent dose will be underestimated by 14%

## ACKNOWLEDGMENTS

This investigation was sponsored by the U.S. Army Test and Evaluation Command Methodology Improvement Program and by the U.S. Army Office of Research and Technology.

## REFERENCES

1. K. Humpherys, W. Quam, C. Rainbolt, R. Durkee, T. Dahlstrom, "Nuclear Radiation Dosimetry Measurements of the White Sands Missile Range Fast Burst Reactor", EG&G Report S-262-R, Edgerton, Germeshausen & Grier Inc., Santa Barbara CA., (1965).

2. J.L. Meason, H.L. Wright, "Neutron Radiation Environment Measuring Techniques", Dosimetry Technical Report Number 70-1, Nuclear Effects Laboratory, White Sands, NM., (1970).

3. J.L.Meason, H.L. Wright, J.C. Hogan, J.T. Harvey, "The Neutron Spectral Distribution From a Godiva Type Critical Assembly", IEEE NS-22, No. 6, (1975).

4. J.T. Harvey, Ph.D Thesis, University of Arkansas, (1977).

5. V.V. Verbinski, N.A. Lurie, V.C. Rogers, "Threshold-Foil Measurements of Reactor Neutron Spectra for Radiation Damage Applications", Nucl. Sci. Eng., 65, (1978).

6. W.A. Rhoades, F.R. Mynatt, "The DOT-III Two-Dimensional Discrete Ordinates Transport Code", Report ORNL-TM-4280, Oak Ridge National Laboratory, Oak Ridge, TN., (1973).

7. J.G. Kelly, D.W. Vehar, "Measurement of Neutron Spectra in Varied Environments by the Foil Activation Method with Arbitrary Trials", SAND87-1330, Sandia National Laboratories, Albuquerque, NM., (1987).

8. J.G. Williams, R.O. Peach, A.M. Ougouag, W.W. Sallee, T.M. Flanders, J.L. Meason, "Neutron Field Characterization for Hardness Testing of Electronics by Least Squares Adjustment Methods", Volume III, 1988 International Reactor Physics Conference, Jackson Hole Wyoming, (1988).

9. W.N. McElroy, S. Berg, T. Crockett, R. Hawkins, "A Computer Automated Iterative Method for Neutron Flux Spectral Determination by Foil Activation", AFWL-TR-67-41, Vol I-IV, Air Force Weapons Laboratory, Kirtland AFB, NM., (1967).

10. F.G. Perey, "Least Squares Dosimetry Unfolding: The Program STAY'SL", ORNL/TM-6062, Oak Ridge National Laboratory, (1977).

11. F.W. Stallman, "LSL-M2: A Computer Program for Least Squares Logarithmic Adjustment of Neutron Spectra", NUREG/CR-4349, Nuclear Regulatory Commission, Washington D.C., (1985).

12 T.M. Flanders, M.H. Sparks, "Monte Carlo Calculations of the Neutron Environment Produced by the White Sands Missile Range Fast Burst Reactor", Nucl. Sci. Eng., 103, 265-275, (1989).

13. M.H. Sparks, T.M. Flanders, J.G. Williams, J.G. Kelly, W.W. Sallee, M. Roknizadeh, J.L. Meason, "Energy Dependence of Neutron Damage in Silicon Bipolar Transistors", IEEE Transactions on Nuclear Science, Vol 36, no. 6, pp. 1904-1911, (1989).

14. LOS ALAMOS TRANSPORT GROUP, "MCNP - A General Monte Carlo Code for Neutron and Photon Transport", Los Alamos, NM., (1981).

15. D.T. Ingersoll, R.W. Roussin, C.Y. Fu, J.E. White, "DABL69: A Broad-Group Neutron/Photon Cross-Section Library for Defense Nuclear Applications", ORNL/TM-10568, Oak Ridge National Laboratory, Oak Ridge, TN., (1989).

16. ASTM, Standard E722-85, 1986 Annual Book of ASTM Standards, Vol 12.02, pp 448-454, American Society for Testing and Materials, (1986).

EFFECTS OF SPECTRAL PERTURBATION ON

CALCULATED DAMAGE COEFFICIENTS

M.H. Sparks and T.M. Flanders

Nuclear Effects Laboratory

White Sands Missile Range, New Mexico, USA

**ABSTRACT**

Calculations have been made for different types of shielding materials using an updated version of the DNA Cross Section data set, DABL69, and a Godiva-type leakage source. ANISN was used to run the transport calculations to predict the change in spectral shape due to the perturbation effects of different materials. One material of particular interest has been iron, due to new cross section evaluations.

**INTRODUCTION**

Fast Burst Reactors (FBR) provide ideal environments for testing material responses to neutron exposure. In particular, they can be used to test the degradation of electronic devices in severe conditions. Silicon devices have been proposed as fluence monitors. However, difficulties in analysis and other limitations restrict their use[1,2] This is particularly true for strongly perturbed neutron fields such as those encountered in shielding applications. For these cases, transport analysis of the environment provides significant benefits for dosimetry purposes. A large selection of processed cross section libraries are available.

The Defense Applications Broad-Group Library (DABL69)[3] has been released for use. This 46-neutron and 23-gamma ray group library replaces the well known DLC-31/FEWG1 Library[4]. DABL69 was designed to improve deep penetration calculations such as those encountered in shielding problems. Direct comparisons can be made between the two libraries because of similarities in energy group structure. We have modeled several shields using both libraries to determine what differences, if any, might be encountered due to the change in cross sections.

In addition, there have been new silicon neutron displacement kerma calculations using the NJOY[5] cross section processing code. For a 14 MeV source, there is a significant difference between the NJOY and

ASTM predicted damage. This paper considers whether this difference extends to the wide range of spectra which the FBR may produce in a perturbed environment.

## SPECTRAL PARAMETERS

The parameter that characterizes the significant portion of the spectrum is the spectral index:

$$S.I. = \frac{\Phi > 10 kev}{\Phi > 3 MeV} \tag{1}$$

In addition to the spectral index, an important parameter for this reactor is the hardness parameter (HP). The HP is a measure of the damage efficiency of the experimental spectrum relative to the damage produced by a reference monoenergetic source. This value is obtained by folding the silicon displacement kerma with the experimental spectrum. Neutrons with energy below 10 keV cause an insignificant amount of damage in silicon for fission spectra and the lower energy threshold is set at 10 keV in most calculations[6]. If there is significant moderation of the neutron spectrum, the lower limit for damage may need to be adjusted downward. Although the selection of the reference energy is arbitrary, intercomparison of data requires adjustment to the same kerma reference value. The normal reference energy is 1 MeV. The hardness parameter is defined as:

$$H.P. = \frac{\int_{10 keV}^{E_{max}} \Phi(E)\, K_D(E)\, dE}{K_D(E_0) \int_{10 keV}^{E_{max}} \Phi(E)\, dE} \tag{2}$$

where:

| | |
|---|---|
| $K_D(E)$ | is the silicon neutron displacement kerma as a function of energy |
| $\Phi(E)$ | neutron fluence |
| $K_D(E_0)$ | average value of the neutron displacement kerma near 1 MeV. |

At White Sands Missile Range (WSMR), sulfur is used as the routine monitor for fluence greater than 3 MeV. In order to obtain the total fluence greater than 10 keV, the sulfur fluence is multiplied by the spectral index. Multiplying the total fluence by the hardness parameter gives the 1 MeV equivalent silicon displacement dose.

## CALCULATIONS

The shielding runs were not designed to simulate real exposure criteria, but rather to compare the FEWG1 and DABL69 cross section data sets. The source for these calculations was obtained from a previous calculation using MCNP[7]. The source was a Monte Carlo calculated leakage spectrum from a Godiva-type assembly[8] transported to 61 cm (24"). This source is a slightly degraded fission spectrum with wall return[9] from the WSMR exposure cell. Previous calculations have shown the neutron population below 10 keV is less than 1 percent of the total number of neutrons leaking from the core[10] The shielding calculations then transport the source through the materials with the detectors positioned several centimeters behind the shield. Of interest in this work is the way in which different materials and different combinations of materials affect the spectrum.

In these calculations, varying thicknesses of each material were used to determine the significant spectral changes. After calculating the spectral perturbation due to single types of material, the effects of layered combinations of different materials were also calculated. As with the single type of material, varying thicknesses were used in the combined material calculations.

ANISN[11] a one dimensional discrete ordinates code, was used for the shielding calculations. Slab geometry was used in all cases. Two cross section libraries were used for the calculations. The most recent of the two libraries was the DABL69 cross section set which was generated from the VITAMIN-E[12] cross section library. It replaced the older DLC-31 library. The newer library utilizes a 46-neutron and 23 gamma-ray energy group structure. The new energy group boundaries cover the region from 0.2 to 1.8 MeV much better with seven new energy boundaries. Also, there is a new boundary at 34.307 keV to better define the iron total cross section. With the exception of the boundary at 12.523 MeV, the newer energy structure is a superset of the DLC-31 energy group structure. Therefore, direct comparisons can be made. The DABL69 group cross sections are based on the ENDF/B-V files and in a few cases, even more recent evaluations.

## SUMMARY

For typical test conditions, the free field spectral index at 61 cm. (24") from the core centerline has been experimentally determined by Sallee[13] to be 7.3 and the HP is 1.01(ASTM) or 0.917(ENDF B-VI). The spectrum calculated by transport methods predicts 6% higher total fluence than the experimental free field value. Figure 1 shows the 46 group source spectrum used for these calculations. Table I provides the spectral index for a sample of the shielding calculations in this study.

Calculated spectra were folded with various silicon displacement kerma functions to predict the dose received by a test object. For the calculation of the hardness parameter, several silicon displacement kermas were used. The kerma factor curve has not been characterized adequately by experimental means. However, there are several theoretical tabulations of this data. The theoretical tabulations do not all agree, in detail, on the shape of the response function curve. This lack of agreement on the shape of the function has a direct impact on establishing the equivalent damage of neutrons of one energy relative to another.

The function designated DNA is a local modification of the neutron displacement kerma from Reference 4. The original kerma values did not agree with other tabulated kerma values below 100 keV. The kerma values below this energy were adjusted to agree more closely with accepted kerma standard values from Reference 6.

The ASTM recommended response function is based on silicon data which was available in 1975[14]. The current 1 MeV normalization is based on 95 MeV-mb. The ASTM E722 Standard[15] is undergoing revision and the silicon displacement kerma table will be replaced.

**TABLE I**

Spectral Index for Various Shielding Material using 37 Group(DLC-31/FEWG) and 46 Group(DLC-130/DABL69) Shielding Libraries.

| Shield | Thickness | Spectral | Index |
|---|---|---|---|
| | (cm) | FEWG1 | DABL69 |
| Al | 2.54 | 7.71 | 7.78 |
| | 5.08 | 9.07 | 9.17 |
| | 15.24 | 16.5 | 16.8 |
| Fe | 2.54 | 11.79 | 11.29 |
| | 5.08 | 17.15 | 15.83 |
| | 15.24 | 83.16 | 67.37 |
| Cu | 2.54 | 17.8 | 18.4 |
| | 5.08 | 41.1 | 43.0 |
| | 15.24 | 703 | 823 |
| Plywood | 2.54 | 6.88 | 6.94 |
| | 5.08 | 7.13 | 7.27 |
| | 15.24 | 5.91 | 6.10 |
| 5%-Borated Polyethylene | 2.54 | 6.24 | 7.18 |
| | 5.08 | 6.02 | 6.80 |
| | 15.24 | 4.53 | 4.39 |

The NJOY B-VI[1] kerma function is based on ENDF/B-VI cross section data and the NJOY cross section processing code. This calculated kerma function is based on the latest available cross section data. It has been proposed as the replacement response function for the ASTM Standard E722 which is undergoing revision.

The 1 MeV equivalent response function supplied with the DABL69 library was also folded with the 46 group spectra. This response function is based on two separate evaluations of the kerma[14,1

[1] We would like to thank P.J. Griffin of Sandia National Laboratories for kindly supplying a draft version of the new response function in 46 group format.

However, it is normalized to 82 MeV-mb instead of the recommended value of 95 MeV-mb, and the values for the HP average 10% higher than the ASTM response function values.

Note that while the spectral index indicates the spectra changes only very slightly depending on which cross sections are used, the spread in calculated hardness parameters can be quite large depending on which response function is used. This is particularly true when the response function is based on a different normalization value. This is apparent in Table II.

**TABLE II**

Calculated Hardness Parameters Using Various Silicon Displacement Response Functions. Hardness Parameters Are Based on 46 Group Spectra.

| Shield | Thick | H.P. | | | |
|---|---|---|---|---|---|
| | cm. | DNA | ASTM | ENDF B-VI | DABL69 |
| Al | 2.54 | 1.01 | 1.00 | 0.909 | 1.10 |
| | 5.08 | 0.978 | 0.966 | 0.877 | 1.05 |
| | 15.34 | 0.815 | 0.791 | 0.724 | 0.863 |
| Cu | 2.54 | 0.888 | 0.849 | 0.794 | 0.934 |
| | 5.08 | 0.768 | 0.718 | 0.680 | 0.790 |
| | 15.34 | 0.511 | 0.468 | 0.454 | 0.517 |
| Fe | 2.54 | 0.971 | 0.947 | 0.871 | 1.04 |
| | 5.08 | 0.924 | 0.889 | 0.824 | 0.977 |
| | 15.34 | 0.739 | 0.685 | 0.645 | 0.754 |
| Plywood | 2.54 | 0.956 | 0.958 | 0.865 | 1.05 |
| | 5.08 | 0.908 | 0.909 | 0.822 | 0.997 |
| | 15.34 | 0.929 | 0.939 | 0.841 | 1.03 |
| 5% | 2.54 | 0.902 | 0.902 | 0.817 | 0.990 |
| Borated | 5.08 | 0.896 | 0.898 | 0.812 | 0.986 |
| Poly | 15.34 | 1.02 | 1.03 | 0.917 | 1.13 |

Figure 1 is a plot of the source spectrum in 46 groups. Also in the figure is the spectrum after penetrating an iron shield 15.34 cm thick using the 46 group and the 37 group cross sections. What does not show in the 37 group plot of DLC-31 is the finer group structure from 0.2 MeV to 1 MeV. The anti-resonance peak at 24keV shows in both group structures. The spectrum softening by this thick shield is expected.

Figure 2 is the spectrum resulting from three shields (2.54, 5.08 and 15.34 cm) built of 5% borated polyethylene. As the shields of borated-polyethylene becomes thicker, the spectrum noticeably hardens.

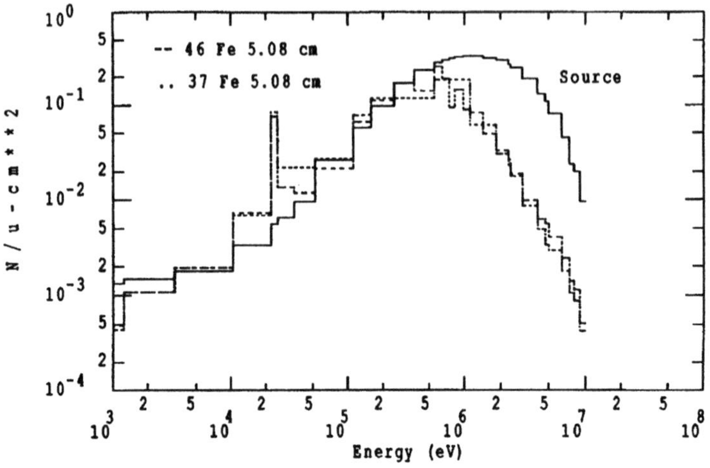


**Figure 1**

Source Spectrum and Spectra Transported Through Iron. Spectra are Plotted as Flux per Unit Lethargy.

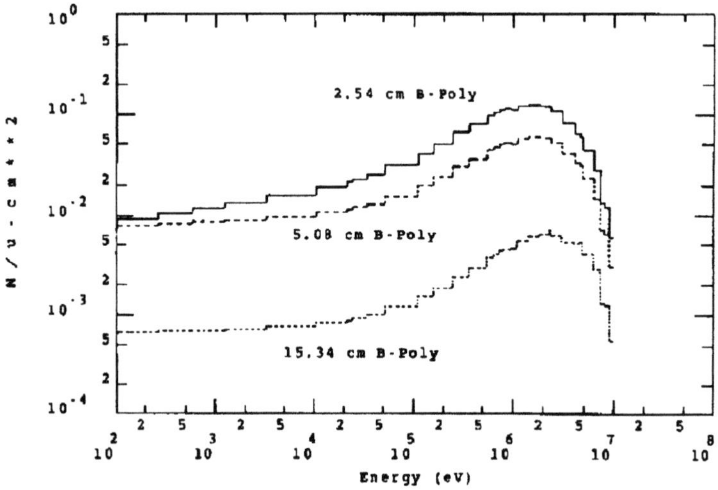


**Figure 2**

Spectra After Penetration Through a Variety of Thicknesses of 5% Borated Polyethylene

## CONCLUSIONS

Shielding calculations using the DLC-130/DABL69 cross section library were compared to identical calculations using the DLC-31/FEWG1 cross section library. The finer resolution of DABL69 in the area from 0.2 to 1.8 MeV is of particular usefulness for deep penetration shielding problems involving iron.

The DABL69 silicon displacement response function gives 1 MeV equivalent fluences that are 10% higher than those obtained using the E722 standard response function. The DABL69 response function is normalized to 82 Mev-mb. When renormalized to 95 MeV-mb the response function gives 1 MeV equivalent fluences that are 5% lower than those given by E722.

The ENDF/B-VI based silicon displacement response function ranges from 3% to 10% lower than the E722 response function for equivalent spectra. The difference in 1 MeV equivalent fluences calculated, using these two response functions, shows a dependence on spectral shape.

There is an abundance of silicon displacement damage cross sections. Shielding calculations very often have used response functions which are different from those used in experimental determinations. It is our recommendation that the same silicon displacement response function be used in both calculation and measurement.

## ACKNOWLEDGEMENT

This investigation was sponsored by the U.S. Army Test and Evaluation Command Methodology Improvement Program.

## REFERENCES

1. M.H. Sparks, T.M. Flanders, J.G. Williams, J.G. Kelly, W.W. Sallee, M. Roknizadeh, J.L. Meason, "Energy Dependence of Neutron Damage in Silicon Bipolar Transistors", IEEE Transactions on Nuclear Science, Vol 36, no. 6, pp. 1904-1911, (1989).

2. J.G. Kelly, Personal communications

3. D.T. Ingersoll, R.W. Roussin, C.Y. Fu, and J.E. White, "DABL69: A Broad-Group Neutron/Photon Cross-Section Library for Defense Nuclear Applications", ORNL/TM-10568, Oak Ridge National Laboratory, Oak Ridge, TN., (1989)

4. D.E. Bartine, J.R. Knight, J.V. Pace III, and R.W. Roussin, "Production and Testing of the DNA Few-Group Coupled Neutron-Gamma Cross-Section Library", ORNL/TM-4840, (1977).

5. R.E. MacFarlane, "NJOY. A Code System for Producing Pointwise and Multigroup Neutron and Photon Cross Sections from ENDF/B-IV and -V Evaluated Nuclear Data", RSIC, PSR-118, Los Alamos National Laboratory.

6. V.V. VERBINSKI, N.A. LURIE, V.C. ROGERS, "Threshold-Foil Measurements of Reactor Neutron Spectra for Radiation Damage Applications", Nucl. Sci. Eng., 65, (1978).

7. LOS ALAMOS TRANSPORT GROUP, "MCNP - A General Monte Carlo Code for Neutron and Photon Transport", Los Alamos, NM., (1981).

8. T.M. Flanders, M.H. Sparks, "Calculated Neutron Spectra of Fast Pulsed Reactors", this conference, (1990).

9. T.M. Flanders, M.H. Sparks, "Monte Carlo Calculations of the Neutron Environment Produced by the White Sands Missile Range Fast Burst Reactor", Nucl. Sci. Eng., 103, 265-275, (1989).

10. T.M. Flanders, M.H. Sparks and J.L. Meason, "A Theoretical Calculation of the Radiation Environment Produced by the White Sands Missile Range Fast Burst Reactor", TR-NEL-015-86, (1986)

11. W.W. Engle Jr, "A Users Manual for ANISN, A One Dimensional Discrete Ordinates Transport Code with Anisotropic Scattering." USAEC Report K-1693, Union Carbide Corporation, (1967)

12. C.W. Weisbin, et. al., "Vitamin E: An ENDF/B-V Multigroup Cross Section Library for LMFBR Core and Shield, LWR Shield, Dosimetry, and Fusion Blanket Technology", ORNL-5505, (1979).

13. W.W. Sallee, Personal Communication and "Adjusted Spectra from Three FBR's", this conference, (1990).

14. V. Rogers, L. Harris, D. Steinman and D. Bryan, "Silicon Ionization and Displacement Kerma for Neutrons from Thermal to 20 MeV", IEEE Transaction on Nuclear Science, NS-22, No. 6, (1975), Erratum (1976)

15. ASTM, Standard E722-85, 1989 Annual Book of ASTM Standards, Vol 12.02, pp 448-454, American Society for Testing and Materials, (1986).

16. T.A. Gabriel, J.D. Amburgey, and N.M. Greene, "Radiation-Damage Calculations: Primary Recoil Spectra, Displacement Rates, and Gas-Production Rates", ORNL/TM-5160, (1976)

# DOSIMETRY IN NEUTRON AND GAMMA ENVIRONMENT ISSUED FROM THE FAST BURST REACTOR"CALIBAN"

J. Dorlet, M. Fourgoux, J.C. Jauréguy*, G. Ledanois, J. Morin, J.C. Nimal, P. Zyromski.

C.E.A.-Centre d'Etudes de VALDUC - 21120 IS-SUR-TILLE - France.

* ETCA-Centre d'Etudes du Bouchet - 94114 ARCUEIL CEDEX-France.

## I - INTRODUCTION

This work has been performed to characterize neutron fluences and gamma dose rates inside converters near a pulsed reactor. It is used to study materials and electronic components behaviour under pulsed irradiation.

## II - "CALIBAN" REACTOR AND ITS ENVIRONMENT

"CALIBAN" is a fast reactor (metallic fuel) designed for pulsed neutron flux generation when operating above prompt critical state.

The reactor core consists of 10% molybdenum alloyed enriched uranium. This device is a vertical axis, critical assembly. It is made of two approximately identical parts, one fixed, and the other one moved by a fast scram mechanism (safety system). Minimum neutron burst duration at half-height is 50 µs, and the integrated fluence for a burst in the cavity (inner cylindrical hole centered in the core h=200 mm Ø= 20 mm) reaches $3.10^{14}$ $n.cm^{-2}$.

The elements to be irradiated are put either in the cavity or outside the reactor.

This study[1] concerns two assemblies, called neutron-gamma converters. They have different dimensions and are made of polyethylene with 20 % in weight cadmium content.

The following figures show the relative converters and reactor positions. The converters may be set at different distances from the reactor, in order to modulate neutron fluences to gamma dose rates ratio.

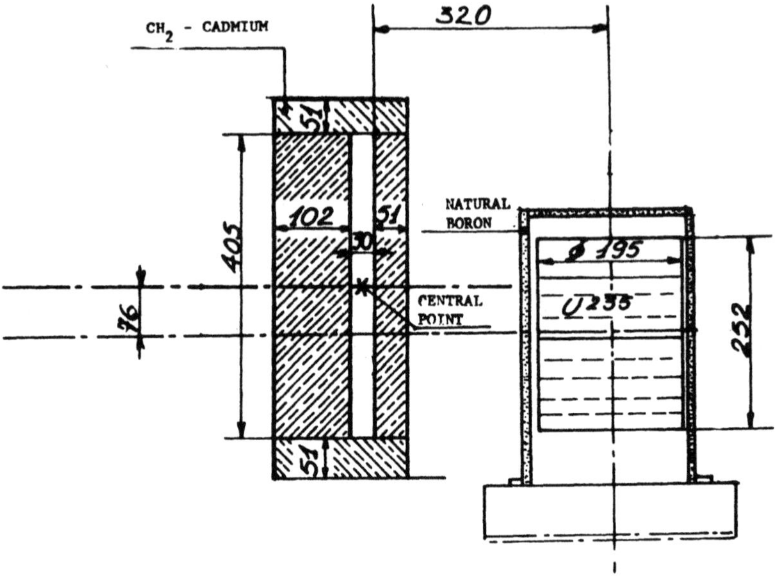


FIG. 1 SMALL CONVERTER

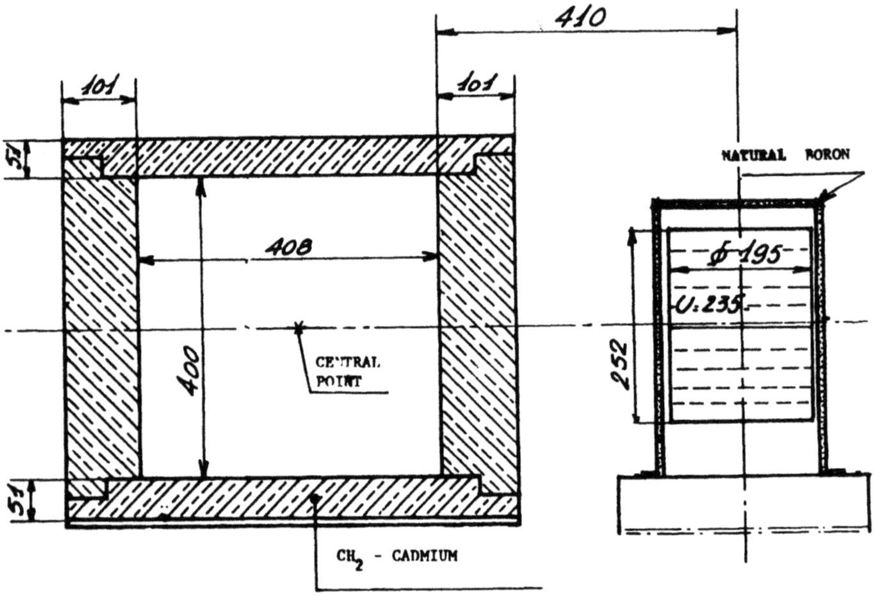


FIG. 2 LARGE CONVERTER

all units in millimeters

## III - DETERMINATION OF THE 1 MeV EQUIVALENT FLUENCE FOR THE SILICON DISPLACEMENT DAMAGE

Neutron effects are directly bound to the 1 MeV equivalent fluence for silicon displacement by the following relation :

$$\Phi_{eq\ 1MeV\ Si} = \frac{\int_0^{E_{max}} \Phi(E)\ K(E)\ dE}{K\ (1\ MeV)}$$

$\Phi(E)$ = differential fluence.
$K(E)$ = displacement kerma.

$\Phi(E)$ is calculated from several detectors reaction rates.

$K(E)$ is a datum. Up to 10 keV, it is a silicon displacement damage function (cf. Verbinski[2] ). Beyond 10 keV, it is a function given by the ASTM E722-85 standard[3].

### Activation detector method[4]

The detectors are either activation detectors or fission detectors.

Reactions used with activation detectors are :

$$^{103}Rh(n,n')^{103m}Rh\ ;\ ^{115}In(n,n')^{115m}In\ ;\ ^{58}Ni(n,p)^{58}Co\ ;$$
$$^{56}Fe(n,p)^{56}Mn \text{ and } ^{27}Al(n,\alpha)^{24}Na.$$

Reactions used with fission detectors are :

$^{235}U(n,f)$ ; $^{235}U_{Cd}(n,f)$ ; $^{238}U(n,f)$ ; $^{237}Np(n,f)$ and $^{239}Pu(n,f)$

Reaction rate for a "i-order" detector :

$$A_i = N_i \int_0^{E_{max}} \sigma_i(E)\ \Phi(E)\ dE$$

$N_i$ is the nuclei number of the i-order detector,
$\sigma_i(E)$ the concerned reaction cross-section,
$\Phi(E)$ the differential fluence rate,
$E_0$ the concerned threshold energy reaction,
$E_{max}$ the maximum energy of the studied spectrum.

If there are n detectors, there will be n integral equations, there are several spectra satisfying such a system. It is convenient to find the best "appropriate" one.

## IV - INVESTIGATION OF THE "APPROPRIATE" SOLUTION CALCULATION OF INDUCED DAMAGES IN SILICON

The methods used are called perturbation methods, because the mathematical process consists of a modification of the input spectrum, until the best agreement between measured and calculated activities is obtained.

Two codes are used to calculate the induced damages in silicon and the differential fluences.

They use :
- the cross-sections in 48 groups (ENDF/B-IV and B-V)
- the measured reaction rates
- the displacement damage function in silicon.

The 1 MeV equivalent fluence for silicon displacement is calculated from the displacement damage :

$$\Phi_{eq.1MeV\ Si} = \frac{\text{displacement damage}}{\text{1 MeV displacement damage} = 95\ \text{MeV.mb}}$$

The DOSELIN code[5] uses a linear programmation method. It allows us to obtain directly a damage interval, and to calculate the differential fluence. The input spectrum is arbitrary (in this case, it is flat).

The CORRENS code[6] is based on the "bayesian" correlations method. It enables us to calculate the differential fluence, the damage, and the accuracy on this damage. The "most likely" input spectrum is a linear combination of 4 components :
- a thermal component,
- an epithermal component,
- a slowing-down spectrum
- and a fission spectrum.

This formalism allows us to solve cross-sections step by step. The multigroup representation, with 48 groups, has been maintained.

## V - RESULTS

The 1 MeV equivalent fluence for silicon displacement has been determined for several points in the converters.

1 MeV Si equivalent fluence (CORRENS code)
(standardization $2.10^{14}$ $n.cm^{-2}$ in the reactor cavity)

| | Small converter Central point | Large converter Central point |
|---|---|---|
| 1 MeV displacement damage for Si | $2.00 \times 10^{14}$ | $2.43 \times 10^{13}$ |
| 1 MeV Si equivalent fluence ($n.cm^{-2}$) | $2.10 \times 10^{12}$ | $2.56 \times 10^{11}$ |

N.B.: figures 1 and 2 show the positions of the converters and of the reactor.

Distance of the internal front face from the reactor axis :
- 320 mm for the small converter.
- 410 mm for the large one.

## VI - GAMMA DOSIMETRY

The transient effects induced by gamma-rays in electronic circuits irradiated in a neutron-gamma converter near the CALIBAN pulsed reactor are not negligible. Therefore, it is necessary to determine the reactor irradiation specifications for experimental conditions that could be very diversified.

The method used to determine the gamma dose rate consists in measuring the total integrated dose with passive dosimeters, and deducing the dose rate, taking into account the time pulse shape.

The passive dosimeters are $Al_2O_3$ thermoluminescent powders, calibrated with $^{60}Co$ sources. The absorbed dose is a function of the material nature.

The gamma dose has been determined for the same converters points as the 1 MeV neutron equivalent fluence.

Results given below correspond to the central point of each configuration.

| | Small converter | Large converter |
|---|---|---|
| Gamma dose (Gy.Si) | 90 | 16.4 |

## VII - COMPARISON BETWEEN MEASUREMENT AND CALCULATION

Simultaneously, computations have been done with the TRIPOLI-2 code. It is a 3-D code which solves neutron and gamma transport problem by a Monte-Carlo method. The constants used in this code are taken from the ENDF-B/IV library for neutrons, and from the Los Alamos Scientific Laboratory for gamma-rays.

Equivalent fluences and gamma dose
determined by measurement and computation
(standardization $2.10^{14}$ $n.cm^{-2}$ in the reactor cavity)

| | Measurement (CORRENS code) | Computation (TRIPOLI code) |
|---|---|---|
| 1 MeV Si eq. fluence ($n.cm^{-2}$) | $2.10 \times 10^{12}$ | $1.90 \times 10^{12}$ |
| Gamma dose (Gy.Si) | 90 | 54.4 |

## VIII - CONCLUSION

We have determined by computation and by experiment the 1 MeV Si equivalent fluences and the gamma doses in silicon inside neutron-gamma converters near the pulsed reactor CALIBAN.

Results obtained for neutrons with the TRIPOLI code and by experiment are in good agreement. Nevertheless, it is necessary to improve our measurement , by an optimalization of our method and a better nuclear data choice.

However, there is a discrepancy between measured and calculated results for gamma-rays.

The treatment of the neutrons to gamma-rays conversion in cadmium is not satisfying. We intend to realize complementary studies to solve this problem.

We also intend to compare our results and those obtained by A.H. Kazi[7] with the APRF reactor.

## REFERENCES


1. Etude de convertisseurs neutrons-gamma auprès du réacteur CALIBAN. (Internal report).

2. Threshold foil measurement of reactor neutron spectra for radiation damage applications.
V. Verbinski, N. Lurie and V. Rogers, Nucl.Sci.Eng. 65, 316 (1978).

3. Characterizing neutron energy fluence spectra in terms of equivalent monoenergetic neutron fluence for radiation hardnes testing of electronics.
ASTM-E 722-85.

4. Détermination de spectres appropriés de neutrons des assemblages critiques rapides PROSPERO et CALIBAN, par la méthode des détecteurs à seuil.
C. Butaye, G. Simonet and J. Morin. (Internal report).

5. Applications des méthodes de la programmation linéaire au dépouillement des résultats de mesures par activation.
J. Dorlet, Knoxville (Oct. 1977).

6. Solution d'un problème de dosimétrie par la méthode des corrélations bayesiennes.
G. Ledanois (Private communication).

7. Characteristics of pulsed fast neutron to gamma ray converter
A.H. Kazi, T.A. Dunn, R.C. Harrisson and D.O. Williams, Nucl. Technol., 25, 450 (1975).

# THE ADVANCED NEUTRON SOURCE REACTOR: AN OVERVIEW*

Colin D. West

Oak Ridge National Laboratory

Oak Ridge, Tennessee 37831-8218, USA

## ABSTRACT

The Advanced Neutron Source (ANS) will be a new user facility for all kinds of neutron research, including neutron scattering, materials testing, materials analysis, isotope production, and nuclear physics experiments. The centerpiece of the facility is to be the world's highest flux beam reactor. There will be beams of hot, cold, and thermal neutrons for more than 40 simultaneous scattering and nuclear physics experiments. In addition, there will be irradiation positions and rabbit tubes for in-pile experiments, testing, and isotope production (including transuranium isotopes).

To reduce technical risks and to minimize safety issues, the reactor design is based on technology already employed in existing research reactors. The fuel elements are annular assemblies of aluminum-clad involute fuel plates, similar to the design of the High Flux Isotope Reactor (HFIR) at Oak Ridge and the Institut Laue-Langevin (ILL) Reactor in Grenoble. As is common with many other research reactors, the core is cooled, moderated, and reflected by heavy water. The preferred fuel is $U_3Si_2$, a high-density fuel form developed by Argonne National Laboratory and Babcock and Wilcox that has been extensively tested in reactors in the United States, Europe, and Japan.

*Based on work performed at Oak Ridge National Laboratory, operated for the U.S. Department of Energy under contract DE-AC05-840R21400 with the Martin Marietta Energy Systems, Inc.

A high flux beam reactor must have a compact core, and the consequential high power density requires particular attention to safety questions. From the beginning of the preconceptual design phase of the project, careful attention has been given to safety questions. One example of this has been the development of a preconceptual design probabilistic risk assessment (PRA) that has already influenced the design concept. Several inherent aspects of the ANS contribute desirable safety features: the relatively low power level and fission product inventory (<10% of a typical power reactor), the large containment dome needed to house experiments around the reactor, design constraints that limit the coolant temperature to <100°C to eliminate any flashing into steam on depressurization, and the small core with little stored energy. In addition, the design of the primary coolant system minimizes the potential for a pipe break to lead to loss-of-coolant, and the choice of an upflow coolant direction in the core enhances transition from forced- to natural-convection cooling.

Several features of the new facility will pose particular challenges in dosimetry. The experiments close to the core will experience very high neutron and gamma fluxes and very steep flux gradients, with spectra that change significantly during the 14-d core life. Totally internally reflecting neutron guides will be used to transport very intense beams of neutrons up to 100 m from the reactor and into an experimental area that can house 40 or more experiments and their users; the beams are 10 times as intense as those at existing reactors. Neutron-capture gamma background radiation from these intense beams can change significantly as the experimental samples are moved or replaced, but protection of the users must be maintained at all times. The ANS Project would welcome early debate of these and other issues of dosimetry in a high flux research reactor facility.

---

## INTRODUCTION

The Advanced Neutron Source (ANS) is a new facility for all kinds of neutron research, planned for construction at the U.S. Department of Energy's Oak Ridge National Laboratory. The major technical objectives of the project are

(1) to design and construct the world's highest flux research reactor for neutron scattering;

(2) to provide isotope production facilities that are as good as, or better than, the High Flux Isotope Reactor (HFIR); and

(3) to provide materials irradiation facilities that are as good as, or better than, the HFIR.

These technical objectives set the performance goals for the research reactor that will provide the source of neutrons (Table 1). A National Steering Committee for an Advanced Neutron Source (NSCANS), composed of neutron users from all fields of science, is also defining the needs in other areas, including materials analysis and nuclear physics. The reactor is only a part of the ANS facility, and NSCANS has also devoted a good deal of time to planning the neutron beam lines and instrument space that will be needed in the new facility (Figs. 1 and 2).

Table 1. Some quantitative performance goals for the ANS reactor

| Parameter | Goal |
|---|---|
| Neutron beams | |
| Peak thermal flux in reflector, $m^{-2} \cdot s^{-1}$ | $5\text{-}10 \times 10^{19}$ |
| Thermal/fast flux ratio | ≥80 |
| Materials irradiation | |
| Fast flux at irradiation positions, $m^{-2} \cdot s^{-1}$ | $\geq 1.4 \times 10^{19}$ |
| Fast/thermal flux ratio | ≥0.5 |
| Transuranium production | |
| $^{252}$Cf production rate, g/year | 1.5 |
| $^{254}$Es production rate, μg/year | 40 |

The peak thermal neutron flux in the reflector is an important measure of performance of a research reactor for beam work and helps to define the sensitivity and resolution of the research experiments that can be performed. The number of beams and the total number of neutrons delivered are a rough measure of the number of experiments that can be carried out, and they give some general indication of the scientific productivity that can be expected from a facility. In this respect, the plans proposed by our National Steering Committee will lead to a truly outstanding research capability (Table 2).

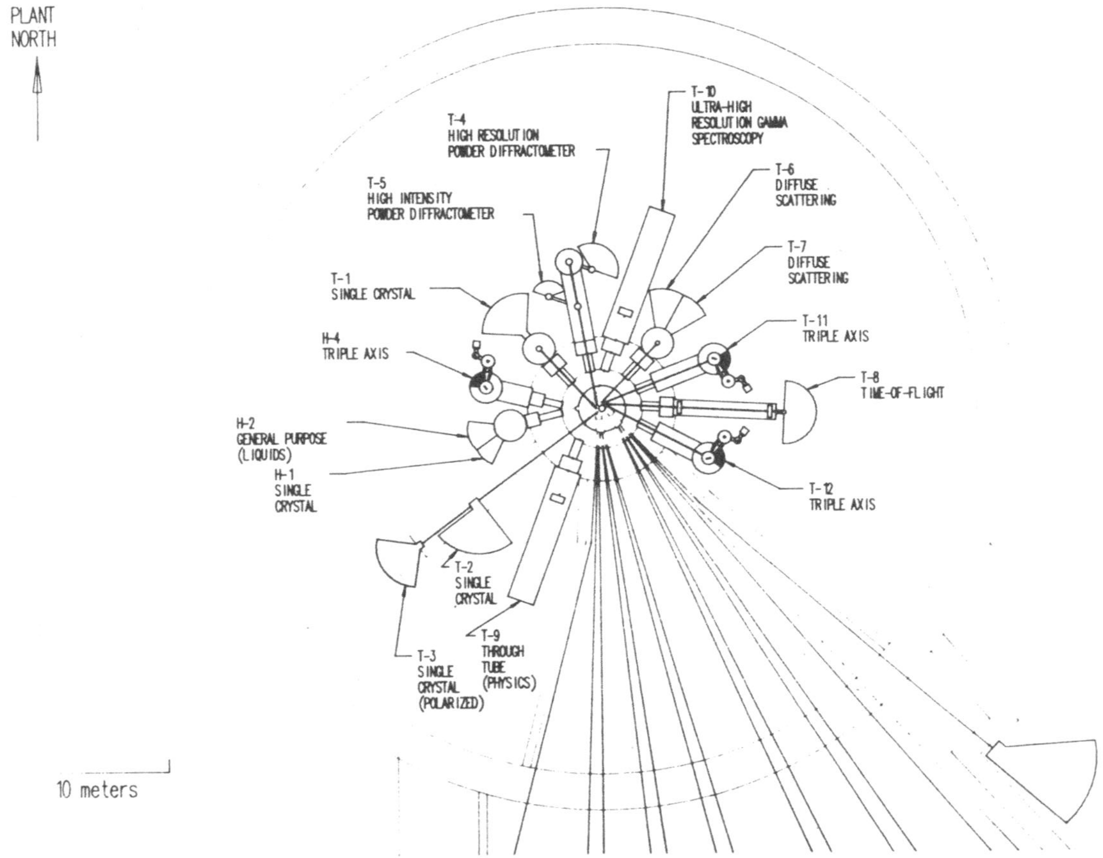


Fig. 1. Instrument layout on the thermal and hot neutron beam lines.

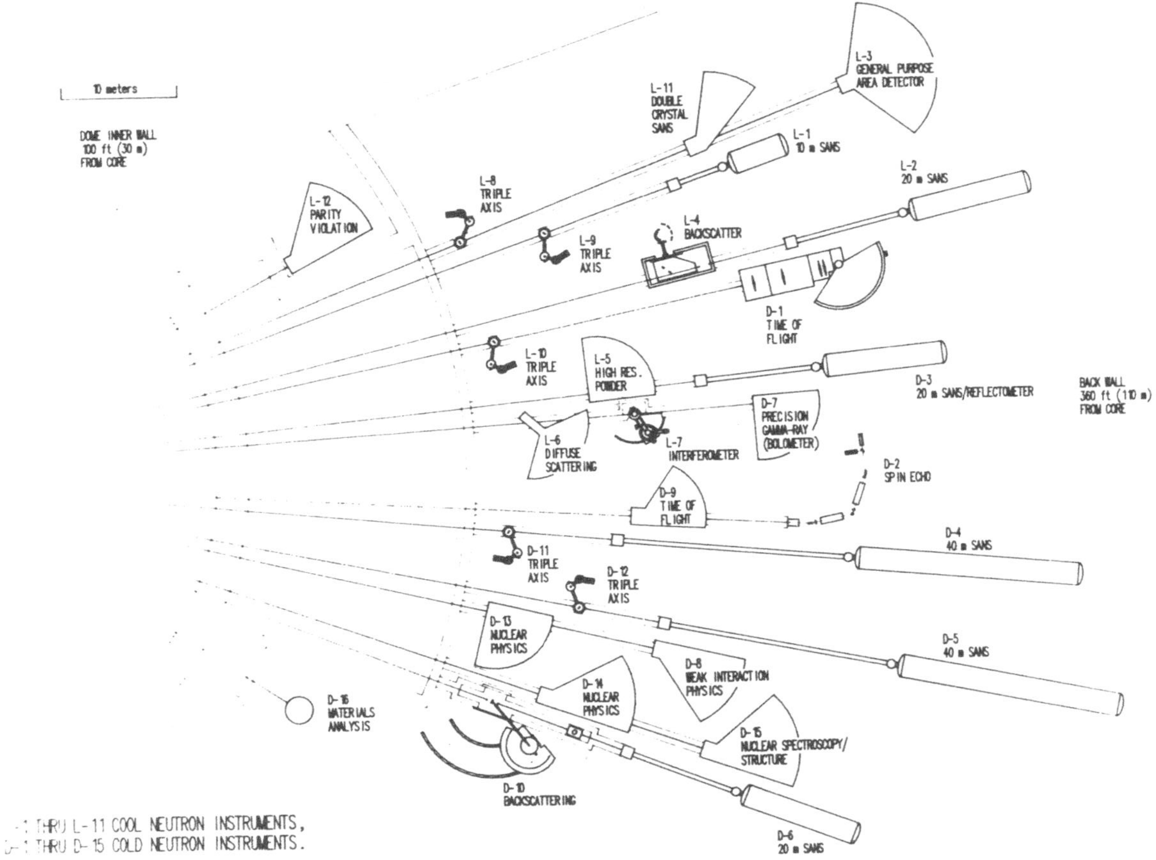


Fig. 2. Instrument layout on the cold neutron guides.

Table 2. Research capabilities
(effective neutrons per second delivered to users)[a]

| Neutrons | HFBR[b] | HFIR[c] | ILL[d] | ANS[e] |
|---|---|---|---|---|
| Hot | 0[f] | 0[f] | 450 | 450 |
| Thermal | 350 | 400 | 1,000 | 11,500 |
| Cold | 30 | 0[f] | 700 | 16,000 |
| Very cold | 0[f] | 0[f] | 40 | 250 |
| Ultra cold | 0[f] | 0[f] | 450 | 2,800 |

[a]A number of beams times useful area of each beam times peak flux in reflector ($10^{19}s^{-1}$).

[b]40-MW operation.

[c]85-MW operation.

[d]Beams available to users are overdemanded by a factor of ~4.

[e]Minimum requirements assessed by the scientific community through four national workshops and by NSCANS.

[f]No sources for these beams are included in these reactors.

## Reactor Design

To meet the performance criteria agreed upon by the scientific community particularly for a thermal neutron flux available for neutron beams 5 to 10 times higher than the best existing facilities, we need a higher power and a higher power density than present research reactors. The essential requirement is a high thermal flux in the reflector tank (i.e., outside the core) where it is accessible to beam tubes and guides for neutron scattering. The basis for the design is therefore a high power — to produce many neutrons — in a small, undermoderated core that will permit most of the fission neutrons to escape before they are thermalized. Thermalization takes place in a reflector surrounding the core. The coolant and the reflector should have low neutron absorption and should have a reasonably long slowing-down length: in the case of the ANS reactor and many other research reactors, this is achieved by selecting heavy water as the coolant and reflector.

The combination of high power and a small core demands a design with a large cooling surface in a small volume and a fuel with high density and high thermal conductivity. The conceptual design for the ANS uses the thin, aluminum-clad, involute fuel plates that have proven so successful in, for example, the Institut Laue-Langevin (ILL) reactor and the HFIR. The reference fuel form is highly enriched $U_3Si_2$ (for such a high flux reactor, low enriched fuel is not viable), with $UAl_x$ as a fallback alternative.

A new core geometry has been devised that is comprised of two coaxial annular fuel elements with different diameters, separated axially (Fig. 3). The arrangement combines the short heated length of the HFIR design with the greater "neutronic length" of the ILL core and offers a number of advantages. One advantage is that there is space around and inside the core with an appropriate spectrum for transuranium isotope production and materials irradiation capsules (Fig. 4).

In the reference design for ANS, the core is enclosed in a section of the primary coolant piping called the core pressure boundary tube (CPBT), which

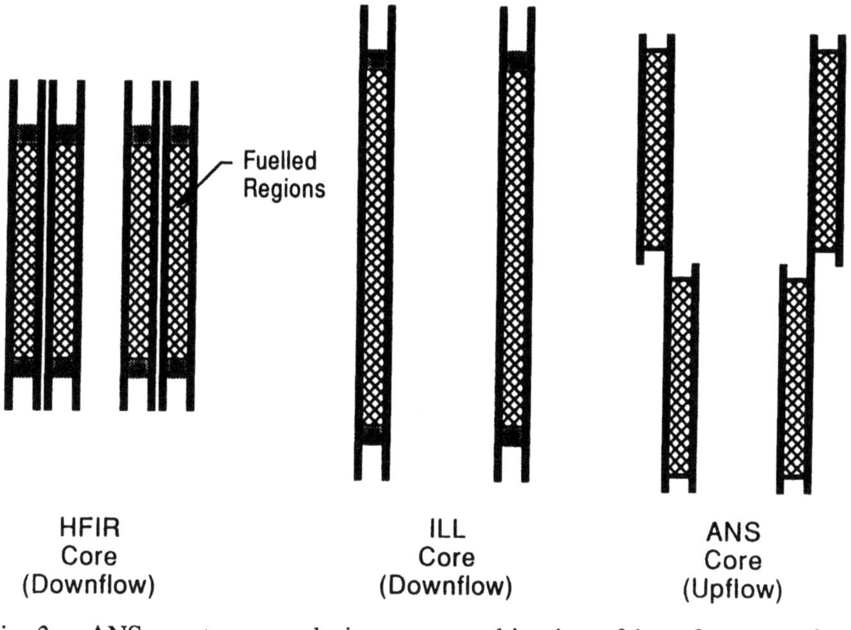


Fig. 3. ANS reactor core design as a combination of best features of earlier designs.

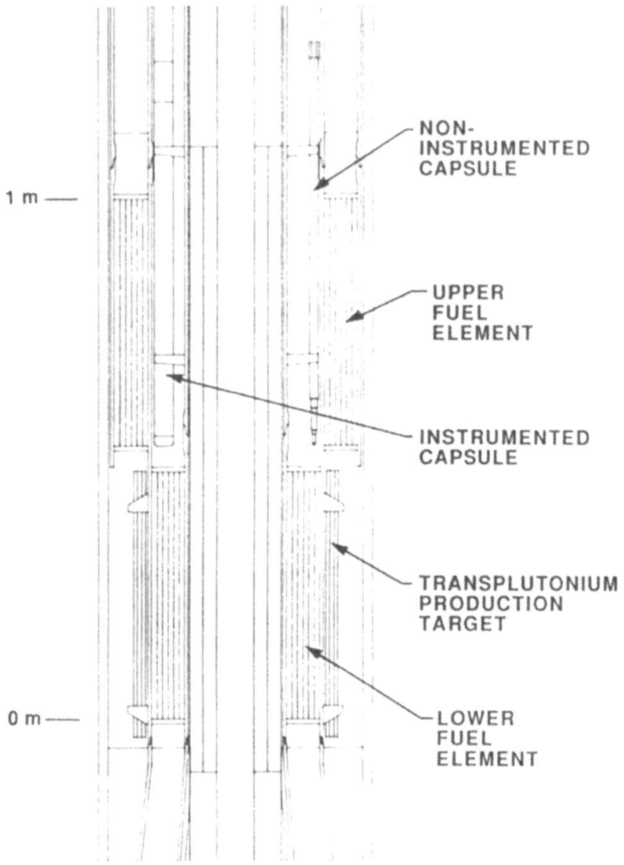


Fig. 4. ANS reactor core design showing materials irradiation capsules and transplutonium production target positions.

corresponds to the pressure vessel in a power reactor. There are two shutdown systems: one, which also provides control, is inside the core, and the other is located in the reflector region outside the pressure boundary (Fig. 5).

The CPBT is surrounded by a large reflector tank of heavy water with space for beam tubes and two large, liquid deuterium cold neutron sources (Fig. 6). The cold neutrons are conducted to the guide hall area by means of neutron guides.

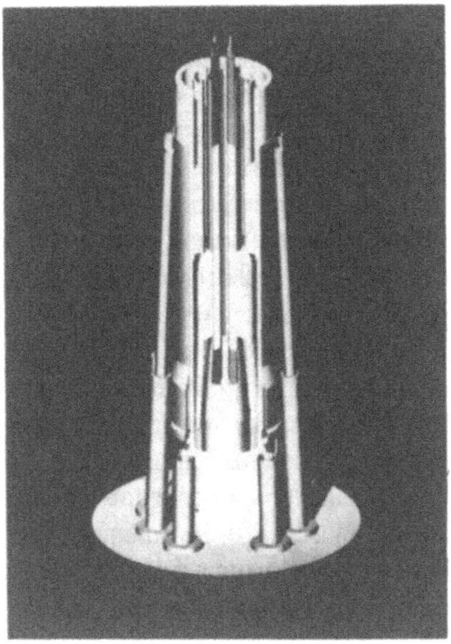

REFERENCE CORE DESIGN

- The Project Has Devised An Improved Research Reactor Core Arrangement, With Two Annular Fuel Elements Of Different Diameters
  - The Arrangement Offers Many Safety And Performance Advantages
  - The Design Continues To Use The Well-proven Fuel Assembly Technology From Existing Reactors
- Two Separate, Diverse, Fast Shutdown Systems Are Incorporated: One (Also Used For Control) Inside The Core And One Outside The Primary Coolant System In The Reflector Tank
  - Either System Alone Can Safely Shut Down The Reactor, Even With One Rod Stuck

  This Arrangement Leads To Reduced Probability Of Anticipated Transient Without Scram (ATWS)

Fig. 5. Reference core design.

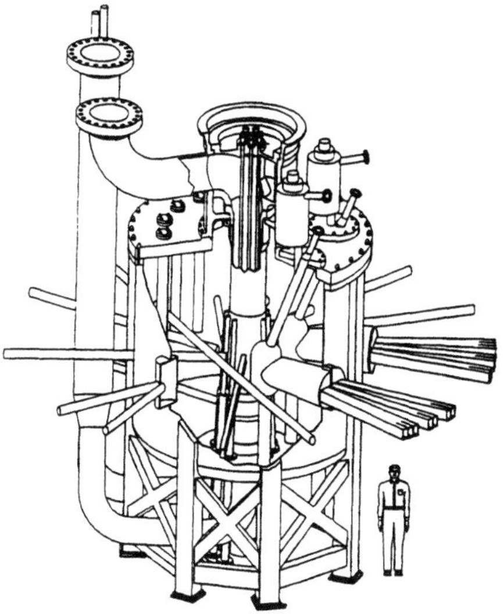

Fig. 6. The reactor core surrounded by a large, low-pressure tank of heavy water containing beam tubes, cold sources, and isotope production facilities.

## Cooling System Design

The reactor cooling systems are designed to make maximum use of passive features for enhanced safety. There are four identical primary loops of deliberately simple design with only one high point and one low point. In normal operation, three loops are in use. The design emphasizes natural convection for decay heat removal; for example, an upflow coolant direction through the core was chosen mainly to avoid any flow reversal during the transition from forced to natural convection. In addition, each of the main heat exchangers is followed by an emergency heat exchanger, immersed in a pool, for decay heat removal. The secondary side of the emergency heat exchanger is cooled by natural convection of the pool water (Fig. 7). Normally, primary water entering the emergency heat exchanger from the main heat exchanger is cool, and there is little driving force for natural convection through its secondary side; but if cooling in the main heat exchanger were lost (although this is guarded against by the provision of battery-powered pony motors in addition to the main motors, on each pump), the emergency heat exchangers would automatically begin to provide passive cooling. Likewise, the cooling towers are elevated above the heat exchangers, so that in the event of a loss of pumping, natural convection would be available with the water in the seismically hardened cooling tower basins as a heat sink. There is an independent secondary circuit for each of the primary loops.

To prevent loss-of-coolant inventory in the event of a break or major leak in the primary circuit, most of the components are located under (light) water pools. The remaining components are housed in limited volume dry cells; each primary loop has a gas-pressurized accumulator containing 5 $m^3$ of heavy water, enough to flood any one of the limited volume cells. The accumulator also slows down the rate of depressurization following a major leak or break, thus further protecting the core.

The design process has deliberately been organized and planned to emphasize safety. A key part of this approach has been the use of probabilistic risk analysis (PRA) techniques, even during the preconceptual design phase. In the absence of a detailed design for which quantitative calculations of risk can be made, the PRA approach provides guidance on the components and event sequences that can, through early design attention, lead to significant risk reduction. Some of the fruits of this powerful technique can be perceived in the previous description of the cooling system reference design, and there are many other examples.

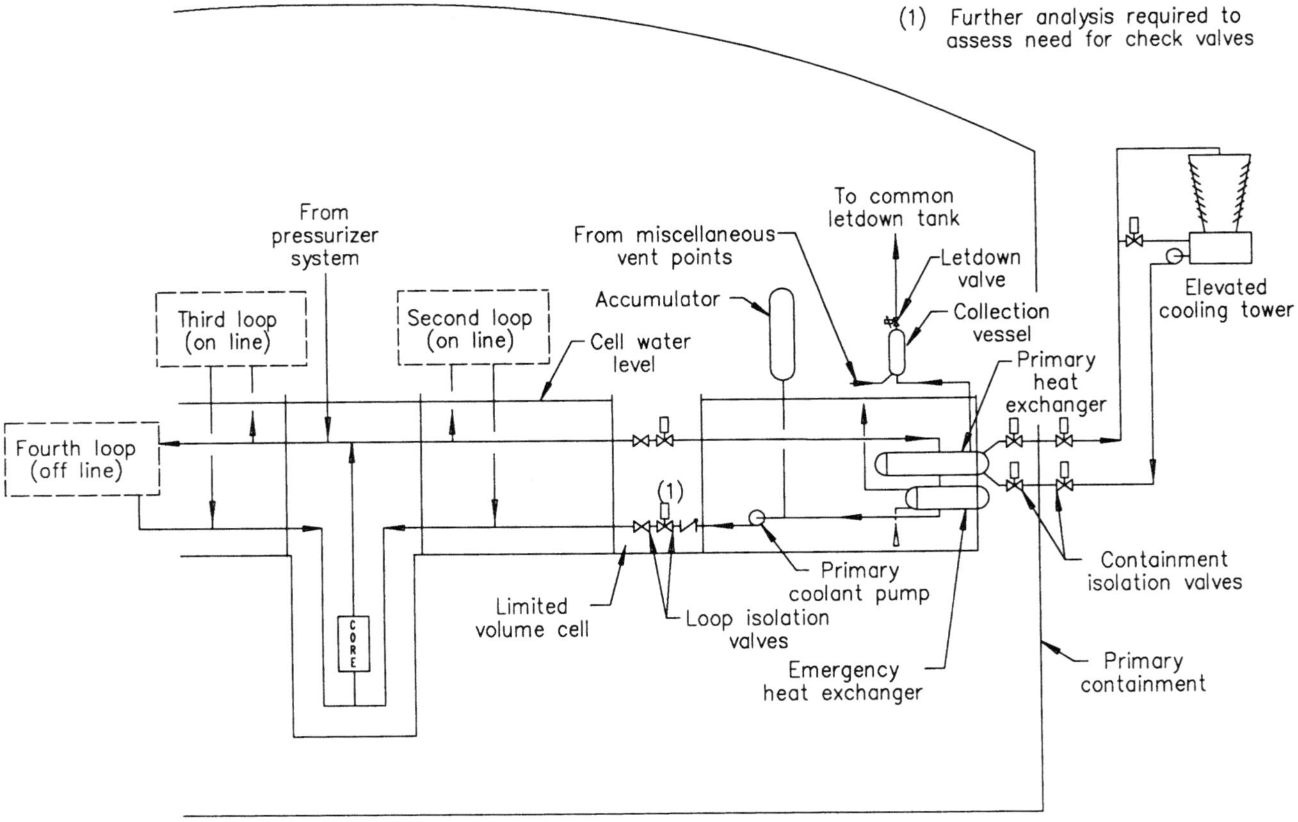


Fig. 7. Four separate cooling circuits designed to optimize reactor safety. Three are in use during normal operation.

It is only within the last several months that the reference or baseline design features described earlier have been fixed. The next steps in the design process include trade-off studies to seek improvements in safety, performance, or economics of the ANS.

## Dosimetry Issues

A research reactor facility, and especially one designed as a user facility, poses particular challenges in dosimetry. Among the more obvious technical ones is the accurate measurement of the neutron flux and spectrum — both of which will vary during the operating cycle — at the in-core irradiation positions.

The ANS is intended as a user facility; we expect more than 1,000 users a year to come and carry out experiments, mainly beam experiments, when the ANS is fully operational. Many of these users will have had little or no training in a reactor environment, so radiation protection arrangements must be designed accordingly. The neutron guide tubes that are so essential to the research capabilities of the ANS also transport intense beams of neutrons up to 100 m or more away from the reactor core. Normally, few neutrons escape from the guide, but when an experimental sample is placed in the beam, then neutrons are captured or scattered; indeed, the main purpose of most experiments is to measure the direction and energy of the scattered neutrons. Thus, the guide hall, which must be accessible to users, is a place of potentially significant background radiation that can change from day to day or even from hour to hour. These effects are found at existing research reactor facilities, of course, and appropriate measures have been taken. However, the beams at the ANS will be from 10 to 100 times more intense than at any present-day facilities and may require new approaches. The control, measurement, and administration of radiation in the guide hall and other experimental areas deserve attention early in the design and planning of the ANS.

## CONCLUSION

The Advanced Neutron Source facility will be a world class facility for neutron research, producing neutron beams an order or magnitude more intense than present facilities. A research reactor based on present-day technology, and designed from the beginning with an emphasis on safety, will be the source of neutrons. To operate such a reactor as part of a user facility, while providing the necessary radiation protection and measurement capabilities, will require early attention to dosimetry issues.

# CALCULATION AND MEASUREMENT OF NEUTRON AND GAMMA-RAY FLUXES IN AND AROUND REACTORS

Yoshinori Sakurai, Shusuke Fujihara*, Itsuro Kimura
Department of Nuclear Engineering, Kyoto University
Yoshidahonmachi, Sakyo-ku, Kyoto 606 Japan

Katsuhei Kobayashi, Seiji Shiroya and Keiji Kanda
Research Reactor Institute, Kyoto University
Kumatori-cho, Sennan-gun, Osaka 590-04 Japan

Neutron and gamma-ray fluxes in and around the core of a critical assembly (KUCA) were simultaneously calculated by a transport code and were measured with gold wires and thermoluminescent dosimeters(TLD) of magnesium orthosilicate $Mg_2SiO_4$(Tb) (MSO). The calculated distributions of the neutron and gamma-ray fluxes were compared with those measured. The same comparison was carried out for a low power reactor (UTR-KINKI).

Calculation was performed by one or two dimensional transport code, ANISN-JR or TWOTRAN. For this calculation, neutron and gamma-ray combined cross sections were generated by RADHEAT-V3 on the basis of the ENDF/B-IV and POPOP libraries.

The MSO powder was packed in thin polyethylene or aluminum tubes and was irradiated in a reactor. After irradiation, we measured absorbed dose of each MSO powder sample with a TLD recorder.

The calculated neutron flux distribution agreed well with that measured with gold wires at the KUCA B-core (highly enriched uranium fuel, polyethylene moderated and reflected). The calculated absorbed dose distribution was 10 % lower than that measured in the core region with MSO at the KUCA, when the contribution of neutron absorbed dose and that of delayed gamma-rays from fission products were taken into account. For this comparison, we normalized the calculated gold wire activities to those measured at the core center.

Although the measured neutron flux distributions in void and in three filter boxes of bismuth, iron and polyethylene agreed well with the calculated ones in an internal reflector of the UTR-KINKI, the calculated absorbed doses were considerably lower (about 35 %) in average than those measured.

---

* Present address: Kansai Electric Power Co. Ltd., Osaka, Japan.

## INTRODUCTION

The importance of evaluation of gamma-ray flux distribution in a reactor has risen recently. Especially the estimated gamma-ray fluxes at experimental facilities in a research reactor, at therapeutic facilities in a medical reactor or in reactor shields are of great importance. For fusion reactors, it is more important to estimate gamma-ray flux distributions in their blanket.

Transport calculation codes by both discrete ordinate method and Monte Carlo method have been developed and applied to obtain simultaneously neutron and gamma-ray flux distributions in and around core of nuclear reactors. However, not many works have been reported so far, in which measured neutron and gamma-ray fluxes are compared with the calculated ones with these codes.

In this work, we have tried to compare the measured both fluxes in two reactors, Kyoto University Critical Assembly (KUCA) and Kinki University Reactor (UTR-KINKI), with those calculated by one or two dimensional Sn code, ANISN-JR or TWOTRAN. For this calculation, neutron and gamma-ray combined cross sections were generated by RADHEAT-V3.

Hashizume et al.[1] showed the properties of magnesium orthosilicate (MSO) activated with terbium, $Mg_2SiO_4$(Tb), as an excellent thermoluminescence dosimeter (TLD) and applied it to wide range gamma-ray dosimetry. Sato and Ono[2] of Research Reactor Institute, Kyoto University (KURRI) tried to use MSO for gamma-ray dosimetry in experimental facilities of research reactors, mainly the Kyoto University Reactor, KUR. But few works have been reported to measure in-core gamma-ray fluxes with MSO detectors so far. We have adopted MSO as a detector to measure gamma-ray flux, or for absorbed dose in a strict sense, in and around two reactors, the KUCA and the UTR-KINKI.

## EXPERIMENTAL PROCEDURES

### 1) MSO Detectors and TLD Recorder

The MSO powder of about 10 mg manufactured by Kasei Optonics Co. Ltd. was weighed, and was packed in polyethylene tubes ( 1 mm of inner diameter, 0.5 mm thick and 1 cm long) or aluminum tubes ( 1 mm of inner diameter, 0.2 mm thick and 1 cm long). We sealed each tube to make a MSO detector. The MSO detectors were irradiated on an aluminum holder for the KUCA and on a paper holder for the UTR-KINKI. A few days, at least, after irradiation for cooling, the MSO powder was taken out from its tube and was scattered on a sample heater of a thermoluminescence recorder which had been designed and manufactured by Sato[3]. An integral value of the thermoluminescence glow curve was digitally printed out for each MSO powder sample. The relation between this value and absorbed dose in Grey(Gy) unit had been calibrated with a $^{60}Co$ source by Sato, and we recalibrated this relation again by ourselves.

The response of the MSO detector to slow neutrons was experimentally obtained using a supermirror neutron guide tube of the KUR. As a result, the obtained value is about $4.9 \times 10^{-14}$ Gy $cm^2$. The details of this measurement will be published elsewhere[4].

2) Configuration of Reactors

We measured neutron and gamma-ray fluxes in the B-core of KUCA and in filtered neutron fields at the UTR-KINKI. At the B-core of KUCA, two of the present authors (S. Shiroya and K. Kanda) have been carrying out a series of high burn-up and high conversion reactor experiments with low moderation[5]. Thereupon we used one of the undermoderated cores for neutron and gamma-ray flux measurements. The horizontal and vertical views of the B-core of KUCA are shown in Fig. 1. Gold wires or MSO detectors were irradiated at the corner of aluminum sheath for fuel rod, of which position is shown in the same figure.

The UTR-KINKI is a training and research reactor of 1 W and has two divided core regions with highly enriched uranium aluminum alloy fuel plates and water surrounded by a large graphite reflector. The configuration of the UTR-KINKI is given in Fig. 2. At the middle of the divided two cores, a part of inner reflector (graphite blocks) can be withdrawn and a neutron filter box can be inserted into that position. Neutron and gamma-ray fluxes were measured in the void and three filter boxes made of bismuth, iron and polyethylene, respectively. The thickness of each filter was 5 cm. The inside of these filter boxes, especially that of bismuth one, has been often used for the irradiation of biological samples[6].

## CALCULATION AND CORRECTION

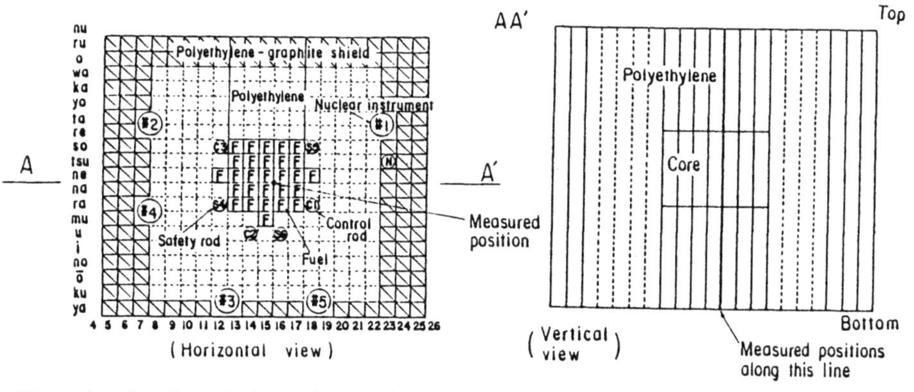


Fig. 1 Horizontal and vertical views of the B-core of KUCA. Neutron and gamma-ray fluxes with gold wires and MSO detectors were measured along the corner of four aluminum fuel sheaths as shown in the figure.

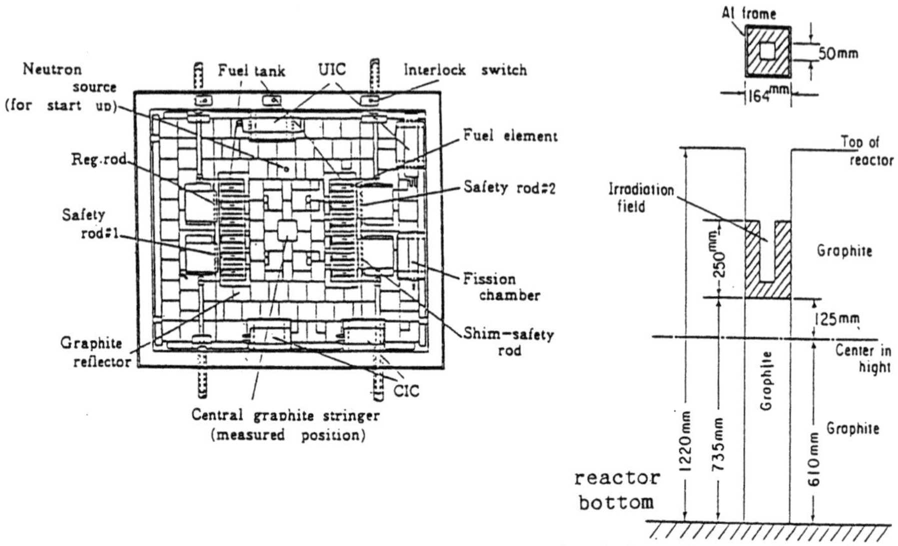


Fig. 2 Cross section view of UTR-KINKI(left) and an enlarged view of its irradiation field with a void or three filter boxes where neutron and gamma-ray fluxes were measured (right).

1) Calculation of Neutron and Gamma-Ray Transport

We assumed an infinite slab geometry with a core sandwiched by two polyethylene reflectors for the B-core of KUCA and a finite cylindrical geometry with the filter, the inner reflector (graphite), the core and the outer reflector (graphite) from inside to outside for the UTR-KINKI.

Neutron and gamma-ray combined cross sections for each region were generated by RADHEAT-V3[7] on the basis of the ENDF/B-IV and POPOP libraries. At first we took 100 groups and 20 groups for neutrons and gamma-rays, respectively, and then we collapsed the number of energy groups for both.

For the transport calculations at the KUCA and the UTR-KINKI, we used one dimensional code, ANISN-JR and two dimensional one, TWOTRAN, respectively. For both codes, calculations were performed with $P_3S_8$.

2) Absorbed Dose Calculation

Using the obtained neutron and gamma-ray fluxes, we calculated the absorbed dose D of a MSO detector by

$$D = D_g + D_n$$

$$= c\int_{E_{min}}^{E_{max}} E\,(\mu_{en}/\rho)\,\phi_g\,dE \; + \; c'\int_{E'_{min}}^{E'_{max}} K_n\,\phi_n\,dE'$$

where

$D_g$ : absorbed dose by gamma-rays,
$D_n$ : absorbed dose by neutrons,
$\phi_g$ : gamma-ray flux,
$\phi_n$ : neutron flux,
$(\mu_{en}/\rho)$ : mass energy absorption coefficient of gamma-rays,
$K_n$ : kerma factor of neutrons,
$E_{min}$, $E_{max}$ : minimum and maximum energies of gamma-rays,
$E'_{min}$, $E'_{max}$ : minimum and maximum energies of neutrons,
c, c' : conversion factors.

The $(\mu_{en}/\rho)$ and $K_n$ values for MSO were obtained by using the tables in the references[8,9].

3) Corrections

The contribution of delayed gamma-rays from fission products was estimated using the classical Way-Wigner's formula. We assumed that the delayed gamma-ray spectrum was similar to that of the fission gamma-rays.

No cavity theory correction was considered, because the equivalent atomic number (11.1) of MSO is close to that of aluminum. The gamma-ray energy dependence on the TLD response of the MSO powder in the aluminum or polyethylene tube was not taken into account.

## RESULTS AND DISCUSSION

The measured and calculated distributions of the neutron flux and those of the absorbed dose in the B-core of KUCA are shown in Figs. 3 and 4, respectively. All data are normalized by the gold wire activation at the core center.

The calculated distribution of the neutron flux or gold activation agrees well with the measured one, except the positions close to the boundary between the core and the reflector, although we took a considerably rough assumption for the geometry in the calculation.

Little difference is observed between the absorbed dose data with the MSO detectors with the two different tubes made of aluminum and polyethylene. The absorbed doses calculated with the prompt gamma-rays alone are about a half comparing with the measured ones, but the corrected values, in which the neutron kerma (absorbed dose) and the contribution of the delayed gamma-rays from fission products are considered, become much closer (lower by 10 %) to the measurements.

The measured and calculated neutron fluxes and absorbed doses in the void and in three filter boxes at the center of the internal graphite reflector of the UTR-KINKI are shown in Figs. 5 and 6, respectively. The normalization similar to the case of KUCA has been performed at 81 cm above the reactor bottom.

It can be seen that (1) the filtering effect to neutron and gamma-ray fluxes by means of wall material is very clear, (2) the

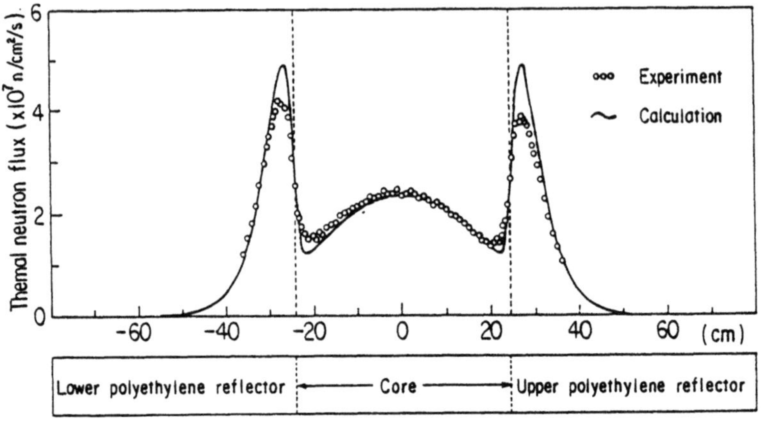


Fig. 3 Measured and calculated neutron flux (gold activation) distribution in the B-core of KUCA. The measured values at the core center is normalized to the calculated value.

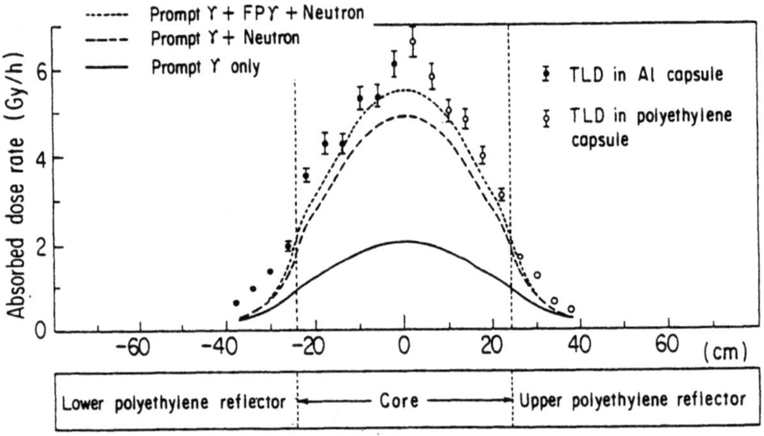


Fig. 4 Measured and calculated absorbed dose of MSO in the B-core of KUCA under the condition of the same normalization given in Fig. 3.

agreement between the measured and calculated neutron flux distributions is excellent, but (3) the calculated absorbed doses are about 35 % lower in average than the measured ones. This discrepancy may be caused by no data correction for this case.

## CONCLUSION

The measured distributions of neutron flux or gold activation agree well with those calculated for both cases, one is the B-core of KUCA and another in four neutron fields ( a void and three filter boxes) of the UTR-KINKI, when we normalize the calculated gold activity to the measured one.

Under the condition of the above normalization, the calculated absorbed dose distributions with MSO is 10 % lower than the measured for the B-core of KUCA, when the contribution of neutron absorbed dose and that of delayed gamma-rays from fission products are taken into account.

Under the similar normalization condition, all of the calculated absorbed doses in void and in three filter boxes at the center of UTR-KINKI are considerably lower (about 35 %) in average than the measured ones, when no correction is done.

More precise data corrections and/or more sophisticated calculation methods may be required to get better agreement between measured and calculated absorbed doses in and around reactors.


## ACKNOWLEDGEMENT

The authors are deeply thankful to Mr. T. Sato for his guidance to use MSO detectors and his TLD recorder.


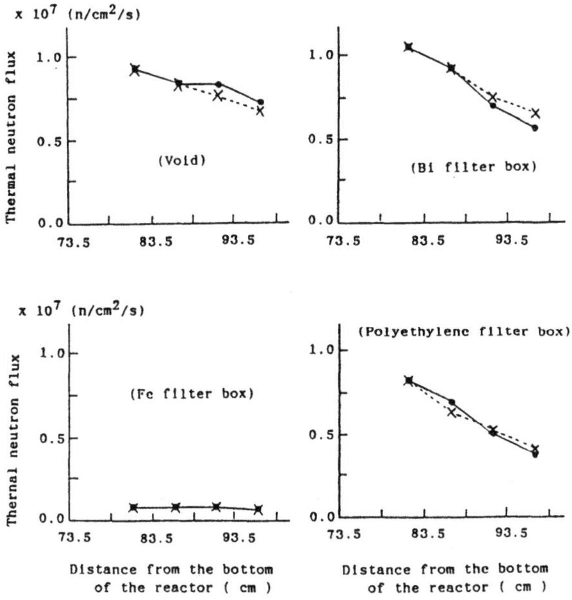


——— : Experiment, - - - - - : Calculation

Fig. 5 Measured and calculated distributions of neutron fluxes (gold activation) in void and in three boxes located in the inner reflector between divided cores of UTR-KINKI. The measured value is normalized to the calculated one at 81 cm above the reactor bottom for each case.

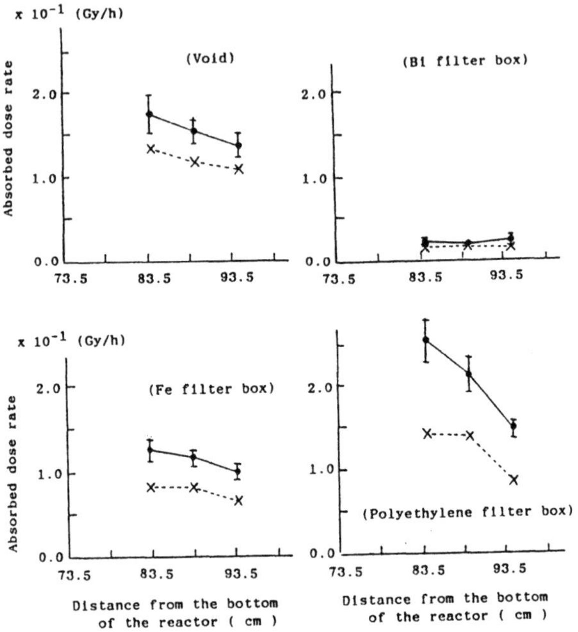


——— : Experiment, ----- : Calculation

Fig. 6 Measured and calculated distributions of absorbed doses of MSO in void and in three filter boxes in the inner reflector between two boxes divided cores of UTR-KINKI under the condition of the same normalization as Fig. 5.

A part of the present works was achieved by the cooperative use program of the UTR-KINKI. Thanks are also due to Prof. R. Miki and Mr. T. Itoh of Kinki University for their cooperation to use the UTR-KINKI.

REFERENCES

1. T. Hashizume et al., Proc. IAEA Symp. on Adv. Rad. Detectors, SM-143/11, p.91 (1971).
2. T. Sato and K. Ono, unpublished.
3. T. Sato, unpublished.
4. Y. Sakurai and I. Kimura, to be published.
5. S. Shiroya, K. Kanda et al., Proc. Int. Conf. on Phys. Reactors: Operation, Design & Computation, Marseille, France, Vol.3, PI-1-3 (1990).
6. Joint Research Program Reports of UTR-KINKI. Osaka University (1988).
7. K. Koyama et al., JAERI-M 5794 (1974).
8. J. H. Hubbell, Photon Cross Sections, Attenuation Coefficients and Energy Absorption Coefficients from 10 keV to 100 GeV, NSRDS-NBS 29 (1969).
9. ICRU Report 26, Neutron Dosimetry for Biology and Medicine, (1977).

# SIMULATION FIDELITY IN REACTOR IRRADIATION OF ELECTRONICS*

T. F. Luera, P. J. Griffin, J. G. Kelly, M. S. Lazo

Sandia National Laboratories & University of New Mexico

Albuquerque, New Mexico, USA

## ABSTRACT

Fast Burst Reactors (FBRs), and pool-type reactors, such as TRIGAs and the Sandia Annular Core Research Reactor (ACRR), are used as sources of neutrons and gamma rays in radiation effects testing of electronics. The applicability of test results to other test or operational radiation environments depends on adequate measurement and control of the test environment, and on the use of a valid damage equivalence methodology. Specific considerations which affect this applicability (simulation fidelity) are discussed. Damage equivalence methodology is discussed in the context of ASTM Standards.

## BACKGROUND

Nuclear reactors are used in the neutron and gamma ray irradiation testing of electronic devices to qualify them for use in radiation environments. In general, radiation environments obtained at test or research reactors differ from the operational environments in which the electronics must perform their intended function. Additionally, all electronic parts which make up a system, and all part lots, may not be tested at the same reactor facility. In order to correctly interpret the test data and assess device performance in the operational environment, it is necessary to: (i) have a basic understanding of the damage mechanisms relating to the operational and test environments; (ii) adequately control the test environment; and (iii) employ a validated damage equivalence methodology. The fidelity of the test simulation need not rely on precise duplication of the operational environment if these three considerations are properly addressed. It is tacitly assumed in the above discussion that required electrical tests are properly performed. ASTM 10.07 is examining the feasibility of creating a matrix standard to show the relationships among the various dosimetric, electrical, and other standards which should be employed to assure test validity.

*Prepared by Sandia National Laboratories, Albuquerque, New Mexico 87185 for the United States Department of Energy under Contract DE-AC04-76DP000789.

## FACTORS AFFECTING SIMULATION FIDELITY

It is essential that the experimenter understand the dominant device-damage mechanisms and adequately control the test environment. A silicon bipolar transistor which is susceptible to total dose as well as to displacement damage (Either results in gain degradation in the transistor.) provides a good illustration of both these points. A free-field irradiation at a fast burst reactor may provide a valid measure of neutron sensitivity because the neutron-to-gamma ratio is high at this type of facility. On the other hand at a pool-type reactor such as a TRIGA, the gamma-ray contribution to damage may have to be subtracted, or the gamma fluence reduced with lead shielding. Another consideration is that if the TRIGA is operated at high sustained power immediately prior to the transistor irradiation, the gamma contribution to damage may be significantly higher than normal because of the activation of surrounding materials. Thus it is necessary to be aware of the total dose sensitivity of the test device and to control unwanted contributors to the damage.

The following are examples of problems that have been encountered because the damage mechanisms were not well understood or the test environments were not adequately controlled:

1. Memory circuits functioned perfectly after exposure to neutrons at the SPR III fast burst reactor; however, the circuits failed when subjected to an equivalent silicon damage exposure at the ACRR, a pool-type reactor. It was later found that the failures resulted from fission product ions traversing sensitive regions of the device. Fissions took place in trace amounts of uranium in the alumina lids of the devices. When gold-kovar lids were substituted no failures occurred. These events were not detected in the SPR III irradiations because there are very few neutrons with energies below 10 keV to produce fissions in that environment. (The fission cross section for uranium-235 increases rapidly with decreasing neutron energy.)
2. Silicon parts may be damaged by thermal neutrons through interactions with dopants such as boron. Further, it is desirable to minimize activation of the semiconductor parts so that they may be handled soon after exposure. For these reasons it is common practice at pool-type reactors such as TRIGAs to surround the exposure volume with cadmium. Unfortunately, in such a neutron environment the gamma dose received by the part may be enhanced by as much as a factor of three through neutron capture in the cadmium. If the part is sensitive to ionization damage, the measured damage may be erroneously attributed to fast neutron effects.
3. Other potential sources of error are:
    a. A changing neutron spectrum as a result of fuel burnup or refueling. Neutron poisons buildup as the reactor is operated.
    b. The neutron/gamma ratio can be altered as a result of the operating history of the reactor. Activation of structural materials and background radiation levels can change significantly depending on operational history of the reactor.

c. In pool-type reactors, the presence of water significantly softens the photon spectrum and dose-enhancement effects may need to be considered. Dose enhancement effects occur at the interface between high and low atomic number materials when photon energies fall below about 200 keV. The absorbed dose in a low atomic number material such as silicon dioxide next to a high atomic number material such as gold can be much higher than indicated by typical dosimeters.

e. Temperature during and after the irradiation may be an important consideration. Damage in silicon anneals slowly at room temperature, but more rapidly at higher temperatures. Damage measured at the same time after otherwise equivalent irradiations at a TRIGA and at a fast burst reactor facility may differ if the temperature of the parts is not controlled.

In summary, when simulation fidelity is an important consideration in a test, the experimenter must know the sensitivity of his test items to the important components of the reactor environment. These may include neutron fluence and spectrum (thermal neutrons may cause problems), ionizing dose from photons and perhaps electrons and neutrons, photon spectrum, dose rate, and temperature. Reactor environments can often be controlled or altered (given enough space) to minimize or eliminate undesirable effects.

## DAMAGE EQUIVALENCE

### Ionization Damage

As noted previously solid state electronics is susceptible to ionization as well as displacement damage. The damage in bipolar silicon devices correlates well with total dose. However, in MOS (metal-oxide-semiconductor) devices, the net damage may be a function of dose rate, and the time between the irradiation and the electrical measurements, because the rates of interface state formation and the annealing of oxide-trapped charge determine the net damage as a function of time. In addition dose enhancement may have to be considered to determine the dose in the sensitive region of the device. The key ASTM dosimetry-related standards applicable to reactor-produced ionization damage are listed in Table 1.

E668 Standard practice for the Application of Thermoluminescence-Dosimetry (TLD) Systems for determining Absorbed Dose in Radiation-Hardness Testing of Electronic Devices

E1249 Practice for Minimizing Dosimetry Errors in Radiation Hardness Testing of Silicon Electronic Devices Using Co-60 Sources

E1250 Methods for Application of Ionization Chambers to Assess the Low Energy Gamma Component of Cobalt-60 Irradiators Used in Radiation-Hardness Testing of Silicon Electronic Devices

**Table 1. Standards Relating To Total Dose Testing**

ASTM Committee F1.11 addresses standard practices for electrical measurements in both total dose and neutron testing.

Reactors may be used as ionizing radiation sources, with or without gamma enhancing material such as CdO-loaded polyethylene, to provide ionizing pulses with durations from 40 $\mu$s up to a few seconds. These pulse widths fill in the range between steady-state isotopic sources such as $^{60}Co$ and the flash X-ray machines and LINACs which provide pulse widths up to 10 $\mu$s in duration.

Displacement Damage

In general if the damage produced in the test environment can be made equal to the damage in the operational environment, as a result of the same damage mechanism, the fidelity of the test is assured. For a single damage mechanism this can be written as:

$$\int D(E)\ \phi_{sim}(E)\ d(E) = \int D(E)\ \phi_{op}(E)\ d(E)$$

The appropriate damage function, D(E), could be the fission cross section, the boron neutron-absorption cross section, or the displacement kerma for silicon, etc. The displacement kerma is proportional to the number of displacements per atom (DPA) times the displacement threshold energy. As can be seen from this fundamental equation of damage equivalence, normalization of the damage function or the neutron spectrum has no effect as long as those who write the test specifications (the right side of the equation) use the same normalization and damage function as the testers (the left side of the equation). The damage equivalence methodology is described in ASTM Standards E720, E721 and E722 (Table 2.) which address the selection of activation foil sets for neutron spectrum determination, the use of adjustment codes and activities for spectrum determination, and the process of establishing damage equivalence given a neutron spectrum. The reactor test community strives to keep errors in delivered displacement kerma below 10% because of the contractual nature of many requirements and associated economic penalties to system developers that may result from failure to meet system specifications.

E720 Selection of a set of neutron-activation foils for determining neutron spectra used in radiation-hardness testing of electronics.

E721 Determining neutron energy spectra with neutron-activation foils for radiation hardness testing of electronics.

E722 Characterizing energy fluence spectra in terms of an equivalent monoenergetic neutron fluence for radiation hardness testing of electronics.

**Table 2. Standards for Neutron Damage Equivalence**

Despite the widespread use of reactors for testing of other semiconductors, the neutron damage equivalence formalism in ASTM Standard E722 (which implies a linear dependence of damage on the displacement kerma) has been validated only for silicon devices.

Joint Sandia National Laboratories/University of New Mexico work to evaluate the applicability of this methodology to other semiconductors such as GaAs has led to proposed improvements to these standards.

## IMPROVEMENTS

An improved calculational model (NJOY89 [1]) and updated cross sections (ENDF/B-V) have been used to generate a new damage function for silicon. The new damage function has been validated with some carefully controlled experiments in reactor and 14-MeV neutron environments. (See papers by P. Griffin and J. Kelly, this conference.) It is proposed to update the silicon damage kerma table in E722-86 with this new damage function. The precision required for comparing damage functions has led to an improved foil and cross section set for E720. (See paper by J. Kelly and P. Griffin, this conference.) The option of using an outer iteration technique for use with SAND II provides significant improvements to E721. Much of our work has also been evaluated and verified by the Dosimetry Laboratory at White Sands Missile Range and their contractor, the University of Illinois [2].

## WORK IN PROGRESS

We have examined the pre-release ENDF/B-VI $^{28}Si$ cross section [3] and found the displacement kerma obtained with this cross section to be consistent with the ENDF/B-V values. Contributions from other isotops are negligible.

At present the NJOY code uses the Lindhard energy partition function which is based on the Thomas-Fermi interaction potential. Work is underway to assess the impact of using the more accurate Universal (Bierzack-Ziegler) potential. Although this step may not be justified for silicon, it appears essential that the methodology be as accurate as practical in order to properly investigate the application to other semiconductor materials.

Only about 1% of the initial neutron damage (Frenkel pairs) remains after neutron irradiation and annealing at room temperature. Because of the importance of initial damage distributions on the annealing process, the relationship between deposited energy and stable damage may not be linear for materials other than silicon. For example, we have shown that carrier removal in GaAs displays a dependence on primary-knock-on-atom (PKA) spectrum for constant displacement energy.[4] A cooperative effort is underway with the University of New Mexico to examine this annealing process from first principles with the molecular dynamics code, MOLDY-CASK [5], developed at Lawrence Livermore Laboratories. A many-body interatomic potential and extensive experimental data already exist for Si, and this method will first be used as a benchmark for these calculations before application to GaAs.

Sandia's investigation of the applicability of the ASTM E722 neutron damage-equivalence methodology to GaAs ran into difficulties immediately. Firstly, the basic neutron cross section data for arsenic was incomplete and inaccurate. The gallium cross section data was not in much better shape. Secondly, careful measurements of carrier removal in GaAs in reactor and 14-MeV environments showed significant deviations from linear dependence on displacement energy. Examination of the available cross-section data and uncertainties has confirmed that the departure from linearity cannot be explained solely on the basis of uncertainty in the cross sections. This behavior of GaAs is similar to that reported for some metals [6] and may depend on damage cascade characteristics. If this behavior is confirmed in other important electronic parameters, neutron simulation fidelity may require development of a GaAs neutron damage equivalence standard based upon a set of empirical damage-weighted displacement-energy response parameters.[7] On the other hand, the feasibility of calculating stable damage efficiencies in semiconductors as a function of primary knock-on atom energy is also being investigated by SNL/UNM.

These efforts are focused on using Monte-Carlo-generated particle distributions as input to molecular dynamics calculations. This method could lead to a damage-equivalence standard based on a theoretical damage-efficiency-weighted displacement-kerma damage function.

With the help of Los Alamos National Laboratory [8] a new evaluation of the $^{69}Ga$ cross section has been performed at the University of New Mexico. Basic cross-section data for an important isotopes of germanium $^{72}Ge$ is lacking. Support for evaluating the $^{72}Ge$ cross section is currently being sought because studies of damage equivalence in this material should also help us to understand damage in GaAs.

Kerma calculations have been performed for a wide variety of materials including: HgCdTe, $CaF_2(Mn)$, $B_4C$, $SiO_2$, epoxy-glass insulators, and some high temperature superconductors.

## CONCLUSIONS

A great deal of work remains to be done to assure simulation fidelity in the reactor irradiation of semiconductors other than silicon. Basic neutron cross-section data is lacking. Improved techniques for the modeling of damage need to be developed, or empirical damage functions generated. Finally, the resulting methodology for non-silicon semiconductors must be validated through experimental confirmation using fission and 14-MeV neutron sources.

## REFERENCES

1. R. E. MacFarlane, "NJOY. A Code System for Producing Pointwise and Multigroup Neutron and Photon Cross Sections from ENDF/B-IV and -V Evaluated Nuclear Data," RSIC, PSR-118, Los Alamos National Laboratory, March 1979.
2. M. H. Sparks, T. M. Flanders, J. G. Williams, J. G. Kelly, W. W. Sallee, M. Rokhizadeh and J. L. Meason, "Energy Dependence of Neutron Damage in Silicon Bipolar Transistors," IEEE Trans. Nucl. Sci., Vol. NS-36, No. 6, December 1989.
3. S. Epperson, D. Larson, D. Hetrick, Y. Fu, private communication from Oak Ridge National Laboratory, September 1989. This data will be released as part of the ENDF/B-VI cross section library release 2. An Oak Ridge National Laboratory report will be published describing the evaluation techniques used.
4. T. F. Luera, J. G. Kelly, H. J. Stein, M. S. Lazo, C. E. Lee, and L. R. Dawson, "Neutron Damage Equivalence for Silicon, Silicon Dioxide and Gallium Arsenide," IEEE Trans. Nucl. Sci., Vol. NS-34, No. 6, December 1987.
5. "MOLDY6 - A Molecular Dynamics Program Simulation of Pure Metals," M. W. Finnis, United Kingdom Atomic Energy Authority, Theoretical Physics Division, Harwell Laboratory, Oxford, England, AERE-R-13182, July 1988.
6. R. S. Averback, R. Benedek, K. L. Merkle, "Ion-irradiation Studies of the Damage Function of Copper and Silver," Phys. Rev. B18, p. 4156, October 1978.
7. J. G. Williams, Univ. of Illinois, private communication, 1989.
8. P. Young and R. MacFarlane, Los Alamos Scientific Laboratory, private communication, 1989.

# BENCHMARKS

# ANALYSIS OF THE VENUS-3 EXPERIMENTS

R. E. Maerker

Oak Ridge National Laboratory

Oak Ridge, Tennessee 37831-6364, USA

P. D'hondt, L. Leenders, A. Fabry

SCK/CEN Mol, Belgium

The results of applying a hybrid superposition-synthesis calculational method to a mockup of a three-dimensional geometry involving a partial length shield assembly at the VENUS-3 facility in Mol, Belgium, are described. Comparisons of transport calculations using the method and many measurements involving nickel, indium, and aluminum dosimeters indicate agreement usually to within measurement uncertainties estimated at around 5%, if effects of inaccuracies in the dosimeter cross sections are minimized and proper orientation of the coordinate system used in the synthesis procedure is observed. These conclusions suggest a solution to the problem of predicting pressure vessel fluence in reactors modified by these partial-length shield assemblies may already exist.

## INTRODUCTION AND BACKGROUND

Over the past several years, pressurized thermal shock (PTS) programs have become operational at many reactors in the United States. Since PTS-induced pressure vessel damage is dependent on the accumulated fluence, these programs are mainly concerned with reducing fluence rates incident on critical locations in the pressure vessel (PV) of pressurized water reactors (PWR's). At least two independent methods have been used, and both involve altering the core-source distribution. The first changes the refueling pattern to an in-in-out rotation at the critical azimuths so that twice-burned fuel now appears in the peripheral assemblies at those azimuths for which accumulated PV fluences have been the highest. This revised fuel loading scheme typically reduces the core leakage at these azimuths by about a factor of two. For many reactors of moderate age, this reduction is adequate to meet the additional requirements of plant-life extension. A second method is being used in several of the older reactors where fluence-rate reduction factors greater that those afforded by the first method have been deemed necessary. This procedure partially replaces the oxide fuel with metallic rods for those peripheral assemblies again located at the critical orientations. This substitution extends axially over a region that depends on the specific plant design, but covers in particular the most critical weld locations which are subject to PTS phenomena. The reduction in the fluence rates at the critcal locations for the latter case is of the order of a factor of ten, and is due principally to the complete removal of the fission source in the critical peripheral assemblies.

One older reactor that has been subjected to both methods of core-leakage reduction has been H.B. Robinson Unit 2, which underwent a revised fuel shuffling scheme for cycle 9 and the introduction of partial-length shield assemblies (PLSA's) consisting of steel rods for cycle 10 and beyond. The analysis of cycle 9 using the LEPRICON procedure appears in the literature and is relatively straightforward(1). However, the use of the PLSA's introduces a geometrical complication that has not existed heretofore in conventional PV transport analysis, and a procedure should first be sought that adequately approximates the results of this three-dimensional transport with calculations involving only the one- and two-dimensional codes already in use.

Accordingly, an agreement was struck between the Nuclear Regulatory Commission (USNRC) and Mol which served the needs of both parties: The USNRC, with the approval of the Electric Power Research Institute, made available LEPRICON software for Mol to use in the dosimetry analysis of their DOEL reactors, while Mol agreed to perform experiments involving a PLSA mockup in their VENUS facility to serve to benchmark the calculational methods. The USNRC, through a contractual agreement with Oak Ridge National Laboratory (ORNL), also agreed to provide the results of an introductory analysis of these measurements.

## THE VENUS-3 FACILITY AND THE MEASUREMENTS ANALYZED

The conclusions drawn from a meeting between ORNL and Mol scientists in January 1987 resulted in a design for VENUS-3 in which steel rods were substituted for the fuel over the complete lower half of the pins in the outermost five rows of two diametrically opposite arms of the cross. In Fig. 1 the PLSA modifications

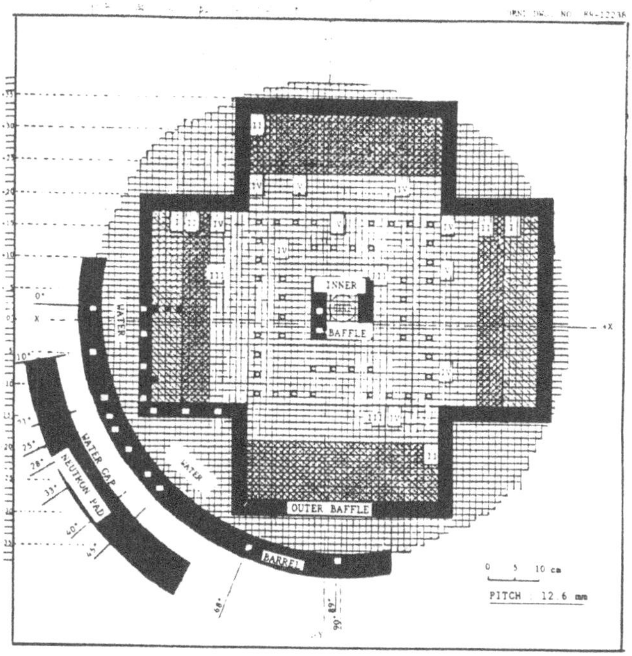


Fig. 1. Plan view of the VENUS-3 facility showing the location of the nickel dosimeters used in the early run of March 1988 as the white squares and the four black squares in the PLSA.

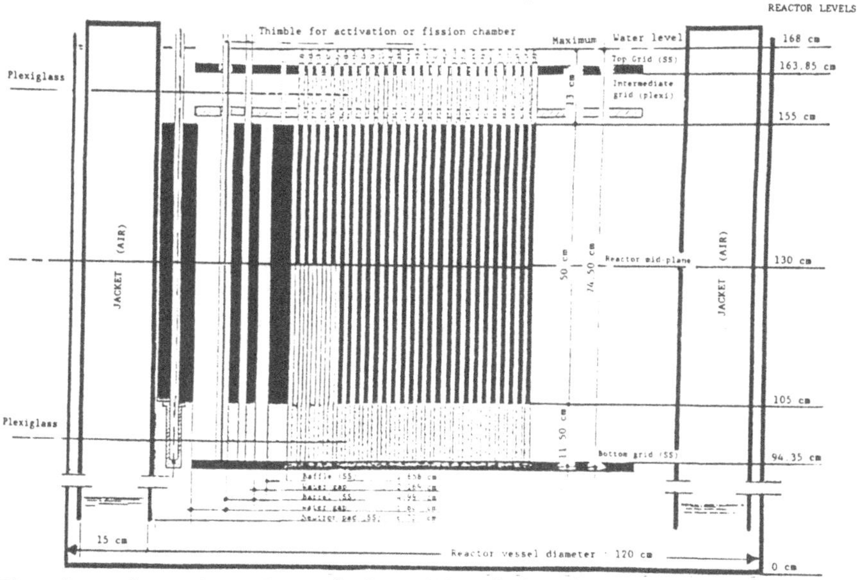


Fig. 2. Elevation view of the VENUS-3 facility showing the substitution of stainless steel rods for oxide fuel in the lower halves of five peripheral pins.

are indicated as the left- and right-hand arms. An elevation view of the full arrangement is shown in Fig. 2. The upper half of these assemblies as well as all the remaining pins contain fuel without modification. The core, however, was modified to employ 4% enriched 235U in steel clad pins in the inner zone and 3.3% enriched 235U in zircalloy clad pins in the outer zone. Criticality was achieved by adjusting a different number and arrangement of pyrex rods surrounding the core inner baffle than those in earlier VENUS configurations.

Since the focus of the VENUS-3 experiments is essentially on providing data to test transport methods for calculating ex-core fluence rates, the source description used in the calculations was taken from measurements in order to reduce this potential source of uncertainty as much as possible. The measurements consisted of obtaining a relative pin-by-pin, three-dimensional distribution of the fission source over one-quarter of the core using a gamma-ray scanning technique, and then following up with a normalization factor that would link this relative distribution to the absolute distribution existing during the runs involving the dosimetry exposures. The fuel-pin scans were performed at fine enough spatial intervals that interpolation between measured values was estimated to be accurate to within 5%; the relative individual measurements were estimated to be accurate to within 2%. Absolute measurements of the core fission rates at several locations using four different methods yielded an averaged value of $8.60 \times 10^9 \pm 2.3\%$ (2) as the multiplicative factor for converting the source, arbitrarily normalized in the calculations to 1 fission/s per pin averaged over the core, to the source at full power. Measured reaction rates available for the present analysis involved the 58Ni(n,p), 115In(n,n'), and 27Al(n,$\alpha$) dosimeters at various axial levels and several horizontal locations throughout the configuration. These measurements are believed to have an accuracy of better than 3%. In this paper, explicit comparisons between calculated and measured values will be limited to those locations indicated by the white squares in the core barrel in Fig. 1 for the nickel dosimeters used in the early run of March 1988. This choice covers a large range of locations, starting at 0° behind the PLSA mockup and continuing around to 90° behind the unmodified assembly.

## DESCRIPTION OF THE CALCULATIONS

The method chosen for the calculations is a modification of a fluence-rate synthesis procedure often used in more conventional geometry(3,4). By decomposing the usual synthesis prescription into two components, one contribution arising from sources (A) lying above the PLSA region and the other from sources (B) lying within the levels defined by the PLSA region, one can write the superposition-synthesis equation as

$$\phi_g(X,Y,Z) = \frac{\phi_g^A(X,Y)\phi_g^A(X,Z)}{\phi_g^A(X)} + \frac{\phi_g^B(X,Y)\phi_g^B(X,Z)}{\phi_g^B(X)}, \tag{1}$$

where X is the principal direction of transport, Y is the transverse direction, and Z is the axial direction.(5)

The sources to be used in the one- and two-dimensional discrete ordinates calculations are

$$S_g^A(X,Y) = \int_{Z_{UL}}^{\infty} S_g(X,Y,Z)dZ, \tag{2}$$

$$S_g^B(X,Y) = \int_{-\infty}^{Z_{UL}} S_g(X,Y,Z)dZ, \tag{3}$$

$$S_g^A(X,Z) = \int_{-\infty}^{\infty} S_g(X,Y,Z \geq Z_{UL})dY, \tag{4}$$

$$S_g^B(X,Z) = \int_{-\infty}^{\infty} S_g(X,Y,Z \leq Z_{UL})dY, \tag{5}$$

$$S_g^A(X) = \int_{-\infty}^{\infty} dY \int_{Z_{UL}}^{\infty} S_g(X,Y,Z)dZ, \tag{6}$$

$$S_g^B(X) = \int_{-\infty}^{\infty} dY \int_{-\infty}^{Z_{UL}} S_g(X,Y,Z)dZ, \tag{7}$$

where $Z_{UL}$ is in general the upper level to which the metal rods extend. For VENUS-3, this is the midplane.

The condition that should be met for validity of the method is that the fluence rate due to each component be separable in Y and Z over the spatial region where that component dominates. In the transport geometry using sources defined by Eqs (2) and (6), the core periphery is composed entirely of active pins, while in the geometry defined by Eqs (3) and (7) the outer five rows (in VENUS-3) contain only steel-filled pins.

## COMPARISON OF CALCULATED AND MEASURED RESULTS

The measurements and calculations were originally expressed as reaction rates in units of reactions/s per target nucleus at full power. Comparisons are more meaningful when expressed in terms of equivalent fission fluence rates in the present application, however, for they remove some of the dependence of the calculated results on the dosimeter cross sections and of the measured results on reaction-specific parameters such as branching ratios. Thus, these resulting comparisons tend to become better measures of the accuracy of the calculational method itself, which is the primary goal of the VENUS-3 program. These equivalent fission fluence rates are reaction dependent, and are defined for reaction i as

$$\phi_i = \frac{\int_0^{\infty} \phi(E)\sigma_i(E)dE}{\int_0^{\infty} {}^{235}\chi(E)\sigma_i(E)dE}, \tag{8}$$

where the numerator is the conventional reaction rate and the denominator is the spectrum-averaged reaction cross section in a 235U fission field. For all three reactions analyzed, calculated values of the denominator using the cross-section library adopted for this analysis(6) underestimate the measured values used by Mol by 3 to 5%.(7)

The calculations described so far involve an XY run, an XZ run, and an X run, all for each of the two source components, for a total of six discrete ordinates calculations. It is to be expected, however, that the two measurements in the core barrel at 68° and 89° (or any such similar locations requiring a considerable re-orientation of the coordinate system defined in Eqs (1) through (7) and Fig. 1) may not be well calculated. As pointed out earlier, the significance of the X and Y axes in these equations is in their interpretation as parallel and perpendicular to the principal direction of flow to the detector, respectively. This direction can be approximated in the present geometry for ex-core detectors by a line joining the core center and the detector. For most of the detectors shown in Fig. 1, both in-core and ex-core, the XY coordinate system shown in Fig. 1 is adequate. For detector locations lying at angles greater than 45° with the X-axis, X and Y may be interchanged in Eqs (1) through (7) and the resulting sources resummed. This has no effect on the XY transport calculations if X and Y are now reinterpreted as the old Y and X respectively, and the original XY results are still valid. The XZ and X transport calculations must be redone, however, since their source terms are now different. This results in additional XZ and X runs for each of the two source components. A grand total of ten discrete ordinates calculations were thus made. The reaction rates at 68° and 89° that were calculated using the re-oriented coordinate system provide good reference points for confirming the accuracy of the synthesis procedure in the vicinity of an unmodified assembly, where no consideration of separate source components (i.e., superposition) is necessary.

Some comparisons of the nickel fluence rates in the core barrel are shown in Table 1. Values of C/E at similar locations to those in Table 1 for the indium and aluminum dosimeters are virtually the same, thus extending verification of the method to higher and lower energies. Comparisons at other locations, both in-core and ex-core, some of which are indicated as additional white squares along with a few black squares in Fig. 1, further substantiate the general applicability of the method. The comparisons at most locations agree to within the estimated standard deviation of around 5% which is a composite of all measurement uncertainties.

Table 1. Some Absolute Comparisons of the Nickel Fluence Rates in the Core Barrel for the March 1988 Run

| Z (cm)[a] | Angle[b] | $\phi_I$ | $\phi_C$ | $\phi_C/\phi_I$ | Angle[b] | $\phi_I$ | $\phi_C$ | $\phi_C/\phi_I$ |
|---|---|---|---|---|---|---|---|---|
| 106.5 | 0 deg | 1.70+7[c] | 1.64+7[d] | 0.96 | 21 deg | 1.41+7 | 1.40+7[d] | 0.99 |
| 114.5 | | 2.83+7 | 2.80+7 | 0.99 | | 2.39+7 | 2.37+7 | 0.99 |
| 122.5 | | 4.39+7 | 4.07+7 | 0.93 | | 3.56+7 | 3.46+7 | 0.97 |
| 128.5 | | 5.54+7 | 5.31+7 | 0.96 | | 4.54+7 | 4.48+7 | 0.99 |
| 131.5 | | 6.09+7 | 5.64+7 | 0.93 | | 5.03+7 | 4.80+7 | 0.95 |
| 134.5 | | 6.45+7 | 6.06+7 | 0.94 | | 5.41+7 | 5.24+7 | 0.97 |
| 141.5 | | 6.44+7 | 6.05+7 | 0.94 | | 5.35+7 | 5.16+7 | 0.96 |
| 145.5 | | 5.87+7 | 5.55+7 | 0.95 | | 4.78+7 | 4.67+7 | 0.98 |
| 149.5 | | 4.94+7 | 4.76+7 | 0.96 | | 4.05+7 | 3.96+7 | 0.98 |
| 153.5 | | 3.76+7 | 3.62+7 | 0.96 | | 3.07+7 | 2.95+7 | 0.96 |
| 106.5 | 45 deg | 1.75+7 | 1.77+7[e] | 1.01 | 68 deg | 3.42+7 | 3.24+7[e] | 0.95 |
| 114.5 | | 2.74+7 | 2.96+7 | 1.08 | | 5.72+7 | 5.50+7 | 0.96 |
| 122.5 | | 3.53+7 | 3.87+7 | 1.10 | | 7.09+7 | 7.00+7 | 0.99 |
| 128.5 | | 3.84+7 | 4.30+7 | 1.12 | | 7.51+7 | 7.51+7 | 1.00 |
| 131.5 | | 3.86+7 | 4.53+7 | 1.17 | | 7.68+7 | 7.53+7 | 0.98 |
| 134.5 | | 3.83+7 | 4.53+7 | 1.18 | | 7.50+7 | 7.40+7 | 0.99 |
| 141.5 | | 3.50+7 | 4.02+7 | 1.15 | | 6.64+7 | 6.50+7 | 0.98 |
| 145.5 | | 3.09+7 | 3.48+7 | 1.13 | | 5.72+7 | 5.65+7 | 0.99 |
| 149.5 | | 2.54+7 | 2.81+7 | 1.11 | | 4.67+7 | 4.67+7 | 1.00 |
| 153.5 | | 1.97+7 | 2.09+7 | 1.06 | | 3.45+7 | 3.43+7 | 0.99 |
| 106.5 | 89 deg | 4.32+7 | 4.60+7[d] | 1.06 | 89 deg | 4.32+7 | 4.23+7[e] | 0.98 |
| 114.5 | | 7.08+7 | 7.82+7 | 1.10 | | 7.08+7 | 7.00+7 | 0.99 |
| 122.5 | | 8.94+7 | 1.02+8 | 1.14 | | 8.94+7 | 8.85+7 | 0.99 |
| 128.5 | | 9.54+7 | 1.09+8 | 1.14 | | 9.54+7 | 9.46+7 | 0.99 |
| 131.5 | | 9.63+7 | 1.09+8 | 1.13 | | 9.63+7 | 9.55+7 | 0.99 |
| 134.5 | | 9.45+7 | 1.07+9 | 1.13 | | 9.45+7 | 9.26+7 | 0.98 |
| 141.5 | | 8.21+7 | 9.36+7 | 1.14 | | 8.21+7 | 8.14+7 | 0.99 |
| 145.5 | | 7.17+7 | 8.05+7 | 1.12 | | 7.17+7 | 7.14+7 | 1.00 |
| 149.5 | | 5.81+7 | 6.49+7 | 1.12 | | 5.81+7 | 5.92+7 | 1.02 |
| 153.5 | | 4.32+7 | 4.78+7 | 1.11 | | 4.32+7 | 4.39+7 | 1.02 |
| 114.5 | 10 deg | 2.48+7 | 2.44+7[d] | 0.98 | 25 deg | 2.18+7 | 2.17+7[d] | 1.00 |
| 131.5 | | 5.21+7 | 4.86+7 | 0.93 | | 4.35+7 | 4.25+7 | 0.98 |
| 145.5 | | 5.04+7 | 4.76+7 | 0.94 | | 4.09+7 | 4.07+7 | 1.00 |
| 114.5 | 33 deg | 2.88+7 | 2.31+7[d] | 0.80 | 40 deg | 2.53+7 | 2.38+7[d] | 0.94 |
| 131.5 | | 4.48+7 | 4.22+7 | 0.94 | | 3.76+7 | 4.34+7 | 1.15 |
| 145.5 | | 3.93+7 | 3.80+7 | 0.97 | | 3.06+7 | 3.62+7 | 1.18 |

[a]The pins extend from 105 to 155 cm; the PLSA extends from 105 to 130 cm. See Fig 2.
[b]See Fig. 1.
[c]Read as 1.70 x $10^7$ n/cm$^2$•s at full power.
[d]Based on orientation of X and Y axes as shown in Fig. 1.
[e]Based on orientation of X and Y axes rotated 90 deg counterclockwise to those of Fig. 1.

## DISCUSSION OF THE RESULTS AND CONCLUSIONS

Success in the introduction of a source decomposition and superposition technique and combining it with a fluence-rate synthesis procedure is conclusively demonstrated by the axial comparisons at the 0°, 10°, 21°, 25°, 28°, and 33° locations in the core barrel as well as four PLSA locations, the water gap, and all seven locations in the outer baffle. The axial profiles at these locations are influenced by the PLSA and are asymmetric, as contrasted with the nearly perfect symmetry of the profiles located well inside the core and in the vicinity of the 68° and 89° core barrel locations. A legitimate conclusion deduced from the comparisons near 0° and 89° is that, at least for VENUS-3, Eqs (1) through (7) are as accurate in the treatment of PLSA-affected regions as earlier equations without source decomposition are in the unaffected ones. It is only at the two locations in the vicinity of 45° in the core barrel that any significant error is incurred with the proposed technique, and presumably this error could be reduced by reorienting the X and Y axes in Eqs (2) through (7) to lie at 45° to the directions shown in Fig. 1, were it really necessary. The error near 45° is still not large, however, being at most 18% (see Table 1). The conclusion that the method provides acceptable accuracy when applied to VENUS-3 does not necessarily imply that comparable accuracies should be realized when applied to PWR's. Differences in geometry between VENUS-3 and H.B. Robinson-2 are still considerable. Nevertheless, the accuracy of the proposed method when applied to VENUS-3 is very encouraging, and suggests that a relatively simple modification to the method, if necessary, could be found for application to PWR's.

## REFERENCES


1. R. E. Maerker, "LEPRICON Analysis of Pressure Vessel Surveillance Dosimetry Inserted into H.B. Robinson-2 During Cycle 9," Nucl. Sci. & Eng. 96, 4,263 (1987).

2. L. Verbruggen, LWR-PVS Benchmark Experiment VENUS-3. Part 3. Absolute Core Power, FCP/VEN/03, Mol, (December 1989).

3. G. P. Cavanaugh, Summary Meeting on PCA Blind Test, National Bureau of Standards (1983).

4. R. E. Maerker, M. L. Williams, and B. L. Broadhead, "Accounting for Changing Source Distributions in Light Water Reactor Surveillance Dosimetry Analysis," Nucl. Sci. & Eng. 94, 4,291 (1986).

5. M. L. Williams, et al., H.B. Robinson Fluence Reduction Analysis for the PLSA Concept, TEC Report R-83-030, Technology for Energy Corporation and Carolina Power and Light Company (1983).

6. M. L. Williams, et al., The ELXSIR Cross-Section Library for LWR Pressure Vessel Irradiation Studies: Part of the LEPRICON Computer Code System, EPRI NP-3654 (September 1984).

7. R. E. Maerker, Analysis of the VENUS-3 Experiments, Oak Ridge National Laboratory, NUREG/CR-5338, ORNL/TM-11106, p.54 (1989).

# BENCHMARK EXPERIMENTS WITH 14 MEV NEUTRONS TRANSMITTED THROUGH PB, FE, NI AND AL.

M.TICHÝ, M. KRÁLÍK, J. PŮLPÁN, T.NOVOTNÝ

Institute of Radiation Dosimetry CSAV,

CS-180 86 Prague 8, Czechoslovakia

## ABSTRACT

Spectra of 14 MeV neutrons transmitted through 7.5 cm of Pb, Fe, Ni and Al in spherical geometry are presented. Measured pulse height spectra (detector was NE-213 scintillator Ø 5.08 cm*5.08 cm) was processed by codes DIFBAS and FORIST (both using matrix method). Their results are compared and the complete covariance matrix gained by the first code is discussed. The unfolding procedure is satisfactory but requires an improvement of input data. Calculations were carried out by one dimensional BLANK (Monte-Carlo) and ANISN (discrete ordinates) codes using ENDF/B-IV, ENDL-83 and Vitamin-C cross section libraries. Influence of used codes and libraries is discussed. Remarkable differences between the experiment and calculations confirmed already published discrepancies in libraries. A detailed investigation of them showed the necessity of utilizing three-dimensional codes for calculations and an improvement in the analysis of the experimental data.

---

## 1.INTRODUCTION

Iron, nickel, aluminum and lead are materials proposed for utilization in the fusion reactor blanket as parts of construction or for ability to multiply the number of neutrons escaping the reactor. To check their cross sections, benchmark experiments i.e.

Program FORIST [4] uses Gramm-Schmidt ortogonalization process for generalized inversion $(A^T W A)^{-1} A^T W$ of the response matrix **A**. The weight matrix **W** is the inverse of diagonal covariance matrix of the p.h. spectrum (Poisson variances). The resulting spectrum must be smoothed because of frequent oscillations. According to previous experience and recommendation of authors [4] we have chosen window widths equal to FWHM recommended by [5] for a similar spectrometer and 3% was used as desired accuracy.

To gain a complete covariance matrix of the spectrum the DIFBAS code was developed [6]. It is based on Bayes' theorem and utilizes the covariance filter method. An a priori neutron spectrum is gained by differentiation of the p. h. spectrum and its uncertainties are guessed and positive correlations generated by a built in Gaussian procedure (see [6] and the next section). Variances of p.h. spectrum follow the Poisson statistics (the same as for FORIST) and correlations of it are not considered.

For comparison of experimental spectra (results of FORIST and DIFBAS) among themselves and with the calculated ones all of them were condensed to a supergroup scale with dividing points 1, 5, 10 and 16 MeV. Therefore, whenever differences will appear in the following text it means differences of group fluences in the above supergroup scale or differences of the total fluences i.e. integrals of fluence from 1 to 16 MeV. All fluences are normalized to 1 neutron emitted from the target to $4\pi$ solid angle.

In this coarse structure, leakage spectra from all four materials unfolded by the two codes differ less than 2% except interval 5-10 MeV where the difference is from 3% to 8% for Al, Fe and Pb and 21% for Ni. It corresponds to a higher a posteriori uncertainty in this region and can be explained by the instability of the background subtracting because there is a very low number of neutrons there (less than 6% of the total fluence). These differences are approximately equal to two "sigma" of the a posteriori spectrum but this comparison can be partly misleading because of a strong influence of several unsure parameters to the a posteriori variance (see next section and [6]). Comparing it with differences from calculation, *results of two unfolding codes were taken as identical* and results of the second code are mentioned as measured or experimental ones in the following sections.

Plots of spectra escaping the Ni & Pb spheres are on Figs. 3 & 4 as examples of all. Two peaks at approximately 1.8 and 3 MeV appeared in the results of both codes for all materials (for Ni they are the most pronounced). From the absence of them in spectra in [12,13] we concluded that they are caused by a mismatch of the real response of the scintillator and the response matrix used.

measurements and calculation of transmitted spectra in a simple arrangement have been already carried out e.g. [1,11-13]. Possible discrepancies between measurement and calculation can indicate errors in used libraries even if cross sections cannot be corrected in this way. The presented paper contains results of such a benchmark. After an outline of geometrical arrangement the results of measurements and their comparison with calculations are presented.

## 2. Experimental Arrangement

An accelerator using energy of deuterons 120 keV with a tritium-titanium target 1.75mg/cm$^2$ thick was the source of neutrons. Emission of neutrons was absolutely calibrated by means of the associated particles method and routine monitoring was carried out by a NE-213 scintillator placed above the target [10]. The target was surrounded by a sphere made from the tested material with thickness 7.5 cm (outer diameter 24 cm). It corresponds to 0.8, 1.7, 1.8 and 1.4 mean free paths for the tested materials - Al, Fe, Ni, and Pb. The detector of the neutron spectrometer - also a NE-213 scintillator with dimensions Ø 5.08cm *5.08cm - was located at the distance 3 m from the center of the sphere. Background was subtracted by means of an iron cone 1 m long. The arrangement in the hall and the geometry of the experiment are on Figs.1 & 2.

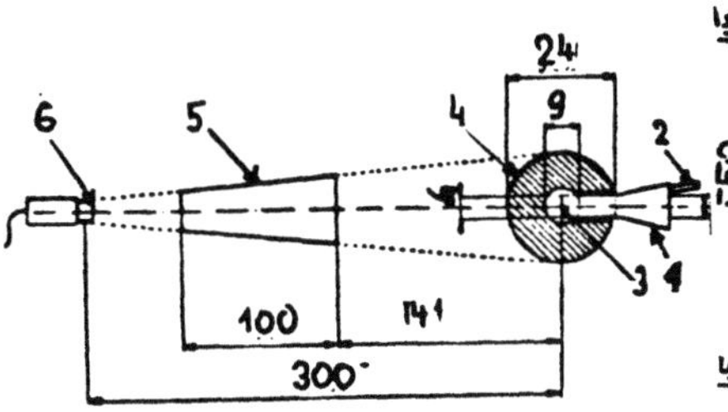


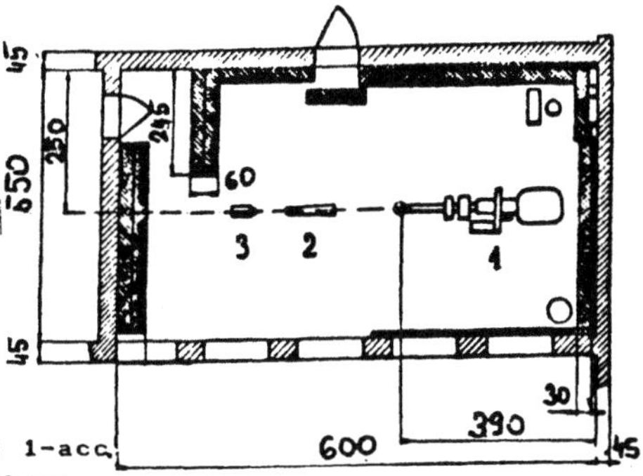


Fig.1: Geometry of the experiment: 1-acc. tube, 2-α detector, 3-target, 4-sphere, 5-iron cone, 6-neutron detector.

Fig.2: Experimental hall: 1-generator, 2-iron cone, 3-detector, borated polyethylene shielding, concrete.

## 3. Measured Spectra

A detailed description of the spectrometer is in [2]. Pulse height (p.h.) spectra from neutron induced reactions were processed independently by codes FORIST and DIFBAS which are based on the matrix method. The so called "Illinois response matrix" [3] was used for unfolding.

## 4. ANALYSIS OF EXPERIMENTAL SPECTRA COVARIANCES

At the present state of the development of the spectrometer and the unfolding procedures the covariance of the a posteriori spectrum is composed of three parts:

- The variance of the p.h. spectrum which is supposed to be well defined by the measurement (Poisson statistics, correlations not used).
- The covariance of the a priori neutron spectrum which is guessed according to the experience of the experimenter. For all materials the following values, acc. to [6], were chosen: variation coefficients of group fluences equal to 50% resp. 100% and 30% in 1-5 MeV resp. 5-12 MeV and 12-16 MeV ranges and the optimized parameters for Gaussian correlations.
- The uncertainty of a normalization factor (to unit emission of the source) was estimated in [10] to 3.7%. This uncertainty was composed of several components. Lack of information about the component concerning the distribution of tritium in the target which influences the correction for d-t reaction anisotropy [10] causes that it can be even 5%.

The prevailing influence of the last item on a posteriori covariance can be seen in Tab.1. A priori and a posteriori covariance matrices of the spectrum from Al sphere are presented condensed to the above supergroup scale with not included and included normalization uncertainty 5% which is greater than the estimated value (3.7%). All other spectra had similar results. A minor role plays a subjective guess of a priori variation coefficients: a decrease of the a priori value to one half caused the decrease of the a posteriori value to about one tenth [6]. The covariances presented in Tab.1 do not include covariances of the response matrix due to non-availability of it and limited possibilities of

| Ener. range [MeV] | a priori covariance matrix [%] | | | | a posteriori covariance matrix [%] normal. not incl. | | | | a posteriori covariance matrix [%] normal. included | | | |
|---|---|---|---|---|---|---|---|---|---|---|---|---|
| 1- 5 | 24.91 | 100 | | | 0.91 | 100 | | | 5.08 | 100 | | |
| 5-10 | 45.62 | 1 | 100 | | 3.87 | -53 | 100 | | 6.32 | 72 | 100 | |
| 10-16 | 17.20 | 0 | 0 | 100 | 0.35 | -11 | -46 | 100 | 5.01 | 98 | 77 | 100 |
| 1-16 | 13.75 | | | | 0.21 | | | | 5.0 | | | |

Tab.1: The a priori and a posteriori covariance matrices of spectrum from Al influenced by normalization uncertainty. Variances and correlations are separated (both in %).

the DIFBAS code [6]. They can add a substantial part to presented numbers.

The above analysis of experimental data has a limited validity. The response matrix used has been taken from FORIST code package (RSIC, PSR-92). Its applicability to any spectrometer is merely not guaranteed and it can cause systematic shifts much greater than the uncertainties presented in Tab.1 (see sections 3 & 6). A special response matrix measured or validated for the spectrometer in question should be used.

## 5. CALCULATIONS

Two one dimensional codes were used for calculations: Monte-Carlo code BLANK [7] and discrete ordinates code ANISN [8]. For calculation by the first one, performed in IAE Moscow, [9] two libraries were used: ENDF/B-IV and ENDL-83. ANISN code using $P_3S_{16}$ approximation and optimized mesh scale was run with group cross section library Vitamin-C (originating also in ENDF/B-IV). To take into account the angular dependence of energy of neutrons from d-t reaction into calculation by one-dimensional codes, the spectrum of the source for both codes was defined as proportional to anisotropy correction factor [10] between 13.5 MeV and 14.9 MeV.

Influence of the code and library used is shown in Tab.2. Total (to $4\pi$ angle) leakage spectrum from the Al, Ni and Fe spheres calculated by ANISN is presented there together with data indicating how results of the BLANK code differ from those. *The choice of library influences mainly Fe and Al* leakage spectra in 1-5 MeV range; other differences are negligible. A comparable *good agreement of results of both codes using the same pointwise library for Ni and Fe and disagreement for Al* can be explained by different ways that the cross sections from the same library are grouped. Greater differences in 5-10 MeV region are probably caused by a low statistics of Monte-Carlo calculation in this region.

| | Aluminum | | | Nickel | | | Iron | | |
|---|---|---|---|---|---|---|---|---|---|
| Ener. range [MeV] | ANISN Vit-C | BLANK ENDF [%] | BLANK ENDL [%] | ANISN Vit-C | BLANK ENDF [%] | BLANK ENDL [%] | ANISN Vit-C | BLANK ENDF [%] | BLANK ENDL [%] |
| 1- 5 | 0.168 | 7.2 | 11.3 | 0.190 | 1.3 | 1.4 | 0.284 | 1.8 | -4.1 |
| 5-10 | 0.054 | 15.2 | 15.3 | 0.022 | 26.7 | 30.1 | 0.037 | 9.8 | 27.6 |
| 10-16 | 0.607 | -11.5 | -12.1 | 0.366 | -0.7 | -0.1 | 0.394 | -.5 | -2.4 |
| 1-16 | 0.830 | -6.0 | -5.6 | 0.578 | 1.0 | 1.6 | 0.716 | .9 | -1.5 |

Tab.2: Comparison of results of ANISN and BLANK codes.

To estimate *the influence of the source definition*, additional tests were carried out with the ANISN code. Three other (in addition to the above described) sources have been tried: "polyenergetic" source in same interval with a "sinusoidal" shape and two "monoenergetic" sources (in 14.2-14.6 MeV and 14.6-14.9 MeV groups). Differences of leakage spectra depend mainly on shape of $\sigma_t$ in the range of the source and in most cases do not exceed 2%; the maximum is 7% for 5-10 MeV supergroup for Fe sphere. Results of the two polyenergetic sources differ less than 1.3%.

## 6.Comparison of Measured and Calculated Spectra

Lack of information concerning anisotropy of the spectrum leaking from a sphere complicates calculation of total leakage from the spectrum measured at a certain angle. Therefore the experimental spectrum was compared with the current streaming from the sphere calculated by ANISN for the "monoenergetic" source in the group 14.6-14.9 MeV (neutron energy at $0^\circ$ and for 120 keV deuterons is between 14.6 and 14.7 MeV depending on target properties [10]).

Results of the comparison are in Tab. 3 and Figs. 3 & 4. The underestimation of the leakage by calculation observed here for all materials confirmed results for Pb in [1], for Fe and Pb in [12] and Fe and Ni in [13]. A rather large size of the differences between the calculation and measurement (C/E) and their nearly same size for the total fluence for all four materials indicated a possible error. That was not found in the normalization procedure: an additional measurement confirmed the estimated uncertainty of the normalization (3.7%). The influence of a non-total absorption of the shadow cone, which was estimated by means of the ANISN code, is less than 0.1% only. "Bad" definition of the source for calculation considered in the previous section (~2%) does not also play the most important role. Therefore the differences may be caused by three other possibilities:

- The response matrix used for unfolding was unsuitable.
- The calculation by one-dimensional codes is not sufficiently truthful.
- There are actual errors in cross sections used for calculation e.g. the anizotropy of the cross sections have a great influence on the calculated results.

At this stage of development we are unable to eliminate the first two possibilities to confirm discrepancies in the cross section library. Conclusion that experimental errors (the first item also mentioned in sec. 3.) are not a major part of the C/E differences can be made from the comparison of Fig.4 and a corresponding very similar plot in [12]. An attempt to eliminate a "systematic" error in the response matrix i.e. error of the total fluence, was done by a renormalization to unit total fluence: The corresponding "relative" differences are compared also in Tab.3.

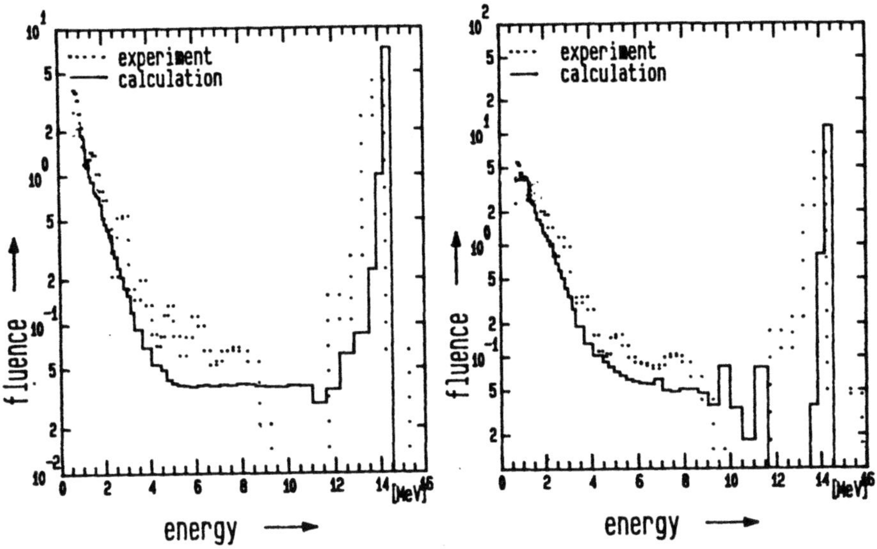


Fig.3 & 4: Leakage spectra from Ni (left) and Pb (right) spheres.

| Ener. range [MeV] | Aluminium Exp. | Aluminium Dif.of Calc.[%] abs. | Aluminium Dif.of Calc.[%] relat. | Nickel Exp. | Nickel Dif.of Calc.[%] abs. | Nickel Dif.of Calc.[%] relat. |
|---|---|---|---|---|---|---|
| 1 - 5 | 2.315 | -36.0 | -23.0 | 2.413 | -28.1 | -13.5 |
| 5 -10 | 0.522 | - 8.1 | 10.5 | 0.334 | -41.9 | -30.2 |
| 10 -16 | 5.982 | -10.2 | 8.0 | 3.504 | - 6.7 | 12.2 |
| 1 -16 | 8.819 | -16.8 | | 6.250 | -16.8 | |

| Ener. range [MeV] | Iron Exp. | Iron Dif.of Calc.[%] abs. | Iron Dif.of Calc.[%] relat. | Lead Exp. | Lead Dif.of Calc.[%] abs. | Lead Dif.of Calc.[%] relat. |
|---|---|---|---|---|---|---|
| 1 - 5 | 3.345 | -20.5 | - 5.9 | 5.617 | -32.6 | -16.7 |
| 5 -10 | 0.458 | -27.6 | -14.4 | 0.447 | -38.1 | -23.5 |
| 10 -16 | 3.885 | - 9.8 | 6.7 | 4.681 | - 0.9 | 22.4 |
| 1 -16 | 7.689 | -15.5 | | 1.075 | -19.0 | |

Tab. 3: Measured fluxes at 3 m from the source multiplied by 1.E+7 (condensed to supergroup scale) and absolute and relative (total flux=1) differences from ANISN calculations.

The smallest differences for Fe and Al confirm conclusions of Takahashi [11]. The necessity to use 3-dim. codes for calculation can be seen from the comparison of results of [12,13]; C/E differences of the second one are much smaller probably due to the use of MNCP code.

## 7. Conclusions

Leakage spectra from the spherical assemblies made of Al, Ni, Fe and Pb were measured and calculated. The new code DIFBAS used for unfolding is a useful instrument but the lack of data or unreliability of them lowers the worth of the presented a posteriori covariances. The comparison of two calculational codes shown, besides their general agreement, the influence of codes used for grouping of cross sections. Differences of cross sections from ENDF/B-IV and ENDL-83 libraries were found mainly for Al and Fe.

Comparison of the presented differences between the measurement and calculation with those from similar benchmarks confirmed our conclusion that calculation with ENDF/B-IV library underestimates the leakage for all tested materials. The detailed analysis of the shown results indicated possible errors in the measurement or calculation. Therefore energetic distribution of emitted neutrons should be taken into account in calculation and it cannot be fulfilled by a one-dimensional code. The reliability of measured spectra has to be improved by utilization of the response matrix generated for the just used detector and by measurements and monitoring at the angle $90^{\circ}$ from the deuteron beam direction to minimize the influence of the unknown parameters of the target.

## 8. References


[1] Elfurth T. et al: Atomkernenergie-Kerntechnik, 49,1987,121

[2] Pulpan J., Kralik M., Tichy M.: NE-213 Neutron Spectrometer for Experiments with 14 MeV Neutrons, IRD 313/90, Prague 1990

[3] Ingersoll D.T. et al: ORNL/RSIC-40, Oak Ridge 1976
Johnson R.H. et al: Nucl. Instr. & Meth. 145,1977,337

[4] Johnson R.H.: PNE-75-107, W. Lafayette 1975

[5] Tschechanski A., Shani G.: Nucl. Sci. & Eng. 87,1984,189

[6] Tichy M.: Bayesian unfolding of Pulse Height Spectra, contribution to this Symposium

[7] Marin S.V. et al.:Preprint IAE-2832, Moscow 1977

[8] CCC-514/ANISN/PC Code Package, ORNL-RSIC, Oak Ridge

[9] Borisov A.A. et al: Proceedings of Symposium on Nuclear Processes in Fusion Reactors, Gaussig, GDR, November 1989

[10] Pulpan J., Kralik M., Tichy M.: Monitoring of a 14 MeV Neutron Yield from a Thick Target, IRD 314/90, Prague 1990

[11] Takahashi A.: Proceedings of Specialists' Meeting on Nuclear Data for Fusion Neutronics, JAERI-M-86-029, Tokai, 1986

[12] Yamamoto J.: JAERI-M-86-029, p.149, Tokai, 1986

[13] Hashikura H. et al: JAERI-M-86-029, p.160, Tokai, 1986

# CALCULS DE FLUENCE SUR LES CUVES DES R.E.P.

# INTERPRETATION DE DEUX BENCHMARKS

# TRANSPORT COUPLE NEUTRON-GAMMA

**T.VERGNAUD* - J.C.NIMAL* - S.H.ZHENG****

* CEA/IRDI/DRN/DMT/SERMA/LEPP
CENTRE D'ETUDES NUCLEAIRES DE SACLAY
91191 GIF SUR YVETTE CEDEX
** THESE préparée dans le même laboratoire

## RESUME

Le calcul des flux de neutrons rapides et de gamma sur la cuve et sur les structures hors coeur des réacteurs à eau pressurisée français est effectué par le code de Monte Carlo TRIPOLI-2. L'interprétation par ce code de deux benchmarks expérimentaux du programme NESDIP permet de valider ce schéma de calcul et plus particulièrement le transport couplé neutron-gamma.

## INTRODUCTION

Afin d'estimer avec précision les taux de dommage et l'échauffement dans l'acier des cuves de R.E.P. et dans les structures hors-coeur (cloisonnement et enveloppe du coeur, écran thermique), il est nécessaire de bien connaître les flux de neutrons et de gamma; ceux-ci sont calculés avec le code de Monte-Carlo TRIPOLI.2 (1)

Le programme expérimental NESDIP, implanté à Winfrith, comporte plusieurs configurations représentatives de la géométrie d'un REP : alternance de plaques d'acier et d'eau,placées derrière une plaque de fission dont la distribution de puissance est connue avec précision. Différents détecteurs permettent d'explorer le spectre neutronique jusqu'à l'énergie thermique et l'exposition due aux gamma est mesurée avec des TLD et des chambres à ionisation..

## METHODE DE CALCUL

Le code TRIPOLI-2 étudie le transport des neutrons et des gamma dans des géométries tridimensionnelles par la méthode de Monte Carlo. L'énergie de la particule est suivie de façon continue et les lois du choc sur chacun des nuclides sont différenciées, l'anisotropie de la collision étant décrite par 32 plages d'égale probabilité pour les neutrons et 64 plages pour les gamma.

Les sections efficaces neutroniques proviennent de la bibliothèque ENDF/B4 et sont décrites sur 315 intervalles en énergie entre 19 MeV et 0. La mise en groupe s'effectue en utilisant un spectre de fission pour les énergies supérieures à 1 MeV et une fonction en $1/E.\Sigma_t$ de 1 MeV à 5 eV, où $\Sigma_t$ est la section totale du matériau contenant le nuclide. Pour des énergies inférieures 5 eV, la thermalisation des neutrons est traitée exactement en prenant en compte les remontées en énergie.

Les sections photoniques proviennent de la bibliothèque de LOS ALAMOS et sont moyennées sur 75 intervalles en énergie entre 15 MeV et 10 KeV.

Le transport des neutrons et des gamma est couplé de façon automatique par le biais de matrices de production gamma, par choc inélastique ou capture, issues de VITAMIN-C.

Pour les neutrons les fonctions réponses : S32(n,p)P32, Rh103(n,n')Rh103m, In115(n,n')In115m, Mn55(n,$\gamma$)Mn56 sont extraites du fichier IRDF/85 et décrites sur le découpage à 315 groupes. Pour les gamma nous avons utilisé la fonction de débit d'équivalent de dose de gamma compilation ANS/77.

## BENCHMARK "REPLICA"

Nous avons d'abord étudié l'expérience REPLICA (2) qui simule la succession des matériaux rencontrés dans un REP : eau, écran thermique en acier inox, eau, cuve en acier doux, intervalle cuve-béton primaire.(voir fig.1) Ce benchmark ne concerne que les neutrons rapides qui sont mesurés par des détecteurs à activation (S32, Rh103, In115) et par un spectromètre à protons de recul.

Le tableau 1 donne la comparaison entre les valeurs mesurées et les valeurs calculées par TRIPOLI; nous avons porté également les résultats obtenus avec le code MCBEND. La figure 2 montre le spectre des neutrons à la position T/4 dans l'acier doux simulant la cuve.

On constate une légère sous-estimation globale des différentes réponses de détecteur : dans l'acier doux elle concerne surtout le Soufre (environ -10%), et elle est de l'ordre de -20% pour le Rhodium dans la boite vide. La comparaison des spectres montrent une sous estimation des résultats de calcul par rapport aux valeurs expérimentales dans le domaine 3 à 5 MeV, à la fois pour TRIPOLI et pour MCBEND.

Malgré ces remarques la comparaison calcul/expérience est satisfaisante et qualifie globalement le traitement des sections efficaces et le calcul du flux supérieur à 1 MeV dans de telles configurations.

**BENCHMARK EXPERIMENTAL POUR LA VALIDATION DU TRANSPORT COUPLE NEUTRON-GAMMA DANS DES LAMELLAGES FER-EAU.**

## DISPOSITIF EXPERIMENTAL

Comme REPLICA, ce deuxième dispositif (3) fait partie du programme Nesdip (Nestor Shielding and Dosimetry Improvement Program). La source de neutrons est une plaque de fission constituée d'un alliage d'aluminium et d'uranium enrichi à 93%, placée derrière un massif de graphite, qui thermalise les neutrons provenant du coeur de Nestor, et suivie d'une plaque de boral de 0.5 cm.La puissance de la plaque de fission est de 3.11 $10^{-4}$ ± 4%watts pour une puissance de 1 watt du réacteur Nestor. La distribution spatiale de la source est donnée en coordonnées cartésiennes.er décrite de la même façon dans TRIPOLI

Ce dispositif est constitué d'une succession de plaques d'acier inox et d'acier doux, de section carrée 177.8x177.8 cm2, plongées dans l'eau. Deux jeux d'eau de 18.31 cm et 19.80 cm sont ainsi ménagés.(voir figure 3)

Des détecteurs de soufre, de rhodium et de manganèse sont placés dans l'eau le long de l'axe du dispositif, ainsi que sur une verticale dans le premier jeu d'eau.

L'énergie déposée par les gamma est mesurée par deux procédés différents utilisant des détecteurs thermoluminescents au LiF et au Be0, étalonnés par rapport à des sources de gamma de $Cs^{137}$ et de $Co^{60}$ et des chambres à ionisation. Ces dernières, de taille plus importante que les TLD et nécessitant des corrections plus complexes (gamma créés dans l'enveloppe en aluminium, déplacement d'eau par le tube guide), n'ont été placées qu'en quatre positions. Les deux procédés de mesure donnent des résultats cohérents.

## CALCULS EFFECTUES PAR TRIPOLI

La géométrie de l'expérience a été modélisée de façon exacte. Les valeurs sur l'axe correspondent à une moyenne sur des disques de 10 cm de rayon où les flux peuvent être considérés comme plats.

Le flux de gamma comporte deux composantes qui ont été séparemment étudiées :

- la contribution essentielle est celle des gamma créés par capture et choc inélastique sur le fer, le chrome et le nickel de l'acier et sur l'hydrogène de l'eau. Cette contribution est évaluée par un calcul couplé neutron-gamma
- la contribution des gamma prompts de fission sur l'U235 de la plaque de fission est étudiée par un transport gamma classique par TRIPOLI. Cette contribution n'a d'importance que sur les premiers points dans l'acier.

Les tableaux 2 et 3 donnent la comparaison calcul/expérience pour les détecteurs de Soufre et de Rhodium.Les incertitudes σ affichées sur les valeurs expérimentales tiennent compte de l'incertitude sur le système (4% pour le soufre et 3% pour le rhodium) et de l'incertitude statistique. Les incertitudes affichées sur les valeurs calculées prennent en compte l'incertitude sur la source (4%) et l'incertitude stochastique due à la méthode de Monte Carlo. Elles sont ajoutées quadratiquement.

Le tableau 4 correspond à la réponse des détecteurs de Manganèse nus et sous Cadmium, pour ce dernier la qualité statistique des résultats est moins bonne vu les résonnances étroites de la réponse au-dessus de la coupure du Cadmium et donc difficiles à calculer par une méthode de Monte Carlo L'incertitude sur les valeurs expérimentales est uniquement d'origine statistique.

Les tableaux 5, 6, 7 donnent les valeurs de l'exposition gamma exprimée en Cs137-Roetgen/10Kw.h, le long de l'axe et le long de deux radiales dans l'acier et dans l'eau. Les valeurs calculées sont comparées avec les mesures effectuées avec des TLD au LiF, corrigées du bruit de fond des gamma provenant du coeur de Nestor et de la contribution des neutrons thermiques de la plaque de fission. L'incertitude sur les valeurs expérimentales est uniquement d'origine statistique. On constate un bon accord, sauf pour les deux premiers points dans l'acier.

## CONCLUSION

De même que pour REPLICA, on constate une certaine sous-estimation des réponses rapides : environ 10% sur le Soufre et 20% sur le Rhodium, dans le deuxième massif d'acier. Cette sous-estimation s'observe également sur la réponse du Mn nu mais ne dépasse pas 25% dans le deuxième jeu d'eau. Par contre, la réponse du Mn/Cd présente une surestimation .

Les valeurs de l'exposition gamma calculées par TRIPOLI présentent un accord satisfaisant avec l'expérience : le rapport C/E est compris dans les barres d'erreur sauf pour les premiers points dans l'acier.

L'accord observé sur le calcul du flux gamma pour ce benchmark teste le calcul du transport couple neutron-gamma effectué dans TRIPOLI et valide les calculs d'échauffement effectués dans les Réacteurs.

## REFERENCES


(1) J.C.NIMAL and al.
TRIPOLI-2 Code de Monte Carlo tridimensionnel polycinétique
ORNL-OLS 80-110 Rapport CEA à paraître

(2) J.BUTLER - M.D.CARTER - I.J.CURL - M.R.MARCH
A.K.McCRACKEN - M.F.MURPHY - A.PACKWOOD
The PCA REPLICA experiment (part1); Winfrith measurements and calculations
AEEW - R1736

(3) M.D.CARTER - I.J.CURL - F.G.ELTHAM - C.J.HOARE
A benchmark experiment to validate coupled neutron/gamma ray transports method for water/steel arrays
AEE Winfrith june 1987

Tableau 1 REPLICA EXPERIMENT

| | d(cm) | Rh103(n,n')Rh103m | In115(n,n')In115m | S32(n,p)P32 |
|---|---|---|---|---|
| before 3.9% shield | 1.91 thermal | E $1.69\ 10^{-20}$ 6.0%<br>T $1.70\ 10^{-20}$ 3.9%<br>T/E 1.00 | $3.86\ 10^{-21}$ 3.9% | $1.20\ 10^{-21}$ |
| before thermal 4.0% | 14.01 shield | E $1.27\ 10^{-21}$ 6.0%<br>T $1.06\ 10^{-21}$ 4.1%<br>T/E 0.83 | $2.46\ 10^{-22}$ 4.0% | 8.21 10-23 |
| behind thermal shield | 19.91 | E $4.23\ 10^{-22}$ 6.0%<br>T $4.02\ 10^{-22}$ 4.1%<br>T/E 0.95 | $8.62\ 10^{-23}$ 4.1% | $2.51\ 10^{-23}$ 4.2% |
| mild steel (T/4) | 39.01 | E $2.07\ 10^{-23}$ 4.0%<br>T $2.03\ 10^{-23}$ 5.1%<br>T/E 0.98<br>M/E 1.01 | $3.93\ 10^{-24}$ 2.9%<br>$3.88\ 10^{-24}$ 2.9%<br>0.99<br>1.01 | $1.08\ 10^{-24}$ 5.5%<br>$9.61\ 10^{-25}$ 4.8%<br>0.90<br>1.03 |
| mild steel (3T/4) | 49.61 | E $5.53\ 10^{-24}$ 4.9%<br>T $5.20\ 10^{-24}$ 5.1%<br>T/E 0.94<br>M/E 0.97 | $8.23\ 10^{-25}$ 3.4%<br>$7.66\ 10^{-25}$ 4.8%<br>0.93<br>0.88 | $1.46\ 10^{-25}$ 5.9%<br>$1.35\ 10^{-25}$ 4.9%<br>0.92<br>1.02 |
| void box | 58.61 | E $1.80\ 10^{-24}$ 4.6%<br>T $1.47\ 10^{-24}$ 4.2%<br>T/E 0.81<br>M/E 0.85 | $2.31\ 10^{-25}$ 4.6%<br>$2.03\ 10^{-25}$ 4.2%<br>0.88<br>0.83 | $3.73\ 10^{-26}$ 5.3%<br>$3.55\ 10^{-26}$ 4.8%<br>0.95<br>1.00 |

d : distance à la plaque de fission  E : valeurs mesurées
T : résultats TRIPOLI  M : résultats MCBEND

Tableau 2 : (n,γ) EXPERIMENT
Taux de réaction S32 (n,p) P32 (Bq.atom$^{-1}$/Nestor.Watt)

| | d(cm) | E | σ(%) | C | σ(%) | C/E |
|---|---|---|---|---|---|---|
| acier | 5.1 | 2.92 $10^{-22}$ | 4.1 | 2.51 $10^{-22}$ | 5.9 | 0.86 |
| | 10.18 | 6.82 $10^{-23}$ | 4.1 | 6.86 $10^{-23}$ | 4.8 | 1.01 |
| | 15.96 | 2.34 $10^{-23}$ | 4.1 | 2.36 $10^{-23}$ | 4.8 | 1.01 |
| | 21.74 | 8 32 $10^{-24}$ | 4.1 | 8.64 $10^{-24}$ | 4.9 | 1.04 |
| eau | 25.44 | | | 4.11 $10^{-24}$ | 4.7 | |
| | 29.94 | 1.78 $10^{-24}$ | 4.5 | 1.48 $10^{-24}$ | 4.7 | 0.83 |
| | 32.94 | | | 8.46 $10^{-25}$ | 5.0 | |
| | 34.44 | 8.14 $10^{-25}$ | 4.5 | 6.46 $10^{-25}$ | 4.8 | 0.79 |
| | 40.04 | | | 2.48 $10^{-25}$ | 5.1 | |
| acier | 46.75 | 1.10 $10^{-25}$ | 5.0 | 1.02 $10^{-25}$ | 5.4 | 0.93 |
| | 52.53 | 4.50 $10^{-26}$ | 6.4 | 3.96 $10^{-26}$ | 5.3 | 0.88 |

d : distance à la plaque de fission

Tableau 3 : (n,γ) EXPERIMENT
Taux de réaction Rh103 (n,n') Rh103m (Bq.atom$^{-1}$/Nestor.Watt)

| | d(cm) | E | σ(%) | C | σ(%) | C/E |
|---|---|---|---|---|---|---|
| acier | 5.1 | 5.30 $10^{-21}$ | 3.2 | 4.85 $10^{-21}$ | 4.9 | 0.92 |
| | 10.18 | 2.34 $10^{-21}$ | 3.3 | 2.29 $10^{-21}$ | 4.5 | 0.98 |
| | 15.96 | 1.25 $10^{-21}$ | 3.2 | 1.18 $10^{-21}$ | 4.7 | 0.94 |
| | 21.74 | 6.74 $10^{-22}$ | 3.2 | 6.02 $10^{-22}$ | 4.8 | 0.89 |
| eau | 25.44 | 3.53 $10^{-22}$ | 3.6 | 3.10 $10^{-22}$ | 4.7 | 0.88 |
| | 29.94 | | | 5.35 $10^{-23}$ | 4.4 | |
| | 32.94 | 2.76 $10^{-23}$ | 4.2 | 2.26 $10^{-23}$ | 4.5 | 0.82 |
| | 34.44 | | | 1.56 $10^{-23}$ | 4.9 | |
| | 40.04 | 5.42 $10^{-24}$ | 6.7 | 4.08 $10^{-24}$ | 4.7 | 0.75 |
| acier | 46.75 | 2.56 $10^{-24}$ | 7.6 | 2.16 $10^{-24}$ | 5.1 | 0.84 |
| | 52.53 | 1.35 $10^{-24}$ | 12.4 | 1.02 $10^{-24}$ | 5.9 | 0.76 |

d : distance à la plaque de fission

Tableau 4

| | | Mn nus (Bq.atom$^{-1}$/Nestor.Watt) | | | | | Mn/Cd (Bq.atom$^{-1}$/Nestor.Watt) | | | | |
|---|---|---|---|---|---|---|---|---|---|---|---|
| | d(cm) | E | σ(%) | C | σ(%) | C/E | E | s(%) | C | s(%) | C/E |
| acier | 5.1 | $4.04\ 10^{-21}$ | 1 | $3.11\ 10^{-21}$ | 11.3 | 0.77 | $3.63\ 10^{-21}$ | 1 | $3.10\ 10^{-21}$ | 11.3 | 0.85 |
| | 10.18 | $2.34\ 10^{-21}$ | 1 | $2.25\ 10^{-21}$ | 11.5 | 0.96 | $2.15\ 10^{-21}$ | 1 | $2.24\ 10^{-21}$ | 11.6 | 1.04 |
| | 15.96 | $1.68\ 10^{-21}$ | 1 | $1.78\ 10^{-21}$ | 13.2 | 1.06 | $1.51\ 10^{-21}$ | 1 | $1.71\ 10^{-21}$ | 13.5 | 1.13 |
| | 21.74 | $2.36\ 10^{-21}$ | 1 | $2.48\ 10^{-22}$ | 9.7 | 1.05 | $1.10\ 10^{-21}$ | 1 | $1.27\ 10^{-21}$ | 14.8 | 1.16 |
| eau | 25.44 | $1.42\ 10^{-20}$ | 1 | $1.18\ 10^{-22}$ | 6.6 | 0.83 | $1.31\ 10^{-21}$ | 1 | $1.52\ 10^{-21}$ | 8.4 | 1.16 |
| | 27.44 | $3.92\ 10^{-20}$ | 1 | $3.65\ 10^{-22}$ | 5.6 | 0.93 | $1.22\ 10^{-21}$ | 1 | $1.62\ 10^{-21}$ | 8.2 | 1.33 |
| | 29.44 | $3.93\ 10^{-20}$ | 1 | $3.62\ 10^{-22}$ | 5.3 | 0.92 | $6.69\ 10^{-22}$ | 1 | $8.88\ 10^{-22}$ | 8.9 | 1.33 |
| | 29.94 | $3.71\ 10^{-20}$ | 1 | $3.51\ 10^{-22}$ | 5.2 | 0.95 | $5.54\ 10^{-22}$ | 1 | $7.05\ 10^{-22}$ | 7.5 | 1.27 |
| | 34.44 | $1.74\ 10^{-20}$ | 1 | $1.39\ 10^{-23}$ | 5.3 | 0.80 | $1.22\ 10^{-22}$ | 1 | $1.83\ 10^{-22}$ | 31.5 | 1.50 |
| | 38.94 | $4.94\ 10^{-21}$ | 1 | $3.98\ 10^{-22}$ | 6.1 | 0.81 | $2.64\ 10^{-23}$ | 2 | $2.35\ 10^{-23}$ | 19.6 | 0.89 |
| | 40.04 | $3.49\ 10^{-21}$ | 1 | $2.73\ 10^{-24}$ | 6.4 | 0.78 | $1.80\ 10^{-23}$ | 3 | $2.10\ 10^{-23}$ | 31.8 | 1.17 |
| | 43.44 | $8.05\ 10^{-22}$ | 2 | $5.05\ 10^{-22}$ | 8.7 | 0.63 | $7.91\ 10^{-24}$ | 5 | $9.07\ 10^{-22}$ | 22.1 | 1.15 |
| acier | 46.75 | $4.00\ 10^{-23}$ | 1 | $3.34\ 10^{-24}$ | 6.6 | 0.84 | $3.63\ 10^{-24}$ | 1 | $6.15\ 10^{-24}$ | 23.9 | 1.69 |
| | 52.53 | $5.46\ 10^{-24}$ | 2 | $4.05\ 10^{-24}$ | 12.6 | 0.74 | $1.08\ 10^{-24}$ | 2 | $1.52\ 10^{-24}$ | 27.4 | 0.84 |
| eau | 55.73 | $1.69\ 10^{-23}$ | 1 | $1.27\ 10^{-22}$ | 9.3 | 0.75 | $1.55\ 10^{-24}$ | 1 | $2.52\ 10^{-24}$ | 9.3 | 1.63 |
| | 57.33 | $4.11\ 10^{-23}$ | 1 | $3.43\ 10^{-22}$ | 7.8 | 0.83 | $1.35\ 10^{-24}$ | 1 | $1.79\ 10^{-24}$ | 7.8 | 1.34 |
| | 59.73 | $4.40\ 10^{-23}$ | 1 | $3.66\ 10^{-22}$ | 7.2 | 0.83 | $7.76\ 10^{-25}$ | 2 | $9.31\ 10^{-25}$ | 7.2 | 1.20 |
| | 63.73 | $2.53\ 10^{-23}$ | 1 | $2.00\ 10^{-22}$ | 7.1 | 0.79 | $2.53\ 10^{-25}$ | 3 | $1.71\ 10^{-25}$ | 7.1 | 0.68 |
| | 65.83 | $1.61\ 10^{-23}$ | 1 | $1.31\ 10^{-23}$ | 7.7 | 0.82 | $1.33\ 10^{-25}$ | 4 | $1.37\ 10^{-25}$ | 7.7 | 1.03 |
| | 71.43 | $4.38\ 10^{-24}$ | 2 | $3.18\ 10^{-22}$ | 8.9 | 0.73 | $3.60\ 10^{-26}$ | 8 | $5.98\ 10^{-26}$ | 8.9 | 1.66 |
| | 74.53 | $1.44\ 10^{-24}$ | 3 | $1.01\ 10^{-24}$ | 9.6 | 0.68 | $2.75\ 10^{-26}$ | 11 | $2.83\ 10^{-26}$ | 9.6 | 1.05 |

d : distance à la plaque de fission

Tableau 5 : (n,γ) EXPERIMENT
Exposition gamma Cs137 R/10Kw-h

| | d(cm) | E | σ(%) | C | σ(%) | C/E |
|---|---|---|---|---|---|---|
| acier | 10.18 | 45.1 | 3.5 | 25.7 | 7.7 | 0.65 |
| | 15.96 | 22.8 | 1.3 | 16.4 | 10.3 | 0.72 |
| | 21.74 | 31.7 | 1.9 | 30.3 | 9.2 | 0.95 |
| eau | 25.84 | 45.0 | 3.3 | 48.2 | 8.0 | 1.07 |
| | 30.44 | 37.7 | 2.7 | 42.7 | 7.0 | 1.13 |
| | 34.74 | 27.0 | 1.5 | 33.8 | 6.4 | 1.25 |
| | 39.14 | 19.4 | 1.0 | 22.2 | 6.0 | 1.15 |
| | 43.44 | 14.0 | 1.4 | 16.3 | 6.3 | 1.16 |
| acier | 46.75 | 5.62 | 1.4 | 5.44 | 6.1 | 0.97 |
| | 52.53 | 1.16 | 1.7 | 1.12 | 6.3 | 0.97 |
| eau | 56.83 | 0.57 | 8.8 | 0.56 | 6.2 | 0.98 |
| | 65.53 | 0.36 | 2.8 | 0.39 | 6.3 | 1.08 |

d : distance à la plaque de fission

Tableau 6 : (n,γ) EXPERIMENT
Exposition gamma dans l'acier à 21.74 cm de la plaque de fission, dans l'acier, direction verticale

| d(cm) | E | σ(%) | C | σ(%) | C/E |
|---|---|---|---|---|---|
| +75. | 5.1 | 2.0 | 2.7 | 21.2 | 0.53 |
| +50. | 14.3 | 4.2 | 12.9 | 12.9 | 0.90 |
| +25. | 25.4 | 2.4 | 21.1 | 8.5 | 0.83 |
| 0. | 31.7 | 1.9 | 29.9 | 9.3 | 0.94 |
| -25. | 25.0 | 2.8 | 18.6 | 7.8 | 0.75 |
| -50. | 12.9 | 3.8 | 9.6 | 12.9 | 0.74 |
| -75. | 4.7 | 4.3 | 3.6 | 21.0 | 0.76 |

Tableau 7: (n,γ) EXPERIMENT
Exposition gamma dans l'eau à 39.24cm de la plaque de fission, dans l'eau, direction verticale

| d(cm) | E | σ(%) | C | σ(%) | C/E |
|---|---|---|---|---|---|
| +75. | 4.2 | 2.0 | 4.1 | 9.8 | 0.98 |
| +50. | 11.2 | 4.2 | 11.7 | 7.2 | 1.05 |
| +25. | 16.0 | 2.4 | 19.9 | 6.1 | 1.24 |
| 0. | 20.3 | 1.9 | 22.0 | 6.0 | 1.09 |
| -25. | 15.9 | 2.8 | 17.7 | 5.7 | 1.11 |
| -50. | 9.0 | 3.8 | 10.4 | 7.3 | 1.05 |
| -75. | 3.8 | 4.3 | 3.4 | 9.8 | 0.90 |

# DOSIMETRY TECHNIQUES AND METHODS USED TO OBTAIN REACTOR CAVITY DOSIMETRY BENCHMARK DATA FOR VESSEL FLUENCE ANALYSIS AT DAVIS-BESSE 1

By

Gary S. Carter

B&W Nuclear Service Company
P.O. Box 10935
Lynchburg, Virginia 24506-0935
U.S.A.

A Paper for

Seventh ASTM-EURATOM Symposium on Reactor Dosimetry
27-31 August 1990
Strasbourg, France

## Abstract

B & W Nuclear Service Company (BWNS), an operating company of B&W Nuclear Technologies (BWNT), headquartered in Lynchburg, Virginia, USA, and a partner with Framatome, have in conjunction with BWNT reactor owners, undertaken a Reactor Cavity Benchmark Dosimetry Program. Over the past seven years, the B&W Owners Group (BWOG) Materials Committee has funded the planning, development, design, fabrication, installation, and removal of reactor cavity dosimetry and hardware for both the temporary benchmark and permanent (replaceable) measurement of neutron fluence in the Davis-Besse 1 reactor cavity. The data obtained will be used to indirectly correlate calculated vessel wall fluence with the measured values of dosimeter activities taken both inside and outside the reactor vessel. Cavity dosimetry, both solid state track recorders (SSTRs) and radiometric, along with the additional radiometric dosimetry from inside the vessel, will provide the means to benchmark analytical techniques on Davis Besse. Reactor vessel cavity fluence data for all but one of the BWNT 177-fuel assembly operating commercial power reactors is now being collected. The benchmark data obtained will form the basis of current and future analytical techniques involving the accurate determination of individual operating plant vessel fluence.

## 1.0 BACKGROUND AND DESCRIPTION

In the 1983 and 1984 time period, Babcock & Wilcox, now B&W Nuclear Technologies (BWNT), and the B&W Owners Group (BWOG) Materials Committee began to explore means and methods to install reactor cavity dosimetry in individual B&W designed commercial nuclear power reactors. It was realized early in the formative stages of this program that a need existed for a benchmark experiment at a typical reactor site. The benchmark data was necessary to indirectly correlate calculated vessel wall fluence with measured dosimetry values taken both inside and outside the reactor vessel.

The BWOG Integrated Reactor Vessel Surveillance Program (IRVSP) provided the vehicle for obtaining a single fuel cycle in-vessel irradiation of two previously unirradiated TMI-2 spare surveillance specimen capsules. These capsules contained basic radiometric dosimetry as specified for the IRVSP. The capsules were positioned inside a standard B&W Surveillance Specimen Holder Tube (SSHT) at the Davis-Besse 1 (DB-1) host reactor. SSHTs are attached to the thermal shield, and the capsules are thus positioned in the downcomer region of the vessel and radially inboard of the inside reactor vessel wall. Two TMI-2 capsules were irradiated at the peak azimuthal flux location in the Yz quadrant (10.88 degrees clockwise of the reactor "Y" axis in the "z" axis direction where "X" axis is zero-degrees reactor azimuth).

An 11- 1/2 degree temporary or benchmark dosimetry assembly was positioned in near direct radial alignment with the two TMI-2 capsules. The first cycle of replaceable or permanent dosimetry and associated hardware containing two permanent dosimetry capsules, and a single SSTR dosimetry orientation test specimen were also installed and irradiated at this peak flux location in the Yz quadrant. However, these permanent dosimetry capsules were positioned further radially from core center than the 11- 1/2 degree temporary or SSHT capsule dosimetry. Figure 1 is a two-dimensional view of the arrangement of the specimen capsules and basic dosimetry assemblies, including the azimuthal locations of five stainless steel chain dosimetry assemblies.

Both SSHT specimen capsules and two of the typical 11- 1/2 degree temporary dosimetry holder assemblies, plus the two permanent cavity dosimetry capsules, were placed so the upper and lower capsules or holder assemblies were equal distances from the cold core midplane (560 feet- 4.75 inches plant elevation). The remaining temporary assemblies located at 26- 1/2 and 42- 1/2 degrees off the "Y" axis in the Yz reactor quadrant, and the 11 and 6 degree assemblies located in the Wx and Wz quadrants, respectively, also had dosimetry holders centered about core midplane elevation. The 11- 1/2 degree temporary dosimetry assembly also had dosimetry holders at elevations above mid-plane.

## 2.0 CONCEPTION AND DECISION

In August 1984, after design evaluation of the BWOG reactor plant configurations for the most promising benchmark reactor cavity, a generic plan was developed. The plan permitted the selection of the best benchmark cavity site, with an eye towards the future for optimum methods of installing plant-specific cavity dosimetry. Although the plants had many similarities with regard to cavity design and hardware, each individual reactor cavity had significant variations in as-built hardware. Common elements in all plant configurations, with the exception of DB-1, were nuclear instrument (NI) source check tubes and cavity neutron shield plugs.

Initially, source check tubes were discounted as practical means of access for plant-specific dosimetry due to their proximity to the cavity walls, and in most cases, heavily lead-shielded nuclear instrumentation. Source calibration tubes also varied in material composition and diameter. Usually they were positioned at un-appealing azimuthal locations. At DB-1, there is for instance, no real tube, but only a penetration through a seal plate support concrete ledge that leads into the reactor cavity. Also, there was evidence of plugged drain holes in the bottom of at least one site-specific group of source tubes. Boron residue was believed to have accumulated after the cover seal leaked around the nuclear instrument well. This leakage could have resulted in standing water, acting as a shield around installed dosimetry. Therefore, even source tubes were not considered a desirable or universal way to access mid-reactor cavity areas in 1984.

However, this view changed substantially after analytical work, by BWNS on dosimetry location sensitivity, was completed in August 1986. This additional analysis confirmed the source tubes as a technically feasible method of locating plant-specific dosimetry in reactor cavity core mid-plane regions.

The other element in BWOG plant configuration design (excepting DB-1) was the presence of reactor cavity neutron shield plugs at the vessel seal surface elevations. However, these plugs all varied in their design, size, materials, and methods of fabrication. Usually plugs interlocked with each other in radial or radial and circumferential directions. Some plants had plugs that were also mechanically fastened to their support structure. Replacing one of these plugs was initially viewed as the best technical alternative for ideal azimuthal, radial and elevation placement of dosimetry. This was in fact accomplished at Oconee Unit 2 with the initial plant specific-dosimetry hardware design.

The Oconee 2 design was completed before BWNS decided to use source calibration tubes for locating plant-specific dosimetry and support hardware. Plant-specific dosimetry and support hardware for Oconee 2, using the modified shield plug concept, has now been installed and is currently accumulating fluence data.

At Rancho Seco, before the decision to use source check tubes for dosimetry placement, serious consideration was being given to penetrating the five-foot-thick concrete and steel reinforced biological shield cavity wall. This route was considered because of the extremely complicated neutron shield plug design at the top of the reactor cavity and an extremely hostile environment under and around the vessel itself. A physical inspection underneath the vessel, in the vessel cavity and also inside the mirror insulation around the vessel, established the impractical nature for hardware design, installation and servicing during periodic replacement of plant-specific dosimetry. As a result, an initial decision was made to penetrate the biological shield wall at core mid-plane elevation. However, the 1986 source tube utilization decision curtailed this approach.

The conditions in and around Rancho Seco's reactor cavity area were typical of all the B&W designed 177-fuel assembly skirt supported reactor vessels. Radiation levels at the bottom of the vessel cavity were in the range of 1 R/hr or greater. Access was not practical, difficult or dangerous. Lighting and ventilation were usually non-existent. Space constraints were always prohibitive. In contrast, DB-1 is the only B&W designed nozzle supported 177-fuel assembly reactor. The plant does not have neutron shield plugs incorporated in the upper cavity design. However, because DB-1 was one of the initial three host reactors for the IRVSP, it was considered a potential top choice as host reactor for the dosimetry benchmark experiment.

Inspection of the DB-1 physical plant revealed excellent under vessel and cavity access with lower than expected ( < 450 mR/Hr ) radiation levels. Lighting and ventilation were available, and space beneath the vessel permitted good walking and working conditions. A small equipment room within a few feet of the cavity area beneath the vessel provided an extremely low-level radiation area ( < 5 mR/Hr ) for staging equipment or other technical activities that could be performed out of the higher radiation area. The arrangement of the cavity permitted simultaneous access to the cavity from both the top and the bottom of the reactor vessel. Therefore, the BWOG decision on the top candidate for the benchmark dosimetry experiment host reactor was DB-1.

## 3.0 METHODS AND CONSIDERATIONS

A close inspection at DB-1 showed that full length cables could be extended from the top to the bottom of the vessel cavity. Dosimetry holders, accurately attached to a pair of cables (designated as a specific azimuthal assembly), at specified elevations, could be located at all significant azimuth and radial positions. This was accomplished, when necessary, by skewing and/or radially slanting individual cable assembly top and bottom anchor bracket radial and azimuthal positions. In some cases this "skewing" was also necessary to avoid interferences with existing cavity equipment or support structures.

The top anchor position for each benchmark cable assembly was obtained by using existing cavity shield plate studs, which helped locate and secured the top anchor brackets of each cable assembly. These top brackets were positioned to within +/- 1/8-inch accuracy in three dimensions by using templates and the installed reactor vessel (RV) closure head studs as survey register points. The closure studs are located very accurately within vessel geometry, and are fabricated with controlled diameters. The azimuthal and radial position for the permanent source tube dosimetry was similarly obtained. The bottom anchor for the permanent dosimetry was plumbed from the top of its target source tube, and then anchored in concrete under the vessel.

Locating the bottom anchor positions for the benchmark cable assemblies was more difficult because they were underneath the reactor vessel and in higher radiation fields. Under vessel survey and layout work was required to accurately position brackets on the bottom head support steel of the mirror vessel insulation. The lower bracket positions were established, again, through the use of templates and installed RV closure head studs. Templates registering off the studs at plant elevation 578 feet (vessel seal surface plant elevation) were used to drop plumb lines to the concrete floor beneath the vessel at the 537 foot plant elevation level. The plumb lines established two plumb points 180 degrees apart at this elevation. Thus, the axial or vertical centerline of the RV was established at the 537' elevation. The as-built elevations (544'- 2- 1/8") of the outer circular section of support steel for the reactor vessel bottom head insulation was also verified. Using the two plumb points, and verified survey techniques, the locations for all benchmark dosimetry assembly lower brackets was then transferred to the insulation support steel. Lower brackets were then located and temporarily attached.

As part of the survey work underneath the reactor vessel, target points for critical individual dosimetry holders were established on the 537 foot elevation floor. These points were calculated, then surveyed and established to define register points at the 537 foot elevation during installation. Looped plumb lines were raised and then dropped from the erected benchmark dosimetry holders to verify spatial location. Deviations from target, other than accepted norms, triggered various corrective responses. All essential holders were thereby verified as within established and acceptable limits.

Materials used in fabricating the dosimetry support and sealing hardware were compatible with the encapsulated dosimetry. This included metallic seals, usually pure aluminum, and holder or capsule bodies, fabricated from 6061-T6 aluminum. Fasteners, with the exception of one test case using 300 series stainless material, were also 6061-T6 or 2024 aluminum. Even fastener lubricant and the aluminum anodize and sealing process were compatible with proximate dosimetry. Small stainless locking helicoils, and metallic O-ring seals were used in the permanent dosimetry capsule hardware. However, their extremely small mass and proximity to irradiated dosimetry was evaluated and any possible secondary irradiation effects were considered insignificant.

6061-T6 aluminum proved to be a good fabrication material for the dosimetry support hardware due to its low activation. Low hardware radiation levels were exhibited at the time of removal or exchange of dosimetry. Maximum measured radiation levels, at near contact upon removal and cleaning, after a one fuel cycle irradiation, were on the order of 50 m/R gamma/beta.

Specific and uniform SSTR dosimetry orientation was maintained throughout the benchmark experiment. This criterion was applied to both the temporary benchmark and the permanent dosimetry SSTRs. With the exception of a single permanent hardware canister of SSTR dosimetry, which was orientated at 90 degrees to all others, SSTR chips were positioned so that the plane of each chip was perpendicular to a theoretical flat radial flux line emanating from the core axial or vertical centerline. Although classical physics reportedly indicates no significant predictable variation in SSTR fluence measurement attributable to SSTR orientation, BWNT took the position of maintaining uniformity in orientation of SSTRs for the benchmark experiment, and for all plant-specific SSTR dosimetry. The single, duplicate test canister was commissioned to position an SSTR sample so that only the edge of the chip, and not the plane of the chip, faced a theoretical flat radial flux line. Currently, no results or conclusions have been drawn from the SSTR orientation mini-experiment, since the SSTR analysis is not yet complete.

At DB-1, a total of six nuclear instrument (NI) source check access penetrations were available for potential use during the benchmark experiment, and only one of these access penetrations was utilized for installing permanent dosimetry hardware. It was therefore decided to install four additional vertical dosimeters, or stainless steel ball chains. One stainless steel ball chain was installed in each azimuthal quadrant of the reactor cavity. A stainless steel ball chain supplied by Arkansas Tech University was also installed in direct radial and vertical alignment with the 11-degree dosimetry assembly.

Each of the four quadrant chains was secured at the top of the NI source tube by collapsible bracket to permit freeing the chain when pulled from under the vessel at the 537 foot elevation. This arrangement allowed the chains to be removed without entering the NI wells surrounding the vessel. The chains thus ran full length in the cavity from plant elevation 576 feet-10 1/2 inches to the 537 foot elevation floor beneath the vessel. The diameters of the chain balls was approximately 0.187-inches, with typically five balls per inch of chain. Pre-irradiation evaluation of sample ball material constituency from the chains will aid in determining fluence levels after irradiation at an almost infinite number of elevation levels within the four quadrants of the reactor cavity.

Finally, the design for all benchmark and permanent plant-specific dosimetry hardware considers the necessity of excluding potential water contamination from the dosimetry cases. The aluminum or gadolinium cases themselves afford some measure of protection to the dosimetry from possible temporary water immersion. Water contact, especially for a long period of time, diminishes SSTR accuracy. Water, as unplanned shielding around dosimetry, could also result in non-conservative dosimetry readings. Water seals have been incorporated as primary or secondary seal boundaries in permanent dosimetry hardware through the use of aluminum or stainless all metallic O-rings. Metallic O-ring seals have been incorporated as primary seals in the top or connecting ends of permanent dosimetry capsules subject to water leakage or flooding from refueling canal water through NI well covers, or other sources of water contamination such as control rod drive mechanism cooling water leaks or canal decontamination activities. At DB-1, dosimetry holder assemblies did receive some water contact from refueling activities which was of no concern because of incorporated seals in the hardware.

## 4.0 CONCLUSION

**BWNS, the design organization responsible for implementing the benchmark and plant-specific reactor cavity dosimetry program, has successfully completed the data accumulation portion of the task. This work included the design, accurate positioning and retrieval of the Davis-Besse 1 Benchmark experimental dosimetry. All BWOG plants have committed to plant-specific dosimetry and dosimetry hardware installation. With the scheduled installation of dosimetry and permanent dosimetry hardware in Arkansas Nuclear One in October or November of 1990, and at Crystal River Unit 3 during refueling outage eight in early 1992, all BWOG utility reactors or reactor sites will have plant-specific dosimetry and permanent dosimetry hardware installed as part of the BWOG Cavity Dosimetry Program. Planned analysis of benchmark experimental data will lead to a methodology for accurately determining vessel wall fluence. This benchmarked methodology will be applied to all BWNT-designed and operating 177 fuel assembly plants and adjusted to account for various plant-specific cavity configurations and cavity equipment. Data accumulated from easily replaceable, permanent dosimetry, which includes SSTRs, will provide continued monitoring of operating fluence levels to reactor pressure vessel wall materials.**

FIGURE 1
R.V. CAVITY DOSIMETRY PLAN VIEW AND SURVEY INFORMATION

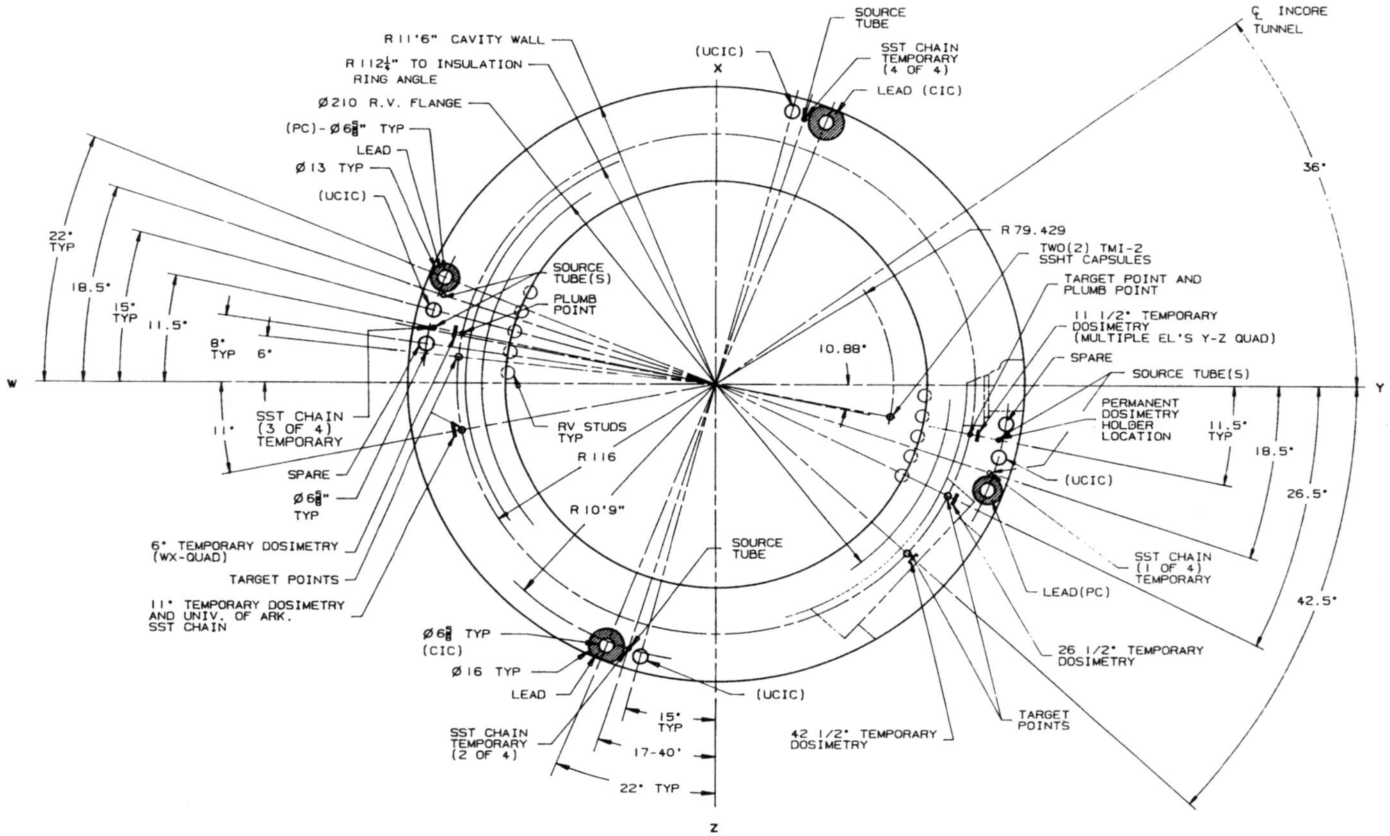

# CALIBRATION AND INTERCOMPARISON EFFORTS

# FOR NEUTRON FLUENCE MONITORING[1]

David W. Vehar

Sandia National Laboratories

P.O. Box 5800, Albuquerque, NM 87185

## ABSTRACT

Three elements of a quality assurance program for neutron fluence measurements as implemented by the Radiation Dosimetry Laboratory at Sandia National Laboratories are discussed. These elements are calibrations, routine checks of system performance, and interlaboratory comparisons. Historical data are presented for interlaboratory comparisons which illustrate nominal agreement within 10 percent among several laboratories for the period from 1981 to 1990. Recommendations for continued work and methods of data reporting are given.

## INTRODUCTION

For many years, the $^{32}S(n,p)^{32}P$ activation reaction has been the preferred reaction for routine monitoring of neutron fluence in reactors used for radiation hardness testing of electronics. This is primarily due to the availability of sulfur in high purity, ease of forming the element into pellets, reasonably high reaction cross section, and convenient half life. In addition, the reaction product is a beta emitter, which can be reliably counted with gas proportional counters. Because the activation reaction has a threshold of about 3 MeV, it is well-suited to measurements in a near-fission spectrum such as that obtained from a fast-burst reactor. Sulfur can also be used in conjunction with a lower-energy-threshold foil to provide information about the hardness of the neutron energy spectrum. Finally, the effective threshold for the $^{32}S(n,p)^{32}P$ reaction is relatively insensitive to spectral hardness, so that errors in spectrum-averaged cross sections due to uncertainties in the neutron energy spectrum are minimized.

---

[1]This work was performed at Sandia National Laboratories, which is operated for the U.S. Department of Energy under contract number DE-AC04-76D000789.

Any system applied to neutron fluence measurements must be used carefully to ensure that reported fluences are valid. This is especially important when an experiment is performed at more than one facility, or when similar experiments at several facilities are compared. This is frequently the case with radiation hardness testing of electronics. Proper use of a measurement system consists of three main elements: 1) calibration of the system against national standards, 2) routine checks of system performance, and 3) interlaboratory comparisons. All three of these elements are essential components of a dosimetry quality assurance program.

This paper addresses these three elements of neutron fluence monitoring as implemented by the Radiation Dosimetry Laboratory (RDL) at Sandia National Laboratories. Special emphasis is placed on the intercomparison program with other laboratories whose main interest is radiation hardness testing.

## CALIBRATION

Calibration of a sulfur system against national standards can be accomplished in several ways, but most commonly is done by irradiating samples in a well-characterized and certified neutron field. The source used by the RDL for these irradiations is a $^{252}Cf$ spontaneous fission source facility operated by the U.S. National Institute of Standards and Technology (NIST) **[1,2]**. The samples are precisely positioned in a fixture surrounding the source and irradiated to several predetermined fluence levels, spanning as much as possible of the range of expected use of the detector system. Corrections, typically a few percent, are made for fluence gradients and neutron scattering in the samples **[3]**. The samples are then counted with the detector system for which the counting efficiency is desired. This efficiency is specific for a given detector system, counting geometry, and sample size or preparation.

The efficiency, $\epsilon$, of the detector system is given by:

$$\epsilon = \frac{C\, f_\tau\, \lambda\, t_i\, e^{\lambda t_d}}{N\, \bar{\sigma}_s\, \Phi\, (1-e^{-\lambda t_c})(1-e^{-\lambda t_i})}$$

where

$C$ = counts recorded in detector, less background;
$f_\tau$ = correction for coincidence losses;
$\lambda$ = $^{32}P$ decay constant;
$t_d$ = decay time;
$t_c$ = count time;
$t_i$ = duration of irradiation;
$N$ = number of $^{32}S$ atoms in sample;
$\bar{\sigma}_s$ = spectrum-averaged cross section for the $^{32}S(n,p)^{32}P$ reaction in the calibration neutron field; and
$\Phi$ = neutron fluence.

$\bar{\sigma}_s$ is determined by folding the energy-dependent cross section for $^{32}S(n,p)^{32}P$ reaction with the (known) energy spectrum for the calibration neutron field:

$$\bar{\sigma}_s = \frac{\int_0^\infty \sigma(E)\ \Phi(E)\ dE}{\int_0^\infty \Phi(E)\ dE}$$

Once the efficiency of the detector system has been determined, the system can be used to monitor other neutron fields, with sample activity given by:

$$A_0 = \frac{C\ f_\tau\ \lambda\ e^{\lambda t_d}}{\epsilon\ (\ 1 - e^{-\lambda t_c}\ )}$$

where $A_0$ is the sample activity adjusted to the end of the irradiation.

The neutron fluence for the field being measured can be determined from the $^{32}P$ activity, providing the $^{32}S(n,p)^{32}P$ spectrum-averaged cross section, $\bar{\sigma}_x$, is known:

$$\Phi = \frac{A_0\ t_i}{N\ \bar{\sigma}_x\ (\ 1 - e^{-\lambda t_i}\ )}$$

or

$$\Phi = \frac{A_0}{N\ \bar{\sigma}_x\ \lambda}$$

for a pulse rather than an extended irradiation.

Note that the value of $\bar{\sigma}_x$ must be calculated using the same set of cross sections (e.g. ENDF-B/V) as is used to calculate the value of $\bar{\sigma}_s$. Large errors can result from using different cross section sets for the two calculations **[4]**.

## ROUTINE SYSTEM CHECKS

Routine system checks are most easily implemented by periodic counting of a long-lived radioactive standard. When performed on a daily basis, these checks can ensure that the system calibration is still valid and can help detect and identify possible sources of fluctuations in counting efficiency. They also provide a measure of intrinsic system precision and long-term stability. In order to minimize the possibility that small changes in counting geometry or absorbers could either go unnoticed or falsely indicate a problem, the source should emit beta particles having end point energies comparable to the 1710-keV beta particles from $^{32}P$. The isotope should also have a half life sufficiently long that decay is negligible or easily corrected for, and that the source will be usable for several years. The RDL uses $^{210}Bi$, which produces beta particles having a maximum energy of 1161 keV. It is obtained in equilibrium with its parent nuclide, $^{210}Pb$, and has an effective half life of 22.3 years.

Figure 1 illustrates the long-term monitoring of the detector system in use by the RDL.

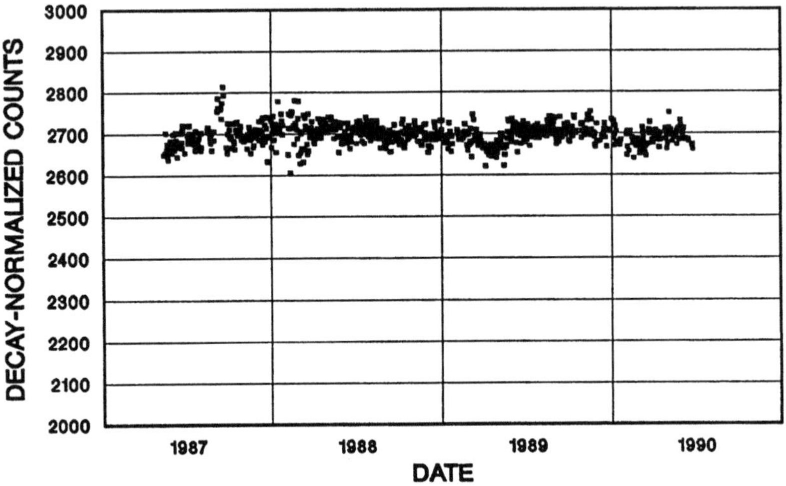


**Figure 1. Bi Reference Counts**

Each data point represents a daily count of the $^{210}Bi$ reference source, corrected for source decay. Counting statistics are not shown, but are smaller than the fluctuations in the data. The figure demonstrates: 1) excellent long-term stability consistent with the decay of the source; 2) data fluctuations consistent with the reproducibility of sample positioning, i.e. about 2 percent; and 3) indications of shifts in detector sensitivity small enough to go unnoticed if not monitored, but still possibly affecting the validity of the dosimetry system. A shift of about 4 percent during the second half of 1987 was traced to a slight misalignment of the reference source in its holder, and thus did not affect any dosimetry measurements. This example illustrates the dangers of using daily reference counts to normalize detector efficiencies, as is sometimes done. Smaller dips of less than 2 percent occurred in early 1989 and again in early 1990. Neither of these were large enough to be of concern unless the trends had continued. The dips were attributed either to variations in counting gas or to contamination of gas lines during changing of gas cylinders, and prompted replacement of the gas cylinders and thorough purging of the lines.

## INTERLABORATORY COMPARISONS

Interlaboratory comparisons are often neglected because many laboratories assume that having calibrations traceable to national standards is sufficient. However, traceability alone does not guarantee accurate dosimetry, which can be affected by factors such as geometric effects, neutron scattering, and spectral hardness, that may not be readily apparent or that may vary among different neutron fields. A comparison among independent measurement systems provides the final assurance that accurate dosimetry is being provided and helps identify possible areas of improvement.

The RDL provides radiation dosimetry services to the Sandia Pulse Reactor (SPR) facility **[5]** at Sandia National Laboratories. This facility consists of two reactors, SPR-II and SPR-III, which are fast-burst reactors used for radiation effects testing. The RDL collaborates with two other laboratories involved in radiation effects testing in intercomparison efforts aimed at identifying possible discrepancies in fluence measurements and improving the quality of dosimetry. One is the Nuclear Effects Laboratory (NEL), which furnishes neutron and gamma-ray dosimetry to the Fast Burst Reactor (FBR) facility **[6]** at White Sands Missile Range. The second is the Army Pulse Radiation Facility (APRF) **[7]** at Aberdeen Proving Ground. Both of these reactors are similar in design to the SPR-II. These intercomparison efforts involve the simultaneous irradiation of sulfur dosimetry at one of the three reactor facilities and subsequent evaluation of the dosimetry at the respective laboratories. The neutron fluences reported by the several laboratories are then compared.

One difficulty that arises when such comparisons are made is the use of inconsistent cross sections among the several laboratories. This manifests itself in two ways. First, cross section sets may differ among laboratories. For example, one laboratory may be using ENDF-B/IV as a basis for calculating spectrum-averaged cross sections, while another may be using ENDF-B/V. Second, laboratories may not have available the spectrum-averaged cross section for the facility at which the irradiations are done. As was illustrated by the equations in the section on calibration, these factors can introduce significant errors in the neutron fluences reported, masking perhaps the inconsistencies that are being investigated. One possible solution to this is reporting neutron fluence data used for comparison purposes as if the irradiations were done in a $^{252}$Cf fission spectrum. This removes errors due to uncertainties in the energy spectrum, and therefore the spectrum-averaged cross section for the reactor spectrum. It also removes ambiguities in the reported data, since fluences are reported in a consistent manner. Finally, effects of updating cross sections are minimized. Although new cross-section evaluations may change radically, thus changing the value for a particular spectrum-averaged cross section, there is negligible change in the ratios of spectrum-averaged cross sections for fission-like fields to that for a $^{252}$Cf fission spectrum **[4]**.

Table 1 lists intercomparison data accumulated since 1981, except for the period from 1982-1984 for which data are not available. The first line of each entry in the table gives the neutron fluences reported by the respective laboratories participating in a given comparison. For a leakage irradiation, the location column shows the distance from the reactor centerline at which the irradiation was performed, or "LEA" if exact placement is not known. The notation "CAV" indicates that the irradiation was done in the reactor cavity. Each value in subsequent columns represents the mean and sample standard

**Table 1. Sulfur Intercomparison Data**

| DATE | LOCATION | RDL | NEL | APRF |
|---|---|---|---|---|
| 09-JUN-81 | SPR3/LEA | 2.33E12 ± 3% | 2.28E12 ± 3%<br>-2.0% | |
| 16-JUN-81 | SPR3/12" | 8.49E12 ± 3% | 8.33E12 ± 4%<br>-1.9% | |
| | | 4.44E13 ± 3% | 4.46E13 ± 3%<br>0.5% | |
| 25-JUL-85 | SPR3/CAV | 1.46E13 ± 8% | 1.36E13 ± 3%<br>-6.6% | 1.34E13 ± 3%<br>-8.1% |
| | SPR3/17" | 7.67E11 ± 4% | 7.30E11 ± 2%<br>-4.8% | 7.13E11 ± 8%<br>-7.1% |
| | SPR3/72" | 4.31E10 ± 2% | 3.94E10 ± 1%<br>-8.6% | 4.02E10 ± 5%<br>-6.8% |
| MAY-86 | SPR/LEA | 1.35E11 ± 3% | 1.34E11 ± 3%<br>-0.5% | |
| | | 6.39E10 ± 3% | 6.35E10 ± 1%<br>-0.6% | |
| | | 2.81E11 ± 7% | 2.80E11 ± 5%<br>-0.5% | |
| 14-OCT-87 | FBR/12" | 5.69E12 ± 2% | 6.02E12 ± 1%<br>5.8% | |
| | FBR/24" | 1.41E12 ± 1% | 1.49E12 ± 2%<br>6.3% | |
| 30-OCT-87 | SPR3/CAV | 1.64E13 ± 3% | 1.71E13 ± 2%<br>4.1% | |
| | SPR3/17" | 2.49E12 ± 3% | 2.61E12 ± 2%<br>4.7% | |
| 28-MAR-88 | SPR3/LEA | 2.37E12 ± 3% | 2.65E12 ± 1%<br>11.7% | |
| 11-APR-88 | FBR/10" | 3.92E12 ± 3% | 4.07E12 ± 2%<br>3.8% | |
| 06-MAY-88 | APRF/LEA | 1.83E12 ± 2% | 1.94E12 ± 2%<br>5.7% | 1.78E12 ± 2%<br>-3.0% |
| 12-FEB-90 | SPR3/CAV | 2.20E13 ± 2% | 2.35E13 ± 3%<br>7.0% | 2.17E13 ± 1%<br>-1.5% |
| | SPR3/19" | 1.01E12 ± 2% | 1.11E12 ± 1%<br>10.3% | 1.05E12 ± 1%<br>4.6% |

Notes: Fluences reported as mean and standard deviation ($1\sigma$) for $^{252}$Cf.
Deviations reported relative to RDL values.
CAV = Cavity; LEA = Leakage.

deviations of several sulfur dosimeters. Multiple entries for a given test date indicate multiple sets of dosimetry. For the reasons discussed above, all data are given in terms of neutron fluence > 3 MeV for a $^{252}$Cf fission spectrum. The second line of each entry gives deviations relative to the fluence reported by the RDL. In general, agreement among the three laboratories is within 10 percent, with most disagreements below 6 percent. Uncertainties are not shown, but are typically reported to be between 5 and 10 percent for all three laboratories. The agreement is consistent with reported uncertainties. However, it is believed that discrepancies of less than 5% should be routinely achievable.

## RECOMMENDATIONS

All three elements of system quality assurance are essential. The most difficult and time consuming is a laboratory intercomparison program, and it is that aspect which is most often neglected. Interlaboratory comparisons must be done on a regular basis, and should involve at least three laboratories. With only two laboratories participating, as was the case for much of the data presented here, specific improvement areas cannot be identified. Furthermore, these intercomparisons should be done more frequently, at intervals of less than 6 months. This would allow a more timely response should problems arise. Finally, in order to avoid ambiguities, data must be reported in a consistent manner. This should include not only fluences determined by a laboratory in its usual fashion, but also fluences in terms of equivalent $^{252}$Cf neutron fluence. Pertinent information such as the spectrum-averaged cross sections used in the calculations should also be provided.

## REFERENCES


1. G. P. Lamaze and J. A. Grundl, "Activation Foil Irradiation with Californium Fission Source," NBS SP-250.13, U. S. Department of Commerce, March 1988.

2. J. A. Grundl and C. M. Eisenhauer, Compendium of Benchmark Neutron Fields for Reactor Dosimetry, NBSIR-3151, U. S. Department of Commerce, January 1986.

3. C. M. Eisenhauer, "Corrections for Fluence Gradient and Neutron Scattering in Calibration of AFRRI Sulfur Sensors at NBS $^{252}$Cf Fission Neutron Facility", Memorandum to Record, National Bureau of Standards, November 6, 1986.

4. J. G. Kelly and D. W. Vehar, Measurement of Neutron Spectra in Varied Environments by the Foil-Activation Method With Arbitrary Trials, pp 133-140, SAND87-1330, Sandia National Laboratories, Albuquerque, NM, December 1987.

5. L. M. Choate, T. R. Schmidt and G. A. Zawadzkas, Editors, Sandia National Laboratories - Radiation Facilities, pp 10-15, SAND89-1399, Sandia National Laboratories, Albuquerque, NM, December 1989.

6. R. L. Long and D. L. Welch, "Operating Experience with the WSMR Fast Burst Reactor-II," in Proceedings of the Army Material Command Research Reactor Symposium, White Sands Missile Range, NM, February 1973.

7. A. H. Kazi, C. R. Heimbach, R. C. Harrison and A. H. Robitaille, "Comparison of Measured and Calculated Radiation Transport in Air-Over-Ground Geometry to 1.6 km from a Fission Source," Nuclear Science and Engineering, Vol. 85, pp 371-386, 1983.

## ACKNOWLEDGEMENTS

The author wishes to thank T. M. Flanders and W. Sallee at the NEL at White Sands Missile Range, and C. Heimbach at the APRF at Aberdeen Proving Ground for their cooperative efforts in these intercomparisons, and for providing much of the data presented herein.

# Effect of ENDF/B-VI Cross Sections on Neutron Dosimetry

**P. J. Griffin, J. G. Kelly,**
**T. F. Luera**

**Sandia National Laboratories**

**P. O. Box 5800, Albuquerque, NM 87185**

## ABSTRACT

ENDF/B-VI cross sections were released to the testing community in January 1990. Work at Sandia National Laboratories, with pre-released versions of the new cross sections indicates that changes in the neutron-induced charged-particle reactions will significantly affect 14-MeV neutron dosimetry. Reactions that are important for fission reactor dosimetry were examined and most did not change significantly.

## INTRODUCTION

While validating the ENDF/B-VI [1] silicon displacement kerma [2], the authors observed that use of the latest ENDF/B-VI $^{58}$Ni(n,p)$^{58}$Co cross section as a fluence monitor for the 14-MeV neutron irradiations led to significant differences in the dosimetry. Further investigation showed that incorporating ENDF/B-VI updated cross sections, one at a time, into the Sandia Dosimetry Laboratory spectral unfolding and fluence monitoring reports may not be the best procedure. That is, there were indications that correlations exist between the individual dosimetry cross section files. It was decided that the ENDF/B-VI cross sections should be incorporated into the Sandia Dosimetry Laboratory set of dosimetry cross sections and validated as a whole rather than on a reaction-by-reaction basis. This paper reports on some of the steps taken to validate the use of the new ENDF/B-VI dosimetry cross sections for the purpose of spectral characterization and on the conclusions drawn from the validation.

This work was performed at Sandia National Laboratories, which is operated for the U.S. Department of Energy under contract DE-AC04-76D00789.

## CROSS SECTIONS

Pre-released versions of the ENDF/B-VI cross sections were used for this study. Since some important dosimetry reactions were not yet updated within the ENDF evaluation, other compilations of dosimetry cross sections were drawn upon to supplement this library of reactions. The GLUCS [3], IRDF-82 [4], and some private communications [5] were judged to represent improvements over the ENDF/B-V data (for dosimetry purposes) and were used within this analysis to verify the consistency of the "updated post-ENDF/B-V" cross sections. Thirty seven dosimetry cross sections were studied. Fourteen of them were validated in the region of 14 MeV. Twenty six of the cross sections were validated for use in the unfold of fission spectra. In the case of some important dosimetry reactions, no update to the ENDF/B-V data could be found.

There are significant differences between the ENDF/B-V and ENDF/B-VI dosimetry reaction cross sections. Many dosimetry reactions were found to differ in the region above 7 MeV. The $^{63}$Cu(n,$\alpha$)$^{60}$ Co reaction is just one example. In this case, folding the cross section in with a fission reactor spectrum results in very little (2-5%) difference in the resulting activity. For most dosimetry reactions, the ENDF/B-V and ENDF/B-VI cross sections were in very good agreement below 7 MeV. The $^{58}$Fe(n,$\gamma$)$^{59}$ Fe reaction was an exception. When this cross section is folded with a typical fission reactor spectrum, the resulting calculated activities from ENDF/B-V and ENDF/B-VI cross sections differ by 30-60%.

## 14 MeV DOSIMETRY

### Energy Determination

The ASTM E-496 Standard Method [6] deals with the determination of the energy and fluence from a $^{3}$H(d,n)$^{4}$He neutron source. This standard recommends use of ratios of the $^{27}$Al(n,$\alpha$)$^{24}$Na, $^{27}$Al(n,p)$^{27}$Mg, $^{58}$Ni(n,p)$^{58}$Co, $^{58}$Ni(n,2n)$^{57}$Ni, $^{63}$Cu(n,2n)$^{62}$Cu, and $^{65}$Cu(n,2n)$^{64}$Cu cross sections to determine the average neutron energy. All but the $^{27}$Al reactions are available in the current ENDF/B-VI release. A December 1989 Oak Ridge National Laboratory (ORNL) version of the GLUCS evaluations [7] was used for $^{27}$Al. This is believed to be a valid and consistent choice since the ENDF/B-VI $^{58}$Ni and $^{63,65}$Cu cross section evaluations were performed by ORNL and were initially drawn from the GLUCS analysis.

Figure1 shows how two of the ASTM E-496 recommended reaction ratios changed in the energy region applicable to $^{3}$H(d,n)$^{4}$He neutron sources. A trend is seen in all of the updated cross sections. Cross sections that are increasing with energy in the 14-MeV region are higher in the new evaluations and cross sections that are decreasing with energy in the 14-MeV region are lower in the new evaluation. The only exception to this is the $^{63}$Cu(n,2n)$^{62}$Cu reaction which stayed the same. In this case, since this reaction was not available from the ENDF/B-V dosimetry tape, the current ASTM E-496 recommended cross section was from the histogram representations in the IRDF-82 library. The IRDF-82 library already represents an update on the ENDF/B-V library of dosimetry reactions and, in the case of the $^{63}$Cu(n,2n)$^{62}$Cu reaction, agrees very well with the latest ENDF/B-VI evaluation. When the E-496 recommended ratios are formed from the updated cross sections, a net decrease of approximately 0.3 MeV in the $^{3}H(d,n)^{4}He$ average neutron energy results. This decrease is independent of which cross section ratio is used.

If the $^{58}Ni(n,p)^{58}Co$ reaction is used as a fluence monitor foil, there is a 15% increase between the ENDF/B-V and ENDF/B-VI fluence determination from a 14.8- MeV neutron source. The 0.3 MeV shift in the energy that results from the application of the ENDF/B-VI cross section ratios results in a compensating decrease of 9% in the fluence. This suggests that the "vintage" of the cross section is very important and that mixing cross sections for the energy and fluence determination of a DT neutron source can result in significant errors. In order to assure ourselves that this compensation effect in the fluence determination is not an anomaly, but reflects a correlation in the cross sections, an experiment was conducted to gather data for 17 dosimetry reactions produced by the Sandia $^{3}H(d,n)^{4}He$ neutron generator.

## Experimental Confirmation

Fourteen of the foils/reactions were used to characterize the spectrum/fluence of the Sandia DT neutron generator. Column 2 of Table 1 shows the slope (the cross section rise over the run expressed as a percent of the14- MeV cross section) of the foil cross sections near 14 MeV. Applicable reactions were chosen from this table to form a "post-ENDF/B-V" cross section ratio to be used in determining the DT source energy. Figure 2 shows the scatter in the energy estimates for the best ratios of "post-ENDF/B-V" cross sections. The bars indicate an upper bound for the error in the energy determination. This error was obtained by adding the experimental and theoretical standard deviations (from Table 1, columns 5 and 6) and taking into account the slope for the relevant cross section ratio. The ratios of the nine reactions with negative slopes to the steeply increasing $^{58}Ni(n,2n)^{57}Ni$ reaction result in very tightly clustered energy estimates. The same nine reactions, when ratioed to the less steeply sloped $^{65}Cu(n,2n)^{64}Cu$ cross sections, result in a moderate energy spread. The ratios to the $^{Nat}Ti(n,X)^{47}$ Sc cross sections are fairly well clustered, but have very large error bars due to the uncertainty of the $^{48}Ti(n,np)^{47}Sc$ cross section.

Figure 2 indicates that cross section ratios with the $^{58}Ni(n,2n)^{57}Ni$ reaction as the increasing function will form the most consistent measure of the DT neutron energy and have the smallest uncertainty. Recommended reactions for the decreasing functions, based on the steepness of the function, are $^{54}Fe(n,p)^{54}Mn$, $^{58}Ni(n,p)^{58}Co$, or $^{32}S(n,p)^{32}P$. The resulting error is estimated to be 2.5%, or 0.36 MeV. The precision or reproducibility of this energy, within the range of decreasing functions shown in Table 1, is better than 1%, or 0.15 MeV.

When ENDF/B-V cross sections are used in an identical manner to analyze the data from this experiment, the data are seen to be less tightly clustered. In addition, the energy determination resulting from a choice of any one of the three increasing functions and averaging over the results for the nine decreasing functions shows an upward shift, relative to the ENDF/B-VI analysis, of about 0.3 MeV for the DT neutron energy. The DT neutron energy, as determined from the ENDF/B-V cross section ratios, is 14.8 MeV with an accuracy of 3.5% and a precision of 1.3%.

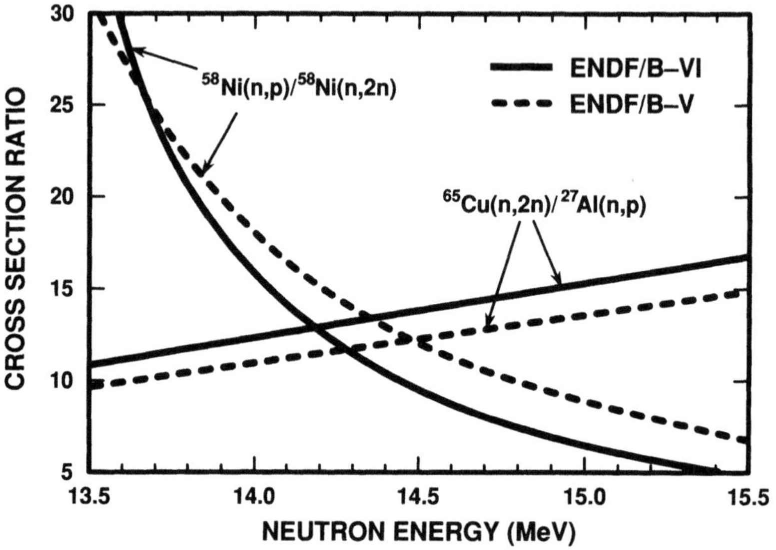


FIG. 1 --- Energy Dependence of Cross Section Ratios Recommended by ASTM E-496

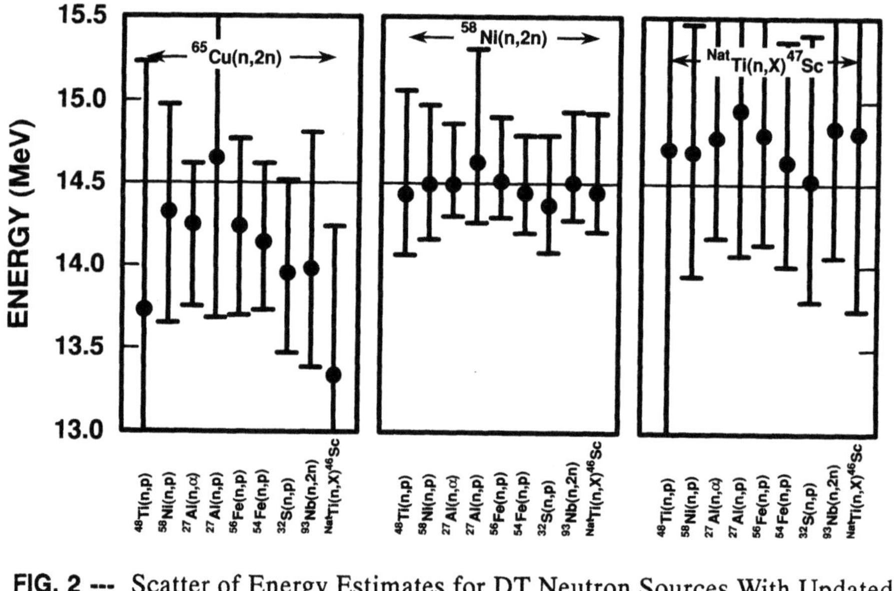


FIG. 2 --- Scatter of Energy Estimates for DT Neutron Sources With Updated Cross Sections

**TABLE 1 ---** Consistency of Updated Cross Sections for 14.5 MeV DT Neutron Fluence Determination

| Reaction | Slope Near 14 MeV | Cross Section (b) | Fluence ($n/cm^2$) ($x10^{12}$) | Theor. Std. Dev. | Expt. Std. Dev. | Comment |
|---|---|---|---|---|---|---|
| **Increasing** | | | | | | |
| $^{58}Ni(n,2n)^{57}Ni$ | 55.0% | 3.42E-2 | 3.33 | 11.7% | 3.0% | Best Inc. Fnc. for Ratio |
| $^{90}Zr(n,2n)^{89}Zr$ | 40.0% | 7.59E-1 | 3.63 | 2.2% | 3.5% | No Recent Eval. |
| $^{Nat}Ti(n,x)^{47}Sc$ | 34.5% | 2.04E-1 | 3.86 | 22.0% | 5.5% | Large Theor. Std. Dev. |
| $^{63}Cu(n,2n)^{62}Cu$ | 26.4% | 5.15E-1 | --- | 1.5% | --- | Short Half-life, 9.74 min. |
| $^{65}Cu(n,2n)^{64}Cu$ | 9.9% | 9.49E-1 | 3.11 | 1.7% | 4.0% | |
| **Decreasing** | | | | | | |
| $^{58}Ni(n,p)^{58}Co$ | -29.9% | 3.22E-1 | 3.11 | 17.3% | 3.5% | Good Dec. Fnc. for Ratio |
| $^{54}Fe(n,p)^{54}Mn$ | -23.5% | 3.07E-1 | 3.51 | 3.7% | 6.5% | Best Dec. Fnc. for Ratio |
| $^{32}S(n,p)^{32}P$ | -23.2% | 2.25E-1 | 3.80 | 8.0% | 7.0% | Thick Target in Expt. |
| $^{63}Cu(n,\alpha)^{60}Co$ | -12.1% | 4.02E-2 | --- | 6.9% | --- | Not Eval. in 14 MeV Expt. |
| $^{27}Al(n,p)^{27}Mg$ | -12.0% | 6.92E-2 | 3.02 | 7.2% | 10.0% | |
| $^{56}Fe(n,p)^{56}Mn$ | -9.7% | 1.11E-1 | 3.28 | 1.9% | 3.0% | |
| $^{27}Al(n,\alpha)^{24}Na$ | -9.1% | 1.16E-1 | 3.33 | 0.7% | 3.0% | |
| $^{64}Zn(n,p)^{64}Cu$ | -9.1% | 1.68E-1 | 3.12 | 4.9% | 5.0% | No Recent Eval. |
| $^{24}Mg(n,p)^{24}Na$ | -8.7% | 1.86E-1 | 3.06 | 1.0% | 3.0% | |
| **Flat** | | | | | | |
| $^{93}Nb(n,2n)^{92m}Nb$ | 1.7% | 4.60E-1 | 3.31 | 5.5% | 3.5% | Best Fnc. for Fluence Det. |
| $^{48}Ti(n,p)^{48}Sc$ | -3.4% | 6.02E-2 | 3.49 | 10.8% | 2.5% | |
| $^{Nat}Ti(n,x)^{46}Sc$ | 5.4% | 2.90E-1 | 3.44 | 22.0% | 3.0% | Large Theor. Std. Dev. |

**Average Fluence Estimate = $3.36x10^{12}$ $n/cm^2$ with a std. dev. of 7.8%**

## Fluence Determination

Table 1 shows the resulting fluence estimates when the latest available cross sections are used and the average neutron energy is specified as 14.5 MeV. The average fluence determined with equal weighting of all available reactions results in a fluence of $3.36x10^{12}$ $n/cm^2$ and a standard deviation of 7.8%. If the reactions considered are restricted to the subset of actual ENDF/B-VI evaluations, the clustering of the fluence is greatly improved resulting in an average fluence of $3.28x10^{12}$ $n/cm^2$ +/- 4.6%. When one considers the slope of the reaction in the 14-MeV energy region and the error associated with the various dosimetry cross sections, the $^{93}Nb(n,2n)^{92m}Nb$ reaction stands out as the optimum choice for use as a DT neutron fluence monitor. This reaction results in a fluence of $3.31x10^{12}$ $n/cm^2$ for this experiment, which is in very good agreement with the average value for all of the foils.

If an identical analysis is performed with the ENDF/B-V cross sections and an energy of 14.8 MeV, an average fluence of $3.34 \times 10^{12}$ n/cm$^2$ with a standard deviation of 8.0% results. Restricting the reactions to the subset for which ENDF/B-VI data are available (in order to allow an identical comparison with the reaction list considered above) results in the same fluence estimate of $3.34 \times 10^{12}$ n/cm$^2$ with a standard deviation of 6.5%. The DOSCROS84 [8] version of the $^{93}$Nb(n,2n)$^{92m}$Nb reaction (which uses an adjustment factor to separate the $^{92m}$Nb metastable state from the $^{92}$Nb ground state activity [9]) gives a fluence of $3.09 \times 10^{12}$ n/cm$^2$.

## Proposed Updates to ASTM E-496

In light of the above results, the authors recommend that the ASTM E-496 Standard be updated. The update should incorporate the ENDF/B-VI cross sections and stress the importance of using the same "vintage" of cross section for both the fluence and energy determination of DT neutrons. The new standard should be modified to recommend that a flat cross section, such as $^{93}$Nb(n,2n)$^{92m}$Nb, be used for the fluence determination. The energy should be determined from a cross section ratio between steeply increasing and decreasing functions. The current ASTM energy error estimate of 0.3 MeV (2.5%) is valid only if the increasing function is the $^{58}$Ni(n,2n)$^{57}$Ni cross section. Ratios to the $^{65}$Cu(n,2n)$^{64}$Cu cross section result in an error estimate double this value. There also is a definite bias in the energy estimate associated with the choice of the increasing function used in the cross section ratio.

## FISSION SPECTRUM DETERMINATION

Spectrum adjustment at Sandia National Laboratories is done primarily with an enhanced version of the SAND-II methodology [10,11]. When a new cross section library was assembled, 15 previously determined spectra [12] were re-analyzed. Integral parameters, such as the spectral index (SI) and hardness parameter (HP) (defined in reference 11), were not significantly altered by the updated cross sections. The quality of the fits, in terms of smoothness and standard deviation of experimental versus calculated activity, was slightly worse for the updated cross sections, but acceptable. Some problems were seen with resonances in specific reactions. This may be due to the fact that experience has led to the exclusion of certain ENDF/B-V dosimetry reactions that do not fit a large body of data. This process of "weeding out" poorly defined reactions is just beginning for the ENDF/B-VI cross sections.

Initial work indicates that the $^{47}$Ti(n,p) reaction, which had previously been excluded from the Sandia foil set based on a poor match with ENDF/B-V data for almost all reactor spectra, is now in acceptable agreement with the other reactions in that energy region. However, the resonances in the 0.3 to 2.0 keV energy region for the $^{55}$Mn(n,$\gamma$)$^{56}$Mn reaction that previously were consistent with other reactions in this region, produced anomalous "bumps" in the spectrum. This happened in each of the spectra studied. In 14 of the 15 spectra the conflict could not be resolved through the elimination of only one of the other foils. The "bumps" were introduced in the very first iteration of the spectral fit. This indicates that the poor agreement is not due to the tendency of iterative unfolding methods to "over-fit" the data, but rather is a problem with the cross sections. In one spectrum a conflict was seen with the $^{98}Mo(n,\gamma)^{99}Mo$ reaction. This is an ENDF/B-V reaction (no update was available) that previously fit well. Many of the 14 spectra discussed here did not include the molybdenum foil, so the authors are still investigating the nature and extent of this foil conflict.

## CONCLUSIONS

This paper has examined the effect of the ENDF/B-VI cross sections on neutron dosimetry and compared the consistency of the most recent cross section evaluations with previous results based on ENDF/B-V data. The studies documented in this paper support the following conclusions:

1) ENDF/B-VI cross section evaluations show significant differences (>10%) at energies above 7 MeV for some reactions important to the dosimetry community.
2) 14-MeV fluence and energy determinations depend upon consistent application of a given cross section set.
3) Use of the most recent ENDF/B-VI and GLUCS cross sections results in a decrease of about 0.3 MeV in the neutron energy from DT sources when compared to previous results based on ENDF/B-V data.
4) DT neutron fluence estimates from recent cross section evaluations are at least as tightly clustered as those from previous ENDF/B-V data. Results based solely on ENDF/B-VI data are more tightly clustered than ENDF/B-V results.
5) Fission reactor dosimetry is not significantly affected by the changes in the ENDF/B-VI cross sections.
6) Spectrum unfolding indicates that changes in the ENDF/B-VI $^{55}Mn(n,\gamma)^{56}$ Mn cross section resonances in the 0.3 - 2.0 keV energy range are in conflict with other activation foil data in this region.
7) ENDF/B-VI silicon displacement kerma has been validated for use as a dosimetry sensor [12].

In view of the work reported here, specific changes to the ASTM E-496 Standard are recommended. Dosimetry comparisons and spectrum unfolding studies with the ENDF/B-VI cross sections should continue to confirm the work reported here and to extend it over a broader range of neutron spectra and dosimetry reactions.

# REFERENCES


[1] R. Kinsey, Ed., "ENDF-102, Data Formats and Procedures for Evaluated Nuclear Data File, ENDF," Brookhaven National Laboratory, BNL-NCS-50496 (ENDF-102) 2nd Edition (ENDF/B-V), October 1979,; BNL-NCS-44945 3rd Edition (ENDF/B-VI) July, 1990.

[2] J. G. Kelly, P. J. Griffin, "Measured Silicon Displacement Damage Ratios Support the Need to Replace the ASTM E-722 Damage Function with the NJOY Calculated Function," presented at the Seventh ASTM-Euratom Symposium on Reactor Dosimetry, held in Strasbourg, France on August 27-31, 1990.

[3] C. Y. Fu, D. M. Hetrick, "Experience in Using the Covariances of Some ENDF/B-V Dosimetry Cross Sections: Proposed Improvements and Additions of Cross-Reaction Covariances," Oak Ridge National Laboratory, Proceedings of the 4th ASTM-Euratom Symposium on Reactor Dosimetry, Vol. 2, pp. 877-887, 1982.

[4] D. E. Cullen, N. Kocherov, P. M. McLaughlin, "The International Reactor Dosimetry File (IRDF-82)," IAEA-NDS-41/R, January 1982.

[5] W. Mannhart, D. Smith, J. Meadows, "The Discrepancy Between Differential and Integral Data on Ti-47(n,p)," in the American Nuclear Society Proceedings on a Topical Meeting on Advances in Nuclear Engineering Computation and Radiation Shielding, Santa Fe, New Mexico, April 9-13, 1989, University of New Mexico Publications, Ed. M. L. Hall, pp. 577-580.

[6] The 1988 ASTM Annual Book of ASTM Standards, Volume 12.02, American Society for Testing and Materials.

[7] D. Hetrick, private communication from Oak Ridge National Laboratory containing the current version of the GLUCS library, December 13, 1989.

[8] W. L. Zijp, H. J. Nolthenius, G. C.H.M. Berhaag, "Cross Section Library DOSCROS84 (in a 640 Group Structure of the SAND-II Type)," ECN-160, October 1984.

[9] S. Blow, "Cross Sections at 14 MeV of the $^{93}Nb(n,2n)^{92}Nb$ Reaction and its Relevance to Fusion Reactor Technology," AERE-R6540, UKAEA, Harwell, January 1971.

[10] W. N. McElroy, S. Berg, T. Crockett, and R. Hawkins, "A Computer-Automated Iterative Method for Neutron Flux Spectral Determination by Foil Activation," AFWL-TR-67-41, Vol. I, Air Force Weapons Laboratory, Kirtland, New Mexico, July 1967.

[11] J. G. Kelly, D. W. Vehar, "Measurement of Neutron Spectra in Varied Environments by the Foil-Activation Method With Arbitrary Trials," Sandia National Laboratories, SAND87-1330, December 1987.

[12] P. J. Griffin, J. G. Kelly, and T. F. Luera, "Effect of ENDF/B-VI Cross Sections on E-722 Silicon Displacement Kerma," presented at the ASTM E10.07 subcommittee meeting held at Las Vegas, Nevada on January 25, 1990.

Learnings from a joint Italian-Belgian neutronic characterization of the TAPIRO source reactor

A. Fabry, H. Aït Abderrahim, P. D'hondt, S. De Leeuw, G. De Leeuw-Gierts, G. Minsart
SCK/CEN, Mol, Belgium

P. Moioli, M. Angelone, M. Carta
ENEA C.R.E. Casaccia, Roma, Italy

ABSTRACT: In the scope of an Interlaboratory Consulting agreement between the ENEA and SCK/CEN Laboratories, an extensive neutronic characterization of the Tapiro source reactor at Casaccia, Italy, has been carried-out. This does encompass the cylindrical enriched uranium core, its surrounding copper reflector and, for some sensors, part of the outer concrete shield. A large array of equivalent fission flux and reaction rate measurements have been performed and experimentally normalized to a unit core-integrated neutron source strength; this covers the entire energy range of practical interest, from $\sim$ 100 eV to $\sim$ 10 MeV.

Tapiro provides a family of neutron spectra of extremely variable hardness, most suitable to many reactor metrology applications, and amenable to good geometrical modeling.

The accuracy of a transport theory evaluation of these reference neutron fields primarily depends on the quality of copper neutronic cross-sections. To supplement the Tapiro experimental data bank in this last respect, a supportive copper shell benchmark has been built at Mol within a $\Sigma\Sigma$-type natural uranium driver source shell at the centre of the one-meter spherical cavity in BR1 graphite thermal column.

This overview highlights the Tapiro neutronic characterization results, with emphasis on learnings deemed to be of broader interest to reactor physics and neutron dosimetry concerns.

KEY WORDS: reactor physics, dosimetry benchmark fields, neutronic nuclear data, deep neutron penetration, kernel methodology, fission neutron spectra.

## INTRODUCTION AND OBJECTIVES

The neutronic characterization of the TAPIRO, enriched-uranium, copper reflected, cylindrical source reactor, is the focus of this paper. The main programmatic objectives of this venture were as follows :

1) To assess detection and calculational techniques for fast, intermediate-energy and epithermal neutrons in nuclear reactors;
2) To establish a reactor dosimetry standardization and referencing capability, featuring permanent radiation fields associated to documented and accurate procedures for long-term, multipurposes utilization.

The Tapiro source reactor provides a rich variety of relevant reference neutron spectra, with reasonably high flux intensities, as indicated by Table 1.

TABLE 1 : Approximate neutron flux intensities and integral neutron spectra available at the TAPIRO source reactor

| RADIUS | NEUTRON FLUX ($cm^{-2}.sec^{-1}$)[(a)] | | |
|---|---|---|---|
| (cm) | > 1 MeV | > 100 keV | TOTAL |
| 0 | 9.6E + 11 | 2.1E + 12 | 2.2E + 12 |
| 10.4 | 1.1E + 11 | 5.0E + 11 | 6.0E + 11 |
| 20.6 | 1.2E + 10 | 1.3E + 11 | 1.9E + 11 |
| 30.4 | 1.4E + 09 | 3.6E + 10 | 7.0E + 10 |

[(a)] At the full reactor power of 5 kW; the total core integrated neutron source strength is of ∿ $3.10^{14}$ n/sec.

The core centre neutron spectrum does not deviate much from the one of fission spectrum neutron sources above 1 MeV. The large array of available spectral shapes is illustrated by Figure 1; shown here is the radial distribution of relative reaction rates for a few integral neutron detectors, ranging from cobalt capture - mostly sensitive to neutrons of energy near 130 eV - to threshold reactions, sensitive above ∿ 0.5 MeV. Turning such an impressive family of neutron spectra into pedigreed benchmark neutron fields has been the long-term thrust of the Tapiro characterization program.

This work is extensively documented in a final report [1] (under print). Only some aspects, not yet covered by recent, related publications [2][3][4], and deemed the more interesting to a Reactor Dosimetry audience, are highlighted here.

## OUTLINE OF FACILITIES AND EXPERIMENTAL APPROACH

Figure 2, a horizontal sketch of the Tapiro reactor at core midplane, differs from published descriptions (for ex.,[5]) only by the fact that uniformity of the concrete shield around the copper reflector has been achieved for all the measurements of this program. The experimental channels displayed on the Figure have all been investigated very thoroughly, as well as an horizontal channel tangential to the core outer radius at its upper edge boundary. Extreme care has been devoted at assessing all causes for possible neutronic field distorsions with respect to an idealized $^{235}U$, copper reflected system : control rod effects, local heterogeneities (cavities, steel structures, copper-concrete interface), instrumental perturbations, ... . Excluding some zones perturbed for traceable and quantified reasons, the Tapiro reflector has been found to display almost perfect spherical neutronic symmetry beyond distances of $\sim$ 13 cm to the core centre : generally, the location of an experimental sensor can be expressed by an effective distance to the centre (labelled "equivalent radius" outside of the reactor midplane).

The Mol Copper Shell Benchmark in support of Tapiro, Figure 3, is derived from the MOL-ΣΣ concept [6] and is similar to the previously documented Iron Shell Benchmark [7]. The natural uranium source shell (driven by the purely thermalized neutron field of the BR1 vertical graphite column) has been kept here as thin as possible (1 cm). Four configurations of increasing copper thickness (up to 18.25 cm ; source completely filled) have been investigated. Detailed $^{6}Li(n,\alpha)$ neutron spectrometry measurements have been done [8], covering the energy range $\sim$ 25 keV to $\sim$ 5 MeV, and agree well with the integral measurements outlined in this paper. Five threshold sensors - miniature $^{237}Np$ and $^{238}U$ fission chambers, radiometric foils $^{115}In(n,n')$ $^{58}Ni(n,p)$ and $^{27}Al(n,\alpha)$ - serve to mutually validate the measurements and calculations linking the Tapiro and its supportive Copper Shell benchmark. The validation encompasses spatial traverses, and as shown by Figure 4, this has required to address (among other perturbing features) the source non-uniformity stemming from the deriving thermal neutron flux gradient.

All integral data in the two facilities have been carefully normalized to the respective integrated neutron source strengths (neutron/sec), as derived from absolute fission rate measurements, appropriate spatial integration and sealing by $\bar{\nu}$ (neutron emission per fission). For all sensors, absolute reaction rates have been measured and are validated by sustained interlaboratory comparisons; this includes participation to a state-of-the-art international confrontation for key sensors [2].

In the threshold detection range, the experimental and analytical emphasis has been placed on equivalent fission fluxes (rather than on reaction rates). In practice, this was implemented by means of dedicated exposures in the MOL Uranium-235 Fission Spectrum Standard Neutron Field; traceability to the NBS (now NIST) $^{252}Cf$ Standard Source has been recently verified. The outstanding relevance of such approach is briefly illustrated below.

## OUTLINE OF RESULTS

### Facility-to-facility Consistency

The inherent complexities of Tapiro can be insured to be properly understood if consistency with nominally simpler, albeit representative assemblies is observed, in terms of deviation (or agreement) patterns with respect to compatible transport theory evaluations. The most critical energy range for such consistency confrontation : > 1 MeV. Indeed, multiple collisions of neutrons with reactor materials do largely mitigate the sensitivity of the neutron field (space, angle) to local material heterogeneities in the slowing-down energy range - i.e., before absorption effects become significant ($\gtrsim$ 10 keV).

In this context, referring equivalent fission fluxes to first-collision or to "removal" kernels has been found to be a powerful methodology in the following instances :

1) The numerical accuracy of discrete-ordinates transport theory can be verified exactly. For ex., above $\sim$ 2 MeV, convergence within 1-2 % at deep Tapiro reflector penetration requires S32 quadrature and 5 mm spatial mesh in the diagonal transport approximation for one-dimensional spherical modeling (Fig. 5).
2) Neutron transport effects can be separated from the heterogeneities of the source. For ex., vertical traverses in the MOL Shell Benchmark (Fig. 4) can be recounciled (Fig. 6) with horizontal ones - unaffected by the source non-uniformity. Subtle perturbation effects are revealed (not illustrated here : for ex., access hole for fission chambers on Fig. 3; etc...).

The inverse source-detector geometries of the Tapiro and Mol Shell Benchmark are actually so different[+] that even fast flux-to-kernel ratios are not displaying exactly the same space variation; the largest difference is recorded for $^{237}Np(n,f)$ (see Fig. 7).

---

[+]Spatial variation of fast neutron flux versus copper "penetration" is order of magnitudes less steep in the Mol Benchmark (Fig. 4) than in Tapiro (Fig. 1).

The flux-to-kernel ratio is a measure of the importance of non-elastic collisions and thus should be used as the abscissa for proper facility-to-facility confrontation; the Tapiro and Mol Benchmark neutronic data in the most critical energy range are entirely consistent within experimental uncertainties, once displayed in such format (Fig. 8).

## Comparison to Transport Theory

Figure 9 presents a synthesis of calculation-to-experiment (C/E) ratios for Tapiro. The measurements feature uncertainties within the range of $\pm$ 3-5 % (1 $\sigma$). No distinction is made here among the various reactor channels (Fig. 2), as the systematic differences are generally very small in the unperturbed zones selected for the final characterization. The two-dimensional S16 (diagonal) transport theory calculations illustrated by the Figure 9 are based on the MOL-BR2 40 group cross section library, which is identical for copper to the old Russian ABBN set above 2.15 eV. Group cross sections for all sensors have been processed from the ENDF/B5 dosimetry file (with space-dependent weighting). A methodology has been developed to address neutron self-shielding in environments departing from 1/E slowing-down conditions. This was applied to two situations : copper-63 in bulk natural copper (Fig. 9c); cobalt capture in thick dosimetry foils (Fig. 9d).

Up to a radius of $\sim$ 35 cm, and even for fast neutron attenuations as large as $\sim$ 10000, most integral observations are predicted to $\sim \pm$ 20-25 %, except above $\sim$ 6-7 MeV (numerical uncertainties do not exceed 2-3 %). The deviations in the intermediate- and fast energy ranges can all be quantitatively traced to small biases of copper inelastic cross sections above 5 MeV and near the threshold of the first discrete levels. Some inadequacies of copper absorption cross sections play a role below $\sim$ 10 keV. More recent data libraries, such as ENDF/B3 and B4, lead to patterns of discrepancy with experiment in copper by far worse.

In concrete, theory fails catastrophically at all neutron energies.

## Benchmark Field Referencing of Threshold Sensors

The use of a simple difference equation allows to establish an integrally-consistent spectral definition in the neutron energy range $\sim$ 0.5 - 14 MeV, directly using equivalent fission fluxes measured by means of benchmark-field referencing. The method has been incorporated in a computer code, Fispec, and validated by application to the core centre of Bigten, the Los Alamos 10 %-enriched critical cylinder; coarse resolution neutron spectra derived from integral data agree exceedingly well with Li(n,$\alpha$) neutron spectrometry results [9].

Conceptually, the Fispec approach defines a ratio between two neutron spectra on the basis of measured integral reaction rate ratios. Another example is illustrated by Figures 10 and 11 : here, the uranium-235 fission neutron spectrum (full line on Fig. 11) is accurately and unchallengeably linked to the Cf spontaneous fission neutron spectrum by the measurements displayed on Figure 10.

Equivalent fission fluxes, if properly measured and interpreted, allow to separate neutron transport effects from both source and detector cross-section inadequacies. Only in this way could one obtain firm evidence (Fig. 12) of a significant bias of the ENDF/B-IV nuclear data of U and U near 2 MeV (probably : inelastic scattering cross sections). Also indicated is a likely deficiency of the Np fission cross section in the ENDF/B-V dosimetry file (Fig. 13). These findings are extensively documented elsewhere [1].


## ACKNOWLEDGMENTS

The authors want to thank Mr. A. PALAMIDESSI (ENEA), Messrs G. STIENNON and J. DEBRUE (SCK/CEN) for their support and encouragement. They are indebted to the reactor operation teams, under the leadership of G. ROSSI at Tapiro and R. MENIL at BR1, for their competent and dedicated help. They appreciate the editorial attention and care of D. LANGELA and S. LENAERTS.


## REFERENCES


[1] A. FABRY et al., "Neutronic Characterization of the Tapiro Fast-Neutron Source Reactor," Final report (in print).

[2] W. SCHOLTYSSEK et al., Proc. Int. Reactor Physics Conf., 1988, Jackson Hole, WY, USA, Vol. 2, p.271.

[3] M. ANGELONE, P. D'HONDT, Nucl. Instr. Meth. in Physics Research, A286 (1990) 301-305.

[4] M. ANGELONE et al., "Radiometric Measurement of Cumulative Mass Chain Fission Yields for Uranium-235 and Plutonium-239 from Hard Fast Reactor Core Spectrum to Reflector/Blanket Degraded Neutron Spectra," This Symposium.

[5] A. D'ANGELO et al., Energia Nucleare (Milan) 20,11,612(1973).

[6] A. FABRY et al., Nucl. Techn. 25, 349 (1975).

[7] G. DE LEEUW-GIERTS, S. DE LEEUW, Proc. Third ASTM-Euratom Symposium on Reactor Dosimetry, EUR 6813 EN-FR, Vol.II, 971 (1980).

[8] G. DE LEEUW-GIERTS et al., "Comparison of theoretical 40 gr. DTF-IV and experimental Li neutron spectra and deduced threshold reaction rates in the centre of two spherical Cu-configurations," Final report (in print).

[9] G. DE LEEUW-GIERTS et al., ibid [7], EUR 6813 EN-FR, Vol.II, p.998 (1980).

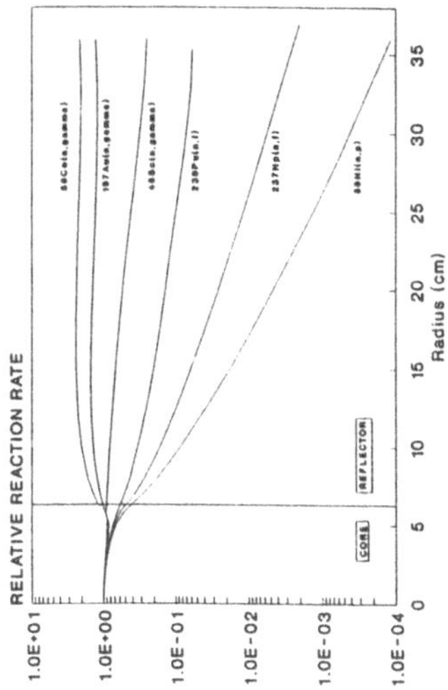


Fig.1. TAPIRO REACTION RATE TRAVERSES
INTEGRAL DOSIMETRY SENSORS

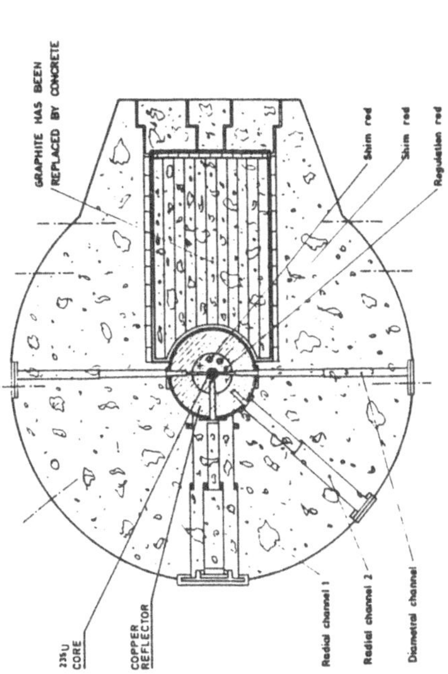


Fig. 2. HORIZONTAL SECTION OF TAPIRO REACTOR AT CORE MIDPLANE : GENERAL CONFIGURATION FOR THE NEUTRONIC CHARACTERIZATION PROGRAM PLAN

Fig.3. MOL COPPER SHELL BENCHMARK : Vertical sketch.

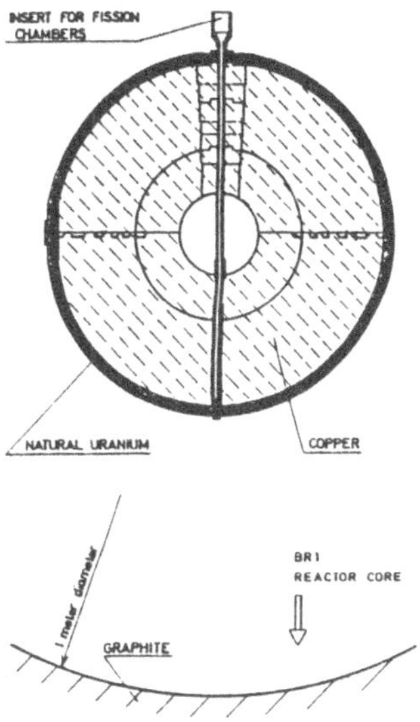


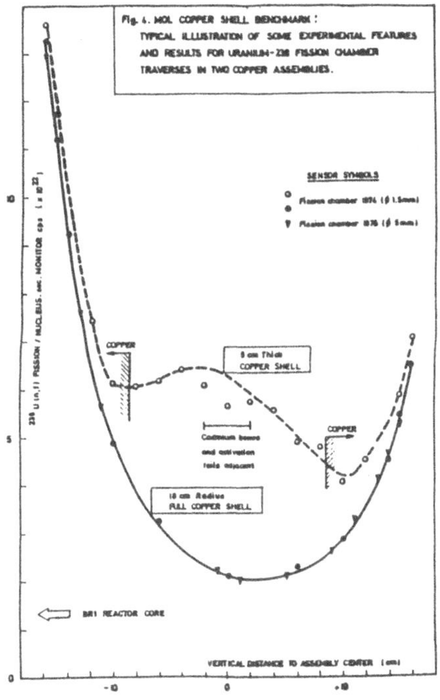


Fig. 4. MOL COPPER SHELL BENCHMARK : TYPICAL ILLUSTRATION OF SOME EXPERIMENTAL FEATURES AND RESULTS FOR URANIUM-238 FISSION CHAMBER TRAVERSES IN TWO COPPER ASSEMBLIES.

Fig.5. CONVERGENCE OF ONE-DIMENSIONAL DISCRETE-ORDINATES TRANSPORT THEORY ABOVE 6 MeV IN A TAPIRO-TYPE REFLECTOR

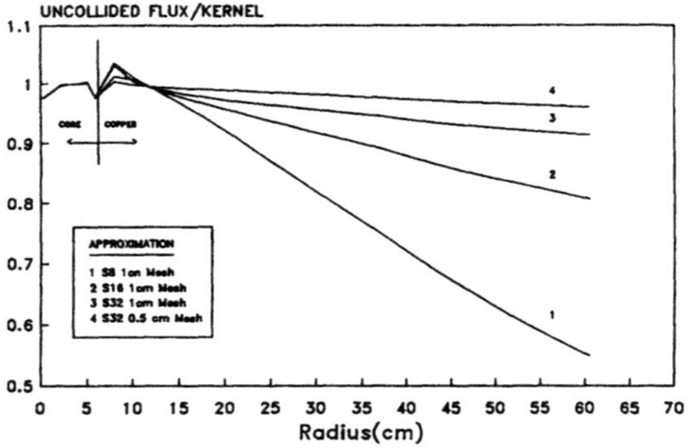


Fig. 6. REMOVING THREE-DIMENSIONALITY OF MOL SHELL BENCHMARK
Unperturbed 115In(n,n') Traverses

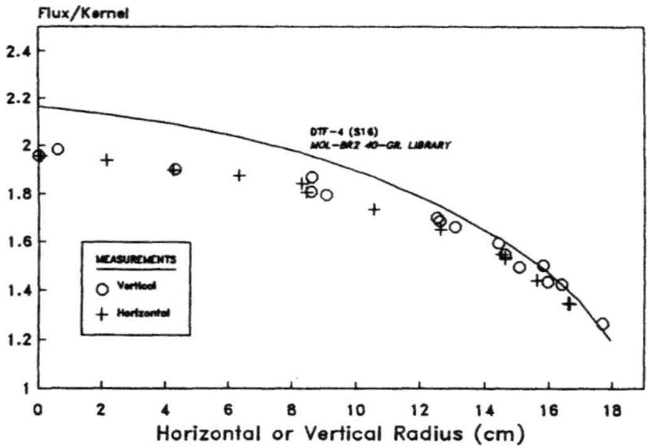


Fig. 7. SELECTED FLUX-TO-KERNEL RATIOS: TAPIRO vs. MOL SHELL

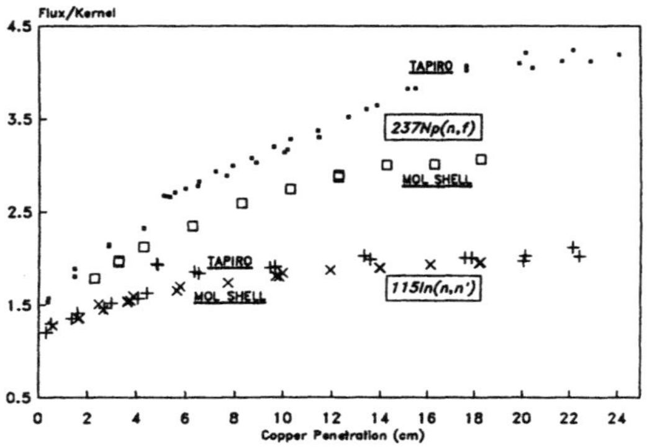


Fig. 8. TAPIRO AND MOL COPPER SHELL : C/E
*Threshold Range: Physically-grounded Interpretation*

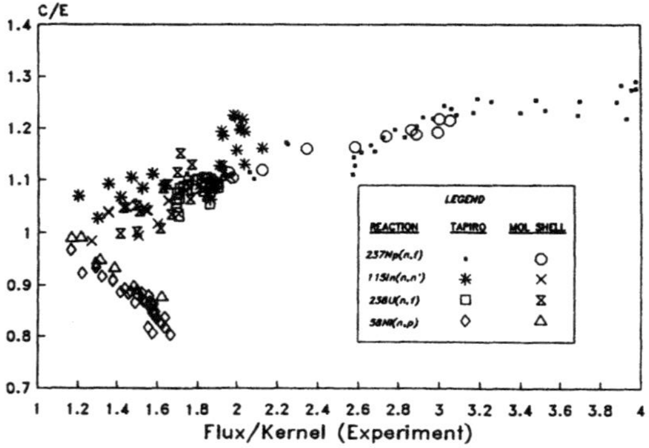

Fig. 9. TAPIRO: SUMMARY OF C/E RATIOS FOR INTEGRAL SENSORS

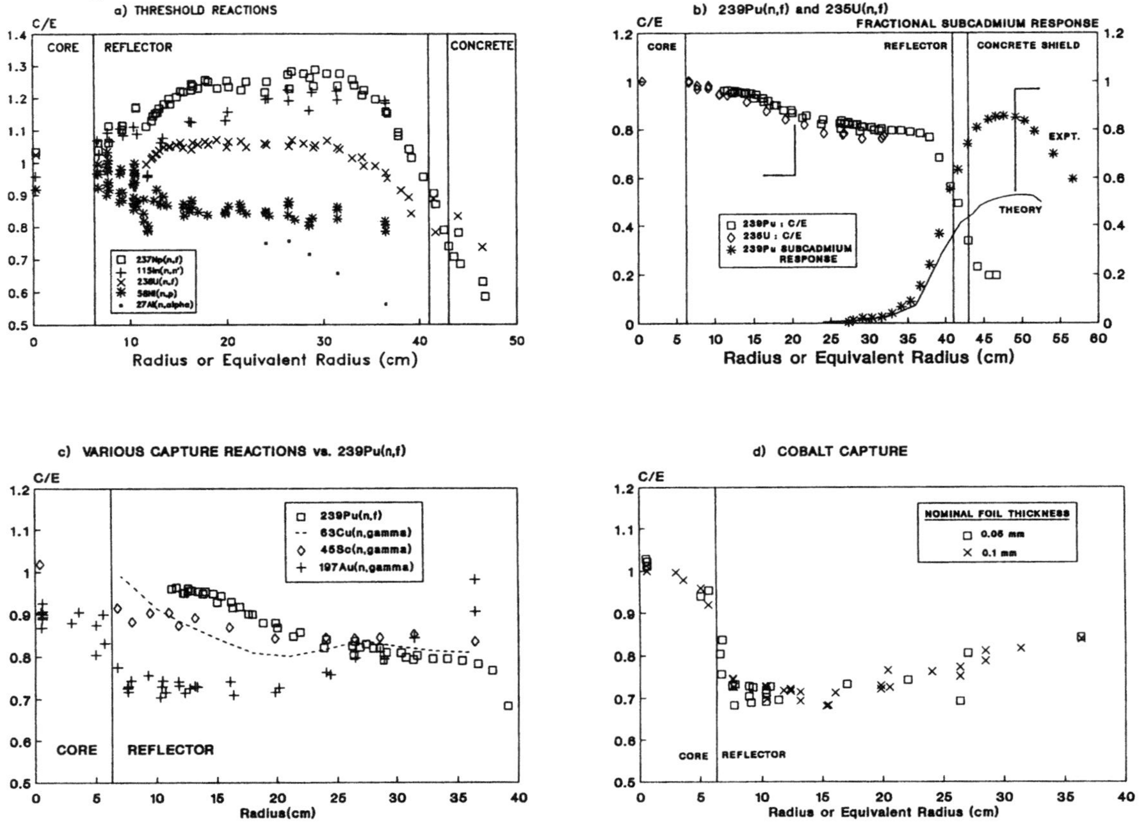

**Fig. 10. 252Cf AND 235U(n,f) FISSION NEUTRON SPECTRA**
RATIO 252Cf/235U

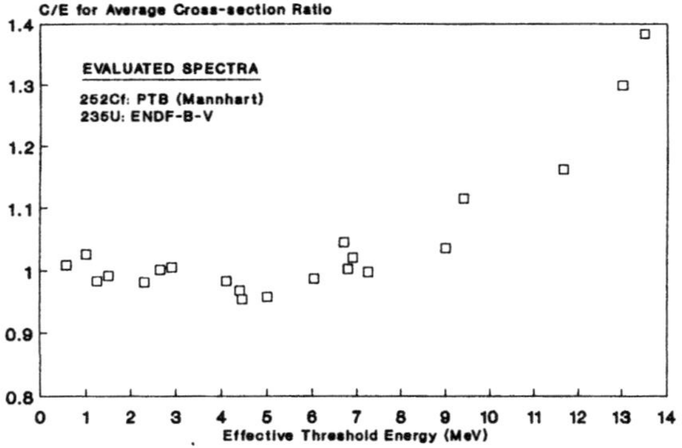


**Fig. 11. EVALUATION OF 235U THERMAL-NEUTRON INDUCED FISSION NEUTRON SPECTRUM**

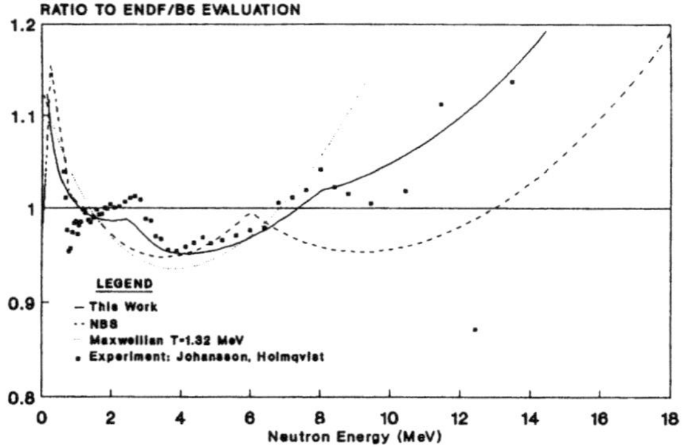


**Fig. 12. EVALUATION OF TAPIRO AND BIGTEN**
HIGH-ENERGY SPECTRAL TRENDS AT CORE CENTRE

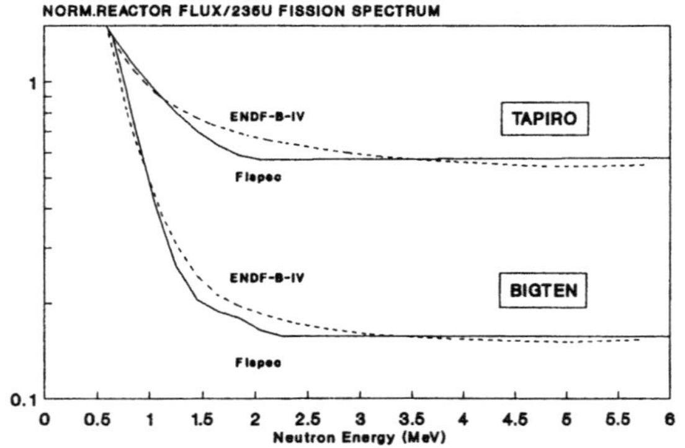


**Fig. 13. 237Np FISSION CROSS-SECTION**

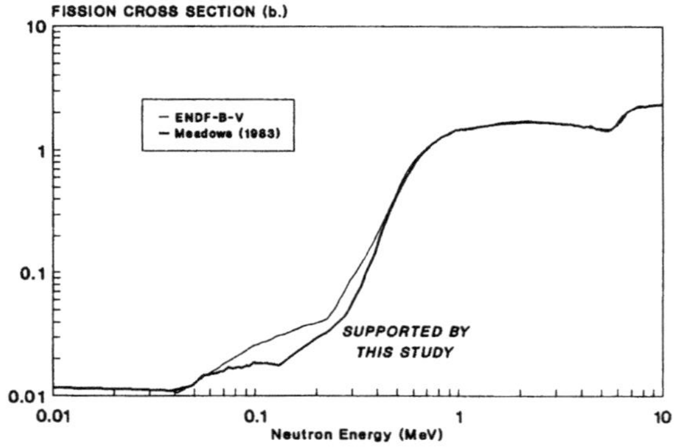

# DAMAGE CORRELATIONS

# GAMMA-RAY INDUCED DISPLACEMENT IN D2O REACTORS

N. P. Baumann

Westinghouse Savannah River Company
Savannah River Laboratory
Aiken, South Carolina, USA

## ABSTRACT

Gamma-ray damage to tank walls is typically more severe in D2O than in H2O moderated lattices because of the much higher ratios of slow-to-fast neutron flux. To estimate this effect it was first necessary to develop energy dependent gamma-ray displacement cross sections for iron. These, along with coupled neutron-gamma-ray transport computations, provided a measure of displacement damage from this source in SRS reactor tank walls. Gamma-ray displacements originating from high energy gammas from neutron capture in and near the tank wall exceeded those from gamma rays created in the reactor core. The displacements from the combined gamma sources ranged from 13% to 16% of that due to iron atom recoil following neutron capture.

## INTRODUCTION

Irradiation of metals by neutrons, gamma rays, or charged particles generally results in damage such as embrittlement and swelling. The damage is due to displacement of metal atoms. To equate damage from the various mechanisms, it is necessary to have a common scale. The concept of Displacements Per Atom (DPA) has been developed, in part, for this purpose (c.f., M. T. Robinson [1]).

High energy gamma rays cause damage in metals from the electrons set in motion by Compton scattering and pair-formation interactions. In light-water power reactors damage from high energy neutrons dominates, primarily because of the high ratio of fast-to-slow neutron flux at the pressure vessel wall. In heavy water reactors, however, this ratio may be reversed and it may be necessary to include the effect of gamma rays along with thermal neutron $(n,\gamma)$ reactions.

The concept of a "displacement cross section" has been developed for various neutron interactions [2]. The displacement cross section for a material is simply the product of a conventional neutron reaction cross section times the average number of displacements per reaction, summed over all relevant reactions. This paper develops a similar cross section for gamma-ray interaction with iron and applies it to SRS (Savannah River Site) reactor tank-wall conditions.

## DISCUSSION

A recent article by Gold, Roberts, and Doran [3] treats the subject of gamma-ray damage. In this paper, the cross section is evaluated in terms of initial beta energy in order to apply it to data from a solid-state detector. For our application, the starting point is a computed gamma-ray spectrum. It was thus necessary to repeat and extend Gold et al., results to obtain the desired gamma-ray displacement cross sections. Their analysis procedure is followed in the present paper except that cross sections are evaluated for one orbital iron electron rather than for one iron atom. A factor of 26 relates the two cross sections for iron. For common stainless steels the average per atom displacement cross section is closely proportional to the average number of orbital electrons. The following steps were employed.

1. Compute the differential rate of energy loss, dE/dX, as a function of electron energy for electrons in iron.

2. Obtain the total number of iron atom displacements as a function of initial electron energy by taking tabulated differential displacement cross sections for electrons and integrating over the distribution of step 1.

3. Compute the energy distribution of scattered electrons and the total cross section for Compton events as a function of gamma-ray energy.

4. Compute the energy distribution of electrons and positrons and the total cross section for pair formation as a function of gamma-ray energy.

5. Compute the number of displacements per Compton event as a function of gamma-ray energy by convoluting steps 2 and 3.

6. Compute the number of displacements per pair formed as a function of gamma-ray energy by convoluting steps 2 and 4.

7. Obtain the displacement cross section per electron (of the iron) as a function of gamma-ray energy by multiplying the results of step 5 by the total Compton cross section of step 3 and summing this with the corresponding product (steps 6 and 4) for pair formation.

(Step 1) Stopping Power for Electrons and Positrons in Iron

A recent compilation of stopping powers for electrons and positrons is given by Seltzer and Berger [4]. These are given in terms of energy dependent coefficients which can be manipulated to derive conventional dE/dX values.

(Step 2) Atom Displacements in Iron by Electrons

The concept of "effective displacement threshold energy" is used in estimating damage due to atom displacements. A value of 40 eV as the threshold energy for iron and stainless steels is now generally used. A threshold of 28 eV was also included to illustrate sensitivity to this parameter.

Differential displacement cross sections, sigma(E), for electrons have been evaluated be Oen [5] for a broad range of elements and threshold energies. The energy grid given by Oen is quite irregular so that graphical interpolation was necessary to perform the numerical integration of the expression:

$$n(E_b) = \int N_a \text{Sigma}(E) dE/(-dE/dX) \qquad (1)$$

where $n(E_b)$ is the total number of iron displacements for an electron of initial energy $E_b$, $N_a$ is the density (atoms/cm$^3$)of iron. Sigma (E) is Oen's displacement cross section, and (dE/dX) is the stopping power.

The results are plotted in Figure 1 for 40 eV and 28 eV thresholds. They are shown both for primary displacements from direct electron interaction and including secondary displacements from the motion of the displaced iron atom.

(Steps 3 and 5) Displacements per Compton Electron

Compton scattering is well understood and is discussed in many standard text books (c.f., [6]). The equation for the differential cross section is a straightforward algebraic equation with the energy of the scattered electron as the variable. If this differential cross section is divided by the integrated total cross section one obtains the conditional probability function

$$P_c(E_g,E_b) = \text{Sigma}(E_g,E_b)/\text{Sigma(total)} \qquad (2)$$

where $P_c$(Eg,Eb) is the probability that if a gamma ray with initial energy $E_g$ undergoes Compton scattering the scattered electron will have energy $E_b$.

The average number of displacements, $N_c(E_g)$, per Compton event with an initial gamma-ray energy $E_g$ is then given by the integral of the products of equations 1 and 2.

$$N_c(E_g) = \int P_c(E_g,E_b)n(E_b)dE_b \tag{3}$$

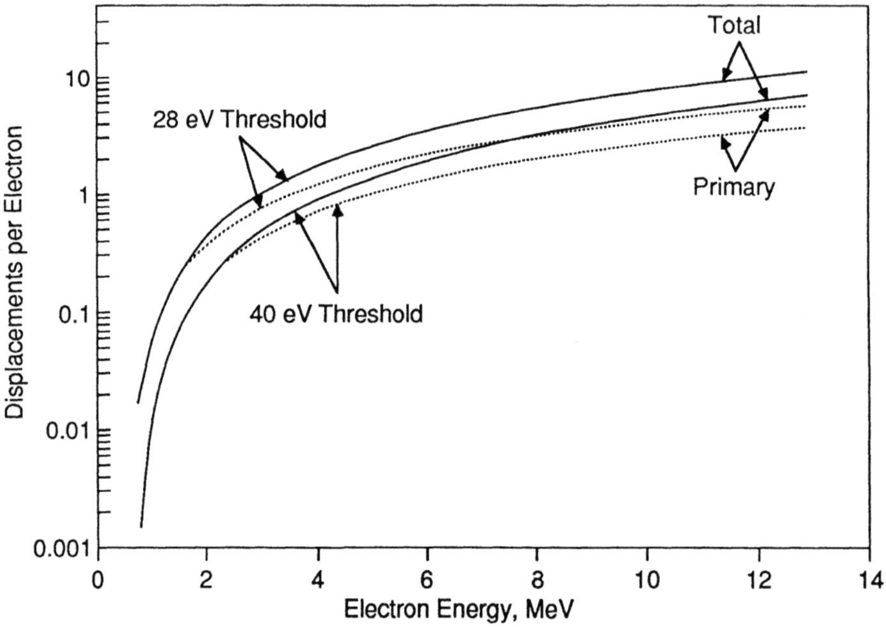


**Figure 1. Electron-Induced Displacements in Iron**

M91feb028.01
02-21-91

(Steps 4 and 6) Displacements per Pair Formed

In contrast to Compton scattering, pair formation is strongly dependent on the nucleus with which the electron is associated. Except for the energy of 1.022 MeV required to form the pair, the entire energy of the gamma ray goes into kinetic energy of the electron and positron. This energy is not equally divided, but covers the entire range of partition. A rigorous theoretical treatment has not been developed for pair formation, but approximate formulae are given by Hough [7] which just cover the region of interest here. A probability function can be defined as for Compton events as

$$P_p(E_g,E_{fract}) = Sigma(E_g,E_{fract})/Sigma(total) \quad (4)$$

where $E_{fract}$ is the fraction of total kinetic energy taken up by the positron and the total sigmas are obtained by numerical integration over the allowed range of positron energies, or more accurately by a separate formula given by Hough [7]. The average number of displacements, $N_p(E_g)$, per pair-formation event with gamma energy $E_g$, is obtained by the integration of the product of equations 1 and 4.

$$N_p(E_g) = \int 2P_p(E_g,E_{fract})n(E_b)dE_b \quad (5)$$

Here the integration is carried out from zero to the gamma-ray energy minus 1.022 MeV. The factor of two accounts for both electron and positron. No distinction is made for the small difference in the two stopping powers. The results are included in Table I.

Table I
Gamma-ray Interaction Cross Sections on Iron

| Eg, MeV | Reaction Xsection* Pair | Reaction Xsection* Compton | Displacements per Event** Pair | Displacements per Event** Compton | Total Displacement Xsection* |
|---|---|---|---|---|---|
| 0.5 | 0.0 | 289.3 | 0.0000 | 0.0000 | 0.0 |
| 1.0 | 0.0 | 211.3 | 0.0000 | 0.0010 | 0.2 |
| 1.5 | 1.1 | 171.6 | 0.0000 | 0.0171 | 2.9 |
| 2.0 | 4.6 | 146.4 | 0.0039 | 0.0568 | 8.3 |
| 2.5 | 8.8 | 128.6 | 0.0320 | 0.1203 | 15.8 |
| 3.0 | 13.1 | 115.1 | 0.0900 | 0.2408 | 24.8 |
| 3.5 | 17.4 | 104.6 | 0.1763 | 0.3080 | 35.3 |
| 4.0 | 21.3 | 96.0 | 0.2872 | 0.4281 | 47.2 |
| 4.5 | 25.1 | 88.9 | 0.4189 | 0.5633 | 60.6 |
| 5.0 | 28.6 | 82.9 | 0.5715 | 0.7121 | 75.4 |
| 5.5 | 31.9 | 77.7 | 0.7404 | 0.8732 | 91.5 |
| 6.0 | 35.0 | 73.3 | 0.9249 | 1.0448 | 109.0 |
| 6.5 | 38.0 | 69.3 | 1.1233 | 1.2261 | 127.6 |
| 7.0 | 40.7 | 65.8 | 1.3342 | 1.4164 | 147.6 |
| 7.5 | 43.3 | 62.7 | 1.5565 | 1.6134 | 168.6 |
| 8.0 | 46.0 | 59.9 | 1.7891 | 1.8178 | 191.2 |
| 8.5 | 48.5 | 57.4 | 2.0309 | 2.0281 | 214.8 |
| 9.0 | 50.8 | 55.1 | 2.2811 | 2.2451 | 239.5 |
| 9.5 | 53.0 | 53.0 | 2.5390 | 2.4675 | 265.4 |
| 10.0 | 55.2 | 51.0 | 2.8042 | 2.6961 | 292.3 |

* All cross sections are in millibarns ($10^{-31}$ $m^2$) per orbital iron electron.

** Displacements are for 40 eV threshold.

### (Step 7) Total Displacement Cross Section for Gamma Rays

The displacement cross section for a gamma ray undergoing Compton scattering is simply the total Compton cross section times the number of displacements per event. A similar product is made for pair-formation events. These factors, for 40 eV, are included in Table I. The sum of the two products is then the total displacement cross section for iron as a function of gamma-ray energy. The sum for 40 eV is the final column of Table I. Figure 2 graphically displays the total displacement cross section for both thresholds.

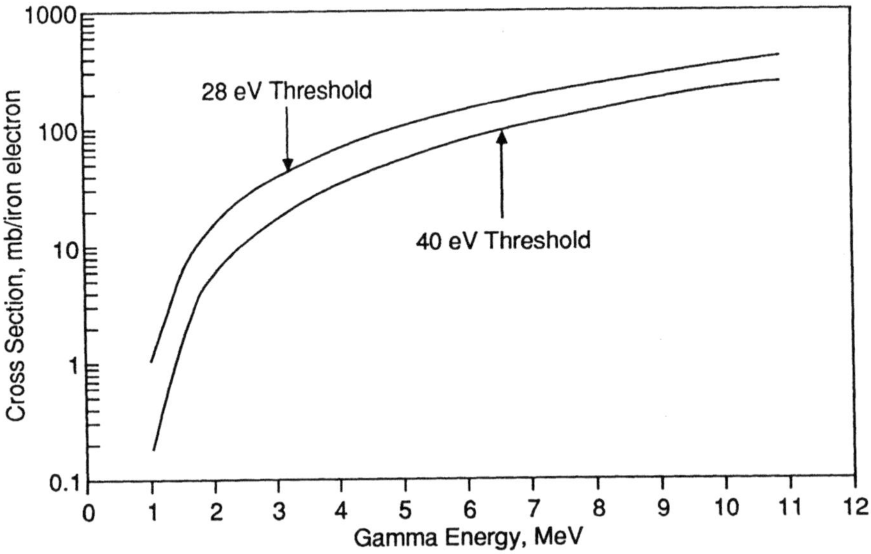


**Figure 2. Gamma-Ray Displacement Cross Section in Iron**

M91feb032.01
02-21-91

### Gamma-Ray Fields in SRS Reactor Tank Walls

A rigorous computation of gamma-ray transport from the core of the reactor to the tank wall requires a code which couples neutrons and gamma rays in an eigenvalue solution. At the present, SRS has no analytic transport code with this capability. The SRS version of ANISN [8] has a coupled 22-group neutron/gamma ray set of cross sections, but performs only "shielding" calculations, which include only first generation fission and capture events. In order to use this code to estimate neutron and gamma-ray transport, the energy and space distribution of the neutron and gamma fluxes in the core must be known a priori. Various SRS codes can calculate the space distribution of the neutron flux fairly accurately. In typical SRS lattices, gamma rays can be assumed to have the same distribution over the core.

For the present study, the lattice selected was a homogenized tritium producing lattice of $D_2O$, U-235, Li-6, and Al with (or without) a homogenized blanket of $D_2O$, Li-6, and Al. The blanket, when present, replaces fuel in the outermost ring of lattice sites. The radial distribution of the source was taken to be flat out to, but not including, the last ring of fuel for both blanketed and unblanketed cases with a reduced source strength in the outer ring. With this distribution, the ANISN computed ratio of thermal neutron flux at the tank wall to that in the core (Table II) agreed well with values computed by more sophisticated reactor transport codes (8.2% and 1.97% vs. 7% and 2%). The reference thermal neutron displacement cross section in Table II is that for iron atom recoil following neutron capture and gamma emission.

Table II
Thermal Neutron and Gamma Fluxes in Reactor Tank Wall
(Normalized to thermal neutron flux in core)

| Group | No Blanket Normal | No Blanket Poisoned | With Blanket Normal | With Blanket Poisoned | Displ. Xsctn Barns |
|---|---|---|---|---|---|
| Thermal Neutrons | .08200 | .000952 | .01972 | .0001112 | 11.9 |
| Gammas | | | | | |
| 6-10 MeV | .02444 | .00315 | .00638 | .00152 | 3.1 |
| 5-6 | .00639 | .00159 | .00188 | .00080 | 2.4 |
| 4-5 | .00639 | .00375 | .00239 | .00185 | 1.6 |
| 3-4 | .00973 | .00717 | .00387 | .00340 | 1.0 |
| 2-3 | .02126 | .01116 | .00721 | .00509 | .41 |
| 0.9-2 | .07033 | .05626 | .02521 | .02272 | .07 |
| Neutron Disp'ments | .976 | .011 | .229 | .0013 | |
| Gamma Disp'ments | .1245 | .0352 | .0367 | .0176 | |

In Table II, the two columns labelled "poisoned" were special computational cases used to separate the effect of core gamma rays from that of capture gamma rays originating at or near the tank wall. For these cases, strong Li-6 absorber, with no capture gamma rays, was added in the vicinity of the tank and shield walls. This virtually eliminated neutron capture in the tank and shield walls without significantly affecting the direct gamma transport from the core. The poisoned and unpoisoned cases can then be algebraically manipulated to separate the neutron capture and transport components.

Normalized to displacements from gamma recoil, the contribution for the full-core lattice (unblanketed) is 9.3% from neutron capture in the wall and shield and 3.4% from core transport. The corresponding values for the blanketed lattice are 8.4 % and 7.8% respectively. The contribution from neutron capture in the wall and shield, about 9%, is nearly independent of core configuration. The component for gamma rays transported from the core does show a strong dependence on lattice configuration. The value of 7.8% for the blanketed tritium producer lattice is near maximum for the various historic lattices at SRS. Although the damage from gamma displacements will never be dominant, in portions of the SRS tank walls it can exceed 10% of the total. This is just large enough to be included in damage estimates. A generally conservative, but reasonable, inclusion of gamma ray damage in SRS reactors can be made by simply augmenting the gamma recoil cross section for thermal neutrons by 15%.

## ACKNOWLEDGEMENT

The information contained in the document was developed during the course of work under Contract No. DE-AC09-89SR18035 with the U.S. Department of Energy. By acceptance of this paper, the publisher and/or recipient acknowledges the U.S. Government's right to retain a nonexclusive, royalty-free license in and to any copyright covering this paper along with the right to reproduce, and to authorize others to reproduce all or part of the copyrighted document.

## REFERENCES

[1] M. T. Robinson, "The Theory of Radiation Induced Defect Production", International Conference on Radiation Damage in Metals, Gatlinburg, TN, (CONF-751006-15), October 1975.

[2] D. G. Doran, "Neutron Displacement Cross Sections for Stainless Steel and Tantalum Based on a Lindhard Model", Nucl. Sci. Eng.,49, pp 130-144, 1972.

[3] R. Gold, J. H. Roberts, and D. G. Doran, "Determination of Gamma-Ray-Induced Displacement Rates", Reactor Dosimetry: Methods, Applications, and Standardization, ASTM STP 1001, Harry Farrar and E. P. Lippincott, eds, American Society for Testing and Materials, Philadelphia, pp 603-613, 1989.

[4] S. M. Seltzer and M. J. Berger, "Evaluation of the Collision Stopping Power of Elements and Compounds for Electrons and Positrons", Int. J. Appl. Radia. Isot. 33, pp 1189-1218, 1982.

[5] O. S. Oen, "Cross Sections for Atomic Displacements in Solids by Fast Electrons", U.S.A.E.C. Report ORNL-4897, 1973.

[6] E. Segre, "Experimental Nuclear Physics", Vol I, pp 317-325, 1953, John Wiley and Sons, Inc, New York.

[7] P. V. C. Hough, "Low Energy Pair Formation", Phys. Rev. 74, 574, 1948.

[8] W. W. Engle, Jr., "A User's Manual for ANISN", ORNL Report K-1693, March 30, 1967.

# DPA VERSUS FLUENCE (E > 1 MeV) FOR REACTOR VESSEL WALL ATTENUATION

A. L. Lowe, Jr. and N. L. Snidow

B&W Nuclear Service Company

Engineering & Plant Services Division

Lynchburg, Virginia, U.S.A.

## ABSTRACT

In assessing the integrity of reactor pressure vessels, the attenuation of the neutron flux through the vessel wall establishes the relative fracture toughness at the various locations within the wall. Traditionally, the damage profile has been based on the attenuation of the flux with energies (E > 1 MeV). However, it has been suggested that the correct approach should use displacement per atom (dpa). The data from available experiments are evaluated. These data show that the exposure parameter of dpa is the most conservative and that the fluence of E > 1 MeV is a reasonable approximation of the attenuation of the properties. Data from the Gundremmingen trepans are compared with the calculated fluence, dpa, and the results based on Regulatory Guide 1.99. The attenuation according to Regulatory Guide 1.99, Revision 2, but based on the calculated fluence is the best representation of the data. The available experimental evidence suggest the damage attenuation in the reactor vessel wall be based on the calculated fast fluence (E > 1 MeV).

---

## 1. INTRODUCTION

One of the important considerations in the study of radiation embrittlement in reactor vessels is the rate at which the radiation damage attenuates going through the wall of the reactor vessel. It is reasonable to assume that the embrittlement caused by neutrons is dependent on the energy of the neutrons. The damage is not only a function of the number of neutrons that enter the vessel, but it is also a function of the energy distribution of the neutrons, that is, the energy spectrum. Measurements of embrittlement in specimens irradiated in surveillance capsules have been obtained in very similar spectra. Furthermore, the spectra in surveillance capsules are similar to those at the inside surface of the reactor vessels. The measured data can, therefore, be applied to the inside surface using any parameter for the exposure index. The situation is quite different, however, inside the reactor vessel wall. The neutron spectrum shifts with penetration into the wall and the measured data from surveillance capsules can only be applied properly if the relationship between damage in different spectra is understood.

In this report, the several aspects of neutron embrittlement and attenuation are reviewed. The attenuation through the wall of a typical B&W reactor vessel of several variables is examined in Section 2. Next, the attenuation specified in Regulatory Guide 1.99, Rev. 2, (1) is discussed in Section 3. Then various attenuation models are compared with experimental data in Section 4. Finally, in Section 5, a damage cross-section based on a model by Nichols is reviewed and a modification to Nichols' cross-section is discussed.

## 2. ATTENUATION OF VARIABLES THROUGH THE REACTOR VESSEL WALL

The attenuation through the reactor vessel wall of several variables related to embrittlement is examined in this section. The calculated values of neutron flux and dpa rate are typical for B&W reactor vessels. The dpa rate was calculated using the displacement cross-section for iron given in ASTM Standard Practice E693-79 (2) and where dpa rate issued without an energy given it means total dpa, that is, the value includes the contributions to dpa from all neutrons. The energy limit is specified when dpa is obtained from neutrons above and specified energy.

Figure 1 shows the attenuation of flux with E > 1 MeV; flux with E > 0.5 MeV, total dpa rate, dpa rate from neutrons with E > 0.5 MeV; and $e^{-0.24x}$ where x is the distance into the vessel wall in inches. (See Section 5 for selection of 0.5 MeV.) Each variable is normalized to 1.0 at the inside surface. Figure 1 shows that flux with E > 1 MeV and dpa rate from neutrons with E > 0.5 MeV fall off at approximately the same rate. The other three are grouped together and fall off at a slower rate. At 3T/4, flux with E > 1 MeV and dpa rate from neutrons with E > 0.5 MeV are approximately 37% lower than the other three.

Some insight into the difference can be gained by examining Figures 2 and 3. Figure 2 shows the neutron flux in three energy ranges: E > 1.0 MeV, 0.1 MeV < E < 1.0 MeV, and 0.00325 MeV < E <

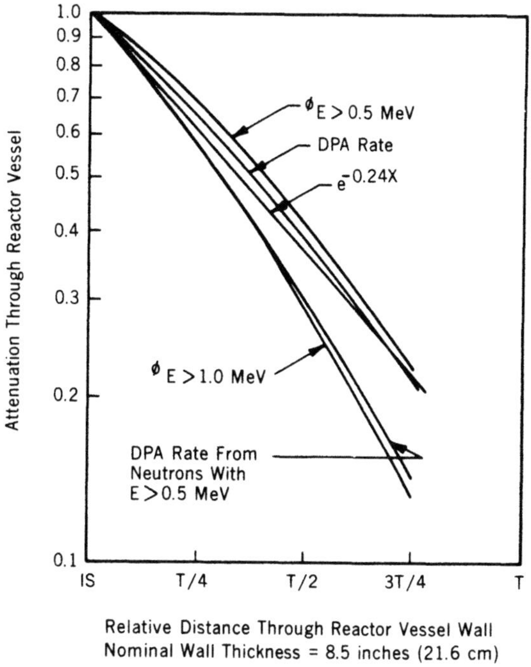


Figure 1. Calculation of the Flux and DPA Rate Attenuation Through a Reactor Vessel Wall

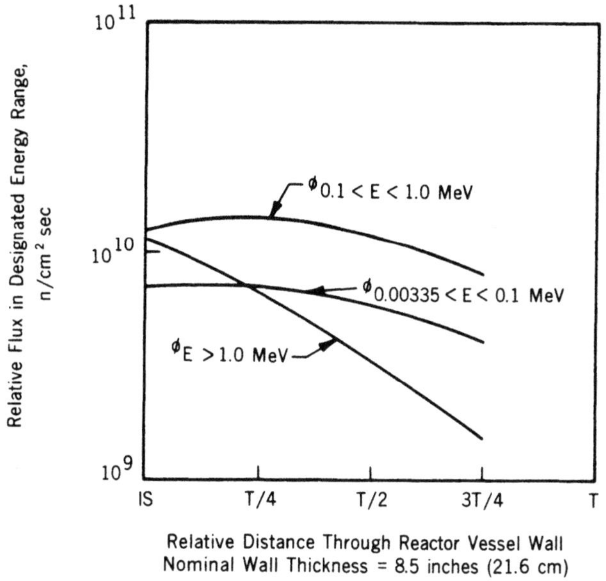


Figure 2. Calculation of the Attenuation of Neutron Fluxes of Different Energy Ranges Through a Reactor Vessel Wall

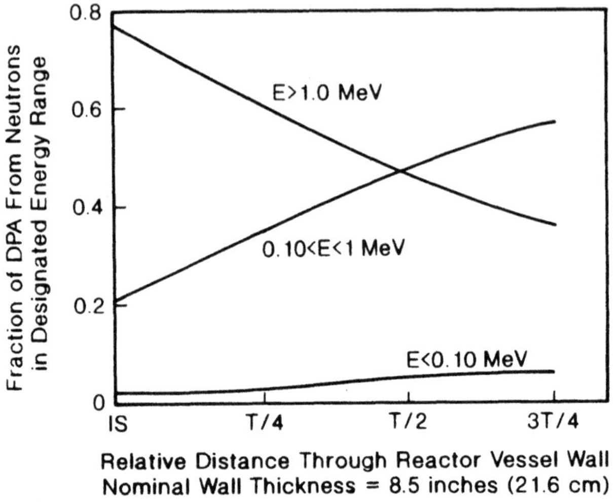


Figure 3. Calculation of the DPA Rate & Contributions to DPA Rate from 3 Energy Ranges As Attenuated Through Reactor Vessel Wall

0.1 MeV. The flux above 1 MeV attenuates at a much faster rate than does the flux in the other two energy ranges. The flux above a threshold energy such as 1 MeV, 0.5 MeV, or 0.1 MeV, therefore, attenuates faster the higher the threshold. The contributions to

The dpa rate from neutrons in the same three energy groups as defined above are shown as a fraction of the dpa rate through the vessel wall in Figure 3.

The contribution to dpa rate from neutrons below 1 MeV is considerable. The fraction of the dpa from neutrons in the 0.1 MeV to 1.0 MeV range increases from 21% at the inside surface to 57% at the 3T/4 point. It is easy to see from this, why DPA rate attenuates at a slower rate than flux with E > 1 MeV.

Approximate uncertainties in fluence and dpa as a function of distance through the vessel wall are illustrated in Figure 4. It is assumed that

$$\phi_{E > 1}(x) = g(x)\, \phi_{E > 1}(o) \tag{1}$$

the uncertainty in $\phi_{E > 1}$ (o) is taken as ± 20%, and the uncertainty in g (x) as 0% at x = 0 and ± 10% at x = 3T/4. Combining the uncertainties as the square root of the sum of squares gives an uncertainty in $\phi_{E > 1}$ at x = 3T/4 of ± 22%. The ratio of DPA rate to flux may be written as

$$\frac{\text{DPA Rate}}{\phi_{E > 1}}(x) = \frac{\text{DPA Rate}}{\phi_{E > 1}}(o)\; q\,(x) \tag{2}$$

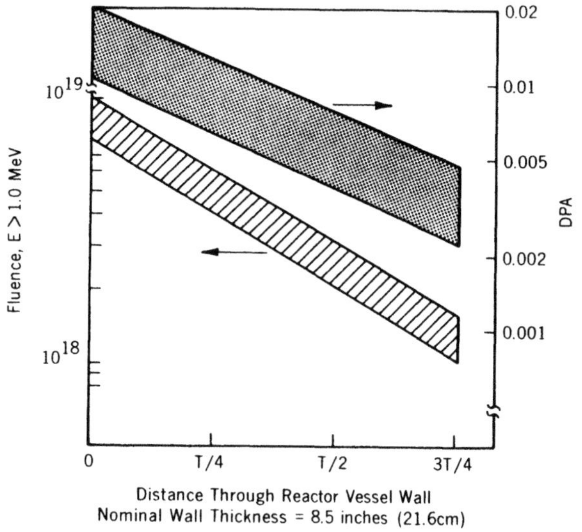


Figure 4. Calculation of Uncertainties in Both Fluence and DPA As a Function of Distance Through a Reactor Vessel Wall

It was assumed that the uncertainty in dpa rate/$\phi_{E>1}$ at x = o is ± 35%. This value is due primarily to the uncertainty in the energy required per displacement. The uncertainty in the spatial term, q (x) was taken as 0% at x = o and ± 10% at x = 3T/4. Combining the above uncertainties statistically gives an uncertainty in dpa rate (o) of ± 40% and in dpa rate (3T/4) of 43%.

Uncertainties in the ratio of flux at x to the value at x = o is just the uncertainty in g (x) and is shown in the lower part of Figure 5. The uncertainty in the ratio of dpa rate at x to dpa rate at x = o is obtained from the uncertainty in the product of g (x) and q (x) and is 0% at x = o and ± 14% at x = 3T/4. Uncertainty bands such as those in Figure 5 are appropriate for the curves in Figure 1.

## 3. ATTENUATION IN REGULATORY GUIDE 1.99, REV. 2

Two measures of radiation embrittlement are considered in Regulatory Guide 1.99: Revision 2 the adjustment of the reference temperature, $\Delta RT_{NDT}$, and the decrease in the Charpy upper-shelf energy level. The attenuation of radiation embrittlement through the reactor vessel will be considered here with respect to the adjustment in reference temperature only. The adjusted reference temperature, ART, is given by Regulatory Guide 1.99 as

$$ART = \text{Initial } \Delta RT_{NDT} + \Delta RT_{NDT} + \text{Margin} \tag{3}$$

$\Delta RT_{NDT}$ is the mean value of the adjustment in reference temperature

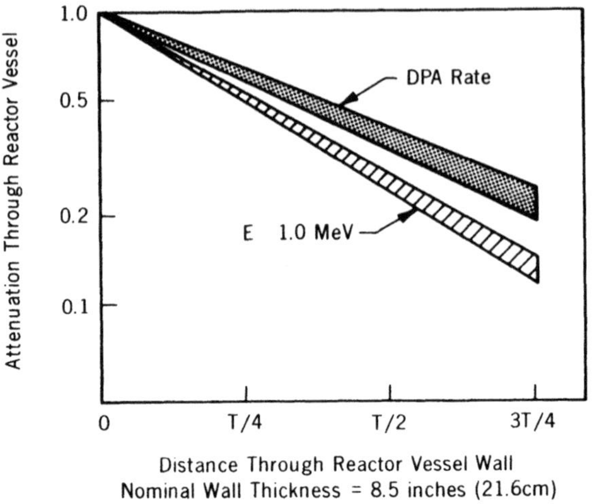


Figure 5. Calculation of Uncertainties in Attenuation of Fluence and DPA As a Function of Distance Through Reactor Vessel Wall

caused by irradiation and according to the Regulatory Guide should be calculated by

$$\Delta RT_{NDT} = CF\ f^{(0.28\ -\ 0.10\ \log f)} \tag{4}$$

where CF is a chemistry factor dependent on the copper and nickel content and f is the neutron fluence with E > 1.0 MeV in units of $10^{19}$ $cm^{-2}$. Equation (4) is basically a fit to surveillance data from commercial power reactors and applies specifically in the capsules and to the inside surface of the reactor vessel to the extent that the spectrum is the same as in the capsules. Since the spectrum is significantly different in the reactor vessel wall from the spectrum at the vessel surface, Equation (4) with f representing fluence with E > 1 MeV does not apply in general. Equation (4) can be made to apply in the wall by changing the meaning of f. Suppose D is an exposure index which is also a perfect damage monitor, that is, when the exposure index D has a given value, the damage is the same regardless of the spectrum that existed during the exposure. Now, let

$$f\ (x) = f_{surf}\ \frac{D\ (x)}{D_{surf}} \tag{5}$$

where D (x) is the value of the perfect damage exposure index at x inches into the wall, $D_{surf}$ is the value of D at the inside surface; and $f_{surf}$ is the fluence with E > 1 MeV at the inside surface. Note that f (x) in Equation (5) is not the fluence with E > 1 MeV at x. Substituting f (x) from Equation (5) into Equation (4) makes $\Delta RT_{NDT}$ a function of x and gives the correct value for $\Delta RT_{NDT}$ within the wall.

Unfortunately, a perfect damage monitor has not been found. The attenuation in Regulatory Guide 1.99, Rev. 2 is based on assuming that dpa is that perfect monitor. Equation (2) becomes for this case

$$f\ (x) = f_{surf} \frac{DPA\ (x)}{DPA_{surf}} \tag{6}$$

Calculations of dpa through commercial power reactor vessel walls has shown that approximately

$$\frac{DPA\ (x)}{DPA_{surf}} = e^{-0.24x} \tag{7}$$

where x is the distance into the wall in inches.

With this, Equation (6) becomes

$$f(x) = f_{surf}\ e^{-0.24x} \tag{8}$$

Regulatory Guide 1.99, Rev. 2 accepts f (x) given by either Equation (6) or Equation (8).

In the next section, experimental data for $\Delta RT_{NDT}$ (x) at various depths in ferritic steel will be compared with calculated values of $\Delta RT_{NDT}$ (x) from Equations (1) and (2) using various exposure indexes for D in an attempt to determine which exposure index fits the data best.

## 4. COMPARISON OF MEASURED $\Delta RT_{NDT}$ WITH VARIOUS ATTENUATION MODELS

The variation in $\Delta RT_{NDT}$ with x determined by Equations (4) and (5) is dependent on the damage exposure index used in Equation (5). There are two sets of measured data that can be compared with predictions from Equations (4) and (5) using various parameters as the exposure index.

One set of measured data is from the Gundremmingen A (KRB-A) reactor pressure vessel that was shut down in 1977 after ten years of operation (Reference 3). Trepans were taken from the vessel and Charpy impact tests were performed on specimens from various depths in the wall. Since the attenuation is being examined in this report and not the chemistry factor, it is convenient to plot the data in the form of the ratio of the reference temperature shift at depth x to reference temperature shift at the inside surface of the vessel. The experimental data from trepans G and C plotted in Figure 6 were taken from Figure 23 of Reference 3. The ratio of $\Delta RT_{NDT}$ (x)/ $RT_{NDT}$ (o) was determined from Equations (4) and (5) using three

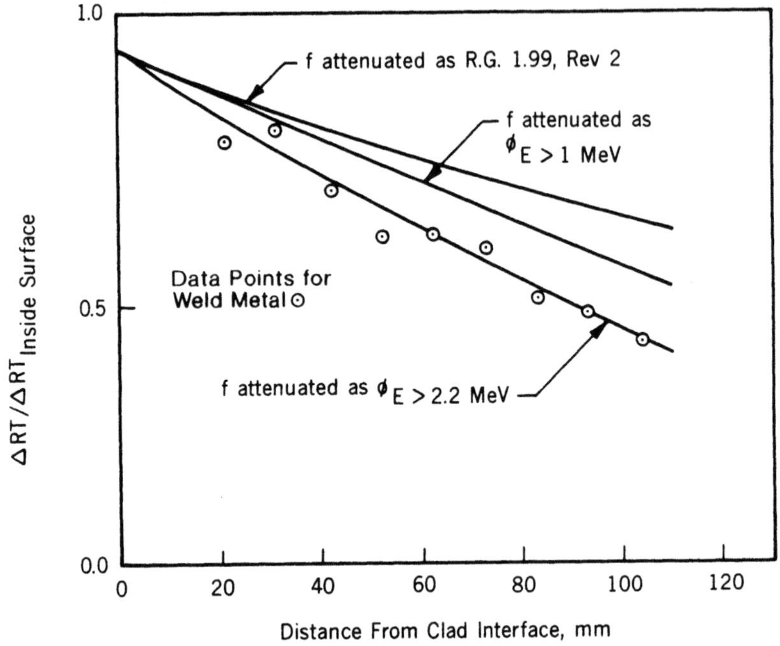


Figure 6. Comparison of Attenuation Methods With the Relative Change in Weld Metal Transition Temperature Properties Through Gundremmingen KRB/A Vessel Wall

models for $D(x)/D_{surf}$ in Equation (5) and the resulting curves are also plotted in Figure 6. The three forms for $D(x)/D_{surf}$ are $e^{-0.24x}$ were x is in inches (from Regulatory Guide 1.99, Rev. 2), $\phi_{E > 1}(x)/\phi_{E > 1}(o)$, and $\phi_{E > 2.2\ MeV}(x)/\phi_{E > 2.2\ MeV}(o)$. The ratio of flux values with E > 1 MeV and E > 2.2 MeV were determined from the spectra in the Gundremmingen vessel calculated by Prillinger (4). The fluence with E > 1 MeV for Trepan C and G at the clad/vessel interface was obtained from Figures 5 and 6 in Reference 4 as $0.309 \times 10^{19}$ $cm^{-2}$.

The attenuation model based on $\phi_{E > 1\ MeV}$ comes closer to matching the experimental data than does the attenuation model based on dpa. Both are conservative with the dpa model being ultra conservative. Attenuation models based on flux greater than a threshold value other than 1 MeV were also examined. The threshold was changed until a match with the experimental data was obtained.

It was necessary to raise the threshold to 2.2 MeV to obtain a match with the experimental data.

Another set of measured data is available for various penetrations through reactor vessel steels from an experiment at the

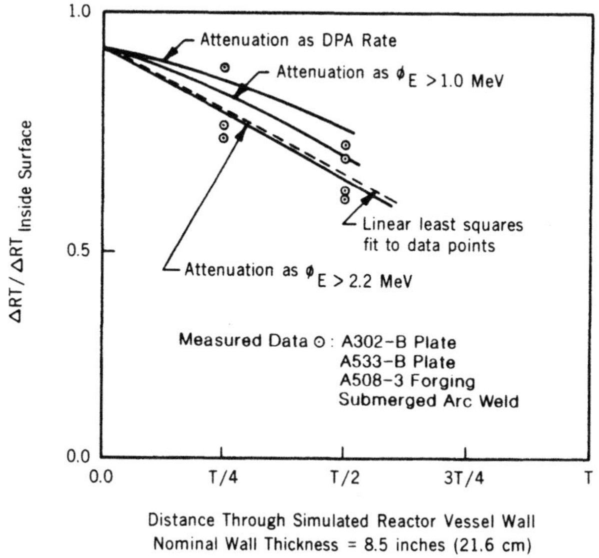


Figure 7. Comparison of Attenuation Methods With the Relative Change in Material Transition Temperatures Through the Simulated Reactor Vessel in the Poolside Experiment

Poolside Facility at Oak Ridge (5). The data from these measurements are plotted in Figure 7 in the same way as for Gundremmingen data. That is, the ratio of temperature shift within the wall to the temperature shift at the surface is plotted versus depth into the wall. The measured data were obtained from the consensus values in Table 3.1, page 3.0-3 of Reference 5. The data plotted in Figure 7 are for the materials listed. Data for 22NiMoCr37 forging and a submerged-arc weld were omitted since their material characteristics are significantly different from the materials used to fabricate U.S. reactor vessels.

The $\Delta RT\,(x) / \Delta RT_{inside\ surface}$ values from Equations (4) and (5) were determined for three attenuation models and were plotted in Figure 7 to compare with the experimental data. Calculated fluence and dpa values were obtained from the Oak Ridge National Laboratory (ORNL) values listed in Table 2.2.2 of Reference 5. The average value for the Charpy specimens of each material type at 0T, 1/4T and 1/2T were used. The data were extended to fluence with E > 2.2 MeV using ratios determined by integrating the spectra listed in Table 2.1.3 in Reference 5 at 0T, 1/4T and 1/2T over energy ranges from 1 MeV up and from 2.2 MeV up. There is considerable spread in the data points: however, it again appears that the attenuation model based on $\phi_{E\,>\,1\ MeV}$ matches the data better than does the model based on dpa rate. One of the eight data points falls above the curve based on dpa and two fall above the curve based on $\phi_{E\,>\,1\ MeV}$. A least squares fit to the data points is shown in Figure 7 and again a model based on $\phi_E$ > 2.2 MeV appears to be closest to the best fit.

## 5. CONSIDERATION OF DAMAGE MECHANISMS

The exposure indexes considered in Section 4 were, for the most part, not based on physical models of a mechanism for producing damage in the vessel material. The dpa index might be considered an exception. However, even if we assume that the physical phenomena considered in calculating the number of displacements per atom have been treated perfectly, the connection between number of displacements and damage has not been made. It nevertheless does seem reasonable, however, that some relationship between displacements and damage probably exist.

The ASTM dpa cross section for iron is based on a calculation of the energy imparted to iron atoms in a lattice from the neutron irradiation.[2] First, the various neutron reactions are considered that can give kinetic energy to the iron atoms. The energy spectra of the iron atoms resulting from direct interaction with the neutrons, called primary knock-on atoms (PKA) were determined for each of the reactions. The number of displacements is the total energy available to displace atoms divided by the energy required per displacement. There is considerable uncertainty in the value of the latter quantity but the total energy going into displacements is probably good.

Nichols proposed a damage model in 1978 that involves more than just the number of displacements (6). He considered the formation of vacancies and interstitial pairs in the cascade of reactions resulting from each PKA and then the number of pairs left after a rapid microannealing process takes place. According to Nichols, this process creates a defect consisting of clusters of vacancies with interstitials relatively far from each other and the vacancies. In the Nichols model the radiation hardening results from such defects acting as barriers to the movement of dislocations. He estimates the size of the defects that are both stable and effective dislocation barriers and concludes that the PKA generating the defect must have at least 15 KeV in energy.

A cross section for the generation of PKA with energy of at least 15 KeV was estimated by integrating the PKA spectra for iron in Table 14 of Reference 7 from 15 KeV up for each neutron energy. This cross-section is shown in Figure 8 where it is compared with the general shape of the dpa cross-section in the 1 to 10 MeV range. The shape of the PKA with $E > 15$ KeV cross section is similar to a response function with a threshold of about 0.4 to 0.5 MeV and a constant value above the threshold except the PKA with $E > 15$ KeV cross-section has the corner rounded off. The attenuation, ds in Figure 1 for a typical B&W reactor spectra, was obtained using the PKA with $E > 15$ KeV cross-section from Figure 8 and the attenuation was found to be similar to that for $E > 0.5$ MeV and DPA. Therefore, it does not appear that the Nichols model without modification will give a better fit to the experimental data in Section 4 then does dpa.

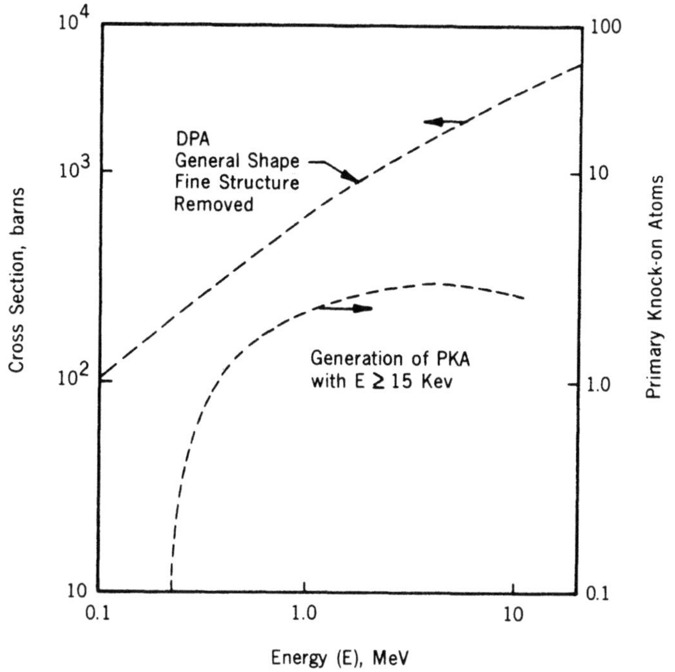


Figure 8. Cross Section for Generation of PKAs with E>15KeV Compared with DPA Cross Section

The PKA with E > 15 KeV cross-section gives a value proportional to the number of PKAs above the threshold. It considers that as long as the energy is above 15 KeV that all PKAs, regardless of energy are equally effective in hindering dislocation movement. An interesting cross-section can be generated retaining the threshold idea of the Nichols model but assuming that the effectiveness of the defects is a function of the size of the defect or say, above the threshold, is proportional to the energy available to create displacements. This cross-section would give an attenuation similar to dpa from neutrons with energy above 0.5 MeV. Shifting the threshold up in energy a bit would give an even better fit to the experimental data in Section 4.

It would appear that this modification of the Nichols model could have considerable promise as a damage monitor.

## 6. CONCLUSIONS

Comparison of the attenuation of $\Delta RT_{NDT}$ through reactor vessel wall as determined by Regulatory Guide 1.99, Rev. 2, with experimental data indicates that the Regulatory Guide, which is based on dpa, is overly conservative. It appears that a damage cross-section combining a threshold for the production of PKA with E > ~15 KeV from the Nichols model with the shape of dpa above the threshold will give a better fit to the experimental data. This is probably related to the fact that dpa is derived for an idealized iron lattice whereas the unirradiated reactor vessel steel already contains many displacements and interstitials from the alloying atoms and material processing. This can also be viewed as requiring an "efficiency factor" as a function of energy when making the transition from displacements to damage.

## 7. ACKNOWLEDGEMENT

This paper reports accomplishments of the Babcock & Wilcox Owners Group Reactor Vessel Integrity Program. The authors acknowledge the group's support and express appreciation for permission to report the program results.

## 8. REFERENCES

1. Regulatory Guide 1.99, Rev. 2, Radiation Embrittlement of Reactor Vessel Materials, U.S. Nuclear Regulatory Commission, May 1988.

2. ASTM E693 Practice for Characterizing Neutron Exposures in Ferritic Steels in Terms of Displacements Per Atom, American Society for Testing and Materials, 1916 Race Street, Philadelphia, PA 19103.

3. K. Kussmaul, et al., Assurance of the Pressure Vessel Integrity With Respect to Irradiation Embrittlement, IAEA Specialist's Meeting on Irradiation Embrittlement and Aging of Reactor Pressure Vessels, Philadelphia, PA, May 27-29, 1987.

4. G. Prillinger, Neutron Spectrum Calculation for the Gundremmingen KRB/A Reactor, NUREG/CR-4791, November 1986.

5. W. N. McElroy, Editor, LWR Pressure Vessel Surveillance Dosimetry Improvement Program: PSF Experiments Summary and Blind Test Results, NUREG/CR-3320, Volume 1, HEDL-TME 86-8, R5, Hanford Engineering Development Laboratory, July 1986.

6. F. A. Nichols, How Does One Predict and Measure Radiation Damage?, Nuclear Technology, 40, August 1978.

7. D. G. Doran and N. J. Graves, Displacement Cross Sections and PKA Spectra: Tables and Applications, HEDL-TME 76-70, Hanford Engineering Development Laboratory, December 1976.

# COMPARISON OF MEASURED SILICON DISPLACEMENT DAMAGE RATIOS WITH ASTM E-722 AND NJOY CALCULATED DAMAGE

J. G. Kelly and P. J. Griffin

Sandia National Laboratories

P.O. Box 5800, Albuquerque, NM, USA 87185

## ABSTRACT

Experiments with silicon bipolar transistors exposed to 14 MeV neutrons and Sandia Pulsed Reactor (SPR III) fission type neutrons demonstrate that the NJOY silicon damage function should replace the one listed in ASTM E-722(86). An annealing procedure applied between exposures ensured accurate damage comparisons between the two environments. Re-evaluation of the dosimetry foil cross sections and fluence determination methodology was also necessary to ensure that damage per neutron/cm$^2$ could be evaluated properly.

---

Accurate neutron energy spectra and a well-established silicon displacement damage function are the important quantities needed to relate simulator induced displacement damage in electronic devices to operational environments. This relationship is made through the concept of 1 MeV neutron equivalent damage established by Verbinski et al., in 1979. [1] In recent years, however, there has been evidence [2,3,4] that differences in the damage rate observed in different laboratories and environments cannot be blamed solely on spectrum errors but must also be attributed in part to the published ASTM E-722(86) [5] damage function. The experiments described herein clearly establish better agreement with the NJOY [6] calculated damage function.

When neutrons displace atoms in the silicon lattice, damage centers are formed whose first effect on the crystal electronic behavior is a reduction in the carrier lifetime in the bulk material. For bipolar devices this translates directly into a reduction in the collector current gain as predicted by Messenger's [1] equation,

---

*This work was performed at Sandia National Laboratories, which is operated for the U.S. Department of Energy under contract number DE-AC04-76DP00789.

$$\frac{1}{h_f} - \frac{1}{h_i} = t_B \bar{K}_D \Phi \ , \qquad (1)$$

where $h_i$ and $h_f$ are the gains before and after exposure, and $t_B$ is the base transit time. $\bar{K}_D$ is the damage constant for the spectrum appropriate for the fluence, $\Phi$, and is proportional to the integral of the damage function, $K_D(E)$, and the neutron energy spectrum. The important point here is that the relative accuracy of $K_D(E)$ functions can be tested by comparing ratios of measured changes in the reciprocal gain of devices (observed from differing spectra) to calculated ratios. We wished especially to compare the damage induced by the neutrons in the Sandia Pulsed Reactor (SPR III) cavity and 14 MeV neutrons, because the ASTM E-722(86) [5] and the NJOY damage functions exhibit their largest differences above 7 MeV. From Eq. (1), the appropriate damage- per-unit-fluence ratio to be measured is

$$R_m = \frac{h_i(14) - h_f(14)}{h_i(SPR) - h_f(SPR)} \cdot \frac{h_f(SPR) \cdot h_i(SPR)}{h_f(14) \cdot h_i(14)} \cdot \frac{\Phi(SPR)}{\Phi(14)} \ . \qquad (2)$$

The experiments therefore require measurements of h and $\Phi$.

## GAIN MEASUREMENTS

The procedures followed below helped ensure accurate damage measurements. Define $\Delta(1/h) \equiv (1/h_f - 1/h_i)$. 1) Environments were chosen to minimize the gamma ray contributions. 2) Exposure levels used ensured $\Delta(1/h)$ errors < 2%. 3) The same transistors were used at both facilities to ensure agreement among damage ratios of ≈ 2%. (This implies gain measurement uncertainty <<2%.) 4) The transistors were annealed after each exposure to induce approximately 80% gain recovery and to stabilize $t_B$. 5) Temperature induced shifts in the gain readings were corrected for with control transistors. 6) Gains were determined at the same time after each exposure to compensate for room temperature gain recovery. The gain recovers as Frenkel defects recombine and displaced atoms return to proper lattice positions. This effect is illustrated in Figure 1a which shows the reciprocal of the gain vs. time after each of two exposures. At $t_2+\Delta t$, the damage is composed of two groups of damage centers each annealing at different rates. One must carefully measure the anneal curve, (curve 1), of these devices and record all relevant time intervals so that at $t_2+\Delta t$ curve 1 can be subtracted from $1/h_2$ to determine the damage at $\Delta t$ from the second exposure. This is the procedure that was used in the past. [2,3,4] However accurate determination of the anneal curve and proper documentation of all the time intervals adds greatly to experimental uncertainties.

Accuracy is improved by annealing the devices at 180°C for 24 hours after each exposure. This stabilizes the gain to 0.5% for months. Thus, as is shown in Figure 1b, the now flat baseline is easily subtracted from the 1/h of the next exposure. It is no longer necessary to accurately record the time intervals between the exposures and peel off in sequence the damage for each earlier exposure.

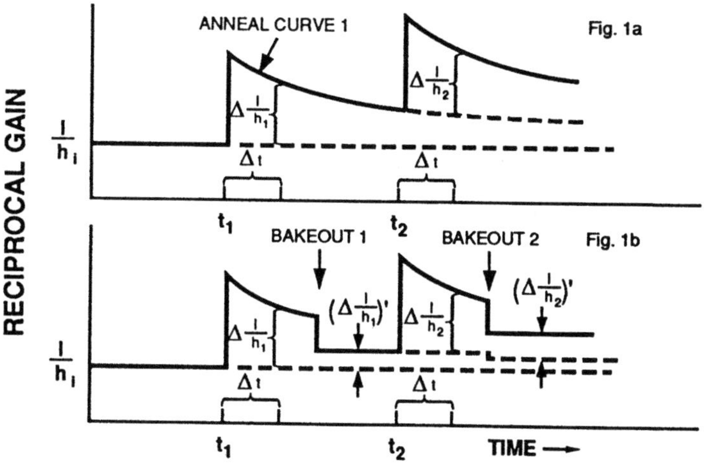


**Figure 1 - Reciprocal Transistor Gain vs. Time**

The quantities needed from Figure 1b are $\Delta 1/h_1$ and $\Delta 1/h_2$ for room temperature damage at $\Delta t$, but the damage after each annealing procedure (bakeout) is proportional to $(\Delta 1/h_1)'$ and $(\Delta 1/h_2)'$. Notice however that on the second bakeout the first group anneals further. Hence, the fractional recovery vs. number of bakeouts for a damage group must also be measured in advance. More is said about after-bakeout damage later.

Define $\alpha_n \equiv d_n''/d_1''$ as the fraction of the damage seen after n bakeouts compared to that seen after 1 bakeout of the same group. Our experiments have shown that $\alpha_2$=0.909, $\alpha_3$=0.864, $\alpha_4$=0.830, $\alpha_5$=0.811, and $\alpha_7$=0.784. For each group, $d_1''$ can be determined from the measurement after the first exposure for that group. It is easily shown that the annealed damage caused by the $n^{th}$ exposure and after the $n^{th}$ bakeout is

$$D_n' = \left\{ \left( \frac{1}{h_n'} - \frac{1}{h_i'} \right) - \sum_{m=1}^{n-1} \left[ \alpha_{n-m} \cdot \Delta \left( \frac{1}{h_m'} \right) \right] \right\} \frac{1}{\Phi_n} \qquad (3)$$

where $h_i'$ is the initial gain, $h_n'$ equals the gain after the $n^{th}$ exposure and bakeout, $\Delta(1/h_m')$ is the change in reciprocal gain after the $m^{th}$ exposure and bakeout for the $m^{th}$ group, and $\Phi_n$ is the fluence during the $n^{th}$ exposure.

## FLUENCE MEASUREMENTS

It is important that all transistors undergo the same exposure. For the 14 MeV exposures where the small source is only a few centimeters from the field point, the transistors and the dosimetry foils were rotated on a disk to ensure that each transistor and foil passed through the same spatial path.

The major concern, however, was in establishing with sufficient accuracy the relative fluence values in the two environments. The $^{58}Ni(n,p)^{58}Co$ reaction alone is not a good dosimeter because at 14 MeV its cross section uncertainty is ±18%. The cross section is falling fairly steeply in that region while the cross section for the $^{58}Ni(n,2n)^{57}Ni$ reaction is rising steeply. This makes the ratio of their activities resulting from a 14 MeV neutron exposure very sensitive to the beam average energy. It then follows that the proper way to determine the fluence in this range is to apply highly energy-sensitive cross section ratios to determine the energy. Then one should use a reaction with a well known and flat response (such as $^{93}Nb(n,2n)^{92}Nb$) to determine the fluence at that energy.

The fluence determination in the fission-like SPR III cavity was made by normalizing the activities of Ni foils (exposed with the transistors) to those obtained during the spectrum determination. The spectrum methodology used here was presented at the ASTM-EURATOM #VI Conference [7]. This methodology is very appropriate for determining the large variety of spectra encountered in simulation experiments. Little knowledge of the experiment environment is required, no transport code has to be run, and an arbitrary trial function can be used.

## EXPERIMENTAL RESULTS

At the 14 MeV source, three foils (Ni, Al, and Cu) and 10 transistors were exposed on the rotating disk. The runs to reach approximately $2 \times 10^{12}$ n/cm$^2$ each required a steady state operation of the Kaman Sciences commercial DT source at Sandia Laboratories of 6 to 7 hours. The five reactions which we were able to count were $^{58}Ni(n,p)^{58}Co$, $^{58}Ni(n,2n)^{57}Ni$, $^{27}Al(n,\alpha)^{24}Na$, $^{27}Al(n,p)^{27}Mg$, and $^{65}Cu(n,2n)^{64}Cu$. The ASTM E 496 standard explains how the values of ratios of the activities of pairs of reactions provide a measure of the average energy of the neutrons. As mentioned before, ratios that change rapidly with energy are most appropriate.

Since at the time of these experiments we did not realize that a flat response foil should be used, fluence values for each of the two exposures were determined by the average value of the 5 reactions. Another paper [8] from this conference provides more details on the 14 MeV fluence measurements.

At fission energies, corresponding to SPR III, the cross sections are well established and are not altered very much in the post ENDF/B-V evaluations. Once the spectrum shape has been obtained and is normalized by the dosimetry foils exposed with the transistors, the fluence $\Phi(E>10 \text{ keV})$ is determined. See ASTM standards E-721 and E-944 [5].

The gain for each of the 10 transistors 50.4 hours after each exposure was measured with an HP 4145A transistor parameter tester. Although individual damage rates differ by almost 60%, the standard deviation of the damage ratios was only 1.2%.

With the precautions taken to make temperature corrections, to use bakeout stabilization, and to ensure adequate changes in the gain on each exposure, the relative uncertainties in the determinations of the average damage for the 10 devices was approximately 1%. As mentioned before, the largest source of uncertainty occurred in the establishment of the fluences. Since the rms deviations of the fluences predicted by 5 reactions at 14 MeV was 3.6%, we have taken that to be a good measure of that uncertainty. These deviations reflect both the random counting errors and any random scatter among the cross sections. Systematic errors in the counting laboratory calibrations would tend to divide out in the ratio determinations.

In the SPR III cavity nickel and/or sulfur activities provide fluence values above 3 MeV with approximately 3% accuracy. This fluence is then multiplied by the spectral index, SI, to establish $\Phi(E>10\text{ keV})$. Although a given set of activities taken during the spectrum measurement usually fixes SI to less than 2%, an independent measurement yielding a new set of activities for the same environment will yield an SI that shows a larger deviation, in the range of 3%. An rms combination of these uncertainties has indicated a combined uncertainty of approximately 6% for the measured damage ratio. This uncertainty is reflected in Table I of the next subsection.

## DISPLACEMENT KERMA AND DAMAGE RATIOS

The methodology for obtaining a theoretical displacement kerma in silicon was guided by the procedures in ASTM E-722 [5] and by section 13 of the ASTM E-521. [5] The NJOY code, [6] version 89.36, was used with the ENDF/B [9] silicon cross sections to calculate the primary knock-on atom (PKA) recoil spectrum, and the Lindhard [10] energy partition function was then applied to calculate the displacement kerma.

Figure 2 compares the silicon displacement kerma currently in ASTM E-722 with the NJOY calculated function derived from the ENDF/B-V natural Si cross section and from a prereleased version [11] of the $^{28}$Si ENDF/B-VI cross section. The ENDF/B-V and ASTM damage comparison has been previously reported [2,3]. The spectrum-averaged ENDF/B-V and ENDF/B-VI silicon kerma from a fission spectrum are virtually identical. However, for 14 MeV neutrons, there is an 8.4% difference in the silicon damage estimates.

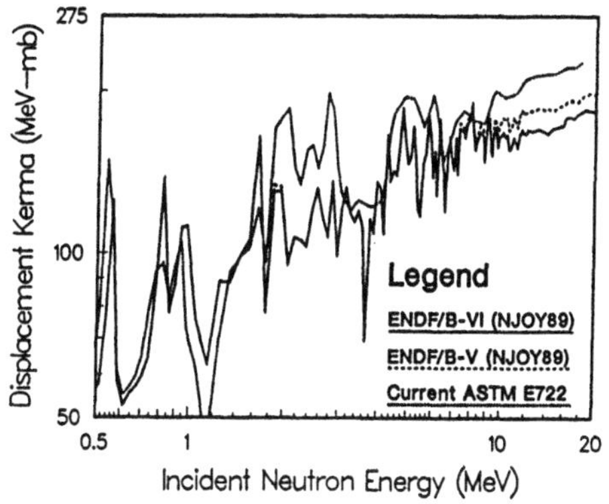


**Figure 2 - Silicon Displacement Kermas**

The NJOY code coupled with the ENDF/B-VI format allows the incorporation of energy/angle distributions to describe the recoil of the PKA. It represents a significant improvement over the application of kinematic constraints that must be employed in conjunction with ENDF/B-V data.

The measured and calculated damage ratios including those from [2,3] are compared in Table I. It is in the current damage ratios between SPR III and 14-MeV spectra that the differences in the experimental and theoretical damage ratios clearly fall beyond the uncertainties in the measurements.

**Table I**

Comparison of Experimental and Theoretical Damage Ratios

| Facility | R(msrd) | R(ASTM) | R(N/E-VI) | R(N/E-V) |
|---|---|---|---|---|
| 14-MeV/SPR | | | | |
| $E_{avg}$=14.5 | 2.17 ± 0.12 | 2.49 | 2.15 | ---- |
| $E_{avg}$=14.8 | 2.21 ± 0.12 | 2.50 | ---- | 2.32 |
| SPR/ACRR | 1.27 ± 0.06 | 1.34 | 1.31 | 1.31 |
| $Cf^{252}$/SPR | 1.28 ± 0.05 | 1.31 | 1.28 | 1.28 |

In the table, ACRR refers to the environment inside a water-cooled, lead-boron bucket in the Sandia Annular Core Research Reactor, a pool-type reactor. Two entries are seen below the 14-MeV/SPR entry, because the use of ENDF/B-V or ENDF/B-VI dosimetry cross sections results in a difference of 0.3 MeV in the characterization of the DT source average neutron energy [8]. The NJOY (R(N/E-VI)) silicon kerma is in better agreement with the experimentally observed silicon damage than is the current ASTM silicon kerma. This data supports the use of the ENDF/B-VI $^{28}Si$ cross section in an update to the ASTM E-722 Standard Practice in determining silicon damage equivalence.

## ANNEALED DAMAGE DEFECT ANALYSIS

In the case of the 14-MeV/SPR exposure, Table I shows that the damage ratio was 2.17. After exposure **and the subsequent anneal procedure**, the damage ratio was 2.51, a 15% increase. This means that a larger fraction of the damage anneals out during anneal of the SPR-induced damage. We believe that this occurs because the cluster size for 14-MeV neutrons is significantly larger and of lower density than from 1-MeV neutrons. Calculations with the binary-collision model code, TRIM, [12] confirm the decreased defect density initially (before any annealing) associated with 14-MeV neutron irradiation.

It is suggested that the silicon damage consists of divacancies, which anneal out from 500-560°K, and more deeply trapped vacancy-oxygen (V-O) defects, which anneal out near 630°K. [13] The density of stable divacancy defects depends upon the motion of the neutral vacancy defects and is likely to be greater in the denser clusters formed by 1-MeV neutrons. On the other hand, the V-O defect distribution is expected to be uniform throughout the damaged area. A large portion of the divacancies anneal out during the anneal sequence while the V-O defects are unaffected. The recombination of the larger fraction of divacancies expected from the SPR irradiation results in a larger 14-MeV to SPR damage ratio after anneal. One ramification of this finding (that the damage structure caused by 14-MeV and SPR-III neutrons differs) is that the 1-MeV equivalent damage may have to be reevaluated for devices subjected to high temperatures.

## DISCUSSION

These studies of silicon damage ratios have led to the important benefits of improved and more accurate methods to use transistors for damage function verification and in improved methodologies for determining neutron spectra and fluences in the fission and 14 MeV energy regions. In addition, we have discovered that silicon itself can be especially valuable in the assurance of simulation fidelity for any material. Because the measured and calculated ratios are in such good agreement, silicon can now be used as a responding material on an equal footing with the other activation foils for the determination of spectra. We have demonstrated that Si provides a response in the critical region from 100 keV to 2 MeV where laboratories without access to fission foils have a paucity of available response functions. In the present situation, without fission foils, spectra often cannot be determined in the sub-1-MeV region with enough precision to establish simulation fidelity.

The procedure is to irradiate a set of silicon transistors in a well characterized environment, measure the damage, and use the new damage function to calibrate their responses. The same devices can be used in the unknown environment with a set of appropriate activation foils to determine the spectrum. Comparisons with spectra in which fission foils have been used have been carried out to delineate the degree to which the transistors act as satisfactory replacements. This technique provides a substantial improvement to the methodology for the determination of neutron spectra and has a potential to significantly advance simulation technology.

## REFERENCES


1. V. V. Verbinski, C. Cassapakis, R. L. Pease, and H. L. Scott, "Transistor Damage Characterization by Neutron Displacement Cross Section in Silicon: Experimental," Nuclear Science and Engineering, Vol. 70, pp. 66-72 (1979).

2. T. F. Luera, J. G. Kelly, H. J. Stein, M. S. Lazo, C. E. Lee, and L. R. Dawson, "Neutron Damage Equivalence for Silicon, Silicon Dioxide, and gallium Arsenide," IEEE Trans. on Nucl. Science, Vol. NS-34, No. 6, pp. 1557-1563, December 1987.

3. J. G. Kelly, T. F. Luera, L. D. Posey, and J. G. Williams, "Simulation Fidelity Issues in Reactor Irradiation of Electronics-Reactor Environments," IEEE Tran. Nucl. Sci., Vol. 35, No. 6, December 1988.

4. M. H. Sparks, T. M. Flanders, J. G. Williams, J. G. Kelly, W. W. Sallee, M. Roknizadeh, and J. L. Meason, "Energy Dependence of Neutron Damage in Silicon Bipolar Transistors," IEEE Tran. Nucl. Sci., Vol. 36, No. 6, December 1989.

5. The 1988 Annual Book of ASTM Standards, Volume 12.02, American Society for Testing and Materials.

6. R. E. MacFarlane and D. W. Muir, "NJOY87, A Code System For Producing Pointwise and Multigroup Neutron and Photon Cross Sections from ENDF/B Evaluated Nuclear Data," PSR-171, Radiation Shielding Information Center, Oak Ridge National Laboratory.

7. J. G. Kelly, "Neutron Spectrum Adjustment with SANDII Using Arbitrary Trial Functions," p. 460 in Reactor Dosimetry: Methods, Applications, and Standardization, ASTM STP1001, H. H. Farrar IV and E. P. Lippincott, editors, American Society for Testing and Materials, Baltimore, MD, 1989.

8. P. J. Griffin, J. G. Kelly, and T. F. Luera, "Effect of ENDF/B-VI Cross Sections on Neutron Dosimetry," presented at the Seventh ASTM-EURATOM Symposium on Reactor Dosimetry, held in Strasbourg France on August 27-31, 1990.

9. R. Kinsey, Ed., "ENDF-102, Data Formats and Procedures for Evaluated Nuclear Data File, ENDF," Brookhaven National Laboratory report BNL-NCS-50496 (ENDF-102) 2nd edition (ENDF/B-V), October 1979; 3rd edition (ENDF/B-VI), 1990.

10. J. Lindhard, V. Nielsen, M. Scharff, and P. V. Thomsen, "Integral Equations Governing Radiation Effects (Notes on Atomic Collisions III)," Matematisk-fysiske Meddelelser-Kongelige Danske Videnskabernes Selskab, KDVSA, Vol. 33, No. 10, pp. 1-42, 1963.

11. S. Epperson, D. Larson, D. Hetrick, and C. Fu, private communication from Oak Ridge National Laboratory, September 1989. This data will be released as part of the 2nd update to the ENDF/B-VI cross section library in late 1990. An Oak Ridge National Laboratory report will be published describing the evaluation techniques used.

12. J. P. Biersack, J. F. Ziegler, and G. Cuomo, "TRIM-88 —— Ions in Solids, Version 4.4," IBM Research, Yorktown Heights, New York, 1988.

13. Ed. B. L. Gregory and G. W. Gwyn, "Radiation Damage in Semiconductor Devices," SC-M-71-0887, p. 5, Sandia National Laboratories, March 1972.

Friedemann W. Stallmann
Oak Ridge National Laboratory
Oak Ridge, Tennessee, USA

Dieter Pachur
Kernforschungsanlage, Jülich, Germany

ABSTRACT

A systematic analysis has been performed of the available instrumented Charpy data from surveillance capsule reports which are collected in the Power Reactor Embrittlement Data Base (PR-EDB). Based on ideas by D. Pachur, fractions of the total impact energy that are due to different segments of the load curve are determined and fit separately to Pachur's model. As a first result, a strong linear relation has been found between the upper shelf values Ea of the energy fractions in segment A and the total upper shelf energy USE. The relation can be written as:

$$Ea = 0.276\ USE + 8.64\ (joule),$$

and holds for all pressure vessel materials, plates, forgings, and welds, both, before and after irradiation in surveillance capsules of commercial power reactors. Based on this relation a separation into energy fractions can be obtained from the total impact energy alone resulting in an improved fit for Charpy impact data. Conversely, reliable values for the upper shelf energy can be obtained from energy fraction Ea, if the impact data in the upper shelf region are missing or erratic. The investigation is being extended to data from materials test reactor experiments.

---

## INTRODUCTION

Records of load versus time are now routinely obtained from the instrumented tup of the testing apparatus for Charpy impact tests. A theoretical analysis of the load curves can be found in monographs about fracture analysis.[1] The main emphasis so far has been on the determination of various stress parameters of the material obtained from the load values at critical points of the curve. The current investigation intends, instead, to use the information from instrumented Charpy data for a more detailed interpretation of the customary impact energy versus test temperature data. The ultimate goal is to reach a better understanding of the relation between Charpy test data and radiation embrittlement and to provide a more accurate basis for trend curve development.

Several stages of deformation and cracking have been identified in the instrumented Charpy curves. The first stage (segment A), the "pre maximum load" segment shows the same pattern as the tensile test consisting of elastic deformation up to the yield point followed by plastic strain to the ultimate load point which coincides with the maximum load of the instrumented curve (Fig. 1). Further plastic deformation of unspecified nature follows (segment B) and is usually identified as the "post maximum load" segment. For impact energies below the upper shelf the segment B is abruptly terminated by a steep fall in the load curve, indicating crack initiation which is followed, for sufficiently high test temperatures, by the segment C that depicts the crack extension. In each of the three segments a specific fraction of the total impact energy is absorbed. These three energy fractions can be fitted separately to a model proposed by D. Pachur[2] which has a shape that is very similar to the more commonly used hyperbolic tangent fit but describes more realistically the thermodynamic basis for the relation between test temperature and impact energy (Fig. 2). The shape of the Pachur curves depend on the upper shelf energy of the corresponding energy fraction, Ea, Eb, and Ec, respectively, and two additional parameters, the "activation energy" needed for the plastic deformation (Ua, Ub, and Uc, respectively, given in eV in Figs. 2, 4, and 5) and a dimensionless parameter Lm (Lma, Lmb, Lmc) which is thought to be related to the defect density. Since each segment has a different transition and upper shelf region, the combination of the three curves gives a more accurate fit to the total impact energy than a single curve. This can have some influence on the determination of the transition temperature at 30 ft-lb which is sometimes part of the transition region of segment A and sometimes of segments B and C. Separate investigation of the three energy fractions may also provide additional information about the dependency of the total upper shelf energy on fluence and chemistry. The segments B and C are difficult to separate in the upper shelf region and appear also to have a common transition region. These two segments are, therefore, not separated in the current investigation but combined and treated as one segment.

## MATERIALS AND METHODS

In what follows we will refer to a "set" of Charpy data the collection of test results for individual specimen that are used

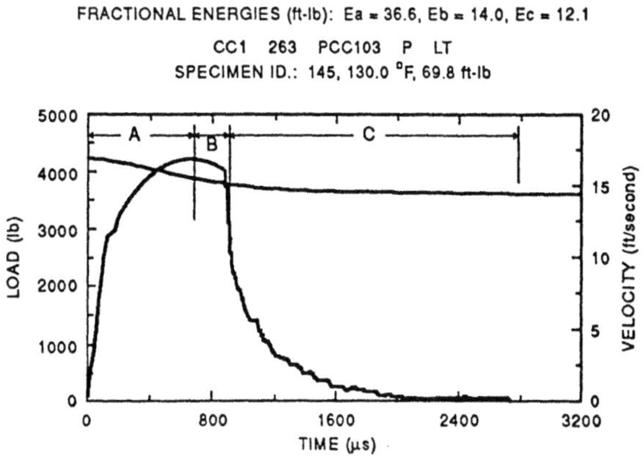


Fig. 1. Typical load curve from an instrumented Charpy test, separated into the three segments A, B, and C. This curve is indicative for the transition region because of the abrupt termination of segment B. In the upper shelf region there is a smooth transition between segments B and C, making their separation difficult. The curve on top of the graph is the velocity of the tup as a function of time. Ea, Eb, and Ec are the fractional energies at test temperature, not the upper shelf values as in the other figures.

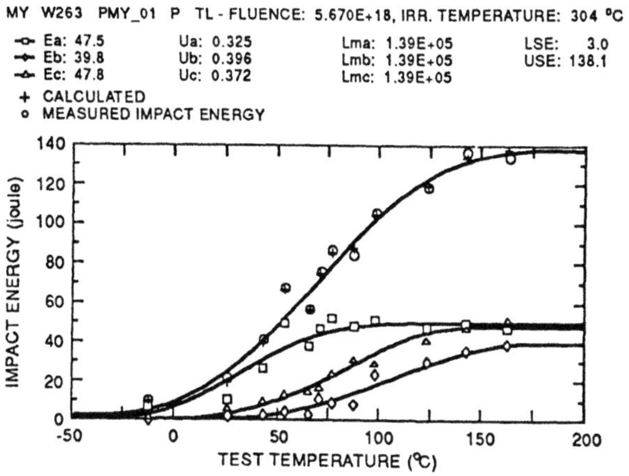


Fig. 2. Separate fitting of the three energy fractions from segments A, B, and C to Pachur's model. Note the close similarity of the fits for B and C which is typical for the fits of power reactor data. LSE is the lower shelf energy which is part of the fit to segment A. The top line contains the code for the reactor, capsule, and material used in the PR-EDB. Fluence is E > 1.0 MeV in neutrons/cm$^2$.

to construct one impact energy versus test temperature curve for the determination of transition temperature and upper shelf energy. Sets containing impact energy versus test temperature for a variety of pressure vessel materials before and after irradiation are collected in the PR-EDB from surveillance reports for commercial power reactors. Instrumented Charpy data are also provided in some surveillance reports either in form of graphical representation of the traces or listings of a variety of load and energy values. Graphic representations of the load values were digitized for 96 sets, each containing between 8 and 12 individual traces. Listings that report the energy fraction of segment A were available for 106 sets with no overlap between the two samples. Each set was fitted with individual curves for two and, where available, three fractions and the results were tabulated. Software was developed which automates the digitizing and fitting procedures. The procedures require no intervention by the evaluator but the traces can be inspected and modified, if appropriate.

Every trace has been separated into the three segments by a computer program which determines first the absolute peak for the separation between segments A and B. This has worked well in almost all cases; in a few cases the peaks were hidden by coordinate lines in the graphic images and unrelated spikes are also sometimes mistaken for the "real" peak. Such cases are easily detected and corrected manually. The separation point between segments B and C was set at the time when the load curve had dropped to 80% of its peak value. This agrees well with the sudden drop of the load in the transition region. However, this separation has not been used in the current investigation. The energy fractions as well as the total impact energy was computed from the load values combined with the information about the initial velocity of the tup and the maximal potential energy of the system. Comparison between the calculated and the dial impact energy shows good relative agreement within each set or groups of sets, but small calibration factors that are common to a set need to be assigned to obtain agreement between calculated and dial energy to within 5% or 3 joules, whichever is larger. In a few cases the traces had been cut off before the completion of the break; the missing amount of the calculated energy has been restored in these cases to obtain agreement with the dial energy.

For each set the energy fractions of segments A and the combined B+C have been fitted separately to Pachur's model resulting in a total impact energy versus test temperature curve as the sum of the two fractions. These fits compare favorably with the customary ones that use only the total impact energy. More importantly, irregularities and data scatter can often be traced to a specific segment which helps in making more reliable determinations of the transition temperature and upper shelf energy. As a first and very surprising result of the fitting a strong linear relation was detected between the total upper shelf value (USE) and that of segment A (Ea) which is discussed below (Fig. 3). To test the validity of this relation 28 sets of power reactor data, obtained from graphic representations of the load curves, were chosen on the basis of well defined upper shelf energies and generally good fits to the energy fractions. The parameters of the linear relation were computed for this set using least squares regression analysis, assuming random errors with normal distribution and the same variance for both energy fractions. This relation was then applied to the remaining fits. According to this test, the linear relation hold for all materials used for power reactor pressure vessels, plates, forgings, and welds, and for unirradiated as well as for irradiated materials, provided the irradiation temperature is in the 288 degree C range. Any significant deviations from the linear relation, i.e., differences between measured and predicted energy fraction A that exceed 20% ($2\sigma$), could be traced to large data scatter or poor fit in the upper shelf region. Details will be discussed in the next section.

Transition temperatures for the different energy fractions were also determined. Its value is not well defined for the segment A since few tests are made in this temperature region and the data scatter is large. The transition temperature for energy fraction in segment (B+C) can be more reliably determined and is, therefore, better suited for trend curve studies. The customary determination of the transition temperature for the total impact energy at 41 and 68 joules (30 and 50 ft-lb) was also performed; since these temperatures are influenced by both segments A and B, many irregularities in their behavior can be traced (and possibly resolved) through the study of the instrumented Charpy data.

For comparison purposes, traces from test reactor data were also considered. Instrumented Charpy data obtained from a variety of PV materials irradiation programs at the Research Center (KFA) in Jülich, Germany have been studied extensively by D. Pachur.[3,4] Data were also obtained from several programs conducted by Materials Engineering Associates (MEA) and from the HSST and HFIR programs at ORNL. Test reactor data are available directly in digitized form as obtained from Nicolet oscilloscopes and have been processed by the same software as for power reactor data after some initial reformatting.

## RESULTS

As mentioned in the previous section, a strong linear relation exists between the upper shelf values of the segments A and B+C, (Ea, and Eb+Ec, respectively) and, by implication, the total upper shelf energy (USE). Based on the coefficients obtained by regression analysis from a selected set the formula can be written as:

$$Ea = 0.276\ USE + 8.64\ (joule) \qquad (1)$$
$$(\ Ea + (Eb+Ec) = USE\ )$$

This relation holds for all PV materials, plates, forgings, and welds, both, before and after irradiation in surveillance capsules of commercial power reactors. It appears also to be true for test reactor data covering similar materials and irradiation conditions. The relation does, however, not hold irradiations at low temperatures.[3] It seems to be also violated for some materials that are outside the specifications used for pressure vessels. It should be noted that small deviations from the formula (1), which may occur for some materials and/or irradiation conditions, cannot be easily detected because of the overall statistical uncertainty for the upper shelf values.

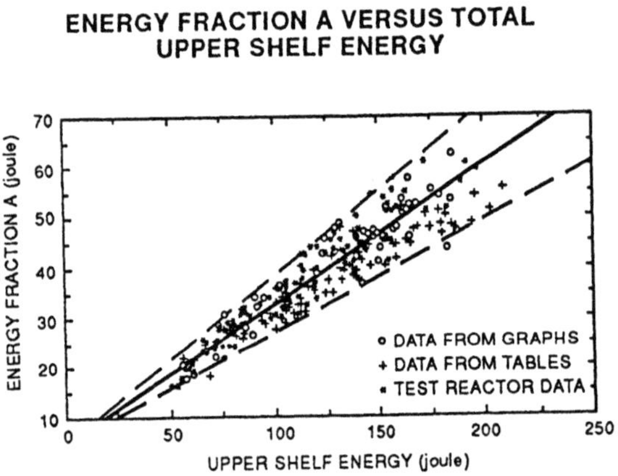


Fig. 3. Linear relation between the upper shelf energy fraction from segment A and the total energy. The dotted line represents the 20% ($2\sigma$) uncertainty bound for determining the upper shelf energies. Excluded from the plot were data with large scatter in the upper shelf region, such as in Fig. 5, or with no data in this region. No data were excluded from test reactor experiments which were of superior quality compared to the surveillance test data.

One consequence of this relation is, that the separate fitting to energy fractions in segments A and B+C can be performed without the information from the load curves. The formula describing the fitting model plus the information about the separation into A and B+C at the upper shelf level suffice in most cases to obtain the separate fits for the energy fractions using only the total impact energy. This has been verified by comparing fits with and without the information from the load curves (Fig. 5). It allows not only the separate determination of transition temperatures for the energy fractions but provides also a somewhat better fit for the total impact energy versus test temperature curve, particularly, if the transition regions for A and B are widely separated. Another application of the linear relation is the possibility to estimate the total upper shelf energy from that of segment A if measurements in the upper shelf region are either missing or unreliable. An example is given in Fig. 5.

As these examples show, the added information from the instrumented Charpy traces in combination with the knowledge about the strong linear relation (1) can provide a better defined and more reliable determination of transition temperature and upper shelf values from the Charpy test data. A detailed study of the energy fractions with their separate transition temperatures and upper shelf energies and the use of more physically based models is likely to give not only a somewhat better predictive power but also a better understanding of the mechanism at work. Physically based embrittlement predictions can then give a more secure footing for regulatory decision than any purely empirical correlation can provide.

Given the wide ranging applicability of relation (1), exceptions from it will be of particular interest. Such exceptions are likely to occur for materials that are outside the narrow range of pressure vessel steels and for irradiations at low temperatures. Some examples of such cases are available but have not been sufficiently analyzed so far. Analysis of these cases will give a better understanding of the general mechanism for radiation embrittlement and may be of interest for damage predictions other than LWR PVs, such as support structures or different types of vessels as in the HFIR.

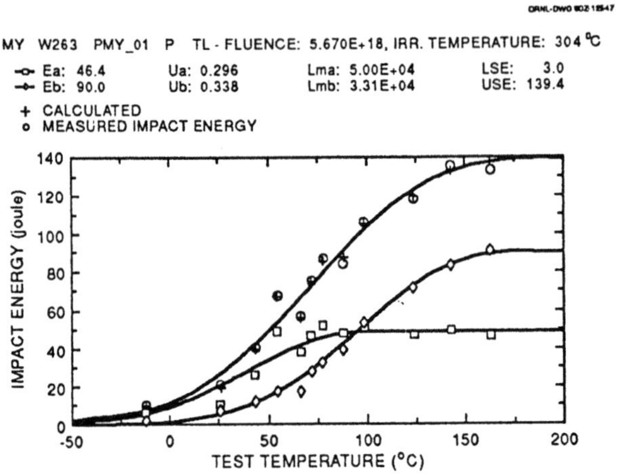


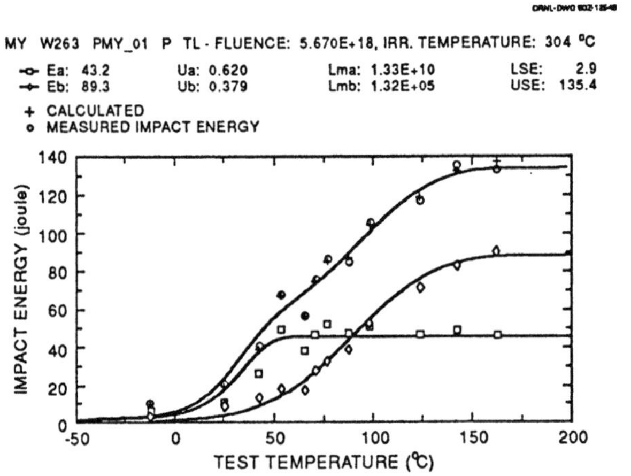


Fig. 4. Comparisons between fits using energy fractions from instrumented Charpy tests and total impact energy in connection with the linear relation between energy fraction Ea and USE. The figure above uses instrumented Charpy data, and the figure below uses only total impact energy. There are slight differences but the goodness of fit is not significantly affected. For the explanation of the symbols see Fig. 2. Eb, Ub, Lmb refers to the combined segments B+C.

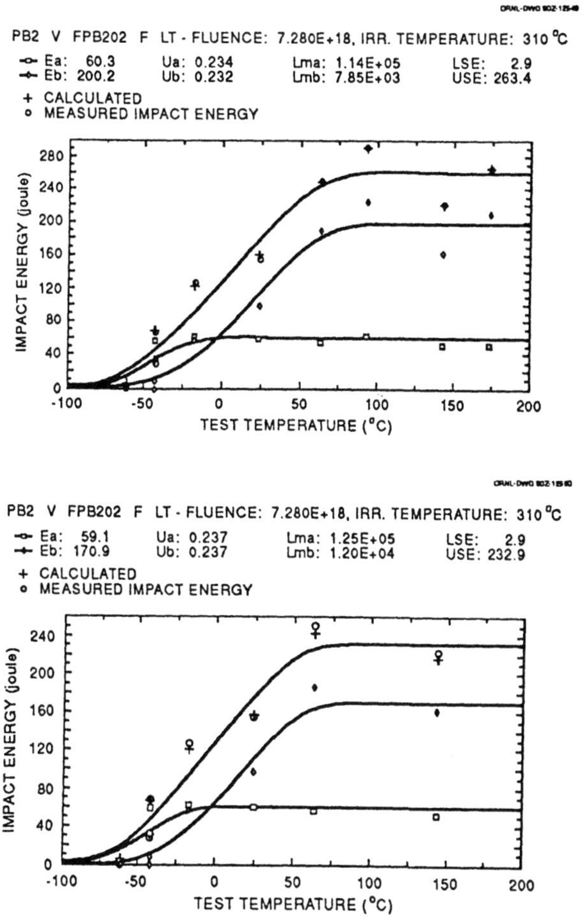


Fig. 5. This is an example for the use of instrumented Charpy data for a more reliable determination of the upper shelf energy. The total impact energy of the set above shows a larger scatter in the upper shelf; however, energy fraction of segment A remains constant, indicating some abnormal behavior of the material during segments B and C. The linear relation indicates that the impact energy for some of the test data is too high and, thus, the upper shelf energy of 263 joule, as determined from the fit. Removing two points with the highest impact energy reduces USE to 233 joule, which is probably still too high, but consistent with the linear relation within uncertainties. The upper shelf energy of the unirradiated material as determined from the same type of fit (without use of instrumented Charpy data, which are not available) is also 233 joule, which is also in better agreement with the lower value. For the explanation of symbols see Figs. 2 and 4.

## ACKNOWLEDGEMENTS

Research sponsored by the Office of Nuclear Regulatory Research, U.S. Nuclear Regulatory Commission, under Interagency Agreement DOE 0551-0551-A1 with the U.S. Department of Energy under contract DE-AC05-84OR21400 with Martin Marietta Energy Systems, Inc.

## REFERENCES

1. A. S. Tetelman and A. J. McEvely, Jr., Fracture of Structural Materials, John Wiles & Sons (1967).

2. Dieter Pachur, "Influence of Radiation Damage Mechanism on the Load Signal and on the Transition Curve of Instrumented Impact Testing, *J. Nucl. Material 160 (1988) 24-33* (see. p. 30).

3. Dieter Pachur, *Ibid. 26-27.*

4. D. Pachur, "Irradiation Embrittlement of Reactor Pressure Vessel Steel Outside the ASTM Specification A508 CL2," *J. Nucl. Material 171 (1990) 128-133.*

# NEUTRON DISPLACEMENT DAMAGE FUNCTIONS FOR IRON

**A.M. Ougouag, M.B. Danjaji, J.G. Williams, and J.F. Stubbins**

University of Illinois at Urbana-Champaign

103 South Goodwin Avenue, Urbana, IL 61801, USA

## ABSTRACT

New calculations of the dpa cross sections for iron have been performed using the NJOY cross section processing code and data from ENDF/B-V. In addition to the total dpa cross sections, the dpa cross sections for production of displacements in cascades of specific size ranges were computed using a modified version of the NJOY code. They are tabulated in the form of a transfer matrix (neutron energy group to displacement cascade size).

## INTRODUCTION

It is a standard practice (ASTM Standard E 693) [1] to characterize neutron exposures in ferritic steels in terms of displacements per atom (dpa) in iron. These can then be used as the starting point for obtaining relative correlations between the neutron exposure and changes in material properties. Computation of the dpa requires a knowledge of the energy dependent neutron fluence and of the displacement cross section as a function of energy. The latter is tabulated in ASTM E 693. The current version of ASTM E 693 contains displacement cross sections calculated by Doran and Graves [2] using ENDF/B-IV [3] cross sections. In the present work, the displacement cross sections for iron have been recalculated using ENDF/B-V [3] cross sections and the cross section processing code NJOY [4]. Results are compared with the previous values. ENDF/B-VI [3] data have been recently released. Computation of the displacement cross sections using these data, which requires the use of a new version of NJOY able to utilize the additional files in ENDF/B-VI, is in progress.

The practice described above intrinsically assumes that the number of displacements effective in causing damage is proportional to the energy expended in atomic collisions. This assumption is necessary if the damage is to be characterized in terms of dpa regardless of the distribution of the damage in displacement cascades of different sizes. Several publications [5,6,7,8] have pointed out that the efficiency for production of stable defects and for production of freely migrating defects each depend on the number of defects initially formed within atomic cascades. Very large cascades have a greater tendency for defect recombination and for formation of stable non-migrating clusters. Very small cascades also may result in more recombination of close Frenkel pairs. In order to account for these effects, a necessary step will be the calculation of the distribution of initial displacements between cascades of different sizes. For the second part of this paper, a modified version of NJOY has been used to compute iron displacement cross sections for production of defects in cascades of specific size ranges. They are tabulated in multigroup form (neutron energy group and cascade energy group), and it is shown how they can be used to obtain the distribution of damage energy between displacement cascades of different sizes.

## METHOD

The computational results obtained in this paper were obtained using the NJOY [4] cross section processing code. The code has been modified to include an improved treatment of the emitted charged particles and to compute the differential displacement kerma cross sections integrated over displacement energy groups. The theory underlying these changes was presented elsewhere [9] and its main points are summarized below.

An energetic charged particle (recoil or emitted particle) slowing down in a material can produce displacements, and hence displacement damage. If the energy of the particle is $E_b$, in the approach of the NJOY code the portion contributing to displacement damage is given by

$$T_{dam} = E_b\ L(E_b) \quad , \qquad (1)$$

where $L(E_b)$ is the Robinson [10] version of the Lindhard [11] partition function for the particular charged particle in the material under consideration. When the above equation is used, the computed quantities are the non-electronic loss kerma cross sections which can be, for large energies ($T_{dam}$ larger than 2.5 $E_d$), translated into a number of displacements $N_D$ via

$$N_D = \beta\ T_{dam} = 0.8\ T_{dam}/2E_d \ , \qquad (2)$$

where $E_d$ is the displacement threshold energy. At lower energies, the modified Kinchin and Pease model predicts no displacement if $T_{dam}$ is less than $E_d$ and only one displacement when $T_{dam}$ is larger than $E_d$ but less than 2.5 $E_d$ [12,13]. This is accounted for in our modified version of the NJOY code by altering the Lindhard partition function accordingly in these two energy ranges. Results obtained using this version of the code are referred to as NJOY-NRT results in this paper.

The theory underlying the computation of displacement kerma cross section matrix elements is based on the representation of a damage response, F, in the following multigroup form:

$$F = \sum_g \Phi_g \sum_j w_j \, {}^jK_{Dg} \quad , \qquad (3)$$

where $w_j$ is a macroeffectiveness factor [9] which quantifies the effectiveness of displacements in cascades of size j for producing a macroscopic effect (e.g. embrittlement); $\Phi_g$ is the multigroup neutron fluence; and ${}^jK_{Dg}$ is the displacement kerma cross section for the production of displacement cascades of a specific size range, indexed by the superscript j, and averaged over the neutron energy group g.

Additional insight and details on the theory presented above can be found in references [9] and [14]. A possible approach to the determination of the macroeffectiveness factors would consist in the discretization of damage efficiency functions, such as those obtained in references [6], [7], and [8] and used in [14]. In this paper the ${}^jK_{Dg}$ matrix elements are computed and tabulated.

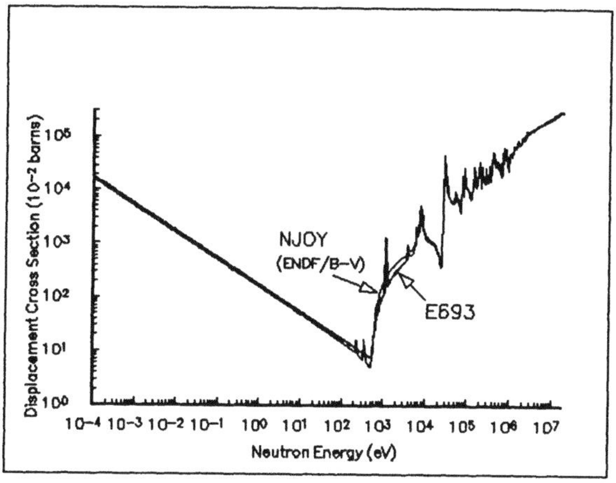


Figure 1: Displacement Cross Section Comparison (NJOY vs E 693)

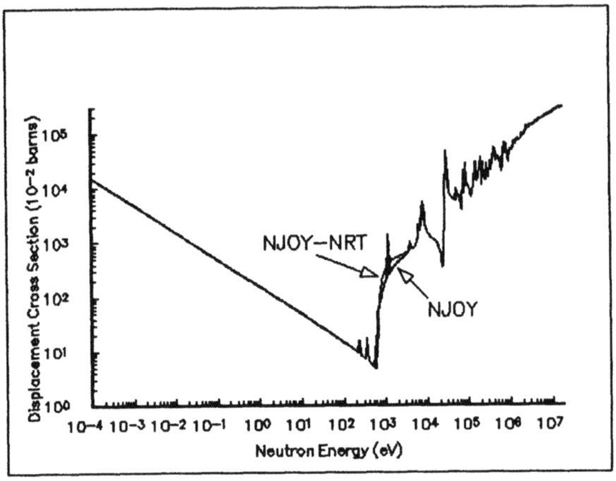


**Figure 2: Displacement Cross Section Comparison (NJOY vs NJOY-NRT)**

RESULTS

A comparison of the NJOY ENDF/B-V iron dpa cross sections with those reproduced in ASTM Standard E 693 is presented in Fig. 1. In both cases the cross sections were obtained using the IAEA recommended [15] conversion of atomic collision energy to displacements, i.e. using a $\beta$ value of 10 $keV^{-1}$, consistent with a value of 40 eV for $E_d$ (see Eq. 2).

In Fig. 2 the same NJOY dpa cross sections are compared with the NJOY-NRT values. The effect of the differences in the three dpa cross section representations, illustrated in Figs. 1 and 2, on the total dpa rates in five different neutron spectra is shown in Table 1. The Table shows that the new NJOY and NJOY-NRT calculations give essentially identical results in these neutron spectra, despite differences in the dpa cross sections near 1 keV (Fig. 2). The ASTM E 693 values differ from these by approximately 10% for thermal neutrons and by less than 5% for the other spectra. In Table 1 and in later tables and figures, ANO-1 T/4 and ANO-1 FPV refer respectively to the quarter thickness position and the front of pressure vessel (internal surface) of Arkansas Nuclear One, Unit 1, which is a Babcock and Wilcox PWR.

The displacement kerma matrix elements ($^{j}K_{Dg}$) for calculating the distribution of displacements between cascades of different sizes are tabulated here (Table 2) in 26 neutron energy groups and 11 cascade size groups. (Similar data in 640 neutron energy groups are available from the authors.) The first cascade size

Table 1: Fe TOTAL DISPLACEMENT CROSS SECTIONS (barns)

| | Method | | |
|---|---|---|---|
| | NJOY | NJOY-NRT | E693 |
| Spectrum | | | |
| THERMAL | 6.213 | 6.293 | 6.961 |
| FISSION | 856.4 | 856.4 | 858.9 |
| ANO-1 T/4 | 349.4 | 349.5 | 335.1 |
| ANO-1 FPV | 308.8 | 308.9 | 311.5 |
| 14 MeV | 2896 | 2893 | 2824 |

Note: The ANO-1 entries for the E 693 method in this table are inferred from dpa values reported in reference [16].

group refers to values of $T_{dam}$ less than 40 eV, the threshold displacement energy, and therefore contributes zero displacements and contains only zeroes for the displacement kerma cross sections when the model referred to as NJOY-NRT is used, as here. For the second cascade size group $T_{dam}$ is in the range 40 eV to 100 eV, resulting in a single Frenkel pair. The remaining cascade size groups correspond to larger values of $T_{dam}$, resulting in the displacement numbers (Frenkel pairs) identified in the second row of the table.

Some of the data from the 640x11 version of the displacement kerma matrix are plotted in Fig. 3. The conversion from displacement kerma cross sections (eV barns) to dpa cross sections (barns) is accomplished by dividing by 100 eV. Some characteristics illustrated in Fig. 3 are: damage energy range 2

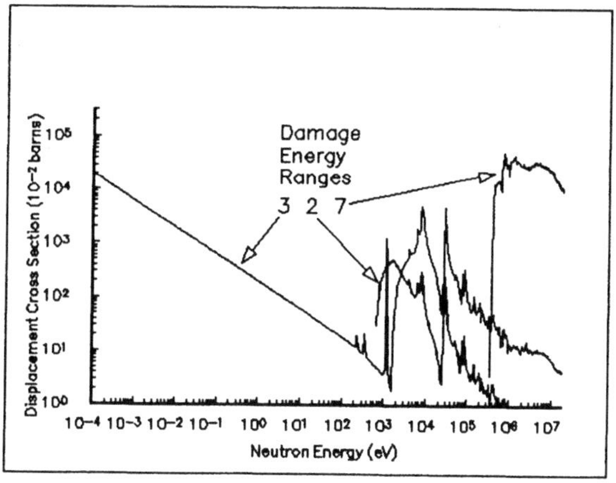


Figure 3: Examples of dpa Matrix Elements

Table 2: Fe DISPLACEMENT KERMA MATRIX (eV.barns)

| | $E_D$/Casc. group | 2 | 3 | 4 | 5 | 6 | 7 | 8 | 9 | 10 | 11 |
|---|---|---|---|---|---|---|---|---|---|---|---|
| | # displ. | 1 | 1-5 | 5-15 | 15-64.7 | 64.7-118 | 118-510 | 510-1430 | 1430-2700 | 2700-3970 | 3970-5240 |
| GRP | range (eV) | | | | | | | | | | |
| 1 | 1.000-5 | 0.000+0 | 1.148+3 | 0.000+0 | 0.000+0 | 0.000+0 | 0.000+0 | 0.000+0 | 0.000+0 | 0.000+0 | 0.000+0 |
| 2 | 1.000-1 | 0.000+0 | 4.418+2 | 0.000+0 | 0.000+0 | 0.000+0 | 0.000+0 | 0.000+0 | 0.000+0 | 0.000+0 | 0.000+0 |
| 3 | 4.140-1 | 0.000+0 | 2.133+2 | 0.000+0 | 0.000+0 | 0.000+0 | 0.000+0 | 0.000+0 | 0.000+0 | 0.000+0 | 0.000+0 |
| 4 | 1.855+0 | 0.000+0 | 6.143+1 | 0.000+0 | 0.000+0 | 0.000+0 | 0.000+0 | 0.000+0 | 0.000+0 | 0.000+0 | 0.000+0 |
| 5 | 1.013+2 | 0.000+0 | 1.331+1 | 0.000+0 | 0.000+0 | 0.000+0 | 0.000+0 | 0.000+0 | 0.000+0 | 0.000+0 | 0.000+0 |
| 6 | 4.540+2 | 2.289+2 | 3.560+2 | 2.337+0 | 0.000+0 | 0.000+0 | 0.000+0 | 0.000+0 | 0.000+0 | 0.000+0 | 0.000+0 |
| 7 | 7.102+3 | 9.503+1 | 1.693+3 | 3.790+2 | 0.000+0 | 0.000+0 | 0.000+0 | 0.000+0 | 0.000+0 | 0.000+0 | 0.000+0 |
| 8 | 1.503+4 | 9.110+0 | 1.855+2 | 5.900+2 | 0.000+0 | 0.000+0 | 0.000+0 | 0.000+0 | 0.000+0 | 0.000+0 | 0.000+0 |
| 9 | 2.418+4 | 4.717+0 | 9.339+1 | 6.616+2 | 5.282+0 | 0.000+0 | 0.000+0 | 0.000+0 | 0.000+0 | 0.000+0 | 0.000+0 |
| 10 | 2.606+4 | 1.510+2 | 2.953+3 | 2.382+4 | 3.400+3 | 0.000+0 | 0.000+0 | 0.000+0 | 0.000+0 | 0.000+0 | 0.000+0 |
| 11 | 3.183+4 | 1.707+1 | 3.246+2 | 2.608+3 | 4.164+3 | 0.000+0 | 0.000+0 | 0.000+0 | 0.000+0 | 0.000+0 | 0.000+0 |
| 12 | 6.738+4 | 8.805+0 | 1.561+2 | 1.148+3 | 1.146+4 | 0.000+0 | 0.000+0 | 0.000+0 | 0.000+0 | 0.000+0 | 0.000+0 |
| 13 | 1.111+5 | 3.500+0 | 6.659+1 | 4.782+2 | 9.662+3 | 4.668+3 | 0.000+0 | 0.000+0 | 0.000+0 | 0.000+0 | 0.000+0 |
| 14 | 1.832+5 | 2.534+0 | 4.997+1 | 3.767+2 | 6.264+3 | 1.445+4 | 0.000+0 | 0.000+0 | 0.000+0 | 0.000+0 | 0.000+0 |
| 15 | 2.972+5 | 1.856+0 | 3.704+1 | 2.830+2 | 3.860+3 | 2.380+4 | 7.130+3 | 0.000+0 | 0.000+0 | 0.000+0 | 0.000+0 |
| 16 | 4.979+5 | 9.795-1 | 2.005+1 | 1.632+2 | 2.542+3 | 1.134+4 | 2.211+4 | 0.000+0 | 0.000+0 | 0.000+0 | 0.000+0 |
| 17 | 7.427+5 | 9.061-1 | 1.860+1 | 1.528+2 | 2.496+3 | 1.160+4 | 3.288+4 | 0.000+0 | 0.000+0 | 0.000+0 | 0.000+0 |
| 18 | 1.003+6 | 6.306-1 | 1.296+1 | 1.070+2 | 1.956+3 | 9.458+3 | 3.819+4 | 1.531+4 | 0.000+0 | 0.000+0 | 0.000+0 |
| 19 | 1.653+6 | 5.621-1 | 1.158+1 | 9.632+1 | 1.825+3 | 8.604+3 | 3.086+4 | 4.671+4 | 0.000+0 | 0.000+0 | 0.000+0 |
| 20 | 2.307+6 | 6.280-1 | 1.293+1 | 1.090+2 | 1.975+3 | 8.906+3 | 2.865+4 | 6.978+4 | 0.000+0 | 0.000+0 | 0.000+0 |
| 21 | 2.466+6 | 5.939-1 | 1.226+1 | 1.054+2 | 1.918+3 | 9.197+3 | 2.801+4 | 8.738+4 | 9.388-4 | 0.000+0 | 0.000+0 |
| 22 | 3.012+6 | 5.177-1 | 1.071+1 | 9.249+1 | 1.685+3 | 8.907+3 | 3.300+4 | 1.013+5 | 3.413+3 | 2.193+0 | 0.000+0 |
| 23 | 4.966+6 | 4.842-1 | 1.004+1 | 8.575+1 | 1.576+3 | 9.821+3 | 3.106+4 | 1.159+5 | 3.109+4 | 3.364+2 | 0.000+0 |
| 24 | 7.408+6 | 4.056-1 | 8.439+0 | 7.166+1 | 1.335+3 | 8.917+3 | 2.621+4 | 1.004+5 | 8.281+4 | 3.183+3 | 1.479+2 |
| 25 | 1.000+7 | 3.112-1 | 6.470+0 | 5.487+1 | 1.024+3 | 7.135+3 | 2.043+4 | 8.291+4 | 1.282+5 | 1.331+4 | 1.513+3 |
| 26 | 1.221+7 | 2.372-1 | 4.927+0 | 4.177+1 | 7.888+2 | 4.333+3 | 1.456+4 | 6.503+4 | 1.594+5 | 3.279+4 | 3.173+3 |
| | 1.733+7 | | | | | | | | | | |

(one Frenkel pair) occurs predominantly in the low keV neutron energy range where neutrons have just enough energy to cause displacements by elastic scattering; damage energy range 3 (more than one, up to five displacements) is produced by recoil from radiative capture, and therefore dominates at the lowest neutron energies; high energy events, exemplified here by damage energy range 7 (118-510 displacements), show threshold behavior requiring high energy neutrons.

The displacement kerma matrix of Table 2 may be folded with neutron fluence spectra to obtain spectrum-averaged values of the displacement kerma cross section, distributed between the 11 cascade sizes. Table 3 shows the results of such calculations for the same five neutron spectra used for the total dpa cross sections of Table 1. The values in Table 3 were also divided by 100 eV so as to obtain displacement, rather than displacement kerma, cross sections. The distribution of displacements between the different cascades size ranges is proportional to the cross section values. The distributions are illustrated, in integral form in Figure 4, which shows that 14-MeV neutrons produce most of their displacements in cascades of more than 510 displacements, the reactor neutrons produce most in cascades of 65 to 1430 displacements, while thermal neutrons produce them in cascades of more than 1 up to 5 displacements.

Table 3: Fe CASCADE-SIZE-DISTRIBUTED DISPLACEMENT CROSS SECTIONS (NJOY-NRT) (barns)

| | Spectrum | | | | |
|---|---|---|---|---|---|
| $E_D$ GRP | THERMAL | FISSION | ANO-1 T/4 | ANO-1 FPV | 14 MeV |
| 1 | 0.000E+00 | 0.000E+00 | 0.000E+00 | 0.000E+00 | 0.000E+00 |
| 2 | 0.000E+00 | 1.214E-02 | 2.047E-01 | 2.263E-01 | 2.273E-03 |
| 3 | 8.724E+00 | 2.298E-01 | 1.194E+00 | 3.123E+00 | 4.721E-02 |
| 4 | 0.000E+00 | 1.757E+00 | 4.074E+00 | 3.364E+00 | 4.002E-01 |
| 5 | 0.000E+00 | 2.456E+01 | 3.206E+01 | 1.819E+01 | 7.603E+00 |
| 6 | 0.000E+00 | 1.044E+02 | 8.095E+01 | 4.881E+01 | 4.131E+01 |
| 7 | 0.000E+00 | 2.816E+02 | 1.142E+02 | 9.200E+01 | 1.421E+02 |
| 8 | 0.000E+00 | 4.195E+02 | 1.026E+02 | 1.255E+02 | 5.893E+02 |
| 9 | 0.000E+00 | 2.644E+01 | 1.458E+01 | 1.847E+01 | 1.681E+03 |
| 10 | 0.000E+00 | 5.572E-01 | 6.644E-01 | 8.314E-01 | 4.145E+02 |
| 11 | 0.000E+00 | 2.957E-02 | 4.960E-02 | 6.445E-02 | 3.300E+01 |

Note: There is an inconsistency in the treatment of emitted photons between the calculations in this table and the total dpa values of Table 1. This has a significant effect only on the thermal data.

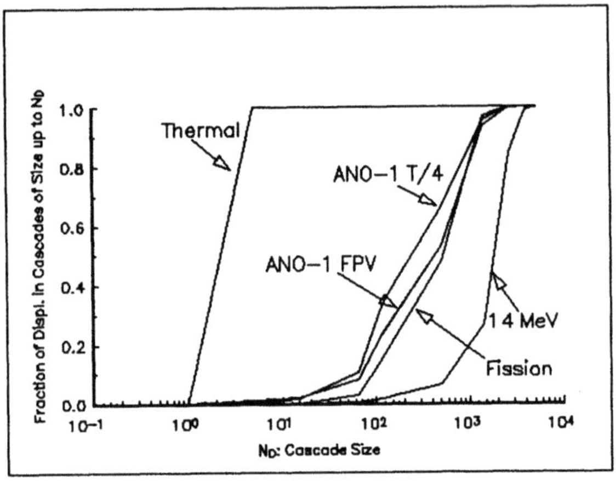


**Figure 4: Cumulative Distribution of Displacement Cascades in Various Spectra**

## CONCLUSIONS

The NJOY and NJOY-NRT calculated cross sections, both based on ENDF/B-V data, give essentially identical results for neutron spectra of interest, despite differences in the dpa cross sections near 1 keV neutron energy. The ASTM E693 values differ slightly from these values, reflecting minor changes in the basic ENDF/B data between versions IV and V and/or in the models used to compute dpa between our work and the earlier work of reference [2]

The cross sections presented here may be used to calculate the size distribution of displacement cascades formed during irradiation by neutrons of any known energy spectrum. Typical fission reactor spectra produce most of their displacements in cascades of 65-1430 displacements, though differences, which may be significant, exist between reactor spectra. Pure neutron spectra produce much smaller cascades and 14-MeV neutrons much larger ones. If the effective macroscopic damage response factor ("macroeffectiveness factor") for different cascade sizes is known, then the effective damage cross section may be found by a weighted sum of the cross sections for different cascade sizes.

## REFERENCES

[1] Annual Book of ASTM Standards, Vol 12.02, Am. Soc. Testing Materials, Philadelphia, 1989, pp. 270-274.

[2] Doran, D. G., and N. J. Graves, "Neutron Displacement Damage Cross Sections for Structural Materials," in ASTM STP 611, Am. Soc. Testing Materials, 1976, pp. 463-482.

[3] Evaluated Nuclear Data File/B, National Neutron Cross Section Center, Brookhaven National Laboratory, Upton, NY.
[4] MacFarlane, R. E., D. W. Muir and F. M. Mann, J. Nucl. Mat. 122 & 123, 1984, pp. 1041-1946.
[5] Averback, R. S., R. Benedek and K. L. Merkle, Phys. Rev. B, 18, 1978, pp. 4156-4171.
[6] Simons, R. L., J. Nucl. Mater., 117, 1986, pp. 70-77.
[7] Muroga, T. Rad. Effects and Defects in Solids, 113, 1990, pp. 119-134.
[8] Wiedersich, H., Rad. Effects and Defects in Solids, 113, 1990, pp. 97-107.
[9] Ougouag, A. M., J. G. Williams, M. B. Danjaji, S.-Y. Yang, and J. L. Meason, "Differential Displacement Kerma Cross Sections for Neutron Interactions in Si and GaAs," to appear in IEEE Transaction on Nuclear Science, Vol. 37, No. 6, Part II, December 1990.
[10] M.T. Robinson, "The Energy Dependence of Neutron Radiation Damage in Solids," in Nuclear Fusion Reactors, proceedings of a B.N.E.S. Nuclear Fusion Reactor Conference at Culham Laboratory, September 1969, pp. 363-377, British Nuclear Energy Society, London, 1970.
[11] J. Lindhard, V. Nielsen, M. Scharff and P. V. Thomsen, "Range concepts and heavy ion ranges (Notes on atomic collisions), Kgl. Dan Vidensk. Selsk. Mat.-Fys. Medd., 33, 1963.
[12] Norgett, M. J., M. T. Robinson, and I. M. Torrens, Nucl. Eng. Design 33, 1974, p. 50.
[13] Biersack, J. P., and L. G. Haggmark, Nucl. Instr. and Methods 174, 1980, pp. 257-269.
[14] Ougouag, A. M., J. G. Williams, J. F. Stubbins, M. B. Danjaji, and M. A. Ghaly, "Defect Production Efficiency in Neutron Irradiation of Iron," presented at the 15th International Symposium on the Effects of Radiation on Matter, June 21, 1990, Nashville, TN, submitted to the ASTM for publication.
[15] "Recommendations for Displacement Calculations for Reactor/Accelerator Studies in Austenitic Steel," Nucl. Eng. Design 33, 1974, p. 91.
[16] Tsoulfanidis, N., et al., EPRI NP-3776, Electric Power Research Institute, Palo Alto, 1984.

## Modified Damage Parameters Applied To a Typical Light Water Reactor's Pressure Vessel Supports

Craig A. Hrabal

U. S. Nuclear Regulatory Commission

Washington, DC, United States

### ABSTRACT

Modified damage parameters from the work of both Heinisch and Greenwood have been used to provide a possible explanation of the observed difference between the neutron damage of the High Flux Isotope Reactor (HFIR) pressure vessel surveillance specimens and the Shippingport neutron shield tank material. These results were used to predict the effects of neutron embrittlement on light water reactor (LWR) pressure vessel supports. Based on these predictions the irradiation effects on LWR pressure vessel supports represent a lesser problem than would be predicted by the HFIR data alone. Therefore, it is likely that the effects of thermal neutrons in the damage process previously have been underrated in cases such as HFIR. However, confirmation of this conclusion is required by experiments specifically designed to test the effects of thermal neutrons.

### INTRODUCTION

Recent results from the work of both Heinisch(1) and Greenwood(2) suggest that the difference in neutron damage experienced by the pressure vessel of the High Flux Isotope Reactor (HFIR) and the neutron shield tank of the Shippingport Reactor can be explained by modified damage parameters. These parameters account for the effects of recombination and, thus, predict neutron damage better than traditionally used correlations with either fluence or displacements-per-atom (dpa). Before 1989, use of published mechanical property correlations led to the conclusion that the damage to reactor pressure vessel (RPV) supports would be negligible. However, correlations with either fluence or dpa became suspect for predicting neutron damage at the reactor vessel supports, because material test results of HFIR surveillance specimens showed greater than expected damage (embrittlement) with low-flux, low-temperature neutron exposure. One proposed explanation for the increased embrittlement was that a "fluence-rate" effect was responsible(3). This explanation might have some basis, but recent test results from the neutron shield tank of the Shippingport reactor, which also experienced low-flux, low-temperature embrittlement, provided contradictory data. Another explanation for the increased embrittlement was that thermal neutrons might produce more damage than can be accounted for by correlations with dpa or fluence(4). This paper attempts to resolve the apparent disagreement by (1) studying previous methods used to predict neutron damage, (2) investigating possible models that explain the difference between HFIR and Shippingport, and (3) applying the results of these investigations to a typical light water reactor's (LWRs) pressure vessel supports.

One common method of predicting damage has used neutron damage correlations with fluence as a parameter. This method, however, does not account for the results of neutron-atom interactions. Some neutrons may be absorbed with subsequent transmutation, release of an energetic sub-atomic particle and concurrent atomic recoil causing an atom displacement; while other neutrons may not interact at all. Therefore, different spectra of neutrons with the same fluence would have different results. This obviously makes fluence a poor choice as a parameter for comparing results of widely different spectra. The problems with using fluence led to the adoption of dpa as a parameter for neutron damage correlations(5).

Dpa predicts neutron damage better than fluence, because the displacement of atoms caused by neutron-atom interactions better models the actual neutron embrittlement process. Atoms are displaced due to the primary knock-on atom (PKA) recoils produced from neutron-atom interactions. To calculate this displacement, a nuclear data processing code, DISCS(6), can be used to calculate the displacement cross section at each neutron energy. The computer code, SPECTER(7), then calculates the dpa from the displacement cross sections and the given neutron energy spectrum.

To calculate the displacement cross sections, DISCS first uses evaluated nuclear data from neutron cross section libaries, such as ENDF/B-V(8), to determine primary knock-on atom (PKA) recoil energy spectra. Then, the Lindhard(9) energy partition model can be used to find the fraction of the recoil energy which will cause nuclear events, referred to as the damage energy. Finally, the damage energy divided by the average energy required to displace an atom multiplied by empirical normalization factors gives the probability of displacing an atom, which is the displacement cross section.

In addition to calculating the displacement cross section, DISCS calculates computer libraries of recoil distributions and other nuclear data. SPECTER, a computer code that was developed to simplify damage calculations, uses these libraries to calculate recoil spectra and spectral-averaged dpa.

Although dpa correlations are better damage predictors than fluence correlations, dpa only gives the number of displacements due to a neutron radiation dose. In reality, displaced atoms can recombine with vacancies, thus annealing the damage caused by the point-defects they produced. Models which take into account the effect of this recombination must be used in order to provide parameters that correlate better with the actually observed neutron damage.

This paper discusses two such models. One model that was both developed for, and applied to, damage calculations by Heinisch is based on computer models of single displacement cascades(1). The other model was based on the results of the formation of stable defects from heavy ion irradiations at the Argonne National Laboratory(10). It involved damage models developed by Wiedersich which were applied to damage calculations by Greenwood(2). Each model was used to modify the displacement cross section input into the SPECTER code for the purpose of calculating a more representative dpa value. The following sections of this paper report (1) further investigations of both models and (2) application of the results to a typical LWRs pressure vessel supports.

## PROCEDURE AND COMPUTATIONS

The displacement cross section at energy E, $\sigma_{dis}(E)$, which is calculated by DISCS, can be expressed mathematically by the following:

$$\sigma_{dis}(E) = \sigma(E)\int_{T_1}^{T_2} K(E,T)\,\nu(T)\,dT \qquad Eqn.1$$

where E = incident neutron energy (lab reference system),
T = recoil nucleus energy (lab reference system),
$\sigma$(E) = total neutron cross section at energy E,
K(E,T) = neutron-atom energy transfer kernel,
and $\nu$(T) = secondary displacement function.

The kernel, or primary recoil distribution, K(E,T), is simply the probability that a neutron of energy E will produce a primary recoil atom of energy T; in other words, a PKA. The quantity $\nu$(T) describes the number of secondary displacement atoms produced per PKA. When the primary recoil distribution includes the nuclear cross section, the equation for $\sigma_{dis}(E)$ can be rewritten as:

$$\sigma_{dis}(E) = \int_{T_1}^{T_2} R(E,T)\,\nu(T)\,dT \qquad Eqn.2$$

where R(E,T) = $\sigma$(E)K(E,T).

Although the displacement cross section is not calculated in SPECTER, Eqn. 2 can be approximated by writing a computer program that integrates the secondary displacement fuction with the recoil spectra calculated by SPECTER. Heinisch used this method to calculate the number of freely migrating defects; in other words, those that escaped the cascade region produced from a neutron-atom interaction. The secondary displacement function calculated from his computer model replaced the usual function that is used to calculate dpa. Because SPECTER omits the contributions due to thermal neutron captures in calculating the recoil spectra, estimates of these contributions were added manually. The effort resulted in calculations of the freely migrating interstitials per atom (fmipa) and freely migrating vacancies per atom (fmvpa).

Greenwood, on the other hand, modified the DISCS code to include the damage efficiency functions of the Wiedersich model. The calculation done in Eqn. 2 was modified to include the damage efficiency for each value of the recoil distribution. Integration over the recoil energy distribution including the damage efficiency times the secondary displacement function led to a modified displacement cross section. The modified displacement cross section was then input into SPECTER to calculate a modified damage parameter called WMOD dpa.

Another way to calculate Eqn. 2 is to modify SPECTER. This paper reports on the results of this approach. The Heinisch and Wiedersich models were both used to calculate their respective damage parameters. Heinisch's model was calculated by changing the secondary displacement function used in the modified SPECTER code. Wiedersich's model was calculated by multiplying the secondary displacement function by the damage efficiency.

## RESULTS

Several neutron spectra were used in the analysis reported here. Greenwood supplied the spectrum for the inner wall of the Shippingport neutron shield tank. Nanstad supplied the spectra at the pressure vessels for the Oak Ridge Research Reactor (ORR) at position A9 and for HFIR at position P6. The speMwrum for the cavity of a typical LWR (McGuire unit 1) was obtained from NUREG/CR-5320.

Before plotting the correlations with damage parameters, experimental property changes from each irradiation were required. Table 1 lists the measured change in 0.2% offset yield stress in MegaPascals (MPa) for A212B vessel steel in the HFIR, ORR, and Shippingport Reactor at several irradiation times(11). In addition, Table 1 lists the weight percent of copper for each of the steels(11).

Table 1. Measured Change in Yield Stress for A212B

| Irradiation time (hours) | Change in 0.2% Yield Stress (MPA) | | |
|---|---|---|---|
| | ORR (0.15 w/% Cu) | HFIR (0.15 w/% Cu) | Shippingport (0.05 w/% Cu) |
| 0.29 | 9 | - | - |
| 0.29 | 5 | - | - |
| 3.65 | 5 | - | - |
| 16.00 | 46 | - | - |
| 16.00 | 35 | - | - |
| 38.00 | 74 | - | - |
| 72.00 | 106 | - | - |
| 72.00 | 128 | - | - |
| 170.00 | 177 | - | - |
| 384.00 | 193 | - | - |
| 384.00 | 160 | - | - |
| 6.67E+04 | - | 53 | - |
| 6.67E+04 | - | 65 | - |
| 6.67E+04 | - | 49 | - |
| 6.67E+04 | - | 51 | - |
| 7.67E+04 | - | 59 | - |
| 7.67E+04 | - | 38 | - |
| 7.67E+04 | - | 57 | - |
| 7.67E+04 | - | 61 | - |
| 1.33E+05 | - | - | 51 |

Calculations were done with SPECTER to obtain the dpa from each spectrum. The results were then used to correlate the change in yield stress with dpa in Fig. 1 for Shippingport, ORR, and HFIR. The HFIR data could not be correlated with the Shippingport and ORR data, thus two best-fit correlations were done, one for the HFIR data alone and one for the rest of the data.

Entering Fig. 1 at the dpa for a typical LWR cavity at 17.53 and 32* Effective Full Power Years (EFPY), the correlation without the HFIR data resulted in a yield stress change of approximately 75 and 100 MPa, respectively. The correlation based on the HFIR data alone, however, resulted in yield stress changes of approximately 180 MPa at 17.53 EFPY and 240 MPa at 32 EFPY.

The Heinisch and Wiedersich models were then used in the modified SPECTER code to calculate modified damage parameters. The results of these calculations along with best-fit curves are shown in Figs. 2 through 4 for fmipa, fmvpa, and WMOD dpa, respectively, assuming that property changes with neutron exposure were negligible for each of the steels studied. Table 2 gives the results predicted for a typical LWR cavity at 17.53 and 32 EFPY.

Table 2. Comparison of Predicted Change in Yield Stress for A212B

| Irradiation time (EFPY) | Change in 0.2% Yield Stress (MPA) | | | | |
|---|---|---|---|---|---|
| | dpa | | fmipa | fmvpa | WMODdpa |
| | HFIR only | without HFIR | | | |
| 17.53 | 180 | 70 | 90 | 85 | 80 |
| 32.00 | 240 | 100 | 135 | 115 | 110 |

## CONCLUSIONS

Based on the results of the correlations with the modified damage parameters, the Heinisch and Greenwood work seems to explain the observed difference between the neutron damage of the HFIR pressure vessel surveillance specimens and the Shippingport neutron shield tank material. The WMOD dpa model appears to provide the best correlation of the data. Based on the above results the irradiation effects on LWR pressure vessel supports represent a lesser problem than would be predicted by the HFIR data alone. Therefore, it is likely that the effects of thermal neutrons in the damage process previously have been underrated in cases such as HFIR. However, confirmation of this conclusion is required by experiments specifically designed to test the effects of thermal neutrons.

* These numbers are based on irradiation times currently under study at the U.S. Nuclear Regulatory Commission.

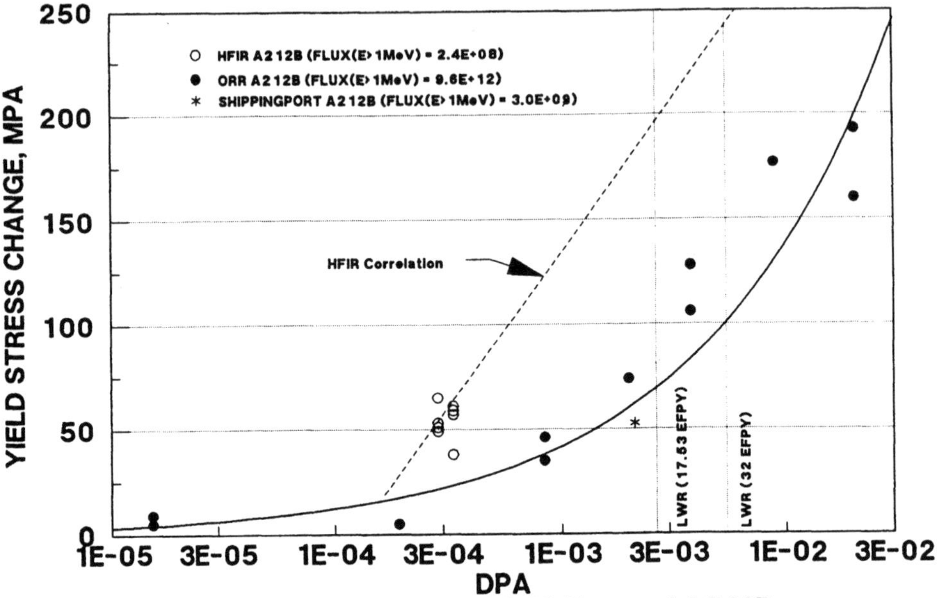


**Figure 1. Change in 0.2% Offset Yield Stress of A212B Pressure Vessel Steels as a Function of the Dpa**

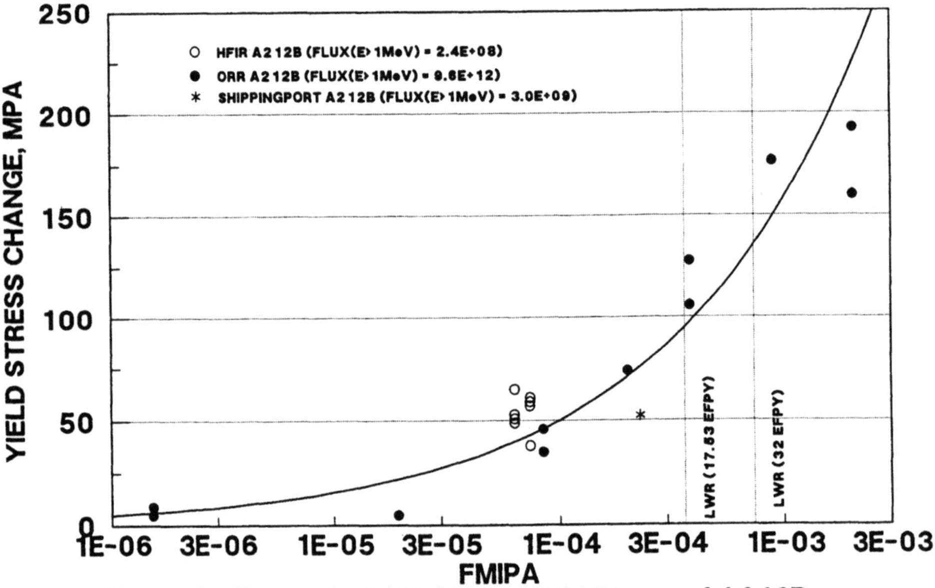


**Figure 2. Change in 0.2% Offset Yield Stress of A212B Pressure Vessel Steels as a Function of the Freely Migrating Interstitials Per Atom (FMIPA)**

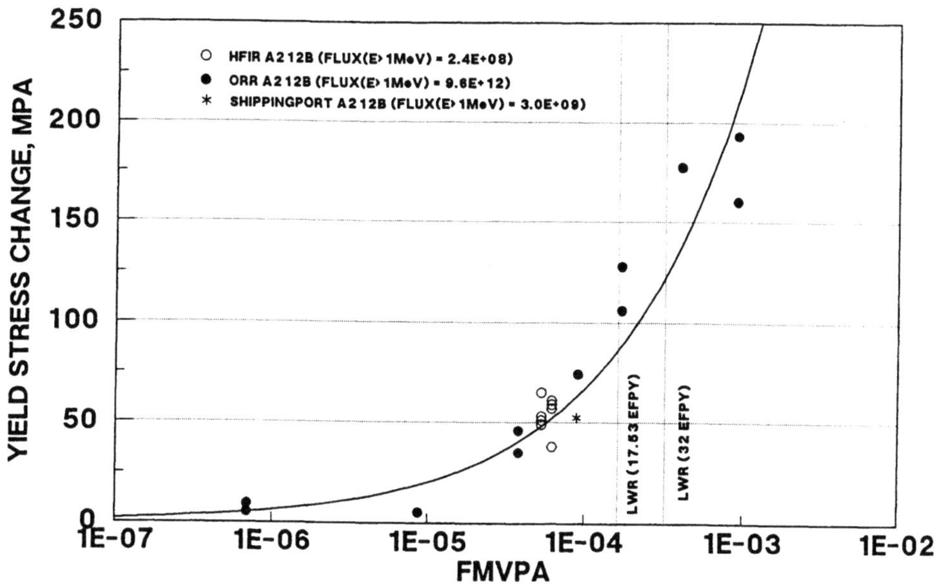


Figure 3. Change in 0.2% Offset Yield Stress of A212B Pressure Vessel Steels as a Function of the Freely Migrating Vacancies Per Atom (FMVPA)

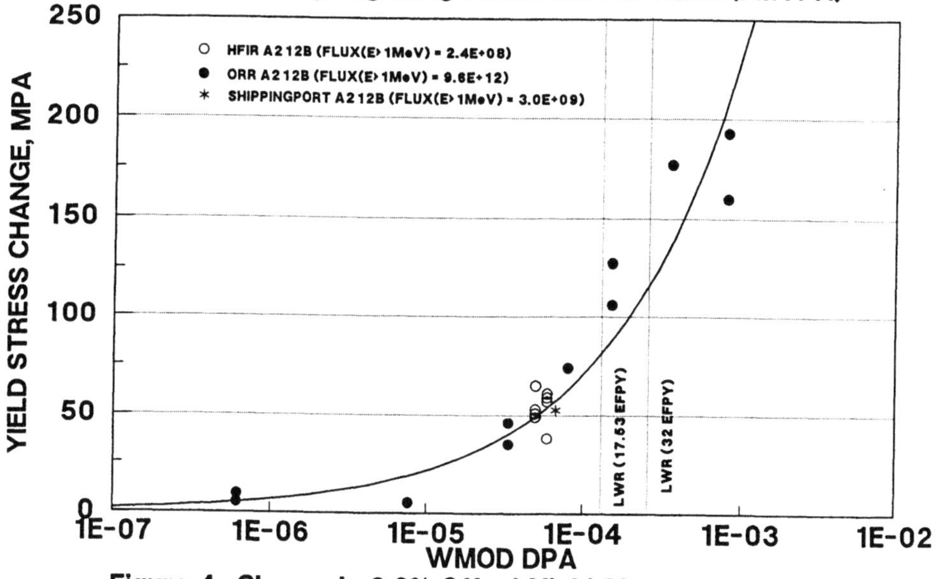


Figure 4. Change in 0.2% Offset Yield Stress of A212B Pressure Vessel Steels as a Function of the Weidiersich Model (WMOD) DPA

## REFERENCES


1. H.L. Heinisch, "Correlation of Mechanical Property Changes in Neutron Irradiated Pressure Vessel Steels on the Basis of Spectral Effects," preprint, Fusion Reactor Materials, Quarterly Progress Report, DOR-ER-0313/6 (1989).

2. L.R. Greenwood, "Damage Efficiency Calculations for Shippingport, HFIR, and ORR," ANL (1989).

3. R.H. Priest, W. Charnock, and B.K. Neale, "Effects of Irradiation on Materials, ASTM-STP-870, American Society for Testing and Materials, Philadelphia, PA, 1985, p. 1150.

4. R.K. Nanstad et al., "Accelerated Neutron Embrittlement of Ferritic Steels at Low Fluence: Flux and Spectrum Effects," J. of Nuc. Matls., 158, pp. 1-6, 1988.

5. ASTM E693, "Standard Practice for Characterizing Neutron Exposures in Ferritic Steels in Terms of Displacements Per Atom (DPA)," Annual Book of ASTM Standards, American Society for Testing and Materials, Philadelphia, PA, 1990 edition.

6. G.R. Odette and D.R. Dorian, Nucl. Technol., 29, p. 346, 1976.

7. L.R. Greenwood and R.K. Smither, "SPECTER: Neutron Damage Calculations for Materials Irradiations," ANL/FPP/TM-197 (1985).

8. Evaluated Neutron Data File, Version V, Part B, National Neutron Cross Section Center, Brookhaven National Laboratory (1979).

9. J. Lindhard, V. Nielsen, M. Scharff, and P.V. Thomsen, Mat. Phys. Medd. Dar. V.D. Selk., 22 (1963)10.

10. H. Wiedersich, "Effects of the Primary Recoil Spectrum on Long Range Migration of Defects," Proc. Workshop on Effects of the Primary Recoil Spectrum, Lugano, Switzerland (1988); to be published in Radiation Effects.

11. "Evaluation of HFIR Pressure-Vessel Integrity Considering Radiation Embrittlement," R.D. Cheverton, J.G. Merkle and R.K. Nanstad, eds., ORNL/TM-10444 (1987).

# TECHNIQUES

# THE APPLICATION OF NIOBIUM FOR RETROSPECTIVE DOSE DETERMINATION IN CANDU REACTORS

M F Banham, A J Fudge, J A Tibbles, B E Sheldon,
AEA Technology, Harwell Laboratory, UK

R Fleck, R A Holt*
Ontario Hydro Power Company of Canada

Samples from zirconium, 2.5 w/o niobium pressure tubes taken from two CANDU reactors and from material irradiated in the DIDO reactor at Harwell have been analysed for $^{93m}Nb$. This is produced by neutrons with energies above 0.1MeV by the reaction $^{93}Nb$ (n,n') $^{93m}Nb$ and has become established as a fast flux dosimeter in test irradiations which normally last for periods of weeks up to a few months. Some of the materials discussed in this paper were irradiated during periods of up to 13 years. Account has been taken of possible burn out by thermal neutron capture of both $^{93}Nb$ and $^{93m}Nb$ and the results are compared with predictions based on the irradiation histories and flux profiles in which the materials were irradiated. Good agreement has been obtained showing the practical value of niobium not only for test irradiation dosimetry but also for retrospective dosimetry in appropriate material after long periods of irradiation.

## INTRODUCTION

In the CANDU reactor system it is important to know how the physical properties of the zirconium alloy used in the fabrication of the pressure tubes change during irradiation with fast neutrons (>1MeV). This is being studied by two methods: 1) an experimental study consisting of the irradiation of material in the Harwell MTR DIDO, and 2) retrospectively on pressure tubes that were removed from two CANDU power reactors for analysis. The neutron

*now AECL, Chalk River, Canada

doses to which these latter samples were exposed had only been obtained by calculation. Material, cut from the tubes, was sent to Harwell so that the doses could be determined retrospectively by measuring the activity induced during the irradiation (1). The material used for the manufacture of the pressure tubes was zirconium alloyed with nominally 2.5 w/o of niobium. Thus analysis of this material for $^{93m}Nb$, the product of the reaction of fast neutrons (the threshold for this reaction is approximately 0.2MeV) was expected to provide the required results. Niobium has become, in recent years, established as a fast neutron dosimeter using the reaction $^{93}Nb$ (n,n') $^{93m}Nb$ (2) but its use retrospectively was still novel. In order to show that the method could be used with confidence, it was also applied to the materials which were irradiated in DIDO and for which the results of iron dosimetry were available.

The irradiation of all these samples took place over several years and doses received were very high. It was therefore necessary during these measurements to show that the results were not affected by the loss of $^{93m}Nb$ by thermal neutron capture, (there are no data on the thermal neutron capture cross-section of $^{93m}Nb$), or by the production of $^{93m}Nb$ by other routes, for example from $^{92}Zr$.

## EXPERIMENTAL

### The CANDU Reactor System and Zirconium Alloy

The CANDU reactor uses natural uranium fuel which is moderated and cooled by heavy water. The fuel bundles are arranged in 389, 390 or 480 horizontal channels. The in-core part of these channels consists of a pressure tube (103mm ID, 4.1mm wall thickness) fabricated from zirconium alloyed with 2.5 w/o niobium. The pressure tubes are concentric with the zircaloy-2 callandria tubes and held in place by spacers in the form of hellical springs wrapped around the pressure tubes. The pressure tubes are 6.3m in length and are irradiated by 12 or 13 fuel bundles over a length of 5.94m.

Details of the samples from CANDU reactors and of the material irradiated in DIDO are given in Table 1. The niobium contents needed for the calculation of the reaction rate or specific activity of the nuclide being measured, i.e $^{93m}Nb$, were obtained from the manufacturers ingot analysis reports and are subject to an uncertainty of ±4% (1 e.s.d). This was one of the major contributions to the uncertainty of the final results.

TABLE 1

Niobium Content of Alloys

| Sample No. | Nb Content (w/o) | Reactor | Irradiation Time (EFPH) |
|---|---|---|---|
| AR-1A | 2.61 | DIDO | 31000 |
| R1-T3 | 2.68 | DIDO | 11300 |
| D1-T4 | 2.50 | DIDO | 11300 |
| B2-P12 | 2.50 | Bruce Generating Station, Unit 2, Channel P12 | 55000 |
| P4-K10 | 2.80 | Pickering Generating Station Unit 4, Channel K10 | 92500 |

## Calculation of doses received by CANDU samples

The integrated power for each fuel bundle location in each channel was obtained from the fuelling records using the fuelling code SORO (3) which is believed to provide power histories with an accuracy of ±5% or better. Fast fluences were then calculated from the integrated bundle powers using a relationship between linear channel power and fast flux derived from the reactor physics code WIMS-CRNL (4,5).

## Reactions of neutrons with niobium

Although elemental niobium is monoisotopic, several different products are created by reactions with neutrons of different energies. In addition to the measurement of fast neutron flux or fluence by means of the (n,n') reaction, thermal neutrons may also be determined by measuring the (n,$\gamma$) product $^{94}$Nb, though because of its very long half-life it is not very sensitive at low fluences. The (n,2n) reaction has a threshold of 10MeV but the half-life of $^{92m}$Nb being only 10.15d results initially in a high specific activity. This may be a possible source of interference with the measurement of $^{93m}$Nb during the first few months after the end of irradiation, due to the emission of zirconium X-rays which are close in energy to the niobium X-rays emitted by $^{93m}$Nb.

$^{93m}$Nb is also produced during the decay of $^{93}$Zr which itself is created by thermal neutron capture of $^{92}$Zr which is one of the naturally occurring isotopes of this element and also the decay product of $^{92m}Nb$ and $^{92}Nb$. However, calculation of the amount of $^{93m}Nb$ created by this route, using the irradiation time for the Pickering 4 sample ($3.331x10^8$s) and the cross-section and nuclear data in reference 6, showed that the contribution to the total $^{93m}Nb$ from this source, was 0.013%.

There are no cross-section data available for neutron reactions with $^{93m}Nb$ and it has been assumed that the thermal neutron reaction cross-section is the same as that for $^{93}Nb$. However, since other metastable nuclides, eg $^{58m}Co$ show very high capture cross-sections this possibility for $^{93m}Nb$ was considered by comparing the $^{94}Nb$ content of the samples with the amounts predicted from calculated thermal fluences.

The reaction cross-section data and nuclear decay data used in this work have been taken from references 6, 7 and 8.

## The Cross-Section for the (n,n') reaction

The reaction cross-section for the production of $^{93m}Nb$ varies with the neutron energy spectrum. The values appropriate to the spectra in which the samples were irradiated were calculated from the differential cross-section (9) and the neutron energy spectra which had been provided (10, 11). The reaction cross-sections for >1MeV neutrons were calculated to be 211mb and 203mb for DIDO and CANDU, respectively.

## Radiochemical Separation of Niobium and Measurement of $^{93m}Nb$

The weighed samples were dissolved in 8M HCl to which a few drops of $HBF_4$ were added to ensure complete dissolution of the niobium. The solutions were weighed and all subsequent aliquoting was carried out by weighing.

The $\gamma$- and X-ray spectra of aliquots of the sample solution showed no emitters present other than the expected nuclides $^{93m}Nb$ and $^{94}Nb$ and a small amount of $^{60}Co$. Nevertheless it was considered necessary to separate the niobium from the bulk zirconium to ensure that the measurement of the $^{93m}Nb$ X-rays was unaffected by factors such as enhancement due to fluorescence caused by the presence of $^{60}Co$ or $^{93}Zr$ or attenuation of the X-rays by the bulk zirconium.

The method chosen was originally developed for the separation of niobium from fission product mixtures (12) and employed the technique of paper chromatography. This method was simple to carry out, quantitative and gave the niobium in a form which could be counted directly and free of any possible interference.

## Counting of $^{93m}$Nb and $^{94}$Nb

The niobium activities were measured on sources prepared on filter papers which were attached to standard aluminium backing trays by means of double sided adhesive tape. The sources obtained after the separation of the niobium were 1cm square. Those made directly from the sample solutions were 9mm discs. Counting was carried out using high purity germanium detectors calibrated for both source geometries. The X- and $\gamma$-emission rates used to calculate the specific activities are summarised in Table 2.

TABLE 2

Niobium Nuclear Decay Data

| Nuclide | Energy KeV | Intensity (%) | $t_{1/2}$ |
|---|---|---|---|
| $^{93m}$Nb | 16.58<br>18.66 | 9.27±0.20<br>1.79±0.10 | 16.13±0.15y |
| $^{94}$Nb | 702.65<br>871.12 | 99.8<br>99.9 | $(2.03\pm0.16)\times10^{4}$y |

When the activities, obtained after separation, were compared with those obtained on the sources made directly from the sample solutions, they were found to be in agreement within the experimental uncertainty.

## RESULTS

The results for the DIDO irradiated samples are shown in Table 3 where they are compared with the results obtained from iron dosimeters. The iron dosimeters were replaced at regular intervals of about one year during the irradiation, which was spread over nearly eight years in the case of sample AR-1A. They were physically very close to the samples whilst in the irradiation rig (2-3mm) and good agreement has been obtained. The amount of $^{93m}$Nb produced by the thermal capture of $^{92}$Zr was calculated using the calculated thermal flux and cross-sections given in Reference 6, and found to be 264Bq $mg^{-1}$ Nb (0.002% of the total) for sample AR-1A. This calculation was not carried out for the other DIDO samples. However, the loss of $^{93}Nb$ by thermal neutron capture has been taken into account in the calculation of the growth-decay factors in all cases. The same cross-section value was also used to allow for the loss of $^{93m}Nb$ by thermal neutron capture since no other value exists. The results are given in SI units.

The results for the CANDU irradiated samples are also shown in Table 3. In the case of sample P4-K10, for which the exposure time was longest, the $^{93m}Nb$ produced from $^{92}Zr$ was calculated to be 0.013% of the total and has been ignored.

TABLE 3

Summary of Results

| | DIDO Irradiations | | | CANDU Irradiations | |
|---|---|---|---|---|---|
| | AR-1A | R1-T3 | D1-T4 | B2-P12 | P4-K10 |
| Flux>1MeV | $7.42x10^{17}$ | $6.27x10^{17}$ | $6.19x10^{17}$ | $2.70x10^{17}$ | $2.15x10^{17}$ |
| Fluence>1MeV | $8.26x10^{25}$ | $2.55x10^{25}$ | $2.51x10^{25}$ | $7.41x10^{25}$ | $7.16x10^{25}$ |
| Fluence>1MeV From Fe Wire | $8.00x10^{25}$ | $2.64x10^{25}$ | $2.67x10^{25}$ | | |
| Fluence>1MeV calculated | | | | $7.00x10^{25}$ | $7.30x10^{25}$ |

The uncertainty associated with the flux and fluence results is composed of the uncertainties of the various parameters used in the calculations. Those which are known are listed in Table 4. The major component is probably the uncertainty of the >1MeV cross-section which could not be calculated, there being no uncertainties given with the neutron spectrum data. There may also be a relatively large uncertainty on the irradiation time, possibly of the order of 5%. If that is assumed the combination in quadrature of the uncertainties listed in Table 4 yields an overall uncertainty of ±7%. The DIDO samples agree within this uncertainty with the iron dosimeters and the CANDU samples with the calculated values.

TABLE 4

Uncertainties of Measured Data

| | Std Dev (%) |
|---|---|
| Sample weighing and sources preparation | ±0.4 |
| Counting | ±2.0 |
| Efficiency of detector | ±2.0 |
| Niobium content of alloy | ±4.0 |
| Irradiation times | ±5.0 |
| Total Uncertainty | ±7.0 |

## DISCUSSION AND CONCLUSIONS

Five samples of irradiated zirconium alloy have been analysed retrospectively to determine the fast (>1MeV) neutron fluence to which they were exposed either in the DIDO reactor or a CANDU power reactor. The results agreed well with iron dosimeters (DIDO irradiations) and with the calculated fluence (CANDU irradiations).

In view of the large doses received it was considered necessary to investigate the effects of possible secondary reactions but these effects were found to be insignificant. These investigations included the possible loss of $^{93m}Nb$ by thermal neutron capture since it was recognised that there are no cross-section data for neutron reactions with $^{93m}Nb$. Although some other metastable nuclides such as $^{58m}Co$ and $^{60m}Co$ do have larger capture cross-sections than the ground state nuclides $^{58}Co$ and $^{60}Co$, the absence of data for $^{93m}Nb$ may itself be an argument for suggesting that this nuclide does not have a large cross-section as it would probably have become apparent. The results obtained on the samples, which have been the subject of this work, support this conclusion and suggest that the thermal neutron capture cross-section for $^{93m}Nb$ is probably no larger than that of $^{93}Nb$.

It has also been shown by calculation that the amount of $^{93m}Nb$ produced by neutron capture reactions with $^{92}Zr$ is negligible even for the sample with the largest fluence. It can therefore be concluded, finally, that the satisfactory use of niobium for retrospective determination of fast neutron fluence has been demonstrated for material exposed over periods of time up to 9.5y and a >1MeV neutron fluence in excess of $7x10^{25}$ $nm^{-2}$ with an acceptable accuracy.

This work forms part of the programme of Corporate Investment Research carried out by AEA Technology and part of a research programme carried out on behalf of Ontario Hydro, Canada.

## REFERENCES

1. M F Banham, A J Fudge and J A Tibbles, Retrospective Dosimetry for Nuclear Reactor Component Materials, A Review, 7th ASTM-Euratom Symposium on Reactor Dosimetry, paper 102, Strasbourg 1990.

2. R McCrohon, M E Plested and A J Fudge, Measurement of $^{93m}Nb$ for Fast Neutron Metrology, Reactor Dosimetry Methods Applications and Standardisation, ASTM STP1001, ASTM, Philadelphia, 1989, pp245-252.

3. C J Schultz, Ontario Hydro, Central Production Services Division, Private Communication (1988).

4. R E Donders, AECL-Research, Chalk River Laboratory, unpublished results, 1988.

5. J V Donelly, AECL Report AECL-8955 January 1986.

6. J W Baard, W L Zijp and H L Nolthenius, Nuclear Data Guide for Reactor Neutron Metrology, Published by Kluwer Academic Publishers, 1989 (Also Rep No EUR12354EN).

7. A Lonery, Nuclear Decay Data for Radionuclides used as Calibration Standards, INDC(NDS-145/GEI) 1983.

8. W Bambynek, Emission Probabilities of Selected X-rays for Radionuclides used as Detector Calibration Sources, GE/R/RN/18/84 (1984).

9. D B Gayther, M F Murphy, K Randle, W H Taylor and C A Uttley, The $^{93}Nb$ (n,n') $^{93m}Nb$ Cross-section, AERE-R12612 (1987).

10. R A Holt, Private Communication (1989).

11. D R Harries, P J Barton and S B Wright, Effects of Neutron Spectrum and Dose-rate on Radiation Hardening and Embrittlement in Steels, Symposium on Radiation Effects and Neutron Dosimetry, Special Technical Publication No.341, ASTM (1962).

12. A J Fudge, The Determination of Radioactivity due to Fission Product Niobium. AERE-C/R1502 (1954).

# NEW REFERENCE MATERIALS PREPARED AT CBNM FOR REACTOR NEUTRON DOSIMETRY

C. Ingelbrecht, F. Peetermans and J. Pauwels

Commission of the European Communities, Joint Research Centre

Central Bureau for Nuclear Measurements
B-2440 Geel, Belgium

## ABSTRACT

A series of reference materials intended for reactor neutron dosimetry have been prepared and certified for impurities likely to cause interference. Twelve different reference materials in the form of alloys, high purity metals or oxides are now available either with EC reference material status, or covered by provisional certificates issued by CBNM. The status of the programme is reviewed. Recently prepared reference materials are rhodium metal foil with iridium and platinum mass fractions to be certified and Al-0.1 % Au alloy intended as a comparator for $k_0$-standardisation of neutron activation analysis. The homogeneity of aluminium alloys containing 0.01 to 1.0 % cobalt was investigated and relative standard deviations of cobalt on 10 mg samples of < 1 % were found. Certification analyses for neptunium-237 dioxide spheres are now complete and a neptunium mass fraction of 873 ± 7 $mg \cdot kg^{-1}$ was calculated.

---

## 1. INTRODUCTION

Reactor neutron dosimetry can only be as good as the dosimeter materials themselves. Reliance on off-the-shelf "nominally pure" metal monitors can lead to a number of pit-falls such as interference from unexpected impurities and inhomogeneous composition or contamination. Extra problems of homogeneity arise when alloy dosimeters are used and small oxide samples can be difficult to handle and package.

In an attempt to solve some of these problems, CBNM, in collaboration with the Euratom Working Group on Reactor Dosimetry, initiated a programme (1,2) to prepare a comprehensive series of neutron dosimetry reference materials systematically certified for the trace elements likely to cause problems for the user. The highest purity starting materials have been used and the certification analyses have been carried out by laboratories of proven expertise using two or more methods in each case. The reference materials : metals, alloys and high density oxide spheres, are prepared in forms suitable for reactor dosimetry with special consideration given to homogeneity on sample sizes appropriate for dosimetry applications, usually 1 to 20 mg.

## 2. STATUS OF THE PROGRAMME

The preparation and characterisation of the reference materials are carried out or coordinated by CBNM. New RMs are generally released for use as soon as sufficient analytical data has been collected to justify this. These candidate RMs are covered at this stage by a provisional certificate issued by CBNM which, for most materials, is up-graded by submission to an EC certification authority. The list of materials currently available with their certified values and physical forms is given in Table 1 and a technical brochure giving more details of the individual RMs has been produced.

Dosimeter encapsulation in miniature quartz ampoules has been successfully demonstrated. Another suitable encapsulation material, producing no long-lived radionuclides and of low neutron capture cross-section is vanadium. Sealing of 0.5 mm diameter $UO_2$ and $NpO_2$ dosimetry spheres in high purity vanadium capsules (Ta < 0.6 $mg \cdot kg^{-1}$) by electron beam welding is being carried out and vanadium tubes suitable for all wire and sphere diameters are being produced.

## 3. NEW REFERENCE MATERIALS

### 3.1 Niobium metal

A long half-life (16.13a) and a low energy threshold (~ 0.6 MeV) make niobium a suitable damage monitor for reactor pressure vessel steel via $^{93}Nb(n,n')^{93m}Nb$. Two niobium RMs, 525 and 526, have now been awarded full EC Reference Material status (3,4). In each case the tantalum mass fraction has been certified (Table 1). This unwanted impurity can be activated by thermal neutrons and interferes by fluoresence with the X-rays produced during the decay of $^{93m}Nb$. EC-NRM 526 is metal of exceptional purity (>99.994 %) produced by electrolysis and electron beam float-zone melting by Max-Planck-Institut Stuttgart

(5). The analytical data, on which the certifications are based, are given in Fig. 1.

| RM | Certified value | | | Form |
|---|---|---|---|---|
| 501 $UO_2$* | U: | 879.4 ± 2.8 | g/kg | 0.5 mm spheres |
| | 238U/U: | 999.9896 ± 0.0005 | g/kg | 1.0 mm spheres |
| | 235U/U: | 10.4 ± 0.5 | mg/kg | |
| 502 $NpO_2$ | Np: | 873 ± 7 | g/kg | 0.5 mm spheres |
| | | | | 0.8 mm spheres |
| 521 Ni* | Co: | < 0.1 | mg/kg | 0.1 mm foil |
| | | | | 0.5 mm wire |
| 522 Cu | Co: | < 0.05 | mg/kg | 0.1 mm foil |
| | Ag: | 0.95 ± 0.04 | mg/kg | 1.0 mm foil |
| | | | | 0.5 mm wire |
| | | | | 1.0 mm wire |
| 523 Al* | Na: | < 0.1 | mg/kg | 0.1 mm foil |
| | | | | 1.0 mm foil |
| | | | | 1.0 mm wire |
| 524 Fe | Co: | < 0.05 | mg/kg | 0.1 mm foil |
| | Mn: | < 0.1 | mg/kg | 0.5 mm wire |
| 525 Nb* | Ta: | 19.6 ± 1.8 | mg/kg | 0.02 mm foil |
| | | | | 0.1 mm foil |
| | | | | 0.5 mm wire |
| 526 Nb* | Ta: | 0.30 ± 0.09 | mg/kg | 0.02 mm foil |
| | | | | 0.1 mm foil |
| | | | | 0.5 mm wire |
| 527 Al-0.1Co | Co: | 1.00 ± 0.02 | g/kg | 0.5 mm wire |
| 528 Al-1.0Co | Co: | 10.0 ± 0.2 | g/kg | 0.5 mm wire |
| 529 Rh | Pt,Ir: | - | - | 0.05 mm foil |
| 530 Al-0.1Au | Au: | 1.00 ± 0.02 | g/kg | 0.1 mm foil |
| | | | | 0.5 mm wire |
| | | | | 1.0 mm wire |

*Full EC certification complete

Table 1 : Reference Materials available

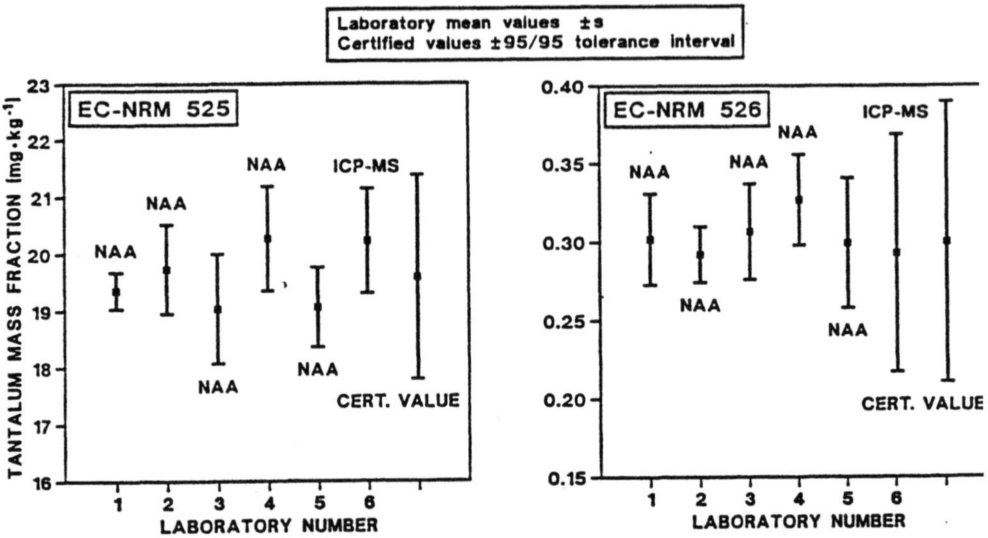


Fig. 1. Certification data for EC-NRMs 525 and 526

### 3.2 Neptunium-237 Dioxide

Neptunium-237 can be used as a fission dosimeter for fast neutrons with an energies > 0.6 MeV. High density dioxide spheres, suitable for easy manipulation and encapsulation were prepared at AEA Technology, Harwell (6). The neptunium mass fraction has been measured by controlled potential coulometry and potentiometric titration. The uncertainty on this value, which is < 1% relative, was assigned on the basis of a homogeneity control by γ-counting on individual spheres (0.9 or 3 mg). The certified value is therefore applicable to single spheres; the smallest unit intended for dosimetry use. The concentration of $^{238}Pu$ was found by α-spectrometry to be ≤ 8 x $10^{-7}$ at % and $^{239+240}Pu$ was ≤ 3.5 x $10^{-5}$ at %. These values are low enough for interference by fission of these plutonium isotopes not to be a problem.

### 3.3 Rhodium Metal

Rhodium finds applications in dosimetry because the neutron-energy response of the $^{103}Rh(n,n')^{103m}Rh$ reaction is similar to that of the fission reaction of uranium-238. Iridium and platinum cause fluoresence interference and these are the elements to be certified. High purity rhodium powder was transformed into buttons by crucible-less induction melting. Rolling was then carried out at 1000°C, firstly with the buttons welded inside stainless steel envelopes, and then under an inert atmosphere using a rolling mill inside a glove box. The final reduction to 50 μm thickness was by cold rolling. Eight laboratories participated in the certification campaign using neutron activation analyses (NAA) and induction-coupled plasma source mass spectrometry (ICP-MS). A platinum mass fraction of $< 5\ mg \cdot kg^{-1}$ and an iridium mass fraction of $24.9 \pm 1.4\ mg \cdot kg^{-1}$ were measured (one-sided and two-sided 95 % confidence intervals respectively). This material appears to be suitable for neutron dosimetry and these values for iridium and platinum mass fractions are likely to be proposed for certification.

### 3.4 Aluminium-Cobalt Alloys

Cobalt is particularly useful as a dosimeter for thermal neutrons via the $^{59}Co(n,\gamma)^{60}Co$ reaction and can also be used for monitoring fast neutrons primarily using the $^{59}Co(n,2n)^{58}Co$ reaction. However, in order to reduce self-shielding effects and because of the high activity of the irradiated dosimeter it is desirable to dilute the cobalt in a non-interfering metallic matrix. Aluminium alloys containing 0.1 and 1.0 wt % cobalt were produced in wire form and certified as CBNM NRMs 527 and 528 respectively. The alloying was carried out by levitation melting, allowing accurate control of the ingot compositions (7,8). The main concern was verification of the homogeneity of the cobalt distribution between 10 mg specimens (9). This was carried out by spectrophotometry (SP), NAA and atomic absorption spectrometry (AAS) and a summary of the data (relative standard deviations of cobalt content) are given in Table 2. The certified values for NRMs 527 and 528 (Table 1) are conservative and take into account any inhomogeneity on a 10 mg scale.

The measurement of high fluence rates and doses has lead to the use of very small cobalt dosimeters (~ 1 mg Al-Co alloy). The certified values of CBNM-NRM 527 and 528 may not be valid for these sample sizes and weighing errors also start to become significant. In response to this, a more dilute alloy, Al-0.01 Co, was prepared and analysed (Table 2). A RSD by NAA slightly larger than that measured for the more concentrated alloys was found. Nevertheless, this alloy appears suitable for dosimetry applications and a larger batch may be prepared for certification.

The cobalt in these alloys is present as $Al_9Co_2$ intermetallic in eutectic form and as larger crystals along the centre line of the casting (9). The segregation of these intermetallic particles within the casting is the cause of inhomogeneity of the cobalt distribution.

| Co content (wt %) | Method | Sample mass (mg) | Co content RSD (%) | Number of measurements |
|---|---|---|---|---|
| 1.0 | SP | 10 | 0.3 | 6 |
| | NAA | 10 | 0.3 | 12 |
| | AAS | 10 | 0.9 | 12 |
| 0.1 | SP | 100 | 0.5 | 6 |
| | NAA | 10 | 0.4 | 12 |
| | AAS | 10 | 0.8 | 14 |
| 0.01 | NAA | 10 | 0.7 | 10 |

Table 2 : Homogeneity of Al-Co alloys by spectrophotometry (SP), neutron activation analysis (NAA) and atomic absorption spectrometry (AAS)

### 3.5 Aluminium-Gold Alloy

The $k_0$-standardisation method is becoming an increasingly useful and widely accepted procedure for neutron activation analysis(10). This makes use of an experimentally determined value (the $k_0$ factor), which is a ratio of nuclear constants. This ratio can be determined relatively accurately and this avoids the use of certain constants ($\gamma$ emission probabilities and neutron cross-sections) for which the published values can be quite inaccurate. The method requires coirradiation with a comparator material. Gold is usually chosen because of the convenient half-life of $^{198}Au$ and because the nuclear constants for gold are well known. Tabulated $k_0$ values based on a gold comparator ($k_{0,Au}$) are published for many nuclides for $(n,\gamma)$ activation analysis (11).

The most convenient form for the comparator is the dilute alloy Al-0.1Au for which self-shielding from thermal and epithermal neutrons is negligible. It is important that this alloy has an accurately known composition and that the homogeneity is good for sample sizes in the range 2 to 60 mg. A batch of 20 g castings of Al-0.1Au was prepared by levitation melting and transformed into foil and wire. The gold is essentially insoluble in the aluminium matrix and is present in the alloy as $AuAl_2$ intermetallic. Metallography on sections through the castings showed that the intermetallic particles were very fine and evenly distributed. Neutron activation analysis on 10 mg samples confirmed that the alloy was extremely homogeneous with a relative standard deviation of gold content of 0.30 %. The gold mass fraction of the alloy has been certified as $1.00 \pm 0.02\ g \cdot kg^{-1}$ and the alloy serves as a reference material for $k_0$-standardisation.

## 4. CONCLUSIONS

The EWGRD reference materials programme has been reviewed. Twelve RMs and candidate RMs are currently available. New additions to the list are rhodium metal foil with low mass fractions of iridium and platinum, and aluminium gold alloy intended as comparator alloy for $k_0$-standardisation of neutron activation analysis.

The programme aims to establish an assured supply of well characterised reference materials to improve accuracy in surveillance and research work and to meet quality assurance needs. It is hoped that many traditional problems associated with availability, purity and homogeneity of monitor materials will be eliminated by the use of this certified stock.

## 5. PARTICIPATING LABORATORIES

The following laboratories have contributed towards the certification analyses for the reference materials in Table 1 :

AEA Technology (UK), Bundesanstalt für Materialforschung und -Prüfung (FRG), Centre National de la Recherche Scientifique (F), Joint Research Centre (EUR), Max-Planck-Institut für Metallforschung (FRG), Rijksuniversiteit Gent (B), Risø National Laboratory (DK), Stichting Energieonderzoek Centrum (NL), Studiecentrum voor Kernenergie/Centre d'Etude de l'Energie Nucléaire (B).

This work was carried out within the European Commission research and development programme.

## 6. REFERENCES

1. J. Van Audenhove, J. Pauwels and A.J. Fudge, "A European Community Source of Reference Materials for Neutron Metrology Requirements" Nucl. Instr. and Meth. in Phys. Res. A236 pp. 600-603 (1985)

2. A.J. Fudge, J. Van Audenhove and J. Pauwels, "Neutron Dosimetry Materials Used in European Community Countries", Reactor Dosimetry Methods, Applications and Standardization, ASTM STP 1001, H. Farrar IV and E.P. Lippincott, Editors, ASTM, Philadelphia, pp. 627-633 (1989)

3. F. Lievens, C. Ingelbrecht and J. Pauwels, "Certification of a Niobium Metal Reference Material for Neutron Dosimetry (EC Nuclear Reference Material 525)" Commission of the European Communities Report EUR 12892 EN (1990)

4. F. Lievens, C. Ingelbrecht, J. Pauwels and K. Schulze, "Certification of a Niobium Metal Reference Material for Neutron Dosimetry (EC Nuclear Reference Material 526)" Commission of the European Communities Report EUR 12894 EN (1990)

5. K. Schulze and M. Krehl, "The Preparation of Pure Niobium for Neutron Dosimetry Purposes" Nucl. Instr. and Meth. in Phys. Res. A236 pp. 609-616 (1985)

6. C.E. Lyon, "The Preparation of Metal Oxide Spheres for Use in Neutron Dosimetry" Nucl. Instr. and Meth. in Phys. Res. A236 pp. 604-608 (1985)

7. J. Van Audenhove and J. Joyeux, "The Preparation by Levitation Melting in Argon of Homogeneous Aluminium Alloys for Neutron Measurements" J. Nucl. Mat. 19 pp. 97-102 (1966)

8. J. Van Audenhove and J. Joyeux, "Sample Preparation by Metallurgical Methods" Nucl. Instr. and Meth. 102 pp. 409-415 (1972)

9. C. Ingelbrecht, F. Peetermans and J. Pauwels, "A Study of the Homogeneity of Aluminium-Cobalt Alloys for Neutron Dosimetry" Nucl. Instr. and Meth. in Phys. Res. A282 pp. 217-222 (1989)

10. F. De Corte, A. Simonits, A. De Wispelaere and J. Hoste, "Accuracy and Applicability of the $k_0$-standardization Method" J. Radioanal. Nucl. Chem. 113, 1, pp. 145-161 (1987)

11. F. De Corte and A. Simonits, "$k_0$-measurements and Related Nuclear Data Compilation for $(n,\gamma)$ Reactor Neutron Activation Analysis" J. Radioanal. Nucl. Chem. 133, 1, pp. 43-130 (1989)

# ELECTRON LEAKAGE MODELS FOR THE EVALUATION OF SELF-POWERED DETECTOR RESPONSES

D.A. Thornton
Nuclear Electric, plc,
Berkeley Nuclear Laboratories,
Berkeley, Gloucestershire, GL13 9PB, UK

## ABSTRACT

This paper discusses methods for the simulation of electron transport in self-powered detectors (SPDs). Previous work has assumed that electron trajectories through the emitters of SPDs are straight and that energy loss can be related to trajectory length. Two new methods have been developed which are more representative of the actual energy losses. SPD response predictions using these methods have been compared with the original work and with Monte Carlo reference calculations. This has shown that a representative electron transport model is crucial to an accurate prediction of SPD responses; the original model can lead to errors of 70% or more.

---

## INTRODUCTION

Self-Powered Radiation Detectors (SPDs) are so named because their response to a radiation field is a current produced without auxiliary power. Their basic design, shown in figure 1, is in the form of a co-axial cable. The emitter and the sheath are conducting electrodes separated by a ceramic insulator. Current is produced by the interactions of radiation resulting in the expulsion of charge from the emitter.

Emitter materials are chosen to produce three basic types of device. These are: delayed SPDs in which the emitter is beta activated by neutrons, prompt SPDs which respond to prompt capture gamma radiation in the emitter, and finally, self-powered gamma detectors (SPGDs) responding to gamma radiation produced outside the device.

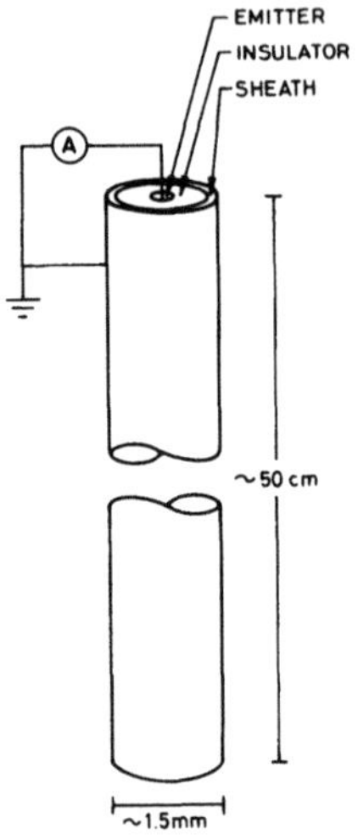


FIG.1. CONSTRUCTION OF A SELF-POWERED DETECTOR.

Although designed to exhibit one of the principal mechanisms described above, all SPDs have response mechanisms other than their principal type. Consequently, behaviour can be difficult to predict and a theoretical understanding of SPDs can make a valuable contribution to their effective deployment.

A full model of SPD responses must account for the reactions of the primary radiations, the production of energetic electrons, the transport of energetic electrons, and the electric field produced by the accumulation of charge in the insulator. A recent study (ref 1) supported by Nuclear Electric plc has developed such work.

A critical area in SPD modelling is predicting the transport of electrons. To date two approaches have been used: Monte Carlo methods (ref 2) which are computationally intensive but represent the physics explicitly, and analytic methods (ref 3) which approximate the net effects of the electron transport process. This paper is primarily concerned with analytic methods and, without loss of generality, is restricted to the discussion of delayed SPDs.

## SPD RESPONSE MODELS

In general form, the response current of a delayed SPD is given by:

$$I = e R Y \varepsilon \qquad [1]$$

where e is the electronic charge, R the neutron capture rate in the emitter, and Y the number of beta particles emitted per neutron

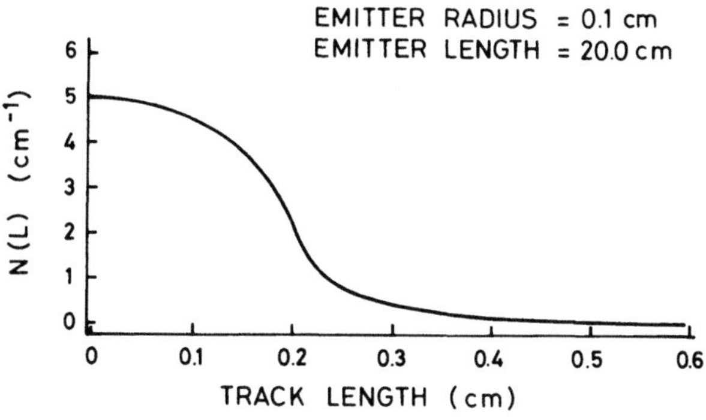


FIG.2. TRACK LENGTH PROBABILITY DISTRIBUTION IN A CYLINDER.

capture. $\varepsilon$ is the fraction of beta particles that contribute current (or the 'current production efficiency'). In the analytic formulation the spectrum and transport of electrons are folded into $\varepsilon$, which is given by:

$$\varepsilon = \int_{EMIN}^{E_\beta} dE \int_{E}^{E_\beta} dE'\, S(E', E) B(E') \qquad [2]$$

where B(E′) is the beta emission spectrum and $E_\beta$ its endpoint. Beta spectra have been generated using a computer code described by Tobias (ref 4). S(E′, E) is called the 'electron transport kernel' and is the differential probability that a beta particle, emitted with kinetic energy E′, leaves the emitter with energy E.

A proportion of the expelled charge comes to rest in the insulator producing a monotonic electric field which has a potential maximum at radius $r_p$. Under the influence of this field, charge drifts away from $r_p$ towards the electrodes. In order to contribute to the SPD's response, an electron leaving the emitter has to have sufficient kinetic energy to cross $r_p$. This cut-off energy is EMIN.

## Evaluation of the Electron Transport Kernel

To derive the electron transport kernel, electron trajectories are represented as straight chords, or tracks, between initialization and the emitter's surface. The tracks originate isotropically and homogeneously throughout the emitter. The probability of a track of length L is found from the distribution N(L) dL, which has been derived by Snidow and Warren (ref 5) for a cylinder and is shown in figure 2. The electron transport kernel is obtained by expressing L in terms of E′ and E.

Original Model. The original model was presented by Warren (ref 2). This model assumes that electron trajectories through the emitter can be approximated by straight lines. Therefore, a track length, L, is equal to the length of trajectory, t, between energies E′ and E:

$$t = R(E') - R(E) \qquad [3]$$

Evaluated in the continuous slowing down approximation (CSDA), the mean electron trajectory length (or range) is given by:

$$R(E) = \int_0^E d\tau \left( -\frac{d\tau}{dx} \right)_\tau^{-1} \qquad [4]$$

where $\left( -\frac{d\tau}{dx} \right)_\tau$ is the electron stopping power in the given material at kinetic energy $\tau$.

Substituting t for L gives the electron transport kernel:

$$S(E', E)dE = N[R(E') - R(E)] \left( -\frac{dE}{dx} \right)_E^{-1} dE \qquad [5]$$

Since electrons do not travel in straight lines, the net displacement experienced by an electron on slowing between E′ and E is less than the corresponding trajectory length. Consequently, this model overpredicts the response of the device. The accuracy attainable with this model is dependent on the validity of the straight line propagation assumption.

Use of Extrapolated Ranges. The extrapolated (or practical range) $R_{ex}(E)$ of electrons is defined as the thickness of material at which the extension of the linearly decreasing region of the number transmission curve becomes zero. This is illustrated in figure 3 which shows a typical electron transmission curve. Tabata et al. (ref 6) have used experimental data for the transmission of electrons through foils to obtain a semiempirical equation for $R_{ex}(E)$. Replacing CSDA ranges in equation 5 gives the electron transport kernel as:

$$S(E', E)dE = N[R_{ex}(E') - R_{ex}(E)] \left( \frac{dR_{ex}}{dE} \right)_E dE \qquad [6]$$

Extrapolated ranges provide practical upper limits on the penetration of electrons. They are not intended to represent the net

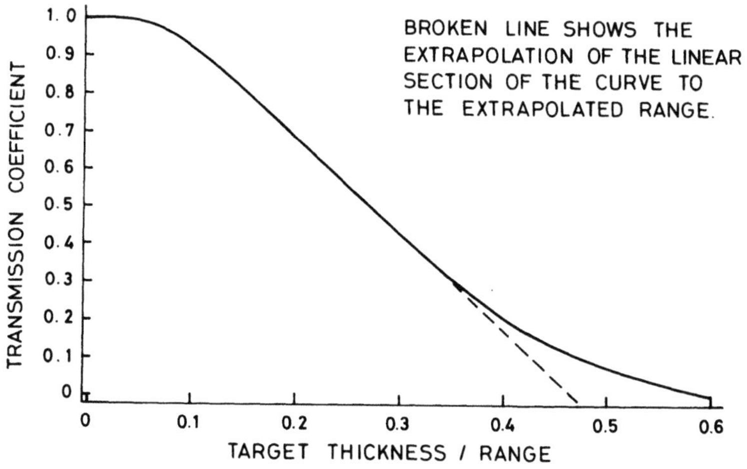


FIG.3. TRANSMISSION COEFFICIENTS OF 2.5 MeV ELECTRONS ON SILVER.

distances traversed between the beginning and end of electron tracks. Their use neglects transverse displacements and overestimates the mean longitudinal displacement. They also neglect backscattered electrons.

These deficiencies imply contradictory trends in the relationship between actual displacements and extrapolated ranges. Consequently, it is difficult to anticipate the likely consequences of using an electron transport kernel evaluated with extrapolated ranges. However, they do provide a kernel that represents electron trajectories more realistically.

Use of Calculated Net Displacements. Bielajew and Rogers (ref 7) have developed an algorithm for the computation of the mean net distance travelled by electrons decelerating between given initial and final energies. The theory involves calculating mean longitudinal and transverse displacements and has been adapted to suit use in the present work. Full details can be found in ref 1.

The mean longitudinal and transverse displacements, $\langle z(t) \rangle$ and $\langle r(t) \rangle$, of an electron after travelling through a trajectory of length $t = R(E') - R(E)$ are:

$$\langle z(t) \rangle = \int_0^t dt' \langle \cos\theta(t') \rangle \qquad [7]$$

$$\langle r(t) \rangle = \frac{1}{2}\, t \langle \sin\theta(t) \rangle \qquad [8]$$

where $\langle \cos\theta(t) \rangle$ and $\langle \sin\theta(t) \rangle$ are the average values of the cosine

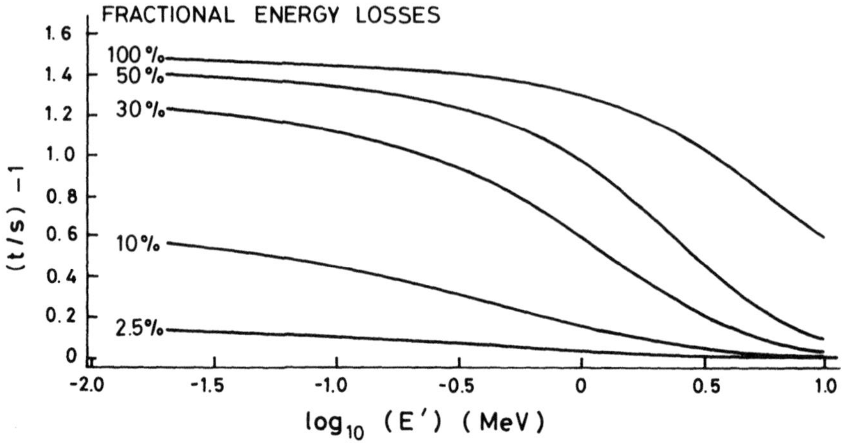


FIG.4. COMPARISON OF ELECTRON DISPLACEMENTS WITH RANGE DIFFERENCES.

and sine of the electron's angle of propagation (with respect to initial motion) after travelling through a trajectory of length t. These averaged trigonometrical functions have been calculated with the electron multiple scattering distribution of Goudsmit and Saunderson which is more generally applicable than the Moliere distribution used by Bielajew and Rogers (ref 7). In this work, the original formulation of the Goudsmit Saunderson distribution (ref 8) has been used, based on the screened Rutherford single scattering cross section.

Combining longitudinal and transverse displacements gives the net displacement, $s(E', E)$. Figure 4 shows a plot of $t/s(E', E) - 1$ against $E'$ in rhodium, for several values of $E/E'$. It can be seen that trajectories can only be represented by straight lines for low energy losses or high kinetic energies. Incorporating $s(E', E)$ into the electron transport kernel gives:

$$S(E', E)\,dE = N[s(E', E)]\left(\frac{\partial s}{\partial E}\right)_{E',\,E} dE \qquad [9]$$

## Monte Carlo Methods

Electron transport calculations have been performed using the Integrated Tiger Series (ITS) of electron-photon transport codes (ref 9). By modelling the transport of beta particles through the emitter and insulator of a SPD, the charge deposition rate can be found in all components of the detector. $\varepsilon$ is equivalent to the charge expelled to outside $r_p$ and is found from the radial integration of the charge deposition rate. For consistency with the analytic model, $r_p$ has been calculated assuming a uniform charge distribution in the insulator.

## RESULTS

Table 1 gives the specifications of the rhodium and vanadium detectors considered in this work. Table 2 gives values of $\varepsilon$ evaluated using the methods discussed in the previous section. For completeness, the values of EMIN used in the analytic calculations are also included. The uncertainties quoted with the Monte Carlo calculations are stochastic.

The data evaluated with the CSDA range kernel are a factor of 1.7 times higher than the data obtained with the more representative methods. It is clear that the assumption of straight line propagation, implicit in the CSDA range kernel, is invalid. The results from the use of extrapolated ranges, net displacements and Monte Carlo methods are consistent to within 10%. This is very encouraging since the individual methods vary considerably in detail and approach.

| Emitter | | Insulator | | | Sheath | |
|---|---|---|---|---|---|---|
| Material | Radius (cm) | Material | Density ($g\ cm^{-3}$) | Thickness (cm) | Material | Thickness (cm) |
| Rh | 0.023 | MgO | 3.58 | 0.031 | inconel | 0.025 |
| V | 0.102 | $Al_2O_3$ | 1.90 | 0.048 | inconel | 0.025 |

**Table 1. Specifications of the detectors considered.**

| Emitter Material | EMIN (MeV) | $\varepsilon$ | | | |
|---|---|---|---|---|---|
| | | Original Model | Extrap. Ranges | Net Disp. | Monte Carlo |
| Rh | 0.343 | 0.609 | 0.378 | 0.386 | $0.358 \pm 0.011$ |
| V | 0.346 | 0.385 | 0.244 | 0.225 | $0.231 \pm 0.003$ |

**Table 2. Current production efficiencies evaluated with different models.**

## CONCLUSIONS

1. Electron transport is a crucially important part of any SPD response model. It is necessary to represent the tortuous nature of electron trajectories in order to calculate the correct proportion of electrons contributing to the detector's response.

2. The original electron transport model, assuming linear propagation, significantly overpredicts the leakage of electrons from the emitter. It has been shown that this method can result in errors of order 70%.

3. The analytic models using extrapolated ranges and calculated net displacements give results consistent to within 10% of the Monte Carlo calculations.

4. The relative accuracy of the extrapolated range model is judged to be fortuitous since the theoretical justification for its use is weak. However, its use in the analytic model is very efficient giving it an advantage over the other analytic models.

## ACKNOWLEDGEMENT

This paper is published by permission of Nuclear Electric plc.

## REFERENCES

1. Thornton D A, 1989, PhD Thesis, Calculating the responses of self-powered radiation detectors
2. Warren H D, 1974, Nuc. Sci. Eng., 48, p.337
3. Hall D S, 1982, IEEE Trans. Nuc. Sci., NS-29 No.1, p.646
4. Tobias A, 1976, CEGB report RD/B/M3740, The average energy of beta spectra
5. Snidow N L and Warren H D, 1967, Nuc. Inst. Meth., 51, p.109
6. Tabata T, Ito R and Okabe S, 1972, Nuc. Inst. Meth., 103, p.85
7. Bielajew A F and Rogers D W O, 1987, Nuc. Inst. Meth., B18, p.165
8. Goudsmit S and Saunderson J L, 1940, Phys. Rev., 57, p.25
9. Halbleib J A and Melhorn T A, 1984, SAND 84-0537, The Integrated Tiger Series of coupled electron-photon Monte Carlo transport codes

# EXPERIENCES WITH THE REACTION $^{93}Nb(n,n')^{93m}Nb$ IN NEUTRON FLUENCE MONITORING

W.E. Freudenreich

Netherlands Energy Research Foundation ECN,
P.O. Box 1, 1755 ZG Petten, The Netherlands

ABSTRACT

First some facts are presented why niobium is of interest in neutron fluence monitoring. Then a description is given of the procedure that was followed with the niobium monitors. This paper shows that the results obtained with this procedure are good with respect to the statistical uncertainties in the $^{93m}$Nb-activity (about 2.5% relative standard deviation). Systematic effects are somewhat larger. One of the systematic effects present in the ratio of the measured specific reaction rates (spectrum index SI) of the reactions $^{93}Nb(n,n')^{93m}Nb$ and $^{54}Fe(n,p)^{54}Mn$ could be identified : the dependence of the SI(Nb/Fe) from the irradiation position in the reactor core (which is qualitatively in good agreement with other observations). However, the distributions of the SI(Nb/Fe) still show that other systematic effects must be present. Up to now it is not known whether they are due to the measurement procedure or due to the irradiation procedure in the reactor.

## 1. INTRODUCTION

The radiation metrology group at ECN provides the neutron fluence monitoring of the material test irradiations in the High Flux Reactor (HFR) at Petten. These irradiations comprise very different materials, e.g. steels, graphite, fuel, ceramics etc. The neutron fluence scoped varies from $1\cdot10^{24}m^{-2}$ to $8\cdot10^{25}m^{-2}$ at a fluence

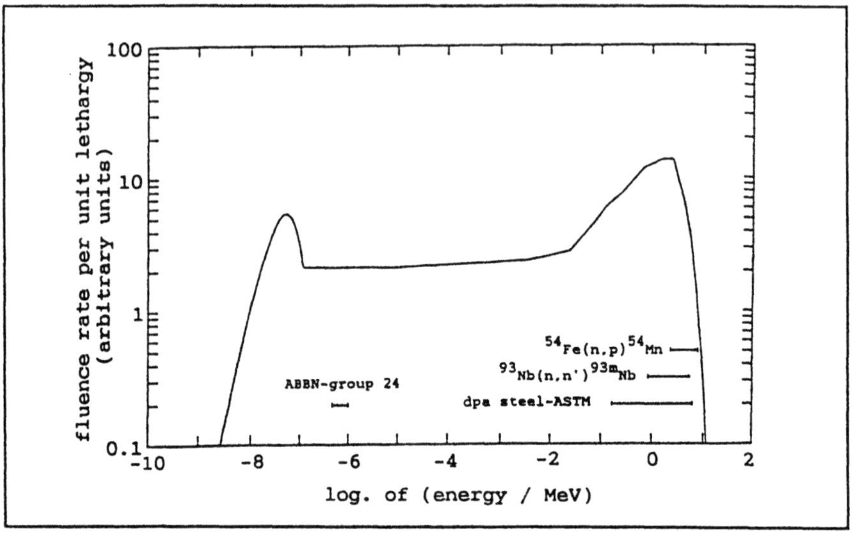


Fig. 1. Typical HFR-spectrum with indication of 90%-response regions and location of ABBN-group 24.

rate level of about $1 \cdot 10^{18} m^{-2} \cdot s^{-1}$.

Niobium has been used for a long time for monitoring of the thermal energy range because of the small decay constant of $^{94}$Nb. Two facts make the (n,n')-reaction of niobium very interesting in neutron fluence monitoring :

- the small decay constant of $^{93m}$Nb, which makes it a good fluence monitor;
- its low threshold of about 100 keV, which gives response in the energy range of about 0.1 to 1.0 MeV that is important for radiation damage.

The latter is shown in fig. 1, where the 90%-response regions of dpa (displacements per atom) in steel-ASTM and of activation of $^{93m}$Nb and $^{54}$Mn in a typical HFR-spectrum are indicated.

In 1986 we started with the measurements of $^{93m}$Nb. At the moment we have results of about 200 irradiated niobium monitors available. The monitors represent about 20 different material irradiations. The irradiations comprised (in some cases) several irradiation channels. Each channel contained about five monitor sets at different vertical positions. The sets comprised among others niobium and iron monitors (see also fig. 2).

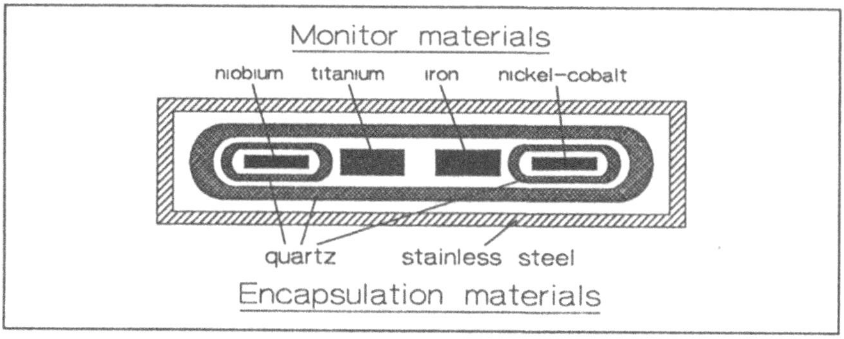


Fig. 2. Schematic view of a typical activation monitor set.

## 2. MEASUREMENT PROCEDURE WITH NIOBIUM MONITORS

The niobium monitors are produced by cutting pieces of about 350 µg from a thin plate or wire of high purity niobium metal (Ta content < 10 ppm). The mass is determined with a microbalance with 1 µg accuracy. Each niobium monitor is put in a first containment of quartz (minitube with l≈5 mm and Ø=0.7/0.2 mm closed by melting). Then these encapsulated niobium monitors are put together with the other monitor materials in a second containment of quartz (minitube with l≈15 mm and Ø=1.4/0.8 mm closed by melting). In all cases, except for an irradiation of graphite, the monitor set is further encapsulated in a containment of stainless steel (tube of l=20 mm and Ø=2.0/1.6 mm closed by welding). Fig. 2 shows a schematic view of a typical monitor set.

After irradiation in the reactor and a waiting time of at least ½ year the monitor set is unpacked. The niobium monitors are kept within their first quartz encapsulation. The measurements of the niobium monitors start with the γ-counting of $^{94}$Nb from the reaction $^{93}$Nb(n,γ). Then the niobium monitor with its quartz containment is dissolved in a mixture of 0.2 ml of 1N HF and 0.2 ml of 1N $HNO_3$. The X-ray counting (16.6 keV and 18.7 keV) of $^{93m}$Nb are done on a high purity germanium X-ray detector. The activity is obtained by comparison with a calibrated reference sample with the same measurement geometry.

## 3. DATA ANALYSIS METHOD

The quality of the results obtained with the (n,n′)-reaction of niobium is tested by an analysis of a spectrum index (SI). The SI is the ratio of the specific reaction rates of two reactions. In this case one of the reactions is the reaction $^{93}$Nb(n,n′)$^{93m}$Nb and as

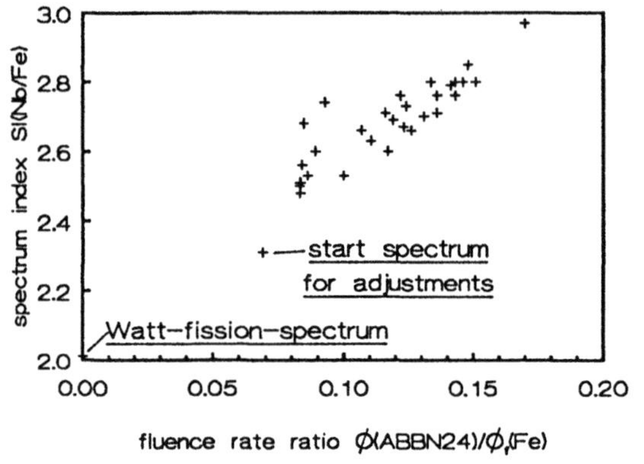


Fig. 3. Dependence of the spectrum index SI(Nb/Fe) from the neutron spectrum.

reference reaction the reaction $^{54}Fe(n,p)^{54}Mn$ is taken. So one gets the SI(Nb/Fe).

Because the response regions of these two reactions are different in a MTR-spectrum (see fig. 1), the value of the SI(Nb/Fe) becomes spectrum dependent. This is shown in fig. 3. This figure shows the SI's(Nb/Fe) for a number of adjusted spectra. The spectrum adjustments have been done with a modified version of the program SAND-II [1] with information of the measurements of about thirty activation and fission reactions for mock-up irradiations in the HFR-Petten (no niobium monitors have been used in these adjustments) and one calculated initial spectrum. The cross-sections for these calculations have been taken from [2].

The characterization of the spectrum is done by the ratio of the fluence rate in ABBN-group 24, Ø(ABBN24), (which is used as a measure for the intermediate fluence rate; see also fig. 1) and the fission equivalent fluence rate obtained with the reaction $^{54}Fe(n,p)^{54}Mn$, $Ø_f(Fe)$.

The values of the SI's(Nb/Fe), obtained from measurements, are analyzed in two ways :
- it is tried to find systematics in the absolute values;
- the spreads in different distributions of the SI's(Nb/Fe) are compared with values for the spreads that one would expect.

## 4. RESULTS

One of the possible systematic dependences of the SI(Nb/Fe) is that of the vertical position in an irradiation channel. Fig. 4 shows the vertical distribution of the SI for two different experiments. The fuel region in the HFR extends from -300 mm to +300 mm. From a number of experiments we found, that the SI-values in the region from about -230 mm to +230 mm show no systematic dependence of the vertical position. Outside this region we find systematically larger values of the SI(Nb/Fe).

Because we could not find any systematic dependence of the SI for the various vertical positions, within the ∓230 mm region, in a particular irradiation channel, the spread within the SI-values for each channel should be purely due to statistics. If this is true, than this spread can be used to estimate the random uncertainty in the measurements of the $^{93m}$Nb-activity. Fig. 5 shows the frequency distribution of the relative spreads found in 32 different irradiation channels. A different number of monitor sets were available for the channels. The average value in the frequency distribution of the spreads per irradiation channel is found to be 2.8%.

This average spread of 2.8% is the result of the random uncertainties in the measurements of the activities of $^{93m}$Nb and of $^{54}$Mn. For the random uncertainty in the activity measurements of $^{54}$Mn we assume a value of 1.5%. We then get a value of 2.4% for the

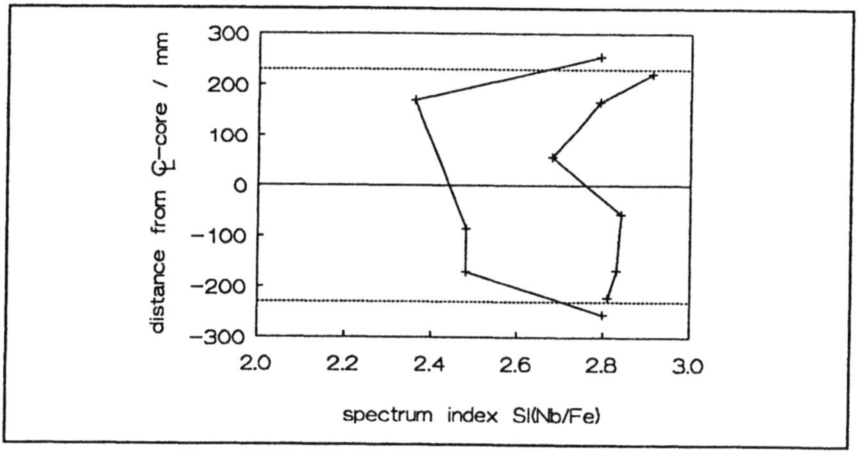


Fig. 4. Dependence of the spectrum index SI(Nb/Fe) from the vertical position in an irradiation channel, illustrated for two different experiments.

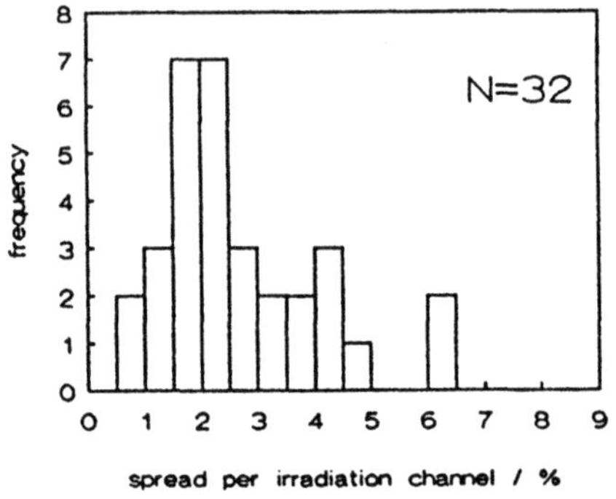


| number of monitors per channel | number of channels |
|---|---|
| 3 | 7 |
| 4 | 7 |
| 5 | 10 |
| 6 | 7 |
| 8 | 1 |

Fig. 5. Frequency distribution of the spreads of the spectrum index SI(Nb/Fe) found in 32 different irradiation channels.

random uncertainty in the activity measurements of $^{93m}Nb$.

For the HFR-core at Petten we roughly distinguish between four irradiation regions : central (C), semi-peripheral (S), peripheral (P) and extreme-peripheral (E). When the SI's for the adjusted spectra, mentioned in chapter 3 (see also fig. 3), are assorted to their appertaining core regions, then we find systematically different average values for the spectrum indices in these regions (see table 1). There is a slight increase of the average SI with the distance from the centre of the reactor core. This is comparable with the increase of the SI with the vertical distance from the centre of the core (see fig. 4).

The same as has been done with the SI's for the adjusted spectra (see above), can be done with the SI's obtained from the measurements in irradiation experiments. The results are listed in table 2. Also the SI's presented in this table show an increase with the distance from the reactor core. However, the values of table 2 are systematically lower by about 4.1% than those of table 1. It is noted that this systematic difference is strongly dependent on the used cross-section evaluation for the (n,n')-reaction of niobium. The evaluation used here [2] gives a fission averaged cross-section of 158 mb. A value that is lower by 15 % is obtained from a preliminary version of the IRDF-90 [3].

When the spreads given in table 2 are compared with the average spread obtained from fig. 5 (2.8%), then it is clear that the SI's for a particular core region are not free from systematic effects (except for the central region with only one channel). At least a part of the larger spread is due to the geometry of the irradiation experiments and the irradiation procedure. Most of the

Table 1. Average values of the spectrum indices SI(Nb/Fe) calculated from a number of adjusted specta in the four HFR irradiation regions.

| region in the HFR-core | number of spectra / channels | SI(Nb/Fe) average | SI(Nb/Fe) spread / % |
|---|---|---|---|
| C | 1 / 1 | 2.50 | - |
| S | 8 / 8 | 2.57 | 3.8 |
| P | 14 / 14 | 2.69 | 2.8 |
| E | 7 / 7 | 2.81 | 3.1 |

Table 2. Average values of the spectrum indices SI(Nb/Fe) calculated from a number of measurements in the four HFR irradiation regions.

| region in the HFR-core | number of monitors / channels | SI(Nb/Fe) average | SI(Nb/Fe) spread / % |
|---|---|---|---|
| C | 4 / 1 | 2.30 | 2.3 |
| S | 61 / 18 | 2.50 | 6.4 |
| P | 41 / 10 | 2.65 | 4.1 |
| E | 74 / 17 | 2.69 | 7.4 |
| total | 180 / 46 | 2.61 | 7.4 |

<u>HFR core regions :</u>

C - central;
S - semi-peripheral;
P - peripheral;
E - extreme-peripheral.

experiments are TRIO-capsules (three irradiation channels at one core position) where the channels have different distances to the centre of the core. Also the position of nearly all experiments is changed in the reactor core during irradiation, although mostly they stay within the same core region. It is not reasonable to assume that the SI has a sharp defined value for a certain core region which also gives a contribution to the spread.

## 5. CONCLUSIONS

The results that have been obtained from the analysis of the SI(Nb/Fe) show that this is a useful method to get information about the quality of the $^{93m}$Nb measurements (if one knows the quality of the $^{54}$Mn measurements).

These results, especially the spreads of the SI per irradiation channel (see fig. 5), also show, that our method of handling the niobium monitors and performing the X-ray countings give good results with respect to the random uncertainties.

One systematic dependence of the SI-data that could be extracted from our data is the dependence of the SI(Nb/Fe) from the neutron spectrum c.q. the irradiation region in the HFR-core. However, as can be seen from the relatively large spread in the SI-data per core region (see table 2) compared with the average spread per irradiation channel (see fig. 5 and chapter 4) there have to be more systematic effects. Up to now we do not know whether the larger spread is completely due to spectrum alterations within a core region or also partly due to other effects, e.g. :
- production of $^{93m}$Nb by high energy photons;
- burn-up of $^{93m}$Nb by thermal neutrons.

## 6. REFERENCES

[1] McElroy, W.N., et al.: "SAND-II neutron flux spectra determination by multiple foil activation iterative method", RSIC computer code collection CCC 112, Oak Ridge National Laboratory, Radiation Shielding Information Centre (May 1969).

[2] Zijp, W.L., Nolthenius, H.J., and Verhaag, G.C.H.M.: "Cross-Section Library DOSCROS84 (in a 640 group structure of the SAND-II type)", Report ECN-160, (ECN, Petten, 1984).

[3] Nolthenius, H.J.: Private communication.

Automated Reader for Solid-state Fission Track Recorders#

C. A. Wemple*, F. M. Clikeman*, and D. W. Vehar**

*School of Nuclear Engineering
Purdue University
W. Lafayette, IN 47907 USA

**Sandia National Laboratories
Albuquerque, NM 87185 USA

## Abstract

An automatic optical track identification/counting system has been developed for counting the total number of fission tracks on a fused quartz solid state track recorder. The system is capable of analyzing up to twenty recorders a day with an operator input of less than two hours. The uncertainty introduced by the counting system is about one percent.

## Keywords

Automatic track counting; fission rate measurements; fused quartz.

## Introduction

Solid state track recorders (SSTR) have been used in various capacities for many years. Conference proceedings such as References 1 and 2, books such as the one by Fleischer, Price and Walker [3] as well as many journal articles describe numerous techniques and applications. Depending on the types of recording media utilized, two methods, spark counting and optical counting, are used to count the tracks in the recorders. Spark counting techniques yield the total number of events recorded. Optical scanning will also give information about the size and shape of the tracks but require the use of a microscope to magnify the track image. The tracks are then either counted manually or automatically. Since the field of view of the microscope is usually much smaller than the size of the track recorder and since most applications require only the

# This work was supported by the United States Department of Energy under Contract DE-AC04-76DP00789.

track density (number of tracks per unit area), most counting procedures simply count the number of tracks in a known field of view, randomly move the microscope platform to a new region and count again. By averaging a number of readings, the track density of the recorder can be established with a reasonable degree of accuracy.

In the Purdue University Fast Breeder Blanket Facility (FBBF), SSTR are used to measure the absolute fission rates of a number of isotopes such as $^{235}U$, $^{239}Pu$, $^{238}U$, $^{232}Th$, and $^{237}Np$. The last three of these isotopes are threshold reactions with low reaction rates. In these measurements, SSTR are sandwiched with a known thin (< 100 $\mu g/cm^2$) deposit of fissionable material and exposed to the neutron flux in the facility. Since the quantity of interest is the absolute fission rate, both the quantity of the isotope used in each experiment and the total number of tracks must be measured.

The use of SSTR for fission rate measurements has a number of advantages. First, in low count rate situations, such as the FBBF facility, the experiments can be left in the facility for long periods (up to 90 days) in order to record enough events to provide meaningful statistics. Electronic counting techniques would place unreasonable requirements on the stability of the electronic system in order to perform the same measurements. In the opposite situation where one encounters a very high reaction rate for a short time such as with a pulsed reactor, techniques using SSTR do not suffer from dead-time losses. The sensitivity of the detector system can be varied by adjusting the area of the fission targets and, to a limited degree, by the thickness of the target. Other advantages of the fission track counting method include the relatively small perturbation introduced by the recorders and the low background when compared to most electronic counting systems.

Two previous automatic counting systems have been developed and used for a number of years at Purdue [4,5,6]. Both of the previous systems were slow; the first system required 24 hours per recorder [4,5] while the second system could analyze up to four recorders per day [6]. New experimental activities require faster analyzing times. The specifications for the new system included analyzing up to twenty recorders a day with a maximum of two hours of operator time per day. This paper describes the hardware, computer software, and performance of the system.

## Track Counting System

**Hardware** A Wild/Leitz TAS Plus system was selected for the automated counting system. The TAS Plus system consists of a Leitz ORTHOPLAN large field microscope with a six-lens rotation nosepiece, swing-out condenser, and FSA binocular tube: a Leitz 100Z lamphousing with a 12V 100W tungsten-halogen lamp; a Bosch TYK9A System Compact Camera with external synchronizer and power supply unit; and the Leitz TAS

Plus Automatic Image Analysis System. The Automatic Image Analysis System consists of the color monitor, keyboard, multi-function control box, operating unit, power supply and PDP 11/03 microcomputer with the RT-11 operating system and various electronic hardware cards for speeding up the computer system and image analysis.

The microscope is equipped with a three-dimensional scanning stage which allows for computer-control motion of the stage. Two drive motors are included with the stage for separate x- and y-direction stage motion. A third computer controlled motor is used for controlling the focus (z-direction) of the system. These drive motors are directly interfaced with the image analysis system controls and can receive control signals either directly from the computer system or from the user via the multi-function control box.

The integration of the Image Analysis System with the camera and traveling stage allows the counting software of the microcomputer to directly control all functions of the automated system hardware, as well as providine a direct user interface with both the system hardware and Image Analysis System. This increases the speed with which counting and track analysis may be performed, which was a main concern in the system development. The electronic hardware cards for image analysis are programmed to perform the area and perimeter measurements for track identification, as well as bitplane digitization and manipulation, and were implemented to increase the analysis speed.

**Operation** The process of track counting with TAS Plus may be broken into three main functions: track identification; recorder scanning; and focusing. Track identification consists of the digitization of the camera image, conversion of the digitized image to a bitplane, parametrization of each potential track in the image, and the discrimination between actual tracks and background. Recorder scanning includes input of the necessary parameters for scanning, and the routines needed for ensuring that the whole recorder is examined during the counting procedure. Focusing entails adjusting the focus during the traverse of the recorder, and the locations of the focus points. These three functions will be more fully described below.

Operator interface with the program(s) is minimal. Several input parameters are required at the beginning of each counting run, as well as several focus and scanning reference points for each of the recorders to be counted. The program output may be as extensive or sparse as the user requires, with many statistical and graphical edits available for detailed study, and minimal results (recorder number, number of stage stops, and number of tracks counted) available for production runs.

**Track Identification** The track identification process begins with the digitization of the camera image. In order to do this, the gray scale discriminator must be set. The gray scale on TAS Plus ranges from 0 (black) to 99 (white), and the user must set a minimum and maximum gray value for discrimination purposes. Each pixel (picture element) in the camera image is then assigned a gray value, and these values are compared with the preset limits on the gray scale values, and the result is written to a bitplane. The bitplane contains 1's where the pixel is within the discrimination limits, and 0's otherwise. This bitplane is used for the remainder of the track discrimination analysis.

The track discrimination analysis determines which sections of the bitplane are eligible to be counted as tracks. Each camera image bitplane is sequentially scanned from top left to bottom right, and for each region of 1's encountered, the area and perimeter are measured, and the shape factor is computed:

$$\frac{4\pi A}{p^2} \tag{1}$$

where A is the area and p is the perimeter of that particular region. For fused quartz, the recorder material used in FBBF experiments, the tracks are nearly circular in cross section. Ideally, all tracks would have identical perimeter and area, and would be perfectly circular (shape factor = 1). However, this is not the case in reality, and thus user-specified ranges for these parameters are required to determine which are actual tracks. These are input at the beginning of each run.

These operations consume the bulk of the running time for the program, and any speed advantages which may be gained are valuable. For this reason hardwired program cards for the area and perimeter calculations, as well as for the image digitization, were installed in the system. Installation of the perimeter board alone reduced the running time by 25%; the other cards also provided a substantial savings in computation time.

**Recorder Scanning** Counting the total number of tracks on a recorder requires not only that the entire recorder be scanned, but also that no track is counted twice. To prevent this, an arbitrary-width border is placed on the bottom and right sides of the image and any tracks which contact this border are removed from the bitplane. Tracks which contact the top and left sides of the border are counted as tracks on that bitplane. This means that the scanning procedure must account not only for the width of the border, but also for the size of potential tracks in contact with this border. Thus, a certain amount of overlap between video images is required to ensure that no track is ignored in the scan.

In order to not waste time counting regions with no tracks, the search pattern for the full recorder scan should match the recorder shape as closely as possible. Since the fission track recorders are circular, a circular search pattern was designed. The search pattern requires user input of the coordinates of the center (x,y coordinates), as well as the coordinates of the top, bottom, and side edges of the recorder. This allows determination of the search radius for the recorder.

The coordinates for the center and edges of the recorder are read from the data file initially entered by the operator. From this information, the platform is moved and an initial image at the extreme top of the recorder is analyzed. The platform is then moved one frame down, giving a new y-coordinate. The x-coordinate of the point farthest to the left of the centerline is calculated from:

$$X = X_c - \sqrt{R^2 - (Y - Y_c)^2} \qquad (2)$$

where:

$R$ = radius of the recorder
$X$ = x-coordinate to be determined
$X_c$ = x-coordinate of the recorder center
$Y$ = y-coordinate of the new image
$Y_c$ = y-coordinate of the recorder center

Note that the coordinate axes of the platform are rotated such that the positive z-direction (focus) is vertically downward, the positive x-direction is to the left, and the positive y-direction is away from the user. The scan continues in the negative x-direction for the new y-coordinate until the radius is exceeded. When this occurs, the platform is again moved one frame down, and the sequence is repeated. This occurs until no further points within the predetermined recorder area remain, and the platform then advances to the next recorder. The system then reads the next set of coordinates and focusing information from the data file and the scan pattern is repeated.

**Focusing** The focusing routines within the scanning/counting software require that the user provide the focus at five predetermined points: the recorder center; top left; top right; bottom left; and bottom right as part of the setup procedure. The coordinates of each of the five positions is calculated within the program to minimize operator requirements. Once these points are provided, the scanning program computes a new focus position for each stage stop using two linear interpolations, one for the x-direction, and one for the y-direction. The data for the focus positions, and the z-coordinates (focus) at these positions, together with the radial positions provided for the circular search, are stored in a data file for use by the scanning routines.

**Operator Input** Operator input for program execution is minimal, and requires approximately two minutes for each recorder to be counted. For each recorder the operator locates the top, bottom and left and right edges of the recorder. The system then determines the center of the recorder and moves the recorder to each of five locations where the operator determines the best focus. These data together with settings for the light intensity, gray scale discriminator, range of areas, and shape factor form the input data file. These data are prompted for during execution of a supplemental routine, which also writes the necessary data files for the counting program.

## System Performance

The ability of the system to identify fission tracks is strongly dependent on both the intensity of the light source and the discrimination level of the gray scale. The light intensity is measured by the system on a relative scale and can vary from 0 to 1000, with zero being dark. Figure 1 shows the variation in the number of tracks identified and counted as a function of the light intensity. Below an intensity value of 413, the count rate sharply decreases. The system parameters are adjusted to operate with the light intensity near but in the region above the maximum of this curve. The variation of the count rate as a function of the discriminator level of the gray scale is shown in Fig. 2. It was found that both the light intensity and the gray scale measurements varied with room temperature and power line voltage and these variations are considered to be the major contributer to the variations in number of tracks counted on a recorder. A longer term effect was also found to be introduced by the aging of the light source.

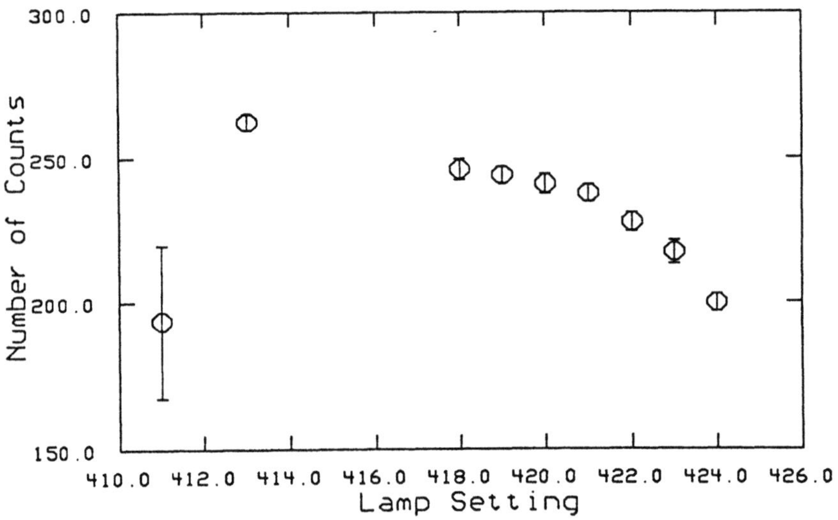


Figure 1. Lamp setting sensitivity study results for a constant discriminator setting. (Relative units)

Repeated counting of one SSTR over a twenty-four hour period yielded a fractional standard deviation of 0.38%. This compares favorably with an experiment reported by Roberts et al. [7] in which a number of Muscovite mica recorders were manually counted by four different observers. This experiment yielded a fractional standard deviation of about 0.4%. When four other SSTR were added to the holder and the experiment repeated with the system cycling through the recorders ten times over a 48 hour period, the fractional standard deviation increased to 0.5%. The standard deviations for the four other recorders were 0.7, 1.1, 0.9, and 0.8 percent. The SSTR with the largest fractional standard deviation had a noticeably greater number of surface imperfections. This experiment with the five SSTR was repeated over a period of several weeks to check on the long term stability of the system. The average standard deviation for the individual runs was 0.78% but the standard deviation for any individual recorder over the period increased to an average of 2.4%. If the average number of tracks counted in each experiment was normalized to the tracks recorded for SSTR #1, then some of the long term variations were removed and the average standard deviation was reduced to 1.4%.

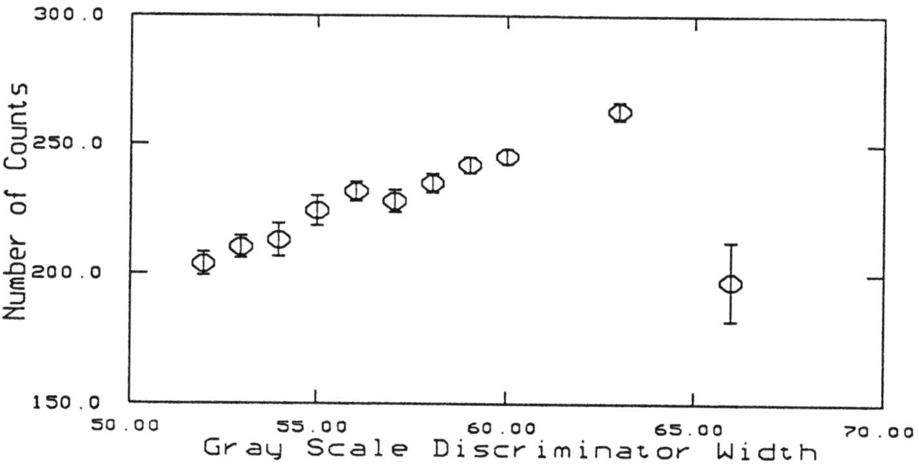


Figure 2. Gray scale discriminator sensitivity study results for a constant lamp setting. (Relative units)

In practice, one "standard" recorder, exposed to a known $^{252}Cf$ fission source, is included with each batch of recorders. This recorder follows the same etch procedure and is counted with the recorders use for the experimental measurements. The reported number of tracks are then corrected to the "standard". By including the standard recorder, the measurements can be corrected for the system efficiency and to some degree for the long term drift.

The track counting system as currently implemented is capable of counting approximately 10,000 fission tracks per hour using the 4X objective lens on the microscope. (Note: The video camera adds additional magnification to the system.) The microscope stage can hold up to 20 SSTR plus one standard in specially designed plates. These 21 SSTR can be counted in 24 hours or less, depending on the track density. With preselected (few surface imperfections) SSTR, the long term standard deviation of the corrected or normalized counts is about one percent which is comparable to the Poisson statistics of one percent for 10,000 tracks. Track overlap and variations in the etching procedure are expected to increase the standard deviation between samples.

## Conclusions

An automatic optical track identification/counting system has been developed for counting the total number of fission tracks on a fused quartz solid state track recorder. The system is capable of analyzing up to twenty recorders a day with an operator input of less than two hours. The uncertainty introduced by the counting system depends to some degree on the quality of the recorders, but with good recorders (few surface imperfections), the standard deviation is about one percent.

## References


[1] *Proceeding of the 11$^{th}$ International Conference on Solid State Nuclear Track Detectors*, Bristol, Sept. 7-12 (1981).

[2] H. Farrar IV and E. P. Lippincott, editors, *Reactor Dosimetry: Methods, Applications, and Standardization*, Sixth ASTM-Euratom Symposium, Jackson Hole, WY, STP 1001 (1987).

[3] R. L. Fleischer, P. B. Price, and R. M. Walker, *Nuclear Tracks in Solids*, University of California Press, Berkeley (1975).

[4] H. P. Chou, *Absolute Fission Rate Measurements in the Fast Breeder Blanket Facility*, Ph.D. Thesis, Purdue University (1981).

[5] H. P. Chou, R. H. Johnson, and F. M. Clikeman, "An Automatic Fission Track Scanner", *Nuclear Inst. & Methods* 1985, 359 (1981).

[6] H. Vital, *Experimental and Calculational Investigation of High Energy Neutron Fluxes and Reaction Rates in the Purdue Fast Breeder Blanket Facility*, Ph.D. Thesis, Purdue University (1985).

[7] J. H. Roberts, R. Gold, and F. H. Ruddy, "Optical Efficiency and Observer Objectivity for fission Track Counting in Muscovite Solid state Track Recorders", *Proceedings of the 11$^{th}$ International conference*, Bristol, 7-12 Sept 1981.

# THE USE OF NEUTRON-SENSITIVE DIODES AS MONITORS FOR 1-MEV EQUIVALENT NEUTRONS

Craig R. Heimbach

Combat Systems Test Activity

Aberdeen Proving Ground, MD 21005 USA

## ABSTRACT

The Harshaw model DN-156 diode has been validated for use as a 1 MeV equivalent neutron monitor. It is small, exposed without bias voltage, not unduly sensitive to gamma rays or thermal neutrons, and has an intrinsic 1 MeV equivalent sensitivity to neutrons. Uniformity of response in a given batch is 5% or better. This is sufficient to use batch calibration, and not individual device calibration. The device is sensitive to temperature, so that either the temperature must be controlled from time of exposure to time of readout, corrections made for variations in temperature, or preheating used to minimize temperature effects. The diode has been used to calibrate a test reactor neutron field to 9%, traceable to a national standard.

---

## INTRODUCTION

Neutron dosimetry can be accomplished by measuring the neutron spectrum and integrating it over a desired response function. For neutron damage to silicon electronic devices the appropriate damage function is the 1 MeV equivalence function defined in Reference 1.

An alternative approach explored here is to use a silicon device as a neutron damage monitor. Since the device is silicon-based, it is by definition damaged in accordance with the silicon damage curve. Thus, a calibrated device can be used directly to measure 1 MeV equivalent fluence. This is similar to previous work using neutron damage to transistors (2,3).

## THE DN-156

The device studied was the DN-156, a PIN diode made commercially by Harshaw(4). It is a 3.5 mm long by 3 mm diameter cylinder with 9 mm long leads. It is used passively.

The read-out procedure used to measure the damage in the DN-156 was to inject a 6 millisec, 25 milliamp, forward current pulse into the diode. The forward voltage drop is measured to the nearest millivolt in the middle of the pulse. Precision resistors are used to set the magnitude of the injected current pulse. An oscilloscope is used to set the pulse width.

Figure 1 shows typical forward voltage readings taken at various times after exposure to neutrons. There is a decline in the voltage for each successive reading. For readings taken soon after exposure, this decline is relatively steep. If the initial readings are taken later, the decline is relatively small. The third reading was used for all results to reduce the variation in measured voltages. Controlling this annealing for various times and temperatures is the main difficulty in the use of the diode. A standard time (24 hours after exposure) was used for readout.

Tests were done to establish readout sensitivity to variations in either pulse height or pulse width. The readout was found to be sensitive to pulse current height, but the use of precision resistors for calibration makes this error less than 0.1%. A slight (less than 0.2%) variation in output was found when varying the pulse width from 4 to 8 msec.

Once the correct readout procedure was determined, the consistency of the production lot could be tested. It was found to be 2% (1 sigma) for each of two lots. This is sufficiently small that calibration of a batch of diodes can be used in place of individual diode calibrations. Production lots of up to several thousand are available with 5% consistency.

## TEMPERATURE DEPENDENCE AT READOUT

The forward voltage is a function of the temperature at the time of readout. The increase in forward voltage ($\Delta V$) varies as

$$\Delta V = \Delta V + 0.0048 * \Delta V * (T-25) \qquad (1)$$

where T is the temperature in degrees centigrade.

This is a small correction for most situations. It amounts to only a 2.5% correction for the range 20 to 30°C. Caution must be used in handling the device, however, since bringing it up to body temperature could cause an error. There is about a 5% increase in forward voltage if the diode is held in the hand for a minute, and the diode recovers to ambient in about 20 sec.

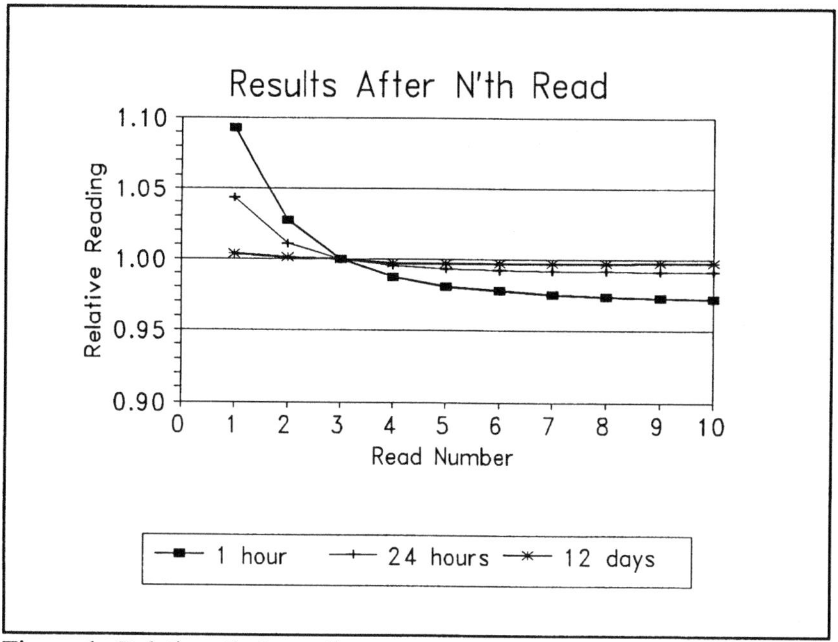


Figure 1. Relative diode reading versus number of read for various times after exposure.

## FADING

### Room Temperature

A fading curve was developed to adjust readings made at various times to 24 hours after exposure. This curve was found not to be sensitive to neutron fluence, but to vary slightly in the two production batches tested.

### Other Temperatures

The fading curve was developed for diodes held at 25°C from exposure to readout. Further testing was done at alternate storage temperatures to see the extent the fading varied. It was found that heat accelerated the fading and that cold retarded it. Fig. 2 shows the results of fading measurements when the diodes were exposed at 25°C, stored for various times at other temperatures, and then read out at 25°C. The result depends strongly on the temperature history of the diode. Storing the diode at high temperatures causes exceptionally high deviations.

Heat treatment of diodes can anneal out much of the shallow damage and make the readings must less sensitive to fading or storage temperature. For example, a readout procedure was tested where the diodes were read out 1 week

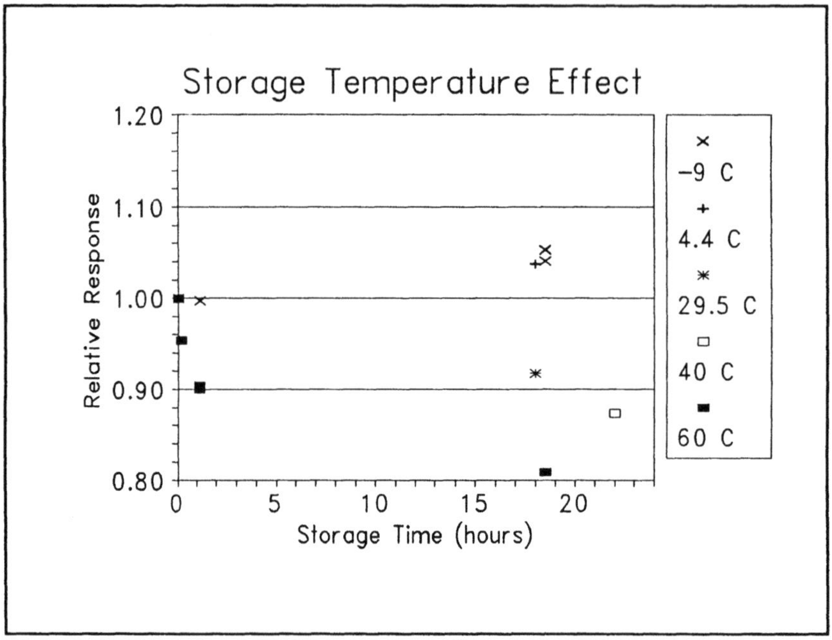


Figure 2. Relative diode response for various storage times and temperatures.

after exposure. Before readout, the diodes were heated for one hour at 100°C. The diodes gave the same result within 2% independent of storage for 24 hours at temperatures ranging from -12 to +50°C. If exposure or storage temperature are problems, such a procedure can reduce uncertainties.

If the irradiation is not performed at room temperature, the heat treatment as specified above should be used. If temperatures above 50°C are applied, the techniques developed here have not been validated.

## CALIBRATION

The diodes were calibrated by exposing several sets to neutrons at the Aberdeen Pulsed Radiation Facility (APRF) reactor. All exposures were at the same distance from the reactor. Only the duration of exposure was changed. The 1 MeV equivalence scale was normalized by exposing three pair of diodes to a Californium-252 source at the National Institute of Standards and Technology (NIST). Thus the shape of the calibration curve was determined by the APRF reactor, and the scale was determined by the NIST exposures. The overall uncertainty of the NIST calibration was about 6%.

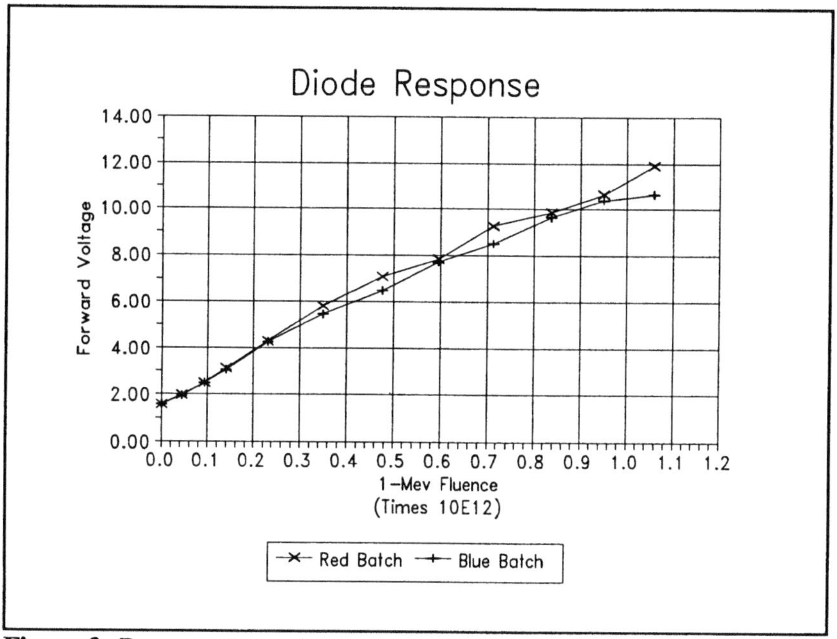


**Figure 3**. Response to neutrons.

Calibrations developed for two different manufacturer's lots differed by as much as 12% in their reading for a given exposure (Figure 3). The calibration curves are not linear. The diode responses saturated at about 1.0E+12 n/cm$^2$ (1 MeV).

## GAMMA-RAY SENSITIVITY

It is desirable for a neutron monitor not to be sensitive to gamma rays. Measurements with a Co-60 source indicated that 122 kRad(Si) gave the same increase in voltage equivalent to 2.6E+9 n/cm**2 (1 MeV), whether delivered before or after the neutron exposure. This is a small response.

## NEUTRON RESPONSE

The sensitivity of the diodes to neutrons of different energies was measured. This may be considered a calibration of the diode. It may also be considered as a verification of the 1 MeV equivalent response curve, since the diodes are silicon devices and the response curve is intended to predict damage to silicon devices.

The National Institute of Standards and Technology (NIST) was used for calibrations with thermal neutrons, in 2, 25, and 144 keV neutron beams, and with bare and D2O-moderated Cf-252 sources. A 14 MeV neutron generator located at the US Naval Academy was used for 14.6 MeV calibration.

Table 1 shows the results from diode exposure in these neutron fields. For many of the fields, the diode voltage changes were quite small, both because of the low intensity of many of the sources and because of the small silicon damage constant for low-energy neutrons. This led to a large uncertainty in some of the results.

The uncertainties in the three lowest-energy readings in Table 1 are quite large due to the low diode response to these neutrons. The values obtained are consistent with zero response to neutrons below 25 keV. Alternatively, the actual response could be a factor of two higher than measured.

The measurement at 144 keV should be within 20% of the correct value. The fact that it is not is probably due to the fact that 144 keV corresponds exactly to a sharp minimum in the neutron cross section in silicon. A silicon filter is used to produce the beam. The ASTM reported value is averaged over a small range of energies, so it might be expected to be higher than the measured value.

The damage measured at 14.6 MeV is 10% higher than would be expected from the damage function. The discrepancy is even larger for more recent evaluations of the damage function of silicon. However, the uncertainty in the measured result is 30% due primarily to uncertainties in the monitoring of the 14.6 MeV neutrons.

Table 1. Neutron Response to Various Sources.

| Neutron Energy (keV) | Measured Damage (MeV-mb) | ASTM Damage (MeV-mb) |
|---|---|---|
| thermal | 0.6(100%) | (a) |
| 2 | 0.2(100%) | (a) |
| 25 | 1.0(100%) | 1.9 |
| 144 | 1.7(20%) | 2.9 |
| 14600 | 237. (30%) | 215.1 |
| D2O | 27.8 (8%) | 28.7(b) |
| Cf-252 | 112. (c) | 112. |

(a) ASTM reports damage values only above 10 keV.
(b) Total non-thermal fluence used.
(c) Cf-252 used for calibration. Measured damage equals ASTM damage by definition.

## CALIBRATION OF THE APRF

The intended use of the diodes is to calibrate the neutron fields in test facilities. Two exposure fields were calibrated at the APRF: 80 cm from core center on the Experiment Exposure Table (EET), and inside the glory hole. Both of these locations had 1 MeV fluence to sulfur (fluence greater than 3 MeV) ratios measured previously (6,7). These were 8.2 and 8.4 for the EET and glory hole, respectively.

Diode measurements were performed simultaneously with sulfur exposures. Using the diodes to determine the 1 MeV equivalent fluences, 1 MeV to sulfur ratios of 8.6 and 8.3 were found for the EET and glory hole. These results are consistent, considering 10% uncertainty in the original measurements and 9% in the diode results.

## SUMMARY AND CONCLUSIONS

The DN-156 diode has been evaluated for use as a 1 MeV equivalent neutron monitor in neutron fields produced at the APRF reactor. The values found with the diode are sensitive to the time after exposure and to the temperature of the readout; however, with care the diode can be used to measure 1 MeV equivalent fluence to 9%, traceable to NIST Cf-252 sources.

## ACKNOWLEDGEMENTS

This work was sponsored by the US Army Test and Evaluation Command methodology program. The author would like to thank Gary Ilko of Harshaw for his cooperation in this investigation. Also, Prof. Martin Nelson of the US Naval Academy and Robert Schwartz, Dale McGarry and David Gilliam of NIST provided expert guidance in the use of their facilities for calibration.

## REFERENCES

1. ASTM E722-85, "Standard Practice for Characterizing Neutron Energy Fluence Spectra in Terms of an Equivalent Monoenergetic Neutron Fluence for Radiation-Hardness Testing of Electronics."

2. J.G. Williams and C.F. Hsun, "The Use of Silicon Transistors as Damage Monitors in Reactor Neutron Metrology" in Reactor Dosimetry, Genthon and Röttger (Ed.) (1985).

3. DeLeeuw et. al., in NUREG/CR-1861 (1981).

4. See DN-156 data sheet, available from Harshaw, for physical and electrical specifications. The use of trade names in this report does not constitute an official endorsement or approval of the use of such commercial hardware. This report may not be cited for purposes of advertisement.

5. R.S. Schwartz, "Neutron Personnel Dosimetry," NBS Special Publication 250-12, 1987.

6. C.R. Heimbach, "Recalibration of APRF Neutron Dosimetry," Proc. of the Fast Burst Reactor Workshop, Albuquerque, NM, SAND87-0098, Vol. 2, 1987

7. E.D. McGarry, A.H. Kazi, G.S. Davis, D.M. Gilliam, "Absolute Neutron-Flux Measurements at Fast Pulse Reactors with Calibration against Californium-252," IEEE Trans. Nucl. Sci., NS-23, NO. 6, pp 2002-2006, 1976

# PLANT LIFE EXTENSION RELATED APPLICATIONS OF SOLID STATE TRACK RECORDER NEUTRON DOSIMETRY

F. H. Ruddy and J. G. Seidel

Westinghouse Science & Technology Center

Pittsburgh, PA 15235-5098

## ABSTRACT

New needs for neutron dosimetry are arising in the context of nuclear plant life extension. Solid State Track Recorder (SSTR) neutron dosimeters provide an excellent means for obtaining these data. Pressure vessel surveillance using SSTR neutron dosimeters deployed in the reactor cavity region is now being carried out on a cycle-by-cycle basis at several operating commercial nuclear power reactors. The status of these dosimetry exposures are summarized and new plant life extension related SSTR neutron dosimetry initiatives currently underway are also discussed. Because of the enormous range of sensitivity of SSTR neutron dosimeters, innumerable other potential plant life extension related applications of SSTR neutron dosimeters exist. Selected applications compatible with the use of SSTRs are described.

---

## INTRODUCTION

As many nuclear power plants approach their 40 year design life, issues related to plant lifetime assurance and plant lifetime extension become of increasing importance. Initial studies of nuclear plant life extension were begun in the late 1970's, (1,2) and aging research as it relates to nuclear plant life extension is now an active field of investigation (3).

The implementation of nuclear plant life extension programs places increased emphasis on the identification and monitoring of age-related degradation mechanisms for key plant components. In those cases where the degradation mechanism is influenced by radiation, a knowledge of the irradiation environment of the component becomes important. In many instances neutron exposure is a factor in the degradation mechanism, resulting in a need for neutron dosimetry for these components.

Some interesting neutron dosimetry aspects have arisen recently in the context of nuclear plant life extension. In addition to the established need for neutron dosimetry for reactor pressure vessel surveillance, the integrity of other key components and materials must be ensured. These include, for example, reactor vessel supports, core components, control rod drive mechanisms, and electrical cable, motors, pumps, etc. used inside reactor containment.

Solid state track recorder (SSTR) neutron dosimeters provide a convenient method for passive monitoring of reactor components. Neutron-induced embrittlement of the reactor pressure vessel is probably the most important factor in reactor plant life extension. SSTRs have proven very useful in pressure vessel neutron dosimetry applications (4-6). The status of SSTR pressure vessel neutron dosimetry measurements carried out to date is reviewed below.

New plant life extension related areas where SSTRs are now being applied are discussed, and other possible areas of neutron dosimetry concern and the applicability of SSTR neutron dosimetry as a means to carry out related measurements are described below.

## SSTR NEUTRON DOSIMETRY

An SSTR neutron dosimeter consists of a thin film of fissionable material pressed in firm contact with a track recording material or SSTR such as mica. Neutron-induced fissions in the fissionable layer result in the formation of latent fission tracks in the neighboring SSTR. These latent tracks are developed to a suitable size by etching and are counted to determine the number of fissions that has occurred in the fissionable deposit. The known deposit mass can be used to calculate the fission rate in fissions per atom. Fission rates from a set of fissionable deposits (for example, $^{235}U$, $^{237}Np$, $^{238}U$, and $^{232}Th$) can be used as input to a spectral unfolding program to determine the neutron spectrum and flux at the location where the dosimeter was deployed.

SSTR neutron dosimetry techniques applied to reactor dosimetry have been described in detail at previous meetings in this series (4,7,8). SSTR neutron dosimeters have an extremely wide range of sensitivity. Present techniques can be shown to be applicable with high accuracy up to at least $5 \times 10^{18}$ n/cm$^2$. At the other extreme, SSTR neutron dosimeters have been used to monitor a flux of less than $2 \times 10^{-3}$ n/cm$^2$/sec (approximately two times the cosmic ray neutron background) in measurements at Three Mile Island Unit 2 (9). Measurements can be carried out reliably and with high accuracy over the entire sensitivity range.

The enormous range of sensitivity of SSTR neutron dosimetry results from the available ranges of fissionable deposit masses and track densities that can be used. Fissionable deposits with thicknesses between $4 \times 10^{-3}$ gms/cm$^2$ and $1 \times 10^{-13}$ g/cm$^2$ have been used successfully for neutron dosimetry (6), and track densities ranging from about 100 tracks/cm$^2$ to almost $10^6$ tracks/cm$^2$ can be used to obtain accurate fission rate results (7).

For extremely low flux neutron dosimetry carried out in fields at or close to the cosmic ray neutron background, thick or self-supporting fissionable foils are used. These so-called asymptotic foils (10) have effective thicknesses that are determined by the range of a fission fragment in the material. The effective thickness for uranium, for example, is $(4.522 \pm 0.070)$ mg/cm$^2$ (11).

Fissionable deposits in the 10-100 $\mu$g/cm$^2$ range can be made routinely and are useful for monitoring neutron fluences in the range between about $10^6$ and $10^{16}$ n/cm$^2$. SSTR neutron dosimetry has been shown to be a highly accurate technique in this fluence range, and, for those measurements not limited by track counting statistics, absolute uncertainties approaching 1% are obtainable (10).

The recent development of ultra low-mass fissionable deposit preparation techniques (12) has extended the range of applicability of SSTR neutron dosimeters to much higher fluences. Measurements are now carried out routinely in fluences of approximately $5 \times 10^{16}$ n/cm$^2$ for pressure vessel surveillance dosimetry at commercial power reactors (4-7). At higher neutron fluences, the techniques currently applied to pressure vessel dosimetry appear to be capable of extension to fluences of at least $5 \times 10^{18}$ n/cm$^2$ (4), while still maintaining overall uncertainties better than 5%. At higher fluences, track scanning limitations lead to higher uncertainties. However, based on the results obtained in Ref. 7, uncertainties of about 5-10% appear to be attainable using established SSTR techniques in the (5-10) $\times 10^{18}$ n/cm$^2$ fluence range and uncertainties of 10-20% in the (1-3) $\times 10^{19}$ n/cm$^2$ range.

Extension to higher fluences will require the development of methods for accurately scanning extremely high track densities (greater than $3 \times 10^6$ tracks/cm$^2$) and/or the extension of ultra low-mass fissionable deposit preparation techniques to even lower masses. Efforts presently in progress at the Westinghouse Science & Technology Center show promise for achieving the latter alternative.

## SSTR APPLICATIONS

SSTRs are a key component of reactor cavity neutron dosimetry sets used for pressure vessel surveillance at commercial power reactors (13,14). In addition to these pressure vessel surveillance dosimetry measurements, a number of other potential applications at high, intermediate, and low neutron fluences are presently under way or are contemplated.

### SSTR Pressure Vessel Surveillance Dosimetry

SSTR pressure vessel surveillance neutron dosimetry has been actively carried out at several Westinghouse pressurized water reactors (PWRs) over the past six years. SSTR neutron dosimetry sets are deployed in the reactor cavity (annular gap) region of the reactor during an outage and left in place for one or more operating cycles. Following exposure, the SSTR neutron dosimeters are analyzed to determine neutron-induced fission rates which are used to deduce the neutron spectra at the dosimetry locations. These neutron spectra are extrapolated into the pressure vessel and used to calculate damage-related parameters such as displacements per atom at key pressure vessel locations (13). Cycle-by-cycle measurements can be used both to verify effects of fuel management on vessel fluence (15) and to track the cumulative neutron fluence at key vessel locations by summing the results from each dosimetry cycle. To date, more than 574 SSTR neutron dosimeters have been analyzed from cavity dosimetry exposures in Westinghouse reactors, and more than 281 additional dosimeters are currently deployed. Cavity dosimetry measurements which make use of SSTR neutron dosimeters are also under way at Babcock and Wilcox reactors (16).

The results of over 574 SSTR cavity neutron dosimeters analyzed to date indicate that, when proper quality assurance procedures are observed, absolute uncertainties in the 2-5% range are obtained for the measured fission reaction rates.

### Other Plant Life Extension Related SSTR Neutron Dosimetry Applications

The SSTR neutron dosimetry techniques developed for reactor cavity pressure vessel surveillance applications can be applied without modification to several other neutron monitoring measurements of relevance to nuclear plant life extension related goals.

Recent material studies indicate the possible presence of "dose-rate" or flux dependent effects on the extent of irradiation induced materials damage (17,18). A possible result of these observations is that materials damage effects at low fluxes may be of consequence at fluences as low as $10^{17}$ n/cm$^2$. Reactor vessel support structures are exposed to much lower fluxes than the pressure vessel itself but may accumulate fluences greater than $10^{17}$ n/cm$^2$ over the lifetime of the reactor, particularly if the reactor is a candidate for plant life extension. Little data is available on the neutron fluxes to which support structures are exposed, and these doses will vary with nuclear plant design. However, SSTR neutron dosimetry is an ideal method for measuring fluxes in the expected range of interest from $10^6$ to $10^9$ n/cm$^2$/sec. For SSTR neutron dosimeters exposed for one reactor operating cycle, total fluences in the range from 3 x $10^{13}$ to 5 x $10^{16}$ result. This fluence range can be completely covered and absolute uncertainties approaching 1% (10) should be obtainable on the low end of the range. SSTR neutron dosimetry measurements at key support structure locations can be used with materials damage studies to assess the magnitude of problems, if any, related to support structure irradiation damage. SSTR dosimetry sets are currently deployed at a US PWR to quantify the neutron fluences and spectra at key vessel support locations.

SSTRs have also found plant life extension related applications in extremely low neutron fluxes. Degradation due to aging and service wear limits the life of many reactor components located within containment. The useful life of pumps, motors, valves, cables, etc. can be shortened by aging effects that are generally accelerated by increased temperature and the irradiation environment to which the component is exposed. Equipment and components are often subjected to controlled neutron exposures as part of pre-use qualification testing (19). SSTR neutron dosimetry provides a means to provide data to verify that equipment qualifications are not exceeded during either routine or extended life service.

At present, SSTR neutron dosimeters are deployed for equipment qualification related neutron surveillance at two commercial nuclear power plants. The expected neutron fluxes are generally less than about 2 x $10^6$ n/cm$^2$/sec. SSTR neutron dosimetry measurements can easily be made at this flux level and extended down to any flux level which is likely to be of concern in terms of accumulated neutron fluence over the service lifetime of a component. In fact, SSTR neutron dosimeters can even provide data in fields which are of concern to personnel dosimetry (8).

Several potential applications of SSTRs in extremely high neutron fluences also exist. Recently, irradiation effects on core structural elements have become a subject of concern (20). Irradiation-assisted stress corrosion cracking can potentially affect the grid plates, core plates, and top guides, control rods, and drive assemblies. Because the materials involved have not been studied nearly as extensively as have pressure vessel steels, increased dosimetry in conjunction with materials studies are likely to be applied to this area. The neutron fluxes delivered to the areas of interest are generally in the range from $10^{11}$ to $10^{12}$n/cm$^2$/sec. (21). If SSTR neutron dosimetry packets are inserted at key locations during an outage and removed at the end of an operating cycle, fluences in the range from 3 X $10^{18}$ to 5 X $10^{19}$ n/cm$^2$ will result at the dosimetry locations. Presently available SSTR techniques are usable on the lower end of this fluence range and, with modest extension, could be used to cover the entire range. The SSTR dosimeters would be exposed to temperatures of about 300°C. However, mica SSTRs are usable at temperatures approaching 600°C (22), and quartz crystal SSTRs at temperatures approaching 800°C (22,23). SSTR neutron dosimetry measurements from individual cycle measurements can provide a basis for assessing the materials effects expected over the life of key components.

An additional high fluence application of SSTR neutron dosimeters is for replacement surveillance capsule dosimetry. However, since the anticipated fluxes at surveillance capsule locations range from 1 to 5 X $10^{18}$ n/cm$^2$/yr, extension of present SSTR techniques would be needed for high accuracy dosimetry measurements for surveillance capsules left in place for more than about six years. New methods for extending ultra low-mass fissionable deposit preparation techniques to even lower masses are currently under development at the Westinghouse Science & Technology Center, and these methods show promise for extending the sensitivity range of SSTRs to higher fluences.

## SUMMARY AND CONCLUSIONS

SSTR neutron dosimetry techniques are presently being extended from reactor pressure vessel surveillance to other plant life extension related applications. A summary of the status of plant life extension related SSTR neutron dosimetry carried out to date is contained in Table 1. At present, more than 900 SSTRs have been deployed during a total of thirty operating cycles at 17 different nuclear power reactors.

SSTRs provide a convenient method for passive integral neutron exposure monitoring. The accuracy of SSTR neutron dosimetry techniques has been demonstrated through irradiations in National Institute of Standards & Technology standard neutron fields (24,25). These irradiations, which serve as a validation

Table 1

Summary of Plant Life Extension Related Deployments of SSTR Neutron Dosimetry

| Application | Number of Plants | Number of Cycles | Number of SSTR Neutron Dosimeters | | |
|---|---|---|---|---|---|
| | | | Analyzed | Currently Deployed | Total |
| Pressure Vessel Surveillance (Reactor Cavity Dosimetry) | 14 | 27 | 574 | 281 | 855 |
| Pressure Vessel Support Structure Surveillance | 1 | 1 | --- | 24 | 24 |
| Equipment Qualification Neutron Dosimetry | 2 | 2 | --- | 36 | 36 |
| Total | 17 | 30 | 574 | 341 | 915 |

of the calibrations of the response of SSTR neutron dosimeters are described in a companion paper (25).

Increasing emphasis is being placed on nuclear plant life extension and neutron dosimetry issues have arisen and will continue to be identified in this context. SSTR neutron dosimeters have many advantages for providing the needed data to address these issues.[6] SSTRs have a proven record in pressure vessel surveillance neutron dosimetry (4,5), and the versatility and wide sensitivity range of SSTR neutron dosimetry makes it an ideal candidate for additional plant life extension related neutron dosimetry.

## REFERENCES


1. C. Negin, et al., "Planning Study and Economic Feasibility for Extended Life Operation of Light Water Reactor Plants," EPRI-TPS-78-788, Electric Power Research Institute, Palo Alto, CA (September 1979).

2. C. Negin, et al., "Extended Life Operation of Light Water Reactors: Economic and Technological Review," EPRI-NP-2418, Electric Power Research Institute, Palo Alto, CA (June 1982).

3. D. F. Ross, Jr., "Aging Research - Needs and Status," Nuclear Power Plant Life Extension, American Nuclear Society, ISBN 0-89448-139-8, p. 29 (1988).

4. F. H. Ruddy and J. G. Seidel, "Solid State Track Recorder Pressure Vessel Surveillance Neutron Dosimetry at Commercial Light Water Reactors," Reactor Dosimetry: Methods, Applications, and Standardization, ASTM STP 1001, American Society for Testing and Materials, Philadelphia, PA, p. 175 (1989).

5. F. H. Ruddy and J. G. Seidel, "Pressure Vessel Surveillance at Commercial Light Water Reactors Using Solid State Track Recorder Neutron Dosimetry," Nucl. Tracks Radiat. Meas. 15, 491 (1988).

6. F. H. Ruddy, "The Use of Solid State Track Recorders for Plant Life Extension Related Neutron Dosimetry," Nuclear Power Plant Life Extension, American Nuclear Society, ISBN 0-89448-139-8, p. 386 (1988).

7. F. H. Ruddy, A. E. Manhardt, and J. G. Seidel, "Analysis of Solid State Track Recorders with the Westinghouse Research and Development Center Automated Track Counter," Reactor Dosimetry: Methods, Applications, and Standardization ASTM-STP 1001, American Society for Testing And Materials, Philadelphia, PA, p. 676 (1989).

8. F. H. Ruddy, "Personnel Neutron Dosimetry Area Monitoring Using Solid State Track Recorders, Reactor Dosimetry: Methods, Applications, and Standardization, ASTM STP 1001, American Society for Testing and Materials, Philadelphia, PA, p. 756 (1989).

9. F. H. Ruddy, J. H. Roberts, R. Gold, and C. C. Preston, "Applications of Solid State Track Recorder Neutron Dosimetry for Fuel Debris Location in the Three Mile Island Unit 2 Makeup and Purification Demineralizers," Nucl. Tracks Radiat. Meas. 8, 473 (1984).

10. R. Gold, R. J. Armani, and J. H. Roberts, "Absolute Fission Rate Measurements with Solid State Track Recorders," Nuclear Science and Engineering 34, 13 (1968).

11. ASTM Standard E854, "Standard Method for Application and Analysis of Solid State Track Recorder (SSTR) Monitors for Reactor Surveillance," (latest version 1988).

12. F. H. Ruddy, "Neutron Dosimetry in High Intensity Fields: A Review", Nucl. Tracks Radiat. Meas. 11, 197 (1986).

13. F. H. Ruddy, W. N. McElroy, E. P. Lippincott, L. S. Kellogg, R. Gold, J. H. Roberts, C. C. Preston, J. A. Grundl, E. D. McGarry, H. Farrar IV, B. M. Oliver, and S. L. Anderson, "Standardized Physics-Dosimetry for US Pressure Vessel Cavity Surveillance Programs," Reactor Dosimetry: Dosimetry Methods for Fuels, Cladding, and Structural Materials, J. P. Genthon and H. Rottger (Ed.), D. Reidel Publishing Co., Dordrecht, p. 473 (1985).

14. S. Q. King, "Description of the Babcock and Wilcox Owner's Group Cavity Dosimetry Benchmark Experiment," in Reactor Dosimetry: Methods, Applications, and Standardization, ASTM STP 1001, American Society for Testing and Materials, Philadelphia, PA, p. 155 (1989).

15. S. P. Grant, "US Utility's Experience in Surveillance Dosimetry," in Reactor Dosimetry: Methods, Applications, and Standardization, ASTM STP 1001, American Society for Testing and Materials, Philadelphia, PA, p. 80 (1989).

16. G. S. Carter, "Dosimetry Techniques and Methods Used to Obtain Reactor Cavity Dosimetry Benchmark Data for Vessel Fluence Analysis at Davis-Besse I" (this Conference).

17. K. Kussmaul, J. Fohl, and T. Weissenberg, "Assurance of the Pressure Vessel Integrity with Respect to Irradiation Embrittlement-Activities in the Federal Republic of Germany," IAEA Specialist's Meeting on Irradiation Embrittlement and Aging of Reactor Pressure Vessels, Philadelphia, PA, ASTM STP (in press) (May 1987).

18. R. D. Cheverton, W. E. Pennell, G. C. Robinson, and R. K. Nanstad, "Impact of Radiation Embrittlement on Integrity of Pressure Vessel Supports for Two PWR Plants," NUREG/CR-5320, ORNL/TM-10966 (January 1989).

19. See, for example, J. M. Snider, "Equipment Qualification Test Report: Target Rock Isolation Solenoid Valve Environmental and Seismic Testing," WCAP Suppl. 2-H10A, Rev. 2 (1985).

20. C. Z. Serpan, "NRC Research in Aging and Extended Life for LWR Primary Systems," Trans. ANS 55, 266 (1987).

21. S. L. Anderson, Private communication (April 1988).

22. J. H. Roberts, R. Gold, and F. H. Ruddy, "Thermal Annealing Studies in Muscovite and in Quartz," Nucl. Tracks Radiat. Meas., Supplement No. 2, 177 (1980).

23. F. H. Ruddy and J. H. Roberts, "Solid State Track Recorder Fission Rate Measurements at High Neutron Fluence and High Temperature," Nucl. Tracks Radiat. Meas., 11, 193 (1986).

24. F. H. Ruddy and E. D. McGarry, "Benchmark Referencing of Ultra Low-mass Solid State Track Recorder Neutron Dosimetry in NBS Standard Neutron Fields" International Atomic Energy Agency Advisory Group Meeting on Status and Requirements of Nuclear Data for Radiation Damage and Related Safety Aspects, Vienna (proceedings in preparation) (September 19-22, 1989).

25. F. H. Ruddy and E. D. McGarry, "Benchmark Referencing of Solid State Track Recorder Neutron Dosimeters in Standard Neutron Fields" (this Conference).

# A REVIEW OF THE DEVELOPMENT OF A LUGGAGE EXPLOSIVE DETECTION SYSTEM

J. Bartko and F. H. Ruddy

Westinghouse Science & Technology Center

Pittsburgh, PA 15235-5098, U.S.A.

## ABSTRACT

In the late 1960's and early 1970's there was a rash of airplane hijacking and bomb threats. This led one of the authors, John Bartko, to speculate on the possibility of using neutron activation analysis for the detection of explosives in baggage to reduce the threat. The initial research and development of the concept was conducted in a Westinghouse internally funded study. This work led to a patent, U.S. Patent 3,832,545 in 1973. The potential of this approach was recognized by the Federal Aviation Administration (FAA) which supported two successive programs. The first was a feasibility study to examine the potential of discriminating explosives from potential false alarms involving suitcases packed with various combinations of explosives and "false alarm" items. The second program investigated the detection rate and false alarm rate of the technique on real baggage. This work has formed the basis for systems currently deployed at several airports.

In this paper we will trace the development of the system from the selection of critical components, to the solution of potential problems to the airport tests.

---

## INITIAL CONSIDERATIONS

Nuclear activation technology has some powerful advantages over other technologies for determining the composition of concealed material. To a great extent this is due to the penetrating power of nuclear radiation, particularly, neutrons and gamma rays and the unique properties of their reactions with nuclei which enables a precise determination of constituent elements. The first step, then, in the evolution of a detection system was to identify an element or elements which would act as an indicator(s) of explosives. The element selected was nitrogen which is found in all common explosives in abundances ranging from ~ 15% to ~ 36%. Oxygen and carbon are also there in some abundance, but these are much more common and are more difficult to detect. Nitrogen, as we shall see, is an ingredient in some other materials, but it is considerably more rare in other materials.

There are three contending reactions that could be employed for detecting nitrogen which involve neutrons and gamma rays: $^{14}N(n,\gamma)^{15}N$, $^{14}N(n,2n)^{13}N$, and $^{14}N(n,n'\gamma)^{14}N$. The first was selected because it is initiated by thermal neutrons (easily produced from fast neutrons), its cross section is better (80 millibarns) than those of its competitors, and the resultant gamma radiation has some unique features. In ~ 15% of the reactions a 10.8 MeV gamma ray is produced. Gamma rays of this and higher energy are rare in other elemental neutron reactions and their production probability is considerably less than 15%.

Since the 10.8 MeV gamma rays are emitted promptly, a detection system based on this reaction must incorporate the means of producing the required thermal neutron flux, as well as the gamma ray detection apparatus. In routine neutron activation analysis studies, this is generally accomplished by constructing a cavity in neutron moderating material with a fast neutron source and detector(s) in close proximity to the cavity. A simple system is shown in Figure 1. The moderating material thermalizes the fast neutrons and deflects a fraction of them into the cavity creating a thermal neutron "field" therein. An item to be examined is inserted into the cavity and the resulting gamma ray activity is detected for elemental analysis. The sensitivity and practicality of this technique is dependent upon the achievable ratio of thermal neutron flux to fast neutron flux, the gamma ray detection efficiency and the capability of discriminating against potential false alarm materials. To optimize these aspects one must select the components, which are very interactive, with great care. These components are the neutron source, the moderator, the cavity, the gamma ray detection system and the shielding. Their selection or development will be discussed in the following section.

## SELECTION OF SYSTEM COMPONENTS

### Neutron Source

Fast neutron sources can be of the machine variety or isotope variety. Machine sources electrically powered D-D or D-T

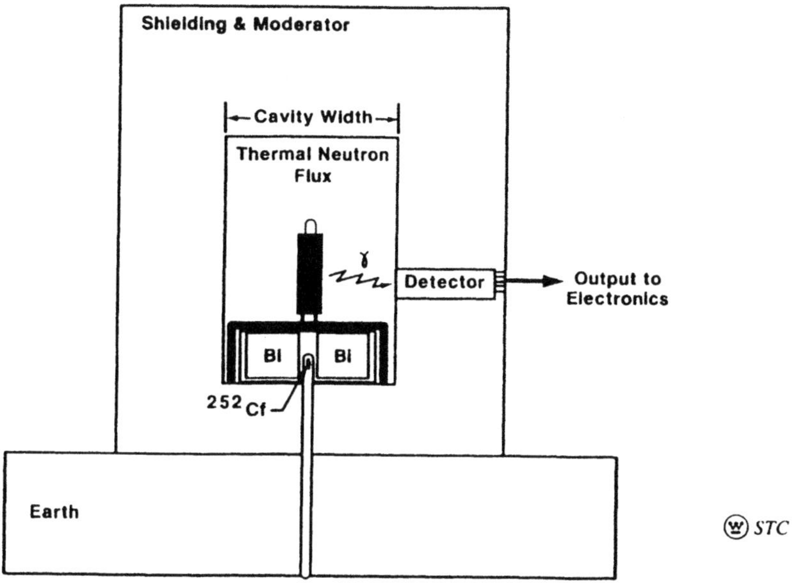


Figure 1 A Prompt Neutron Activation System

generators can be turned off but they are somewhat bulky and the periodic replacement of generator tubes makes them more expensive. The isotopic sources, e.g., $^{252}Cf$, are smaller and lighter, but because they emit neutrons and gamma rays continuously, they must be shielded continuously. We selected $^{252}Cf$ as our source because of its convenient small size, ~ 1 cm diameter and ~ 3.8 cm length, and its low cost. The source intensity was ~ 1 x $10^9$ n/sec.

## Moderator Material

For moderation of the fast neutrons, hydrogenous materials are preferred. The thermal neutrons will be scattered back into the cavity where they will become part of the "thermal neutron gas cloud". In our early work paraffin was used but, because of its potential fire hazard, we shifted to polyethylene. Only 5 cm of polyethylene was needed to produce the maximum thermal neutron flux in the cavity.

## The Cavity

The dimensions of the cavity were prescribed by the FAA. The size was selected which would accommodate most of the baggage items stored in airplanes. The dimensions were ~ 40 cm wide, ~ 60 high and ~ 80 cm long.

### The Gamma Ray Detector

There are three types of gamma ray detectors that are preferred for TNA: Ge detectors, NaI detectors and plastic scintillator detectors. Ge detectors have the best energy resolution, NaI detectors are the most efficient and the plastic scintillator detectors are the fastest. The selection of a particular one for a particular application is a complex process because of the many operational factors that must be considered. Additionally, one must also make decisions on the number and arrangement of the detectors.

We selected plastic scintillator detector principally because of their speed, ease of obtaining different geometries and their low cost. Because the selection of these detectors was key to the detection system feasibility, i.e., it enabled us to succeed where others had failed, we will devote the next section to a more detailed discussion.

### The Shielding

We mentioned earlier that only 5 cm of polyethylene was required for moderation. Thus, the size of the basic system is relatively compact. However, because the neutron and gamma ray radiation must be reduced to safe levels, a considerable amount of shielding was required, and in our studies the primary shielding material was polyethylene. Boron doped polyethylene was used to reduce the number of 2.2 MeV gamma rays resulting from neutron capture by hydrogen. The neutron capture cross section in boron is very high in a normal percentage of the captures and an easily shielded low energy gamma ray is produced. Sixty cm of borated polyethylene adjacent to the moderator produced safe levels. Finally, the $^{252}$Cf source gamma rays were shielded by surrounding the source with 15 cm of Bi.

## THE GAMMA RAY DETECTION SYSTEM

The $^{252}$Cf neutron source is also an intense source of gamma rays. The capture of neutrons by the hydrogenous moderator and shield material is an additional source of gamma ray background. Thus, the selection of $^{252}$Cf for our neutron source and the selection of hydrogenous material for moderation and shielding presented us with a significant problem vis-a-vis the selection of a gamma ray detector. That problem was to find a detector which could handle the high counting rate without producing pileup pulses (overlap of two pulses) due to the intense background, which would interfere with the detection of the 10.8 MeV gamma rays from nitrogen capture. The solution was to select plastic scintillators which produce light pulses $<$ 10 nsec in duration compared to the 250 nsec of the more efficient and widely used NaI detectors.

The poor energy resolution of a plastic scintillator detector was not a problem because of the exceptionally high energy 10.8 MeV gamma ray of the nitrogen capture gamma ray. There are no interfering gamma rays at that energy or higher.

The poor intrinsic efficiency of these detectors could be a problem if a high baggage throughput rate is desired. Because these detectors are inexpensive the solution was to simply increase the volume of the detector. The specific way in which the volume was increased was dictated by the solution of another problem - the problem of distinguishing between explosives and other nitrogenous materials normally found in suitcases, e.g., wool, nylon, orlon, silk and leather.

The key to the solution was the fact that although the nitrogen composition in these materials might be high, e.g., 26% for orlon, the nitrogen densities are significantly lower than that for explosives. Thus, an array of detectors, sensitive primarily to the nitrogen content immediately before it, could distinguish between explosives and other nitrogenous materials. To make a detector sensitive to the nitrogen content before it we took advantage of some aspects of the Compton interaction which is the governing interaction in plastic scintillators for 10.8 MeV gamma rays. The high energy electrons of the Compton tail are directed in a forward direction close to the axis of the incoming gamma ray. Thus, if one made a detector long and narrow, one would: 1) greatly increase the probability of interaction for a gamma ray entering from the front compared to the side and 2) because of the forward direction of the high energy electron enhance its probability of depositing all of its energy within the scintillator. This would provide the detection directionality desired. Indeed, a Monte Carlo computer calculation of the counts and distribution of counts expected for an array of 5 cm diameter by 25 cm long plastic scintillator detectors monitoring a wool sweater and four sticks of dynamite, shown in Figure 2 verified this. Thus, by increasing the length we improved the efficiency and provided self-collimation.

## TEST PROGRAM

### The Laboratory System

With the development of a gamma ray detection system we were in a position to conduct a laboratory test program of the feasibility of this approach with the eventual goal of development of an airport system.

A test system consisted basically of a 25 cm polyethylene cavity shell surrounded by paraffin was designed and built (Figure 3). Along the right-hand wall can be seen the front surfaces of the six detectors (black circular areas) used in our

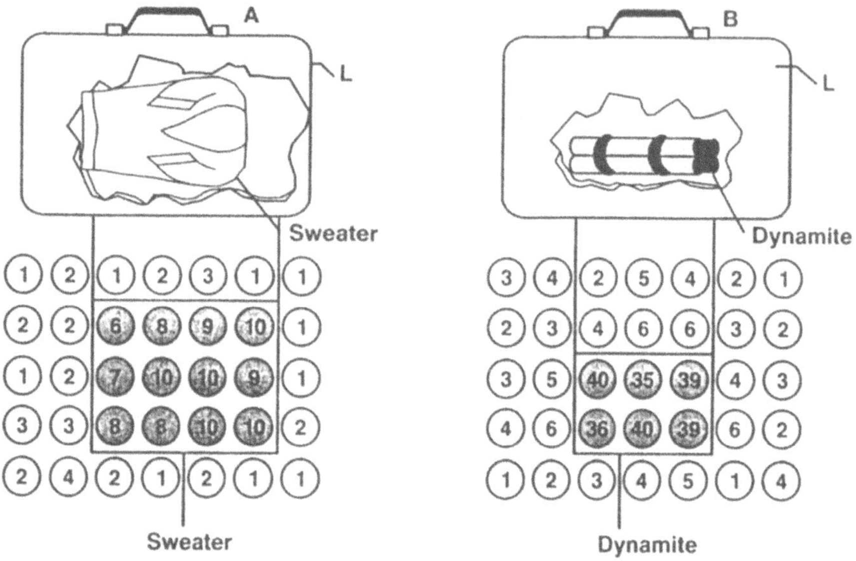


Figure 2 Comparison of the Gamma Ray Count Distribution Produced in An Array of Long Narrow Detectors by a Sweater and by Dynamite

study. The detectors used in the study were 5 cm diameter x 80 cm long NE102 plastic scintillator cylinders. The figure shows the cavity containing a Pullman-sized suitcase on a board used for positioning the suitcases to different areas within the cavity.

## Data-Taking and Processing

After positioning the explosive and/or nitrogenous material within the suitcase, the suitcase was inserted into the cavity on the locating board shown in Figure 3. After taking 200 second counts (recorded on a computer), the suitcase was moved successively to different positions until one half of the suitcase had been scanned.

The FAA required extensive testing of differently sized suitcases filled with wool and orlon sweaters to various capacities, large pairs of shoes and/or simulated explosives. The key items to be determined by the tests were whether the system would be capable of detecting and discriminating the defined explosives against a background of other nitrogenous materials and the neutron source strength required to enable detection at a throughput rate of 10 bags per minute.

Figure 3 View of the Thermal Neutron Activation Cavity with a Large Suitcase Inside

## RESULTS AND DISCUSSION

The test results were grouped into three categories: (1) explosives alone in baggage, (2) false alarm tests, and (3) detection tests. It is obviously impossible to present all of the data from all of the tests. Instead, we will select the most revealing data for presentation in the paper.

### Detection of Explosives Alone in Baggage

In the first run the package of eight sticks of simulated dynamite packed in a cylindrical array was placed on the floor of a large suitcase. Figure 4 is a computer printout of the results and shows the distribution of the counts in slightly more than half of the suitcase. Note that the counts increase toward the bottom-center of the suitcase. The dynamite package is sketched in at its location during the test.

Based on this run as well as the other explosive runs, we selected a threshold of 211 counts for obtaining an "image" of the explosives. The white numbers in Figure 4 represent this image. The dynamite is clearly delineated by a solid block of high counts covering its area.

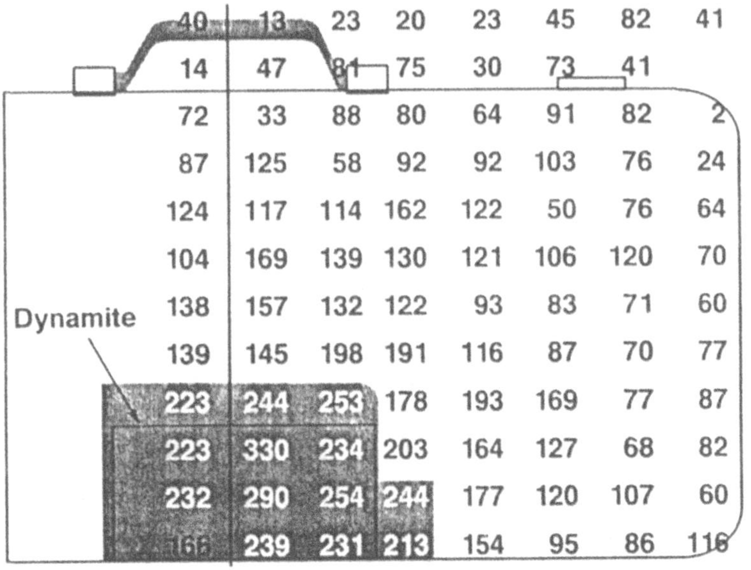


Figure 4 Count Distribution Produced by Dynamite in a Large Suitcase

## False Alarm Tests

The results for 70% packing of wool (three medium and two large sweaters) in the large suitcase are shown in Figure 5. While there are many detector positions exhibiting counts above the threshold, there is no consistent pattern which would lead to a false alarm on the basis of the aforementioned criteria.

## CONCLUSION

Based on the favorable results of the laboratory tests, the FAA initiated a follow-on program to examine real baggage. For this a detection system was designed and constructed in a trailer. The trailer was transported to four different airports and tests were conducted on a sizeable number of baggage or cargo items. The FAA would add simulated explosives to selected items to provide information on detection rates and false alarm rates. The results of the last two tests are shown in Table 1. The results pointed to the feasibility of moving to the prototype stage.

As a result of the Westinghouse development program, TNA explosive detectors are beginning to play an important role in airport security. Presently, there are four TNA systems established at airports and three others are planned. Improvements vis-a-vis detection of plastic explosives continue to be made.

Table I Results of Airport Tests

| Airport | Number of Bags | Type | Detection Rate | False Alarm Rate |
|---|---|---|---|---|
| Chicago | 1,403 | Cargo | 99.2% | 1.6% |
| Philadelphia | 815 | Checked | 96.3% | 4.1% |

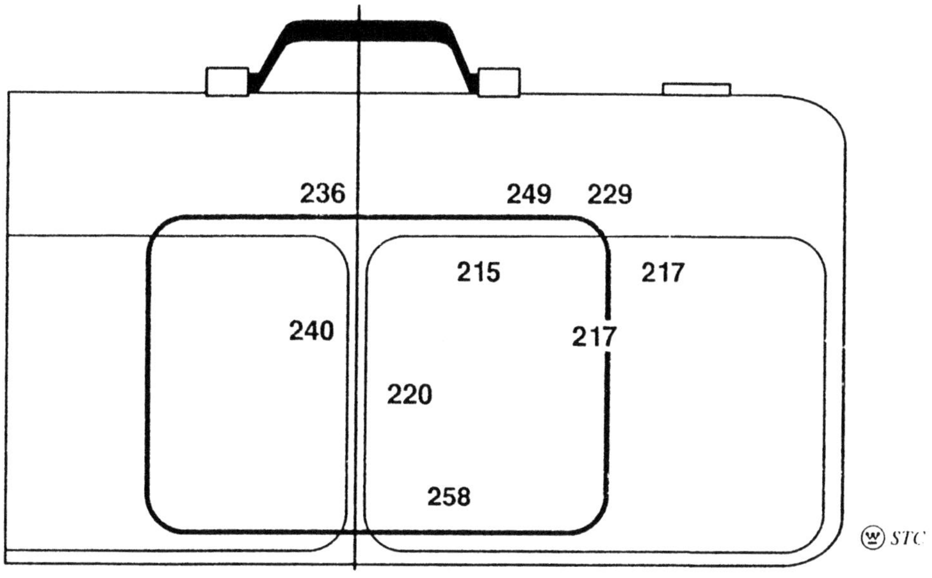


Figure 5 Woolen Sweaters Packed to 70% of Capacity in a Large Suitcase

# A NOVEL ON-LINE NEUTRON DOSIMETER BASED ON A CONDUCTING POLYMER

J. Bartko, F. H. Ruddy, K. F. Schoch and T. V. Congedo
Westinghouse Science & Technology Center
Pittsburgh, PA 15235-5098

E. D. McGarry
National Institute of Standards & Technology
Gaithersburg, MD 20899

S. L. Anderson
Westinghouse Nuclear & Advanced Technology Division
Pittsburgh, PA 15230

## ABSTRACT

A new neutron dosimeter has been conceived and is being tested. The dosimeter consists of a fissile layer placed adjacent to a polymeric material which becomes conducting as a result of irradiation by energetic heavy ions such as those resulting from neutron-induced fission. The increase in conductivity (reduced resistance) of the conducting polymer that results from irradiation of the conducting polymer with neutron-induced fission fragments is monitored as a measure of the fission rate, and with appropriate calibrations can be used to determine the total neutron fluence that has impinged on the detector. The results of preliminary calibration experiments on prototype detectors are described.

## INTRODUCTION

Many polymers, when irradiated with energetic ions, exhibit significant, permanent, and stable increases in conductivity of 10 to 15 orders of magnitude or more.[1] Furthermore, there is a power law relationship between the conductivity and ion dose so that conductivity is extremely sensitive to increases in ion dose. In the case of the polymer poly(phenylene sulfide), PPS, the observed conductivity is a predictable function of the ion stopping power, dE/dx, and the ion dose divided by threshold dose required to produce conductivity.[2]

A novel neutron dosimeter has been proposed[3] that is based on observing conductivity changes brought about in polymers as a result of their exposure to neutron-induced fission fragments.

Based on the behavior of PPS irradiated with known energetic ions, the response of PPS irradiated with fission fragments can be anticipated. Using a representative dE/dx for fission fragments and the predictions of reference 2, the increase in conductivity can be calculated as a function of fission fragment dose. If the fission fragments are being produced by neutron irradiation of a fission foil of known mass, the increase in conductivity can be related to the dose of neutrons delivered to the fissile foil.

The basic dosimeter is composed of two layers. One layer contains fissile material similar to the layers used in Solid State Track Recorder (SSTR) neutron dosimeters. A thin film of PPS forms the second layer. The operation of the dosimeter is as follows. Neutrons captured by the fissile material result in the production of high energy fission fragments which are highly charged ions. The irradiation of the PPS by these ions gives rise to an increase in conductivity. The value of the conductivity is dependent on the ion dose, which in turn is dependent on the neutron fluence.

This dosimeter can be employed either as an active on-line or as a passive detector. In the initial tests, the latter alternative was chosen to simplify interpretation of the detector response.

## EXPERIMENTAL

Layers of PPS were prepared by melt casting PPS powder (Phillips Ryton V-1) onto a 0.60 mm thick aluminum backing plate. After drawing the melted powder into a film, the sample was immediately plunged into an ice/water bath to quench the polymer in the amorphous state. A 0.127 mm 93% enriched $^{235}U$ foil was sandwiched between two 1.19 mm thick aluminum spacers which contained an 11.1 mm circular aperture. Each of the spacers were in turn in contact with an aluminum backed PPS layer as shown in Figure 1. The contents of each dosimetry packet is listed in Table 1. In the case of packet PPS-1, both PPS layers (PPS-1 and PPS-1′) were facing the uranium foil. In the other three cases, PPS-2, PPS-3, and PPS-4 were facing the uranium, whereas PPS-2′, PPS-3′, and PPS-4′ had the PPS layers facing away from the uranium to serve as blanks on the effects of neutron irradiation on the PPS. Each dosimeter packet was wrapped in 0.01 mm aluminum foil, an SSTR neutron dosimetry packet was placed adjacent to the PPS backing (as shown in Figure 1), and a second aluminum wrapping was applied. Each packet was individually heat-sealed in polyethylene prior to irradiation in the National Institute of Standards and Technology (NIST - formerly NBS) Research Reactor. A listing of the contents of the SSTR neutron dosimetry packets is contained in Table 2. The nominal neutron doses for the irradiations are listed in Table 3.

Dwg. 9417A55

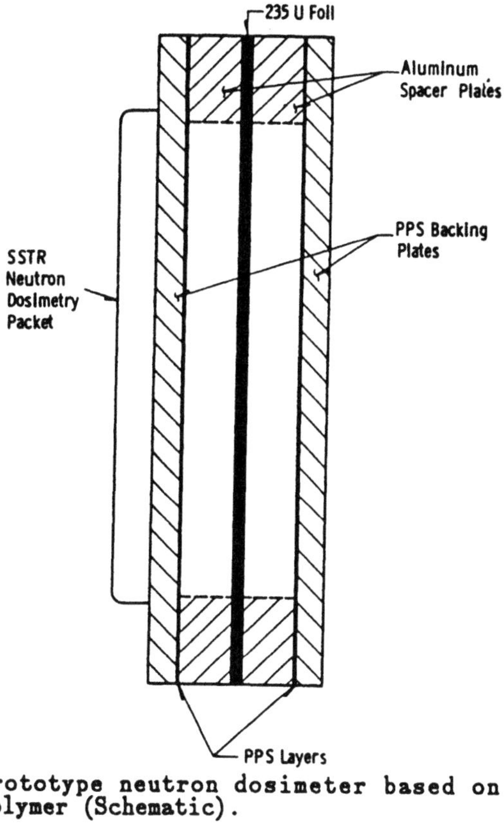


Figure 1. Prototype neutron dosimeter based on a conducting polymer (Schematic).

TABLE 1. CONTENTS OF PPS NEUTRON DOSIMETRY PACKETS

| PACKET LABEL | CONDUCTING POLYMER LAYER LABELS | |
|---|---|---|
| | FRONT | BACK |
| PPS-1 | PPS-1 | PPS-1′ |
| PPS-2 | PPS-2 | PPS-2′ * |
| PPS-3 | PPS-3 | PPS-3′ * |
| PPS-4 | PPS-4 | PPS-4′ * |

*In these cases, the conducting polymer layer was facing away from and therefore shielded from the uranium layer to serve as a blank on the effects of neutron irradiation on the conducting polymer.

TABLE 2. CONTENTS OF SSTR NEUTRON DOSIMETRY PACKETS

| DOSIMETER NUMBER | FISSIONABLE DEPOSIT | MASS (nanograms)* LABEL | SSTR LABEL |
|---|---|---|---|
| PPS-1 | W-76 | 0.349 (± 0.06%) | NBS-111 |
| PPS-2 | W-457 | 0.0647 (± 2.49%) | NBS-112 |
| PPS-3 | W-460 | 0.0362 (± 1.18%) | NBS-113 |
| PPS-4 | W-461 | 0.0167 (± 1.50%) | NBS-114 |

*The number in parentheses represents the relative standard deviation of replicate mass calibrations of the same fissionable deposit.

TABLE 3. RESULTS OF CONDUCTIVITY MEASUREMENTS

| DOSIMETER LABEL | NEUTRON FLUENCE (neutrons/cm²) | RESISTANCE (Ohms) | CONDUCTIVITY[a] (S/cm) |
|---|---|---|---|
| PPS-1 | $3 \times 10^{13}$ | $> 5 \times 10^{12}$ | $< 1 \times 10^{-16}$ |
| PPS-2 | $8 \times 10^{13}$ | $> 5 \times 10^{12}$ | $< 1 \times 10^{-16}$ |
| PPS-3 | $1.5 \times 10^{14}$ | $> 5 \times 10^{12}$ | $< 1 \times 10^{-16}$ |
| PPS-4 | $5 \times 10^{14}$ | $2.9 \times 10^{12}$ | $1.9 \times 10^{-16}$ |

[a]Sheet conductivities were calculated from the four-point probe resistance measurements by assuming an average range of 4 micrometers for the degraded, isotropic fission fragments from the thick $^{235}U$ foils.

After irradiation at NIST, the packets were returned to the Westinghouse Science & Technology Center for analyses. During disassembly, the as-built contents of each packet were confirmed. The uranium foils were generally free from oxidation and the overall condition of the packets was excellent. The irradiated PPS layers are shown in Figure 2. The darker exposed layers

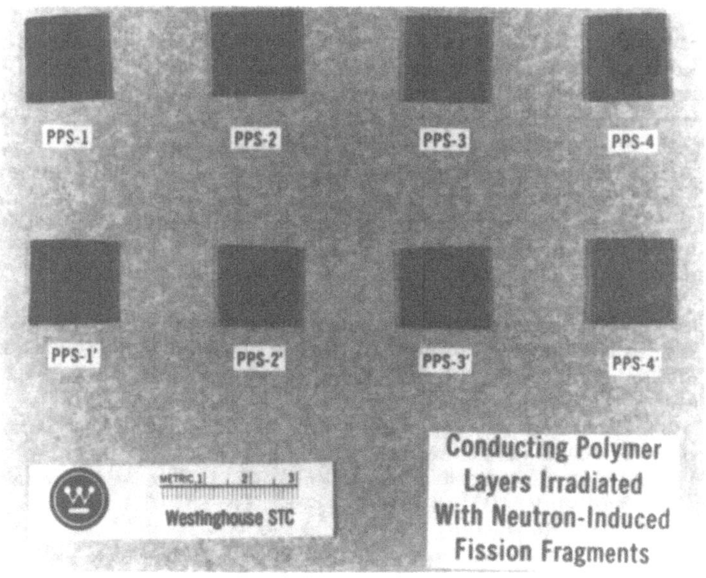


Figure 2. Poly(Phenylene Sulfide) layers after exposure to neutron-induced fission fragments. The darker areas correspond to higher neutron doses.

correspond to the higher neutron fluences. The PPS layers that were exposed to neutrons but turned away from the $^{235}U$ foil show no darkening and appear to be unaffected by neutron or radiation damage encountered during the irradiation.

The conductivities of the PPS layers were measured with a four-point probe as described in reference 4. The results are contained in Table 3.

## RESULTS AND CONCLUSIONS

The results of the four-point probe conductivity measurements are contained in Table 3. In all but one case the resistance was too high to be measurable with our present equipment. However, the dosimeter exposed to the highest neutron fluence had a measurable resistance which corresponds roughly to a conductivity of $1.9 \times 10^{-16}$ S/cm. Additional measurements will be undertaken with more sensitive measuring equipment.

Based on this measurement result it can be concluded that the threshold ion dose for enhanced conductivity has been exceeded for PPS at a neutron fluence of $5 \times 10^{14}$ neutrons/cm$^2$ which corresponds to an ion dose of $2.7 \times 10^{12}$ fission fragments per square centimeter. The average stopping power for fission fragments is about 60 MeV/(mg/cm$^2$) which, according to the predictions of reference 2, would correspond to a threshold for the onset of enhanced conductivity of about $4 \times 10^{11}$ ions/cm$^2$. However, the ions in the present case originate in a thick uranium foil and emerge isotropically from the surface with a degraded spectrum of energies and, since fission fragment stopping power decreases with energy, correspondingly lower stopping powers. The threshold for the onset of conductivity increases as the stopping power decreases.[2] Therefore, the resultant effective conductivity threshold for a thick foil will be shifted to higher doses consistent with the results of the present work.

The conducting polymer dosimeter has been proposed for use in reactor cavity pressure vessel surveillance applications.[3] In this context, the dose of neutrons of energy sufficient to produce displacement damage is of importance and fissionable foils of $^{237}Np$ or $^{238}U$ would be used as sources of energetic ions. In a typical reactor cavity, an ion dose of $2.7 \times 10^{12}$ would be reached at fast neutron fluences ($E > 1$ MeV) of $1.9 \times 10^{17}$ neutrons/cm$^2$ for $^{237}Np$ and $8 \times 10^{17}$ neutrons/cm$^2$ for $^{238}U$, assuming thick sources of both nuclides. Since these doses will be exceeded in approximately 4 and 16 years, respectively, in a typical reactor cavity, the conducting polymer dosimeter would appear to have potential usefulness for pressure vessel surveillance. After the threshold dose has been exceeded, the ion-induced conductivity is expected to increase rapidly with increasing neutron dose.

According to reference 2, an increase in PPS conductivity of four orders of magnitude should result if the threshold dose is exceeded by a factor of 10.

In addition to its potential usefulness in reactor cavity pressure vessel surveillance dosimetry, the conducting polymer neutron dosimeter should find other applications where measurements of neutron fluences commensurate with the onset of materials damage are a concern.

# REFERENCES


1. J. Bartko, B. O. Hall, and K. F. Schoch, Jr., "High Conductivity Poly(phenylene sulfide) Prepared by High Energy Ion Irradiation," J. Appl. Phys. 59, 111 (1986).

2. F. H. Ruddy, J. Bartko, and K. F. Schoch, Jr., "Ion-induced Conductivity in Poly(phenylene sulfide)," J. Matl. Res. 3, 1253 (1988).

3. J. Bartko, K. F. Schoch, Jr., T. V. Congedo, and S. L. Anderson, United States Patent Number 4,857,259 (April 1987).

4. F. M. Smits, "Measurement of Sheet Resistivities with the Four-Point Probe," Bell Sys. Tech. J. 37, 711 (1958).

# BENCHMARK REFERENCING OF SOLID STATE TRACK RECORDER NEUTRON DOSIMETERS IN STANDARD NEUTRON FIELDS


Frank H. Ruddy

Westinghouse Science & Technology Center

Pittsburgh, PA 15235-5098

and

E. D. McGarry

National Institute of Standards & Technology

Gaithersburg, MD 20899



## ABSTRACT

Ultra low-mass fissionable deposit fabrication techniques developed for light water reactor pressure vessel surveillance dosimetry with solid state track recorders have been verified through irradiations in standard neutron fields at the National Institute of Standards and Technology (formerly NBS). Excellent agreement has been obtained between independent absolute calibrations of the fissionable deposits based on ultra low-mass radiochemical spiking techniques and on the response to a known neutron fluence in the standard neutron field. The status of and the results of standard irradiations carried out to date are described.

## INTRODUCTION

Solid state track recorder (SSTR) neutron dosimeters have been deployed for pressure vessel surveillance in the reactor cavity regions of 14 commercial nuclear power reactors during a total of 26 operating cycles.[1] To date, more than 574 SSTR neutron dosimeters have been analyzed or are currently deployed.

A key to the successful use of SSTR neutron dosimetry for reactor pressure vessel surveillance has been the development of techniques for producing and calibrating the ultra low-mass fissionable deposits required for reactor cavity dosimetry.[2,3] As a result of the high neutron fluences encountered in a typical reactor cavity during a reactor operating cycle, the masses of the fissionable deposits used must be kept much less than those conventionally used in SSTR neutron dosimetry. The fissionable deposit masses required for reactor cavity neutron dosimetry range from $1.3 \times 10^{-13}$g/cm$^2$ for $^{239}$Pu to $3.2 \times 10^{-9}$g/cm$^2$ for $^{238}$U. The $^{239}$Pu mass corresponds to only 3 atoms per square micron of surface area.

To verify the ultra low-mass deposit preparation and calibration techniques used for reactor cavity neutron dosimetry, selected deposits have been subjected to an independent calibration by irradiating them in either a thermal neutron field or a $^{235}$U standard neutron fission spectrum at the National Institute of Standards and Technology (NIST - formerly NBS) as part of a collaborative NIST/Westinghouse program to improve and validate methods for power reactor dosimetry. The results of these benchmark irradiations, which also serve as an independent check on the processing and track counting methods for SSTRs,[4] are described below.

## EXPERIMENTAL

Standard neutron fields (or benchmark fields) are maintained by NIST to supply calibration irradiations and certified fluence standards. Two different types of irradiations were carried out at NIST. Isotopes with threshold neutron responses ($^{237}$Np and $^{238}$U) were irradiated in the NIST Cavity Fission Source (the $^{235}$U fission spectrum), and isotopes with primarily thermal neutron responses ($^{235}$U and $^{239}$Pu) were irradiated in a thermal neutron field.

### $^{235}$U Fission Spectrum Irradiations

The NIST Cavity Fission Source (see Figure 1) provides a $^{235}$U thermal-neutron induced fission spectrum in a 30-cm diameter cavity of the NIST Research Reactor. The foils to be irradiated are positioned inside of a cadmium pill box which is sandwiched between two $^{235}U$ driver discs. When the reactor is operated at maximum power (20 megawatts), the $^{235}U$ spectrum fluence rate is $4 \times 10^{10}$ n/cm$^2$-s. Fluences are based upon the known source strength of a standard $^{252}Cf$ field relative to the $^{235}U$ field by comparisons of irradiated nickel foils. First the nickel foils are irradiated in the californium spectrum to obtain a calibration factor for their activity. The nickel foils are then included in each Cavity Fission Source irradiation as fluence monitors. The reactor power level is normally very constant (± 2%), but it is monitored during each $^{235}U$ irradiation by a separate fission chamber, and when necessary power-time variations are taken into account. Certified fluences are normally reported to an accuracy of 1.5-2%.[5]

Fission neutron return spectra for spherical cavities in graphite have been studied extensively by means of neutron transport calculations and by experiments accomplished to develop standard neutron fields in spherical geometries.[6,7] For the 30-cm diameter cavity at NIST, the response of threshold fission detectors to neutrons returning from the cavity walls is less than 0.3% of the uncollided fission neutron fluence from the fission source disks; for detectors with thresholds above 1 MeV, the response to the wall-returned neutrons is less than 0.1%. The Cavity Fission Source is generally not used to irradiate low-energy-sensitive integral dosimeters such as $^{235}U$ and $^{239}Pu$.

The $^{235}U$ fission disks and the internal structure of the cadmium pill box are shown in the upper half of Figure 1. Small perturbations to the $^{235}U$ fission spectrum (which result in corrections of less than 5%) come from scattering in the structure of the Cavity Fission Source. The scattered neutrons may be separated into two categories:

(1) Neutrons scatter into the dosimeters from the fission disks, the cadmium pill box, and the aluminum top-hat-shaped assembly. This inscattering is essentially independent of position of the dosimeter in the stack. These scattered neutrons may produce more activity in the dosimeters than would occur in an unperturbed fission spectrum. Because the fluences reported by NIST are for an unperturbed $^{235}U$ fission spectrum at the locations of the SSTRs or nickel fluence monitors, scattering corrections are applied to each measured reaction rate.

(2) Neutrons scatter in other foils in the dosimeter stack. The effects are both a function of position and the number of foils in the stack. Nearest neighbors are most significant and

Dwg. 9417A75

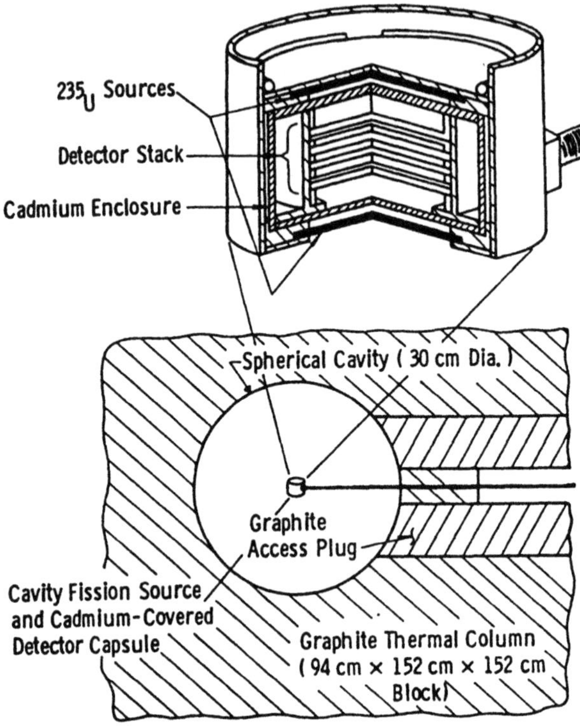


Figure 1. $^{235}U$ Cavity fission source. Upper view: relative positions of neutron sensor foils, fission disks and cadmium enclosure. Lower view: the assembly in the thermal column of the NBS reactor.

the correction factors are larger for the lower energy thresholds. Depending upon the relative positions and type of threshold, the integrated correction factors can be greater than or less than unity. These corrections are determined by Monte Carlo neutron transport calculations.

Total net corrections for these two types of scattering and the particular irradiation geometry have been applied for all irradiated SSTR deposits. These corrections include the wall-return spectrum correction factors of 0.997 for $^{237}Np$ and 0.999 for $^{238}U$ and the $^{58}Ni(n,p)^{58}$ Co reaction.

## SSTR Thermal Irradiations

SSTR dosimeters with low-energy neutron response were not exposed in the Cavity Fission Source but in a thermal neutron field at the center of the 30-cm graphite cavity in the NIST Reactor thermal column. For thermal neutron irradiations, $^{235}U$ fission rates are specified. This was accomplished by irradiating an SSTR package back-to-back with a $^{235}U$ fission deposit of known mass, 95.6 nanograms ($\pm$ 1.5%). The deposit is an active electrode in a fission chamber which looks much like the cadmium pill box mounted in the aluminum holder in the Cavity Fission Source. The irradiations were carried out by placing the fission chamber containing the monitor deposit and one SSTR package into the cavity before the boral curtain at the thermal column was raised. The fission chamber was then made operational on a predetermined time cycle and the curtain was raised. By monitoring the fission rate, the irradiation time history and the duration were determined. Halfway through the irradiation, the fission chamber is rotated 180° to ensure equivalence of the active chamber and SSTR irradiations. The sum of the monitored fissions and the deposit mass provide the basis for determination of the fission density.

## RESULTS

In the cavity source irradiations, radiometric analyses of the $^{58}Ni(n,p)^{58}Co$ reaction in the nickel foils were used to determine the neutron fluence rates at the locations of the nickel foils. The magnitudes of the fluence rates were plotted as a function of foil position in order to determine the fluence rates at the locations of the SSTR neutron dosimeters by interpolation. Figure 2 is a plot of typical neutron fluence rate results for an irradiation in the NIST cavity neutron source. The SSTR fluence rates were corrected for the radial gradient (0.500" diameter nickel foils and 0.250" diameter fissionable deposits were used) and for neutron scattering as discussed previously. After the irradiation, the SSTRs were etched in 49% HF for one hour at 22.0°C to reveal the tracks which were counted both manually by two independent scanners and with the Westinghouse Science & Technology Center automated track counter.[4] The known irradiation time was used to calculate the total fast neutron fluence, which was then used with the number of fissions derived from the track count to derive a deposit mass. The neutron-induced fission cross sections used were 312 mb ($\pm$ 2.3%) for $^{238}U$ and 1359 mb ($\pm$ 2.1%) for $^{237}Np$.

The deposit masses derived from the NIST irradiations are compared with the masses obtained from radiometric spiking techniques in Table 1. In general, the overall agreement of the masses derived from the NIST benchmark irradiations and those derived from the independent radiometric mass calibrations is good. A slight bias (5%) exists between the NIST and radiometric mass scales. The cause for this discrepancy is being investigated, although its overall impact on the 2-5% uncertainty requirement for reactor cavity dosimetry is minimal if the average of the two mass determinations is used.

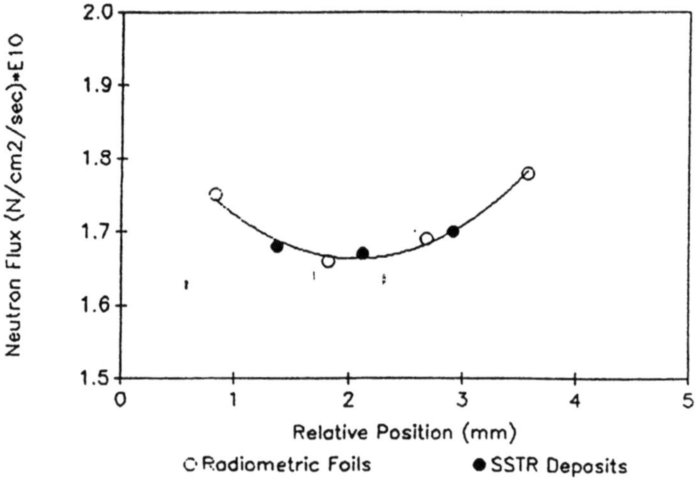


Figure 2. Neutron fluxes as derived from Ni-$^{58}$Co flux foils plotted as a function of foil location. The fluxes obtained at the locations of the SSTR neutron dosimeters by interpolation are also plotted.

For the thermal neutron irradiations SSTR fission rates were calculated from the number of fissions obtained from scanning the SSTR divided by the fissionable deposit mass as determined radiometrically. A direct comparison between the SSTR fission rates and the corresponding fission rates obtained in the active half of the NIST fission chamber is shown in Table 2. In the case of the $^{235}$U fission rates, the agreement between the NIST fission chamber and SSTRs is excellent. The $^{239}$Pu reaction rate ratio differs from one by one standard deviation. More $^{239}$Pu data are needed to determine if this is a real discrepancy. Further thermal neutron benchmark irradiations are currently in progress.

Table 1

Overall Comparison of $^{237}Np$ and $^{238}U$ Fissionable Deposit Masses Derived From NIST Benchmark Neutron Irradiations and By Ultra Low-Mass Radiometric Spiking Techniques

| Isotope | Spike Solution[a] | Mass (nanograms) RM[b] | Mass (nanograms) NIST[c] | Mass Ratio RM/NIST |
|---|---|---|---|---|
| $^{238}U$ | HEDL-12 | 8.01 | 7.52 | 1.065 |
| $^{238}U$ | HEDL-12 | 4.34 | 4.19 | 1.035 |
| $^{238}U$ | HEDL-K1 | 1.63 | 1.51 | 1.077 |
| $^{238}U$ | HEDL-13 | 1.58 | 1.57 | 1.005 |
| $^{238}U$ | HEDL-13 | 1.64 | 1.59 | 1.030 |
| $^{238}U$ | HEDL-13 | 2.06 | 2.14 | 0.964 |
| $^{238}U$ | W R&D-5 | 17.07 | 15.98 | 1.068 |
| $^{237}Np$ | HEDL-8 | 3.63 | 3.34 | 1.086 |
| $^{237}Np$ | HEDL-7 | 0.315 | 0.303 | 1.040 |
| $^{237}Np$ | W R&D-7 | 4.88 | 4.54 | 1.074 |
| $^{238}U$ | HEDL-12 | 2.88 | 2.71 | 1.063 |
| $^{237}Np$ | HEDL-8 | 0.667 | 0.620 | 1.076 |
| $^{238}U$ | W R&D-9 | 1.73 | 1.60 | 1.080 |
| $^{237}Np$ | W R&D-8 | 7.48 | 7.19 | 1.040 |
| | | | Average | 1.050 ± 0.033 |

[a] HEDL spike solutions were prepared and calibrated at the Hanford Engineering Development Laboratory. W R&D spike solutions were prepared and calibrated at the Westinghouse Science & Technology Center.

[b] Mass determined by ultra low-mass radiometric spiking techniques. Overall uncertainties in the mass determinations are less than 2%.

[c] Mass determined from fissions produced in an irradiation in an NBS standard neutron field. The combined precision due to contributions from nickel counting, gradient corrections, and scattering corrections is 0.6%. An additional uncertainty of 0.7% results from the track counting process. Statistical uncertainties related to the total number of fission tracks observed range from 2.5% to less than 1%.

Table 2

Overall Comparison of $^{235}U$ and $^{239}Pu$ Fission Rates
Derived from NIST Standard Fission Chamber Irradiations and
By Ultra Low-Mass Radiometric Spiking Techniques

| Isotope | Spike Solution[a] | Fission Rate[b] (fissions/nanogram) WSTC[c] | Fission Rate[b] (fissions/nanogram) NIST[d] | Ratio (WSTC/NIST) |
|---|---|---|---|---|
| $^{235}U$ | HEDL-13 | $6.479 \times 10^5$ ($\pm$ 2.5%) | $6.492 \times 10^5$ ($\pm$ 2.5%) | 0.9980 ($\pm$ 3.6%) |
| $^{235}U$ | HEDL-14 | $1.909 \times 10^5$ ($\pm$ 2.6%) | $1.91 \times 10^5$ ($\pm$ 2.5%) | 0.9997 ($\pm$ 3.6%) |
| $^{235}U$ | HEDL-16 | $6.603 \times 10^5$ ($\pm$ 3.0%) | $6.492 \times 10^5$ ($\pm$ 2.5%) | 1.0172 ($\pm$ 3.9%) |
| | | | Average | 1.0050 ($\pm$ 0.86%) |
| $^{239}Pu$ | WR&D-2 | $6.758 \times 10^5$ ($\pm$ 3.0%) | $6.503 \times 10^5$ ($\pm$ 2.5%) | 1.0392 ($\pm$ 3.9%) |

[a] HEDL spike solutions were prepared and calibrated at the Hanford Engineering Development Laboratory. W R&D spike solutions were prepared and calibrated at the Westinghouse Science & Technology Center.

[b] Fission rates are in units of fissions per nanogram of $^{235}U$. Plutonium fission rates are converted to equivalent uranium fission rates by using the factor 1.397 ($\pm$ 1.7%) $^{239}Pu$ fissions per $^{235}U$ fission.

[c] Mass determined by ultra low-mass radiometric spiking techniques. The quoted uncertainty in parentheses includes uncertainties due to track counting of the SSTR from the benchmark irradiation in addition to uncertainties due to the mass calibrations.

[d] Fission rate determined by the number of fissions produced in a NIST standard double fission chamber containing a standardized $^{235}U$ deposit.

## CONCLUSIONS

The excellent agreement between the independent absolute radiometric mass calibrations (Westinghouse STC) and the masses derived by benchmark field neutron irradiations (NIST) is encouraging and serves as an overall verification of the radiometric spiking methods[3] used to calibrate the ultra low-mass fissionable deposits used in SSTR neutron dosimetry and of the track scanning techniques[4] used to analyze the exposed SSTRs.

## REFERENCES


1. F. H. Ruddy and J. G. Seidel, "Plant Life Extension Related Applications of Solid State Track Recorder Neutron Dosimetry," Seventh ASTM-Euratom Symposium on Reactor Dosimetry, Strasbourg, France, August 27-31, 1990 (Proceedings to be published).

2. F. H. Ruddy, "Neutron Dosimetry in High Intensity Fields: A Review," Nuclear Tracks and Radiation Measurements 11, 197 (1986).

3. F. H. Ruddy and J. G. Seidel, "Solid State Track Recorder Pressure Vessel Surveillance Neutron Dosimetry at Commercial Light Water Reactors," in Reactor Dosimetry: Methods, Applications, and Standardization, ASTM STP 1001, American Society for Testing and Materials, Philadelphia, PA, p. 175 (1989).

4. F. H. Ruddy, A. E. Manhardt, and J. G. Seidel, "Analysis of Solid State Track Recorders with the Westinghouse Research and Development Center Automated Track Counter," in Reactor Dosimetry: Methods, Applications, and Standardization, ASTM STP 1001, American Society for Testing and Materials, Philadelphia, PA, p. 676 (1989).

5. G. P. Lamaze and J. A. Grundl, "Activation Foil Irradiation by Reactor Cavity Fission Sources," NBS Special Publication 250-14, U.S. Department of Commerce's National Institute of Standards and Technology, Gaithersburg, MD (March 1988).

6. J. A. Grundl, "Cavity Fission Spectra," Trans. Am. Nucl. Soc. 5, 381 (1962).

proportions 2-1-0.2] and limited by two electrodes made of phenolic resin (bakelite) with volume resistivity of the order of $10^{11}$ ohm*cm (Fig.1). The electrodes are lined with a very thin layer (0.1-0.2 mg/$cm^2$) of linseed oil, in order to smooth out surface defects which could lead to accidental discharges (3).

The power is carried by a conductive graphite-based coating. Electric field strength is such as to match the conditions required for streamer formation and propagation. The discharge, some nanoseconds of duration, affects only a small volume surrounding the ionizing particle track. As a consequence, the charge-depleted electrode area is small and, even if the recharging time is long, the inactive volume fraction is negligible.

The detector readout is performed by means of insulated metallic pick-up strips, so that the counter can be read directly with low-voltage signal electronics. When a streamer takes place, an electric charge is induced on a number of adjacent strips. The charge distribution is unambiguously related to the interaction-point coordinate in the direction orthogonal to the strips. Therefore, collecting the charge on strips placed orthogonally on either side of the detector will lead to bidimensional position determination.

Fast time response, negligible dead time and low cost per unit area make in principle this detector ideal for mapping of radiation fields. Scope of this work was :

- to investigate the applicability of RPC counters to gamma-ray and, by means of a suitable converter, thermal neutron field mapping;
- to develop the signal pick-up and data processing electronics for a monodimensional detector.

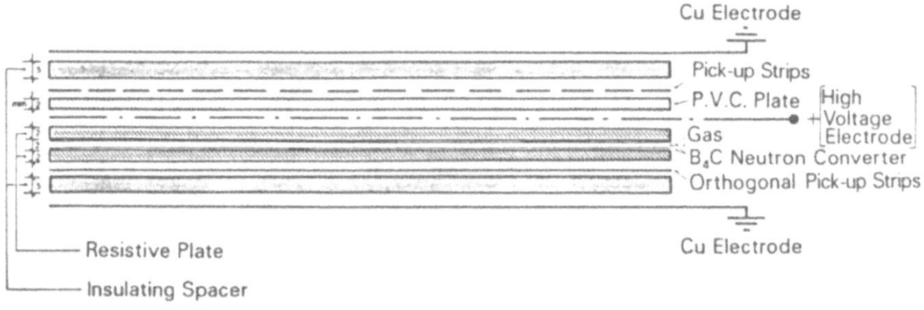


Fig.1 - The Bidimensional Resistive Plate Counter.

## THERMAL NEUTRON DETECTOR

The RPC detector has very high sensitivity (near 100%) to charged particles which penetrate its active volume. Neutral particles must be converted into charged ones.

For gamma rays the converter is naturally provided by the walls of the counter itself. For thermal neutrons an appropriate converter must be provided. Since the available reactions are limited in practice to B-10(n,alpha)Li-7 and Li-6(n,alpha)T, which produce low-energy heavy charged particles, the converter must be lined on the detector walls inside the active volume. This means that the converting layer, in addition to the usual conditions (conversion efficiency as high as possible, transparency to the charged particles produced, good uniformity and mechanical strength, etc.), must satisfy the additional one of having the high resistivity required for the correct counter operation. Moreover, even small defects on the converter surface must be avoided, since they may become very important sources of noise because of the high electric field present in the active gap.

A mixture of Boron Carbide and Epoxy Resin was selected as a neutron converter material. The energy release in the B-10 (n,alpha) reaction is 2.79 MeV, of which only 2.31 MeV are available in 94% of cases as kinetic energy of the reaction products. The energy partition is 0.84 MeV to Li-7 and 1.47 MeV to the alpha particle, so that a $B_4C$ converter layer 1 mg/cm$^2$ thick is adequate to give good thermal-neutron sensitivity.

Dow Chemicals DER 736, $B_4C$ powder (grain diameter:about 1 $\mu$m) and Metyl-Isobutyl-Ketone (MIBK) in the proportions 1:2:60 by weight were carefully blended with the help of an ultrasonic bath. Sedimentation was preferred to spraying as a deposition technique. The surface to be lined, suitably confined, was covered with a quantity of the mixture to give a layer of the desired thickness. After complete MIBK evaporation, which under normal conditions took about 24 h, a fairly uniform and mechanically resistant layer was obtained. It was also verified that such a layer did not alter significantly the electrode resistivity.

The neutron conversion efficiency was measured by exposing to a thermal neutron flux a gas ionization chamber, whose high-voltage electrode was lined with a converting layer 1.8 mg/cm$^2$ thick. Fig.2 shows the pulse spectrum obtained, displaying two broad peaks that can be ascribed to Li-7 and alpha particle, respectively. The efficiency value resulting from these measurements was $6.13*10^{-3}$ counts/incident neutron in good agreement with a theoretic datum, $5.82*10^{-3}$, obtained for an "infinite" thickness layer with an evenly distributed charged particle production.

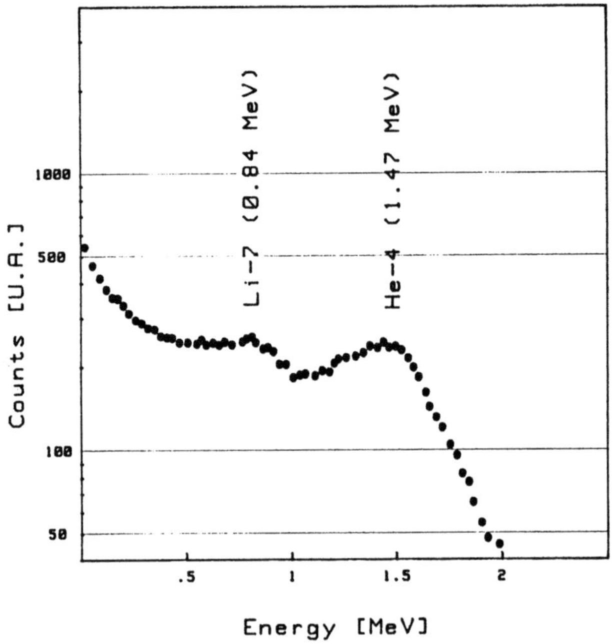


Fig.2 - Pulse Spectrum from an Ionization Chamber with Neutron Converter Exposed to a Thermal Neutron Flux.

SIGNAL READOUT

The strip width affects both the induced charge distribution and the detector spatial resolution. Experimental tests showed that 1.5-cm wide strips are adequate for both criteria. With this choice, about 50 pC are collected in the "hit" strip over 50-ns time gate. The typical pulse at the strip end connected to a 50-ohm resistor has 2-ns rise time and 10-ns full-width-half-max..

The counter resolution can be made finer than the strip width by analysing the size distribution of the pulses on adjacent strips. Since high count rates are expected, this analysis is not carried out in software, but implemented in the hardware. This is done by comparing the digitized values of the amounts of charge collected on the strips with pre-computed values identifying the possible across-the-strip locations of the interaction point. The complete interaction-point coordinate is then obtained by adding a base number, which corresponds to the coordinate of the "hit" strip, to an offset showing the across-the-strip location. This procedure effectively increased resolution by a factor of four.

Finally, if two or more particles interact within the time gate during which signals are integrated, the event is rejected because it would lead to ambiguity about the two interaction points in a bidimensional detector.

## ACQUISITION ELECTRONICS

The acquisition electronics consists of an analog/digital unit and a real-time data handling computer.

The analog/digital unit is required to :
- pick up, amplify and discriminate signals from the detector;
- check whether the signal pattern is consistent with those characteristic of the interaction;
- compute the coordinate of the "hit" strip and, using the pre-assessed amplitude distribution, the across-the-strip coordinate of the point from which the streamer propagated.

Physically, the module is made of three boards : one pre-amplifier board, consisting of four current-sensitive hybrid technology quad preamplifiers, placed close to the detector, and two boards for acquisition and control that can be placed several meters from the detector in a suitably radiation-shielded location. Fig.3 shows the flowsheet of the analog/digital unit for the 16-channel system adopted in the monodimensional configuration.

Amplified signals are fed into passive splitters through line drivers. From splitters onward, signals are processed concurrently by two logic blocks. The former is composed of sixteen free-running integrators followed by three sixteen-input analog multiplexers. The latter consists of sixteen comparators which perform the first signal discrimination against a fixed threshold.

The 16-bit word coming out from the discriminators is fed into a priority encoder, which provides the address of the left-most bit set to 1 (i.e., of the left-most strip whose signal was above threshold), and into two majority logic units, that check whether more than one cluster is present and/or more than a fixed number (3 in this case) of contiguous lines are active. If any of the two conditions is met, the acquisition sequence is reset, since either the event is not unambiguously reconstructable (more than one cluster) or the cluster is too wide.

If the event is accepted, the time synchronization unit activates the three analog multiplexers. Multiplexing addresses are provided by the priority encoder in such way that MUX 1 receives address N, MUX 2 address N+1 and MUX 3 address N+2, N being the address of the left-most active strip. The outputs of the three selected integrators are then multiplexed and converted by three 8-bit flash ADCs. The 4 most significant bits from each ADC are assembled to make a 12-bit pattern, which is the image of the original charge distribution across the strips.

The 12-bit word is used as an address to point into the RAM table, that provides a 4-bit wide across-the-strip coordinate. This half byte is then assembled with the 4-bit address of the cluster, making an 8-bit wide coordinate which is stored in a FIFO I/O buffer interfaced to the external VME I/O unit.

Afterwards the real-time data handling unit :
- acquires the digitized coordinate values from the analog/digital unit;
- generates and updates the event map;
- displays and stores the map.

To avoid bottle necks, the word width is kept constant

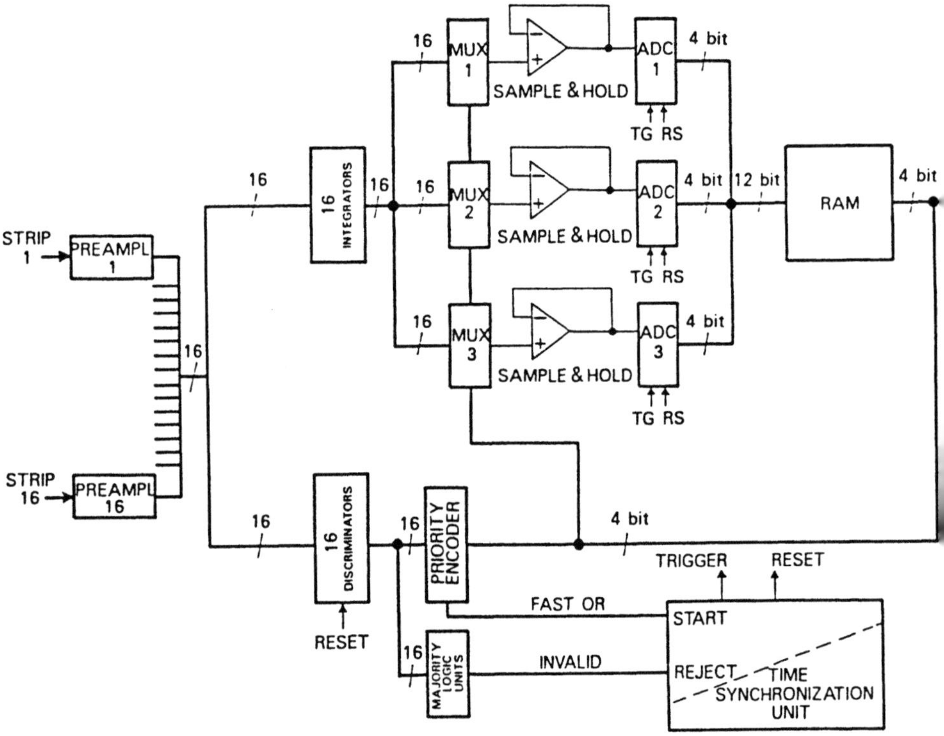


Fig.3 - Flowsheet of the Analog/Digital Unit for a 16-Channel Readout System in the Monodimensional Configuration.

throughout the digital system : a 16-bit wide path was selected. Considerations of speed, flexibility and commercial availability of hardware led to choose a VMEbus architecture for this subsystem. It consists of a parallel I/O unit, a CPU, a RAM board, and a graphic unit with high-resolution monitor.

## RESULTS

A 16-strip counter, having both surfaces facing the 2-mm wide gap lined with a $B_4C$ layer about 1 mg/cm$^2$ thick, and a modular 16-channel readout system for a monodimensional configuration were realized.

The scope was to measure gamma-ray and thermal neutron detection efficiency, to verify the performance of the adopted acquisition principle and evaluate the spatial resolution and the maximum acquisition rate attainable. Anyway, it must be reminded that the exact performance of the counter depends on the details of construction and use, so no absolute value should be attributed to our results.

The counter efficiencies were measured for thermal neutrons, and for Cs-137 and Am-241 gamma rays as a function of the operating voltage, within the 7.3-7.8 kV range. Typical values (e.g., at 7.5 kV) were of the order of $10^{-3}$ counts/incident particle for thermal neutrons, $10^{-5}$ counts/particle for 660-keV gamma rays, and $10^{-6}$ counts/particle for 60-keV gamma rays.

The substantial loss in neutron efficiency with respect to the optimal one measured in the ionization chamber is probably due to the presence of the linseed oil layer. Even so, the thermal neutron efficiency is higher by about a factor of 100 than the 660-keV gamma-ray efficiency, almost independently of the operating voltage. Our counter had a noise intensity of 0.15 Hz/cm$^2$, so that it could be used to measure thermal neutron and gamma-ray flux densities higher than some $10^3$ n/cm$^2$*s or some $10^5$ gammas/cm$^2$*s. The maximum acquisition rate turned out to be 2 MHz.

Although in principle the signal readout system can operate on three contiguous strips for the reconstruction of the event coordinate, tests with a gamma-ray source indicated that the spatial resolution could be increased by a factor of 4 (to 0.4 cm) by analysing the relative pulse heights of only two adjacent strips.

## CONCLUSIONS

The RPC detection unit seems very promising for use as a

position-sensitive detector for thermal neutrons and gamma rays.

It offers fast time response (a few nanoseconds), very short dead time (about 500 nanoseconds), and large active surfaces with low cost per unit area.

Spatial resolution of 0.4 cm was attained with the present 16-channel counter and readout system. However, in principle the 1.5-cm wide strips and the 4-bit wide offset allow a 1.5/16 $\approx$0.1 cm resolution, well below the value attainable for physical reasons (e.g., beam spread due to scattering).

On the other hand, if 1.5-cm wide strips are used, a 4-bit wide word for the cluster coordinate allows single detection units of only 1.5*16=24 cm. Anyway, electronics is designed to be modular, so that more units can be connected to obtain a 16-bit wide coordinate (by adding a board coordinate to the in-board coordinate). This way, an 8-bit wide word for the board coordinate would allow in principle the handling of up to 256* 0.24=61-m wide detectors.

Although our RPC detection unit has been tested till now only for monodimensional operation, the extension to bidimensional data handling is straightforward, being only a question of synchronization of signals coming from two sets of orthogonal strips. The dedicated software for this is already available.

At present, the main drawbacks seem to be a relatively low reproducibility from counter to counter and a strong dependence of the output signal on the composition of the flowing gas mixture. The last drawback can probably be overcome by flowing an appropriate gas mixture in a closed circuit.

Our main interest was focused on detection units to be used in the low-energy field for radiation mapping purposes, but we think that some of the solutions adopted for the system readout have a general applicability.

## REFERENCES


1. Yu.N. Pestov and G.V. Fedotovich, Izv. Akad. Nauk SSSR, ser.Fiz.(SLAC Translation 184, 1979).

2. R.Cardarelli and R. Santonico, Nucl. Instr. and Meth. 187, 377 (1981).

3. R. Cardarelli, R. Santonico, A. Di Biagio, and A. Lucci, "Progress in Resistive Plate Counters Technique," Nucl. Instr. and Meth. A263, 20 (1988).

# INDUCED X-RAY EMISSION IN NIOBIUM FOIL DOSIMETERS

T. G. Williamson* and A. C. Chubb
Department of Nuclear Engineering and
Engineering Physics
University of Virginia
Charlottesville, VA 22901

## ABSTRACT

A procedure has been developed to account for induced X-ray emission by impurity activities in niobium activation detectors. The X-ray fluorescence contribution has been measured in activated niobium foils which contain tantalum and tungsten impurities. The contribution of beta rays to the fluorescence was measured independently with external beta sources. Both experiments were modeled with Monte Carlo calculations. An apparent $^{93m}$Nb activity / $^{182}$Ta activity of 2.0 (resp. 2.5) in a 0.127 mm (resp. 0.254 mm) thick niobium foil was measured. These values have been used with reproducible results for foils irradiated in a power reactor environment.

---

## INTRODUCTION

The reaction $^{93}$Nb(n,n')$^{93m}$Nb is useful for fast neutron dosimetry in the pressure vessel environment. The radioisotope $^{93m}$Nb decays by internal conversion from a 30 keV excited state so that the primary emissions are niobium X-rays. A difficulty in using niobium activation detectors results from the fact that niobium usually contains trace quantities of tantalum which activate readily. The isotope $^{182}$Ta decays with a 115 day half-life and emits a wide spectrum of beta and gamma rays which can induce X-ray fluorescence in niobium and interfere with the determination of the $^{93m}$Nb activity. This problem has been addressed by making measurements of the tantalum interference and modeling the experiments with Monte Carlo calculations.

---

* Present address: Westinghouse Savannah River Laboratory, Aiken, SC.

Two sets of experiments were done to measure fluorescence in niobium from other isotopes. In one set of experiments niobium foils with tantalum as an impurity were irradiated in the University of Virginia Reactor (UVAR) with covers designed to produce different tantalum activities. The irradiation produced the isotopes $^{93m}$Nb, $^{182}$Ta, $^{187}$W and $^{92}$Nb. The activities of these isotopes and the X-ray emission rates were measured over a period of several months to follow the decay of the various isotopes. In the second set of experiments $^{99}$Tc, $^{38}$Cl and $^{210}$Bi beta-ray sources were placed in contact with unirradiated niobium foils and the production of niobium X-rays measured. This experiment was done because of the uncertainty in the calculation of the production of X-rays by isotopic beta sources. Monte Carlo routines were written to model each experiment.

For the irradiation experiment two sets of 12.7-mm diameter niobium foils were irradiated in the reflector of the research reactor. One set contained foils 0.127 mm thick with a nominal tantalum concentration of 0.075 % and the other set contained foils 0.254 mm thick with a tantalum concentration of 0.017 %. Each set contained five foils, one irradiated with no neutron absorbing cover, one with a 0.25-mm cadmium cover, one with a 0.51-mm cadmium cover, one with a 0.51-mm cadmium and 0.13-mm tantalum cover and one with 0.51-mm cadmium and 0.25-mm tantalum cover. The covered foils were packaged together and displaced about 60 mm from the bare foils. The foil set was irradiated to a fast neutron fluence of 4 X $10^{16}$ $cm^{-2}$.

The $^{182}$Ta, $^{187}$W and $^{92}$Nb activities of each foil were measured with a germanium detector and the emission rate of 16.6 keV X-rays measured with a planar germanium detector. The X-ray emission rate was followed over a period of several months to track the fluorescence contribution from the tantalum decay. For the tantalum analysis, data was not included for the first two weeks to allow for decay of the $^{187}$W and $^{92}$Nb isotopes. For each foil a plot was made of the X-ray emission rate against the tantalum activity. The slope of this line is the ratio of the X-ray production rate to the tantalum activity and the intercept at zero tantalum activity is the $^{93m}$Nb activity. A correction was made for each data point for $^{93m}$Nb decay.

After correcting for the fluorescence contribution from the tantalum isotope, the decay data taken during the first several days was used to determine the effect of the short lived tungsten isotope. The fluorescence contribution from 92Nb was too small to measure from this data.

In the second set of experiments, commercially purchased $^{99}Tc$, $^{38}Cl$ and $^{210}Bi$ beta-ray sources were placed in contact with unirradiated niobium foil material and the production of niobium X-rays measured with a planar germanium detector. The beta-ray emission rate of each source was measured with a proportional counter and each source was radiographed to determine the location and extent of the active region. Each source was placed in intimate contact with niobium foil material of two thicknesses, 0.127 mm and 0.254 mm.

Monte Carlo routines were written to simulate two source geometries, one in which the sources are uniformly distributed throughout the foil volume, and one in which the source is external to the foil but in intimate contact with one surface. In both cases the foil is modeled as a cylinder with a flat surface parallel to the detector plane. The code calculates the location of the first interaction, produces a 16.6 keV X-ray if that location is within the foil volume, and then follows the X-ray to determine if it strikes the detector. Multiple Thomson scattering is allowed for the X-ray.

For source photons the total attenuation coefficient is used to calculate the initial path length. Each interaction is weighted by the photon abundance, the ratio of the photoelectric to the total cross section, the number of K-shell vacancies per photoelectric absorption, the fluorescence yield and the k$\alpha$ X-ray transition probability. The program does not trace the primary photon beyond the first interaction.

For beta-ray sources and conversion electrons an attenuation coefficient is calculated from

$$\mu = \rho \cdot 4.47 \text{ X } 10^4 \cdot (\text{Emax})^{-1.14} \quad \text{cm}^{-1},$$

where

$\rho$ = density of foil material, g · $cm^{-3}$,
Emax = beta spectrum maximum energy - keV.

This attenuation coefficient is used to calculate the location of the electron interaction and to generate an X-ray at that location. Each interaction is weighted by the charged particle decay abundance and the probability of the charged particle producing an X-ray. The number of X-rays produced per beta interaction is

$$I_p = n_K \cdot g_{K\alpha} \cdot \omega_K,$$

where

$n_K$ = K-shell vacancies per $\beta$ particle absorbed,
$g_{K\alpha}$ = K$\alpha$ x-rays produced per K-shell vacancy, and
$\omega_K$ = fluorescence yield.

The number of K-shell vacancies per electron, $n_K$, is

$n_K$ = $n_{Kdir} + n_{Kind}$
$n_{Kdir}$ = direct K-shell ionizations per electron and
$n_{Kind}$ = indirect (i.e., secondary particle such as bremstrahlung) K-shell ionizations per electron.

The number of direct K-shell ionizations per electron is found from

$$n_{Kdir} = R \cdot (A/N_{av}) \cdot \int_{E_K}^{E_O} (Q_K/S)\, dE,$$

$$Q_K = .76 \cdot \pi \cdot e^4 \cdot z_K \cdot \ln(U) / (U\, E_K^2) \quad (1),$$

$$S = (3.458 \times 10^{10} / E_{avg}) \ln(.002106 \cdot E_{avg}) \quad (2)$$

$$E_{avg} = (E_O + E_K) / 2, \text{ and}$$

$$n_{Kind} = 1.46 \times 10^{-8} \cdot (Z-2)^2 \cdot Z \cdot (U_O \ln(U_O) - U_O + 1) \quad (3),$$

where

$R$ = backscatter correction factor,
$A$ = atomic weight,
$N_{av}$ = Avogadro's number,
$Q_K$ = K-shell ionization cross section, $cm^2$,
$S$ = stopping power, eV $cm^2$ $g^{-1}$,
$E_0$ = initial electron energy, eV,
$E_K$ = binding energy of the K-shell electron, eV,
$E$ = instantaneous electron energy, eV,
$z_k$ = number of orbital electrons in the K-shell,
$U$ = overvoltage $E/E_k$, and
$U_o$ = initial overvoltage $E_0/E_K$.

These equations have been integrated numerically. A backscatter correction factor is applied for beta sources external to the foil to take into account the loss of ionizations due to the backscatter of electrons out of the target. (4)

The following constants were used for the $^{93m}$Nb decay: Half life - 16.13 years (5), fraction of photoelectric interactions which result in a K-shell vacancy - 0.85, fluorescence yield - 0.748, Kα X-rays per K-shell X-ray - .833, K-shell X-rays per $^{93m}$Nb disintegration - 0.1112, (6) and Kα X-rays per $^{93m}$Nb disintegration - 0.0926.

## RESULTS AND DISCUSSION

The results for measurements made with external beta sources are shown in Table I which lists the calculated and measured count rates of 16.6 keV X-rays. The agreement between the two values is not as satisfactory as one would like. With the thinner foils the ratio of calculated to measured values is within 12% but for the thicker material the agreement is no better than 25% and the calculated values are consistently lower than the measured values. Some reasons for the discrepancies could be an improper treatment of the calculation of the site of the production of the X-ray, improper treatment of the reflection coefficient, wrong calculation of the production rate of X-rays by betas, and measurement inaccuracies, such as the knowledge of the number of beta particles entering the foil volume. It should be noted that the location of the site of the beta interaction and the reflection coefficient are not as important for sources uniformly distributed throughout the volume as for external sources.

TABLE I
FLUORESCENCE FROM EXTERNAL BETA SOURCES

| Source | Avg. Beta energy keV | Foil Size mm | Calculated cps | Measured cps | Ration C/M |
|---|---|---|---|---|---|
| Tc-99 | 85 | .127 | .0174 | .0156 | 1.12 |
| | | .254 | .00232 | .00314 | 0.074 |
| Cl-36 | 251 | ,127 | .0665 | .0714 | 0.94 |
| | | .254 | .00961 | .0120 | 0.80 |
| Bi-210 | 389 | .127 | .131 | .124 | 1.06 |
| | | .254 | .0276 | .0281 | 0.98 |

The results for the measurements made in the irradiated foils are shown in Table II. This table lists measured and calculated values of apparent Niobium 93m activity per tantalum and tungsten activity in niobium foils of two thicknesses. The ratio of the calculated to measured values for tantalum is within one percent for the thinner foils and 15% for the thicker foils. For tungsten the agreement is not as good. For the tungsten measurement, corrections were made to the measured X-ray disintegration rate to account for the tantalum contribution. This procedure introduced an uncertainty in the tungsten measurements which has not been quantified. Tables III and IV list the calculated values of the fluorescence contribution for each photon and charged particle included in this analysis. In some cases photons or charged particles of similar energy were combined into one energy group with an equivalent energy and intensity. Also tabulated is the fraction of the fluorescence from the photons, the beta decay groups and the conversion electrons. For tantalum the majority of fluorescence is caused by the photons, particulary the low energy gamma rays. For tungsten 64% of the fluorescence is caused by photons and 36 % by charged particles. The uncertainty in calculating the charged particle effect contributes to the discrepancy between the measured and calculated values for tungsten. It is also apparent from Tables III and IV that fluorescence from the niobium isotopes is much less important than fluorescence from tantalum.

TABLE II
NIOBIUM FLUORESCENCE FROM TANTALUM DECAY

| FOIL | | NB ACTIVITY/ TA ACTIVITY | NB ACTIVITY/ W ACTIVITY |
|---|---|---|---|
| FOIL THICKNESS .127 MM | | | |
| E-5 | | 1.99 | .742 |
| G-5 | | 1.95 | .758 |
| H-5 | | 1.94 | .782 |
| J-5 | | 2.19 | .858 |
| B-5 | | 2.07 | .768 |
| | Average | 2.03 | .782 |
| | Calculated | 2.04 | 1.32 |

Niobium foils and wires which had been irradiated in power reactor surveillance capsules have been analyzed. Each sample was counted over a period of several months to observe a change in the tantalum induced contribution to the measured activity. For some of the samples the tantalum induced X-ray emission was greater than 50% of the total. Results from a typical foil are shown in Table V. In this case the foil was counted over a period of approximately one half life of the tantalum decay and the contribution to the measured 93mNb activity from tantalum fluorescence ranged between 20 and 40%. The corrected activity listed in the fifth column is the activity of 93mNb on September 1. One standard deviation from the mean of the five measurements is bettei than 2%. The uncertainty assigned to each individual measurement decreased from 7% for the first measurement to 4% for the last. In the uncertainty calculation a 10% uncertainty assigned to the fluorescence correction factor was combined quadratically with the various counting, foil attenuation and standard source uncertainties. Reproducibility of measurements in the range of one to three per cent have been attained for foils in which the tantalum correction is less than 50%.

TABLE III
NIOBIUM FLUORESCENCE - TA-182

| PARTICLE TYPE | ENERGY KEV | NB ACTIVITY PER DISINTEGRATION | | | |
|---|---|---|---|---|---|
| | | 0.127 MM | | 0.254 MM | |
| TA-182 | | RATIO | % | RATIO | % |
| γ | 34 | 0.43 | 87 | .050 | 91 |
| | 58 | .63 | | .88 | |
| | 67 | .18 | | .27 | |
| | 68 | .74 | | 1.08 | |
| | 85 | .029 | | .042 | |
| | 100 | .11 | | .18 | |
| | 163 | .036 | | .058 | |
| | 222 | .009 | | .014 | |
| | 1177 | .002 | | .004 | |
| β | 263 | .022 | 9 | .023 | 7 |
| | 441 | .043 | | .046 | |
| | 527 | .12 | | .12 | |
| e- | 34 | .002 | 4 | .002 | 2 |
| | 60 | .009 | | .009 | |
| | 90 | .047 | | .047 | |
| | 153 | .016 | | .016 | |
| TOTAL | | 2.04 | | 2.84 | |

TABLE IV
NIOBIUM FLUORESCENCE - NIOBIUM ISOTOPES
0.0127 MM FOIL

| PARTICLE TYPE | ENERGY KEV | NB ACTIVITY PER DISINTEGRATION | | | |
|---|---|---|---|---|---|
| | | 0.127 MM | | 0.254 MM | |
| NB-94 | | RATIO | % | RATIO | % |
| γ | 702 | .0065 | 4 | .011 | 7 |
| β | 871 | .0042 | | .007 | |
| | 431 | .24 | 96 | .25 | 93 |
| TOTAL | | .25 | | .27 | |
| NB-92 | | | | | |
| γ | 934 | .004 | 100 | .006 | 100 |
| TOTAL | | .004 | | .006 | |
| NB-95 | | | | | |
| γ | 766 | .005 | 23 | .009 | 33 |
| β | 160 | .019 | 77 | .019 | 67 |
| TOTAL | | .024 | | .028 | |

In summary, the production of niobium X-rays by fluorescence in niobium dosimetry foils has been measured and calculated by Monte Carlo techniques. The agreement between the measured and calculated values provides reasonable confidence in the calculations. The techniques have been applied to dosimetry foils and wires from operating power plants.

TABLE V
PWR IRRADIATED FOIL ACTIVITY

FOIL THICKNESS 0.127 MM

| DATE | MEASURED CPS | TA ACT BQ | APPARENT ACTIVITY BQ | CORRECTED ACTIVITY BQ | TA % |
|---|---|---|---|---|---|
| SEP 8 | 2.90 | 733 | 3844 | 2365 | 39 |
| SEP 17 | 2.93 | 694 | 3880 | 2482 | 36 |
| OCT 6 | 2.77 | 619 | 3670 | 2436 | 34 |
| OCT 28 | 2.65 | 542 | 3515 | 2438 | 31 |
| JAN 11 | 2.32 | 344 | 3073 | 2415 | 21 |
| | | | AVERAGE ACT | 2427 | |
| | | | ONE STND DEV | 38 (1.6%) | |

ACKNOWLEDGEMENT

Much of this work was done by A.C. Chubb as part of a Masters thesis in Nuclear Engineering. The support of Virginia Power and Babcock and Wilcox is gratefully acknowledged.

REFERENCES

1. Powell, C. J., Rev. Mod. Phys., 48 (1976) 33.

2. Poole, D. M., and P. M. Thomas, J. Inst. Metals, 90 (1961-62) 228.

3. Green, M. and V. E. Cosslet, Proc. Phys. Soc. London 78 (1961) 1206.

4. Duncumb, P. and S. J. B. Reed, in "Quantative Electron Probe Microanalysis", Ed. K. F. J. Heinrich, NBS Special Publ. 298, National Bureau of Standards, US Dept of Commerce, Washington, D.C. (1968) 133.

5. Vaninbroukx, R., Int. J, of Appl. Rad. and Iso., 34 (1983) 1211-1214.

6. Coursey, B. M., R. Vaninbroukx, D. Reher, J. M. R. Hutchinson, P. A. Mullen, K. Debertin, R. J.Gerke, J. W. Rogers, L. D. Koeppen and D. Smith, Nucl. Inst. and Meth. A290 (1990) 537-550.

# THE RADFET SYSTEM FOR REAL-TIME DOSIMETRY IN NUCLEAR FACILITIES

by Andrew Holmes-Siedle*, Leonard Adams**, J. Stephen Leffler and Steven R. Lindgren***

* R.E.M, 64A Acre End St.,Eynsham, Oxford 0X8 1PD, England
** ESA/ESTEC, POB 299, 2200AG Noordwijk zh, The Netherlands.
*** General Atomics, POB 85608, San Diego, CA 92186, USA

The RADFET is a solid state, integrating dosimeter for ionising radiation. Dose is determined by the measurement of long-lived trapped charge, generated by radiation in a silicon dioxide layer grown on silicon. The principle and construction of the RADFET and the design of reader circuits for various applications are introduced, with emphasis on nuclear facilities. Radiation dose may be read out nondestructively, during or after exposure, by means of low power dc electronics. The available dose range covers seven orders of magnitude from 1 Rad (10 mGy) to 100 MRad (1 MGy) with very little fading after exposure. The small size, ruggedness and low cost of the device make it suitable for severe environments. RADFETs have been shown to respond to gamma-rays, hard and soft x-rays, electrons, protons and neutrons. In radiation effects testing of electronic devices, the RADFET is suitable for integrating the ionising component of dose to be expected in a semiconductor component. In nuclear facilities, a RADFET damage monitor system can be fitted to Class 1E instruments, robots, area monitoring arrays and in reactor dosimetry capsules.

## 1. INTRODUCTION

A silicon dioxide film, grown by thermal oxidation on the surface of silicon, is an integral part of semiconductor devices and a key element of metal-oxide-semiconductor field-effect transistors (MOSFETs). Early in the history of MOSFET development, the movement and trapping of charge during irradiation was observed. This effect created a positive charge sheet which degraded the operation of the device. It was then realised that the effect could be controlled and could be of major

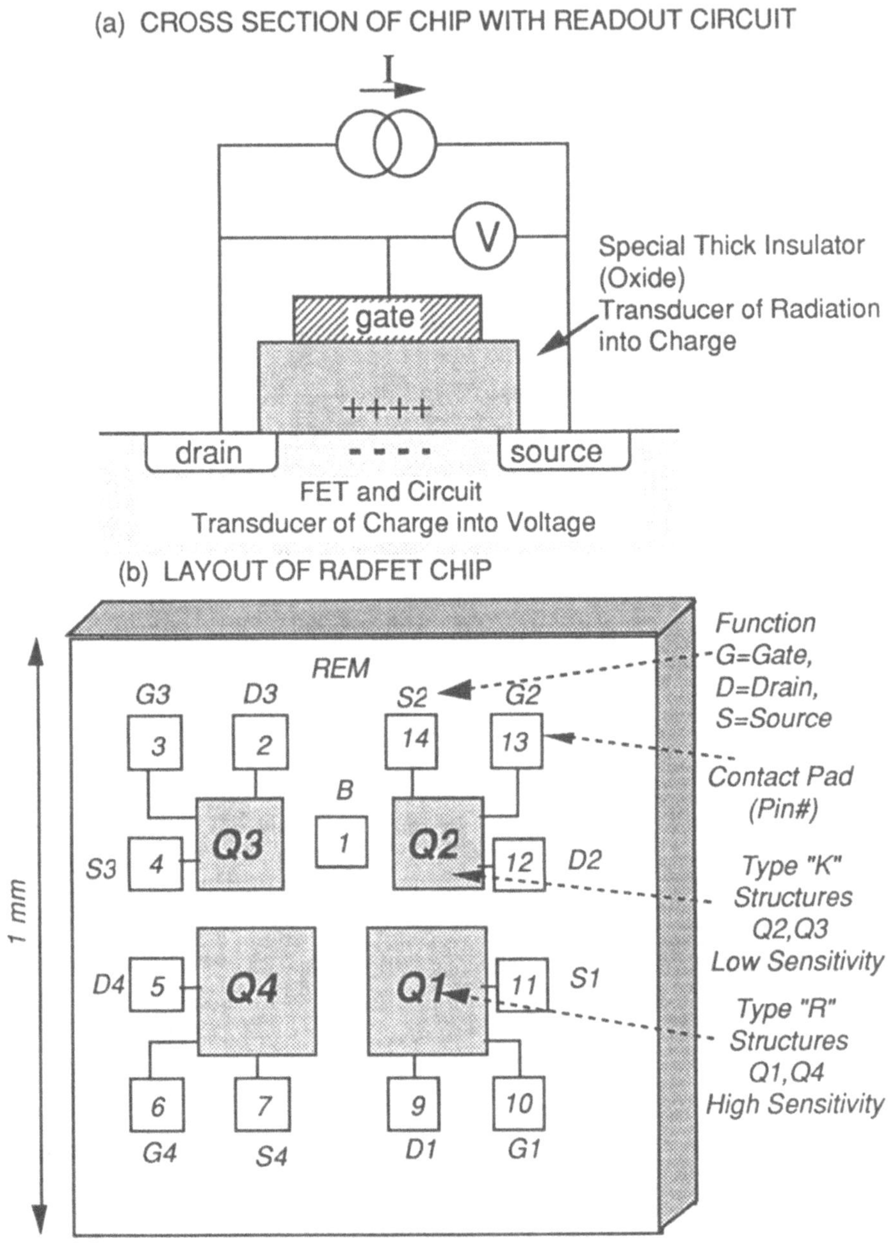


Figure 1. Principle of RADFET.(a) Microscopic cross section. (b) Chip layout.

use in dosimetry. The fact that the stored space charge lies in close proximity to a semiconductor offers an ideal method of transducing a radiation-induced charge into a simple low-voltage DC electrical signal. The system was named "Space-Charge Dosimetry" by Holmes-Siedle in 1974 and the silicon devices used were later called "Radiation-Sensitive Field-Effect Transistors" (RADFETs) by Hughes (1985). This method bears little similarity to that of existing dosimeters. The collecting volume is uniquely small, and the transduction system is not used elsewhere in radiation detectors. Radiation dose may be read out nondestructively, during or after exposure, by means of low power dc electronics. The available dose range covers seven orders of magnitude from 1 Rad (10 mGy) to 100 MRad (1 MGy) with very little fading after exposure. The small size, ruggedness and low cost of the device make it employable autonomously or as part of a hard-wired area monitoring array. Several applications will be be discussed after a brief introduction to RADFET operation.

## 2. DESCRIPTION OF RADFET SYSTEM

### 2.1 RADFET Operation

The RADFET principle was first described in detail by Holmes-Siedle in 1974 and a full description is to be found in a review by Holmes-Siedle and Adams, written in 1986. When exposed to radiation, the RADFET exhibits a shift in its current-voltage characteristics. The change is caused by the buildup of positive charge. The yield of charge is the dosimetric information. It is stored in the MOS structure itself and can be "read" at any time. Figure 1 summarises the principle.

An electronic system designed to perform bench-top tracking and control of a RADFET is often called a "RADFET Reader". The system in fact does more than read: the threshold voltage is measured ("V(T) tracker") and displayed; exposure voltages are applied to the gate; and the RADFET is checked for "health" and for its identity. In the active mode, a voltage is applied to the device during irradiation. In the passive mode, no voltage is used.

### 2.2 The Type TOT500 RADFET

The RADFETs developed by REM are based on Al-gate p-channel MOS technology with extra thick oxides and very thin metal gates. The TOT500 contains four pMOS FETs (see Fig.1). One pair of the FETs shown, called Type "R", carries a "thick" gate insulator, suitable for measuring low doses ("R" stands for "rad" doses). The other pair, Type "K", carries an oxide which is relatively "thin", suitable for moderate to high doses ("K" stands for "kilorad" doses). Fig. 2 summarises the ranges over which RADFETs with various oxide thickness values can be used. Work is in progress on mounting the chip at the end of a small probe with thin encapsulation.

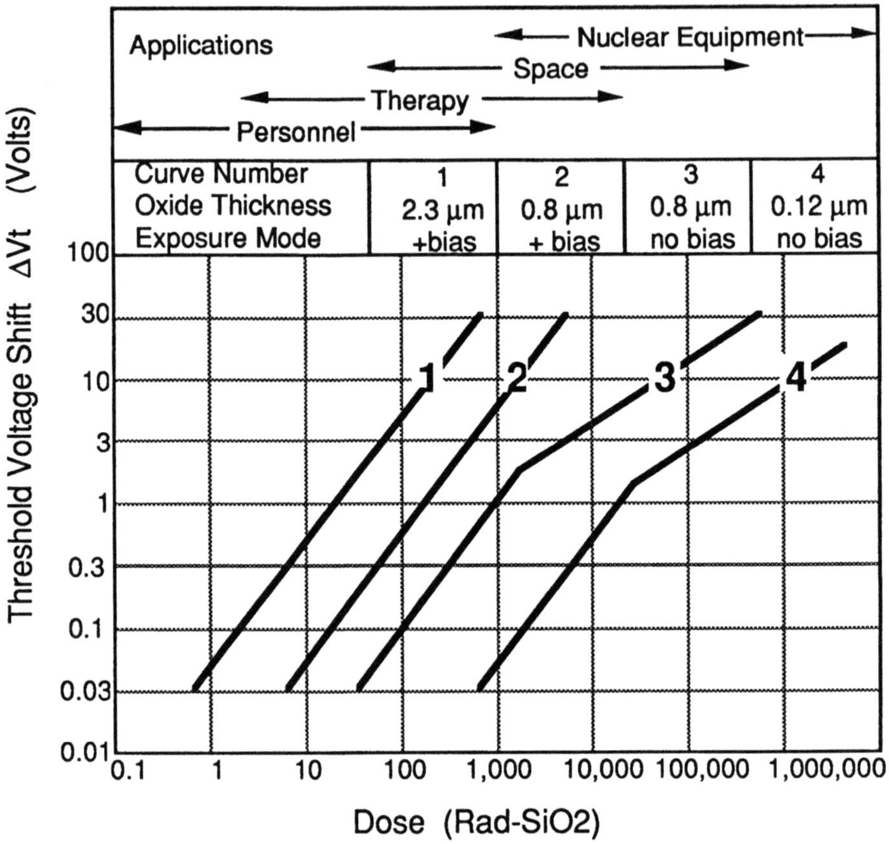


Figure 2. Model of the threshold voltage response vs. dose for several types of RADFET, shown in relation to the dose ranges encountered in various applications. Two exposure modes and three values of gate oxide thickness are shown. "+" indicates positive 2-20 volts bias to gate during irradiation.

2.3 Simulation of Devices under Test

In the radiation testing of electronic devices, the RADFET is suitable as a dosimeter for integrated circuits (IC) under test. The RADFET element is very similar in construction and device action to the complementary metal-oxide-semiconductor logic switching transistor used in microprocessors and memories. Because the TOT500 RADFET chip is smaller than an IC chip, it can be inserted and read in situ in any IC package using exactly the same lids, wiring and die attach alloys as for the IC under test. This cannot be done with other dosimeters such as plastic foils or thermoluminescent chips. RADFETs have been tested under gamma-rays, hard and soft x-rays, electrons and protons, neutrons and fission fragments

(Holmes-Siedle and Adams 1986 ; Blamires et al.; Ensell et al 1988; Stapor et al 1985). In summary, the RADFET is an excellent dosimeter for cross-referencing doses from radiation test sources.

## 3. APPLICATIONS TO DATE

The first use of the RADFET and its tracker was the successful application of the RADFET in space (Adams and Holmes-Siedle, 1978). Laboratory beam mapping and field tests have been carried out with a bench reader. Because the RADFET registers dose when unconnected, a very small "dosimeter pill" can be made. This has many of the advantages of the thermoluminescent dosimeter chip and few of the disadvantages. Some proven applications are described briefly below.

(a) END OF LIFE MONITORING: Equipment, when exposed to radiation, has to be replaced after reaching a specified dose level. Dosimetry on unmanned satellites in high radiation orbits is one proven example of this application. An "ESA Modular Dosimeter" has been designed (Ward 1989). Several tests of RADFETS have taken place on US military satellites. A similar need is developing in the nuclear industry and is discussed further in a following section.

(b) THERAPY SYSTEMS: The RADFET may be used in large arrays (imaging) and in very small physiological probes. A "dosimeter pill" can be swallowed or implanted and read later. A miniature RADFET probe and tracker have been used in cancer treatment by Hughes and others (1988) and further work using the TOT500 chip is reported by Gladstone (1990).

(c) ENVIRONMENTAL AND PERSONNEL BADGES: Type TOT501 RADFET can be used for an "Accident Dosimeter"(ANSI 1981). The same applies to troops who require "Tactical Dosimetry". The "zero bias" mode is useful, because a "pill" can be left in any situation as an environmental dosimeter.

(d) NUCLEAR BEAM MAPPING : Single RADFETs or arrays can be used to map the profile and intensity of radiation beams. This may be used in beam quality control, radiation hardness testing, nuclear physics, semiconductor photolithography and industrial radiography, sterilization and materials processing. A specially designed RADFET reader board has been used to monitor radiation fluxes near the General Atomics DIII-D Tokamak in San Diego (nuclear fusion research device) by Leffler et al (1989). The use of MOS devices to monitor short-wavelength UV and soft X-ray beams has also been demonstrated.

## 4. RADFETS IN NUCLEAR FACILITIES

### 4.1 Radiation Damage

In a nuclear reactor facility, semiconductor devices are sensitive to two types of radiation damage and the effects occur in two clearly distinguishable regions of the semiconductor chip. Firstly, radiation-induced ionisation by photons and particles, leads to deleterious interface effects, often termed "Surface Damage". Secondly, transfer of kinetic energy from particles to atoms, causes atomic displacements, often referred to as "Bulk Damage". The environment in a nuclear facility produces a mixture of these effects, due mainly to neutrons and gamma rays. In general, neutrons will produce "Bulk Damage" and the gamma rays will cause "Surface Damage". In silicon microelectronics, these combine to cause electrical degradation.

### 4.2 Typical Radiation Levels in Nuclear Facilities

Johnson et al (1983) have surveyed the doses and dose rates to electronics in nuclear facilities under both normal and accident conditions. Gamma doses vary from 1,000 rads to 100 Megarads over 40 years in normal operation. In accident conditions, many instruments will receive total gamma doses as high as 20 megarads in a much shorter time. Holmes-Siedle (1984) estimated the total doses in several nuclear facility locations and came to broadly the same conclusions. In spent fuel handling and processing equipment; lifetime doses well above 100 Megarads are possible.

### 4.3 Radiation Testing and Hardening

In the design of electronic equipment for nuclear facilities, plans are in hand to use design methods to increase tolerance to radiation. This will include the following: analysis and prediction of semiconductor and optics degradation and their effects at system level; followed up by precise and practical design solutions such as part substitution of the commercial microelectronics by more tolerant forms of device; shielding; and a policy of frequent replacement. This analysis usually requires extensive radiation testing of components. This in turn requires the calibration of the radiation fields used to simulate the working environment. As already mentioned in sections 2 and 3, the RADFET has been found useful in this application.

### 4.4 Reactor Dosimetry

At a workshop in the present conference (J.A. Mason, Workshop on Gamma Dosimetry), the need to measure gamma rays within reactor pressure vessels was discussed. Special TLDs are used by Aït and Fabry; King and Gilliam and others (see relevant papers in this conference). The use of RADFETS as an alternative was discussed and is being pursued. The RADFET shown as "4" in Figure 2 is seen to be compatible with the required dynamic range. It will withstand fast neutron fluences of the order of $10^{16}$ $cm^{-2}$.

## 4.5 Equipment Monitoring

Radiation levels in nuclear facilities affect both equipment lifetime and personnel safety. At present, equipment is replaced when estimated exposure exceeds an established maximum. Area radiation monitors are distributed throughout a nuclear facility to estimate the radiation dose and dose rates likely to be suffered by personnel and equipment. These estimates can be correlated by personnel dosimetry badges and individual equipment monitors in order to establish the credibility of the radiation field model. The levels of interest for personnel health protection are lower than the the present level of RADFET sensitivity (see Fig. 3). However, RADFET sensitivity levels are very well matched to the equipment monitoring requirement. Since actual exposure levels at all locations are not known, equipment must be replaced according to conservative dose estimates long before maximum allowable exposure levels are reached. Monitoring the actual dose levels with a Radiation Damage Monitor could extend equipment life and hence reduce replacement costs. For equipment which is mobile in non-uniform radiation fields, individual Radiation Damage Monitors represent the only way to monitor dose.

Equipment monitoring based on RADFETs can be accomplished in a variety of ways, depending on the sophistication desired:

(1) In the simplest and least intrusive scheme, small, unbiased RADFET "patches" , as shown in Fig. 3, are attached to the equipment during preparation and remain attached throughout its operation. The patch has a unique bar coded label. At intervals, the the dose is read out. This can be done by collecting the patch or visiting the instrument with a portable reader. Alternatively, the patch can be wired to a central reader and the operator could inititate the readout simply by reading the barcode with a light pen. A portable, computer-controlled patch and reader system has been developed at General Atomics. It is capable of storing the exposure history (dose, date, time and temperature) of hundreds of patches.

(2) A more sophisticated monitoring scheme would involve embedding RADFET devices into the active electronics. Since RADFETs are mounted in standard integrated circuit packages, they can be included in printed circuits, and could be incorporated in standard Class 1E circuit designs.

(3) An area monitoring system employing "stand-alone" RADFETs could fulfil a role in both personnel protection and equipment life extension. Being a low-cost device, hundreds of RADFETs could be incorporated in a given structure, and be addressed by a low-voltage data bus. The resulting map of doses built up in the area would complement the dose rate data received from a smaller number of fixed counters or survey meters, many of which are more bulky, more expensive and require high-voltage supplies.

(4) Remote handling in radioactive environments is likely to lead to some special requirements for equipment dosimetry. A prime example of this requirement is the use of mobile robots in dismantling reactor vessels during decommissioning and after a reactor accident.

## 5. CONCLUSION

In nuclear facilities, there is a need for dosimeters which measure dose in Gy(Silicon). These include (a) dosimeters which signal the need for the replacement of electronic equipment (including Class 1E instruments); (b) dosimeters for the calibration of radiation fields used in the testing of electronics (including mixed neutron-gamma fields). There is also a general need for a small, inexpensive, low-power system which reads cumulative environmental doses in "hot" areas of a reactor or isotope storage building.

RADFET dosimeters meet both of these needs and should be an ideal dosimeter choice for the vast majority of microelectronic circuits used in the nuclear industry, since they incorporate similar structures. Commercial prototype control units, already designed for scanning RADFETs in space vehicles and radiographic facilities will be adapted in the near future to the uses listed here. Other types of damage monitors can be accomodated within the system.

One advantage of the RADFET is the very wide dynamic range of doses possible. The low dose limit is a few milliGray(Si) (a few hundred millirad(Si)); the upper limit is near the instrument dose tolerance recommendations developed after the Three Mile Island event (many MegaGrays). This range is made possible by some special features of the RADFET such as non-linear response at high dose and the variety of device structures which have been developed by one of the authors and co-workers.

There are many other advantages of the RADFET system: The semiconductor chip can be read out without change (nondestructive readout). The low-cost, low-voltage, rugged sensor can be read at the end of a long cable or "piggybacked" on a nuclear instrument or robot. The technology is suitable for close-spaced arrays and integration with a microprocessor. Finally, power drain is small compared with other automatic dosimeter readers. All of these advantages reinforce the suitablility of the system for extending equipment life in a radiation environment.

## 6. REFERENCES


1) ANSI N13.3.1969(R1981), "Dosimetry for Criticality Accidents", American National Standards Institute, Washington D.C., USA 1981.

2) L. Adams and A. Holmes-Siedle, "The Development of an MOS Dosimetry Unit for Use in Space", IEEE Transactions on Nuclear Science, **NS-25** (6) 1607(1978).

3) L.S. August and R.R. Circle, "Advantages of Using a pMOS FET Dosimeter in High Dose Radiation Testing", IEEE Trans. Nucl. Sci., **NS-31** (6) 1113(1984).

4) N.G. Blamires, D.H.J. Totterdell, A.G. Holmes-Siedle and L.Adams, "pMOS Dosimeters: Long-Term Annealing and Neutron Response", IEEE Trans. Nucl. Sci., **NS-33** (6) 1310(1986).

5) G. Ensell, A. Holmes-Siedle and L. Adams, "Thick Oxide pMOSFET Dosimeters for High Energy Radiation", Nucl. Instr. Meth., **A269** 655(1988).

6) D.J. Gladstone, L.M. Chin, J. Hansen, H.F. Bowman, J. Humm and G.K. Svensson, "A Radiation Dosimetry System Using Miniature MOSFET Probes", Proc. Annual Mtg. Rad. Res. Soc./ North Am. Hyperthermia Group, April. 1990.

7) A.G. Holmes-Siedle, "The Space Charge Dosimeter-General Principles of a New Method of Radiation Dosimetry", Nucl. Instr. Meth. **121** 169(1974).

8) A. Holmes-Siedle, "Robotics Which Tolerate the Nuclear Environment" in I. D. Salter, "Nuclear Industry Requirement for Robots", Contract Report, Directorate General XII, Commission of the European Communities" (Sep. 1984).

9) A. Holmes-Siedle, L. Adams, S. Marsden and B. Pauly, "Calibration and Flight Testing of a Low-Field MOS Dosimeter", IEEE Trans. Nucl. Sci., **NS-32** (6) 4425(1985).

10) A. Holmes Siedle and L. Adams, "RADFETs: A Review of the Use of Metal-Oxide-Silicon Devices as Integrating Dosimeters", Radiation Physics and Chemistry, **28**(2) 235(1986).

11) R.C.Hughes, D. Huffman, J.V.Snelling, T.E.Zipperian, A.J.Ricco and C.A.Kelsey, "Miniature Radiation Dosimeter for *In Vivo* Radiation Measurements", Int. J. Radiation Oncology Biol. Phys, **14**, 963(1988).

12) IEEE Standard 323-1974, "Qualifying Class 1E Equipment for Nuclear Power Generating Stations", IEEE, New York 1974.

13) R.T. Johnson, F.V. Thome and C.M. Craft, "A Survey of Aging of Electronics with Application to Nuclear Power Plant Instrumentation", IEEE Trans. Nucl. Sci., **NS-30** (6) 4358-4362 (Dec 1983).

14) J.S.Leffler, S.R.Lindgren, A.G.Holmes-Siedle, R. Stav and S. Williams, "Initial Results and Applications of RADFET Dosimeters in the Fusion Radiation Environment", American Physical Society Meeting, Anaheim, CA, 1989.

15) J.S.Leffler, S.R.Lindgren, and A.G.Holmes-Siedle, "The Application of RADFET Dosimeters To Equipment Radiation Qualification and Monitoring", Trans. 1989 Winter Meeting of the ANS, San Francisco, TANSAO **60** 534 (1989).

16) W.J. Stapor, L.S. August, D. H. Wilson, T.R. Oldham and K.M. Murray, "Proton and Heavy Ion Radiation Damage Studies in MOS Transistors", IEEE Trans. Nucl. Sci., **NS-32** (6) 4399 (1985).

17) A.K. Ward, N. Blower, L. Adams, J. Doutreleau, A. Holmes-Siedle, M. Pignol, J. Berneron and M. Mehlen, "The METEOSAT-P2 Radiation Effects Experiment", 40th Congress of the International Astronautical Federation, October 1989, Malaga, Paper No. IAF-89-055; Astronautica Acta, in press.

# FAST NEUTRON DOSIMETRY OF THE REACTOR PRESSURE VESSEL BY MEANS OF THE SCRAPING SAMPLING METHOD. MEASUREMENT OF THE NIOBIUM MASS BY TXRF SPECTROMETRY.

F.Hegedüs
Paul Scherrer Institute
CH-5232 Villigen PSI

ABSTRACT

At the Fifth ASTM-Euratom Symposium on Reactor Dosimetry,a paper was presented on the fast neutron dosimetry by means of the scraping sampling method (1). The principal problem of this method is the determination of the niobium weight of the final sample in which the 93m Nb activity is counted.

After chemical separation of the main components of steel: iron, nickel and chromium, as well as the tantalum, the final solution contains mostly niobium. This niobium concentration is not well known, because the Nb percentage in steel (nominal value 0.6%) and the niobium lost in the separation procedure can vary from one sample to another.

The Totally Reflected X-Ray Fluorescence (TXRF) spectrometer is well suited to measure low concentrations in a small mass of sample material, therefore this method has been used for the measurement of the niobium concentration in the final solution.

## 1. INTRODUCTION

The fast neutron fluence on the reactor pressure vessel inner wall is usually not measured, but only calculated. In order to improve the reliability of these calculated fluence values,neutron fluence measurements were done by means of the scraping sampling method. The principle (1) is simple: during the shut down period, on the RPV inner wall splinter samples (typically a few hundred of mg) were taken. As the inner plating of the RPV contains Fe, Ni and Nb, by measuring the activities of 54Mn, 58Co and 93mNb, the fast neutron fluence can be determined.

The evaluation of the fast neutron fluence from the 54Mn and 58Co activities was done by using the procedure given in the DIN norm (2,3). In the case of the 93mNb activity the DIN norm (4), which is applicable for pure niobium detectors, can not be used.

The niobium was chemically extracted from the steel as described in (1). The main source of uncertainty in this method was the niobium weight determination in the final thin sample. A considerable improvement on the niobium weight determination was achieved by using the Totally Reflected X-Ray Fluorescence (TXRF) spectroscopy.

## 2. NIOBIUM WEIGHT MEASUREMENT

In contrast to iron and nickel, the niobium concentration in the scraped steel samples was not constant: it varied between 0.55% and 0.65%. Furthermore the niobium lost in the chemical extraction procedure can vary from one sample to another.

After the chemical extraction of other elements, the final solution contained mostly niobium. The thin counting sample was made by evaporation of the final solution. Before that, the concentration of the niobium in the final solution has to be measured. Conventional analytical methods were not successful as the mass of niobium to be determined was so small and the solution was chemically very agressive.

The Totally Reflected X-Ray Fluorescence (TXRF) spectrometer is well suited to measure low concentration of trace elements in small mass of sample materials (5), therefore it was used to determine the niobium concentration in the final solution. The final solution (2 ml) contained about 100-200 ug of niobium. The samples for TXRF measurement were prepared by pipetting 2 ul of solution on a quartz plate.

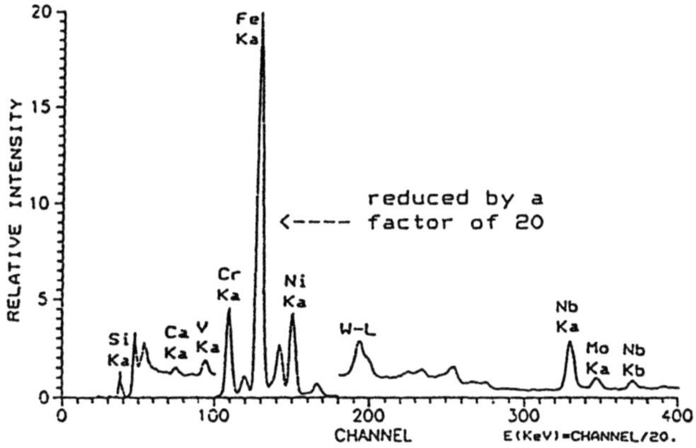


FIG.1. X-Ray spectrum of the reactor inner plating steel before chemical extraction. Sample weight: 57 μg. Internal standard: 0.1 μg Mo. W X-ray tube, 55 kV, 5 mA.

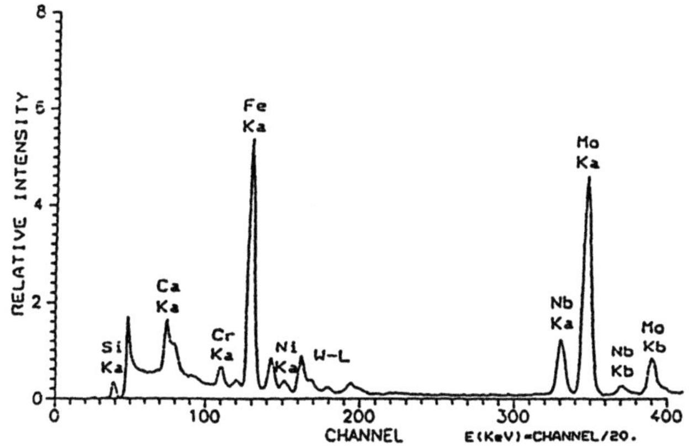


FIG.2. X-Ray spectrum of the reactor inner plating steel after chemical extraction. Sample weight: 13.6 μg. Internal standard: 0.2 μg Mo. W X-ray tube, 55 kV, 5 mA.

TABLE 1. Measured X-K-ray intensities of the steel samples.

| element | Ka+Kb intensity (cps/μg steel) | | |
|---|---|---|---|
| | before chem.sep. | after chem.sep. | ratio |
| Cr | 6.55 | 0.030 | 0.0046 |
| Fe | 31.7 | 0.700 | 0.0221 |
| Ni | 6.00 | 0.013 | 0.0022 |
| Nb | 0.43 | 0.250 | 0.5880 |

3. RESULTS

Fig.1. and 2. show the fluorescence X-ray spectra of a dissolved steel sample before and after chemical separation respectively.By comparing Fig.1. to Fig.2.,the efficiency o the chemical extraction was determined. Table 1. shows that the matrix elements: Cr, Fe and Ni have been almost removed meanwhile the niobium loss was relatively low : 40%.

The extraction efficiency of the tantalum trace element was measured by means of gamma spectroscopy. It was found that the initial concentration of a few hundred ppm of tantalum in steel was reduced by a factor of >100.

The niobium weight in the final thin sample was determined by comparing the Nb Ka peak intensity to the Ka peak intensity of the Mo internal standard. The error on the Nb weigl determination was less than +/- 6%.

REFERENCES:


1. F.Hegedüs: Fast Neutron Dosimetry by means of Scraping Sampling Method.
Proceedings of the Fifth ASTM-Euratom Symposium on Reactor Dosimetry. Edited by J.P.Genthon and H.Röttger,1985.

2. Neutron Fluence Measurement: Determination of the fast Neutron Fluence with Iron Activation Detectors.
DIN Norm 25456/2.

3. Neutron Fluence Measurement: Determination of the fast Neutron Fluence with Nickel Activation Detectors.
DIN Norm 25456/3.

4. Neutron Fluence Measurement: Determination of the fast Neutron Fluence with Niobium Activation Detectors.
DIN Norm 25456/4.

5. F.Hegedüs et al.:TXRF Spectrometer for Trace Element Detection. Advances in X-Ray Analysis. Vol.33.

# RADIATION DOSIMETRY WITH FIBER-OPTIC SENSORS

H. Bueker, F.W. Haesing

Research Center Juelich, P.O. Box 1913, D-5170 Juelich
Fed. Rep. of Germany

## ABSTRACT

A survey is given of measurement methods and fields of applications for fiber-optic dosimeters. Advantages and disadvantages of the different methods are discussed. First results with dosimeter systems under development for medical applications and the surveillance of nuclear facilities are given.

---

## INTRODUCTION

If optical fibers are used as dosimeters the geometrical dimensions (diameter 10-3000 µm, great length) and the material properties (high flexibility, homogeneity, low light attenuation, chemical stability) of optical fibers open up considerably extended fields of applications for solid-state dosimetry in general as compared with conventional dosimeters. A fiber-optic dosimeter can be adapted to the respective requirements with regard to geometry, dose range and environmental conditions. Reviews of papers related to the effects of ionizing radiation in fibers and fiber-optic dosimetry are given by Seger /1/, Lyons /2/ and Friebele /3, 4/.

The optical properties of solid materials used for the production of optical fibers are changed by the interaction of nuclear radiation with the fiber material. The interaction of particle radiation ($\alpha$, $\beta$, n) can cause dislocations of atoms and volume defects in the lattice of the solid or changes in the electronic structure of the solid. In the case of $\gamma$-radiation, lattice defects are produced which cause electron-traps (colour-centers). The energy gap between the induced electronic states or the energy gap between

Another field of application is the ex-vivo or in-vivo real-time monitoring of the radiation load of patients in radiotherapy /7/. Fig. 2 shows an example of the design of a fiber-optic dosimeter for in-vivo application. This dosimeter allows the measurement of the dose in the immediate vicinity of a tumor. A short length (10-50 mm) of a radiation-sensitive lead glass fiber is coupled to a radiation-resistant fiber in a capillary. The total diameter of the probe including the biocompatible catheter is about 0.5 mm. These dimensions of the dosimeter make it possible to place the sensor in the patient's body either via a cannula or during a surgical operation. The measurement of the radiation-induced absorption averages the exposure of the tissue over the length of the lead glass fiber. The small diameter of the fiber (0.1mm) causes a high local resolution perpendicular to the fiber axis and therefore allows a precise positioning of the radiation field inside the body. The doses used in radiotherapy are in the range of 1 to 10 Gy for a single treatment. During the therapy the total dose may accummulate up to 100 Gy. The fiber-optic dosimeter system set-up (Fig. 2) allows the sensor to be disconnected from the electronic evaluation

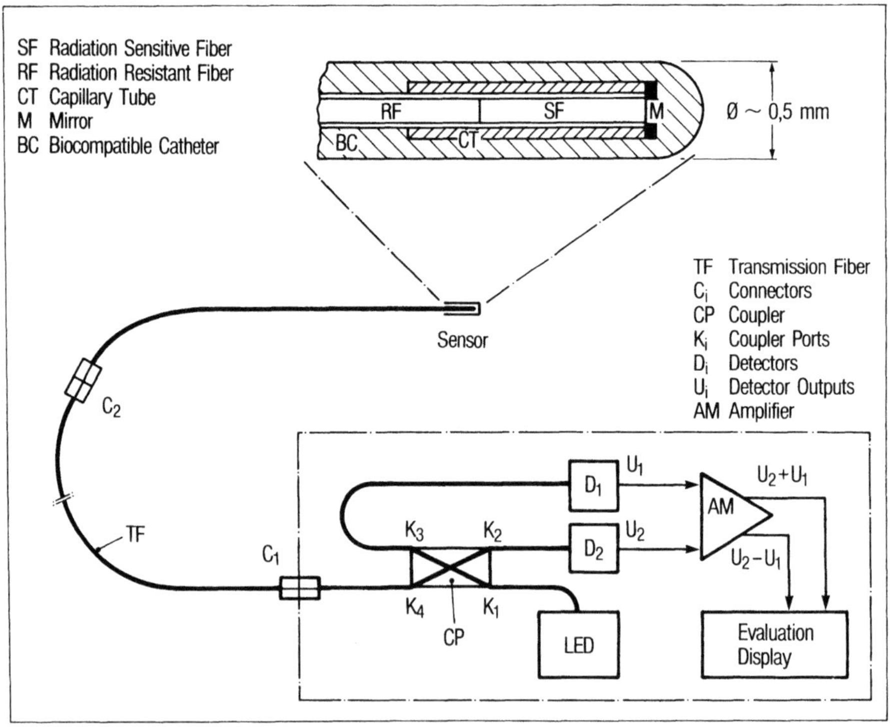


Fig. 2: Experimental Set-up of a Fiber-optic Dosimeter System (FOD) /7/

system between treatments without any disturbance of the dosimeter reading. Good reproducibility and linearity of the dosimeter signal up to a total dose of 120 Gy were observed. Other applications in radiotherapy of this dosimeter may be the on-line surveillance of dose loads of highly sensitive tissue (e.g. eyes) or the determination of the total absorbed dose during the treatment. In this case the fiber will be placed outside the body. A local distribution of dose loads may be measured by using the fiber as a distributed sensor, described below.

The application in radiotherapy is one of the most important fields for fiber-optic dosimetry. Therefore some other fiber-optic dosimeters have also been developed for this application. Most of the systems use the scintillation effect. A scintillation crystal or a short length of a scintillating solid or liquid fiber is attached to an optical fiber, which is used for the transmission of the scintillation photons to the detector /11,12,13/. Due to the short wavelengths of the photons the absorption of the transmission fiber is high, which restricts the system to short lengths of transmission fibers.

For fiber-optic radiophotoluminescence measurements a special phosphorus crystal was developed, which can be excited to fluorescence (wavelength 470 nm) by a Nd-YAG laser at a wavelength of 1060 nm /8/.

There are two other fields where optical fibers are used or have been proposed as radiation detectors. The first one is the determination of the traces of elementary particles by bundles of scintillating fibers at accelerator facilities. Such a system was installed at CERN /10/. The other one is the surveillance of nuclear facilities or contaminated areas.

For monitoring applications in accidental situations (radioactive fall-out, power failure) Kronenberg /6/ developed a fiber-optic dosimeter based on a liquid lightguide. A capillary tube is filled with a liquid dye. The refractive index and the absorption coefficient of the dye is changed in the visible spectral range by the impinging ionizing radiation. The induced refractive index in the liquid core is lower than the refractive index of the capillary tube, leading to the loss of the light guided. This dosimeter may be used independently of any artifical light source.

Also complex structures like nuclear reactors or nuclear plants may be monitored for escape of radioactive substances. In this application especially the method of distributed sensors may be of great interest /14/. Another method for fiber-optic dosimetry, with local resolution was proposed by Gaebler /9/. Different segments of an irradiated fiber may be heated (Fig. 1) and the light of thermoluminescence will be observed at one end of the fiber.

## RADIATION DOSIMETRY WITH DISTRIBUTED SENSORS

Optical time domain reflectometry (OTDR) is a well known technique in the diagnostics of fiber-optic networks. Part of the power of an injected light pulse is scattered at imperfections of the fiber core. The reflected power as a function of time can be used to analyze the optical properties as a function of local position in the fiber. The deriative of the function gives the local attenuation of the fiber. The radiation-induced loss may be measured by the OTDR method and from this the distribution of radiation fields in complex geometries or extended areas may be calculated. An OTDR system has been used by Lyons /2/ in the laboratory for detecting the influence of low radiation doses (< $10^{-2}$ Gy/d). First experiments with a commercially available OTDR system ($\lambda$ = 850 nm) show that changes in dose loads of 0.1 Gy can be determined at a distance of 1 km from the detector system with a local resolution of 3 m, using a Ge-P-doped fiber. The sensitivity for the detection of nuclear radiation can be significantly increased by using light of shorter wavelengths. In order to use the OTDR method for the radiation surveillance of nuclear facilities, an arrangement has therefore been conceived which is shown in Fig. 3. The light source used is a diode-pumped Q-switched Nd-AYG laser which is frequency doubled.

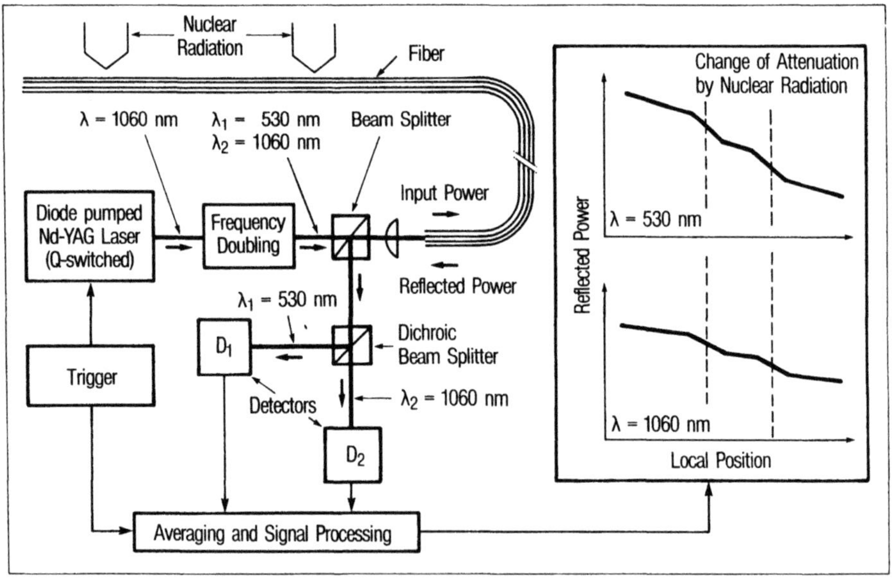


Fig. 3: Scheme of a Two-wavelength OTDR Dosimeter System /14/

Both wavelengths, $\lambda_1$ = 530 nm and $\lambda_2$ = 1060 nm, are coupled into the fiber. Behind a beam splitter the backscattered light is split into the two wavelengths, $\lambda_1$ and $\lambda_2$, using a dichroic beam splitter and detected by the detectors $D_1$ and $D_2$. Since the induced loss due to nuclear radiation is highly dependent on wavelength, the measurement at wavelength $\lambda_2$ can be used as the reference measurement. The sensitivity and the local resolution may be improved further by optimizing the geometrical arrangement of the fiber.

## SUMMARY

An evaluation of the different fiber-optic methods is given in Table 2. The most important advantage of the method of measuring the radiation-induced losses is the option of compensation for cable and connector losses and the option of using commercially

| Methods | Advantages | Disadvantages |
|---|---|---|
| Induced Loss | - Free Choice of Measurement Wavelength and Fiber Material<br>- Option of a Distributed Sensor for Local Resolution<br>- Compensation for Cable and Connector Losses | - Only Medium Sensitivity<br>- Fading Effects<br>- No Measure of Dose Rate |
| Radio-photo-lumines-cence | - High Sensitivity<br>- Partial Compensation for Cable and Connector Losses<br>- Measurement of Dose Rate | - Non-linearity at Medium and High Dose Rates<br>- Expensive<br>- No Option for Distri-buted Sensors |
| Thermo-lumines-cence | - High Sensitivity | - No Real-Time Monitoring<br>- Fading Effects<br>- No Compensation for Cable and Connector Losses |
| Scintil-lation | - High Sensitivity<br>- No External Light Sources<br>- Measurement of Dose Rates | - No Compensation for Cable and Connector Losses<br>- Restriction to Short Fiber Lengths<br>- No Options for Distri-buted Sensors |

Tab. 2: Evaluation of some Fiber-optic Methods for Nuclear Radiation Dosimetry

available fibers as distributed radiation detectors over great distances. The radiophotoluminescence and scintillation systems have higher sensitivities but up to now they have been restricted to short distances and they are sensitive to cable and connector losses. The thermoluminescence method is not a purely optical method and requires heating of the irradiated fiber.

A lot of work has been done in testing the radiation sensitivity of optical fibers for the development of radiation-resistant fibers for fiber-optic data transmission. Also first developmental work for the use of optical fibers as dosimeters has been undertaken. Remaining problems for the development of the fiber-optic dosimeters are: Compensation for cable and connector losses as well as reduction of crosstalks of environmental parameters and the dependence of radiation sensitivities on environmental parameters.

## REFERENCES


/1/ Seger, G., BMFT-Bericht (02U5311) (1982)
/2/ Lyons, P., Proc. SPIE Vol. 566 (1985), p. 166
/3/ Friebele, E.J. et al., Nucl. Instr. Methods Phy. Res. Sect. B Vol. 1, (1984), p. 355
/4/ Friebele, E.J. et al., Proc. SPIE Vol. 541 (1985), p. 70
/5/ Evans, B.D. et al., IEEE Trans. Nucl. Sci. Vol. 25, (1978), p. 1619
/6/ Kronenberg, S. et al., ECOM-Report 4545 (1977)
/7/ Bueker, H. et al., Proc. SPIE Vol. 1201 (1990), p. 419
/8/ Soltani, P.K. et al., NUREG/CR-5100 (1989)
/9/ Gaebler, W. et al., Proc. SPIE Vol. 404 (1983), p. 132
/10/ Ansorge, R.E. et al., Nucl. Instr. Methods Phy. Res. Sect. A Vol. 265 (1988), p. 33
/11/ Robillard, J., FR-OS 22 31017
/12/ Caine, S. et al., IEEE Trans. Nucl. Sci. Vol. 25 (1978), p. 539
/13/ Koechner, W., US Patent, Ptno. 4.788.436 (1988)
/14/ Hille, R. et al., Nucl. Instr. Methods Phy. Res. Sect. A Vol. 299 (1990), p.217
/15/ Bueker, H. et al., Intern. Conf. on Optics in Life Science, Garmisch-Partenkirchen, Springer Series of Optical Science (to be published)

# NIOBIUM AS AN EX-VESSEL NEUTRON DOSIMETER FOR PWRS

L.M. Hodgson, S.C. Apple, R.R. Culp
Arkansas Tech University
Russellville, AR 72801

C.O. Cogburn
University of Arkansas
Fayetteville, AR 72703

J.N. Miller
Entergy Inc
Russellville, AR 72801

## ABSTRACT

Since 1983, several niobium foils have been irradiated along with conventional neutron dosimeters in two nuclear reactors at Arkansas Nuclear One. The niobium monitors have been analyzed for $^{93m}$Nb by applying correction factors for self-absorption, and for the more recently irradiated foils, correcting for x-ray fluorescence. The data have been used to compute reaction rates for the $^{93}$Nb(n,n')$^{93m}$Nb reaction. These reaction rates have been compared with reaction rate values determined from fluences measured by conventional dosimetry methods.

## INTRODUCTION

The use of niobium as a neutron monitor has generated considerable interest among those doing reactor pressure vessel dosimetry. The reaction $^{93}$Nb(n,n')$^{93m}$Nb occurs in the presence of neutrons. The reaction product, $^{93m}$Nb, has a half-life of 16.1 years. It decays back to $^{93}$Nb by internal conversion accompanied by the emission of k-series x-rays at energies of 16.6 and 18.6 keV.

Niobium's most siginificant attribute is that its neuton cross-section is similar to the damage cross-section of iron. In addition, niobium monitors may be useful for irradiation periods ranging from half of a fuel cycle to several fuel cycles because of its long half-life and activation sensitivity.

Niobium's disadvantage as a neutron monitor is that $^{93m}$Nb emits weak x-rays. These x-rays are susceptible to being absorbed and scattered within the sample itself, creating the need for self-absorption correction. Another disadvantage is that any higher energy radiation from radioactive isotopes coexisting in the sample can induce x-ray fluorescence in the niobium that is indistinguishable from the x-rays produced from the decay of $^{93m}Nb$. Tantalum impurity in the niobium is particularly troublesome.

Since 1983, several niobium foils have been irradiated along with conventional neutron dosimeters in the two nuclear reactors at Arkansas Nuclear One (1,2). In some of the older foils, $^{182}Ta$ is below minimum detectable activity. In foils from more recent irradiations, the decay of $^{182}Ta$ was studied to determine the contribution of x-ray fluorescence to the total x-ray activity. All foils were analyzed for $^{93m}Nb$, and reaction rates were determined for each. Average neutron fluence rates were calculated from these reaction rates and from the neutron cross-sections for the $^{93}Nb(n,n')^{93m}Nb$ reaction. Table 1 presents the nuclear data used to make the calculations in this study.

## DETERMINATION OF NET PEAK AREA

The foils were counted on a high purity germanium low energy photon detector connected to an Ortec 8000 channel Spectrum Ace Card™ through a spectroscopy amplifier. The detector, an Ortec model GLP-10180/07, had an active diameter of 10 mm and a 0.127 mm beryllium window. The resolution (FWHM) of the detection system at 16.6 keV was measured at 265 eV.

The detection system was calibrated using three different sources. Two were low mass sources traceable to NIST. The other was a well characterized foil obtained from the University of Virginia (3). Table 2 lists the parameters of each of these sources.

Figure 1 shows a typical niobium x-ray spectrum. The spectral components have been stripped out using a combination of mathematical and graphical methods. The spectrum is composed of the full energy photo peaks ($k_{\alpha 1}$, $k_{\alpha 2}$, $k_{\beta 1}$ and $k_{\beta 2}$), continuous background radiation, and compton-scattered photons from higher energy emissions such as $^{182}Ta$ and $^{94}Nb$. The $k_\alpha$ and $k_\beta$ photons scatter off the air, the source holder, and the source, adding scatter peaks to the spectrum.

Figure 2a compares peak shapes as the distance between the source and detector changes. The only noticeable differences appear to be attributable to changes in the amount of scattering which increased as the distance between the source and detector increased. Figure 2b shows a comparison of peak shapes among the three source types. In the foil, the $K_\beta/K_\alpha$ ratio was higher than that of the low mass sources because of the greater self-attenuation of the $K_\alpha$ x-ray. The sources with lighter mass showed more scattering than the foil source since the foil attenuated more of the emissions at the wider

| | |
|---|---|
| Half-life $^{93m}Nb$ | 5881 ± 73 days |
| Half-life $^{182}Ta$ | 115 ± 2 days |
| Nb k x-ray emission probability | .1114 ± 0.020 |
| Nb $K_\beta/K_\alpha$ ratio | 0.189 ± .003 |
| Niobium mass absorption coefficient | 19.52 $cm^2/gm$ |

Table 1. Values for the nuclear data used in the calculations for this study.

| Source | NIST | INEL | UVa FOIL |
|---|---|---|---|
| Calibration date | 23-Aug-85 | 23-Feb-84 | 8-Aug-84 |
| Calib. activity (Bq) | $2.632 \times 10^5$ | $3.280 \times 10^4$ | 1573 |
| Source diameter (mm) | 7.0 | 12.7 | 12.7 |
| Source thickness (mm) | 0.25 | 0.51 | 0.127 |
| Source mass (mg) | 0.0421 | 0.00525 | 130.7 |
| Mounting type | Al annulus | Al plates | None |
| Mounting material | 0.006 cm tape | kapton film | none |

Table 2. Niobium sources used for detector calibration.

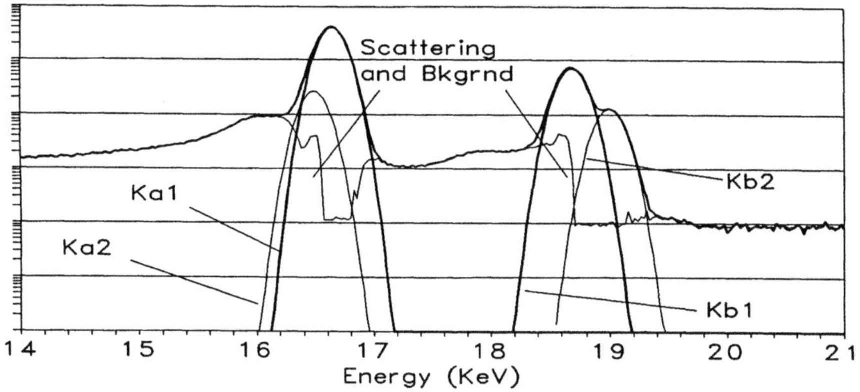


Figure 1. Components of a niobium X-ray spectrum from a low mass niobium source at 5 cm from the face of a high purity germanium low energy photon detector.

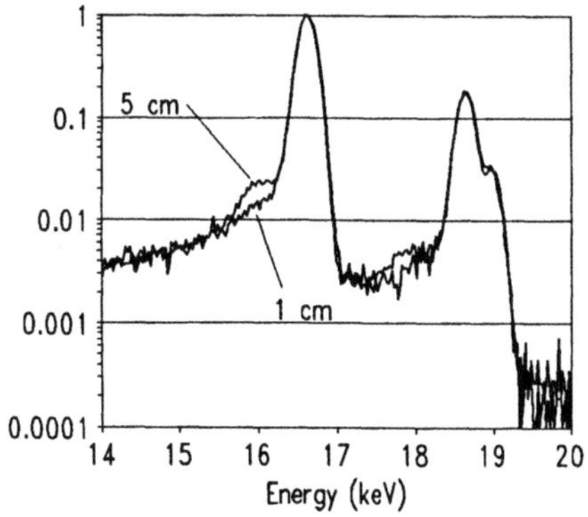


Figure 2a. Comparison of niobium x-ray peak shapes for various source/detector distances.

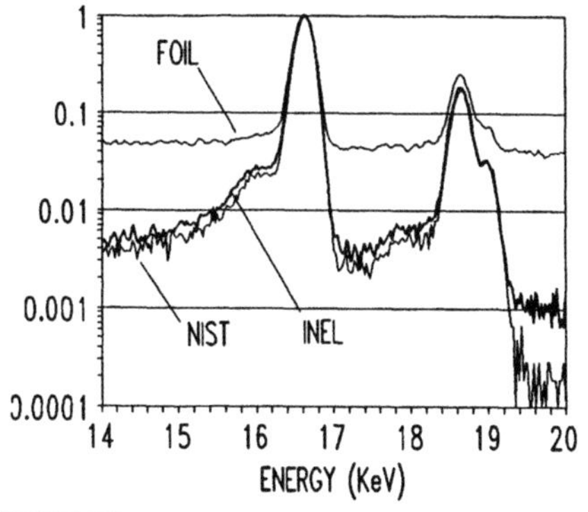


Figure 2b. Comparison of niobium x-ray peak shapes for the three source types at 5 cm.

angles, where scattering into the detector predominated. The INEL source showed more scattering than the NIST source, probably because of the mounting materials that were used.

In the selection of the energy ranges to use for the calculation of net peak area, best results were obtained when both $k_\alpha$ and $k_\beta$ full energy peaks were used. Background was determined by averaging several channels just on either side of each peak. This method of determining NPA was sufficient so long as the scattering components of the two sources being compared was not significantly different.

## Detection Efficiency

A computer program was written to help understand changes in detection efficiency at various source-detector geometries and with various source masses. (The term "detection efficiency" will be used here to include self-attenuation in the source.) The program numerically integrated the absorption probabilities for emissions originating at discrete points within the source and radiating at incremental elevation and azimuthal angles. Table 3 lists both measured and calculated efficiencies for the three sources at four distances from the detector face.

At 5 cm, source geometry differences between the INEL source and the NIST source are small according to calculation, but large according to measurement. While the NIST source's measured efficiency was 2.4% greater than the calculated efficiency, the measured efficiency for the INEL source was 2.1% less. The 5% difference between the measured efficiencies of the two sources was attributed to differences in scattering and absorption in the source and source holder.

Self-absorption in the foil source was determined from the stated activity of the foil and the x-ray emission rate measured using the NIST source for calibration. Self-absorption was also calculated for the foil using the computer program described above. Self-absorption measured at 5 cm was 0.556 (photons absorbed in the volume of the foil per photon emitted). The calculated value for the same distance was 0.571. Self-absorption was calculated for foil masses ranging from 10 to 300 mg. A plot of calculated self-absorption versus foil weight is shown in Figure 3.

| Distance From Detector | Measured Efficiency / Calculated Efficiency | | |
|---|---|---|---|
| | **NIST** | **INEL** | **UVa FOIL** |
| 1 cm | .01443 / .01442 | .01171 / .01381 | .005338 / .006097 |
| 2 cm | .005928 / .005760 | .005226 / .005655 | .002450 / .002500 |
| 3 cm | .003127 / .003064 | .002884 / .003035 | .001312 / .001337 |
| 5 cm | .001317 / .001292 | .001252 / .001286 | .0005668 / .0005620 |

Table 3. Detection efficiencies for three niobium sources at four distances from a germanium LEPS detector compared with values calculated from numerical integration of photon absorption.

## Fluorescence Induced X-rays

Four niobium foils were studied in order to determine the contribution of fluorescence induced x-rays to the total x-ray emission rate. Two of these foils were nominally 140 mg foils containing approximately 750 ppm tantalum. They were irradiated in different reactors. The other two were 290 mg foils containing approximately 170 ppm tantalum. They were irradiated in the same reactor and at the same location. One was covered with cadmium, and the other was bare.

These four foils were counted many times over several $^{182}$Ta half-lives. To determine $^{93m}$Nb activity, it was assumed that the total x-ray emission was composed of a portion that was from the decay of $^{93m}$Nb and a portion that was from fluorescence induced by photons from the decay of $^{182}$Ta. This may be stated mathematically as

$$\dot{X}_{total} = \dot{X}_F + \dot{X}_{Nb} .$$

When counted over time, the components of the total x-ray activity are related by the decay constants and the decay time T, by the equations

$$\dot{X}_{F_2} = \dot{X}_{F_1} \cdot e^{-\lambda_{Ta}T} \quad \text{and} \quad \dot{X}_{Nb_2} = \dot{X}_{Nb_1} \cdot e^{-\lambda_{Nb}T} .$$

Combining the equations, substituting for the desired component, simplifying, and solving yield the following formulae for each portion of the total x-ray emission.

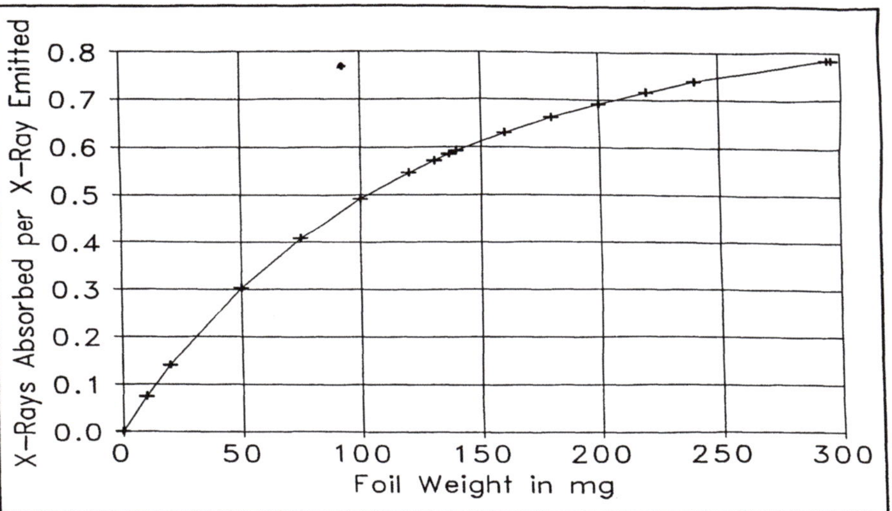


Figure 3. Self absorption for 12.7 mm diameter niobium foils at 55 mm from 10 mm diameter detector for various source masses.

$$\dot{X}_{Nb_2} = \frac{\dot{X}_{Total_1} - \dot{X}_{Total_2} \cdot e^{\lambda_{Ta} T}}{e^{\lambda_{Nb} T} - e^{\lambda_{Ta} T}} \quad \text{and} \quad \dot{X}_{F_2} = \dot{X}_{Total_2} - \dot{X}_{Nb_2} .$$

Results of this analysis for the four foils are presented in Table 4. The data for the two lighter foils, I-15 and II-10, appeared to have a trend in that as the $^{182}Ta$ decayed away, the calculated portion of the x-ray activity due to $^{93m}Nb$ decreased faster than could be accounted for from its decay constant. For the heavier foils, in which the fluorescence contribution still was very high, the results were erratic. The variance on the calculation of $^{93m}Nb$ activity can be determined from the formula

$$\sigma^2_{N_2} = \frac{\sigma^2_{\dot{X}_{Total_1}} \, e^{2\lambda_{Ta} T} + \sigma^2_{\dot{X}_{Total_2}}}{\left(e^{2\lambda_{Nb} T} - e^{2\lambda_{Ta} T}\right)^2} .$$

This error over time continues to decrease as the $^{182}Ta$ decays. However, if the ratio of $^{93m}Nb$ decay x-rays to fluorescence x-rays is small, the error on the calculation of $^{93m}Nb$ activity will be large. Figure 4 shows the coefficient of variation on the calculation over time for different beginning activity ratios. The variance on the x-ray emission rates were estimated from an assumed 2% coefficient of variation.

| Decay Time Days | Measured Total X-rays/s | $^{93m}Nb$ Component X-rays/s | Flourescence Component X-rays/s | $^{182}Ta$ Activity (Bq) | Fluorescence Coefficient |
|---|---|---|---|---|---|
| **ANO Unit II Experiment 10 - Foil Weight 140.9 mg** | | | | | |
| 83.16 | 2070 | 1496 | 574.0 | 2 233 | 0.2570 |
| 230.74 | 1657 | 1404 | 252.7 | 921 | 0.2745 |
| 293.11 | 1542 | 1363 | 179.0 | 630 | 0.2842 |
| 321.65 | 1521 | 1373 | 148.6 | 532 | 0.2793 |
| 414.67 | 1433 | 1347 | 85.73 | 304 | 0.2823 |
| 461.33 | 1445 | 1384 | 61.89 | 229 | 0.2704 |
| **ANO Unit I Experiment 15 - Foil Weight 140.3 mg** | | | | | |
| 74.62 | 673.0 | 171.6 | 501.5 | 1 902 | 0.2636 |
| 166.04 | 451.5 | 158.1 | 293.4 | 1 099 | 0.2669 |
| 302.09 | 289.2 | 160.9 | 128.3 | 484 | 0.2650 |
| 363.23 | 238.2 | 148.0 | 901.4 | 335 | 0.2688 |
| 491.16 | 184.6 | 142.8 | 41.86 | 156 | 0.2684 |
| 532.57 | 181.5 | 149.2 | 32.30 | 212 | 0.2668 |
| **ANO Unit II Experiment 12 (bare) - Foil Weight 296.2 mg** | | | | | |
| 63.66 | 5035 | 365.6 | 4670 | 24018 | 0.1944 |
| 199.08 | 2419 | 352.8 | 2067 | 10645 | 0.1941 |
| 230.45 | 2088 | 386.2 | 1702 | 8778 | 0.1939 |
| **ANO Unit II Experiment 12 (Cd covered) - Foil Weight 294.7 mg** | | | | | |
| 60.72 | 4740 | -97.2 | 4837 | 24257 | 0.1994 |
| 203.09 | 2254 | 331.6 | 1922 | 10432 | 0.1842 |
| 206.56 | 2216 | 334.6 | 1882 | 10061 | 0.1870 |

Table 4. Correction for $^{182}Ta$ induced fluorescence.

# A STUDY OF A PROTON ACCELERATOR SYSTEM AS A SOURCE OF EPITHERMAL AND THERMAL NEUTRONS AT FLUENCE RATES CHARACTERISTIC OF NUCLEAR REACTOR BEAMS

Thomas E. Blue, T.X. Bruce Qu,
C.K. Chris Wang**, James W. Blue*

The Ohio State University
Columbus, Ohio, U.S.A.

## INTRODUCTION

In the early 1980's, researchers at Chalk River Nuclear Laboratories reported design studies for an accelerator-based source of neutrons (CANUTRON) for research and industrial applications.[1,2] The CANUTRON is based on a radiofrequency quadrupole (RFQ) accelerator producing a continuous 10 mA beam of 2.5 MeV protons, which strike a thick liquid lithium target. The protons interact with the $^{7}Li$ via a $^{7}Li(p,n)^{7}Be$ reaction to produce neutrons with a maximum energy of 790 keV. The neutrons are produced at the center of, and are moderated by, a 50 cm radius water sphere which is surrounded by 20 cm of lead shielding. Design calculations predict that the thermal neutron flux at the irradiation point, at the center of the sphere, is $2 \times 10^{11}$ neutrons/cm$^2$•sec. In this paper, we report our extension of the CANUTRON concept to the design of epithermal and thermal neutron sources for the treatment of cancerous tumors by Boron Neutron Capture Therapy (BNCT). As in the CANUTRON design studies, we assume in our design studies that a continuous high current beam of 2.5 MeV protons strikes a thick liquid lithium target. Also, as in the CANUTRON design studies, the neutrons which are produced in the target are moderated in a moderator assembly. However, in our design study, the irradiation point is external to the moderator assembly, and the neutrons emerge from the moderator assembly to strike the patient.

Specifically, this paper reviews the design of a moderator assembly for an accelerator epithermal neutron irradiation facility (AENIF)[3] for the treatment of deep tumors, and contrasts this design with the design of a moderator assembly for an accelerator thermal neutron irradiation facility (ATNIF)[4] for the treatment of

** Kansas State University
* Cleveland Clinic Foundation/NASA-Lewis

superficial tumors. Taken together, the design calculations for the CANUTRON moderator assembly and for our AENIF and ATNIF moderator assemblies represent three different neutron environments which can be achieved with a 2.5 MeV high current proton accelerator.

## MODELS AND METHODS

The AENIF moderator assembly is illustrated in Fig. 1. It consists of a cylinder of BeO, which is 25 cm in diameter and 22.5 cm in height, surrounded by a 30-cm-thick alumina reflector. Also, 0.01 g/cm$^2$ of $^6$Li in the form $^6$LiF is placed at the moderator assembly exit window to reduce thermal neutron contamination at the irradiation point (a point on the centerline of the moderator assembly 3 cm from the exit window).

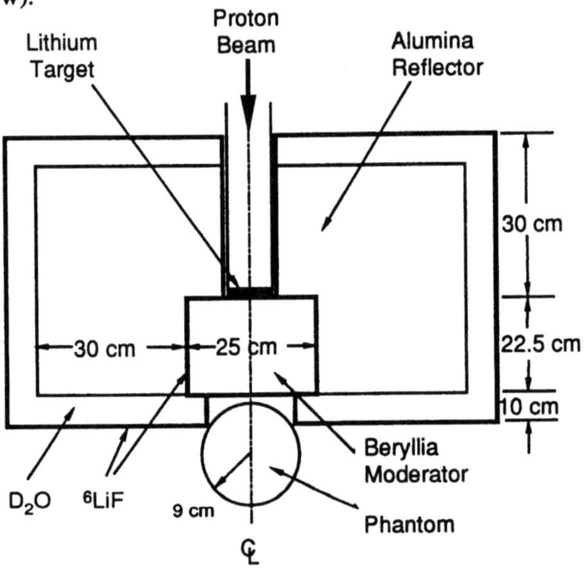


**Figure 1. The Configuration of the AENIF Moderator Assembly**

The ATNIF moderator assembly is illustrated in Fig. 2. It consists of a cylinder of $D_2O$, 50 cm in height and 30 cm in diameter, surrounded on its top and sides by a 40 cm thick graphite reflector.

The design criteria for the AENIF moderator assembly are that the neutron current emerging from the moderator assembly, with energies between 1 eV and 1 keV, is as large as possible and that the contamination of this neutron current by fast neutrons, thermal neutrons, and gamma rays is as low as possible. The design criteria for the ATNIF moderator assembly are that the thermal neutron current emerging from the moderator assembly is as large as possible, and that the contamination of the thermal neutron current by fast and epithermal neutrons and gamma rays is as small as possible.

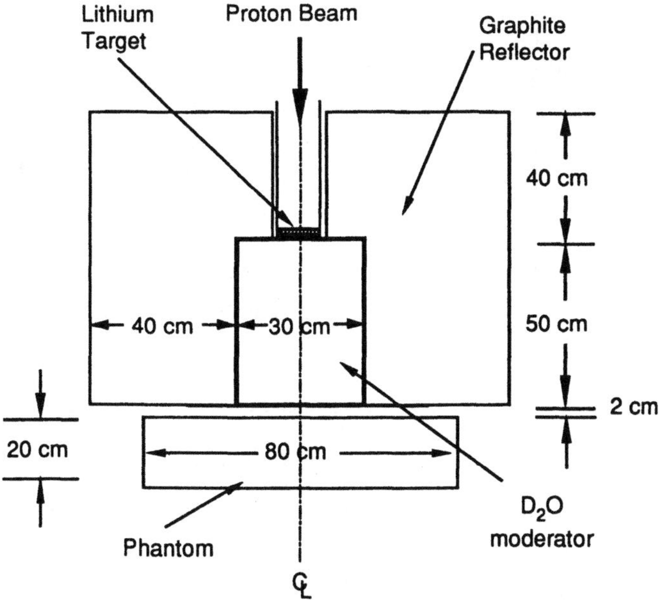


**Figure 2. The Configuration of the ATNIF Moderator Assembly**

In order to assess the suitability of the AENIF neutron current, we have identified two figures of merit: (1) the useful (i.e. E > 1 eV) neutron flux ($\Phi_u$) at the irradiation point in air and (2) the ratio of the neutron kerma to the fluence for a differential volume of tissue at the irradiation point. In order to assess the suitability of the ATNIF neutron field, we have identified two figures of merit: (1) the thermal neutron flux ($\Phi_t$) at the irradiation point in air, and (2) the Ratio of the Thermal neutron flux to the sum of the epithermal and Fast neutron flux (RTF).

Our analysis models two distinct physical processes: (a) neutron generation in the $^7$Li target and (b) neutron transport in the target assembly and head phantom. Neutron generation in the $^7$Li target was calculated by simulating the production of neutrons as protons slow down in the target, using the doubly differential cross section for the $^7Li(p,n)^7Be$ reaction [5] and the stopping power for protons in lithium.[6] Neutron transport in the target assembly and head phantom were modeled using the MORSE Monte Carlo neutron transport code [7] and the BUGLE-80 neutron cross-section library.[8]

## RESULTS AND DISCUSSION

The calculated neutron flux, in units of neutrons per cm$^2$•sec per unit lethargy per 10 mA of proton current, is plotted versus neutron energy in Fig. 3 for the AENIF moderator assembly.

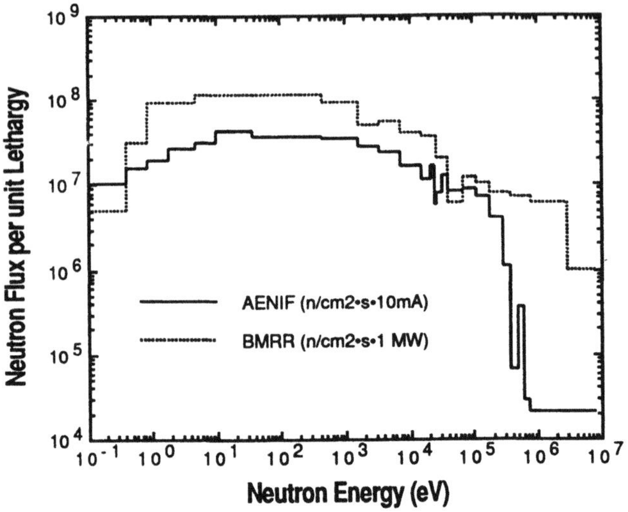


**Figure 3. The Neutron Flux Spectra for the AENIF and BMRR**

For the AENIF moderator assembly with a 30 mA proton beam , $\Phi_u$ = (9.0 $\pm$ 0.3) x $10^8$ neutrons/cm$^2$•sec at the irradiation point in air, with 89% and 84% of $\Phi_u$ due to neutrons with energies less than 10 keV and 1 keV, respectively. For this spectrum the ratio of the neutron kerma to the fluence (average kerma factor) for a differential volume of tissue at the irradiation point is (4.4 $\pm$ 0.5) x $10^{-11}$ cGy•cm$^2$, which is equivalent to the kerma factor for 5 keV neutrons.

The magnitude of $\Phi$u for the AENIF moderator assembly with a 30 mA proton beam is comparable with the magnitude of $\Phi$u for reactor neutron beams of similar spectra, which can be extracted only with special beam filters from nuclear reactors which are in the mega-watt power range. As an example, $\Phi$u for the Brookhaven National Laboratory Medical Research Reactor (BMRR) beam is 9.0 x $10^8$ neutrons/cm$^2$•sec for the BMRR operating at 1.4 MW.

In order to compare in more detail the BMRR beam and the AENIF neutron field, the calculated neutron flux spectra [9] for the BMRR BNCT beam is plotted in Fig. 3 along with the calculated neutron flux spectra for the AENIF moderator assembly. The units of the neutron flux for the BMRR beam are neutrons per cm$^2$•sec per unit lethargy per MW of reactor thermal power. As can be seen from Fig. 3, the neutron flux in the epithermal region (i.e. from 1 eV to 1 keV) is comparable for the AENIF operating at 30 mA and for the BMRR operating at 1 MW. For neutron energies greater than 20 keV but less than 300 keV, the neutron flux for the BMRR operating at 1 MW is less than the neutron flux for the AENIF operating at 30 mA. However, for neutron energies from 300 keV to several MeV, the neutron flux spectrum for the BMRR decreases slowly; while for neutron energies greater than 300 keV, the neutron flux spectrum for the AENIF decreases precipitously, because the $^7$Li(p,n)$^7$Be reaction does not produce neutrons with energies greater than 790 keV for a proton kinetic energy of 2.5 MeV. Despite the differences in the spectra, the average kerma factors are nearly equal for a differential volume of tissue irradiated in the two beams. For the BMRR BNCT beam, the average kerma factor is 4.5 x $10^{-11}$ cGy•cm$^2$ [9], while as stated above, the average kerma factor for the AENIF is (4.4 $\pm$ 0.5) x $10^{-11}$ cGy•cm$^2$.

For the ATNIF moderator assembly with a 30 mA proton beam, $\Phi_t$ at the irradiation point in air equals (1.8 ± 0.1) x $10^9$ neutrons/cm$^2$•sec and the RTF equals 80 ± 20. The value of $\Phi_t$ per milliamp of proton current is less by a factor of about 400 for the ATNIF moderator assembly than for the light water moderator assembly for the CANUTRON accelerator neutron source described in reference 1. However, for the ATNIF the neutron field is less contaminated with fast neutrons and gamma rays. Also, for the ATNIF, $\Phi_t$ is measured at an irradiation point which is external to the moderator assembly, while for the CANUTRON $\Phi_t$ is measured at the center of the moderator assembly.

## CONCLUSION

The design calculations for the CANUTRON moderator assembly and for our AENIF and ATNIF moderator assemblies represent three different neutron environments which can be achieved with a 2.5 MeV high current proton accelerator. The CANUTRON moderator assembly is designed to provide a source of predominately thermal neutrons for industrial applications. The AENIF and ATNIF moderator assemblies are designed, respectively, to provide sources of epithermal and thermal neutrons for BNCT.

For the AENIF moderator assembly with a 30 mA proton beam the useful neutron flux $\Phi_u$ at the irradiation point in air is calculated to be (9.0 ± 0.3) x $10^8$ neutrons/cm$^2$•sec, with an average kerma factor equivalent to that of a 5 keV neutron. For the ATNIF moderator assembly with a 30 mA proton beam the thermal neutron flux $\Phi_t$ at the irradiation point in air is calculated to be (1.8 ± 0.1) x $10^9$ neutrons/cm$^2$•sec with the Ratio of the Thermal neutron flux to the sum of the epithermal and Fast neutron flux (RTF) equal to 80 ± 20.

The calculational methods used for the design of the AENIF and ATNIF moderator assemblies have been confirmed by comparison of calculations with measurements for a test AENIF moderator assembly. The test AENIF moderator assembly is simply a right circular cylindrical tank of $D_2O$ with a diameter of 36 cm and a height of 20 cm. The neutron spectrum emerging from the test moderator assembly was measured at the Ohio State University Van de Graaff accelerator for an approximately 1 μA beam of 2.5 MeV protons bombarding a solid $^7$Li target. Details of the measurements and the measurement methods are described elsewhere [10, 11]. Confirming measurements have not yet been made for the ATNIF moderator assembly.

## ACKNOWLEDGMENT

This work was supported in part by the National Cancer Institute under Grant 1 R01 CA 47298-01 and by the U.S. Department of Energy under contract DE-AC02-76CH00016.

## REFERENCES


1. S.O. Shriber, M.A. Lone, B.G. Chidley, M.S. de Jong, S.A. Kushneriuk, and W.N. Selander, "A Compact Neutron Source for Research and Industrial Applications," IEEE Trans. Nucl. Sci. 30, p. 1668 (1983).

2. B.G. Chidley, G.E. McMichael, and S.O. Shriber, "Canutron-A Clean Accelerator Neutron Generator," Nucl. Instr. and Meth. B10/10, p. 888 (1985).

3. C.K. Wang, T.E. Blue, and R.A. Gahbauer, "A Neutronic Study of an Accelerator-Based Neutron Irradiation Facility for Boron Neutron Capture Therapy," Nucl. Tech. 84, pp 93-107 (1989).

4. T.X. Qu, T.E. Blue, C.K. Wang, R.A. Gahbauer, "A Neutronic Study of an Accelerator-Based Neutron Irradiation Facility for the Treatment of Superficial Melanoma by Boron Neutron Capture Therapy," Nucl. Tech. 91, pp. 404-412 (1990).

5. H. Liskien and A. Paulsen, "Neutron Production Cross Sections and Energies for Reaction for $^{7}Li(p,n)^{7}Be$ and $^{7}Li(p,n)^{7}Be^{*}$," Atomic and Nuclear Data Tables 15, pp. 57-84 (January 1975).

6. H.H. Anderson and J.F. Zeigler, Hydrogen Stopping Powers and Ranges in All Elements (Pergaman Press Ltd., 1977, Vol. 3), pp. 16-22.

7. M.B. Emmet, "The MORSE Monte Carlo Radiation Transport Code System," Oak Ridge National Laboratory report ORNL-4972 (February 1975).

8. R. Roussin, "BUGLE-80, Coupled 47-Neutron, 20-Gamma Ray, P3, Cross Section Library for LWR Shielding Calculation Calculations," DLC-60, Radiation Shielding Information Center, Oak Ridge National Laboratory (1978).

9. F.J. Wheeler, D.K. Parsons, D.W. Nigg, D.E. Wessol, L.G. Miller, R.G. Fairchild, "Physics Design for the Brookhaven Medical Research Reactor Epithermal Neutron Source," in Neutron Beam Design, Development, and Performance for Neutron Capture Therapy, Edited by Otto K. Harling, et al, (Plenum Press, 1990) pp. 83-95.

10. C.K. Chris Wang, T.E. Blue, & J.W. Blue "An Experimental Study of the Moderator Assembly for a Low - Energy Proton Accelerator Neutron Irradiation Facility for BNCT," in Neutron Beam Design, Development, and Performance for Neutron Capture Therapy, Edited by Otto K. Harling, et al, (Plenum Press, 1990) pp. 271-280.

11. C.K. Wang, T. E. Blue and J.W. Blue, "A Neutron Spectrometer for Neutrons with Energies between 1 eV and 10 keV", Nucl. Instr. and Meth. A290, pp. 237-241 (1990).

# EVALUATION OF TWO THRESHOLD FOIL NEUTRON SPECTROMETRY METHODS FOR IN-PHANTOM MEASUREMENTS

VICTOR V. VERBINSKI

SCIENCE APPLICATIONS INTERNATIONAL CORPORATION

SAN DIEGO, CALIFORNIA, U.S.A.

## ABSTRACT

A simple method of in-phantom neutron dosimetry is presented that takes into account changes of spectral shape by means of five threshold-activation foils. It employs a perturbation technique that utilizes ratios of activations of these five foils for a standard spectrum, comparing them with activation ratios for an unknown spectrum. Changes in these ratios denote a change of spectral shape and provide a corresponding shape-change correction. The accuracy of the method is evaluated by comparisons with the more complex SAND II spectrum-unfolding method. The SAND II unfolding code was tested, in this investigation, against National Institute of Standards and Technology (NIST) standard fission spectrum. The evaluation of the simple method included a dosimetry determination for a free field reactor spectrum, and one that was significantly altered above about 1 MeV by means of a 5.08cm thick lead wall. The simple method agreed with the SAND II results within a standard deviation of 6% in neutron dose.

The method can be extended to neutron damage measurements of reactor pressure vessels or electronics by weighting each spectral segment (Table 1) by the respective spectral damage function instead of KERMA values.

# I. INTRODUCTION

In the course of a program designed to evaluate both calculational and day-to-day standard-dosimetry techniques for in-phantom dosimetry, a series of in-phantom measurements was carried out with 13 neutron threshold activation foils (12 epicadmium), utilizing the Armed Forces Radiobiology Research Institute (AFRRI) TRIGA reactor facility. The SAND II unfolding code (1-3) was utilized to unfold the in-phantom spectra, with a trial spectrum, $\phi_{tr}$, obtained from the ANISN calculation (4,5). The trial spectrum provided only the starting-spectrum shape, which was modified only slightly in shape by the SAND II code to agree with the foil activations in absolute spectral fluence.

The threshold-foil spectrometry method, which utilized the ENDF/BV cross section library, was in turn evaluated by means of an exposure in the NIST standard-fission-spectrum assembly, utilizing for $\phi_{tr}$ the spectrum shape from the NIST - provided, evaluated fission spectrum for $^{235}U$ in 45 - group format. The fast neutron flux provided by NIST and the SAND II results fortuitously agreed to within 2.5% with an overall probable error of 3 to 4%.

# II. MEASUREMENTS AND CALCULATIONS

Measurements were taken at three locations in a monkey phantom located in exposure room ER1, behind a 5.08 cm thick lead wall, and in the same in-phantom locations in exposure room ER2, no lead wall ("free field"), for a total of six measurements. One of the spectral measurements was taken as a standard of reference, for use as a "base point" in the perturbation method whereby only five foils are used to measure the neutron dose in each of the remaining five in-phantom measurements. This "base point" spectrum was measured in mid-thorax, in ER1, and the corresponding data are presented in Table 1 in a format designed for use in the 5-foil perturbation method of neutron dosimetry. All six spectra were unfolded with the SAND II code and all but the base-point spectrum were compared with the simple 5-foil perturbation method for measuring neutron dose.

The data of Table 1 were for a mid-thorax exposure in a monkey phantom located in exposure room ER1 at 130cm from the reactor "dimple" (the half-cylinder air-water interface), for a 4187.5kW-Min exposure yielding a fast neutron tissue dose of 9296 Gy. The monkey-phantom thorax was a 10.2cm diameter (upright) cylinder, 34cm high, with midpoint coincident with the reactor centerline. The head measurement was at the center of an 8cm diameter spherical head, whose center was

25.5cm above the mid-thorax point and along the extended thorax axis. The hip measurement was 5cm above the bottom of the thorax and displaced laterally from the centerline within 1.7cm of the thorax outer wall. All phantom parts consisted of thin-walled lucite vessels filled with body-tissue - equivalent liquid, and the neutron threshold activation foil stack, in cadmium, was suspended in the liquid with a thin lucite rod. When the sub-cadmium foil stack exposure was conducted at one location (mid-thorax, say), a bare gold foil was exposed at an alternate location (mid-head, say), along with an extra nickel foil for cross referencing. However, only sub-cadmium data are of interest here because the thermal neutron dosimetry measurement is quite independent, consisting of a cadmium-difference measurement with gold or manganese foils.

For a monkey phantom exposed in exposure room ER2 to a free-field AFRRI-TRIGA reactor spectrum, the spectral shapes measured for the three locations inside the phantom are almost identical. Therefore, another three in-phantom measurements were made, behind a 5.08cm thick lead wall, in exposure ER1. The lead wall produced a significantly large spectral change (softening) above the canonical slowing-down spectral shape (See Figure 1), thus providing a meaningful test for the calculational method, the day-to-day dosimetry method, and the simplified 5-foil (for epithermal neutrons) perturbation method presented in Section III. There is no spectral change, however, (See Figure 1) in the canonical slowing-down region below about 1MeV in the highly hydrogenous phantom, and this enhances the accuracy of the 5-foil perturbation method because practically all the foil thresholds are above the slowing-down region.

Normalization of the reactor spectral flux output to the day-to-day reactor power monitor is totally inadequate without a reactor core flux plot and a reactor core neutron-transport calculation. Even with that, the uncertainty in the water gap between the movable TRIGA reactor core and the exposure room would seriously compromise the accuracy of the reactor-output fluence rate determination. Instead, a nickel-foil flux monitor, used for <u>scaling</u> the spectral fluence, was place on the "dimple" (i.e., the half-cylindrical aluminum water-air interface between the swimming pool reactor and the walk-in exposure room) at the mid-core centerline, and the reactor spectral-flux output was determined via the Vitamin-E ENDF/B-V activation cross sections for the $^{58}$Ni (n,p)$^{58}$Co reaction. The resulting reactor flux output was converted to neutron tissue KERMA factor values (Rad tissue or Gy per Kw-Min), and these values were 78.6% and 85.8% of the AFRRI monitor values taken at the same locations but on different day. The calculated (4.5) KERMA factor values from neutrons above 10keV agreed with the values derived from the measured data within a standard deviation of 4% for the mid-thorax in-phantom location (6), where comparison with the 1-D calculation was of greatest validity (because the mid-thorax was located at the reactor center line).

## III. SIMPLIFIED 5-FOIL PERTURBATION METHOD

This method is in essence a two-step method. In the first step, the first order neutron dose $D_1(Ni)$ is simply equal to the (epithermal) dose for the " standard base point" spectrum, $D_s$, for which the foil activity data are listed in Table 1, column 4 (i.e., the $(N_o\lambda)_s$ values, where $N_o$ = the probability of activating one atom and $\lambda$ is the decay constant in $sec^{-1}$), multiplied by the ratio, $R_s$, of nickel-foil activations; i.e.,

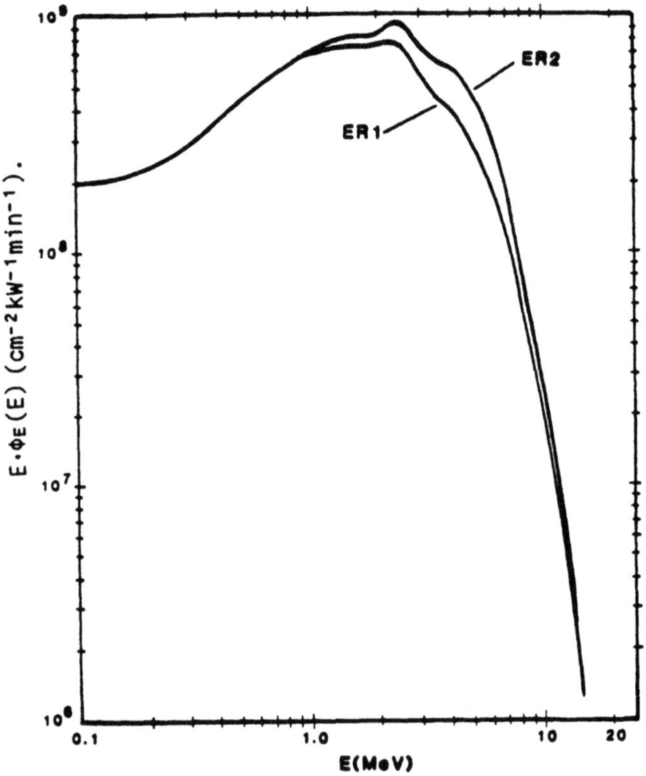


Figure 1. Spectrum shape comparison

$$D_1 = D_s * (N_o\lambda)_x(Ni)/(N_o\lambda)_s(Ni). \quad (1)$$

Here, $D_s$ = 9296 rad (Table 1). If there is no change of spectral <u>shape</u> between the standard (baseline) spectrum $\phi_s$ and the new spectrum $\phi_x$, then Equation 1 yields the correct dose value $D_x$ (i.e., $D_1 = D_x$), which is defined as the second order (final) dose value for spectrum $\phi_x$. In this case, the nickel-normalized foil activation ratios $R_s$ in Table 1 will not have changed; namely, $R_s = R_x$ for all non-nickel foils.

If, on the other hand $R_x \neq R_s$, then the spectral shape has changed (i.e., $\phi_x \neq K\phi_s$), and the ratios $R_x/R_s$ for the i-th non-nickel foil, defined as $A_i$, yield the second-order dose value $D'_x$ by means of the following "four-group shape-correction relationship":

$$D'_x = D_1 \left( (\Sigma_i A_i F_i) - 1 \right) \quad (2)$$

where $D_x$ is now the sum of $D_1$ (Eq 1) and $D_x$ (Eq 2). The $F_i$ values, given in Table 1, are the (non-nickel) foil weighing values to convert the change-of-shape data to second-order-dose-correction data. The $F_1$ represented the fraction $\Delta D_s$ of the non-nickel dose (See Table 1) in the energy region $\Delta E$. Here, $\Delta E = E_{L+1} - E_L$ in Table 1, listed in order of increasing lower-energy limit (or threshold value).

Table 2 presents the values of $A_i$ for all six in-phantom dosimetry measurements (e.g., $A_4 = R_x/R_s$ (Mg)), the $D_x$ values (second-order dose determination) derived from these data, as well as the total (epithermal) dose $D_s$ for the standard run(i.e., the mid-thorax run in experimental room ER1, at 130 cm from the reactor "dimple" and behind a 5.08cm thick lead wall). For runs other than $\phi_s$ (i.e., ER1, mid-thorax), the $D_x$ and D (SAND II) data agreed within a standard deviation of 6%. Thus, the simplified threshold-foil dosimetry method appears to yield about twice the uncertainty as the rigorous SAND II results, but is appreciably better than the 27% and 17% overestimate by the AFRRI dose-monitor compared to the SAND II results.

Table 1. Five-foil parameters for standard exposure (AFRRI, ER1) 130 cm behind lead, 9296 Gy (tissue) mid-thorax epithermal dose per 4187.5 kW.Min.

| Foil | Lower Energy (MeV) | $F_i$ | Activations Per Nucleus $N_o\lambda$ ($sec^{-1}$) | $R_s = N_o\lambda/N_o\lambda(Ni)$ | (Rad Tissue) $\Delta D$ |
|---|---|---|---|---|---|
| $^{197}Au$ | 4.14 - 7 | 0.20 | 6.68 - 16 | 2.09 + 4 | 1408 |
| $^{115}In$ | 0.55 | 0.60 | 2.25 - 17 | 7.03 + 2 | 4386 |
| $^{54}Fe$ | 2.31 | 0.15 | 5.87 - 21 | 1.83 - 1 | 1074 |
| $^{58}Ni$ | 3.01 | | 3.20 - 20 | 1.00 + 0 | 2066 |
| $^{24}Mg$ | 6.38 | 0.05 | 6.08 - 20 | 1.90 + 0 | 362 |

Table 2. Five-foil epithermal dose values via Equations 1 and 2, compared to D from SAND II with 5 foils, in Gy (tissue) per kw.Min.

| | ER1 (5.08 cm lead) | | | ER2 (Free Field) | | |
|---|---|---|---|---|---|---|
| | Mid-Thorax | Head | Hip | Mid-Thorax | Head | Hip |
| D (SAND II) | 2.22 | 2.150 | 2.810 | 6.060 | 5.130 | 7.210 |
| $R_x/R_s$(Au) | 1.00 | 0.970 | 0.954 | 0.746 | 0.780 | 0.778 |
| $R_x/R_s$(In) | 1.00 | 0.986 | 1.067 | 0.867 | 0.900 | 0.917 |
| $R_x/R_s$(Fe) | 1.00 | 1.027 | 1.041 | 0.993 | 0.991 | 0.953 |
| $R_x/R_s$(Mg) | 1.00 | 1.011 | 0.941 | 1.075 | 1.047 | 1.074 |
| $R_x/R_s$(Ni) | 1.00 | 0.980 | 1.190 | 3.380 | 3.750 | 3.720 |
| $D_1$ | 2.22 | 2.100 | 2.640 | 7.490 | 6.110 | 8.260 |
| $D_x$ | 2.22 | 2.080 | 2.600 | 6.530 | 5.480 | 7.450 |
| $\Delta(\%)(D_x-D_s)$ | | -3.4 | -8.0 | +7.7 | +6.8 | +3.4 |

## IV. EXTENSION TO OTHER APPLICATIONS.

Table 1 presents a breakdown of the epicadmium neutron dose for each segment of the neutron spectrum assigned to a respective neutron threshold activation foil. This was done for one of the neutron spectra selected as a standard of reference. The neutron dose in each segment is, of course, that segment of the spectrum weighted by the KERMA values. If, instead of the KERMA values, one used the reactor pressure vessel damage function values (or the silicon displacement values), one would thereby have extended the method to reactor lifetime measurements (or silicon-transistor radiation damage measurements).

## REFERENCES


1. ASTM Standards Derived from References 2,3; E 720-80, E 721-80, E 722-80 and E 763-80.

2. V.V. Verbinski, et al, "Threshold-Foil Measurements of Reactor Spectra for Radiation Damage Applications," Nuclear Science and Engineering <u>65</u>, 316 (1978).

3. V.V. Verbinski, et al, "Transistor Damage Characterization by Neutron Displacement Cross Section in Silicon: Experimental," Nuclear Science and Engineering <u>70</u>, 66 (1979).

4. W.W. Engle, Jr., "A User's Manual for ANISN," K-1963 (RSIC-CCC-254) Union Carbide Corporation, 30 March, 1967, rev. 6 June, 1973.

5. D.C. Kaul, et al, "Phantom Dosimetry Calculations for Use in Radiation Effect Correlations," DNA-TR-83-51, Defense Nuclear Agency, 30 July, 1984

6. V.V. Verbinski and C.G. Cassapakis, "Neutron Field Measurements in Phantom with Foil Activation Methods," SAIC 85/1090R, Science Applications International Corporation, November 1986

# PULSED SPHERE EXPERIMENT APPLIED TO NEUTRON TRANSPORT IN LIQUID NITROGEN

AUTHORS : J-B. POIRE, M. FOURGOUX, C. FONTAINE

C.E.A./VALDUC 21120 IS-SUR-TILLE . FRANCE

**ABSTRACT**

*The interaction of 14 MeV and fission neutrons with a thick layer of air has been studied by the time-of-flight method (TOF). Atmospheric air is simulated by a liquid nitrogen sphere (79% $N_2$ in the air). Different shields (steel, $CH_2$, $CH_2+B$) surround the sphere to study the neutron transmission. Measurements (neutron and gamma time spectra) and Monte-Carlo calculations are compared.*

## INTRODUCTION

We used the pulsed sphere technique to study the interaction of neutrons with air. This technique is based on the time-of-flight spectrometry with a pulsed fast neutron source. In these measurements, the sphere is irradiated by a neutron source at its center, the most isotropic in fluence and in energy, generated by the $T(d,n)^4He$ reaction.

The neutrons and secondary gamma leaking the sphere are measured by detectors located at an adequate long distance.

The main point is the neutron arrival time spectrum on the detector; the gamma pulse height spectrum is typical of a direct dose measurement.

## EXPERIMENTAL APPARATUS AND PROCEDURE

### 1. 14 MEV NEUTRONS

The present measurements were performed using an electrostatic accelerator, a 140 keV deuteron beam and a tritium-loaded titanium target. The target (2 cm in diameter), mounted at the end of a pipe, is positioned at the center of the nitrogen sphere.

The 14 MeV neutron production at the tritium target is monitored by counting the associated alpha particle with two silicon surface-barrier detectors positioned at 179° from the incident deuteron direction. The two alpha-diodes are calibrated by two little fission chamber detectors placed close to the neutron source.

To reduce the background and to be able to analyze the neutron component and the gamma component, the accelerator is used in a pulsed mode; the ion beam is deflected and bunched. At a repetition rate of 400 kHz, the delivered bursts of 14 MeV neutron have a gaussian shape of 2 ns Full Width at Half Maximum.

## 2. 14 MEV AND FISSION NEUTRONS

Such a source is made of a multiplying assembly placed around the accelerator target. This assembly is a 93,5% enriched $^{235}U$ sphere (weight=17,6 kg) surrounded by a cadmium layer and a protective steel sheath.

## 3. LIQUID NITROGEN SPHERE

The experimental set-up is shown in figure 1.

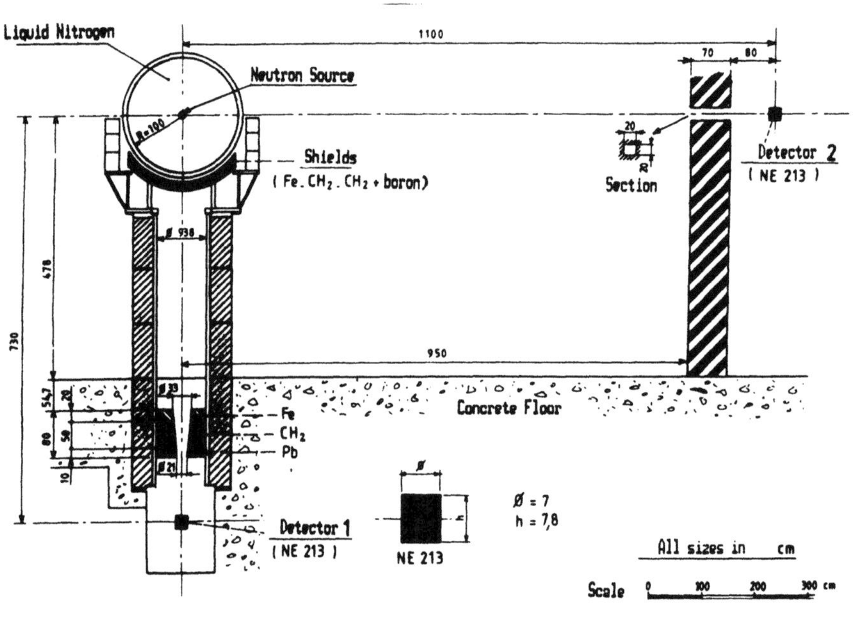


*Figure 1 : EXPERIMENTAL SET-UP*

The sphere contains liquid nitrogen (density = 0,81 $g.cm^{-3}$) corresponding to six mean free path for 14 MeV neutrons. The sphere internal radius is 1 m. The insulator material between the two stainless steel walls (thickness = 2*3 mm) is 9,7 cm thick polyurethane; its density is 0,028 $g.cm^{-3}$. Although the insulation is not perfect, the nitrogen thickness remains constant in the detector directions during each measurement. It is equal to 96 cm, that is simulating 650 m air at atmospheric pressure.

## 4. SHIELDS

They are in contact with the external wall of the sphere and have a hemispheric shape. Their specifications are :

| | | | |
|---|---|---|---|
| - steel | : density = 7.8 | , thicknesses = | 5 to 20 cm |
| - $CH_2$ | : " 0.955 | " | 5 to 15 cm |
| - $CH_2$ + B | : " 1.115 | " | 5 cm |

Several geometric combinations were studied.

## 5. DETECTION

Neutron and gamma-rays are detected simultaneously in two perpendicular directions at 7,3 and 11 m from the tritiated target.

The detection system consists of two 7 cm diameter by 7,8 cm thick NE213 liquid scintillators directly coupled to fast photomultiplier tubes.

The detector efficiencies are calculated by a Monte-Carlo code for a 1 MeV neutron threshold. The detectors are calibrated by means of gamma sources and their linearities are measured up to 6,13 MeV. This kind of detector allows an excellent neutron-gamma discrimination due to the different luminous pulse shapes produced by these radiations within the scintillator.

Pulse shape discrimination is used to minimize gamma-ray background during the neutron measurement spectrum.

Two separate monoparametric spectra as a function of time are recorded (the time pitch is 3 ns on the multichannel analyzer).

* The first spectrum is the neutron detection instants between 1 and 14,1 MeV (TOF). They are typical of diffusion and slowing down neutrons inside nitrogen and possible shields (see fig. 2,5,6).

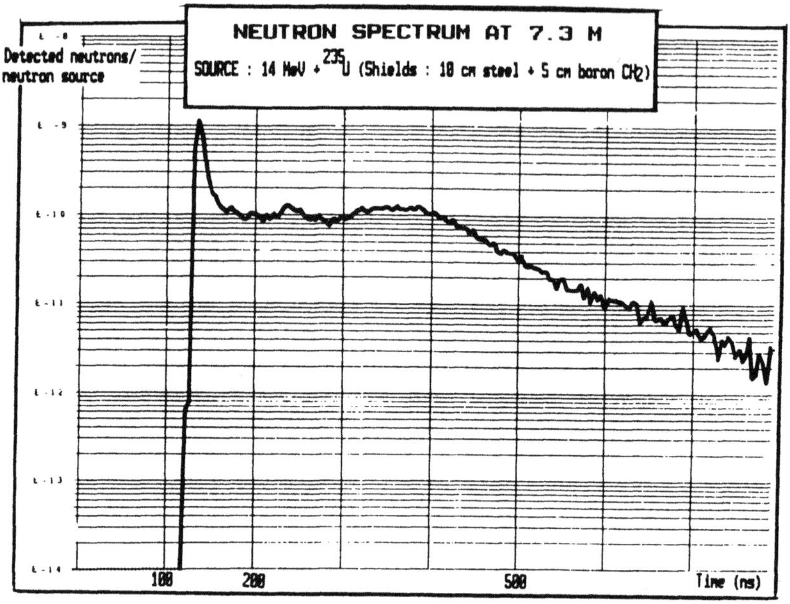


*Figure 2 : EXAMPLE OF AN EXPERIMENTAL NEUTRON TOF SPECTRUM FROM LIQUID NITROGEN AND SHIELDS PULSED WITH 14 MeV NEUTRONS.*

* The second spectrum is significant of the gamma arrival moments into the detectors (see fig. 3a-3b). These gamma radiations are generated by inelastic diffusions inside nitrogen.
Another helpful information about gamma effects is the electron recoil spectrum (ERS) created by the gamma-rays interacting with hydrogen and carbon inside the scintillators (see fig.4). From an energy calibration between the 0,15 MeV threshold up to 10 MeV, the ERS is recorded during a time window on the gamma spectrum (see fig. 3b) before the fast neutronic arrival signal. Its integral is representative of the energy deposition or absorbed dose in a medium with the same composition such as these scintillators.

## DATA REDUCTION AND EXPERIMENTAL RESULTS

Data reduction is initiated with substraction of the background spectra after normalization to one source neutron for each measurement. The background causes are the detector background and the capture gamma-rays coming from highly moderated neutrons.

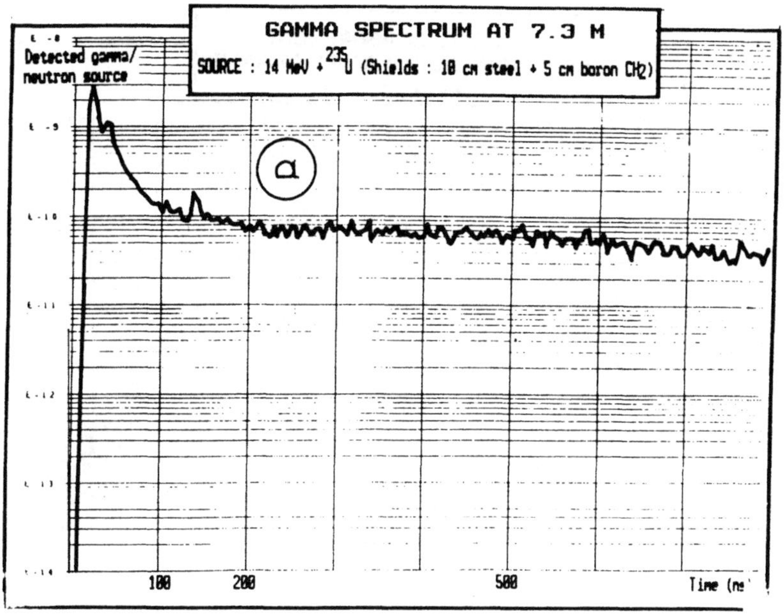


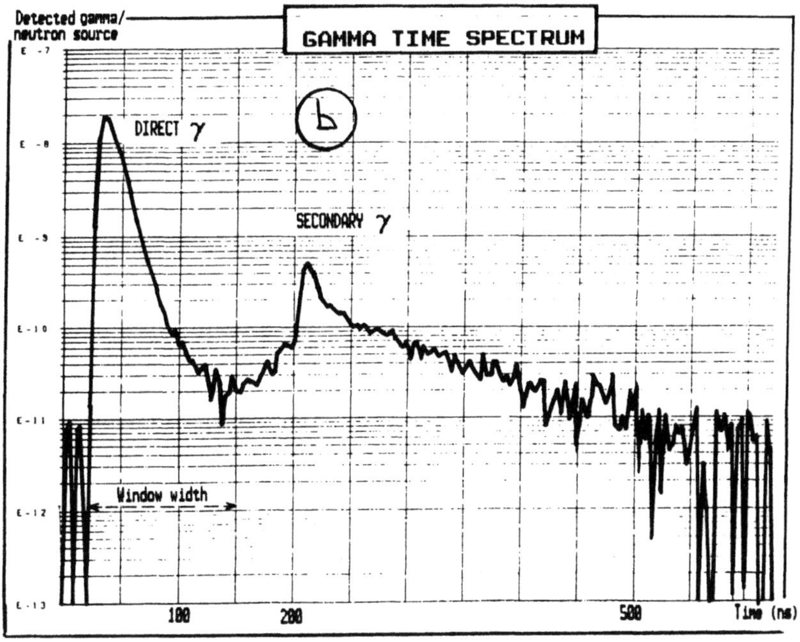


*Figures 3a-3b : EXAMPLES OF EXPERIMENTAL GAMMA SPECTRA*

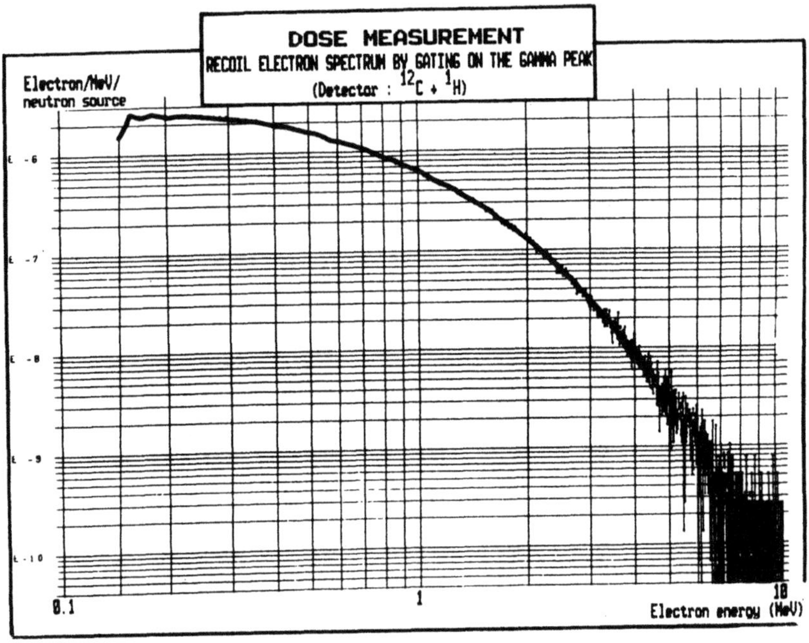


*Figure 4 : EXPERIMENTAL ELECTRON RECOIL SPECTRUM*

The TOF neutron spectra show one component of 14 MeV neutrons transmitted without a shock and a tail characterized by inelastic diffusion peaks inside nitrogen; that tail is also an indication of fissions in $^{235}U$.
We confirmed the effect of iron as a good opportunity shield for high energy neutrons.
The boron enrichment of $CH_2$ has no effect for neutron attenuation.
Gamma spectra show several peaks that we have identified. Gamma-rays are created by neutrons :
- by inelastic scattering into the differents shields, target and its pipe, sphere walls, liquid nitrogen
- or by fission inside the uranium.

In that last case, the fission peak is broadened : it is the temporal distribution image of the fission neutron source.
The boron influence is not determinant on the gamma dose.

## CALCULATIONS

Monte-Carlo calculations of the neutron and gamma time spectra were done with MCNP and SICOMORE codes. In these calculations, an accurate description of the geometries and materials compositions were input into the problem.

Neutrons and gamma-rays emerging from the sphere through the collimators, were tallied at the detector positions. The energy and time of arrival for each particle reaching the detectors were calculated; the detector efficiencies were included.

The cross-sections libraries used, EBDF/B-IV and ENDL, give results in good agreement.

The ERS spectra have not been calculated yet.

## COMPARISONS TO EXPERIMENTS

Two examples are given in figures 5 and 6.

The first comparisons between calculations and experiments show a very sensible discrepancy for the neutron spectra. Discrepancies are reduced when shields are present compared with more singular geometries (e.g. without nitrogen).

These conclusions are similar for the two codes.

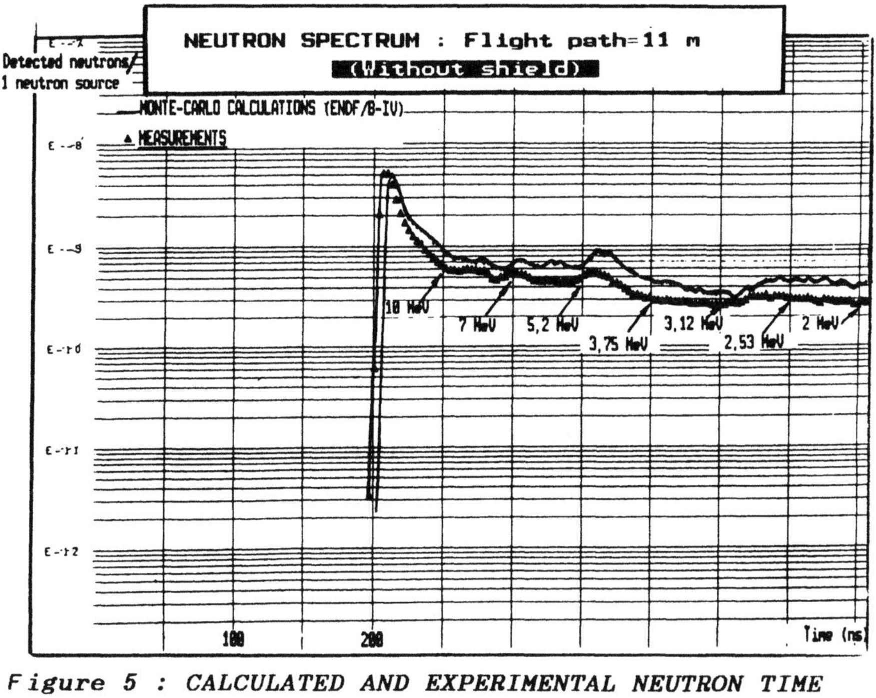


*Figure 5 : CALCULATED AND EXPERIMENTAL NEUTRON TIME SPECTRUM - NO SHIELD IN THIS DIRECTION*

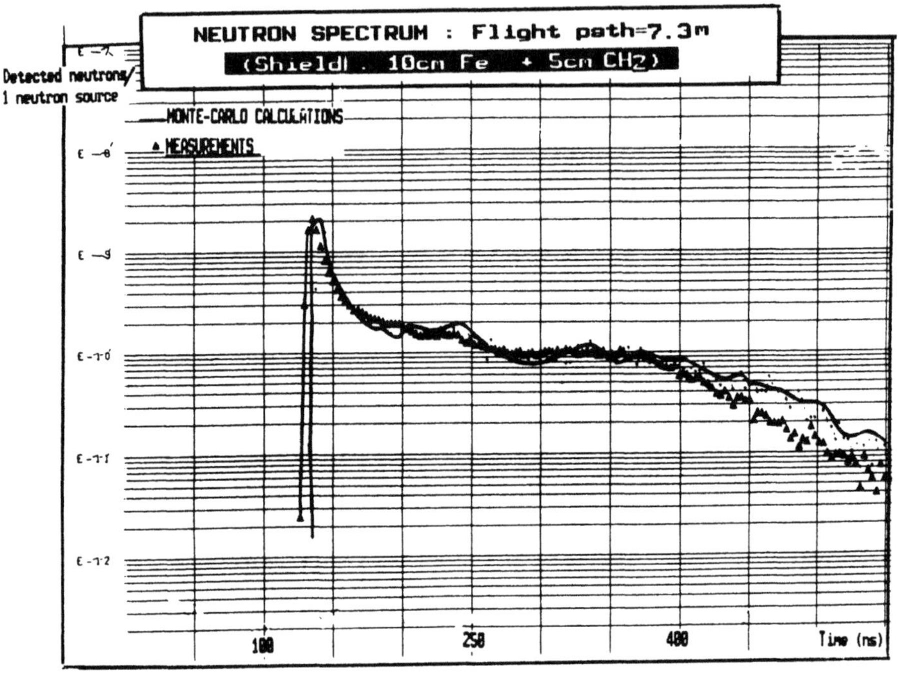


*Figure 6 : CALCULATED AND EXPERIMENTAL NEUTRON TIME SPECTRUM - VARIATIONS ARE ACCEPTABLE FOR THE 300 FIRST NANOSECONDS.*


**SUMMARY**

In case of long neutron ranges studies, the interactions of neutrons with the crossed media are more and more numerous. The errors in cross-sections and in the models are cumulative and can introduce large incertainties in the calculated fluences.

In our study, the most important discrepancy is about 50 % at a given time, between neutron TOF spectra, in experiment and calculation.

Experimental errors don't exceed 10 % .

We conclude that these integral measurement spectra provide a sensitive test of the quality of the nuclear data bases in the 14 MeV neutron energy range.

# PROGRESS IN NEUTRON BEAM DEVELOPMENT AT THE HFR-PETTEN (FEASIBILITY STUDY FOR A BNCT FACILITY)

G.CONSTANTINE**, P.R.D.WATKINS*, C.A.PERKS*, H.J.DELAFIELD*, D.ROSS***, W.P.VOORBRAAK****, A.PAARDEKOOPER****, W.E.FREUDENREICH****, F.STECHER-RASMUSSEN****, R.L.MOSS**.
* AEA TECHNOLOGY, HARWELL, OXFORDSHIRE OX11 ORA, UK,
** JRC-PETTEN, P.O.BOX 2, NL-1755 ZG PETTEN, THE NETHERLANDS,
*** BIRMINGHAM UNIVERSITY, BIRMINGHAM, UK,
**** ECN, P.O. BOX 1, NL-1755 ZG PETTEN, THE NETHERLANDS

ABSTRACT

Boron Neutron Capture Therapy (BNCT), using intermediate energy neutrons to achieve the deep penetration essential for treating brain tumours, can be implemented with a filtered reactor neutron beam. At the Joint Research Centre (JRC) Petten an optimized neutron filter is currently being built for installation into the HB11 beam tube of the High Flux Reactor (HFR). Part of the development leading to this design, has been an extensive experimental and calculational study of broad spectrum, filtered beam performance on the HB7 beam tube facility. The agreement between the calculated and measured spectra has given confidence in the reactor and filter modelling methods used to design the HB11 therapy facility.

## 1. INTRODUCTION

Boron Neutron Capture Therapy (BNCT) offers a possible method for treating tumours, particularly in the brain, which are resistant to conventional attack by surgery, chemotherapy and radiotherapy [1].

In particular the use of intermediate energy neutrons to achieve the deep penetration necessary for treating brain tumours looks very promising. A suitable filter combination which generates a beam of neutrons of the required specification, i.e. of low mean neutron energy and acceptable gamma-ray dose, leads to a significant reduction of the neutron fluence rate. Hence reasonably short therapy periods can only be achieved using reactors with a high power density, such as the HFR at Petten, the Netherlands.

Following the Bremen Conference on NCT [2] in 1988, a "Concerted Action (CA) on BNCT was initiated at Abingdon (UK) at the beginning of 1989. The CA, sponsored by the CEC, coordinates the various activities concerning BNCT in Western Europe and covers such diverse disciplines as: neutron physics, radiation dosimetry, chemistry, radiotherapy, pharmacology and clinical management.

The first priority of the CA is the realisation of a BNCT facility in Petten. Later, other facilities in Europe will be considered . The experience gained elsewhere eg. at the DIDO reactor at Harwell [1] is used in the CA. A self-imposed, tight time schedule to implement BNCT at Petten by late 1991, requires a start of the necessary cell radiobiology and animal irradiations by late 1990. A validation exercise, preliminary to designing and constructing the therapy facility in HB11, was considered essential. Prior commitments for nuclear physics experiments up to August 1990 precluded the use of HB11 itself, but the smaller HB7 beam, which was available, offered several advantages.

This paper gives a survey of the campaign of experiments to validate calculations of the neutron fluence rate and neutron spectrum with the Monte Carlo MCNP code in HB7 and the results of the comparison.

The measurements at HB7, and their subsequent analysis, performed in the period July - November 1989, were a collaborative undertaking by the authors [3], [7], [13].

## 2. THE HB7 FACILITY

The HFR is a 45 MW(th) Materials Testing Reactor (ORR Type), cooled and moderated by light water [4]. The reactor is equipped with 12 horizontal beam tubes of which HB11 and HB12 are foreseen to supply intermediate and thermal neutrons for BNCT, and of which, HB7 was used for this validation exercise. Measurements took place along the beam axis at 7 positions, (I-VII) up to 11.7 m downstream from the reactor vessel edge, (fig. 1).

The HB7 beam tube has a diameter of 180 mm and is currently occupied by an experiment incorporating a "neutron mirror system", comprising a number of tapered slits delineated by 1 mm thick neutron mirrors of a CoFe alloy. The slits are 50 mm high and taper from 2.5 mm wide at the core end to 1.2 mm at the far end and are 2 m in length. For our measurements, the system was operating as a geometrical attenuator. The mirror system focuses the neutrons to a point 9307 mm from position I, after which the beam diverges again. A polyethylene collimator was used to limit the beam diameter to 50 mm.

Three different core configurations were investigated, which differ from each other by the type of assembly loaded in the core position directly at the entrance of the beam: an aluminium assembly (A), a beryllium assembly (B) or a fuel assembly (F).

## 3. MCNP CALCULATIONS

The neutron fluence rate and absorbed dose (neutrons and photons) were calculated at Harwell [5] with the Monte Carlo code, MCNP code [6] for a position 5.0 m from the core edge. Some additional calculations were made at ECN [7].

The influence of different core configurations, different combinations of materials to be used as filter and/or spectrum shifters and different dimensions for the various materials were examined. For validation reasons, extra calculations of both fluence rates and activation responses were performed for positions I and II, in which multi-foil sets were irradiated. Some of the results, for different core and filter configurations are given in Table 1, including preliminary gamma-ray dose rates.

## 4. MEASUREMENTS

Different measurement techniques were used to examine the neutron spectrum, for which the neutron fluence rate varies by more than 6 orders of magnitude in an energy range which extends from fission down to the cadmium cut-off at 0.5 eV.

### 4.1 Activation and Fission Monitor Sets

Foil sets, mostly with a 1 mm Cd shield were irradiated in all positions. The composition of the sets varied from multiple foils (up to 13), mostly of diluted materials in position I, to a single pure gold foil (thickness: 0.1 mm) in position VII. Multi-foil sets, bare and Cd covered, used for neutron spectrum adjustment calculations were irradiated for core configurations A and B (positions I and II) and F (position II). The intermediate energy fluence rate distribution along the beam-axis was measured with diluted gold and/or pure gold foils covered by 1 mm Cd. Both foil materials were irradiated in position II, from which a self-shielding correction in the pure gold foils of 4.0 ($\sigma$ = 5%) was derived.

### 4.2. Proton Recoil Spectrometry [8]

Above 30 keV, proton recoil proportional counters, diameter 40 mm, were used. The counters are filled with hydrogen to pressures of 100, 300 and 1000 kPa which respond to the neutron spectrum in the overlapping energy ranges 10 - 200 keV, 100 - 500 keV and 200 keV - 1.5 MeV respectively. Due to the narrowness of the beam and the limited time available, the counters were positioned one behind the other in positions V, VI and VII. The 300 and 1000 kPa counters in position VI and VII respectively, contained a trace of $^{3}$He for calibration purposes and were covered with Cd (thickness 1.0 mm) to minimise their response to thermal neutrons. Additional count rate measurements were made with the proton recoil counters in line in various orders to allow corrections to be made for intensity fall-off along the beam-line and attenuation by the preceding counters. A reactor power of 500 kW and the attenuation of the mirror system were sufficient to keep the count rate below the required 300 $s^{-1}$.

Two measurements were performed, both for core A, but with different beam filter combinations, i.e. 0.5 mm Cd, 50 mm Al, 1000 mm Ar, with and without 10 mm Ti.

## 4.3. Spectrum Modification Technique [1].

A range of $^{10}B$ plates of various thicknesses and a Ti plate were placed in the beam and the counting rates of two detectors situated in position VI, when thus modified, were measured (see fig. 2). These two counters, (a $BF_3$ counter, diameter 40 mm, and a 38.0 mm radius polyethylene sphere containing a $^{3}He$ proportional counter [9]), have very different energy response functions. Due to the HB7 set-up, the response functions had to be modified due to the non-uniform beam profile. Therefore MCNP calculations were performed for a range of discrete neutron energies in a parallel beam with an experimentally determined radial profile. This distribution was also used to correct the sphere and proton recoil counters to the values they would have shown, had they been irradiated uniformly. Factors of 2.35 ± 0.1 (1σ) and 1.14 ± 0.02 (1σ) respectively were obtained. The measurements were carried out for two cores, A and B, and with filters of 1000 mm argon with, in turn, 300, 150 and 50 mm aluminium; the last with, then without, 100 mm of sulphur. Frequent checks on the reactor power were taken with a 127.0 mm radius sphere located at the rear of position VII. This was otherwise covered with a borated polythene block to minimise the effect of the backscatter of low energy neutrons on the counting during the measurements.

# 5. ANALYSIS

Neutron spectrum adjustment calculations were performed for positions I and II, using an ECN version of the SAND-II code [10]. This code normalises the MCNP spectrum using the measured reaction rates and a cross-section library [11]. The count rates of the different counters were first normalised to the 5000 mm position on the HB7 axis taken for the MCNP calculations, using the results of the cadmium covered gold foils. The spectrum unfolding code SENSAK [12] was used to adjust the calculated spectrum on the basis of the observed count rates and the response functions modified as above [13]. Uncertainties in the response functions and the initial spectrum, together with uncertainties in measurements, and where appropriate covariance matrices, are all accounted for in the SENSAK calculation.

For core configuration A, the total flux derived from the proton recoil measurements was also used, assuming a cross section of zero below 30 keV and a constant value above. Figure 3 shows the spectrum adjustment factors for core B. The average ratios of calculated and measured reaction rates in positions I and II for the 3 cores are given in Table 2. Table 3 gives a summary of the calculated to measured fluence rate ratios for the 2 cores and the 4 filter combinations. More details are given in ref. [13].

# 6. DISCUSSIONS AND CONCLUSION

It is evident when comparing Tables 2 and 3, that the agreement between calculations and experiments is much closer for the beryllium reflected core (B) and configuration (F) with the fuel assembly, than for the aluminium reflected core (A). This is probably because the power distribution appropriate to core B used in each MCNP run is unrealistic for the core when reflected by aluminium. Another shortcoming may be the recognized incorrect modelling of the aluminium reflector assemblies. In conclusion it can be stated that the neutron fluence rate measurements in HB7 have shown reasonably good agreement with calculations, especially for the beryllium reflected core.

A more relaxed time schedule for proton recoil measurements, a more uniform beam in HB11 and the modification of the adjustment programs SAND-II and/or SENSAK for activation monitors will contribute to more accurate characterization of neutron fluence rate and spectrum in HB11. In addition, information from alternative types of counters, including larger polyethylene-spheres, a $^6$Li spectrometer and multi-layer gold foils will be employed.

## 7. LOOKING AHEAD

A filter combination of 1 mm Cd, 150 mm Al, 50 mm S, 10 mm Ti and 1500 mm liquid Ar and an appropriate fast (main) beam shutter with an emergency back-up system was installed in the HB11 beam in the HFR in August 1990. After testing of the liquid argon filter system, a start will be made with the characterisation of the neutron and gamma field using the experience gained at HB7.

## 8. REFERENCES


1. G. Constantine, A. Brenen, P.G.F. Moore and C.A. Perks,: "Spectrum Measurements on Filtered Neutron Beams for Medical Applications", Reactor Dosimetry: Methods, Applications, and Standardization, ASTM STP 1001 ,1989, pp. 699-709

2. Third International Symposium on Neutron Capture Therapy, Bremen 31 May - 3 June 1988; Zeitschrift fur Radioonkologie, Strahlenbiologie, Strahlenphysik, Band 165, Heft 2/3, Februar/ März 1989.

3. F. Stecher-Rasmussen et al.: "Characterization of intermediate energy neutron beams for BNCT at the HFR in Petten" (paper presented at the 4th Int. Symposium on NCT, Sydney, 1990).

4. H. Röttger, A. Tas, P. von der Hardt, W.P. Voorbraak: "High Flux Materials Testing Reactor HFR Petten; Characteristics of facilities and standard irradiation devices", EUR 5700, 1986.

5. P.R.D. Watkins: "Report of Calculations Performed by Harwell for the Petten BNCT Project",(Reactor Physics Note 168, September 1989).

6. J.F. Briesmeister: "MCNP - A General Monte Carlo Code for

Neutron and Photon Transport " LA-7396-M, Revision 2. (Los Alamos National Laboratory, September 1986).

7. W.P. Voorbraak et al.: "Neutron Metrology in the HFR. Neutron fluence rate and gamma dose measurements in HB7" (Contribution to the design of a neutron beam for Boron Neutron Capture Therapy, BNCT), to be published.

8. R. Birch, H.J. Delafield, L.H.J. Peaple, K.G. Harrison.," The Neutron Leakage Spectrum through the Steel Top Plates of two Heavy-water-cooled Research Reactors" Radiat. Prot. Dosim. 12 No. 3, pp. 285-291 (1985)

9. P.M. Thomas, K.G. Harrison, M.C. Scott: "A Multisphere Neutron Spectrometer using a Central $^3$He Detector", Nucl. Instr. Meth. 224, pp. 225-232 (1984).

10. W.N. McElroy et al. :"A Computer Automated Iterative Method for Neutron Flux Spectra Determination by Foil Activation", Vol. I - IV, Report AFWL-TR-67-41 (1967).

11. W.L. Zijp, H.J. Nolthenius, and G.C.H.M. Verhaag: "Cross-section Library DOSCROSS84 (in a 640 group structure of the SAND-II type", Report ECN-160 (October 1984).

12. A.K. McCracken: "Few Channel Unfolding in Shielding - the SENSAK Code "Proc. 3rd ASTM/EURATOM Symp. on Reactor Dosimetry, Ispra, Italy 1979, EUR 6813.

13. G. Constantine, C.A. Perks, H.J. Delafield, D. Ross, P.R.D. Watkins,: "Neutron Spectrum Characterisation of the HB7 Beam in the HFR, Petten", (Reactor Physics Note 169, Harwell Laboratory, December 1989).

## TABLES

Table 1: MCNP results for 2 cores and 3 filter combinations [5]

| Filter | Assembly at beam entrance Aluminium (A) | | | Beryllium (B) | | |
|---|---|---|---|---|---|---|
| liquid Argon (mm) | 1000 | 1000 | 1000 | 1000 | 1000 | 1000 |
| Cadmium (mm) | 1 | 1 | 1 | 1 | 1 | 1 |
| Aluminium (mm) | 300 | 150 | | 300 | 150 | |
| Sulphur (mm) | | | 50 | | | 50 |
| fluence rate($10^{12}$ $m^{-2}\cdot s^{-1}$) | 3.337 | 10.23 | 29.22 | 1.801 | 5.591 | 16.20 |
| uncertainty | 0.025 | 0.023 | 0.022 | 0.034 | 0.033 | 0.032 |
| mean energy (keV) | 11.80 | 12.32 | 20.04 | 10.30 | 10.96 | 18.93 |
| uncertainty | 0.047 | 0.038 | 0.034 | 0.059 | 0.052 | 0.049 |
| Photon Dose (mGy/h) | 34.5 | 109.8 | 264.9 | 36.5 | 118.3 | 289.3 |
| uncertainty | 0.030 | 0.029 | 0.028 | 0.045 | 0.041 | 0.038 |

Table 2: Ratio Measured to Calculated Reaction Rates in Positions I and II (based on activation monitors) [7]

| Core Configuration 1) | A | | B | | F |
|---|---|---|---|---|---|
| Position | I | II | I | II | II |
| (n,γ) - reactions with Cd-cover | 0.52 | 0.69 | 0.79 | 0.83 | 0.81 |
| std deviation (%) 2) | 12 | 22 | 8 | 25 | 24 |
| (n,p) - reactions 3) | 0.65 | 0.90 | 0.94 | 0.95 | 0.96 |
| std deviation (%) | 18 | 7 | 17 | 6 | 5 |

1) A refers to an aluminium reflector, B to a Beryllium reflector and F to a combination Al/Fuel/Al for the core positions directly at the entrance of the beam.
2) Average ratio for all (n,γ) reactions; the number of reactions varies per position.
3) Average ratio for all (n,p) reactions; the number of reactions varies per position.

Table 3: Ratio Measured to Calculated Reaction Rates in Position VI (based on spectrum modification technique and proton recoil spectrometry).

| Filter combination | | | | core configuration 1) | |
|---|---|---|---|---|---|
| Al (mm) | S (mm) | Cd (mm) | Ar (mm) | A | B |
| 50 | 0 | 0.5 | 1000 | 0.58 | 0.86 |
| 300 | 0 | 0.5 | 1000 | 0.65 | 0.97 |
| 50 | 100 | 0.5 | 1000 | 0.60 | 0.88 |
| 150 | 0 | 0.5 | 1000 | 0.59 | 0.88 |
| average | | | | 0.60 | 0.90 |
| std deviation (%) | | | | 5 | 5 |

1) A refers to an Aluminium reflector, B to a Beryllium reflector.

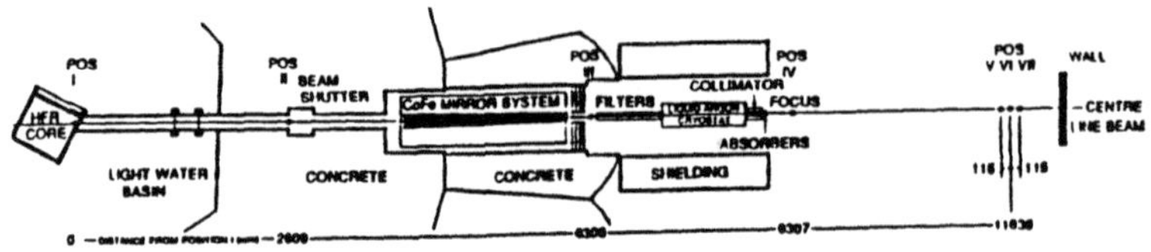

Fig. 1. Measurement postions HB7.

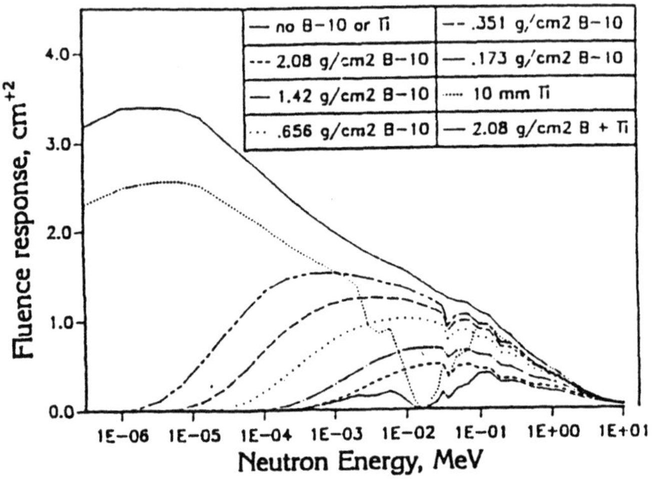


Fig. 2. Fluence response of the 76 mm diameter single sphere without and with covers as indicated, as function of the energy.

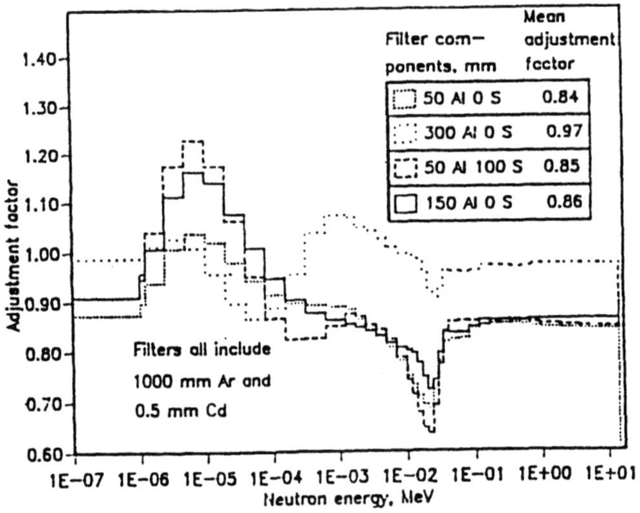


Fig. 3. Spectrum adjustment factors for the Be reflected core.

All foils were also analyzed for $^{182}$Ta activity so that the amount of fluorescence could be equated to the $^{182}$Ta activity. The fluorescence coefficient, which is the ratio of the fluorescence x-ray emission rate due to the $^{182}$Ta, is specific to the source. Foils with different masses require different corrections. Fluorescence correction coefficients for these foils are also listed in Table 4.

## Comparison of Results

Saturated reaction rates were estimated from the $^{93m}$Nb activities, irradiation histories, and the physical properties of the foils. Table 5 tabulates reaction rates for the niobium foils evaluated in this study. Neutron cross-sections for the $^{93}$Nb(n,n')$^{93m}$Nb reaction were determined from reaction cross-sections provided in SAND-II group structure and restructured for the energy group structure used in the transport calculations from which the cross-sections for the conventional dosimetry sets were determined. Reaction rates for the niobium foils were calculated from the $^{93}$Nb(n,n')$^{93m}$Nb reaction cross-sections and the fluences measured by the conventional dosimetry sets. These values are also listed in Table 5.

## Conclusion

The differences in the calculated to estimated reaction rates could be the result of errors in any or all of several measurements. The activity of $^{93m}$Nb was corrected for

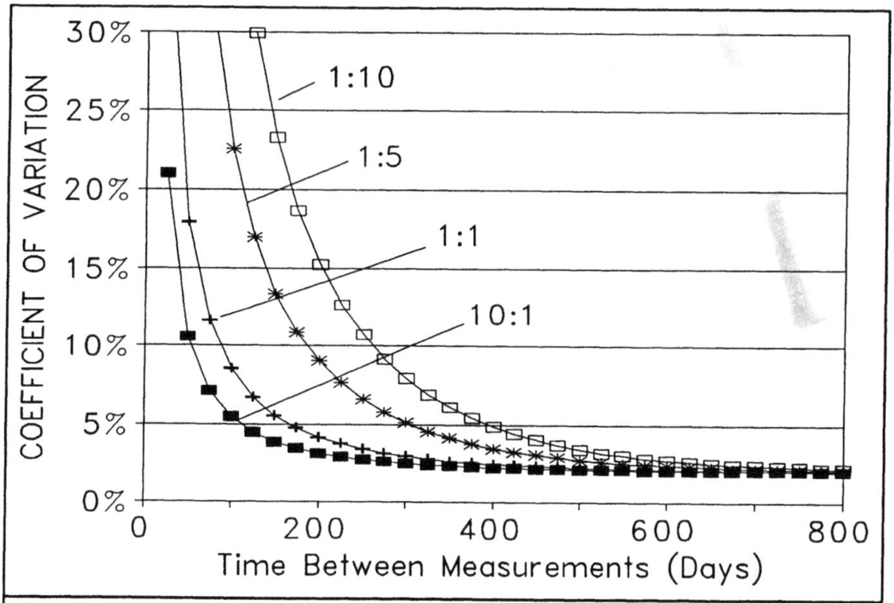


Figure 4. Coefficients of variation as $^{182}$Ta decays for several initial ratios of x-rays produced from the decay of $^{93m}$Nb to x-rays induced by fluorescence.

| FOIL | ANO Reactor Unit - Exp | End of Irradiation | Effective Full-Power Days | Niobium Dosimetry Reaction Rate dps/atom | Convent. Dosimetry Reaction Rate dps/atom | Niobium to Convent. Ratio |
|---|---|---|---|---|---|---|
| I | II - 7 | 26-Sep-83 | 233.9 | 3.878E-16 | 3.361E-16 | 1.15 |
| J | II - 7 | 26-Sep-83 | 233.9 | 3.893E-16 | 3.361E-16 | 1.16 |
| E | II - 10 | 28-Aug-88 | 443.55 | 2.939E-16 | 2.299E-16 | 1.29 |
| A | II - 12 | 25-Sep-89 | 125.29 | 1.214E-16 | 1.884E-16 | 0.63 |
| B | II - 12 | 25-Sep-89 | 125.29 | 1.116E-16 | 1.884E-16 | 0.59 |
| | | | | | | |
| T | I - 11 | 16-Mar-84 | 236.9 | 4.695E-17 | 4.052E-17 | 1.16 |
| U | I - 11 | 16-Mar-84 | 236.9 | 5.174E-17 | 4.052E-17 | 1.28 |
| E' | I - 15 | 28-Aug-88 | 230.07 | 5.914E-17 | 4.618E-17 | 1.28 |

Table 5. Reaction rate measurements and comparisons with values calculated from fluences obtained from conventional dosimetry methods.

self-absorption, and in some cases, fluorescence. These corrections are not expected to account for large portions of the differences, except in the case of the heavier foils. One area that needs to be looked at more closely is the possibility of additional fluorescence induced by the decay of $^{94}Nb$. Other factors that would contribute to the differences are reaction cross-sections and the fluence measurements from the conventional dosimeter sets. Future effort will involve resolving these discrepancies.

**Acknowledgements:**

The authors would like to acknowledge Entergy Operations, Inc for their assistance and support.

**References**

1. C.O. Cogburn and J.G Williams. Pressure Vessel Neutron Dosimetry at Three PWRs. EPRI NP-5733, April 1988.

2. C.O. Cogburn, J.G. Williams, N. Tsoulfanidis. "Pressure Vessel Dosimetry at U.S. PWR Plants". Proceedings 5th ASTM-Euratom Symposium on Reactor Dosimetry, 1985.

3. T.G. Williamson, T.A. Ayers, W.L Hammersten, G.P. Lamaze, and F.J. Schima. "Measurements of the 93Nb(n,n') Fission Spectrum Cross Section". Reactor Dosimetry, Methods, Applications, and Standardization. STP 1001, 1989.

# GENERAL INTEREST

# RADIOIODINE MONITORING OF NUCLEAR POWER PLANT AIRBORNE EMISSIONS UNDER ACCIDENT CONDITIONS

William A. Jester, T.T. Tseng, and B.S. Lee
Penn State University; Atomic Energy Council, ROC; PSU
Radiation Science and Engineering Center, University Park, PA 16802

## ABSTRACT

In testing a patented concept for a real time radioiodine monitor designed for operation under nuclear accident conditions, the authors observed conditions which would render conventional real time radioiodine monitors ineffective. The observed conditions include the following:

1. The loss of radioiodine information due to the passages of larger amounts of radioactive noble gas through the system.

2. The loss in $^{131}I$ detection sensitivity when lead collimators are employed to reduce detector dead time.

3. Interference from $^{88}Rb$ produced by the decay of $^{88}Kr$ within the detector system.

4. The delayed release of $^{132}I$ by the decay of $^{132}Te$ long after the start of the accident, at a time in which both reactor and monitor radioiodine absorbers may have become saturated with water vapor and thus are unable to retain some of the possible chemical forms of this radioisotope.

---

## INTRODUCTION

For the last nine years, a monitor capable of on-line detection of iodine radioisotopes emitted from a nuclear power plant under accident conditions has been developed. This system is designed to measure in real time, trace amounts of radioiodine fission products in the presence of orders of magnitude higher concentrations of noble gases. The system has been tested under simulated accident conditions at three nuclear reactor facilities(1,2,3). In 1985-1986, a second version of the system(4,5,6) and in July 1989 an improved version of this system was tested(7).

In all of these tests, the authors have used both NaI(Tl) and HPGe gamma photon detectors like those commonly used in many commercial radioiodine stack monitors, including those used for routine monitoring and those used for accident conditions. They have observed a number of phenomena which could occur to affect the operations of both types of commercial monitors when operated under reactor accident conditions. This paper contains a description of these phenomena.

## OBSERVED PROBLEMS

### Problem 1

Response of detectors to high concentrations of noble gases passing through the system.

During the 1985-86 ANL tests, conditions existed whereby relatively large amounts of radioactive noble gas passed the system during the intake mode operation, approximately 0.5 curies to 0.9 curies. The major components of this gas were $^{135}Xe$ and $^{88}Kr$. Figures 1 and 2 show two successive NaI(Tl) spectrum measurements (i.e., an intake mode spectrum). In the intake mode, the counting chamber contained about 70 μCi of noble gas while during the purge mode the chamber contained less than 0.1 μCi of noble gas. The purge mode spectrum shows the normal undistorted NaI(Tl) spectrum giving a pulse height analysis dead time of less than 1%. The sample mode spectrum shows the whole spectrum as a continuous curve and had a dead time of about 42%. In this test the system was set to view the 1260 keV photopeak of $^{135}I$. As can be seen from these figures, during the sampling mode, the detector output would give a very large but false value for the $^{135}I$ content. In this example $^{135}I$ was not seen in the purge mode because of its low concentration and the presence of $^{88}Rb$, a problem discussed below, but the resulting false $^{135}I$ purge mode value was at least a factor of 100 less than the sample mode value.

Under the same conditions, the HPGe detector was completely shut down, but about one second after the start of the purge mode, the counting chamber was swept clear of the radioactive noble gas and the detector worked satisfactorily, producing acceptable gamma spectra.

#### Conventional System Ramifications

Conventional radioiodine stack monitors employing NaI(Tl) detectors operated without the benefit of a purge mode. Thus, under conditions of moderate to high amounts of noble gases through the system, these systems will indicate the presence of much higher concentrations of the radioiodine of interest than actually exists in the vent gases. An even more serious problem may exist for those systems employing HPGe

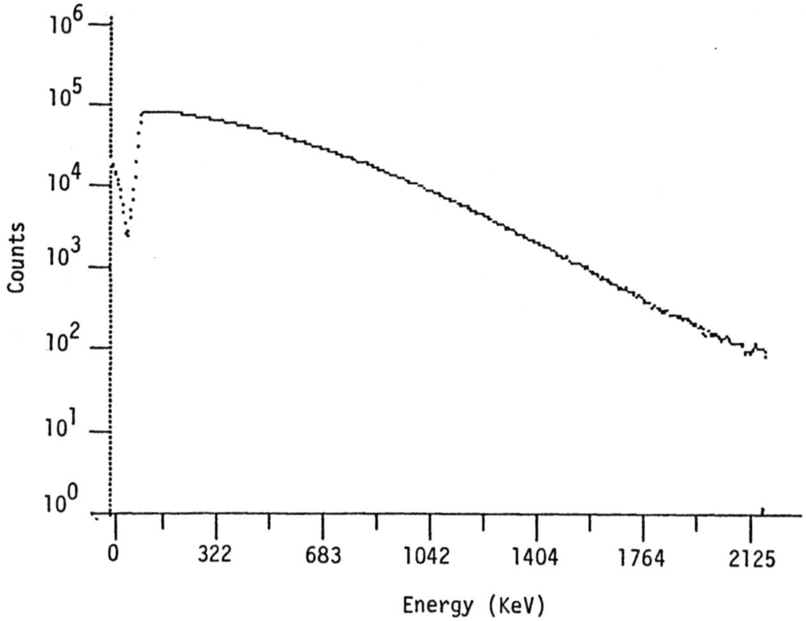


Figure 1. Typical NaI(Tl) Intake Mode Spectrum

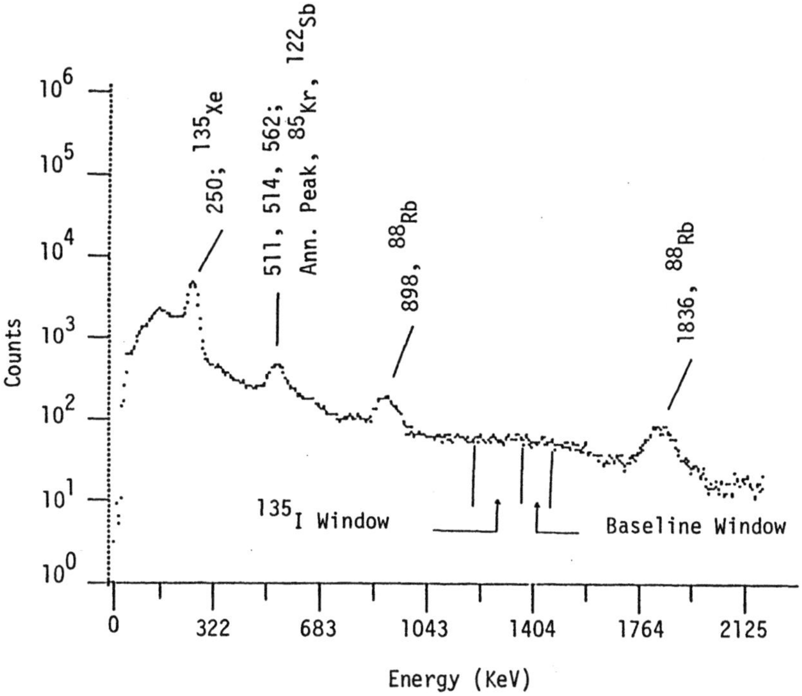


Figure 2. Typical NaI(Tl) Purge Mode Spectrum

detectors, since at high noble gas concentrations these systems may shut down, giving operational personnel no immediate information about iodine release rates.

## Problem 2

### Effect of lead collimators on $^{131}$I detection efficiency

Sometimes in both tests lead collimators were used to reduce dead time losses. Detector efficiencies for both the two 2"x2" iodide and the 50 cc germanium detector were determined using 1" (2.54 cm) thick lead collimators. The collimator for the NaI(Tl) contained a 1/8" hole (0.3175 cm) while the collimator for the HPGe detector contained a 1/4" hole. Figures 3 and 4 show the efficiency curves with and without the collimator, respectively. These figures show that the lead collimators cause a rapid fall-off of efficiency with decreasing photon energy, below 800 keV, due to photoelectric absorption.

### Conventional System Ramifications

The significance of this effect is that most utility iodide monitors are set to monitor the release of iodine-131. Thus, if lead collimators are used to prevent the saturation of the system under accident conditions ,the detection of iodine-131 by its major photopeak (364 keV) will be considerably reduced relative to the detection of radioiodine isotopes having major photopeaks greater than 800 keV (i.e, $^{132}$I, $^{134}$I, and $^{135}$I).

In case a lead collimator is used, then at least for the first 24 hours the system should be set up to detect the 1.14 MeV and/or the 1.28 MeV photopeaks of $^{135}$I and not the 364 keV photopeak of $^{131}$I. After that time, the lead collimator(s) could be removed and the system adjusted to monitor $^{131}$I.

## Problem 3

### Interference from $^{88}$Rb produced by the decay of $^{88}$Kr within the detection system

A surprise result of our 1985-86 test was the discovery that the major radionuclide collected on the silver zeolite filter was $^{88}$Rb (Figure 2). This radionuclide is the short-lived daughter ($t_{1/2}$ = 2.8 hours). Since $^{88}$Rb should exist as a particle, it was initially believed that these particles were penetrating the inlet HEPA filter. Subsequent study showed that all of the $^{88}$Rb collected on the silver zeolite filter could be explained by $^{88}$Kr

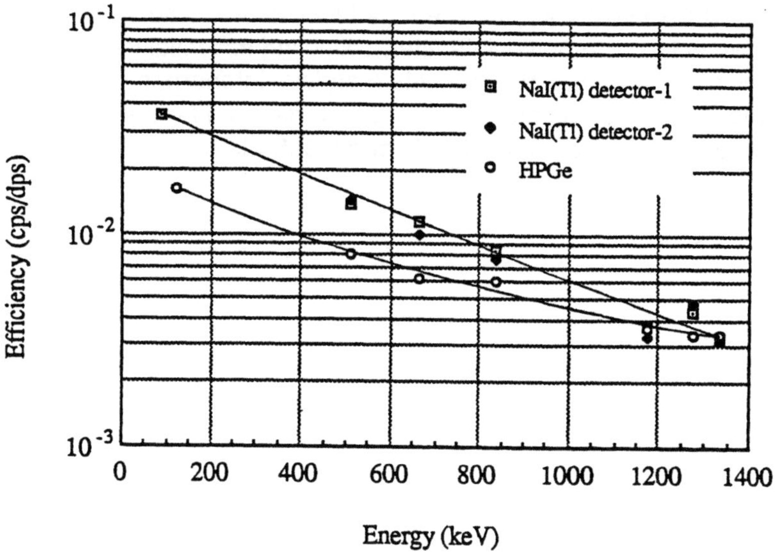


Figure 3. Radioiodine Monitor Detector Efficiencies Without Lead Collimator

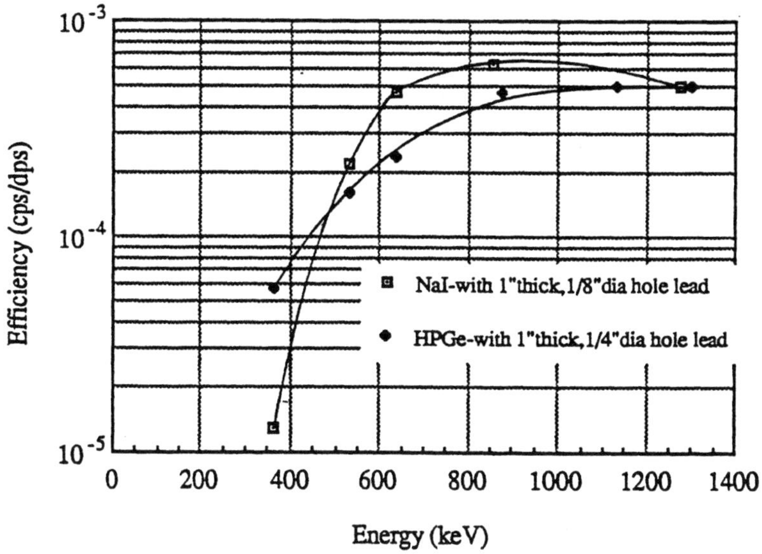


Figure 4. Radioiodine Monitor Detector Efficiencies With Lead Collimators in Place

decay in the sample lines between the two filters(7). The silver zeolite filter then acted as a particulate filter, collecting essentially all of the $^{88}Rb$. As shown in Figure 5, $^{88}Rb$ presents a serious interference for $^{131}I$, $^{133}I$, $^{134}I$, and $^{135}I$. It also presents a serious interference for the major photopeaks of $^{132}I$, i.e., 0.67, 0.773, and 0.995 MeV.

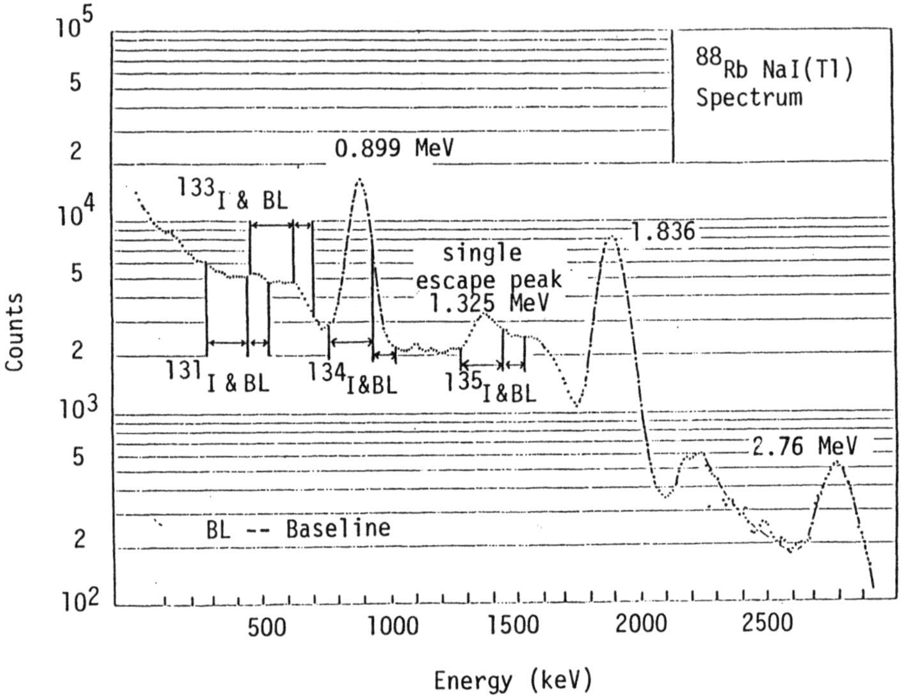


Figure 5. NaI(Tl) Spectrum of $^{88}Rb$ Showing Its Interference with $^{131}I$, $^{133}I$, $^{134}I$ and $^{135}I$

A system modification was made to reduce, by over a factor of 10, the volume of the gas flow line between the HEPA filter and the silver zeolite filters, reducing the magnitude of the problem by the same factor.

## Conventional System Ramifications

There are three noble gas-decay alkali metals which under conditions of very high radioactive noble gas to radioiodine ratio, could contribute to interferences in radioiodine monitoring by conventional systems. For about the first hour after an accident, $^{89}Rb$, from the decay of $^{89}Kr$, could be a significant interference. After that time, $^{88}Rb$ and to a lesser extent $^{138}Cs$ produced by the decay of $^{138}Xe$, could be the major contributor. The collection of these alkali metals on the iodine monitor filter would give false high results on radioiodine releases until the filter is removed and counted on a high resolution detector.

## Problem 4

### The potential late release of iodine-132 and its iinterference with the detection of other iodine radionuclides

To characterize the various airborne radioiodine species under accident conditions, experiments were conducted in 1989 at a radiopharmaceutical production facility's hot cell laboratories. Surprisingly, we found that the dominant form of radioiodine was $^{132}I$ ($t_{1/2}$ = 2.3h) and not $^{133}I$ ($t_{1/2}$ = 20.3h). But while the chemical distribution of $^{131}I$, $^{133}I$, and $^{135}I$ were quite similar, the chemical distribution of $^{132}I$ was quite different, being mostly in the methyl iodine form. This phenomenon was explained as follows. The $^{132}I$ was primarily not that released from the irradiated uranium, as is the case of the other three iodine radioisotopes. Instead it was being released from the waste storage container being produced in the waste by the decay of the fission produce $^{132}Te$ ($t_{1/2}$ = 3.5d).

### Conventional System Ramifications

Under nuclear accident conditions all of the radioiodine entering the reactor containment may not be released at the time of the initial disruption of the fuel. The decay of $^{132}Te$ can continually release $^{132}I$ for many days after an accident. Conditions could exist in which the $^{132}I$ concentration could be high enough that it might mask the $^{131}I$ detection by on-line radioiodine monitors. The delayed $^{132}I$ release will occur under different physical and/or chemical conditions compared to the radioiodine released during the initial stages of the accident, and thus, the delayed released radionuclide may be in a different chemical form than that of the initially released radioiodine. The chemical form difference was observed in the final LOFT test.(8) This delayed release may occur under situations in which both the plant and the monitor charcoal filters are saturated with moisture so that significant amounts of $^{132}I$ might escape from the plant without detection. The solution to this monitoring problem is to replace the radioiodine filters on a regular basis so they do not become saturated with moisture and radioiodine compounds.

## CONCLUSION

Current plant radioiodine monitors designed to detect radioiodine concentration in real time are inadequate to provide data under many types of conceivable accidents. The authors' patented design using alternating sampling and purging modes will overcome many of these problems, but this system will require filter replacement to prevent loss of system sensitivity with time after an accident.

## REFERENCES


1. T.T. Tseng, "Concept Evaluation of a Radioiodine Monitor Capable of Operating in High Levels of Radioactive Noble Gases," M.S. Thesis, The Pennsylvania State University (1982).

2. T.T. Tseng, W.A. Jester, A.J. Baratta, I.B. McMaster, and D.W. Miller, "The Development and Testing of a Prototype On-Line Radioiodine Monitor for Nuclear Power Stations," Health Physics, Vol 50, No. 1, pp. 65-72 (1986).

3. W.A. Jester, I.B. McMaster, and A.J. Baratta, "Reactor Radioactive Emission Monitor," U.S. Patent 4,663,113 (1987).

4. T.T. Tseng, "A Study of the Interferences With the On-Line Radioiodine Measurement Under Nuclear Accident Conditions," Ph.D. Thesis, The Pennsylvania State University (1986).

5. T.T. Tseng and W.A. Jester, "The Penetration of Noble-Gas Decay Produced Alkali Metal Aerosols Through HEPA Filters," Proceedings 4th Internat. Symposium on Particulate Phenomena and Multiphase Transport, ISBN-0-89116-817-6, 4, pp. 163-168 (1988).

6. T.T. Tseng, W.A. Jester, and G.R. Imel, "Interferences to Conventional Radioiodine Air Monitors Caused by Noble-Gas Decayed Alkali Metal Aerosols Under Nuclear Accident Conditions," Proceedings of the ANS First Regional Conference, ISBN-0-89448-130-4, pp. 47-51 (1987).

7. B.S. Lee, W.A. Jester, and J.M. Olynyk, "Iodine Speciation in Hot Cell Effluent Gases," Transaction of ANS 1990 Annual Meeting, ISSN 0003-018X, A1, pp. 41-42 (1990).

8. J.W. Mandler, J.W. Tkachyk, J.T. Case, J.H. McDaniel, and H.K. Peterson, "In-situ Measurements of Efficiency of Silver Zeolite for Removal of Airborne Radioiodine Species," 19th DOE/NRC Nuclear Air Cleaning Conference, August 18-21, Seattle, Washington, U.S. Department of Energy, pp. 159-166 (1986).

# UMMARIES OF WORKSHOP MEETINGS

## SUMMARIES OF WORKSHOP MEETINGS

Nine workshop meetings were held at the symposium to foster an informal exchange of information on topics of major interest to the participants. Each workshop lasted over two hours and ran currently in groups of three. A survey was conducted before the symposium to determine workshop topics and to produce a schedule to allow each participant to attend as many workshops of primary interest as possible.
Each workshop had two chairman who had the responsibility of organizing the workshop, including defining the scope and relating the workshop to papers presented at the symposium. The program schedule was designed so that the workshop discussion would build upon the presentations. After the workshop the chairmen produced summaries of the discussions and conclusions. These summaries are presented in this section.

### Workshop on Adjustment Methods, Cross-section Files, Uncertainties and REAL88

Chairmen: M. Matzke (PTB)
F.W. Stallmann (ORNL)

The workshop was opened by remarks of F.W. Stallmann. He pointed out that presently regulatory and licensing, agencies in the USA as well as in Europe intend to require quantitative statements about calculated fluences that can be validated by dosimetry measurements using statistical means and, perhaps, adjustment procedures. This places a large responsibility on researchers in the field of adjustment procedures to assure that only scientifically valid methods are permitted or required for regulatory purposes.

Subsequently the following topics are discussed:

IRDF90. N.P. Kocherov presented the current status of this file (He had also given a paper concerning IRDF90 at this symposium). He mentioned that some type of covariance files (LB=8) cannot be processed by the current software, but this problem will be rectified soon. In the discussion interest was expressed for inclusion of U235 and, perhaps, Pu239 fission spectra, but at present, only new data for the Cf252 fission spectra are available. M. Matzke gave a brief summary of the evaluations performed by M. Wagner et al. (to be published in Physics Data). These evaluations cover 14 dosimetry cross-sections that are being included in the IRDF90 file. Of particular importance is the new evaluation of the Nb93(n,n') data that show a significant decrease around 3 MeV compared to the old data.

LEPRICON adjustment procedure. This code has been released to the public, as announced by R.E. Maerker at his presentation at this symposium. He presented additional data at this workshop concerning primarily corrections to the iron inelastic cross-section by C.Y. Fu. Further improvement is to be expected from the use of the new ENDF/B-VI cross-sections. The question was raised why European researchers appear to be able to calculate cavity fluences more accurately than their U.S. colleagues. It was mentioned that in Europe older iron cross-sections were used which are closer to the ENDF/B-VI values.

REAL88. E.M. Zsolnay reported on the present status of the REAL88 amplifying on her talk on the same subject at this symposium. A small number of participants have revised the original REAL84 input data to provide a consistent data base for interested users. This data base contains 6 benchmark neutron spectra and can be obtained from the IAEA on request. Also collected were 3 adjustment codes, two of which are derivations from STAYS'L, the third being LSL-M2. It is intended to add LEPRICON to the collection. In the discussion potential problems with updating the REAL88 input data while preserving consistency were mentioned. Further support from the IAEA on that topic was required.

GENERAL DISCUSSION. J. Williams pointed out that for many experiments in test records calculated fluences are difficult to obtain and the neutronics characterization must, therefore, rely primarily on dosimetry. It may be difficult, under these circumstances, to uphold the high standards that are required.

**Workshop on gamma-ray dosimetry**

Chairmen: J.A. Mason (A.N. Technology LTD)
S.Q. King (Nuclear Service Company)

The workshop consisted of several presentations on specific topics of interest to the participants and the presentations were followed by wide ranging discussions on different aspects of gamma-dosimetry.
The first presentation was given by the chairman on behalf of Dr. Ray Gold who was not able to attend the symposium. He has provided notes from which the chairman spoke.
In the presentation Dr. Gold cited the outstanding discrepancy between gamma-ray spectrometry measurements and calculations for the 12/13 configuration of the PCA. The measurements were made in 1981 as part of the LWR PVSD improvement programme. He noted that C/E ratios for Nuclear Research Emulsions (NRE) and gamma-ray spectrometry are approximately 0.5. The issues were discussedat the workshop in Raleigh N.C.
In 1986 the matter was not resolved. Subsequently data from the NESDIP replica benchmark at Winfrith in the U.K. have shed light on the issue. The NRE and SSTR fission rate ratios for the PCA 12/13 configuration are roughly a factor of 2 higher than those observed in the NESDIP replica. This result would appear to explain the 0.5 C/E ratio for the PCA. The conclusion which is drawn from the result is that the 1981 NRE an gamma-ray spectrometry measurements were not made in the 12/13 configuration. The data should therefore be ignored and not used in further comparisons.
Dr. Gold proposed that the status of the gamma-ray spectra characterization of LWR-PV environments should be assessed from the measurements based on the 4/12 SSC configuration. Here TLD dosimetry and gamma-ray dosimetry based on gamma-ray spectrometry agree within the experimental uncertainties for these methods, to better than approximately 10 per cent at all locations of the 4/12 SSC configuration. These 4/12 SSC configuration gamma-ray spectrometry results agree within the uncertainties with calculations performed at CEN/SCK Mol, except in the void box region where inadequacies of the model are suspected.
Dr. Harry Farrar IV reported on an initiative to write standards for dosimetry for radiation processing. Control and measurement of absorbed gamma dose is important in the field of radiation processing.

Radiation processing is a rapidly developing technology with numerous applications in food treatment, sterilization and polymer modification. the effectiveness of the process depends, however, on the proper application of the gamma dose and it's measurement.These aspects are being considered by a wide group of experts from around the world who have joined together to write a comprehensive set of standards for dosimetry for radiation processing. This group has now expanded into a full subcommittee of ASTM, E10-01 and the scope of the standards under development includes dosimetry for all forms and applications of radiation processing. Dr. Farrar described the standardization program, and discussed some of the technical aspects of radiation processing dosimetry . He welcomed participation in the committees activities.

Dr. Norm Baumann described recent work on gamma-ray induced displacements in iron. At the sixth symposium at Jackson Hole, one of the concerns raised was that, all relevant radiation damage mechanisms might not be included in damage estimates. One such mechanism, gamma-ray interaction with iron was shown by R. Gold and W.N. McElroy to be neglectable for a specific LWR for which they had data. They emphasized, however, that the effect could be significant for reactors with high thermal-to-fast fluence rate ratios. N. Baumann has subsequently shown that the effect is indeed significant for tank wall sections in SRS (Savannah River Site) heavy water production reactors where up to 15 per cent of total damage results from this mechanism. For this evaluation, he developed a set of displacement cross-sections for gamma-rays on iron. These cross-section provide a convenient way of including damage estimates from this mechanism via any computation of neutron damage using a coupled neutron/gamma-ray transport code. Dr. Baumann described the work which is also the subject of his paper.

Dr. Peter Heffer reported on the use of TLD's in Nuclear Electric, citing the efforts of T.A. Lewis. He discussed the need for an accurate and reliable method for determining gamma fluence over a wide dynamic range and on the various TLD's that have been considered.

Dr. Albert Fabry described recent discrepancies between calorimetry measurements and calculations in the reflector of the BR2 reactor where the results of the calorimetry measurements are greater by a factor of 4 .

Dr. John Mason noted some of the types of errors which can arise in calorimetry measurements, which were the subject of his poster presentation.

Dr. Andrew Holmes-Siedle revealed in this workshop the need for a reliable gamma fluence dosimeter for use in the 1 krad to 1 Mrad range.

Dr. Holmes-Siedle was prompted by this to describe the principle of the RADFET technique, (Radiation Field Effect Transistor). This device employs the space charge build-up in oxide films on silicon as the gamma energy sensitive property. In the "zero loss " mode , the RADFET has a sublinear response function which allows it to cover a dynamic range from ca. 100 rad to 10 Mrad. This appears to fill the gap in reactor gamma dosimetry field. Neutron damage effects in the silicon can be tolerated up to about 1.E+16 fast neutrons / cm2. Thermal neutrons do not cause a problem.

Dr.Fabrizio Cleri of ENEA described recent gamma-ray energy deposition measurements at the Masurca fast critical assembly. He described the calculation and the use of TLD's in the measurements. He noted a significant and as yet unresolved discrepancy in the C/E ratios for measurements in the plutonium core of the reactor.

Dr. Bill McElroy noted that the use of thermoluminescence dosimeters TLD's for surveillance is not common, but these monitors should be mentioned as a technique for the determination of gamma exposure and heating rates. ASTM standard E668 covers the use of TLD's for the range of 0. to 1.E+6 rad. TLD's were used by Fabry et al. to determine measured values of gamma heating in the VENUS PWR mock-up. BeO TLD's were used by Heffer and Lewis for measuring gamma exposure rates, but their usefulness is limited by their effective range. Gilliam and King have used LiF chip gamma detectors successfully in the measurement of MegaGray range gamma fluence in a B&W 177 reactor cavity. When available, the combined results of the VENUS and the DAVIS BESSE 1 benchmark studies should be carefully reviewed and compared, and consideration should be given to presentation of a new state of the art ASTM LWR surveillance standard covering the use of gamma fluence dosimeters.

**Workshop on Neutron dosimetry with Niobium**

Chairmen: B. Baers (Technical Research Center of Finland)
J.W. Rogers (EG&G)

The workshop was attended by 27 people and there was very good participation with several presentations, discussions and comments on the subject. In general, the status of the use of niobium as fast neutron monitor is apparently expanding based on the papers presented at this meeting. The status of the 93Nb(n,n')93mNb nuclear data was discussed and the K X-ray emission probability, Half-life and integral fission spectrum averaged cross-sections are considered to be well established with no major discrepancies. Currently the "best" values are found to be:

| | |
|---|---|
| .K x-ray emission probability | 0.1112 +/- 0.0022 |
| .Half-life | 16.12 +/- 0.16 years |
| .235U fission spectrum averaged cross-section | 0.145 +/- 0.006 barn |
| .252Cf fission spectrum averaged cross-section | 0.148 +/- 0.007 barn |

The energy dependent cross-section has been recalculated by nuclear models (by Odano) with significant improvements in the agreement with the measurements. Good agreement between the calculated and measured fission spectrum averaged cross-sections exists when the IRDF-90 evaluation is used (by Wagner et al.). It appears that the accuracies are sufficiently good for use in spectrum adjustment applications.

Some users (Lloret) are finding consistent results with other fast neutron monitors which respond in a similar energy range such as 54Fe(n,p) and 238U(n,f). One comment by McGarry (NIST) indicated there have been some discrepancies observed when compared with 237Np(n,f) in benchmark fields with steel and water environments but in the fission spectrum fields the discrepancy does not exist.

The discussion on the preparation of niobium dosimeters centred around the purification chemistry associated with the use of steel material samples and each method needs to be documented and published so that other users have better information on the laboratory details.

Counting sources consist of deposits from dissolved material, solid foils and liquid scintillation solutions. Wires are frequently used but no direct counting of wires has been developed. Sufficiently accurate mass attenuation corrections can apparently be made for "thick" foils.

Counting equipment in use consists of germanium and silicon detector based spectrometer systems with one user (Martin ) saying that acceptable results could be obtained from high activity sources using germanium detectors without Be windows. Liquid-scintillation counting provides some discrimination and correction possibilities against interfering activities.
The possibilities to use laser resonance ionization spectroscopy techniques for the direct measurement of the 93mNb/93Nb ratio have been investigated in Finland with proposals to use several two or three level excitation methods. So far no measurements on radioactive samples have been conducted. A number of applied and potential methods for determination of small masses (at
the micro gram level) of niobium was reported in a paper by Baers. The available equipment, contamination problems, etc. may determine the method to be used (emission and absorption spectrophotometry. mass spectrometry, polarography, x-ray fluorescence in niobium induced by external electron, gamma or x-ray radiation).
The use of the total x-ray reflection method for measurements of niobium down to the ppb level was reported by Hegedus. Together with the sensitive liquid scintillation radioactivity measurement technique these mass determination methods provide the possibilities to make accurate 93mNb specific activity measurements on samples of a large variety of origin.
A new 93mNb radioactivity standard is now being prepared in the UK and will be available in the near future. An interlaboratory comparison of niobium foil counting is being conducted in eastern European laboratories.
A revised DIN standard on the use of niobium has been issued in the FRG (with English version available). A new ASTM standard on the use of niobium (E-1297) has been issued in the US.
Measurement of the 93Nb(n,2n) fission (235U) spectrum averaged cross-section was reported at this symposium (by Willamson). A calculation of the 93mNb(n,g) cross-section by the Japanese indicates that there is no significant transmutation of this product in pressure vessel surveillance applications.
Several papers at this symposium (15) dealt with the use of niobium as a fast neutron monitor in one way or another. Although no specific recommendations came from this workshop it is anticipated that more data and experiences in the use of niobium will be forthcoming at future symposia an in the literature.
Dr. Hegedus should be saluted for initiating the use of niobium (25 years ago) and many others for their efforts to refine and develop the data and techniques.

**Workshop on LWR surveillance dosimetry**

Chairmen: J.J. Wagschal (Racah Institute of Physics)
E.P. Lippincott (Westinghouse)

The LWR surveillance dosimetry workshop was attended by about thirty participants. The workshop had four presentations and each was followed by a lively discussion in which many people participated.
The first presentation was given by E.P. Lippincott who talked about exposure calculations. His presentation reflected Stan Anderson and his experience. He pointed out that "as built " dimensions may not be equal to the nominal design values, that one has to consider changes in fuel composition and water density and that all these call for additional measurements in the cavity. Any work that has to result in a defendable estimate of the damage or fluence , to the PV, has to be a continuous comparison of calculation and measurements.

The second presentation by E.M. Zsolnay described the surveillance program of the four 440 MWe PWR's at Paks in Hungary. She mentioned, among other things, the Monte-Carlo fluence calculations, near capsules in the surveillance position and in the cavity, and the shift to low leakage cores in units 1 and 2. The new capsules all contain extensive dosimetry for the characterization of the neutron spectrum. She came to the conclusion that there is adequate safety in these plants as embrittlement is concerned.
H. Ait Abderrahim discussed the trend of the C/E as a function of penetration depth in benchmarks and in a Belgian PWR. The different trend in the PWR cavity C/E ratios could be explained by errors in the calculation of the backscattering from the concrete walls outside the cavity. The presentation was followed by Ait Abderrahim's proposal for a " concrete benchmark experiment" and by a lively discussion of the relative importance of calculation and measurements. The conclusion is that one has to be careful in taking various effects and uncertainties into account in the calculations, and these have to be anchored by agreement to measurements at surveillance points in an out of the pressure vessel.
The last presentation by H.C. Mehner described the situation of the four 440 MWe PWR's in East Germany. Out of the four units only one is operating now, and two have undergone annealing. Mehner described also a technique to use multi component detectors by using a Ti, Fe, Ni, Co alloy for a surveillance in reactor cavities.

**Workshop on dosimetry for irradiation facilities at test and research reactors**

Chairmen: D. Beretz ( SRS/SAPR-CEN)
J. Williams (Nuclear Reactor Lab. )

The motivation for this workshop topic was to discuss the large variations which exist between reactor facilities in their provision of dosimetry information to users of their facilities.Some users are accustomed to performing dosimetry measurements and neutron calculations, but the majority, especially those involved in other areas than reactor materials, do not have the knowledge or resources needed. It is recognized that facility operators must accept the responsibility to provide information on their irradiation facilities, then what is the extent of that responsibility?

Topics included in the workshop are:
Neutron and photon fluence and spectral measurements, calculations of the same quantities, and perturbations that may change them from the reference value.
Experiences of workshop participants from French CEA (Grenoble and Saclay), from Belgium (BR2) , The Netherlands (Petten) and from the USA (EG&G, Idaho, Sandia NL , White Sands Missile Range and Battelle, NM) reflected differences due to the reactors' characteristics and due to different applications. Common problems were encountered, however in providing calculated spectrum at position outside the reactor cores and in regions which are perturbed by the presence of experiments. In some cases irradiations are performed in positions for which no calculations are available, or for which the calculational accuracy is poor. In such cases the characterization depends heavily on measurements. Problems due to non-linearity of adjustment procedures may arise when large adjustments from starting neutron spectra are needed. The validity of adjusted spectra was questioned in such cases.

Participants from Saclay advocated the use of damage monitors in mock-ups instead of or besides spectrum determination. This was supported by Sandia participants who suggested the use of silicon damage monitors for inclusion on spectrum adjustments for electronics testing irradiations.
Discussion of the problem of perturbations due to experimental assemblies raised the question of the facility operator's responsibility to account for these. In the case of perturbations due to core changes or neighbouring experimental installations, the facility operator's are responsible for controlling and limiting these and providing information to users. It was also suggested that the operators should provide characterization of unperturbed cavities into which experiments are installed, but that users must account for perturbations within their own experiments.
T.F. Luera (Sandia) presented a proposal protocol for reactor testing of electronics materials. A key recommendation was that facility operators must be "willing and able to consistently expend the resources to do the job right and be able to prove it year after year". Specificly, they should supply a spectrum determination for every advertised configuration, and this should be based on experiments that include either fission reactions or damage detectors. S.C. Apple from Arkansas Tech. Univ. provided information that it is currently impossible for non-government laboratories in the US to obtain fission monitors. The lack of standards that provide guidance to facility operators was discussed. Some of those responsible for providing irradiation facilities indicated that they would welcome such guidance.

C. Ertek (IAEA) expressed the opinion that it was urgent to rectify the lack of standards. ASTM E10.07 subcommittee is in process of developing a guide for use in electronics testing, which can be useful to broader applications.

A. Holmes-Siedle (REM, UK) has drafted a standard on the same topic for the BS 9000 series. He will provide a copy to D.Vehar, chairman of E10.07. The situation for characterization of photon dose and spectra was raised. No good solutions to this problem were proposed, and it was pointed out that none of the standard methods applicable to gamma dosimetry at high doses has been validated for use in the presence of reactor neutrons.

A general conclusion of the workshop was that there are major issues which need to be addressed in the area of dosimetry for facilities and that standards development is urgently needed.

**Workshop on Radiation Damage Correlations**

Chairmen: A. Fabry (SCK/CEN)
A.L. Lowe Jr. (Babcock & Wilcox)

The workshop has focused on embrittlement of LWR pressure vessel steels. The issues proposed for discussion at the onset were as follows:

1. "Indexation" of radiation damage
2. Neutron fluence dependency
3. Microstructure
4. Chemistry
5. Environmental effects: dose rate vs. spectral vs. irradiation temperature effects
6. Thermal aging
7. Standards

The last topic has been added upon request of W.N. McElroy. This agenda has been accepted by the audience, after a final comment by R. Dierckx: links with fusion program concerns and related learnings should be covered as appropriate in a perspective of possible cross-fertilization.
Topic 1, "Indexation", was described here as encompassing two related types of questions in the context of the relationship between the steel microstructural (i.e. microscopic) changes and the resulting (macroscopic) changes of mechanical properties:

a) How far can different engineering tests (Tensile, hardness, Charpy-v, compact tensile,...) be considered equivalent to develop improved predictive embrittlement correlations?

b) When interpreting the results of a given test type, can one always adopt engineering/ regulatory definitions while still properly tracing the microstructural causes of fracture? Emphasis was placed on the nil-ductility transition temperature (NDT), i.e. on the risk of brittle fracture. The answer given to the above questions was prudently negative. For ex., it was shown that tensile yield changes do not automatically reflect NDT shifts and that the 30 ft.lb energy level in the Cv-test is as well an inadequate indexation in some cases ( such as: Whenever the upper shelf energy is lower than about 60 ft.lb).

Topic 2, Fluence dependency, opens to questioning the usual practice of adopting power laws that do lump together different damage mechanisms (copper precipitation vs. "matrix damage", for ex.), each characterized by their own kinetics. i.e. fluence dependencies. How does this impact on the assessment of: chemistry correlations, dose rate effects,...?

Topic 3, Microstructure, challenges the neglect of this parameter in current engineering correlations. Examples were discussed in which either the prior-austenite or the ferrite grain size ( depending on the heat treatment) do play a significant role.

Topic 4, Chemistry, encompasses such concerns as how much of a given detrimental element (e.g., copper) is already precipitated in the unirradiated steel (so that correlation to the bulk content is biased). Little interest into this area was expressed by the present audience.

Topic 5, Environmental effects, did by contrast raise much interest, probably because it is the most familiar to the participants. Recent experimental results on dose rate effects by Hawthorne & Hiser (Report NUREG/CR-5493, MEA-2376, 1990, finalizing ASTM-STP-1001 p. 535) are synthesized and amply commented by McElroy. long the same line, Manahan briefly describes his earlier modelling work, which predicted these effects, while Fabry indicates that his previous evaluation (as highlighted in ASTM-STP-1001 p. 17), is being updated for improved fluence dependency and this leads to dose rate effects significantly less pronounced than before. As stressed by Mc.Elroy, the outstanding question remains: How can one quantify better the still lower flux regimes within actual vessels?
Spectral sensitivity does complicate the picture further. The paper by Heffer et. al. relative to Magnox RPV submerged-arc welds present very strong experimental evidence in support of dpa as appropriate exposure unit, but Lowe's evaluation of the Gundremmingen and PSF in-vessel attenuation data suggests that fluence › 1 MeV is more adequate for PWR conditions. According to Hrabal's paper, the effects of thermal neutrons in the damage process seem to been have underrated, which would explain why the "surprise" of anomalously large low temperature embrittlement at HFIR vessel is not found by the Shippingport neutron shield tank examination (thermal-to-fast flux ratio large at HFIR as compared to PWR environments of

interest). Furthermore, irradiation temperature effects are shown to depend on the material fabrication, microstructure, copper content in solid solution, etc.,, Clearly, experiments aimed at specifically isolating all these mutually interacting environmental parameters are a must in such complex context: The paper by Alberman et. al., gives an excellent example of such initiative (In this planned experiment, the ratio dpa / fluence >1MeV is varied by about a factor of 2, while all other factors are fixed).

Topic 6, Thermal aging, was not seriously addressed at this workshop.

On topic 7, Standards, the ASTM guide " Proposed ASTM standard methods for application and analysis of standard reference materials data in nuclear reactor pressure vessel surveillance", prepared by M.P. Nanahan, ( now being submitted for balloting) was discussed. This entails steel correlation monitors (A302 ASTM reference plate, HSST plate 02). It is recommended in particular that an inventory be undertaken to locate the remaining specimens, primarily the un-irradiated ones, the irradiated ones stored without testing, and as feasible, available broken remnants.

## Workshop on reactor life time assessment

Chairmen: A. Alberman (CEA/CEN)
E.D. McGarry (Nat. Inst. of Stand. & Techn.)

The aim of the workshop was to examine the status of international efforts to address reactor life-time assessment and to determine requirements for future research. Discussions revealed that the situation is different in almost every country but that common problems allow identification of three necessary directions:

1) Evaluation of major reactor components to withstand radiation damage during expected lifetime;
2) Better understanding of damage mechanisms to develop suitable irradiation parameters;
3) Improvement of in- and ex-vessel dosimetry.

With respect to these directions, the recommendations were as follows:

1a) It was recommended that more fracture toughness tests should be conducted whenever possible. There appears to be evidence that shifting the Kir curve by the Charpy dt30 is not conservative for some materials, fracture mechanics specimens should be included in future irradiations.

1b) There were numerous recommendations for the improvement in timely and permanent documentation of records of exposures of materials. This includes suggestions for archive specimens of the unirradiated material for future investigations. The recommendations included requests for information on all fuel burns, core designs, and all quality control documentation. In particular, there was a request for pre-irradiation contents of stainless steels.

2a) It was recommended that detailed microstructure evaluations be performed prior to the irradiation of materials, modelling efforts at some laboratories indicate that the initial microstructure condition can significantly influence the irradiation damage response. Furthermore, the microstructure of the copper content of a specimen changes with exposure and this also has effect on the post irradiation properties of ferritic steels. It was also recommended to develop microstructure modelling to predict mechanical behaviour.

2b) There were recommendations for work on dpa damage correlation. Spectrum sensitivity to embrittlement should be further developed through experimental studies.

2c) It was recommended that the question of possible dose rate effects be addressed and resolved before it is time to make decisions on specific plant life extensions. In support of this was further the recommendation to report both fluence and fluence rates associated with each report of metallurgical irradiation damage data.

3a) The discussions of dosimetry efforts centred on recommendations for more and better dosimetry. Advanced methods should be implemented in relationship to evaluation techniques such as scraping, replacement surveillance capsules eventually equipped with irradiated Charpies and tensile specimens reconstituted by laser welding techniques.

3b) Retrospective dosimetry through specific radioactive and stable isotopes created by irradiation could be a promising future technique.

3c) Ex-vessel cavity dosimetry performed on a cycle-by-cycle basis, such as carried out by SSTR programs, should be of major importance for monitoring exposures from low-leakage cores. In conclusion, it appears that life-time prediction could be improved through development of plant-specific trend curves, established through accurate dosimetry. The establishment of these trend curves require damage correlation studies and more experimental work on possible dose rate effects.

**Workshop problems with activation detectors**

Chairmen: W.P. Voorbraak (ECN)
G.C. Martin (General Electric Comp.)

The workshop was attended by 31 people; 30 per cent of them reported on a small inquiry which was set up to prepare the meeting. Our aim was to exchange practical information between users of activation detectors, more than setting recommendations. Therefore the variation in irradiation conditions which were encountered by the attendants was too large.
Concerning preparation of detector sets no special problems were met. It was stated only that Copper and Al-Co detectors must be physically separated.
Cadmium is regularly used as shield for thermal neutrons for temperatures below 300 degrees Celsius, Cadmium oxide and Gadolinium above these temperatures. The shields were used to minimize activation of impurities in the materials by thermal neutrons. The attendants of the workshop encapsulated , if they did, their detectors in quartz, stainless steel or vanadium. Germanium detectors are widely used and have reduced our counting problems. In special cases liquid scintillators, sodium iodine well detectors and ionization chambers are used one should be aware of corrections to be made at low counting geometries because of gamma-gamma coincidences. All laboratories should have a good quality assurance by periodic check with e.g. an europium source. Also interlaboratory comparisons are still very useful.
Because of significant problems with the application of fast fission monitors, namely U238, Np237, and Th232, Nb is becoming the monitor of choice, even in the US, for characterization of phi >1 MeV of neutron spectra. However, we must still cope with problems from fission monitors which are still being irradiated and will be analyzed in the future.

Photonuclear reactions, that is (Gamma, fission) reactions can seriously compete with neutron, fission reactions. For deep water penetrations, specifically at pressure vessel locations where corrections of 30 to 40 per cent are needed for BWR's and 15 to 20 per cent for PWR's. These photo fission effects are, however, much less serious in PWR cavities and at accelerated locations. We must also be aware of build-up and interferences of Pu239, Pu238, and U233 in U,Np, and Th which can cause serious problems at moderately high fluences.
Fortunately, finally , a new Nb93m activity standard will soon be available from Harwell. The reactor power time history does not necessarily reflect the local power at an experimental location due to movement and substitutions of fuel bundles between cycles, and for movement of control rods within a cycle.
Errors up to 30 per cent in fluence rates have been observed from inaccurate power histories.
It is important in many cases to obtain the exposures of adjacent fuel bundles with time and also control rod configuration information from the reactor utility to access this local history. The solution to this problem, also, is not an easy one. Fission monitors, however, can be used as an aid in correcting for local power history through the measurement and normalization of reaction rates of fission products of different half-lives. Experience has shown that impurities in certain monitors and composition of certain alloy wires, especially Co in Al have given problems in the past. Inhomogeneity of alloys can be significantly different from nominal values. Now, quite pure Np237, U238 and Nb, as well as homogeneous alloys of Co, Au, and Ag in Al are available.

## Workshop on dosimetry for fusion applications

Chairmen: V. Sangiust (Politechnico di Milano)
L.R. Greenwood (Pacific Northwest Laboratory)

Agenda

| | | |
|---|---|---|
| 1. | Opening remarks | L.R. Greenwood |
| 2. | Nuclear data for fusion | E.T. Cheng (general discussion) |
| 3. | Calculation of nuclear data | C.Y.Fu (general discussion) |
| 4. | Neutron sources | R.L. Dierckx, F. Hegedus (general discussion) |
| 5. | Radiation damage | R.L. Dierckx, C. Morin (general discussion) |

E. Cheng (TSI Research) presented a list of nuclear reactions needed for fusion reactor dosimetry (see table 1). The list was prepared at the Sixth Coordination Meeting of the Division of Nuclear Data Needs of the Office of Fusion Energy in Athens, Ohio, 19-21 September 1989. (The complete report is available from E. Cheng.) These reactions must be short-lived to provide high specific activities for fusion diagnostics. The point was also made that some reactions are needed with thresholds above 14 MeV in order to monitor neutral beam performance to 17 MeV and to monitor runaway plasma conditions.

Concern was raised that some of these reactions may suffer interference from (gamma,n) reactions - more study is needed to determine these effects.
C.Y. Fu (ORNL) discussed computer codes to calculate nuclear reaction cross-sections. Due to long list of cross-sections required in fusion activation, it would require unreasonable effort and expenses to perform the state-of-the-art calculations,e.g., TNG, GNASH, for each reaction.
Simple codes (like ALICE or THRESH) have thus been used to estimate many reaction cross-sections in order to identify which reactions may be important. Such analyses were sponsored by the IAEA and approximately 256 reactions were identified as needing further study.
This work is now being done by the WHC-US, ECN Petten, and in the U.K. An OECD Task Force was also established following a meeting at ANL, September 1989, to calculate the reaction CO60(n,p)Fe60 (1.5xE06 y).
Eight participants will perform these calculations using the best codes available and the results will then be compared at the International Conference on Nuclear Data for Science and Technology in Julich in May 1991.
There was some discussion about how to calculate cross-sections above 100 MeV for spallation applications. There are several semi-empirical codes; however, the accuracy may only an order of magnitude.
R. Dierckx (JRC Ispra, Italy) discussed possible future sources including (d,t) 14 MeV, Li(d,n), spallation, and beam plasma facilities. Such a source is needed to test materials and nuclear reactions for fusion reactors. However, the prospects are not encouraging that such a source will be constructed. A Li-deuteride converter was also discussed for placement in a fission reactor.
However, concern was raised that the fluence rate might be too low if we want a ratio of 14 MeV / fission neutrons above 1 per cent.
F. Hegedus (PSI, Switzerland) discussed proton irradiation experiments (PIREX) with 590 MeV proton beams of about 20 microA. The experiments will study materials damage from high energy particles in small foils and measure transmutation by synchrotron -induced X-ray fluorescence techniques. Helium will also be measured and it was pointed out that helium production may be very high at energies above 50 MeV for accelerator sources.
C. Morin (CEA Saclay) described radiation damage experiments for Li tritium-breeder compounds in the HFR reactor (fluence rate about 2.E+14 cm-2.s-1 for 40 days). The main damage comes for the Li6(n,alpha)t reaction. Tritium will be measured and compared to results from 2D transport calculations. Preliminary results indicate that tritium calculations are too low (approximately 15 per cent). Damage calculations are difficult for these compounds.
R. Dierckx described radiation damage experiments at the Los Alamos Meson Physics Facility (800 MeV beam stopped in Cu target). A mass spectrometer is being used to measure transmutation products. Of particular note was the production of Ar39 from Nb and V. Such large bubbles could be especially damaging A list of activation reactions needed for spallation sources was mentioned in the Proceedings of the IAEA Advisory Group Meeting on the Status and Requirements of Nuclear Data for Radiation Damage and Related Safety Aspects, 19-22 September 1989, in Vienna.

**Table 1 List of high priority dosimetry cross-sections**

| No. | Material and Reaction | Energy Threshold | Reaction product Half-life | Review Comment |
|---|---|---|---|---|
| 1 | 16O(n,a)13C | ~2.4 MeV | - | 5 |
| 2 | 24Mg(n,p)24Na | ~4.9 MeV | 15.0 h | 1,2 |
| 3 | 27Al(n,p)27Mg | ~1.9 MeV | 9.5 min | 1 |
| 4 | 27Al(n,a)24Na | ~3.2 MeV | 15.0 h | 1 |
| 5 | 28Si(n,p)28Al | ~4.0 MeV | 2.24 min | 4 |
| 6 | 31p(n,p)31Si | ~0.73 MeV | 2.6 h | 1,2 |
| 7 | 34Cl(n,2n)34mCl | ~11.8 MeV | 32 min | 5 |
| 8 | 39K(n,2n)38K | ~13.4 MeV | 7.6 min | 5 |
| 9 | 47Ti(n,p)47Sc | | 3.4 d | 1 ? |
| 10 | 48Ti(n,p)48Sc | ~3.3 MeV | 43.7 h | 1 |
| 11 | 56Fe(n,p)56Mn | ~3.0 MeV | 2.6 h | 1,4 |
| 12 | 58Ni(n,2n)57Ni | ~12.4 MeV | 36 h | 1,2 |
| 13 | 59Co(n,a)56Mn | ~5 MeV | 2.6 h | 1,4 |
| 14 | 63Cu(n,g)64Cu | ~0 MeV | 12.7 h | 1 |
| 15 | 63Cu(n,2n)62Cu | ~11.0 MeV | 9.7 min | 1,2,4 |
| 16 | 64Zn(n,p)64Cu | ~2 MeV | 12.7 h | 1,2 |
| 17 | 64Zn(n,2n)63Zn | ~12 MeV | 38 min | 6 |
| 18 | 85Rb(n,2n)84mRb | ~10.7 MeV | 20.5 min | 6 |
| 19 | 90Zr(n,p)90mY | ~1.5 MeV | 3.2 h | 5 |
| 20 | 90Zr(n,2n)89mZr | ~12.1 MeV | 4.2 min | 1,2 |
| 21 | 103Rh(n,n')103mRh | ~0.04 MeV | 56 min | 1,2 |
| 22 | 115In(n,g)116mIn | ~0 Mev | 54 min | 1 |
| 23 | 115In(n,n')115mIn | ~0.8 MeV | 4.5 h | 1,4 |
| 24 | 197Au(n,g)198Au | ~0 Mev | 2.69 d | 1,4 |
| 25 | 197Au(n,2n)196Au | ~8.1 MeV | 6.2 d | 3 |
| 26 | 199Hg(n.n')199mHg | ~0.5 MeV | 43 min | ? |
| 27 | 235U(n,f) | ~0 Mev | Varied | 1,4 |
| 28 | 237Np(n,f) | ~0.5 MeV | Varied | 1 |
| 29 | 238U(n,n') ? | ~1.5 MeV | Varied | 2,4 |

1. Cross-section in IRDF-82
2. Being updated in Vienna for IRDF
3. New evaluation in Vienna for IRDF
4. New evaluation for ENDF/B-VI
5. Old evaluation, consider dropping
6. Evaluation needed, not in IRDF or ENDF/B.

## PARTICIPANTS OF THE SEVENTH ASTM-EURATOM SYMPOSIUM


Dr. H. AIT ABDERRAHIM
SCK/CEN - Reactor Physics
Boeretang 200
2400 M O L
Belgium
tel. 32-14-311801-2307
fax. 32-14-315021

Mr. A. ALBERMAN
CEA/CEN - SACLAY
DRE/SRO/SEN
91191 GIF SUR YVETTE
France
tel. 33-1-69 08 45 38
fax. 33-1-69 08 90 08

Dr. W.G. ALBERTS
Physikalisch-Technische
Bundesanstalt
Bundesalle 100
W-3300 BRAUNSCHWEIG
Germany
tel. 49-531-592-7300
fax. 49-531-592-4006

Mr. A.V. ALEVRA
Physikalisch Technische
Bundesanstalt
Bundesallee 100
3300 BRAUNSCHWEIG
Germany
tel. 49-531-592 7221
fax. 49-531-592 4006

Dr. M. ANGELONE
ENEA - CRE FRASCATI
Via F. Fermi
00044 FRASCATI-ROME
Italy
tel. 39-6-94005518
fax. 39-6-94005400

Mr. S.C. APPLE
Arkansas Techn. University
Engrg. Dept. - SCA
RUSSELLVILLE, AR 72801
USA
tel. (501) 968-0629
fax. (501) 968-0677

Mr. A. BALLESTEROS
Tecnatom
Km 19 Ctra. N.I. Madrid-Irun,
S.S. de los Reyes
28709 MADRID
Spain
tel. 34-1-6516700
fax. 34-1-6541531

Dr. M.F. BANHAM
AEA Technology, Fuel Services
Process Services Div.,
B 220, HARWELL LABORATORY
OXFORDSHIRE OX11 ORA
United Kingdom
tel. 44-235 821111 4736
fax. 44-235 43 4522

Dr. L.B. BÄRS
Technical Research Centre of
Finland/Reactor Laboratory
Otakaari 3A
SF-02150 ESPOO
Finland
tel. 358-0-4566340
fax. 358-0-4566390

Dr. BARAK BEN-SHAHAR
NRC-NEGEV
PO Box 9001
BEER SHEVA
Israel
tel.
fax.

Dr. P. BATISTONI
ENEA-CRE FRASCATI
Via E. Fermi
00044 FRASCATI-ROME, Italy
tel. 39-6-94005739
fax. 39-6-94005400

Dr. N.P. BAUMANN
Westinghouse Savannah River
Company
P.O.Box 616
AIKEN, SC 29802, USA
tel. 1-803-725 5296
fax.

Mr. D. BERETZ
SRS/SAPR - CEN GRENOBLE
B.P. 85 x
38041 GRENOBLE CEDEX
France
tel. 33-76 88 45 21
fax. 33-76 88 51 78

Mr. A. BEVILACQUA
DRE/SAPR - CEA GRENOBLE
BP 85X
38041 GRENOBLE CEDEX
France
Tel. 33-76 88 45 26
Fax. 33-76 88 51 78

Mr. J.J. BONNIN
SRS/SAPR - CEA
BP 85X
38041 GRENOBLE CEDEX
France
tel. 33 76 88 37 87

Mr. L. BOURDET
SERMA/LEPP - CEA
91191 GIF SUR YVETTE CEDEX
France
tel. 33-1-69-08-53-82
fax. 33-1-69 08 45 72

Mr. G.S. CARTER
Babcock & Wilcox Nuclear
Service Company
PO Box 10935
LYNCHBURG, Virginia
24506-0935
USA
tel. 1-804-385-2095
fax. 1-804-385-3663

Mr. C.A. CASADO
ENUSA - Nuclear Fuel Division
St. Santiago Rusinol, 12
28040 MADRID
Spain
tel. 34-1-3474230
fax. 34-1-3474215

Dr. E.T. CHENG
TSI Research
225 Stevens Ave.,Suite 110
SOLANA BEACH, CA 92075
USA
tel. 1-619-793-3567
fax. 1-619-793-0569

Dr. F. CLERI
ENEA
CRE CASACCIA
PO Box 2400 - Sacco 59
00100 ROMA A.D.
Italy
tel. 39 6 3048 3442
fax. 39 6 3048 6308

Dr. N.C. COURSOL
CEA/DTA/DAMRI
BP 6
91191 GIF SUR YVETTE CEDEX
France
tel. 33-1-69 08 27 78
fax. 33-1-69 08 26 19

Mr. F. CURRAN
Ohio State University
c/o Dr. T. BLUE
2091 Robinson Lab
206 W. 18th Ave
COLOMBUS, OHIO 43210
USA
tel. 1-614-292-0629
fax. 1-614-292-3163

Dr. R. DIERCKX
Commission of the European
Communities/JRC
21020 ISPRA
Italy
tel. 39 332 789438
fax. 39 332 789108

Mr. J.R. DUMAIS
Dept. of Nuclear Engineering
TOHOKU UNIVERSITY
Aramaki-Aza-Aoba, Aoba-ku
SENDAI 980
Japan
tel. 81-222-1800 4674
fax. 81-268-1539

Mr. C. ERTEK
IAEA
Wagramerstrasse 5
PO Box 100
1400 VIENNA
Austria
tel. 43-1-2360-2164
fax. 43-1-234564

Mr. A. FABRY
SCK/CEN Reactor Physics
Boeretang 200
2400 MOL
Belgium
tel. 32-14-311801-2267
fax. 32-14-315021

Dr. H. FARRAR IV
Rockwell International
Rocketdyne Division - NA02
6633 Canoga Ave
CANOGA PARK, CA 91303
USA
tel. 1-818-700-4355
fax. 1-818-700-3790

Dr. T.M. FLANDERS
Stews-NE
Nuclear Effects Laboratory
White Sands Missile Range
NEW MEXICO
USA 88002
tel. 505-679-9478
fax.

Mr. M. FOURGOUX
C.E.A.
21120 IS SUR TILLE
France
tel. 33-80 23 43 44

Mr. W.E. FREUDENREICH
ECN
PO Box 1
1755 ZG PETTEN
The Netherlands
tel. 31-2246-4281
fax. 31-2246-3490

Dr. C.Y. FU
Building CO/0 MS i 356
Oak Ridge National Lab.
OAK RIDGE, TN 37831
U.S.A.
tel. 615 574-6116
fax. 615-574-9619

Dr. A.J. FUDGE
AEA Technology, Fuel Services
Process Services Division
Bldg. 220, HARWELL
OXFORDSHIRE OX 1 ORA
United Kingdom
tel. 44-235-43-4635
fax. 44-235-43-4522

Dr. L.R. GREENWOOD
Battelle Pacific Northwest
Laboratory
MSIN P7-22
P.O. Box 999
RICHLAND, WASHINGTON 99352
USA
tel. 1-509-376-6918
fax. 1-509-376-1517

Dr. P.J. GRIFFIN
Sandia National Laboratories
P.O. Box 5800
ALBUQUERQUE, NM 87185
USA
tel. (505) 845-9121
fax. (505) 844-7154

Dr. HAGGAG
Fracture Mechanics Group
Metals and Ceramics Division
Oak Ridge National Laboratory
Building 45005, MS 6151
PO Box 2008
OAK RIDGE, TN 37831
U.S.A.
tel. (615) 574-7007
fax 624-7007

Dr. F.W. HASING
KFA Jülich
5170 JÜLICH
Germany
tel. 49-2461-613187
fax

Dr. P.J. HEFFER
Nuclear Electric plc
Berkeley Nuclear Laboratories
BERKELEY GLOUCESTERSHIRE ,
GL 13 9PB
United Kingdom
tel. 44-453-810451
fax. 44-453-812529

Dr. F. HEGEDUS
Paul Scherrer Institute
CH-5232 VILLIGEN
Switzerland
tel. 41-56-992473
fax. 41-56-982327 PSI CH

Mr. C.R. HEIMBACH
Combat Systems Test Activity
1603 Samantha CT
FOREST HILL, MD 21050
U.S.A.
tel. (301) 279-4881

Mr. HELM
Dept. of Applied Physics
200 MUDD
Columbia University
NEW YORK, NY 10027
USA

C. HICKMAN
DPHD/SIDR
CEN (FAR)
B.P. 6
92265 FONTENAY-AUX-ROSES
CEDEX
France

Mr. L.M. HODGSON
Arkansas Tech. University
Engineering Department
RUSSELLVILLE, AR 72801
USA
tel. (501) 968-0626
fax. (501) 968-0677

Mr. HOFMANN
Siemans AG UB KWU
Hammersbacher St. 12+14
8520 ERLANGEN
Germany
tel. 49 9131 185192
fax. 49 9131 187263

Dr. A. HOLMES-SIEDLE
R.E.M.
64A Acre End St.
EYNSHAM OXFORD OX8 1PD
United Kingdom
tel. 44 865-880050
fax. 44 865-883773

Dr. P. D'HONDT
SCK/CEN-Reactor Physics
Boeretang 200
2400 MOL
Belgium
tel. 32-14-311801-2200
fax. 32-14-315021

Mr. C.A. HRABAL
US Nuclear Regulatory
Commission
Mail Stop NL/S-302
WASHINGTON, DC 20555
USA
tel. 301 492-3644
fax. 301 443-7804

Dr. C.D. INGELBRECHT
CBNM
Steenweg op Retie
2440 GEEL
Belgium
tel. 32-14-571602
fax. 32-14-584273

Mr. S. IWASAKI
Dept. of Nuclear Engineering
Tohoku University
Aramaki-Aza-Aoba, Aoba-ku
SENDAI 980
Japan
tel. 81-22-222-1800 ext. 4674
fax. 81-22-268-1539

Dr. J. KALLNE
Uppsala University
INF Box 535
S-751 21 UPPSALA
Sweden
tel. 46-18-182394
fax. 46-18-183833

Mr. F.B. KAM
Oak Ridge National Laboratory
PO Box 2008, Bldg 3001
OAK RIDGE, T.N. 37831-6028
USA
tel. 1-615-574-7000
fax.

Mr. S.O. KING
B&W Nuclear Service Company
3315 Old Forest Road
PO Box 10935
LYNCHBURG, Virginia 24506 0935
USA
tel. 1-804-385-2594
fax. 1-804-385-3663

Dr. N.P. KOCHEROV
IAEA
Wagramerstr. 5
P.O. Box 100
A-1400 VIENNA
Austria
tel. 43-1-23 60 17 18
fax. 43-1-23 45 64

Dr. K. KOBAYASHI
Research Reactor Institute
Kyoto University
Kumatori-cho, Sennan-gun
OSAKA 590-04
Japan
tel. 81-724-52-0901
fax. 81-724-53-0360

Dr. G.P. LAMAZE
National Institute of
Standards & Techn.
Bldg. 235
GAITHERSBURG, M.D. 20899
USA
tel. 1-3-1-975-6202
fax. 1-3-1-921-9847

Dr. E.P. LIPPINCOTT
Program Chairman, ASTM
Symposium Committee
Westinghouse Electric
Corporation
Mail Stop 4-17
Westinghouse Energy Centre
PO Box 355
PITTSBURGH, PA 15230-0355
tel. W (412) 374-4191
H (412) 733-4720
fax. (412) 374-4697

Mr. R. LLORET
CEA/SRS/SAPR
B.P. 85 X
38041 GRENOBLE CEDEX
tel. 33 76884763
fax. 33 76885178

Mrs M.C. LOPES
LNETI/ICEN/DEEN
Institute of Nuclear Sciences
& Engineering
Apartado 21, 2686 SACAVEM
Portugal
tel. 351 19550021
fax. 351 19550117

Mr. A.L. LOWE, Jr.
B & W Nuclear Service Company
PO Box 10935
LYNCHBURG, Virginia 24506-0935
USA
tel. 1-804-385-3276
fax. 1-804-385-3663

Dr. T.F. LUERA
Sandia National Laboratories
Division 6452
1515 Eubank S.E.
ALBUQUERQUE, NM 87123
USA
tel. 1-505-844-0049
fax. 1-505-844-7154

Dr. R.E. MAERKER
Oak Ridge National Laboratory
PO Box 2008, Bldg. 6025
OAK RIDGE, T.N. 37831-6364
USA
tel. 1-615-574-6190
fax.

Dr. William N. McELROY
Executive Chairman
ASTM Symposium Committee
President
Consultant & Technology
Services
113 Thayer Drive
RICHLAND, WA 99352
USA
tel. H (509) 946-8139
W (509) 943-5953
fax.

Mr. P. MANAHAN
Penn State University
231 Sacket, University Park
P.A. 16802
USA
tel. 1-814-865-13-41

Mr. A. MARCAULT
CEA/DRE/SRO/SEN
981191 GIF SUR YVETTE
France
tel. 33-69 08 36 46
fax. 33-69 08 90 08

Dr. G.C. MARTIN
General Electric Company
Vallecitos Nuclear Center
PLEASANTON, CA 94566
USA
tel. W (415) 862-4347
H (415) 657-6920
fax. 415 862-4515

Dr. J.A. MASON
A.N. Technology Ltd
Technology Transfer Centre
Solwood Park
ASCOT, BERKSHIRE SL5 7PW
England

Dr. M. MATZKE
Physikalisch-Technische
Bundesanstalt
Bundesallee 100
3300 BRAUNSCHWEIG
Germany
tel. 49-531 592 7320
fax. 49-531 592 4006

Mr. E. Dale McGARRY
Building 235-A155
National Institute of
Standards & Technology
GAITHERSBURG, MD 20899
USA
tel. W (301) 975-6205
(FTS: 879-6205)
H (301) 774-2165
fax.

Mr. MEHNER
Zentralinstitut für
Kernforschung
Rossendorf
PSF 19
O-8051 DRESDEN
Germany
tel. 37-51-5912381
fax. 37-51-136069

Mr. G. MINSART
SCK/CEN Reactor Physics
Boeretang 200
2400 MOL
Belgium
tel. 32 14 311801-2306
fax. 32 14 315021

Mr. C. MORIN
CEA/DRE/SRO/SEN
91191 GIF SUR YVETTE CEDEX
France
tel. 33-69 08 23 68
fax. 33-69 08 90 08

Dr. M. NAJZER
J. Stefan Institute
Jamova 39
61000 LJUBLJANA
Yugoslavia
tel. 38-61-371-321
fax. 38-61-374-919

Mr. H. NOLTHENIUS
ECN
PO Box 1
1755 ZG PETTEN
The Netherlands
tel. 31-2246-4296
fax. 31-2246-3490

Mr. N. ODANO
Dept. of Nuclear Engineering
Tohoku University
Aramaki-Aza-Aoba
SENDAI 980
Japan
tel. 81-22-2221800 etx. 4674
fax. 81-22-2681539

Mr. P.M. ORTEGO
ENUSA - Nuclear Fuel Division
Santiago Rusinol, 12
28040 MADRID
Spain
tel. 34-1-3474234
fax. 34-2-3474215

Mr. R.L. PEREL
Racah Institute of Physics
Hebrew University of
Jerusalem
91904 JERUSALEM
Israel
tel. 972-2-584521
fax. 972-2-666804

Dr. M. PETILLI
ENEA - Casaccia - Sacco 121
CP 2400
00100 ROMA A.D.
Italy
tel. 39-6-3048-3456
fax. 39-6-3048-4786

Mr. L. PETRUSHA
Babcock & Wilcox Nuclear
Service Company
3315 Old Forest Road
PO Box 10935
LYNCHBURG, Virginia
24506-0935
USA
tel. 1-804-385 2584
fax. 1-804-385 3663

Mr. J.B. POIRE
CEA Centre de Valduc (DPEN/CE)
21120 IS SUR TILLE
France
tel. 33-80 23 43 59
fax. 33-80 23 50 27

Mr. A.N. POLLE
ECN
PO Box 1
1755 ZG PETTEN
The Netherlands
tel. 31-2246-4372
fax. 31-2246-3490

Mr. S.W. POWER
Atomic Energy Authority
Winfrith Technology Centre,
Dorchester
DORSET DT2 8DH
United Kingdom
tel. 44-305 251888-2877
fax. 44-305 251140

Dr. K.M. PYTELL
Institute of Atomic Energy
05-400 OTWOCK-SWIERK
Poland
tel. 48-22798359
fax. 48-22793481

Mr. U. QUADE
Interatom GmbH
Postfach 100351
D-5060 BERGISCH-GLADBACH 1
Germany
tel. 49-2204/843527
fax. 49-2204/843045

Mr. C.M.J. DE RAEDT
SCK/CEN Reactor Physics
Boeretang 200
2400 MOL
Belgium
tel. 32-14 311801-2312
fax. 32-14 315021

Mr. I. REMEC
Jozef Stefan Institute
PO Box 100
Jamova 39
61111 LJUBLJANA
Yugoslavia
tel. 38-61-371-321
fax. 38-61-374-919

Mr. C. RIEG
EDF/SEPTEN
12-14 Av. Dutrievoz
69628 VILLEURBANNE CEDEX
France
tel. 33-78 94 47 67
fax. 33-78 94 49 99

Mr. JW ROGERS
EG & G Idaho Inc.
PO Box 1625
IDAHO FALLS, ID 83415
tel. W (208) 526-4252
(FTS: 583-4252)
H (208) 523-0256
fax.

Dr. F.H. RUDDY
Westinghouse Science &
Technology Center
1310 Beulah Road,
Bldg. 301 3B4
PITTSBURGH, PENNSYLVANIA 15235
USA
tel. 1-412-256-1064
fax. 1-412-256-1348

Dr. K.E. SALE
Lawrence Livermore National
Laboratory L-41
PO Box 808
LIVERMORE, CA 94550
USA
tel. (415) 423-0686
fax.

Dr. G. SANDRELLI
ENEL - Thermal & Nuclear
Research Centre
Via Monfalcone 15
20132 MILANO
Italy
tel. 39-2 8847 3424
fax. 39-2 8847 3496

Mr. V. SANGIUST
Politecnico di Milano
Via ponzio N. 34/3 cap.
20133 MILANO
Italy
tel. 39-2-23996390
fax. 39-2-23996309

Mr. T. SEREN
Technical Research Centre of
Finland
Reactor Laboratory
PO Box 200 (Otakaari 3A)
SF-02151 ESPOO
Finland
tel. 358-0-4566340
fax. 358-0-4566390

Mr. W.L. SERVER
EPRI Representative
Senior Project Manager
Tenera Engineering Services
1955 University Ave.
BERKELEY, CA 94704
USA
tel. W (415) 845-5200
H (414) 845-8453
fax.

Dr. N.H. SHAFRIR
Technion Israel Institute of
Technology
Nuclear Engineering
IL-32000 HAIFA Technion City
Israel
tel. 972-4-230068
fax. 972-4-211581

Dr. Mary H. SPARKS
Stews-NE
Nuclear Effects Laboratory
White Sands Missile Range
NEW MEXICO 88002
USA
tel.
fax.

Dr. F.W. STALLMANN
Oak Ridge National Laboratory
Bldg. 3001, MS-028
PO Box 2008
OAK RIDGE, TN 37831-6028
USA
tel. 1-615-574-7076
fax.

Dr. M. TICHY
Institute of Radiation
Dosimetry
Czechoslovak Academy of
Science
Na Truhlarce 39/64,
180 86 PRAGUE 8
Czechoslovakia
tel. 42-2 823341
fax. 42-2 8152171

Dr. G. TSOTRIDIS
Commission of the European
Communities
PO Box 2
1755 ZG PETTEN
The Netherlands
tel. 31-2246-5122
fax. 31-2246-1449

Dr. D.W. VEHAR
Sandia National Laboratories
PO Box 5800
ALBUQUERQUE, N.M. 87185
USA
tel. 1-505-845-8659
fax. 1-505-844-7154

Mrs T. VERGNAUD
CEA/SERMA/LEPP
91191 GIF SUR YVETTE CEDEX
France
tel. 33-69 08 59 40
fax. 33-69 08 45 72

Mr. W.P. VOORBRAAK
ECN
PO Box 1
1755 ZG PETTEN
The Netherlands
tel. 31-2246-4295
fax. 31-2246-3490

Pr. J.J. WAGSCHAL
Racah Institute Hebrew
University of Jerusalem
IL-91904 JERUSALEM
Israel
tel. 972-2-584521
fax. 972-2-666804

Dr. C.D. WEST
Oak Ridge national River
PO Box 2009
FEDC Bldg.
OAK RIDGE, T.N. 37831-8218
USA
tel. 1-615-574-0370
fax. 1-615-676-3041

Dr. Peter WILLE
GKSS-Forschungszentrum
Geesthacht GmbH
Institut für Werkstofforschung
PO Box 1160
D-2054 GEESTHACHT
Germany
tel. 04152 871271

Dr. John G. WILLIAMS
Dept. of Nuclear Engrg.
214 Nuclear Engrg. Lab.
University of Illinois
103 S. Goodwin Ave.
URBANA, IL 61801
USA
tel. W (217) 333 7710 or
(217) 333-0866
H (217) 352 7592
fax. (217) 333-2906

T.G. WILLIAMSON
Westinghouse Savannah
River Lab.
AIKEN S.C. 29801
USA
tel. 803-725-5999 (0)
fax.

Dr. E.M. ZSOLNAY
Technical University of
Budapest
Institute of Nuclear Techn.
Muegyetem RKP 9
H-1521 BUDAPEST
Hungary
tel. 36 1 181 2754
fax. 31 1 166 6808

Mr. P. ZYROMSKI
CEA
21120 IS-SUR-TILLE
France
tel. 33-80-23-51-60

## AUTHORS OF THE SEVENTH ASTM-EURATOM SYMPOSIUM

**Batch number: 10755317**

Printed by Printforce, the Netherlands